NEURAL STEM CELLS AND THERAPY

Edited by **Tao Sun**

Neural Stem Cells and Therapy
Edited by Tao Sun

Published by InTech
Janeza Trdine 9, 51000 Rijeka, Croatia

Publishing Process Manager Vedran Greblo

Technical Editor Teodora Smiljanic

Cover Designer InTech Design Team

Additional hard copies can be obtained from orders@intechopen.com

Neural Stem Cells and Therapy, Edited by Tao Sun
p. cm.
ISBN 978-953-307-958-5

For EU product safety concerns:
IN TECH d.o.o., Prolaz Marije Krucifikse Kozulić 3, 51000 Rijeka, Croatia,
info@intechopen.com or visit our website at intechopen.com.

Meet the editor

Dr Tao Sun is an Associate Professor of Cell and Development Biology at Weill Medical College of Cornell University in New York City, the United States of America (USA). He undertook his PhD studies in Neurobiology at University College London, United Kingdom. After his PhD work, Dr Sun did his postdoctoral training in Neurogenetics at Harvard Medical School in Boston, USA. He became a faculty member at Weill Medical College of Cornell University in 2005. Dr Sun's research interests include neural stem cell development in embryonic and adult brains and spinal cords, noncoding RNA regulation in neural development and function, molecular control of brain asymmetry and cognitive functions, and genetic causes of neurological disorders.

Contents

Preface

We have made exciting progress in understanding the neural stem cells (NSCs) in the past twenty years. We have learned what genes control NSC proliferation and differentiation, discovered how to culture NSCs and trace their lineage in a culture dish, and have even developed methods to either stimulate endogenous NSCs to repair damaged neurons or transplant cultured NSCs to damaged regions in the central nervous system (CNS). Research from neurodevelopmental biologists using various invertebrate and vertebrate models, in particular rodents, has advanced the NSC field and accelerated therapy using NSCs.

The notion of germinal cells in the neurogenic region, such as the ventricular zone (VZ) in embryonic human brains, came very early, in the 1870s. Later on, the advance of labeling techniques, in particular using the DNA replication marker [3H]-thymidine, allowed scientists to visualize dividing progenitors in primate and rodent brains. In embryonic mammalian brains, neuroepithelial cells are the first identified proliferating cells and they are in fact NSCs. These NSCs are then transformed into radial glial cells, which are now known as neural progenitors, and then intermediate progenitors. The proper proliferation of these progenitors is believed to be important for controlling brain size. Similar NSCs are also identified in other regions in the CNS, such as the spinal cord.

The adult brain has long been recognized as a hard-wired system that neither generates new neurons, nor consists of NSCs. However, the observation of new neurons in the song bird brain has changed our view of adult neurogenesis. Using [3H]-thymidine labeling, dividing cells were detected on the wall of the lateral ventricle and, 30 days later, new neurons were detected in the high vocal center (HVC), a region that is believed to be responsible for song production. Furthermore, dividing cells were observed in the SVZ region of adult rodent brains and in the dentate gyrus (DG) region in the hippocampus of rodent and even human brains. Thus, in contrast to the previously held view of the hard-wired adult brain, new neurons are constantly generated in the SVZ and then migrate along the rostral migratory stream (RMS) into the olfactory bulb and the DG of the hippocampus, which may contribute to learning and memory aptitude.

Numerous exciting studies have focused on illuminating the molecular mechanisms that regulate NSC proliferation and survival in both developing and adult brains. Many transcription factors and growth factors have been identified to control NSC proliferation and differentiation into various cell types. In recent years, epigenetic regulation of NSC development has also been revealed. Moreover, the niche that maintains NSCs has been realized. For example, the vascular system in the SVZ of adult brains has been shown to promote NSC proliferation.

Parallel to the growth of our understanding of NSCs at cellular and molecular levels, our attempt to utilize NSCs for repair of damaged neurons in neurodegeneration disorders and injuries has also made significant progress. Cultured NSCs have been transplanted into the brains of stroke models and into the spinal cord after injuries, and significant recoveries have been observed. Moreover, it has been found that ischemia promotes the endogenous NSCs to proliferate and migrate into damaged regions.

Taking the benefit of our knowledge of neurodevelopment and neural stem cell specification, embryonic stem cells (ESCs) have been used to produce NSCs and their progenies in cultures. The induced pluripotent stem cell (IPS) technology allows for NSCs to be generated directly from fibroblasts of patients with neurological disorders. Excitingly, recent studies have shown that fibroblasts can be reprogrammed directly into neurons by skipping the IPS step. Consequently, we are no longer restricted to post-mortem samples of patients with neurological disorders. These new technologies allow scientists to reprogram patient fibroblasts into NSCs or neurons, identify abnormal gene regulations responsible for these disorders, and screen potential drugs for treatment. We still face many challenges, such as the difficulty of producing homogeneous neuronal populations for transplantation, and the strain in leading new neurons to form synaptic connections with exiting neurons. However, there is no doubt that NSCs are becoming a promising means for treatment of neurological diseases and injuries.

The publication of this book is timely. It contains the characterization of embryonic and adult neural stem cells in both invertebrates and vertebrates, and highlights the history and the most advanced discoveries in neural stem cells. This book provides the strategies and challenges of utilizing neural stem cells for therapy of neurological disorders and brain and spinal cord injuries.

I am honored to have had this opportunity to work with over 20 authors on this book. The expertise and scientific contribution from each author has enriched the depth and broadness of the book and I have learned a tremendous amount from each and every one of them. It has been a great pleasure to work with the staff members at InTech Open Access Publisher. In particular, I feel fortunate to have worked closely with Mr. Vedran Greblo, who has coordinated the publication of this book from the beginning to the end. It is his professional insight in publishing, and his patience and encouragement that has made this book possible.

Tao Sun
Weill Medical College of Cornell University
USA

Part 1

Characterization of Neural Stem Cells

1

Neural Stem Cells from Mammalian Brain: Isolation Protocols and Maintenance Conditions

Jorge Oliver-De la Cruz and Angel Ayuso-Sacido
Regenerative Medicine Program, Centro de Investigación Príncipe Felipe, REIG and Ciberned
Spain

1. Introduction

Traditionally, the adult brain has been considered a quiescent organ, lacking the production of new cells, or more exactly, new mature and functional neurons. This dogma has been widely refused in the last decades with the discovery of proliferative cells with stem cell properties in the adult brain.

First evidences come from the demonstration of neurogenesis in non-mammal vertebrates such as birds or lizards (as reviewed in Garcia-Verdugo et al., 2002). Neurogenesis was also confirmed to occur in adult mammals, like mice and rats, and, finally, in primates and humans (for a complete revision see Gil-Perotín et al., 2009). Though the process of neurogenesis in the adult is primarily confined to the subventricular zone (SVZ) and the subgranular zone (SGZ) of the dentate gyrus, glial progenitors exist in other brain regions. These widespread glial progenitors remain quiescent and do not generate mature glial cells, but, in certain situations such as traumatic injury, they may act as true stem cells (Belachew et al., 2003; Rivers et al., 2008).

The terminology of stem cell, progenitor cell and precursor cell has been adapted from others tissues. Basically, a *bona fide* neural stem cell (NSC) must meet all these three features: capacity of self-renewal, capacity to differentiate into the three neural lineages (neuron, astrocyte and oligodendrocyte) and, finally, the ability to regenerate neural tissue. When cells show a limited self-renewal and are already committed toward a specific fate, they are classified as progenitor cells, while the term "precursor" represents intermediate stages.

Neural stem/progenitor cells (NSPc) primary cultures provide the best *in vitro* model to study proliferation and differentiation signaling pathways, a difficult issue to address *in vivo.* Additionally, these cells might be used in future replacement cell therapies, thus motivating the development of protocols aimed to isolate and expand these cells *in vitro.* These protocols display significant variations among them, and the introduction of new technologies has increased drastically their number. The differences in the protocols have rendered different results in terms of stem cell subpopulations, differentiation potential and the amount of cells. The last is especially relevant in the case of human samples because of their low availability.

Therefore, the aim of this chapter is to recapitulate some of these technical differences that could induce variances in the final results. We have analyzed the main isolation protocols from the two canonical neurogenic zones in the adult (subventricular zone and hippocampus), described for both animal models (mouse and rat) and human.

2. Neural stem/progenitor cell isolation

The NSPc isolation procedures follow common steps including tissue dissection, digestion and cell enrichment. However, comparing the different protocols found in the bibliography, it is notable the presence of significant differences between them even when they are consecutive works from the same group. The introduction of new technologies has also increased drastically the number and variety of protocols. Additionally, some tissues like normal human brain are particularly difficult to manage due to their low availability, which requires improvements in the protocol to include modifications that increase the rate of isolated cells. Interestingly, the diversity of isolation procedures results in the obtaining of different stem/progenitor cell subpopulations with distinct differentiation potential, and might be also responsible for the, sometimes contradictory, results observed in the literature.

Although the development of standard protocols would be the best option to assure that results can be easily compared, in the practice, this is almost impossible. Different groups have generated independently alternative procedures for the isolation, dissociation and enrichment of NSPc. Furthermore, the animal model, the specific location of the brain sample, or even the characteristics of the experiment have requirements that would make unmanageable the use of universal procedures. Usually, the same group employs very similar strategies to isolate cells from different samples, independently of their developmental stage or animal/human origin. Nevertheless, it will be interesting to establish flexible guidelines to indicate what can be modified from the standard procedures and how to do so.

The basic scheme followed by NSPc isolation protocols is reflected in figure 1, and we will discuss the specific methodology associated with every step in the following headings

2.1 Tissue dissection methods

The origin of the tissue influences the type of isolated cells as well as their proliferation and differentiation capacity. A number of profound differences have been reported between brain samples from different species (mouse, rat of human) or from different stages of development within a given specie (Gritti et al., 2009; Svendsen et al., 1997). However, the accurate dissection of specific regions of the brain has become more relevant as the knowledge on the NSPc biology and location increases. In fact, regardless of the animal model, one of the main factors that might determine the final results is the specific location of the brain tissue from where NSPc are isolated.

Different regions of the brain have been used as a primary source of NSPc and, consequently, discrepancies in the isolated cells have been reported. In this sense, analyzing the distinct approaches for the tissue dissection might be useful to contextualize such a controversy.

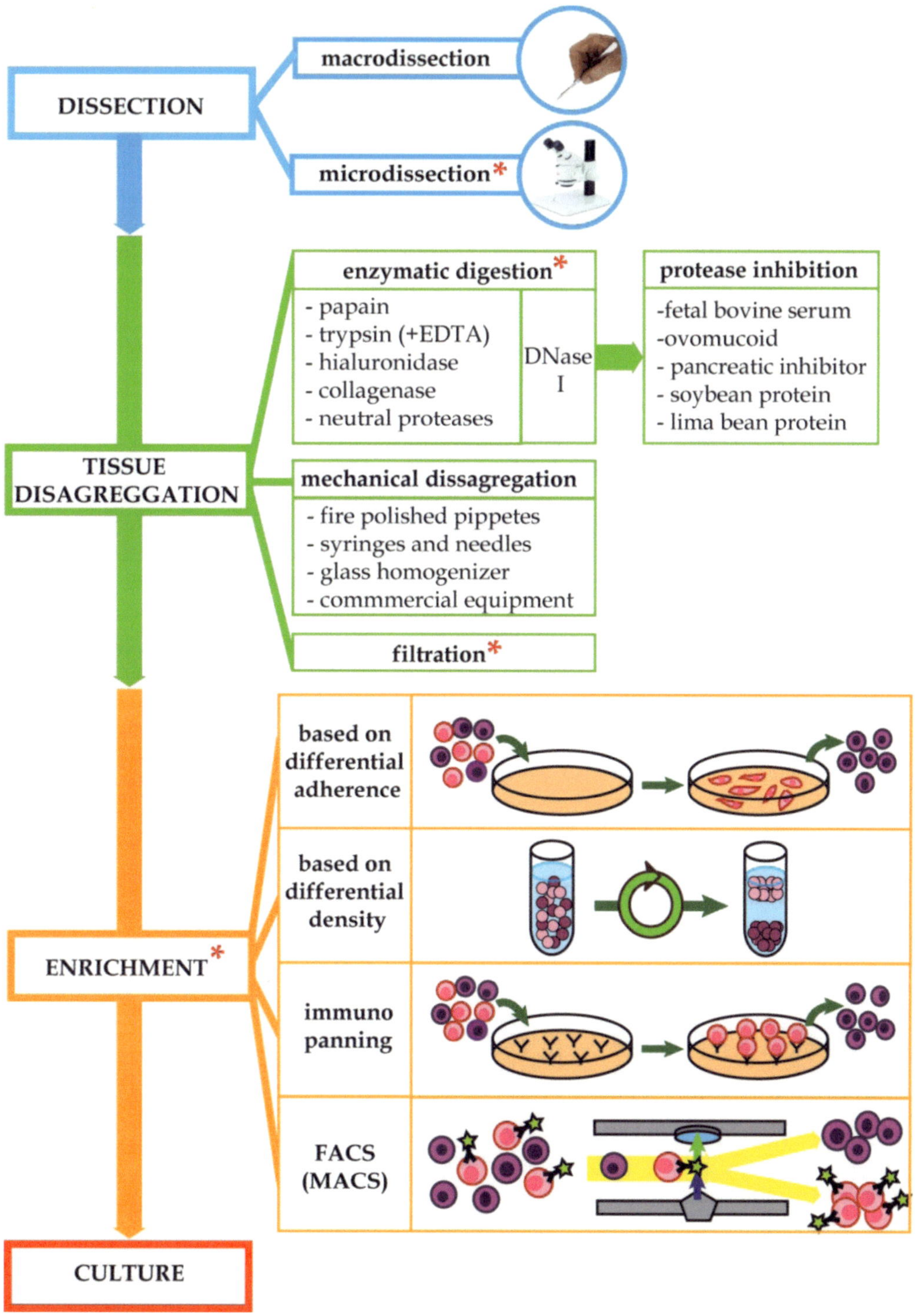

Fig. 1. Diagram depicting the main steps of standard neural stem/progenitor cell isolation protocol. Headings marked with an asterisk are not always included.

We might consider three different levels of dissection according to the amount and location of tissue, ranging from large unselected brain tissue to microdissection. In a first level, a number of works start from whole brain (e.g. Von Visger et al., 1994) or large areas that include heterogeneous regions (e.g. whole human temporal lobe, Kirschenbaum et al., 1994). In these cases, the results can be highly variable, because of the different types of progenitors coming from distinct locations and giving rise to an artefactual impression of cell heterogeneity. An intermediate step of complexity is found in those works that use tissue from specific areas, but without the exclusion of contiguous tissues, i.e. macrodissection. In this regard, some authors reported the presence of multipotent stem cells from different regions of the adult parenchyma that differ from canonical neurogenic zones (SVZ and SGZ), e.g. from striatum (Reynolds & Weiss, 1992). However, these cells might arise from the cross-contamination of adjacent neurogenic regions (Lois & Alvarez-Buylla, 1993). Likewise, as will be discussed later, the existence of real neural stem cells in adult dentate gyrus of hippocampus has become a controversial subject. Some authors claim that there are true stem cells from this zone. However, others state that these isolated cells should be considered progenitors because of their low proliferation *in vitro* and their doubtful multipotentiality. The main argument of these authors is the lack of fine dissection, and the inclusion of neural stem cells from other adjacent tissues, like SVZ. Therefore, considering the current knowledge on NPSc niches location, an exhaustive microdissection is essential to take out the region of interest in a reliable way before starting the isolation procedures. Then, it is highly recommendable the use of thin slices of tissue for the accurate microdissection of different compartments under a dissecting microscope (e.g. Seaberg & van der Kooy, 2002).

Tissue dissection is particularly challenging in the case of human surgical samples, where orientation and anatomical organization is usually altered after surgery, making difficult the recognition of particular zones and, consequently, a good dissection. Alternatively, some authors have demonstrated the isolation of viable cells from postmortem tissue, especially in the case of human samples (e.g. Schwartz et al., 2001). While these procedures might be the only way to access some type of tissues, there might be some logistical inconveniences, the main one being that collection of tissue and cell isolation protocols need to be performed within few hours, because the number of NSPc decreases with time (Leonard et al., 2009; Xu et al., 2003), especially when samples are exposed to environmental temperature instead of 4ºC (Laywell et al., 1999).

2.2 Tissue digestion methods

2.2.1 Enzymatic dissociation

NSPc are surrounded by a highly structured extracellular matrix mainly composed by lecticans, hyaluronic acid, tenascin-C and tenascin-R (Rutka et al., 1988). These molecules interact among them and with membrane molecules on cell surfaces, and can regulate part of their behavior.

Therefore, one of the most successful strategies for removing NSPc from the rest of the tissue implies the use of proteases to degrade this matrix.

The first step, to prepare the tissue for enzymatic digestion, involves the mincing into small pieces (less than 1 mm^3) in order to provide more degradable surface for the action of proteases. In this sense, the use of two different enzymes stands over the rest in the literature: trypsin (examples of its applications in different samples and developmental stages can be read at Kirschenbaum et al., 1994; Kukekov et al., 1997, Reynolds et al., 1992; Reynolds&Weiss, 1992; Svendsen et al., 1998) and papain (Babu et al., 2007; Roy et al., 2000a; Wang et al., 2000; Windrem et al., 2004). Trypsin is the most employed one, and is often combined with ethylenediaminetetraacetic acid (EDTA), a Ca2+ chelating agent that weakens intercellular unions. Regarding the concentration and the incubation time, it is not always possible to compare between different protocols as the enzyme units are not always specified and the incubation time ranges from 10 to 90 minutes. Additionally, other enzymes can be found in the bibliography such as hialuronidase (e.g. Gritti et al., 1995; Weiss et al., 1996), collagenase (e.g. Uchida et al., 2000), and neutral protease (dispase) (e.g. Babu et al., 2007), alone or in combination with others.

Generally, the use of proteases is linked to the utilization of Desoxiribonuclease I (DNase I), usually from bovine origin, in order to eliminate the DNA mucus originated by cell lysis, which could hinder cell survival and further experiments.

In any case, the employment of enzyme specific buffers (with adjusted pH and containing activators) is necessary to allow the action of these enzymes. In some cases, antibiotic/ antimitotic is added to the digestion solution to prevent contamination. At this stage, some authors also include kynurenic acid in order to reduce glutamate excitotoxicity through NMDA receptor channels (e.g. Reynolds&Weiss, 1992). Afterward, the use of protease inhibitors is necessary to stop enzymatic reaction. Papain is usually neutralized with fetal bovine serum, whereas in the case of trypsin, the most employed method includes ovomucoid, although there are commercially available soy, lima bean, and basic pancreatic protein -based inhibitors.

The criterion for the choice of one or another enzyme is not clear, and frequently it has more to do with the previous experience and skills of the group. Nevertheless, as a general rule, embryo and early fetal samples require less amount of enzyme due to its laxity. For this reason, some protocols reduce protease concentrations and/or exposure time (e.g. Svendsen et al., 1998) or even recommend the use of mechanical disaggregation techniques alone (e.g. Ciccolini&Svendsen, 1998; Reynolds&Weiss, 1996).

The enzymatic digestion is a critical step because it affects directly to the NSPc survival rate. In this sense, some studies have been done to compare cell survival after dissociation with different protease. Maric et al., 1998, used murine embryonic tissue to evaluate the efficacy of papain, trypsin, and collagenase treatment, or mechanical disaggregation alone. The results indicate that papain dissociation is optimal, achieving the maximum reproducible cell recovery and viability. On the contrary, trypsin, collagenase, and mechanical dissociations resulted on suboptimal and highly variable yields. Another study, carried out by Panchision et al., 2007 also compared the results obtained for mouse embryonic stem cell isolation when using papain, Tryple™ (a commercial analog of trypsin), or collagenase/neutral protease commercial cocktails (Accutase™ and Liberase-1™). Data also confirmed that mechanic dissociation induced more variability, cell death and more number of aggregates. However, Tryple™ and papain produced more quantity of DNA mucus (but not an increased cell death) and a lower adherence to culture plate after planting. They conclude that the best results were obtained with papain, independently of the exposure time to the enzyme.

Moreover, this work also revealed another important factor to take into account when optimizing protease dissociation: cell surface markers can be altered by these enzymes, inducing false negatives when immunocytochemistry or Fluorescent Activated cell sorting is performed just after isolation. Table 1 includes a list of sensitive markers described in this paper and similar reports. In addition, another work detected that trypsin cleavage can lead to an increased positivity of some tumor -related surface markers, depending on the state of glycosylation (Corver et al., 1995).

	enzyme -sensitive	**very weakly enzyme -sensitive**
Trypsin	hCD133, CD31, O4, CD81, c14, Ca125, BMa180	A2B5, CD15
Papain	PSA-NCAM, CD24, BMP IA, BMP IB	CD15, O4, CD81
Liberase-1™	BMP IA, BMP IB	
Accutase™	none of the studied	
Tryple™	none of the studied	

Table 1. List of enzyme - sensitive markers that are reported in Corver et al., 1995; Panchision et al., 2007; and commercial report by Reiβ et al (Miltenyi). CD133 has been reported to be sensitive to trypsin treatment in human cells, but not in rodent cells.

2.2.2 Mechanical disaggregation

Usually, the enzymatic digestion is not enough to remove the NSPc from the remaining tissue. After or during enzymatic digestion, the tissue must be triturated to break up the digested pieces into a single cell suspension. It is a dramatic process that ends up with an important number of dead cells. However, different strategies have been described in the literature in order to reduce, to some extent, this number. The most common method consists in passing the suspension through fire polished glass pipettes, due to their high availability and lower price (e.g. Ciccolini & Svendsen, 1998; Gage et al., 1995; Reynolds&Weiss, 1992). Moreover, they can be narrowed into different diameters, adapting their thickness to samples of different size. Many protocols include the sequential trituration through pipettes with decreasing diameters in order to disaggregate the tissue in successive steps and reduce cell death (e.g. Wang et al., 2000). Nevertheless, this system also presents some technical problems. First, cells display a relative adherence to glass and might be lost. Furthermore, as glass pipettes are usually prepared specifically for each experiment, their diameter can vary, and therefore, different cell survival rates can be obtained. The cell adherence issues might be partially resolved by coating pipettes with silicone. Alternatively, some commercially available plastic pipettes (Kukekov et al., 1997) are treated to reduce the adherence, but they cannot be fire polished.

Another strategy is based on the utilization of sterile syringes and needles (e.g. Shi et al., 1998). In this sense, a large range of needle gauges is available commercially, ensuring the reproducibility of the technique; however, their edges are too sharp and that results in an increase in cell death. Although less frequent, it is worth mentioning the use of different devices like the glass homogenizer, used for embryonic neural stem cell isolation (Carpenter et al., 1999), and some commercial equipment that appeared in the last years, promising a higher efficiency via the automation of the isolation procedure (Reiβ et al (Miltenyi)).

2.2.3 Filters utility

Some groups, after enzymatic digestion and mechanical disaggregation, include a filtering step to remove the debris from the cell suspension. This additional step might eliminate undissociated tissue pieces as well as avoid the presence of necrotic particles in the final pellet that would potentially induce cell death. However, it also reduces the final number of viable cells trapped into the filter. In any case, the use of filters usually requires a DNase I treatment, to remove the mucus that can difficult the filtering, and it is strongly recommended the dilution of cell suspension in a considerable volume of medium. Regarding the type and size of the filters, some authors describe the use of cell strainers, whereas others prefer sterile gauze (e.g. Kukekov et al., 1997). The mesh size also differs among protocols (40 um (Wang et al., 2000), 70 um (Rietze et al., 2001), 100 um, etc), and should be chosen in accordance with the efficiency of preceding methodology.

2.3 Neural stem/progenitor cells enrichment procedures

The initial protocols for NSPc isolation were designed with the only purpose of isolating and culturing these cells to study their biology *in vitro*. However, as the knowledge on the biology and differentiation potential of NSPc increased, it was evident that cell cultures comprised a number of different subpopulations with different degree of stemness. Consistently with this reality, many authors have recently included separation steps into their NSPc isolation protocols. This separation is usually based on the NSPc phenotypic characteristics closely related to their stem cell features.

In this sense, the first works on NSPc isolation and culture described a selection based on their capacity to proliferate in the chosen medium and growth factors. Obviously, it was not enough to discriminate heterogeneity. Consequently, many technical approaches have been developed since then, for the enrichment of a specific subpopulation. This way, the biological significance behind the molecule chosen to enrich for a specific type of cell and the technology used for the procedure become an important step determining the differentiation potential of the final cell culture. The current techniques for the separation and enrichment of NSPc are described below.

2.3.1 Methods based on differential adherent properties of cells

One of the first methodologies for the enrichment of particular subpopulations was based on the differential attachment of cells to the culture plate due to their particular adhesion molecule patterns. By optimizing some parameters like substrates and time in culture it is possible to distinguish between different types of cells. Astroglial cells show the biggest adherence, even in untreated culture plate, whereas oligodendrocytes can be easily detached through the agitation on a rotary shaker at slow revolutions (200-300 rpm) for 12-20 h. This procedure has demonstrated to be useful, easy and affordable. As a consequence, it has been common in the purification of specific cell types like oligodendrocytes (McCarthy & de Vellis, 1980; Chen et al., 2007b).

Taking advantage of these properties, Lim&Alvarez-Buylla, 1999, reported the isolation of 4 cell fractions using serial streaming of medium or PBS over the surface of poly-D-lysine treated plates, and a final step with trypsin. The first fraction (or fraction 1), which contains the less adherent cells, was enriched in PSA-NCAM and Tuj1 (identified as migrating neuroblasts). On the contrary, cells from the most adherent fraction (fraction 4) were GFAP$^+$ and show characteristics of neural stem cells (type B/C according to the model of SVZ organization (Fig.2). However, it is important to mention that this procedure does not allow the obtaining of high purity cultures.

2.3.2 Differential gradient centrifugation

Another group of technical approaches for NSPc enrichment is based on fractionating cell populations according to their buoyant density. Previously, the cells are dissolved in specific solvents that, after centrifugation, generate a density gradient. The cells distribute in this gradient and can be collected separately. The gradient might be formed by using different types of reagents, being Percoll the most widely used (e.g. Palmer et al., 1999; K. Chen et al., 2007a). It consists of colloidal silica particles coated with a layer of polyvinylpyrrolidone (PVP) that can be used to form solution densities between 1.00 and 1.20 g/ml. A combination of Percoll gradients can be generated in order to separate more subpopulations. Using a discontinuous density gradient, Maric et al., 1998 reported the isolation of 20 different bands and the delimitation of density bands can be facilitated by commercial color-coded density marker beads. While its application has become very common because of its low interaction with cells and low toxicity, it is restricted to research as it may contain variable quantities of endotoxin (PVP). Alternatively, density gradients can be also generated using sucrose solutions (Johansson et al., 1999) and Bovine Serum Albumin (Ericsson, 1977).

2.3.3 Immunopanning

Initial immunopanning applications were essentially directed to eliminate specific cell subpopulations by antibody union and complement-mediated lysis (e.g. Gard&Pfeiffer, 1993). Nevertheless, the present acceptation of the immunopanning technic comprises the purification of a cell population by exploiting their differential binding to the culture dishes previously coated with a cell-surface antibody. Cells expressing this surface antigen are retained on the dish and are thereby separated from the remaining cell population. It has been especially applied to the isolation of oligodendrocyte progenitor cells, using A2B5 or O4 (Barres et al., 1992; Wu et al., 2009; Mayer-Proschel, 2001) as molecular surface markers, but it can also be adapted to segregate immature neurons (PSA-NCAM) (Ben-Hur et al., 1998; Schmandt et al., 2005). Although the use of immunopanning has become less popular with the introduction of Fluorescence-activated cell sorting (FACS) technology, some authors had reported that immunopanning provides a higher survival (Mayer-Proschel, 2001).

2.3.4 Fluorescence activated cell sorting (FACS)

The main improvement in terms of separation and enrichment of specific NSPc comes with the introduction of the FACS technology. As a specialized form of flow citometry, it provides a method for sorting heterogeneous cells based upon the specific union of a fluorophore-labeled antibody to a cell surface maker. In addition to antibodies, other molecules like lectins can be used to recognize the glycosylation state of some membrane epitopes (as reviewed in Kitada et al., 2011).

The main advantage of this procedure is its high sensitivity, reaching values of purity above 95%. Moreover, the possibility of labeling cells with simultaneous antibodies allows the isolation of a particular subset of cells with a combination of membrane markers (e.g. Uchida *et al*, 2000). Moreover, the use of this technology makes possible the sorting of cells according to the expression of either cytoplasmic or nuclear markers. This advantage allowed the design of transgenic animal models that express a given fluorophore under the control of specific promoters. Additionally, the introduction of small DNA molecules can also induce the expression of a fluorescent molecule in both animal and human cells.

Alternatively, magnetic labeled antibodies might be used through a variation known as Magnetic-activated cell sorting (MACS). This technology uses a more reduced and affordable equipment, although it does not allow the labeling of more than one surface marker. Table 2 lists the stem cell markers used in the isolation of NSPc subpopulations by FACS or MACS.

NEURAL STEM CELLS	ANTIBODY	Integrin α1β5	Yoshida et al., 2003
		CD15	Capela et al., 2002; Corti et al., 2005; Panchision et al., 2007
		CD24LOW	Murayama et al., 2002; Rietze et al., 2001
		CD133	Cortie tal., 2005; Panchision et al., 2007; Uchida et al., 2000
		CXCR4	Corti et al., 2005
		EGFR1 (EGF)	Ciccolini et al., 2005; Pastrana et al., 2009
		NOTCH1	Johansson et al., 1999
	LECTINS	PHA-E4	Hamanoue et al., 2008
		WGA	Hamanoue et al., 2008
	FLUOROPHORE UNDER PROMOTOR CONTROL	p/GFAP	Doetsch et al., 1999; Pastrana et al., 2009
		p/MELK	Nakano et al., 2005
		p/MSI1	Keyoung et al., 2001
		p/Nestin	Kawaguchi et al., 2001; Keyoung et al., 2001; Roy et al., 2000a, 200b; Sawamoto et al., 2001; Yoshida et al., 2003
		p/SOX1	Barraud et al.,2005

			p/SOX2	Brazelet al., 2005; Ellis et al., 2004; ; Keyoung et al., 2001; Suh et al., 2007; Wang et al., 2010
LINEAGE RESTRICTED PRECURSOR	glial	ANTIBODY	A2B5	Maric et al., 2003; Nunes et al., 2003; Windrem et al., 2002; Windremet al., 2004; Wright et al., 1997
			CD44	Liu et al., 2004
			GD3	Maric et al., 2003
			NG2	Aguirre et al., 2004
			O1	Duncan et al., 1992
		FLUOROPHORE UNDER PROMOTOR CONTROL	p/CNP	Aguirre et al., 2004; Nunes et al., 2004; Roy et al., 1999; Yuan et al., 2002
	neuronal	ANTIBODY	PSA-NCAM	Panchision et al., 2007:Windrem et al, 2004
		FLUOROPHORE UNDER PROMOTOR CONTROL	P/Tα1	Piper et al., 2001; Roy et al, 2000[a], 2000b; Sawamoto et al., 2001; Wang et al., 2000
			P/Neurogenin2	Thompson et al., 2006
		OTHERS	Cholera toxin	Maric et al., 2003
			Tetanus toxin	Maric et al., 2003

Table 2. Main markers used for NSPc isolation by FACS or MACS.

3. Primary neural stem/progenitor cell culture

3.1 Culture conditions

3.1.1 Culture media

The culture media commonly used to grow NSPc includes two components, the basal media and supplements and the use of growth factors. The basal culture media formulation does not change significantly between different groups. It is based on the use of Dulbecco's modified Eagle's Medium (DMEM), which is composed by a defined mixture of inorganic salts, amino acids and vitamins among other nutrients. DMEM is usually combined (1:1) with Ham's F-12 (F-12), which basically increases the level of some nutrients and provides different inorganic salts. Although less frequent, other alternatives with similar characteristics have been reported as basal media, such as Neurobasal™, or Ex Vivo™ 15 (e.g. Babu et al., 2007).

The basal media is frequently supplemented with N2 or B27 supplements which contain nutrients like insulin, transferrin or putrescine, among others. These supplements cannot be added to basal formulation until they are used because of their short life at 4ºC. Although both of them might be used, even in combination, they have different properties that may influence cell culture behavior. B27 has a more complex composition than N2 supplement and only enhances cell survival during the period immediately following isolation (Svendsen et al., 1995), while N2 offers the same results, at a lower price. Babu et al. (2007) concluded that monolayer cells maintained with N2 supplement generated more neurons after differentiation, whereas B27 supplement promoted proliferation.

3.1.2 Serum and growth factors

Although basal media and supplements are quite similar in most cases, the most important issue in terms of culture media is the use of either specific growth factors or serum. The first works on NSPc isolation and maintenance described the use of serum in their culture media. However, as the knowledge on NSPc biology increased, researchers found that the use of serum, generally fetal bovine serum (FBS), had several disadvantages. As a complex solution of undefined composition that can vary drastically among batches, the use of serum does not contribute to improve our knowledge about trophic signals requirements. Additionally, it is not a physiological condition, since neural stem cells are not exposed directly to serum *in vivo*. Finally, serum includes a combination of different growth factors that are able to maintain stem cell phenotype and also induce differentiation. All these reasons made the authors substitute serum for a specific combination of purified growth factors. The utilization of two main growth factors stands out from the rest: fibroblast growth factor 2 (FGF-2, also called basic FGF or bFGF) and epidermal growth factor (EGF), alone or in combination. Moreover, FGF-2 must be used in combination with heparin, which mediates the binding of the growth factor to its receptor (Yayon et al., 1991).

Initial works (Reynolds, 1992; Reynolds & Weiss, 1992) described the isolation of an EGF-responsive neural stem cell population from striata/lateral ventricle, although some authors reported that similar cell cultures could be also maintained with FGF (Gritti et al., 1995; Vescovi et al., 1993). Similarly, some works also found a synergic effect of both EGF and FGF in proliferation, but only at low cell densities (Svendsen, 1997; Tropepe, 1999). Finally, a series of studies (Martens et al., 2000; Tropepe et al., 1999; Ciccolini, 2001; Maric et al., 2003) demonstrated that FGF- responsive cells arise earlier at development, and then give rise to both EGF/FGF- responsive cells. Moreover, it was revealed that the acquisition of EGF responsiveness is promoted by FGF *in vitro* (Ciccolini & Svendsen, 1998). First isolations could be explained with the discovery of a small autocrine/paracrine FGF production by neural stem cells, allowing the survival of FGF-2 dependent cells without FGF until the acquisition of EGF responsiveness (Maric et al., 2003).

Other growth factors that have been reported to support cell culture are Transforming growth factor alpha (TGF-α) (Reynolds et al., 1992), Leukemia inhibitory factor (LIF) and its equivalent Ciliary neurotrophic factor (CNTF) (Carpenter et al., 1999), or Brain-derived neurotrophic factor (BDNF), although its capacity to enhance later neuronal production and survival has been questioned (Kirschenbaum & Goldman, 1995; Ahmed et al., 1995; Reynolds & Weiss, 1996).

Platelet-derived growth factor alpha (PDGFα) is frequently used in the maintenance media for oligodendrocyte progenitor cells. The signaling pathway through the PDGFα/PDGFRα has different effects depending on the stage of differentiation of these progenitors: it provides signals favoring proliferation and migration in murine and human oligodendrocyte progenitors (Wilson et al., 2003; Calver et al., 1998), whereas later in development is related with cell survival (Gogate et al, 1994). Similarly, FGF promotes proliferation and blocks differentiation of oligodendrocyte progenitors, in part through the modulation of PDGFRα receptors expression (McKinnon et al., 1990).

Finally, some works have attempted to co-culture NSPc in the presence of other supportive cells like astrocytes (Richards et al., 1992; Lim et al., 1999), that seem to favor the NSPc growth by physical contact, or endothelial cells, that also enhance cell proliferation via VEGF production (Sun et al., 2010).

3.1.3 pH and oxygen levels

The metabolic processes undergone by the cells in culture give rise to acidic components that eventually are released to the media, thus decreasing the pH. This alteration, easily followed by the inclusion of a pH indicator like phenol red, has a direct influence in the behavior of the cells. Therefore, buffering agents are commonly added to medium formulation in order to control variations in the pH. In this sense, two main systems are routinely used in the elaboration of the media: sodium bicarbonate buffer, which is dependent on the CO_2 concentration present in the incubator, and HEPES (4-(2-hydroxyethyl)-1-piperazineethanesulfonic acid), independent of atmospheric CO_2. Although HEPES is better at maintaining physiological pH controls, the exposure of HEPES-containing media to light must be reduced as HEPES-containing media generates hydrogen peroxide when exposed to ambient light ({Zigler et al., 1985).

In contrast, the level of O_2 tension remains to be optimized for NSPc cultures. The standard conditions to culture NSPc had included atmospheric levels of O_2 (21%), although physiological levels are much lower (around 3%). Several studies have confirmed that NSPc expansion under low level of O_2 correlates with the expression of stemness markers and higher survival rate both *in vitro* and *in vivo* after engraftment (reviewed in De Filippis & Delia, 2011).

3.2 Monolayer versus neurosphere cultures

To maintain and propagate stem cell cultures different authors have published two alternative methods of NSPc culture and expansion: as free-floating cell clusters (neurospheres) or as adherent cultures forming a monolayer on the plate surface.

The neurosphere assay has been the most extended method to demonstrate the presence of NSPc in culture (Reynolds et al., 1992) and it is still used with different modifications (Rietze& Reynolds, 2006). Some authors claim that each neurosphere represents a microenvironment that recapitulates neurogenic niche and allows survival of stem cells *in vitro* (Bez et al., 2003) through direct cell-to-cell interaction. Nevertheless, a single neurosphere contains only a small percentage of true stem cells, whereas the remaining cells are in different stages of differentiation. Necrotic and apoptotic cells are also present (Lobo et al., 2003). Interestingly, it has been reported that committed progenitors, like oligodendrocyte precursors, can generate cell clusters similar to neurospheres (Chen Y. et al., 2007).

This culture method has also several technical disadvantages. First, when neurospheres become larger, the diffusion of nutrients and growth factors through the neurosphere is compromised (Svendsen et al., 1997b), which makes difficult the interpretation of some experimental results. Second, packed neurospheres do not allow the tracking of individual cells, which also hinders studies relating to differentiation processes. Finally, recent publications demonstrate that neurospheres are not static particles originated from a single cell and isolated from the rest of the neurospheres and cells (Rietze, 2006). On the contrary, they are dynamic structures within the culture, were cells are exchangeable from one to another sphere. This effect, may be circumvented by either using a limiting dilution analysis to obtain a single cell in each well or using semisolid cultures by adding methylcellulose (Gritti et al., 1999; Kukekov et al., 1997) or collagen (Neural Colony-Forming Cell Assay (Louis et al., 2008).

By contrast, monolayer cultures obviate some of these restrictions. They can be used to study the properties of stem cells at individual cell level, although it does not allow cell interaction during differentiation. Moreover, cells are exposed homogeneously to growth factors and serum, with the consequent reduction in cell heterogeneity.

In all, there is not a prevalent method over the other. It has not been demonstrated a total equivalence between both type of cultures and the two methods have advantages and limitations that researchers should take into consideration in the experimental design. The formation of neurospheres may be promoted by following several strategies, being the most common one the use of nonadherent surfaces like poly-2-hydroxyethyl methacrylate (Kukekov et al, 1999). Furthermore, it has been also reported the addition of mercaptoethanol to avoid cell attachment (Kukekov et al., 1997). However, not all attempts to transform an adherent culture into neurospheres have been successful (Walton et al., 2006). Alternatively, cell attachment may be induced by coating the plate surface with charged molecules such as poly-l-ornithine, poly-d-lysine or laminin.

3.3 Cell passaging

Before cells become totally confluent, it is necessary to subculture them after disaggregation of cell clusters into single cell suspensions. Regardless of the type of culture, monolayer or suspension, passaging should be performed before cells achieve their maximum confluence (monolayer) or cell cluster become necrotic (neurospheres) in order to avoid senescence associated with prolonged high cell density. The methodology employed for the disaggregation step depends on the cell type.

Adherent cells are usually detached from the surface of the culture vessel by enzymatic means. Trypsin, alone or in combination with EDTA, has been the most used protease (e.g. Palmer et al., 1997); but in the last years it has been substituted in current protocols by Tryple™, since this commercial product is free of animal- and human- derived components, less damaging to cells, and does not require the use of inhibitors.

In the case of neurospheres, cell disaggregation is performed by using mechanical procedures which involve triturating spheres with fire polished pipettes. However, this is an aggressive method that renders high levels of cell death. Enzymatic digestion can be also used before triturating, however, this may alter the experimental results if FACS assays are conducted right after disaggregation.

An alternative method was reported by Svendsen & ter Borg, 1998 for passaging neurospheres isolated from human fetal tissue. Briefly, neurospheres were cut into 4 pieces instead of standard trituration into single cell suspension. According to their data, this sectioning method reduces cellular trauma and preserves cell interaction, allowing NSPc to proliferate more replication rounds *in vitro.*

3.4 Cryopreservation

Cryopreservation allows the maintenance of NSPc in a suspension mode awaiting for future experiments and saving expensive culture reagents. Considering the low number of cells obtained from each sample, especially in human tissue, increasing the survival ratio after long-term preservation of NSPc becomes a major concern. The main cryopreservation protocols employ dimethylsulfoxyde (DMSO) diluted at 10-20% in culture media to avoid ice crystallization, accompanied by a slow cooling step in isopropanol recipients. Although less popular, glycerol can be used instead of DMSO. Cellular viability can be improved adding animal serum to freezing medium, but it can potentially introduce contaminants, and induce differentiation. In any case, cryopreservation must follow some general rules to ensure the successful preservation of cells. It must be performed during the logarithmic growth phase and high cell density in each ampoule seems to facilitate cell recuperation. Smaller neurospheres survive better than larger, so triturating cells until getting a suspension of small neurospheres improves cell survival.

Recently, a new alternative preservation method, named vitrification, has been adapted for NSPc (Tan et al., 2007). In brief, cells are sequentially submerged in a series of freezing solutions with increasing concentrations of cryoprotectant (ethylene glycol and sucrose), and finally transferred into borosilicate glass capillaries, snap-frozen and stored in liquid nitrogen. The results showed that vitrification offered the best combination of cell viability, multipotency, and preservation of structural integrity of neurospheres.

3.5 Differentiation

After isolation of proliferating cells, it is necessary to confirm the stemness characteristics of the cells, that is, the multipotent and self-renewal capacities. In this sense, cells with lower self-renewal or with potential to generate just one type of cell should be considered as progenitor cells. To evaluate the differentiation capacity, cells are exposed to differentiation signals coming from animal serum or chemically defined compounds.

The use of serum has the same problems highlighted above. Nevertheless, this is still the standard methodology, because the specific signals inducing NSPc differentiation into a specific lineage remains largely unknown. Cells maintained in defined medium tend to differentiate when exposed to serum in a variable concentration (from 1% up to 10%)(e.g. Ciccolini & Svendsen, 1998; Palmer et al., 1999; Roy et al., 2000; Wanget al., 2000), although a preference towards astroglial differentiation has been reported (Palmer et al., 1995). The use of serum is usually accompanied by the addition of molecules such as Poly-L-ornithine, laminin or matrigel to promote adhesion to substrate, which seem to enhance differentiation of neurospheres cultures (Ciccolini & Svendsen, 1998; Reynolds & Weiss, 1996; Tropepe et al., 1999). In some cases, the removal of growth factors in conjunction with an adherent substrate has been also used to differentiate NSPc (Gritti et al., 1996).

Alternatively, media previously exposed to other cell cultures (conditioned medium) may be used to induce differentiation. Probably the most employed one is B104 conditioned medium, which is exposed to a neuroblastoma cell line and induces oligodendroglial differentiation (Young & Levinson, 1997).

Few authors have conducted NSPc differentiation assays by using growth factor cocktails in the absence of serum. Uchida et al., 2000 reported that a combination of BDNF and glial-derived growth factor (GDNF) was enough to differentiate CD133+ cells from human fetal tissue. Ling et al., 1998 reported a more specific differentiation protocol, proving that the combination of Interlekine-1b, Intelkeulin-11 and GDNF promoted the appearance of dopaminergic neurons (tyrosine hydroxylase -positive cells).

Furthermore, a number of chemical signals have been also reported to stimulate the differentiation toward a particular neural lineage.

In the case of neuronal maturation, BDNF, retinoic acid, Neurotrophin (NT3), and Sonic Hedgehog (SHH) have been associated to an enhanced neural obtaining (Babu et al., 2007; Bull & Bartlett, 2005; Dutton et al., 1999; Roy et al., 2000a, 2000b).

Oligodendroglial differentiation can be also enhanced using PDGFa, which promotes their survival (Gogate et al., 1994) in collaboration with NT3 and Triiodothyronine (T3), factors necessary for the correct development of oligodendrocytes and the expression of myelin proteins (Billon et al., 2002; Park et al., 2001).

4. Isolation from neurogenic zones

Neural stem cells seem to reside within specific niches of the adult brain. These regions are located in the subventricular zone of the lateral ventricles and the subgranular zone in the hippocampus. The origin of NSPc in these two areas has been the focus of intense debates in the literature and the isolation procedures of such cells from these specific locations need special attention.

Since the discovery of adult neural stem cells, the isolation procedures have been modified along with the increased knowledge of NSPc biology. Initially, these cells were supposed to be scattered within the brain parenchyma. However, soon after it was restricted to the SVZ, although the individual cell identity is still a source of division among researchers due to the lack of a specific marker to label neural stem cells. The nature and origin of the neural stem cell in the SGZ of the hippocampus has been also a subject of an intense debate, questioning whether they could be considered true neural stem cells or committed progenitors. Additionally, other types of neural progenitors like oligodendrocyte progenitor cells (OPCs) seem to be dispersed through the white matter, and their isolation procedures and characterization have become recently relevant in the context of demyelinating diseases.

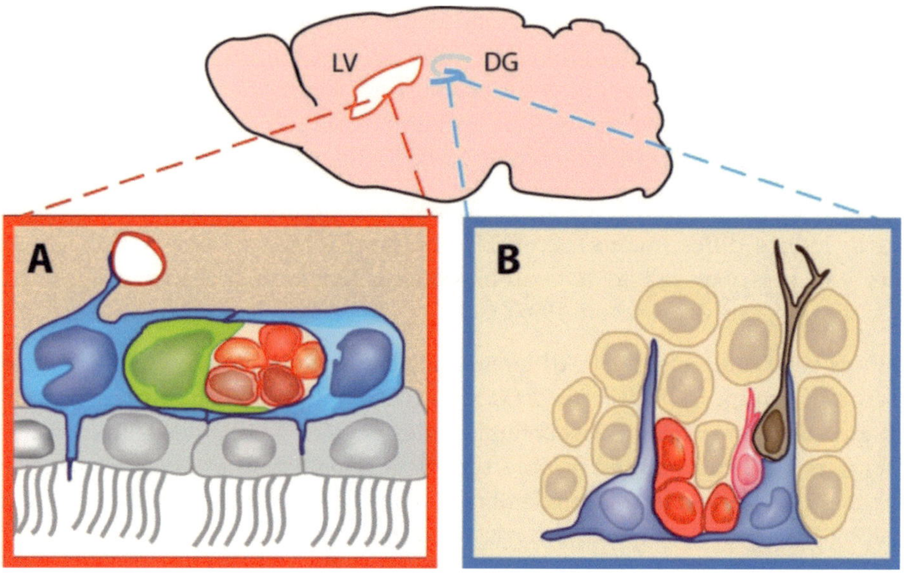

Fig. 2. Schematic representation of both adult neurogenic niches and their location in a sagittal section of rodent brain. A) The subventricular zone (SVZ) of the lateral ventricles (LV) contains an astrocyte-like stem cell population termed as type B1 cells (blue) that, unlike ependymal cells (grey), contact with lumen occasionally, showing a single cilium. Through asymmetrical divisions, Type B1 cells generate rapidly dividing, transit-amplifying cells, termed as Type C (green) which, in turn, give rise to immature neuroblasts or Type A cells (red). Those neuroblasts migrate through an astroglial scaffold toward the olfactory bulb, where they fully differentiate into granular neurons. B) The subgranular zone (SGZ) of the dentate gyrus of hippocampus also harbors an astroglial stem cell subpopulation (green). These cells generate directly immature neuroblasts , which can divide once (Type D1, red), migrate while undergoing differentiation (Type D2, type D3,pink) and integrate as granular neurons (brown).

4.1 Subventricular zone

The first population of adult neural stem cells in mammals is located in a specific niche along the SVZ of the lateral ventricles. Several types of cells can be distinguished within these niches including type B cells (slowly dividing astroglial cells and the *bona fide* adult neural stem cells), type C cells (transit-amplifying cells derived from asymmetric division of Type B cells), type A cells (immature neuroblasts derived from Type C cells) and ependymal cells lining the lumen of the ventricle. Although this is a general model found in rodents, some differences have been described in primates and humans (for a complete revision see Quiñones-Hinojosa et al., 2006). However, the accumulating knowledge on the neural stem cells biology, and their interaction with other elements of the niches, has transformed the description of the neurogenic regions into a dynamic process were the acquisition of new findings changes the model over the time. The evolution of this model has also derived in the inclusion of changes in the NSC isolation and maintenance protocols.

Once the presence of neural stem cells was demonstrated within the adult brain of mammals, the first question to address was the location of such cells and the neurogenic region. Early studies reported the isolation of NSC from the striatal tissue of both rat (Gritti et al., 1995; Reynolds & Weiss, 1992;) and mouse (Richards et al., 1992). Nevertheless, the striatum is a relatively large region that is not consistent with data about NSC biology. Afterwards the investigations in this field confirmed that only SVZ tissue was able to generate cell cultures with stem cell properties, whether via explants (Kirschenbaum et al., 1994; Kirschenbaum & Goldman, 1995; Lois & Alvarez-Buylla, 1993) or neurospheres cultures (Morshead et al., 1994) were used. Subsequent articles which described the isolation of neural stem cells from striatum specified the inclusion of the SVZ region within the dissected tissue (Weiss et al., 1996) and confirmed that the results of neural stem cell isolation exclusively from SVZ were identical to those obtained from extensive anatomical regions containing the SVZ (Kukekov et al., 1997). These initial works emphasized the relevance of a fine orientated dissection for a successful NSC isolation protocol.

Nonetheless, the SVZ comprises a heterogeneous population, and those early reports did not reveal the cellular identity of the NSC. Probably, the first work addressing the NSC identity *in vitro,* was performed by Johansson et al., 1999. Marking ventricular cells with DiI (a lipophilic membrane stain that diffuses laterally to stain the entire cell), they concluded that NSC were actually ciliated ependymal cells. By contrast, Doetsch et al. (1999) reported that only GFAP expressing cells (marked via adenovirus which allowed the expression of the green fluorescent protein (GFP) under the control of glial fibrillary acidic protein (GFAP) promoter) give rise to neurospheres culture. Moreover, it was previously described that those astrocyte-like cells occasionally contacted the ventricle and displayed a single cilium, suggesting that DiI isolated cultures could be originated from them instead of ependymal cells (a more complete model can be consulted at Mirzadeh et al. (2008)). Following previous findings, two new studies described that both types of cells were able to proliferate *in vitro,* but only SVZ astrocytes generated neurospheres with self-renewal and mutipotential capacity (Chiasson et al., 1999; Laywell et al., 2000). These findings also marked the need for an identification and selection step in the NSPc isolation protocol. With the introduction of FACS technology, two phenotypical features of NSC supported the astrocyte-like theory, the isolation of a CD15+ population with stem cell characteristics (Capela & Temple, 2002), a carbohydrate only expressed in astrocytes, and the identification of NSC as $CD24^{low}$ (Rietze et al, 2001), whereas ependymal cells are $CD24^{+}$. Using an opposite strategy to deplete adult $GFAP^{+}$ cells, two independent studies demonstrated that the ablation of adult astrocytes resulted in the loss of multipotent neurosphere formation (Morshead et al, 2003; Imura et al., 2003). However, with the controversy surrounding the identity of NSCs, two later studies reported the isolation of CD133+ cells from adult brain, as previously reported by Uchida et al. (2000) for fetal tissue. Nevertheless, opposite results were obtained from each one. The first one (Corti et al., 2007) concluded that CD133 stained a small number of cells underlying the ependymal layer, and the sorting of those CD133+ cells leads to the isolation of a NSC population. Interestingly, Mirzadeh et al. (2008) found that 29% of the apical processes of B1 cells were positive for CD133. By contrast, Coskun et al. (2008) found that both CD133+ ependymal cells and NSC originate from ependymal cells. In any case, it will be necessary to find new markers in order to improve the identification and selection of either the real adult neural stem cells and the different range of progenitor cells. This will allow the study of their specific biological features and maybe modulate their behavior *in vivo.*

4.2 Subgranular zone

In the adult hippocampus, NSCs are located in the SGZ, a cellular layer found between the granule cell layer and the hilus, in the dentate gyrus. Similarly to SVZ, NSCs have been identified as astrocyte-like cells, with cell bodies located in the SGZ and vertical processes extended through the molecular layer. However, unlike what happens in the SVZ, these astrocytic cells generate an earlier immature neuroblast that divides only once and expresses neuronal markers (PSA-NCAM and doublecourtin) (Type D1 cells). Those cells migrate short distances within the granule cell layer while undergoing morphological changes (type D2 and Type D3 cells) until they reach a final position and differentiate into mature granular neurons.

The first data from adult hippocampal cell cultures were published by Palmer et al., 1995. They isolated a monolayer culture from adult female rat hippocampus. These cells were capable of proliferating in serum free media supplemented with FGF. In Gage et al., 1995, they were also able to derivate a FGF-2 -dependent adherent culture that differentiates into mature neurons when engrafted into adult rat brain. Short after, *in vitro* differentiation was attempted by Palmer et al., 1997. Using several combinations of growth factors, they demonstrated the multipotency of those progenitor cells, and the enhancement of neuronal maturation when BDNF was added (similarly to NT3 and retinoic acid results), whereas serum addition promoted astroglial differentiation.

As described previously for initial reports on SVZ, those studies also isolated cells with similar features from non-neurogenic zones, including the septum of striatum. However a few years later was published the first work that isolated specific hippocampal neural stem cells from adult human samples (Kukekov et al., 1999). They observed the presence of neurospheres when cultured in non-adherent conditions by using mercaptoethanol, similarly to previous studies carried out with SVZ samples (Kukekov et al., 1997). Following the enrichment step based on the expression of GFP under the control of an specific promoter (P/Tα1:hGFP and E/nestin:EGFP), described previously to identify NSC from the SVZ, Roy et al., 2000b, isolated, for the first time, neural stem cells from human hippocampal samples. However, these results did not reproduce in mice and later works criticized the gross microdissection done in these publications. Afterwards, Seaberg & van der Kooy, 2002 tried to generate neurospheres from microdissected mouse dentate gyrus. They were unable to generate neurospheres capable of self-renewal, and multipotency was also compromised. Nevertheless, hippocampus obtained with gross dissection was able to generate neurospheres, suggesting that previous results could be explained if the dissection procedure included contaminating cells from the ventricular layer next to the hippocampus, or even white matter progenitors. They also assayed different culture conditions and dissection techniques, including dentate gyrus microdissection, and were unable to obtain proliferating neurospheres. After a number of publications showing controversial results in this regard Babu et al., 2007, using dentate gyrus microdissection from p/Nestin:EGFP transgenic mouse, were able to obtain monolayer cultures with self-renewal (up to 66 passages) and multipotency characteristics. Moreover, they observed spheres-like colonies when performing a modified neurosphere assay using semisolid medium. Additionally, after trying different media and supplements they concluded that although B27 and EGF promoted a slightly higher proliferation, N2 supplement and bFGF maintained cells differentiate better into mature neurons. Moreover, they reconfirmed that BDNF, NT3, SHH promoted neuronal differentiation, while LIF and Bone morphogenetic protein 2 (BMP-2) promoted glial differentiation

More recent works found that isolated neural progenitor/stem cells display paracrinal production of BMP, and the addition of noggin to culture media favors the formation of multipotential and self-renewal neurospheres (Bonaguidi et al., 2005; Bonaguidi et al., 2008).

The differences, in terms of culture media requirements, between NSC isolated from the SVZ and those from the SGZ might be due to their behavior *in vivo*. NSPc from lateral ventricle are prepared to migrate a larger distance to the olfactory lobes, while new neurons produced from dentate gyrus integrate nearby the stem cell niche and, therefore, are not prepared to maintain their stemness capacity in the absence of the niche signals.

5. Conclusions

Cell isolation and culture provides a powerful tool for the study of neural stem and progenitor cells. Although *in vitro* analysis has several limitations, and results cannot be directly extrapolated to the *in vivo* behavior of the isolated cells, it allows the analysis of their features and potential capacities in a controlled environment that can be modified and monitored more accurately.

Every step of the isolation procedure is likely to be optimized. Any protocol amendment should be tested and not considered trivial, as it can have a high impact on the cell population obtained. Consequently, isolation methods should be planned according to further experimental applications and not based on the routine practices of each research group, especially in the case of adaptation of protocols used previously for tissues collected from different species.

Moreover, it should be considered that the final purpose of most experiments is to improve our knowledge about stem cells and their clinical applications. For this reason, steps in the protocol which include reagents with undefined composition or with the possibility of introducing contaminants, such as serum, must be redesigned, because it is the only way to understand the chemical signals underlying the biological behavior of neural stem or progenitor cells.

6. Acknowledgements

We are especially grateful to Mario Soriano Navarro and Josefa Carrión Navarro for their assistance. Jorge Oliver De La Cruz is a recipient of a Predoctoral Fellowship from The FPU program (AP2008/02823), Ministerio de Educacion y Ciencia, Spain. This work was supported in part by grants from Alicia Koplowitz Foundation (JMGC, AAS, JOC) The Gent x Gent Foundation (AAS) and Fondo de Investigaciones Sanitarias (FIS) del Instituto de Salud Carlos III (PI10/01069)(AAS).

7. References

Aguirre, A. A., Chittajallu, R., Belachew, S., et al. (2004). NG2-expressing cells in the subventricular zone are type C-like cells and contribute to interneuron generation in the postnatal hippocampus. *J Cell Biol,* Vol. 165, No. 4 (May 2004), pp. (575-589), ISSN 0021-9525

Ahmed, S., Reynolds, B. A.&Weiss, S. (1995). BDNF enhances the differentiation but not the survival of CNS stem cell-derived neuronal precursors. *J Neurosci,* Vol. 15, No. 8 (August 1995), pp. (5765-5778), ISSN 0270-6474

Babu, H., Cheung, G., Kettenmann, H., et al. (2007). Enriched monolayer precursor cell cultures from micro-dissected adult mouse dentate gyrus yield functional granule cell-like neurons. *PLoS One,* Vol. 2, No. 4, (2007), pp. (e388), ISSN 1932-6203

Barres, B. A., Hart, I. K., Coles, H. S., et al. (1992). Cell death and control of cell survival in the oligodendrocyte lineage. *Cell,* Vol. 70, No. 1, (July 1992), pp. (31-46), ISSN 0092-8674

Belachew, S., Chittajallu, R., Aguirre, A. A., et al. (2003). Postnatal NG2 proteoglycan-expressing progenitor cells are intrinsically multipotent and generate functional neurons. *J Cell Biol,* Vol. 161, No. 1, (April 2003), pp. (169-186), ISSN 0021-9525

Ben-Hur, T., Rogister, B., Murray, K., et al. (1998). Growth and fate of PSA-NCAM+ precursors of the postnatal brain. *J Neurosci,* Vol. 18, No. 15 (August 1998), pp. (5777-5788), ISSN 0270-6474

Bez, A., Corsini, E., Curti, D., et al. (2003). Neurosphere and neurosphere-forming cells: morphological and ultrastructural characterization. *Brain Res,* Vol. 993, No. 1-2, (December 2003), pp. (18-29), ISSN 0006-8993

Billon, N., Jolicoeur, C., Tokumoto, Y., et al. (2002). Normal timing of oligodendrocyte development depends on thyroid hormone receptor alpha 1 (TRalpha1). *EMBO J,* Vol. 21, No. 23, (December 2002), pp. (6452-6460), ISSN 0261-4189

Bonaguidi, M. A., McGuire, T., Hu, M., et al. (2005). LIF and BMP signaling generate separate and discrete types of GFAP-expressing cells. *Development,* Vol. 132, No. 24, (December 2005), pp. (5503-5514), ISSN 0950-1991

Bonaguidi, M. A., Peng, C. Y., McGuire, T., et al. (2008). Noggin expands neural stem cells in the adult hippocampus. *J Neurosci,* Vol. 28, No. 37, (September 2008), pp. (9194-9204), ISSN 1529-2401

Brazel, C. Y., Limke, T. L., Osborne, J. K., et al. (2005). Sox2 expression defines a heterogeneous population of neurosphere-forming cells in the adult murine brain. *Aging Cell,* Vol. 4, No. 4, (August 2005), pp. (197-207), ISSN 1474-9718

Bull, N. D.&Bartlett, P. F. (2005). The adult mouse hippocampal progenitor is neurogenic but not a stem cell. *J Neurosci,* Vol. 25, No. 47 (November 2005), pp. (10815-10821), ISSN 1529-2401

Calver, A. R., Hall, A. C., Yu, W. P., et al. (1998). Oligodendrocyte population dynamics and the role of PDGF in vivo. *Neuron,* Vol. 20, No. 5, (May 1998), pp. (869-882), ISSN 0896-6273

Capela, A.&Temple, S. (2002). LeX/ssea-1 is expressed by adult mouse CNS stem cells, identifying them as nonependymal. *Neuron,* Vol. 35, No. 5, (August 2002), pp. (865-875), ISSN 0896-6273

Carpenter, M. K., Cui, X., Hu, Z. Y., et al. (1999). In vitro expansion of a multipotent population of human neural progenitor cells. *Exp Neurol,* Vol. 158, No. 2, (August 1999), pp. (265-278), ISSN 0014-4886

Ciccolini, F.&Svendsen, C. N. (1998). Fibroblast growth factor 2 (FGF-2) promotes acquisition of epidermal growth factor (EGF) responsiveness in mouse striatal precursor cells: identification of neural precursors responding to both EGF and FGF-2. *J Neurosci,* Vol. 18, No. 19, (October 1998), pp. (7869-7880), ISSN 0270-6474

Ciccolini, F. (2001). Identification of two distinct types of multipotent neural precursors that appear sequentially during CNS development. *Mol Cell Neurosci,* Vol. 17, No. 5, (May 2001), pp. (895-907), ISSN 1044-7431

Ciccolini, F., Mandl, C., Holzl-Wenig, G., et al. (2005). Prospective isolation of late development multipotent precursors whose migration is promoted by EGFR. *Dev Biol,* Vol. 284, No. 1, (August 2005), pp. (112-125), ISSN 0012-1606

Corti, S., Locatelli, F., Papadimitriou, D., et al. (2005). Multipotentiality, homing properties, and pyramidal neurogenesis of CNS-derived LeX(ssea-1)+/CXCR4+ stem cells. *FASEB J,* Vol. 19, No. 13, (November 2005), pp. (1860-1862), ISSN 1530-6860

Corti, S., Nizzardo, M., Nardini, M., et al. (2007). Isolation and characterization of murine neural stem/progenitor cells based on Prominin-1 expression. *Exp Neurol,* Vol. 205, No. 2, (June 2007), pp. (547-562), ISSN 0014-4886

Corver, W. E., Cornelisse, C. J., Hermans, J., et al. (1995). Limited loss of nine tumor-associated surface antigenic determinants after tryptic cell dissociation. *Cytometry,* Vol. 19, No. 3, (March 1995), pp. (267-272), ISSN 0196-4763

Coskun, V., Wu, H., Blanchi, B., et al. (2008). CD133+ neural stem cells in the ependyma of mammalian postnatal forebrain. *Proc Natl Acad Sci U S A,* Vol. 105, No. 3, (January 22 2008), pp. (1026-1031), ISSN 1091-6490

Chen, K., Hughes, S. M.&Connor, B. (2007). Neural progenitor cells derived from the adult rat subventricular zone: characterization and transplantation. *Cell Transplant,* Vol. 16, No. 8, (2007), pp. (799-810), ISSN 0963-6897

Chen, Y., Balasubramaniyan, V., Peng, J., et al. (2007). Isolation and culture of rat and mouse oligodendrocyte precursor cells. *Nat Protoc,* Vol. 2, No. 5 (2007), pp. (1044-1051), ISSN 1750-2799

Chiasson, B. J., Tropepe, V., Morshead, C. M., et al. (1999). Adult mammalian forebrain ependymal and subependymal cells demonstrate proliferative potential, but only subependymal cells have neural stem cell characteristics. *J Neurosci,* Vol. 19, No. 11, (June 1999), pp. (4462-4471), ISSN 1529-2401

De Filippis, L.&Delia, D. (2011). Hypoxia in the regulation of neural stem cells. *Cell Mol Life Sci,* Vol. 68, No. 17, (September 2011), pp. (2831-2844), ISSN 1420-9071

Doetsch, F., Caille, I., Lim, D. A., et al. (1999). Subventricular zone astrocytes are neural stem cells in the adult mammalian brain. *Cell,* Vol. 97, No. 6, (June 1999), pp. (703-716), ISSN 0092-8674

Duncan, I. D., Paino, C., Archer, D. R., et al. (1992). Functional capacities of transplanted cell-sorted adult oligodendrocytes. *Dev Neurosci,* Vol. 14, No. 2 (1992), pp. (114-122), ISSN 0378-5866

Dutton, R., Yamada, T., Turnley, A., et al. (1999). Sonic hedgehog promotes neuronal differentiation of murine spinal cord precursors and collaborates with neurotrophin 3 to induce Islet-1. *J Neurosci,* Vol. 19, No. 7, (April 1 1999), pp. (2601-2608), ISSN 0270-6474

Ellis, P., Fagan, B. M., Magness, S. T., et al. (2004). SOX2, a persistent marker for multipotential neural stem cells derived from embryonic stem cells, the embryo or the adult. *Dev Neurosci,* Vol. 26, No. 2-4, (March-August 2004), pp. (148-165),ISSN 0378-5866

Ericsson, R. J. (1977). Isolation and storage of progressively motile human sperm. *Andrologia,* Vol. 9, No. 1, (January-March 1977), pp. (111-114),ISSN 0303-4569

Gage, F. H., Coates, P. W., Palmer, T. D., et al. (1995). Survival and differentiation of adult neuronal progenitor cells transplanted to the adult brain. *Proc Natl Acad Sci U S A,* Vol. 92, No. 25, (December 1995), pp. (11879-11883), ISSN 0027-8424

Garcia-Verdugo, J. M., Ferron, S., Flames, N., et al. (2002). The proliferative ventricular zone in adult vertebrates: a comparative study using reptiles, birds, and mammals. *Brain Res Bull,* Vol. 57, No. 6, (April 2002), pp. (765-775) ,ISSN 0361-9230

Gard, A. L.&Pfeiffer, S. E. (1993). Glial cell mitogens bFGF and PDGF differentially regulate development of O4+GalC- oligodendrocyte progenitors. *Dev Biol,* Vol. 159, No. 2, (October 1993), pp. (618-630), ISSN 0012-1606

Gil-Perotin, S., Alvarez-Buylla, A.&Garcia-Verdugo, J. M. (2009). *Identification and characterization of neural progenitor cells in the adult mammalian brain.* Springer, ISBN 3540887180, Berlin, Germany

Gogate, N., Verma, L., Zhou, J. M., et al. (1994). Plasticity in the adult human oligodendrocyte lineage. *J Neurosci,* Vol. 14, No. 8, (August 1994), pp. (4571-4587), ISSN 0270-6474

Gritti, A., Cova, L., Parati, E. A., et al. (1995). Basic fibroblast growth factor supports the proliferation of epidermal growth factor-generated neuronal precursor cells of the adult mouse CNS. *Neurosci Lett,* Vol. 185, No. 3, (February 1995), pp. (151-154), ISSN 0304-3940

Gritti, A., Parati, E. A., Cova, L., et al. (1996). Multipotential stem cells from the adult mouse brain proliferate and self-renew in response to basic fibroblast growth factor. *J Neurosci,* Vol. 16, No. 3, (February 1996), pp. (1091-1100), ISSN 0270-6474

Gritti, A., Frolichsthal-Schoeller, P., Galli, R., et al. (1999). Epidermal and fibroblast growth factors behave as mitogenic regulators for a single multipotent stem cell-like population from the subventricular region of the adult mouse forebrain. *J Neurosci,* Vol. 19, No. 9, (May 1999), pp. (3287-3297), ISSN 0270-6474

Gritti, A., Dal Molin, M., Foroni, C., et al. (2009). Effects of developmental age, brain region, and time in culture on long-term proliferation and multipotency of neural stem cell populations. *J Comp Neurol,* Vol. 517, No. 3 (November 2009), pp. (333-349), ISSN 1096-9861

Hamanoue, M., Sato, K.&Takamatsu, K. (2008). Lectin panning method: the prospective isolation of mouse neural progenitor cells by the attachment of cell surface N-glycans to Phaseolus vulgaris erythroagglutinating lectin-coated dishes. *Neuroscience,* Vol. 157, No. 4, (December 2008), pp. (762-771), ISSN 0306-4522

Imura, T., Kornblum, H. I.&Sofroniew, M. V. (2003). The predominant neural stem cell isolated from postnatal and adult forebrain but not early embryonic forebrain expresses GFAP. *J Neurosci,* Vol. 23, No. 7, (April 2003), pp. (2824-2832), ISSN 1529-2401

Johansson, C. B., Momma, S., Clarke, D. L., et al. (1999). Identification of a neural stem cell in the adult mammalian central nervous system. *Cell,* Vol. 96, No. 1, (January 1999), pp. (25-34), ISSN 0092-8674

Kawaguchi, A., Miyata, T., Sawamoto, K., et al. (2001). Nestin-EGFP transgenic mice: visualization of the self-renewal and multipotency of CNS stem cells. *Mol Cell Neurosci,* Vol. 17, No. 2, (February 2001), pp. (259-273), ISSN 1044-7431

Keyoung, H. M., Roy, N. S., Benraiss, A., et al. (2001). High-yield selection and extraction of two promoter-defined phenotypes of neural stem cells from the fetal human brain. *Nat Biotechnol,* Vol. 19, No. 9, (September 2001), pp. (843-850), ISSN 1087-0156

Kirschenbaum, B., Nedergaard, M., Preuss, A., et al. (1994). In vitro neuronal production and differentiation by precursor cells derived from the adult human forebrain. *Cereb Cortex,* Vol. 4, No. 6, (November-December 1994), pp. (576-589),ISSN 1047-3211

Kirschenbaum, B.&Goldman, S. A. (1995). Brain-derived neurotrophic factor promotes the survival of neurons arising from the adult rat forebrain subependymal zone. *Proc Natl Acad Sci U S A,* Vol. 92, No. 1 (January 1995), pp. (210-214), ISSN 0027-8424

Kitada, M., Kuroda, Y.&Dezawa, M. (2011). Lectins as a tool for detecting neural stem/progenitor cells in the adult mouse brain. *Anat Rec (Hoboken),* Vol. 294, No. 2, (February 2011), pp. (305-321), ISSN 1932-8494

Kukekov, V. G., Laywell, E. D., Thomas, L. B., et al. (1997). A nestin-negative precursor cell from the adult mouse brain gives rise to neurons and glia. *Glia,* Vol. 21, No. 4, (December 1997), pp. (399-407), ISSN 0894-1491

Kukekov, V. G., Laywell, E. D., Suslov, O., et al. (1999). Multipotent stem/progenitor cells with similar properties arise from two neurogenic regions of adult human brain. *Exp Neurol,* Vol. 156, No. 2, (April 1999), pp. (333-344), ISSN 0014-4886

Laywell, E. D., Kukekov, V. G.&Steindler, D. A. (1999). Multipotent neurospheres can be derived from forebrain subependymal zone and spinal cord of adult mice after protracted postmortem intervals. *Exp Neurol,* Vol. 156, No. 2, (April 1999), pp. (430-433), ISSN 0014-4886

Laywell, E. D., Rakic, P., Kukekov, V. G., et al. (2000). Identification of a multipotent astrocytic stem cell in the immature and adult mouse brain. *Proc Natl Acad Sci U S A,* Vol. 97, No. 25, (December 2000), pp. (13883-13888), ISSN 0027-8424

Leonard, B. W., Mastroeni, D., Grover, A., et al. (2009). Subventricular zone neural progenitors from rapid brain autopsies of elderly subjects with and without neurodegenerative disease. *J Comp Neurol,* Vol. 515, No. 3, (July2009), pp. (269-294), ISSN 1096-9861

Lim, D. A.&Alvarez-Buylla, A. (1999). Interaction between astrocytes and adult subventricular zone precursors stimulates neurogenesis. *Proc Natl Acad Sci U S A,* Vol. 96, No. 13, (June 1999), pp. (7526-7531), ISSN 0027-8424

Ling, Z. D., Potter, E. D., Lipton, J. W., et al. (1998). Differentiation of mesencephalic progenitor cells into dopaminergic neurons by cytokines. *Exp Neurol,* Vol. 149, No. 2 (February 1998), pp. (411-423), ISSN 0014-4886

Liu, Y., Han, S. S., Wu, Y., et al. (2004). CD44 expression identifies astrocyte-restricted precursor cells. *Dev Biol,* Vol. 276, No. 1, (December 2004), pp. (31-46), ISSN 0012-1606

Lobo, M. V., Alonso, F. J., Redondo, C., et al. (2003). Cellular characterization of epidermal growth factor-expanded free-floating neurospheres. *J Histochem Cytochem,* Vol. 51, No. 1, (January 2003), pp. (89-103), ISSN 0022-1554

Lois, C.&Alvarez-Buylla, A. (1993). Proliferating subventricular zone cells in the adult mammalian forebrain can differentiate into neurons and glia. *Proc Natl Acad Sci U S A,* Vol. 90, No. 5 (March1993), pp. (2074-2077), ISSN 0027-8424

Louis, S. A.&Reynolds, B. A. (2005). Generation and differentiation of neurospheres from murine embryonic day 14 central nervous system tissue. *Methods Mol Biol,* Vol. 290, (2005) , pp. (265-280), ISSN 1064-3745

Louis, S. A., Rietze, R. L., Deleyrolle, L., et al. (2008). Enumeration of neural stem and progenitor cells in the neural colony-forming cell assay. *Stem Cells,* Vol. 26, No. 4, (April 2008), pp. (988-996), ISSN 1549-4918

Maric, D., Maric, I.&Barker, J. L. (1998). Buoyant density gradient fractionation and flow cytometric analysis of embryonic rat cortical neurons and progenitor cells. *Methods,* Vol. 16, No. 3, (November 1998), pp. (247-259), ISSN 1046-2023

Maric, D., Maric, I., Chang, Y. H., et al. (2003). Prospective cell sorting of embryonic rat neural stem cells and neuronal and glial progenitors reveals selective effects of basic fibroblast growth factor and epidermal growth factor on self-renewal and differentiation. *J Neurosci,* Vol. 23, No. 1, (January 2003), pp. (240-251), ISSN 1529-2401

Martens, D. J., Tropepe, V.&van Der Kooy, D. (2000). Separate proliferation kinetics of fibroblast growth factor-responsive and epidermal growth factor-responsive neural stem cells within the embryonic forebrain germinal zone. *J Neurosci,* Vol. 20, No. 3, (February 2000), pp. (1085-1095),ISSN 1529-2401

Mayer-Proschel, M. (2001). Isolation and generation of oligodendrocytes by immunopanning. *Curr Protoc Neurosci,* Vol. Chapter 3, No. (May 2001), pp. (Unit 3 13),ISSN 1934-8576

McCarthy, K. D.&de Vellis, J. (1980). Preparation of separate astroglial and oligodendroglial cell cultures from rat cerebral tissue. *J Cell Biol,* Vol. 85, No. 3, (June 1980), pp. (890-902), ISSN 0021-9525

McKinnon, R. D., Matsui, T., Dubois-Dalcq, M., et al. (1990). FGF modulates the PDGF-driven pathway of oligodendrocyte development. *Neuron,* Vol. 5, No. 5, (November 1990), pp. (603-614), ISSN 0896-6273

Mirzadeh, Z., Merkle, F. T., Soriano-Navarro, M., et al. (2008). Neural stem cells confer unique pinwheel architecture to the ventricular surface in neurogenic regions of the adult brain. *Cell Stem Cell,* Vol. 3, No. 3, (September 2008), pp. (265-278),ISSN 1875-9777

Morshead, C. M., Reynolds, B. A., Craig, C. G., et al. (1994). Neural stem cells in the adult mammalian forebrain: a relatively quiescent subpopulation of subependymal cells. *Neuron,* Vol. 13, No. 5, (November 1994), pp. (1071-1082), ISSN 0896-6273

Morshead, C. M., Garcia, A. D., Sofroniew, M. V., et al. (2003). The ablation of glial fibrillary acidic protein-positive cells from the adult central nervous system results in the loss of forebrain neural stem cells but not retinal stem cells. *Eur J Neurosci,* Vol. 18, No. 1, (July 2003), pp. (76-84), ISSN 0953-816X

Murayama, A., Matsuzaki, Y., Kawaguchi, A., et al. (2002). Flow cytometric analysis of neural stem cells in the developing and adult mouse brain. *J Neurosci Res,* Vol. 69, No. 6, (September 2002), pp. (837-847), ISSN 0360-4012

Nakano, I., Paucar, A. A., Bajpai, R., et al. (2005). Maternal embryonic leucine zipper kinase (MELK) regulates multipotent neural progenitor proliferation. *J Cell Biol,* Vol. 170, No. 3, (August 2005), pp. (413-427), ISSN 0021-9525

Nunes, M. C., Roy, N. S., Keyoung, H. M., et al. (2003). Identification and isolation of multipotential neural progenitor cells from the subcortical white matter of the adult human brain. *Nat Med,* Vol. 9, No. 4, (April 2003), pp. (439-447), ISSN 1078-8956

Palmer, T. D., Ray, J.&Gage, F. H. (1995). FGF-2-responsive neuronal progenitors reside in proliferative and quiescent regions of the adult rodent brain. *Mol Cell Neurosci,* Vol. 6, No. 5, (October 1995), pp. (474-486), ISSN 1044-7431

Palmer, T. D., Takahashi, J.&Gage, F. H. (1997). The adult rat hippocampus contains primordial neural stem cells. *Mol Cell Neurosci,* Vol. 8, No. 6, (1997), pp. (389-404), ISSN 1044-7431

Palmer, T. D., Markakis, E. A., Willhoite, A. R., et al. (1999). Fibroblast growth factor-2 activates a latent neurogenic program in neural stem cells from diverse regions of the adult CNS. *J Neurosci,* Vol. 19, No. 19, (October 1999), pp. (8487-8497), ISSN 1529-2401

Panchision, D. M., Chen, H. L., Pistollato, F., et al. (2007). Optimized flow cytometric analysis of central nervous system tissue reveals novel functional relationships among cells expressing CD133, CD15, and CD24. *Stem Cells,* Vol. 25, No. 6 (June 2007), pp. (1560-1570), ISSN 1066-5099

Park, S. K., Solomon, D.&Vartanian, T. (2001). Growth factor control of CNS myelination. *Dev Neurosci,* Vol. 23, No. 4-5, (2001), pp. (327-337), ISSN 0378-5866

Pastrana, E., Cheng, L. C.&Doetsch, F. (2009). Simultaneous prospective purification of adult subventricular zone neural stem cells and their progeny. *Proc Natl Acad Sci U S A,* Vol. 106, No. 15, (April 2009), pp. (6387-6392), ISSN 1091-6490

Piper, D. R., Mujtaba, T., Keyoung, H., et al. (2001). Identification and characterization of neuronal precursors and their progeny from human fetal tissue. *J Neurosci Res,* Vol. 66, No. 3, (November 2001), pp. (356-368), ISSN 0360-4012

Quinones-Hinojosa, A., Sanai, N., Soriano-Navarro, M., et al. (2006). Cellular composition and cytoarchitecture of the adult human subventricular zone: a niche of neural stem cells. *J Comp Neurol,* Vol. 494, No. 3, (January 2006), pp. (415-434), ISSN 0021-9967

Reynolds, B. A., Tetzlaff, W.&Weiss, S. (1992). A multipotent EGF-responsive striatal embryonic progenitor cell produces neurons and astrocytes. *J Neurosci,* Vol. 12, No. 11, (November 1992), pp. (4565-4574), ISSN 0270-6474

Reynolds, B. A.&Weiss, S. (1992). Generation of neurons and astrocytes from isolated cells of the adult mammalian central nervous system. *Science,* Vol. 255, No. 5052, (March 1992), pp. (1707-1710), ISSN 0036-8075

Reynolds, B. A.&Weiss, S. (1996). Clonal and population analyses demonstrate that an EGF-responsive mammalian embryonic CNS precursor is a stem cell. *Dev Biol,* Vol. 175, No. 1, (April 1996), pp. (1-13),ISSN 0012-1606

Richards, L. J., Kilpatrick, T. J.&Bartlett, P. F. (1992). De novo generation of neuronal cells from the adult mouse brain. *Proc Natl Acad Sci U S A,* Vol. 89, No. 18, (September 1992), pp. (8591-8595), ISSN 0027-8424

Rietze, R. L., Valcanis, H., Brooker, G. F., et al. (2001). Purification of a pluripotent neural stem cell from the adult mouse brain. *Nature,* Vol. 412, No. 6848, (August 2001), pp. (736-739),ISSN 0028-0836

Rietze, R. L.&Reynolds, B. A. (2006). Neural stem cell isolation and characterization. *Methods Enzymol,* Vol. 419, (2006), pp. (3-23), ISSN 0076-6879

Rivers, L. E., Young, K. M., Rizzi, M., et al. (2008). PDGFRA/NG2 glia generate myelinating oligodendrocytes and piriform projection neurons in adult mice. *Nat Neurosci,* Vol. 11, No. 12, (December 2008), pp. (1392-1401), ISSN 1546-1726

Roy, N. S., Wang, S., Harrison-Restelli, C., et al. (1999). Identification, isolation, and promoter-defined separation of mitotic oligodendrocyte progenitor cells from the adult human subcortical white matter. *J Neurosci,* Vol. 19, No. 22, (November1999), pp. (9986-9995), ISSN 1529-2401

Roy, N. S., Benraiss, A., Wang, S., et al. (2000a). Promoter-targeted selection and isolation of neural progenitor cells from the adult human ventricular zone. *J Neurosci Res,* Vol. 59, No. 3, (February 2000), pp. (321-331), ISSN 0360-4012

Roy, N. S., Wang, S., Jiang, L., et al. (2000b). In vitro neurogenesis by progenitor cells isolated from the adult human hippocampus. *Nat Med,* Vol. 6, No. 3, (March 2000), pp. (271-277), ISSN 1078-8956

Rutka, J. T., Apodaca, G., Stern, R., et al. (1988). The extracellular matrix of the central and peripheral nervous systems: structure and function. *J Neurosurg,* Vol. 69, No. 2 (August 1988), pp. (155-170), ISSN 0022-3085

Sawamoto, K., Yamamoto, A., Kawaguchi, A., et al. (2001). Direct isolation of committed neuronal progenitor cells from transgenic mice coexpressing spectrally distinct fluorescent proteins regulated by stage-specific neural promoters. *J Neurosci Res,* Vol. 65, No. 3, (August 2001), pp. (220-227), ISSN 0360-4012

Schmandt, T., Meents, E., Gossrau, G., et al. (2005). High-purity lineage selection of embryonic stem cell-derived neurons. *Stem Cells Dev,* Vol. 14, No. 1 (February 2005), pp. (55-64), ISSN 1547-3287

Schwartz, P. H., Bryant, P. J., Fuja, T. J., et al. (2003). Isolation and characterization of neural progenitor cells from post-mortem human cortex. *J Neurosci Res,* Vol. 74, No. 6, (December 2003), pp. (838-851), ISSN 0360-4012

Seaberg, R. M.&van der Kooy, D. (2002). Adult rodent neurogenic regions: the ventricular subependyma contains neural stem cells, but the dentate gyrus contains restricted progenitors. *J Neurosci,* Vol. 22, No. 5, (March 2002), pp. (1784-1793), ISSN 1529-2401

Shi, J., Marinovich, A.&Barres, B. A. (1998). Purification and characterization of adult oligodendrocyte precursor cells from the rat optic nerve. *J Neurosci,* Vol. 18, No. 12, (June 15 1998), pp. (4627-4636), ISSN 0270-6474

Suh, H., Consiglio, A., Ray, J., et al. (2007). In vivo fate analysis reveals the multipotent and self-renewal capacities of Sox2+ neural stem cells in the adult hippocampus. *Cell Stem Cell,* Vol. 1, No. 5, (November 2007), pp. (515-528), ISSN 1934-5909

Sun, J., Zhou, W., Ma, D., et al. (2010). Endothelial cells promote neural stem cell proliferation and differentiation associated with VEGF activated Notch and Pten signaling. *Dev Dyn,* Vol. 239, No. 9, (September 2010), pp. (2345-2353), ISSN 1097-0177

Svendsen, C. N., Fawcett, J. W., Bentlage, C., et al. (1995). Increased survival of rat EGF-generated CNS precursor cells using B27 supplemented medium. *Exp Brain Res,* Vol. 102, No. 3, (1995), pp. (407-414), ISSN 0014-4819

Svendsen, C. N., Caldwell, M. A., Shen, J., et al. (1997a). Long-term survival of human central nervous system progenitor cells transplanted into a rat model of Parkinson's disease. *Exp Neurol,* Vol. 148, No. 1, (November 1997), pp. (135-146),I SSN 0014-4886

Svendsen, C. N., Skepper, J., Rosser, A. E., et al. (1997b). Restricted growth potential of rat neural precursors as compared to mouse. *Brain Res Dev Brain Res*, Vol. 99, No. 2, (April 1997), pp. (253-258), ISSN 0165-3806

Svendsen, C. N., ter Borg, M. G., Armstrong, R. J., et al. (1998). A new method for the rapid and long term growth of human neural precursor cells. *J Neurosci Methods*, Vol. 85, No. 2, (December 1998), pp. (141-152), ISSN 0165-0270

Tan, F. C., Lee, K. H., Gouk, S. S., et al. (2007). Optimization of cryopreservation of stem cells cultured as neurospheres: comparison between vitrification, slow-cooling and rapid cooling freezing protocols. *Cryo Letters*, Vol. 28, No. 6, (November-December 2007), pp. (445-460), ISSN 0143-2044

Thompson, L. H., Andersson, E., Jensen, J. B., et al. (2006). Neurogenin2 identifies a transplantable dopamine neuron precursor in the developing ventral mesencephalon. *Exp Neurol*, Vol. 198, No. 1, (March 2006), pp. (183-198), ISSN 0014-4886

Tropepe, V., Sibilia, M., Ciruna, B. G., et al. (1999). Distinct neural stem cells proliferate in response to EGF and FGF in the developing mouse telencephalon. *Dev Biol*, Vol. 208, No. 1, (April 1999), pp. (166-188), ISSN 0012-1606

Uchida, N., Buck, D. W., He, D., et al. (2000). Direct isolation of human central nervous system stem cells. *Proc Natl Acad Sci U S A*, Vol. 97, No. 26, (December2000), pp. (14720-14725), ISSN 0027-8424

Vescovi, A. L., Reynolds, B. A., Fraser, D. D., et al. (1993). bFGF regulates the proliferative fate of unipotent (neuronal) and bipotent (neuronal/astroglial) EGF-generated CNS progenitor cells. *Neuron*, Vol. 11, No. 5, (November 1993), pp. (951-966), ISSN 0896-6273

Von Visger, J. R., Yeon, D. S., Oh, T. H., et al. (1994). Differentiation and maturation of astrocytes derived from neuroepithelial progenitor cells in culture. *Exp Neurol*, Vol. 128, No. 1, (July 1994), pp. (34-40), ISSN 0014-4886

Walton, N. M., Sutter, B. M., Chen, H. X., et al. (2006). Derivation and large-scale expansion of multipotent astroglial neural progenitors from adult human brain. *Development*, Vol. 133, No. 18, (September 2006), pp. (3671-3681), ISSN 0950-1991

Wang, S., Roy, N. S., Benraiss, A., et al. (2000). Promoter-based isolation and fluorescence-activated sorting of mitotic neuronal progenitor cells from the adult mammalian ependymal/subependymal zone. *Dev Neurosci*, Vol. 22, No. 1-2, (2000), pp. (167-176), ISSN 0378-5866

Wang, S., Chandler-Militello, D., Lu, G., et al. (2010). Prospective identification, isolation, and profiling of a telomerase-expressing subpopulation of human neural stem cells, using sox2 enhancer-directed fluorescence-activated cell sorting. *J Neurosci*, Vol. 30, No. 44, (November 2010), pp. (14635-14648), ISSN 1529-2401

Weiss, S., Dunne, C., Hewson, J., et al. (1996). Multipotent CNS stem cells are present in the adult mammalian spinal cord and ventricular neuroaxis. *J Neurosci*, Vol. 16, No. 23, (December 1 1996), pp. (7599-7609), ISSN 0270-6474

Wilson, H. C., Onischke, C.&Raine, C. S. (2003). Human oligodendrocyte precursor cells in vitro: phenotypic analysis and differential response to growth factors. *Glia*, Vol. 44, No. 2, (November 2003), pp. (153-165), ISSN 0894-1491

Windrem, M. S., Roy, N. S., Wang, J., et al. (2002). Progenitor cells derived from the adult human subcortical white matter disperse and differentiate as oligodendrocytes within demyelinated lesions of the rat brain. *J Neurosci Res,* Vol. 69, No. 6, (September 2002), pp. (966-975), ISSN 0360-4012

Windrem, M. S., Nunes, M. C., Rashbaum, W. K., et al. (2004). Fetal and adult human oligodendrocyte progenitor cell isolates myelinate the congenitally dysmyelinated brain. *Nat Med,* Vol. 10, No. 1, (January 2004), pp. (93-97), ISSN 1078-8956

Wright, A. P., Fitzgerald, J. J.&Colello, R. J. (1997). Rapid purification of glial cells using immunomagnetic separation. *J Neurosci Methods,* Vol. 74, No. 1, (June 6 1997), pp. (37-44), ISSN 0165-0270

Wu, C., Chang, A., Smith, M. C., et al. (2009). Beta4 tubulin identifies a primitive cell source for oligodendrocytes in the mammalian brain. *J Neurosci,* Vol. 29, No. 24, (June 2009), pp. (7649-7657),ISSN 1529-2401

Xu, Y., Kimura, K., Matsumoto, N., et al. (2003). Isolation of neural stem cells from the forebrain of deceased early postnatal and adult rats with protracted post-mortem intervals. *J Neurosci Res,* Vol. 74, No. 4, (2003), pp. (533-540), ISSN 1097-4547

Yayon, A., Klagsbrun, M., Esko, J. D., et al. (1991). Cell surface, heparin-like molecules are required for binding of basic fibroblast growth factor to its high affinity receptor. *Cell,* Vol. 64, No. 4, (February 1991), pp. (841-848), ISSN 0092-8674

Yoshida, N., Hishiyama, S., Yamaguchi, M., et al. (2003). Decrease in expression of alpha 5 beta 1 integrin during neuronal differentiation of cortical progenitor cells. *Exp Cell Res,* Vol. 287, No. 2, (July 2003), pp. (262-271), ISSN 0014-4827

Young, G. M.&Levison, S. W. (1997). An improved method for propagating oligodendrocyte progenitors in vitro. *J Neurosci Methods,* Vol. 77, No. 2, (December 1997), pp. (163-168),ISSN 0165-0270

Yuan, X., Chittajallu, R., Belachew, S., et al. (2002). Expression of the green fluorescent protein in the oligodendrocyte lineage: a transgenic mouse for developmental and physiological studies. *J Neurosci Res,* Vol. 70, No. 4, (November 2002), pp. (529-545), ISSN 0360-4012

Zigler, J. S., Jr., Lepe-Zuniga, J. L., Vistica, B., et al. (1985). Analysis of the cytotoxic effects of light-exposed HEPES-containing culture medium. *In Vitro Cell Dev Biol,* Vol. 21, No. 5, (May 1985), pp. (282-287), ISSN 0883-8364

2

Neurogenesis in Adult Hippocampus

Xinhua Zhang and Guohua Jin
Nantong University
China

1. Introduction

Hippocampus as a whole has the shape of a curved tube including CA1-CA4 regions with a single layer of densely packed pyramidal neurons which curl into a tight "U" shape. One edge of the "U", field CA4, is embedded into a backward facing strongly flexed V-shaped cortex, the dentate gyrus (DG) which comprises molecular, granular, subgranular cell layers and poly-morph layer called hilus (Figure 1). The ability to learn or form a memory requires a neuron to translate a transient signal into gene expression changes that have a long-lasting effect on synapse activity and connectivity. There are many neural circuits formed by multi-class neurons in hippocampus. One of them is the trisynaptic circuit (Figure 1) that is made up of three major cell groups: granule cells, CA3 pyramidal neurons, and CA1 pyramidal cells. The axons of layer II neurons in the entorhinal cortex (EC) project to the dentate gyrus through the perforant pathway. The dentate gyrus sends projections to the pyramidal cells in CA3 through mossy fibres. CA3 pyramidal neurons relay the information to CA1 pyramidal neurons through Schaffer collaterals. CA1 pyramidal neurons send back projections into deep-layer neurons of the EC. This kind of circuit is involved in long term potentiation (LTP) mediating learning and memory. CA3 also directly receives the projections from EC layer II neurons through the perforant pathway. CA1 receives direct input from EC layer III neurons through the temporoammonic pathway. The dentate granule cells also project to the mossy cells in the hilus and hilar interneurons, which send excitatory and inhibitory projections, respectively, back to the granule cells. The complicated neural circuits in hippocampus form the foundation of hippocampal functions.

The external relation between hippocampus and other brain regions also plays an important role in cognition and attentional behaviors. Hippocampal afferents are from the septal area, the locus coeruleus, and the raphe nuclei via 3 anatomically distinct pathways, cingular bundle (CB), Fimbria Fornix (FiFx) and a ventral pathway whose exact anatomical location is not well defined but is thought to reach the hippocampus after passing in the vicinity of the amygdalar complex (Cassel et al., 1997; Eckenstein et al., 1988; Gage et al., 1994; Hong & Jang, 2010; Saper, 1984). Afferent fibers via the FiFx and CB provide the hippocampus with cholinergic, extrinsic GABAergic, noradrenergic and serotonergic inputs. A very important projection comes from the medial septal area, which sends cholinergic and GABAergic fibers to all parts of the hippocampus. The inputs from the septal area play a key role in controlling the physiological state of the hippocampus: destruction of the septal area abolishes the hippocampal theta rhythm, and severely impairs certain types of memory. Hippocampal efferents carry fibers from hippocampal pyramidal CA2-CA4 cells projecting to the anterior thalamic nucleus, medial mamillary nucleus, cingular gyrus, and the nucleus basalis of Meynert (Cassel et al., 1997). Cholinergic projections comprise a complex neural network that supports higher brain functions, and the FiFx and CB are the principal cholinergic pathways that communicate between the basal forebrain and hippocampus and cortex.

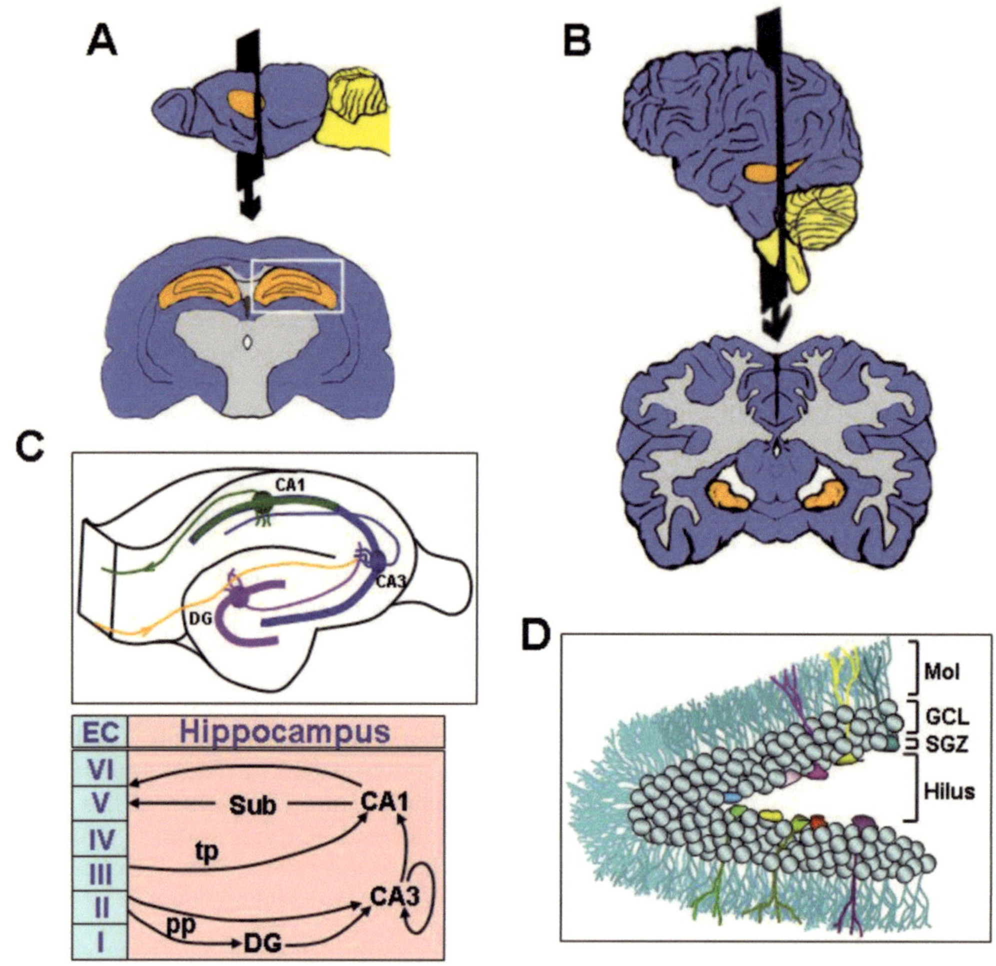

A) Hippocampus (orange region) sits below the surface of the neocortex in rodent brain. The lower is a coronal section through hippocampus. B) Hippocampus (orange region) in human brain is also located under the surface of the neocortex. The lower is a coronal typical section through hippocampus. C) Basic circuit of the hippocampus. Neurons in EC II project to the DG through the perforant pathway (pp). DG sends projections to pyramidal cells in CA3 through mossy fibres. CA3 also receives the projections from EC II neurons through the perforant pathway. CA3 pyramidal neurons send axons to CA1 pyramidal neurons. CA1 also directly receives input from EC III neurons through the temporoammonic pathway (tp). CA1 pyramidal neurons send back projections into deep layers of EC. D) The details of cell layers in rodent DG indicate the neurogenic cells migrate along SGZ and into GCL, and finally form mature granule cells projecting processes into Mol. Abbreviation: DG, dentate gyrus; EC, entorhinal cortex; GCL, granule cell layer; Mol, molecular layer; SGZ, subgranular zone; Sub, subiculum.

Fig. 1. Location and inner structure of the hippocampus.

2. Distribution and fate of neural progenitor cells in hippocampus

Findings of new neurons in the adult brain challenge the dogma that cells of the central nervous system (CNS) are incapable of regeneration. It is well established that the DG in the hippocampus is one of two adult well-accepted regions with continuous addition of new neurons throughout life (Gage, 2000; Kempermann & Gage, 2000). The adult hippocampal neurogenesis is a complex process that originates from proliferation of neural progenitor cells (NPCs) located in the subgranular zone (SGZ), a germinal layer between the granular layer and hilus. The majorities of NPC progenies are specified to become dentate granule cells (DGCs) and go through the initial differentiation and migrate into the inner granule cell layer within a week of their birth. The adult immature DGCs generated from NPCs in SGZ undergo maturation and make important contributions to learning and memory (Deng et al., 2009).

Subventricular zone (SVZ) is another adult region continuously generating new neurons. SVZ NPCs give rise to neuroblasts that migrate in chains to the olfactory bulb through the rostral migratory stream (RMS) where they differentiate into granule and periglomerular neurons (Bovetti et al., 2007; Corotto et al., 1993; Lois & Alvarez-Buylla, 1994; Lois et al., 1996). In the adult DG, new neurons from NPCs are born in the SGZ and migrate a short distance to differentiate into granule cells that project their dendrites into the molecular layer (ML) and axons to the CA3 pyramidal cell layer via the mossy fiber pathway (Markakis & Gage, 1999; Stanfield & Trice, 1988) and establish synaptic connection with local neurons (McDonald & Wojtowicz, 2005).

There are four main cell types in the SVZ: neuroblasts (Type A cells), SVZ astrocytes (Type B cells), immature precursors (Type C cells) and ependymal cells (Doetsch et al., 1997). The neuroblasts (Type A cells) which are from the focal clusters of rapidly dividing precursors (Type C cells) along the SVZ network of chains divide as they migrate as chains through glial tunnels formed by the processes of slowly dividing SVZ astrocytes (Type B cells).

As in the SVZ, there are four types of cells in dentate gyrus: SGZ astrocytes (Type B cells), immature dividing cells (type D cells), granule neurons (type G cells) and endothelial cells. SGZ astrocytes are in close proximity to blood vessels and extend basal processes under the blades of the dentate gyrus and an apical process into the granule cell layer. It is the same as SVZ that SGZ astrocytes are the primary precursors of neurons. The SGZ astrocytes divide to give rise to immature dividing D cells and generate granule neurons. So the type D cells are adjacent to SGZ astrocytes. Neurogenesis in the SGZ occurs in foci formed by these cells suggesting mutual co-regulation between them (Palmer et al., 2000). Endothelial cells are likely an important source of signals for neurogenesis.

Accumulating evidences lead to a detailed classification of the SGZ cells characterized by their properties and specific markers (Figure 2). Adult hippocampal neurons originate from a radial glia-like precursor cell (type-1) which is glial fibrous acid protein (GFAP) positive but negative to S100 beta, doublecortin (DCX) and polysialic acid-neural cell adhesion molecule (PSA-NCAM) in the SGZ of DG through a number of intermediate cell types (type-2, $GFAP^-$, $S100^-$, DCX^+, $PSA\text{-}NCAM^+$ and type 3 with DCX expression). Type-1 cells correspond to type B cells because they have a proliferative capacity and are marked by GFAP (Seri et al., 2004; Suh et al., 2007; Zhao et al., 2006). Nestin, Sox2, and brain lipid-binding protein (BLBP) are also expressed in type-1 cells suggesting their radial glial features and the expression persists into the type-2 cell stages (Steiner et al., 2006). Although type 1 cells have a proliferative capacity, their cycles are much slower than the followed type-2 progenitor cells supposed to be the type D cells (Filippov et al., 2003; Fukuda et al., 2003; Kronenberg et al., 2003; Steiner et al., 2004). Type-2 cell stage marks the transition between cells with astrocytic phenotype (type-2a cells, the early stage of type-2 cells) and cells with early features of the neuronal lineage (type-2b cells, the later stage of type-2 cells). A panel of different markers (Sox2, BLBP, DCX, and NeuroD) discriminates between the type 2a and type 2b cells. Type-2a cells feature, to some degree, properties of radial glia-like cells marked with BLBP and Sox2. NeuroD and DCX, the markers of immature neurons, appear in type-2b cells and persist into postmitotic but immature granule cell precursors with transient Calretinin-expression. That is to say, type-2b cells are committed to the neuronal lineage. The type-3 cells are the terminal postmitotic differentiation of granule cells that exits from the cell cycle (Kempermann et al., 2004; Steiner et al., 2006). Finally, these cells mature into granule cell neurons in the DG that express specifically NeuN, calbindin and Prox1 (Figure 2). These newborn granule cells elongate their dendrites and axons integrating into the DG circuitry (Jessberger & Kempermann, 2003; Song et al., 2005; van Praag et al., 2002).

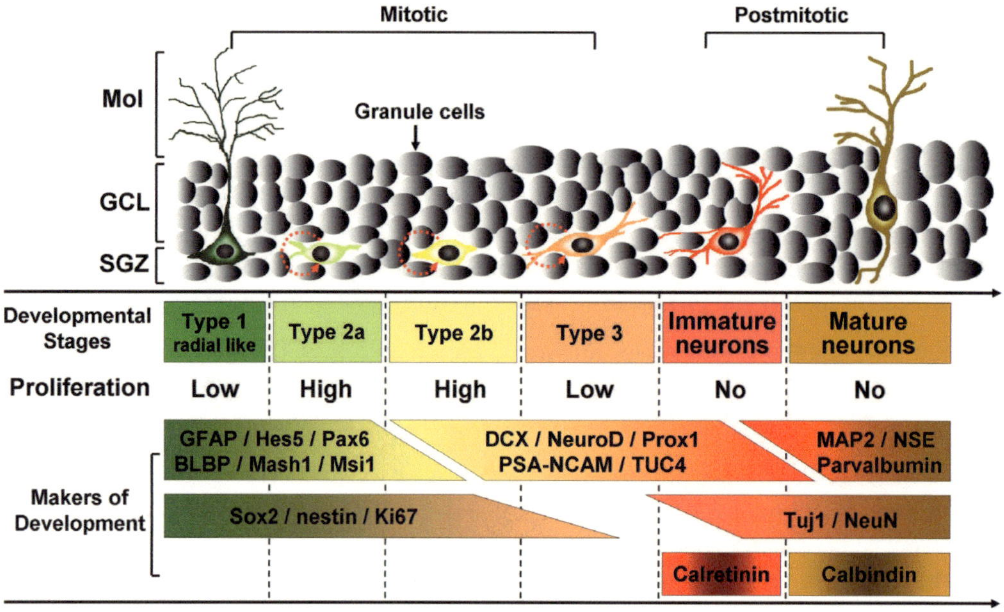

Adult hippocampal neurons originate from type-1 cell with radial glia properties through a number of intermediate type-2 and type 3 cells. Type 2 cells with transit rapid proliferation have two types 2a and 2b. The neuronal determination is at stage type 2b. Type 3 cells gradually exit from the cell cycle and then subsequently form the immature and mature neurons. These newborn granule cells elongate their dendrites and axons integrating into the molecular layer. Cells in different stages of neurogenesis express neural specific markers highlighted in this figure.

Fig. 2. Proposed course of adult hippocampal neurogenesis.

Recent studies in increasing detail showed that a sequence of markers express in the SGZ cells of various stages during the adult hippocampal neurogenesis in mice and rats (Kempermann et al., 2004; Kim et al., 2008; Steiner et al., 2006; Steiner et al., 2008). The stage-specific expressions of neural markers are summarized in Figure 2. In an addition to the putative markers described above, other genes are expressed in different stages of hippocampal neurogenesis. The neuronal marker Hu appears in the GFAP positive intermediate progenitors committed to the neuronal lineage, while Hu is undetectable in primary progenitors and astrocytes, indicating that Hu is a useful marker for discriminating GFAP+ astrocytes and GFAP+ neural progenitors that generate neurons (Liu et al., 2010). The transcription factor Pax6 is expressed not only in precursor cells during embryonic development of the central nervous system but also in the adult SGZ (Sakurai & Osumi, 2008). It plays an important role in the regulation of cell proliferation and neuronal fate determination (Englund et al., 2005; Gotz et al., 1998; Heins et al., 2002). About half of the Pax6-positive cells in the SGZ display a radial glial phenotype which is marked for GFAP, whereas about 30% of the Pax6-positive cells are immunoreactive to PSA-NCAM or DCX (Maekawa et al., 2005; Nacher et al., 2005). In addition, more than 50% of Pax6-positive cells are immunoreactive to NeuroD (Nacher et al., 2005). Thus, Pax6 may represent a suitable marker for type 1 and type 2a cells. The transcription factor NeuroD is expressed in later stages of neuronal commitment (Lee et al., 1995) and during neurogenesis in the adult DG (Kawai et al., 2004). It is important for the proper development of the DG, the proliferation and postnatal differentiation of neuronal progenitors (Liu et al., 2000; Miyata et al., 1999). Thus it could serve as a specific marker. TUC-4 is not only expressed in postmitotic neurons during brain development as they begin their migration but also re-expressed in adult neurogenesis again. Its expression pattern during neurogenesis resembles that of PSA-NCAM and DCX. Thus, TUC-4 can be used as a marker for different stages of adult neurogenesis in the DG. Calretinin is expressed in specific non-pyramidal γ-aminobutyric acid (GABA)-ergic neurons within the adult hippocampus. At late phases of neurogenesis, new neurons express calretinin and doublecortin or NeuN but do not express GABA (Brandt et al., 2003). At later time-points, the newly generated neurons stop expressing calretinin and start to express calbindin, a marker of mature dentate granule cells (Brandt et al., 2003). So that calretinin expression within the DG is restricted to a short postmitotic time window in which axonal and dendritic target their destination regions (Kempermann et al., 2004; Ming & Song, 2005). FABP7 (BLBP) is expressed in the type 1, 2a, and 2b cells, since FABP7 (BLBP) were found in bromodeoxyuridine (BrdU)-positive newly generated cells whereas Tuj1 or PSA-NCAM positive newborn neurons in the vicinity of the astrocytes express none of the FABPs. (Boneva et al., 2011). Musashi1 (Msi1) is a neural RNA binding protein (Sakakibara et al., 1996) that expressed in early-stage NPCs (Kaneko et al., 2000; Sakakibara et al., 1996). The clarity of the development stage-specific markers is not only helpful for gaining further insights into the genesis of new neurons in the hippocampus, but also might be applicable to the development of strategies for therapeutic interventions.

3. Survival and differentiation of grafted NSCs in hippocampus

In CNS the mature neurons lose the ability to undergo cell division once they fully differentiate. Therefore, cell replacement is recognized as a potential strategy to treat neurodegenerative diseases.

The past studies showed that hippocampus is vulnerable to many pathogenic factors or chemical substances. Since that, the hippocampus is preferred as pathological model to investigate the mechanisms and therapies of nervous disorders, such as ischemia, epilepsy, aging and excitotoxicity, all of which disturb the physiological balances in the circuits of hippocampus. For example, cholinergic input plays an important role in cognition and attentional behaviors, and cholinergic dysfunction is a prominent feature of dementias including Alzheimer's disease (AD).

Although the pharmacotherapy, such as acetylcholinesterase inhibitors (Gauthier, 2002), secretase inhibitors (Lanz et al., 2003), transition metal chelators (Gnjec et al., 2002) and Aβ immunization (Ferrer et al., 2004; Heppner et al., 2004), has exerted curative effects to some extent on the amelioration of hippocampal neurodegeneration syndromes, but can not completely rescue or replace the dying neurons. Neuro-transplantation has been proposed recent years as a potential treatment for neurodegenerative disorders (Bachoud-Levi et al., 2000; Gaura et al., 2004). Grafts of neural stem/progenitor cells (NSCs/NPCs) present a potential and innovative strategy for the treatment of many disorders of central nervous system, with the possibility of providing a more permanent remedy than present drug treatments.

Cholinergic projections comprise a complex neural network that supports higher brain functions. FiFx and CB are the principal cholinergic pathways that communicate signals between the basal forebrain and hippocampus and cortex. Lesions of the FiFx plus CB lead to substantially reduced cholinergic innervation (Gage et al., 1994) and produce lasting impairments of spatial learning and memory (Liu et al., 2002), all of which are among the earliest events in the pathogenesis of AD (Geula & Mesulam, 1989; Schliebs & Arendt, 2006; Szenborn, 1993). Selective depletion of cholinergic neurons in the basal forebrain elevated APP immunoreactivity in the cerebral cortex and hippocampus, and increased APP levels correlated with decreased cholinergic activity (Leanza, 1998; Lin et al., 1998). The increased expression of APP after cholinergic lesion can potentially lead to increased Aβ production, thereby possibly causing Aβ accumulation and deposition, which is one of the main pathological features.

In our study [(Zhang et al., 2007) and Figure 3] we transplanted SVZ progenitors directly into the denervated and contralateral hippocampi of the AD rat models and determined the effect of different hippocampal environment on the fate of NPCs. The donor neural progenitors in this study were derived from the neonatal SVZ for their features prior to and following transplantation that make them candidates for cell replacement therapy. The grafted cells survived well even through the longest span, 2 months after implantation, and migrate along the subgranular layer after. The same model was treated through neural stem cell transplantation by Xuan and his colleagues (Xuan et al., 2009). The results indicate that the deafferented hippocampus provided proper microenvironment for the survival and neuronal differentiation of neural progenitors and transplanted NSCs can differentiate into cholinergic neurons and enhance the learning and memory abilities. Another kind of AD model produced by injections of amyloid-β peptide (1-40) ($Aβ_{1-40}$) received neural stem cell transplantation into the hippocampus dentate gyrus. The grafted cells can survive, and differentiate with high yield into immunohistochemically mature glial cells and neurons of diverse neurotransmitter-subtypes. More importantly, transplanted cells demonstrate characteristics of proper synapse formation between host and grafted neural cells (Li et al., 2010).

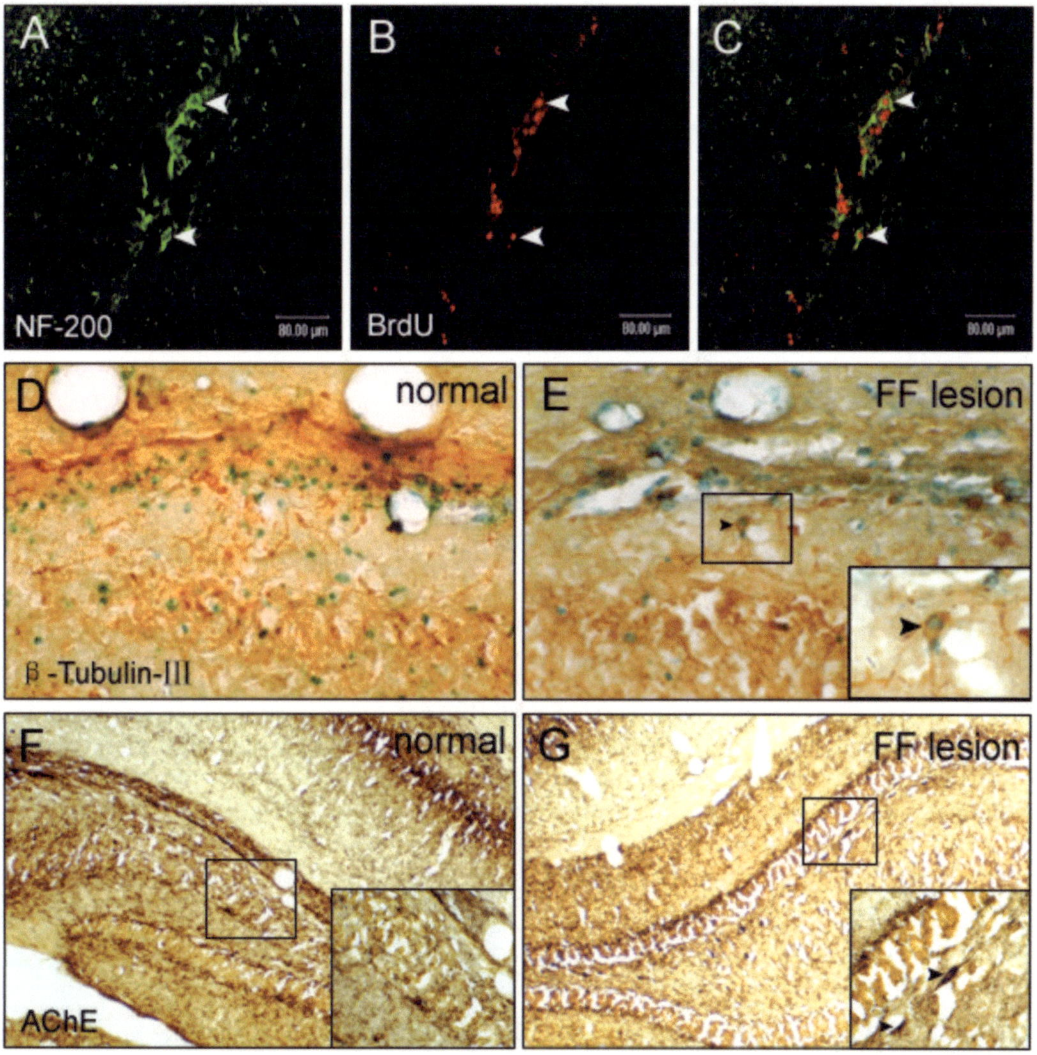

(A–C) Cofocal images of NF-200 positive (green) and BrdU positive (red) neurons in denervated hippocampus at day 30 after transplantation. Arrow showed the neurons double positive to BrdU and NF200. (D and E) β-Tubulin-III (Tuj1, brown) and BrdU (blue) immunohistochemistry on the normal (D) and denervated (E) hippocampus. Arrow showed the β-Tubulin-III and BrdU positive neurons in denervated hippocampus. (F and G) AChE histochemistry on the normal (F) and denervated (G) hippocampus. Arrow showed the AChE positive neurons in denervated hippocampus which may be from differentiation of the grafted cells or endogenous NPCs because there originally are no cholinergic neurons in normal hippocampus.

Fig. 3. Immunodetection to the neuronal differentiation of SVZ NPCs grafted into adult hippocampus.

Prophylactic cranial radiotherapy involves giving radiotherapy to a person's head to prevent or delay the possible spread of cancer cells to the brain, but induces progressive and debilitating declines in cognition that may, in part, be caused by the depletion of the normal neural cells or NSC in hippocampus. Acharya and his colleagues (Acharya et al., 2011) used NSC replacement as a strategy to combat radiation-induced cognitive decline by intrahippocampal transplantation with human neural stem cells (hNSC). Unbiased stereology revealed that 23% and 12% of the engrafted cells survived 1 and 4 months after transplantation, respectively. Engrafted cells migrated extensively, differentiated along glial and neuronal lineages, and expressed the activity-regulated cytoskeleton-associated protein (Arc), suggesting their capability to functionally integrate into the hippocampus. Behaviorally the irradiated animals engrafted with hNSCs showed significantly less decline in cognitive function.

After transplantation if these cells survive the injured and/or degenerative insult(s), they may migrate within damaged areas and promote repair or neuroprotection via cell replacement, integration or neuroprotection. The neuroprotection from grafted NPCs may be the results of in situ release of immunomodulatory molecules (e.g., anti-inflammatory cytokines) and neurotrophic factors [e.g., nerve growth factor (NGF), fibroblast growth factor (FGF)-2, ciliary neurotrophic factor (CNTF) and brain-derived neurotrophic factor (BDNF)] (Martino & Pluchino, 2006; Pluchino et al., 2005). On the other hand, transplanted NPCs may also differentiate into local specific cells to replace the dying cells and integrate within the host neural cells. Thus, we can propose the concept of 'therapeutic plasticity', which can be viewed as the capacity of somatic stem cells to adapt their fate and function(s) to specific environmental needs occurring as a result of different pathological conditions.

It is indicated that NPCs afford a promising strategy for functionally restoring defects induced by hippocampal degenerations or injuries. However, neural transplantation to correct congenital or acquired disorders using multipotent progenitor cells has several major limitations: migration of the transplanted cells is limited; the cells seldom develop into neurons; the limited sources of donor cells and many ethical concerns and political restrictions. Motivating endogenous neural progenitors may be another good strategy for the neurodegenerative disorders.

4. Adult neurogenesis of endogenous NSCs in hippocampus

During the past decade, the progress in the field of stem cells has fueled the hope to cure currently intractable diseases by cell replacement. In regard of ethical concerns and political restrictions that have been raised regarding the use and manipulation of fetal tissue and embryonic stem cells and the limitation of heterogeneous graft, adult endogenous NPCs have been prefered as a cellular source for the treatment of CNS diseases. The use of endogenous sources for cell replacement offer a potential advantage over other cell sources: Immunological reactions are avoided.

After injury or during neurodegenerative processes in restricted brain regions the NPCs frequently reside in niches that regulate their self-renewal, activation and differentiation. Within the niche, both environmental cues and intrinsic genetic programs are two factors required to direct/regulate stem and precursor cell proliferation, differentiation and integration. The adult born functional neurons in the neural networks is believed to experience sequential steps in a highly regulated fashion: proliferation of the NSC, generation of a rapidly amplifying progenitor cell, differentiation into an immature neuron, migration to the final location, growth of axons and dendrites and formation of synapses with other neurons in the circuits, and ultimately maturation into a fully functional neuron. Although these steps are equivalent to the ones that newborn neurons undergo during development, the fundamental difference between the developmental and adult neurogenesis is that new adult neurons undergo these processes in an already mature environment and integrate into preexisting circuits in adult hippocampal neurogenesis. During this period, the newborn neurons undergo dying, surviving, migrating into the granular layer, sending axons to the CA3 region to form mossy fibers and projecting dendrites to the outer molecular layer (Hastings & Gould, 1999; Kempermann et al., 2003; Markakis & Gage, 1999; Seri et al., 2001; van Praag et al., 2002). Simultaneously, the newly generated neurons receive synaptic inputs from the other region within four to six weeks after birth (van Praag et al., 2002). The complexity and density of their dendritic spines have to continuously grow for several months. Thus, the course of neuronal development for granule neurons born in the adult hippocampus appears much more protracted than those generated during embryonic stages.

The endogenous NPCs in the SVZ and SGZ are the source of adult neurogenesis and remodeling which are implicated in responses to multiple insults including ischemia (Arvidsson et al., 2002; Jin et al., 2001; Miles & Kernie, 2008; Nakatomi et al., 2002), trauma (Johansson et al., 1999; Yoshimura et al., 2001), seizure (Parent et al., 1997; Parent & Murphy, 2008) and neurodegeneration (Fallon et al., 2000; Magavi & Macklis, 2002). Adult neurogenesis in hippocampus can be regulated by numerous factors associated with an animal's behavioural and cognitive states. Indeed, an animal's experiences on cognition and mood, including hippocampus-dependent learning, environmental enrichment, voluntary running and chronic treatment with antidepressants, can affect the rate of neurogenesis. The factors enhancing hippocampal neurogenesis are summarized in the following and Figure 5 which also enumerates the factors decreasing adult hippocampal neurogenesis.

4.1 Enriched environment

Gage and his colleagues have demonstrated that mice placed in an enriched environment where there are more social interactions, inanimate objects for play and a wheel for voluntary exercise have an increased rate of neurogenesis relative to mice that are kept in standard cages (Kempermann et al., 1997). Subsequently, the similar experiments have been repeated and proven by other laboratories (Beauquis et al., 2010; Brown et al., 2003; Ehninger & Kempermann, 2003; Kempermann et al., 2002; Kohl et al., 2002; Llorens-Martin et al., 2010; Olson et al., 2006; Steiner et al., 2008). The dual-birthdating analysis used to study two subpopulations of newborn neurons born at the beginning and end of enrichment suggested that enriched environment induces differential effects on distinct subpopulations of newborn neurons depending on the age of the immature cells and on the duration of the enriched environment itself (Llorens-Martin et al., 2010). This work points to a hypothesis that the effects of physical-cognitive activity on neurogenesis depend on the interaction of two critical parameters: the age/differentiation status of the immature neuron plus the time the individual is under the effects of an enriched environment.

4.2 Exercise

Studies of voluntary exercise demonstrate that running on wheel without other components of enriched environment is sufficient to increase proliferation and recruitment of granule cells into the adult DG (van Praag et al., 1999a; van Praag et al., 1999b). Although the exact mechanism underlying the exercise-induced up-regulation of neurogenesis remains unclear, exercise is reported to increase the expression of certain trophic factors, such as BDNF and FGF-2 (Ding et al., 2011; Gomez-Pinilla et al., 1997; Griffin et al., 2011; Russo-Neustadt et al., 1999), which have also been shown to increase neurogenesis during development or in adult brain (Ding et al., 2011; Zigova et al., 1998).

4.3 Psychotropic drugs

Serotonergic antidepressant drugs have been commonly used to treat mood and anxiety disorders. In experimental animals, chronic antidepressant treatments can facilitate neurogenesis in the DG of the adult hippocampus (Dagyte et al., 2010; Kitamura et al., 2011; Malberg et al., 2000; Nasrallah et al., 2011). The adult hippocampal neurogenesis has been implicated in some of the behavioral effects of antidepressants (Airan et al., 2007; Santarelli et al., 2003; Wang et al., 2008). Two molecular mechanisms are possibly involved in the antidepressant drug-induced hippocampal neurogenesis. One is the increased BDNF in hippocampus. Previous studies have demonstrated that repeated antidepressant administration increases the expression of BDNF in hippocampus (Duman et al., 1997; Duman et al., 2000; Lee & Kim, 2010; Pilar-Cuellar et al., 2011; Reus et al., 2011; Rogoz et al., 2008). In contrast, stress decreases BDNF expression in this brain region (de Lima et al., 2011; Murakami et al., 2005) and causes atrophy of hippocampal neurons and decreased neurogenesis (Gould et al., 1998; Yap et al., 2006). All these results have contributed to a neurotrophic hypothesis of depression and antidepressant action. Antidepressant treatment may block or even reverse these effects of stress via increased expression of BDNF. The other is the Notch1 signaling. New evidences indicated that fluoxetine (antidepressant) administration increased mRNA and protein expression of Notch1 signaling components (including Jag1, NICD, Hes1 and Hes5) and simultaneously up-regulated hippocampal cell proliferation and survival, suggesting that activation of Notch1 signaling might partly contribute to increased neurogenesis in hippocampus (Sui et al., 2009). In addition to promotion of neurogenesis, the psychotropic drugs significantly increased the survival of newborn neurons in dorsal hippocampus by approximately 50% (Su et al., 2009). Results from Kobayashi and his colleagues (Kobayashi et al., 2010) showed that serotonergic antidepressants can reverse the established state of neuronal maturation in the adult hippocampus, termed "dematuration" of mature granule cells, and up-regulate 5-HT4 receptor-mediated signaling which may play a critical role in this distinct action of antidepressants. Such reversal of neuronal maturation could affect proper functioning of the mature hippocampal circuit. Together with these results support the hypothesis that antidepressants exert therapeutic effects on neuropsychiatric disease via not only activating the hippocampal neurogenesis but also reinstating neuronal functions of the matured granular cells.

Evidences have not show the confirmed effects on the repeated antipsychotic drug administration because of the contradictory results that Dawirs et al. work (Dawirs et al., 1998) demonstrated granular cell proliferation by chronic administration of haloperidol while Backhouse et al., (Backhouse et al., 1982) reported a decrease in hippocampal cell proliferation. Abuse of drugs including opiates and psychostimulants can influence cognition, learning and memory, which is accompanied by decrease of the proliferation of granule cells in adult rat hippocampus (Eisch et al., 2000).

4.4 Ischemia

Studies have noted that ischemia also produces enhanced neurogenesis in neuroproliferative regions of the adult rodent brain, including the SVZ of the lateral ventricles and SGZ of DG (Burns et al., 2009; Jin et al., 2001; Parent et al., 2002; Yagita et al., 2001). Proliferation induced by transient focal or global ischemia peaks 7 to 10 days after ischemia and returns to baseline levels within several weeks. Some of the new cells die but others survive to adopt a neuronal fate in the ischemic and uninjured dentate gyrus. Newborn cells labeled EGFP retroviral reporter are found to move from the subgranular proliferative zone to the DGC layer, shift from coexpression of immature to mature neuronal markers, and increase in dendritic length (Tanaka et al, 2004), suggesting that newly generated DGCs in the ischemic brain follow a time course of neuronal maturation. A new report from Liu and his colleagues (Wang et al., 2011) indicated that transient brain ischemia initiates a sustained increase in neurogenesis for at least 6 months and promotes the normal development of the newly generated neurons in the adult DG.

4.5 Traumatic Brain Injury (TBI)

The hippocampus, a region responsible for memory and learning, is particularly vulnerable to brain trauma. Learning and memory deficits are the most enduring and devastating consequences following TBI on hippocampus. A slow but significant improvement in cognitive function after TBI indicates that innate mechanisms for repair exist in the brain (Schmidt et al., 1999). Although the mechanisms underlying this innate recovery remain largely unknown, the findings that NSCs persist in the hippocampal DG throughout life (Gage, 2000; Kempermann & Gage, 2000) and exhibit high activation of proliferation and neurogenesis in response to brain trauma (Chirumamilla et al., 2002; Dash et al., 2001; Urrea et al., 2007; Yu et al., 2008) suggest that neurogenesis may contribute to the cognitive recovery observed following TBI.

In our laboratory transection of FiFx plus CB is deemed as a kind of TBI to produce deafferented hippocampus. The denervated hippocampus provided a proper microenvironment for the survival and neuronal differentiation of exogenous neural progenitors (Zhang et al., 2007). Subsequently, we determined the endogenous NPCs in DG of adult hippocampus after denervation trauma. The results showed that traumatic injury by transecting FiFx and CB which carry cholinergic inputs promoted proliferation of the local NPCs and increased the number of newborn neurons in SGZ of hippocampus (Figure 4). Indicating that the changes in the deafferented hippocampus provided a suitable microenvironment for neurogenesis of endogenous progenitors of adult hippocampus. However, Christiana et al. (Cooper-Kuhn et al., 2004) produced a cholinergic depletion model by infusion of the immunotoxin 192IgG-saporin into lateral ventricle to selectively lesion cholinergic neurons of the cholinergic basal forebrain. Oppositely, their results showed a significant declination of neurogenesis in the granule cell layer of the dentate gyrus and olfactory bulb. Furthermore, immunotoxic lesions led to increased numbers of apoptotic cells specifically in the SGZ and the periglomerular layer of the olfactory bulb. The model of TBI created by distinct ways may contribute to the conflict results because the immunotoxin might exert negative effects on the neural progenitors and newborn neurons.

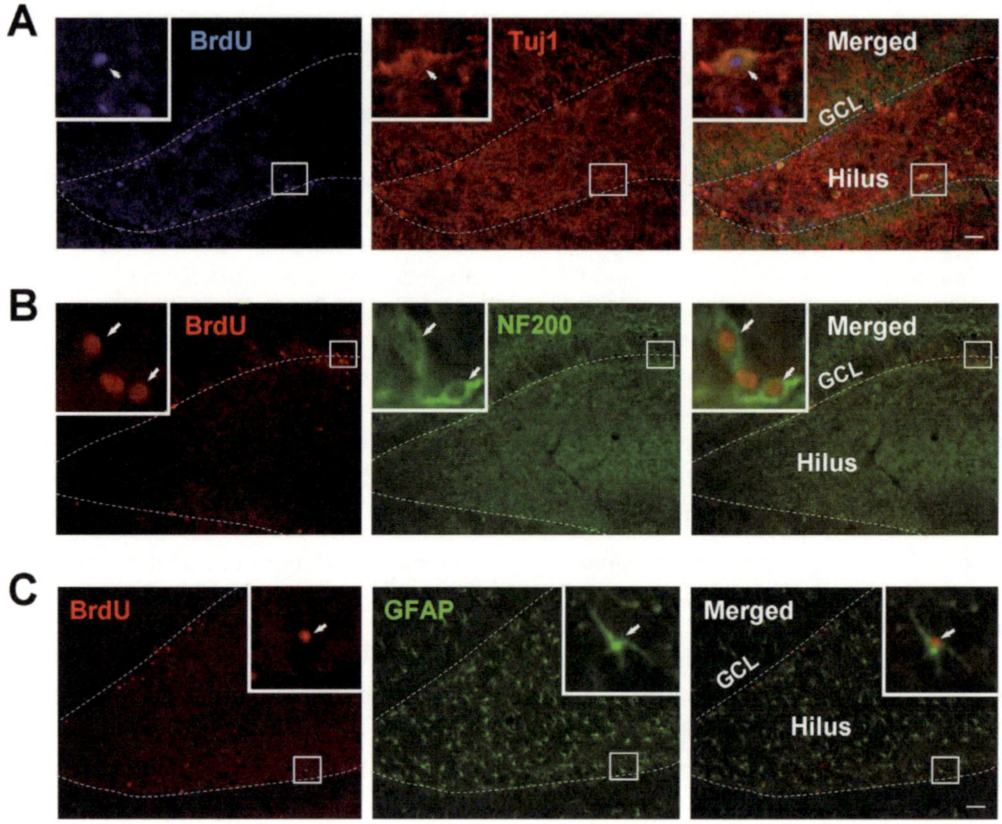

A) Immunofluorescence micrographs of anti–BrdU (Blue), β-tubulin III (Tuj1, red) in coronal sections of the hippocampus at day 35 after denervation operation. Arrows show the cells immunoreactive to BrdU and Tuj1. B) Microscope images of sections through deafferented hippocampus stained by BrdU and NF-200 antibodies on day 42 after transection. Arrows show the BrdU and NF-200 double positive neurons. (C) Microscope image of BrdU positive (red) and GFAP positive (green) astrocytes in denervated hippocampus 28 days after transection. Arrow showed the BrdU and GFAP positive astrocytes.

Fig. 4. Endogenous NPCs labeled with BrdU differentiate into neurons and astrocytes in deafferented hippocampus.

4.6 Seizures

Seizures characterize the periodic and unpredictable occurrences of epilepsy. Accumulating evidences indicate that seizures alter not only the amount, but also the pattern of neurogenesis, though the overall effect depends on the type of seizures. Acute seizures abnormally increase the amount of hippocampal neurogenesis and induce aberrant migration of newly born neurons into the dentate hilus and the dentate molecular layer

(Bengzon et al., 1997; Jessberger et al., 2005; Kralic et al., 2005; Parent et al., 1997). Examination of the hippocampus from young temporal lobe epilepsy patients (<4 years of age) suggested increased cell proliferation of neural precursor cells (Blumcke et al., 2001). However, recurrent spontaneous seizures typically observed in chronic temporal lobe epilepsy lead to a radically waned neurogenesis (Hattiangady et al., 2004; Kralic et al., 2005), which, interestingly, coexists with learning and memory impairments and depression. Heinrich et al. (Heinrich et al. 2006) reported a gradual fall in neurogenesis at 1 week and virtual loss of all neurogenesis by 4-6 weeks after the initial seizure episode. However, a modest increase in neurogenesis was observed even at 2 months post status epilepticus in a lithium-pilocarpine model of epilepsy using postnatal day 20 rats (Cha et al., 2004). It emerges that decreased levels of hippocampal neurogenesis in chronic epilepsy depend on the model and the age of the animal at the time of the initial seizure episode.

4.7 Others

Lithium was noticed to have mood stabilizing properties in the late 1800s when doctors were using it to treat gout. Australian psychiatrist John Cade published the first paper on the use of lithium in the treatment of acute mania. Lithium, as a mood stabilizer, is used as an add-on treatment for clinical depression. Recent reports have described that lithium increases cell proliferation and/or promotion of neuronal differentiation of NPCs (Boku et al., 2011; Chen et al., 2000; Fiorentini et al., 2010; Hanson et al., 2011; Kim et al., 2004; Kitamura et al., 2011; Son et al., 2003; Wexler et al., 2008) and blocks the effects of stress on depression-like behaviors through increasing hippocampal neurogenesis in adult rodent models (Silva et al., 2008). Results of these studies suggest that adult hippocampal neurogenesis plays an important role in the therapeutic action of mood stabilizers as well. Inhibition of GSK-3β and subsequent activation of Wnt/β-catenin signalling may underlie lithium-induced hippocampal neurogenesis and therapeutic effect (Boku et al., 2010; Fiorentini et al., 2010; Wexler et al., 2008).

Acupuncture or electroacupuncture, the ancient Chinese treatments through stimulating the acu-points, can ameliorate syndromes of many illnesses pain, metabolic and pathological brain disease, and even mental disorders, such as major depression. Although the mechanisms underlying treatment of acupuncture on these diseases remain unclear till now, neurogenesis must be considered as a potential one of mechanisms in the process of therapy. It has been reported that acupuncture and electroacupuncture in the acu-points ST36 (*Zusanli*) and GV20 (*Baihui*) increase significantly neurogenesis in the normal DG, while electroacupuncture has greater effects on neuroblast plasticity in the DG than acupuncture (Hwang et al., 2010). In addition to normal status, relieves of illnesses were paralleled with the hippocampal neurogenesis in DG. For example, decreased cell proliferation in the DG of dementia model was improved by *Yiqitiaoxue* and *Fubenpeiyuan* acupuncture (Cheng et al., 2008). In addition, electroacupuncture at GV20 and EX17 increased hippocampal progenitor cell proliferation in adult rats exposed to chronic unpredictable stress (Liu et al., 2007). In ischemic models (Kim et al., 2001) and streptozotocin-induced diabetic models (Kim et al., 2002), acupuncture (ST36)-induced alleviation is paralleled with increased cell proliferation in the DG. Acupuncture at *Tanzhong* (CV17), *Zhongwan* (CV12), *Qihai* (CV6), ST36, and *Xuehai* (SP10) improve spatial memory impairment (Yu et al., 2005), maintain oxidant-antioxidant balance, and regulate cell proliferation in a rodent dementia model (Cheng et al., 2008; Liu et al., 2006).

After comparing the cell proliferation in DG of adult mice fed on hard and soft diet, Yamamoto et al. (Yamamoto et al., 2009) found that sufficient mastication activity enhanced hippocampal neurogenesis since that the total number of BrdU-labeled cells was fewer in the soft-diet group than in the hard-diet group at 3 and 6 months of age.

Additionally, Leuner et al. (Leuner et al., 2010) found that sexual experience that the adult male rats were exposed to a sexually-receptive female increased circulating corticosterone levels and the number of new neurons in the hippocampus and stimulated the growth of dendritic spines and dendritic architecture, suggesting that a rewarding experience actually promotes adult-born neuronal growth.

The persistence of neurogenesis in the adult mammalian forebrain suggests that endogenous precursors provide a potential source of neurons for the replacement of the dying or lost neurons due to brain damage or neurodegeneration. Based on the multiple stimuli inducing hippocampal neurogenesis, strategies that are designed to increase adult hippocampal neurogenesis specifically, by targeting the cell death of adult-born neurons or by other mechanisms, may have therapeutic potential for reversing impairments in pattern separation and DG dysfunction such as those seen during normal ageing.

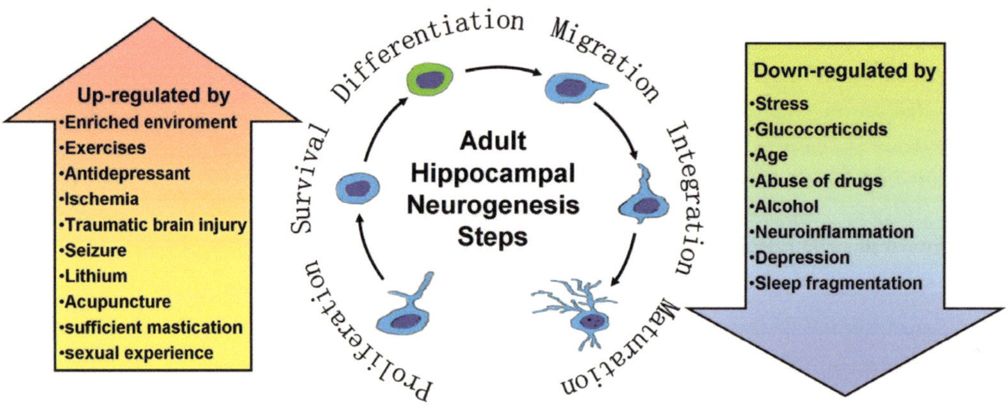

Fig. 5. Adult hippocampal neurogenesis can be up- or down-regulated by various stimuli. This summarizes the sequent steps of adult hippocampal neurogenesis and a variety of stimuli positively or negatively influencing adult hippocampal neurogenesis.

5. Signal pathways involved in hippocampal neurogenesis

Understanding the mechanisms underlying adult neurogenesis and differentiation of NPCs is crucial to delineate the function of NPCs and their progeny and ultimately their therapeutic potential. The initial investigations on environmental niches and intrinsic genetic programs that regulate early and adult neurogenesis have revealed many extrinsic and intrinsic elements playing critical roles in differential phrases of neurogenesis, such as proliferation, migration, differentiation, integration and maturation. The following lists the signal molecules involved in adult hippocampus neurogenesis.

5.1 Wnt (wingless)

Traditionally, Wnt proteins are assumed to act as stem cell growth factors, promoting the maintenance and proliferation of stem cells (Willert et al., 2003) and inducing of neural specification (Muroyama et al., 2002). Interaction of Wnts with their receptors can trigger several signaling pathways, including the β-catenin dependent pathway. Studies of Lie et al. (Lie et al., 2005) show that Wnt signalling components and their receptors were expressed in the adult hippocampal progenitor cells. Overexpression of Wnt3 is sufficient to increase neurogenesis of adult hippocampal progenitors in vitro and in vivo. By contrast, blockade of Wnt signalling reduces neurogenesis of adult hippocampal progenitor cells in vitro and abolishes neurogenesis almost completely in vivo. Evidence also suggests that β-catenin, which is present in neural progenitors and newborn granule neurons, plays an important role in the dendritic development of adult born hippocampal neurons (Gao et al., 2007). These data show that Wnt signalling is a principal regulator of adult hippocampal neurogenesis.

5.2 Notch

Notch (1 - 4 in mammals) signaling pathway is crucial for maintenance of stem cell self renewal, proliferation, and specification of cell fate (Mason et al., 2006). Notch signaling is highly activated in type-B cells of the SVZ of the lateral ventricle and type-1 cells of the SGZ of the DG (Ehm et al., 2010; Imayoshi et al., 2010; Lugert et al., 2010). In postnatal and adult mice, Overexpression of Notch1 in postnatal and adult mice increased hippocampal cell proliferation and maintained GFAP-expressing NSCs, while depletion of Notch signaling led to a decrease in cell proliferation and a shift in the differentiation of newly born cells towards a neuronal lineage suggesting that Notch1 signaling is required to maintain a reservoir of undifferentiated cells and ensure continuity of adult hippocampal neurogenesis (Ables et al., 2010; Breunig et al., 2007). In addition, Notch1 signaling modulates the dendritic morphology of newborn granule cells by increasing dendritic arborization (Breunig et al., 2007). These evidences suggest that Notch1 signaling is involved in the cell proliferation, fate determination, and maturation of adult hippocampal neurogenesis. Pathologically, antidepressant therapy chronic fluoxetine administration increased expression of Notch1 signaling components including Jag1, NICD, Hes1 and Hes5 in the hippocampus, accompanied by cell proliferation and survival (Sui et al., 2009). This indicated that activation of the Notch1 pathway might partly contribute to chronic antidepressant therapy-increased neurogenesis in hippocampus.

5.3 Bone Morphogenetic Protein (BMP)

BMP proteins, the extracellular signaling molecules, regulate cell proliferation and fate commitment throughout development and within the adult SVZ and SGZ neurogenic niches (Bonaguidi et al., 2005; Bonaguidi et al., 2008; Mehler et al., 2000). The cysteine knot proteins noggin, chordin and follistatin regulate BMP actions via competitively binding BMPs in the extracellular space to prevent receptor activation and the downstream signaling activity (Dal-Pra et al., 2006; Ebara & Nakayama, 2002). The inhibition of noggin in vivo by RNA interference decreased hippocampal cell proliferation (Fan et al., 2004). Study of Gobeske et al. indicated that BMP signaling mediates effects of exercise on hippocampal neurogenesis and cognition in mice (Gobeske et al., 2009).

5.4 Sonic hedgehog (Shh)

Shh is reported to be crucial in the expansion and establishment of postnatal hippocampal progenitors (Palma et al., 2005). The Shh receptors patched (Ptc) and smoothened (Smo) are expressed in the dentate gyrus subfield including the neurogenic niche of SGZ and in neural progenitor cells derived from hippocampus (Lai et al., 2003; Traiffort et al., 1998). Recently, it is addressed that Shh signaling regulates adult hippocampal neurogenesis (Han et al., 2008; Lai et al., 2003; Palma et al., 2005). In rats, overexpression of Shh in the DG increased cell proliferation and survival (Lai et al., 2003). On the other hand, inhibition of Shh signaling by injections of inhibitor cyclopamine reduced cell proliferation (Banerjee et al., 2005; Lai et al., 2003). Removal of Shh signaling in these animals resulted in dramatic reduction in number of neural progenitors in both the postnatal SVZ and hippocampus. Consistently, conditional null alleles of hedgehog signaling also resulted in abnormalities in the DG and olfactory bulb (Machold et al., 2003). These studies emphasize the importance of the Shh signaling pathway in adult neurogenesis. Findings from Banerjee et al. (Banerjee et al., 2005) demonstrated that Shh pathway may be involved in electroconvulsive seizure-enhanced adult hippocampal neurogenesis. The primary cilia are important sites of signal transduction which unite the receptors and the signal-transduction components, such as Wnt and Hedgehog (Hh) signaling cascades (Huangfu et al., 2003; Huangfu & Anderson, 2005). It is demonstrated that, in the absence of cilia, there is a dramatic diminution in Shh signaling, decreased early proliferation and a consequent loss of quiescent precursor cell (Breunig et al., 2008).

5.5 PI3K-Akt

PI3K-Akt signalling pathway is the downstream of neurotrophic and growth factor receptors, as well as monoamine receptors (Datta et al., 1999). It is potentially implicated in a number of different functions and especially associated with cell survival by inhibiting the activation of proapoptotic proteins and transcription factors (Aberg et al., 2003). Akt has three different isoforms, Akt1, -2, -3, each encoded by independent genes (Coffer et al., 1998). It was shown that Akt1 and Akt2 knockout mice had lower levels of hippocampal cell proliferation compared to wild type animals (Balu et al., 2008). However, only Akt2KO mice had impairment in the survival of adult born hippocampal progenitors (Balu et al., 2008). Subsequent report also showed the nonredundant roles of Akt in the regulation of hippocampal neurogenesis since that physical exercise activated Akt and three downstream targets, BAD, GSK3b and FOXO1 and inhibition of PI3K-Akt signaling blocks exercise-mediated enhancement of adult neurogenesis and synaptic plasticity in the DG (Bruel-Jungerman et al., 2009).

6. Conclusion

These findings have fuelled the hope of using neurogenesis, exogenous or endogenous, in regenerative medicine for neurological diseases, arguably the most difficult diseases to treat. The proposed regenerative approaches to neurological diseases include (1) cell therapy approaches in which donated cells are delivered by intracerebral injection or infusion through an intravenous or intra-arterial route; (2) stem cell mobilization approaches in which endogenous stem or progenitor cells are activated by cytokines or chemokines; (3) trophic and growth factor support in which the factors, such as BDNF and GDNF, were delivered through grafted stem cells modulated genetically into the brain to support the injured neurons. These approaches may be used together to maximize therapeutic effects. Although the mechanisms underlying these therapeutic processes are still unclear, the neurogenic cells must survival various complicated and difficult barriers from proliferation to maturation. Understanding the factors in NPC niches and intracellular molecules regulating/directing adult neurogenesis will largely speed the steps to make use of exogenous or endogenous NPCs in treatment of neural disorders. The past evidences indicate that cell therapy to the injured tissue and brain may be contributed by several processes including angiogenesis, neurogenesis and trophic or 'chaperone' support.

7. References

Aberg MA, Aberg ND, Palmer TD, Alborn AM, Carlsson-Skwirut C, Bang P, Rosengren LE, Olsson T, Gage FH & Eriksson PS. (2003). IGF-I has a direct proliferative effect in adult hippocampal progenitor cells. Mol Cell Neurosci 24(1):23-40.

Ables JL, Decarolis NA, Johnson MA, Rivera PD, Gao Z, Cooper DC, Radtke F, Hsieh J & Eisch AJ. (2010). Notch1 is required for maintenance of the reservoir of adult hippocampal stem cells. J Neurosci 30(31):10484-92.

Acharya MM, Christie LA, Lan ML, Giedzinski E, Fike JR, Rosi S & Limoli CL. (2011). Human neural stem cell transplantation ameliorates radiation-induced cognitive dysfunction. Cancer Res 71(14):4834-45.

Airan RD, Meltzer LA, Roy M, Gong Y, Chen H & Deisseroth K. (2007). High-speed imaging reveals neurophysiological links to behavior in an animal model of depression. Science 317(5839):819-23.

Arvidsson A, Collin T, Kirik D, Kokaia Z & Lindvall O. (2002). Neuronal replacement from endogenous precursors in the adult brain after stroke. Nat Med 8(9):963-70.

Bachoud-Levi AC, Remy P, Nguyen JP, Brugieres P, Lefaucheur JP, Bourdet C, Baudic S, Gaura V, Maison P, Haddad B and others. (2000). Motor and cognitive improvements in patients with Huntington's disease after neural transplantation. Lancet 356(9246):1975-9.

Backhouse B, Barochovsky O, Malik C, Patel AJ & Lewis PD. (1982). Effects of haloperidol on cell proliferation in the early postnatal rat brain. Neuropathol Appl Neurobiol 8(2):109-16.

Balu DT, Easton RM, Birnbaum MJ & Lucki I. 2008. Deletion of Akt Isoforms Reduce Hippocampal Neurogenesis, Fear Conditioning and Antidepressant Behavioral Responses. Society for Neuroscience. Washington, DC.

Banerjee SB, Rajendran R, Dias BG, Ladiwala U, Tole S & Vaidya VA. (2005). Recruitment of the Sonic hedgehog signalling cascade in electroconvulsive seizure-mediated regulation of adult rat hippocampal neurogenesis. Eur J Neurosci 22(7):1570-80.

Beauquis J, Roig P, De Nicola AF & Saravia F. (2010). Short-term environmental enrichment enhances adult neurogenesis, vascular network and dendritic complexity in the hippocampus of type 1 diabetic mice. PLoS One 5(11):e13993.

Bengzon J, Kokaia Z, Elmer E, Nanobashvili A, Kokaia M & Lindvall O. (1997). Apoptosis and proliferation of dentate gyrus neurons after single and intermittent limbic seizures. Proc Natl Acad Sci U S A 94(19):10432-7.

Blumcke I, Schewe JC, Normann S, Brustle O, Schramm J, Elger CE & Wiestler OD. (2001). Increase of nestin-immunoreactive neural precursor cells in the dentate gyrus of pediatric patients with early-onset temporal lobe epilepsy. Hippocampus 11(3):311-21.

Boku S, Nakagawa S & Koyama T. (2010). Glucocorticoids and lithium in adult hippocampal neurogenesis. Vitam Horm 82:421-31.

Boku S, Nakagawa S, Masuda T, Nishikawa H, Kato A, Toda H, Song N, Kitaichi Y, Inoue T & Koyama T. (2011). Effects of mood stabilizers on adult dentate gyrus-derived neural precursor cells. Prog Neuropsychopharmacol Biol Psychiatry 35(1):111-7.

Bonaguidi MA, McGuire T, Hu M, Kan L, Samanta J & Kessler JA. (2005). LIF and BMP signaling generate separate and discrete types of GFAP-expressing cells. Development 132(24):5503-14.

Bonaguidi MA, Peng CY, McGuire T, Falciglia G, Gobeske KT, Czeisler C & Kessler JA. (2008). Noggin expands neural stem cells in the adult hippocampus. J Neurosci 28(37):9194-204.

Boneva NB, Kaplamadzhiev DB, Sahara S, Kikuchi H, Pyko IV, Kikuchi M, Tonchev AB & Yamashima T. (2011). Expression of fatty acid-binding proteins in adult hippocampal neurogenic niche of postischemic monkeys. Hippocampus 21(2):162-71.

Bovetti S, Bovolin P, Perroteau I & Puche AC. (2007). Subventricular zone-derived neuroblast migration to the olfactory bulb is modulated by matrix remodelling. Eur J Neurosci 25(7):2021-33.

Brandt MD, Jessberger S, Steiner B, Kronenberg G, Reuter K, Bick-Sander A, von der Behrens W & Kempermann G. (2003). Transient calretinin expression defines early postmitotic step of neuronal differentiation in adult hippocampal neurogenesis of mice. Mol Cell Neurosci 24(3):603-13.

Breunig JJ, Silbereis J, Vaccarino FM, Sestan N & Rakic P. (2007). Notch regulates cell fate and dendrite morphology of newborn neurons in the postnatal dentate gyrus. Proc Natl Acad Sci U S A 104(51):20558-63.

Breunig JJ, Sarkisian MR, Arellano JI, Morozov YM, Ayoub AE, Sojitra S, Wang B, Flavell RA, Rakic P & Town T. (2008). Primary cilia regulate hippocampal neurogenesis by mediating sonic hedgehog signaling. Proc Natl Acad Sci U S A 105(35):13127-32.

Brown J, Cooper-Kuhn CM, Kempermann G, Van Praag H, Winkler J, Gage FH & Kuhn HG. (2003). Enriched environment and physical activity stimulate hippocampal but not olfactory bulb neurogenesis. Eur J Neurosci 17(10):2042-6.

Bruel-Jungerman E, Veyrac A, Dufour F, Horwood J, Laroche S & Davis S. (2009). Inhibition of PI3K-Akt signaling blocks exercise-mediated enhancement of adult neurogenesis and synaptic plasticity in the dentate gyrus. PLoS One 4(11):e7901.

Burns TC, Verfaillie CM & Low WC. (2009). Stem cells for ischemic brain injury: a critical review. J Comp Neurol 515(1):125-44.

Cassel JC, Duconseille E, Jeltsch H & Will B. (1997). The fimbria-fornix/cingular bundle pathways: a review of neurochemical and behavioural approaches using lesions and transplantation techniques. Prog Neurobiol 51(6):663-716.

Cha BH, Akman C, Silveira DC, Liu X & Holmes GL. (2004). Spontaneous recurrent seizure following status epilepticus enhances dentate gyrus neurogenesis. Brain Dev 26(6):394-7.

Chen G, Rajkowska G, Du F, Seraji-Bozorgzad N & Manji HK. (2000). Enhancement of hippocampal neurogenesis by lithium. J Neurochem 75(4):1729-34.

Cheng H, Yu J, Jiang Z, Zhang X, Liu C, Peng Y, Chen F, Qu Y, Jia Y, Tian Q and others. (2008). Acupuncture improves cognitive deficits and regulates the brain cell proliferation of SAMP8 mice. Neurosci Lett 432(2):111-6.

Chirumamilla S, Sun D, Bullock MR & Colello RJ. (2002). Traumatic brain injury induced cell proliferation in the adult mammalian central nervous system. J Neurotrauma 19(6):693-703.

Coffer PJ, Jin J & Woodgett JR. (1998). Protein kinase B (c-Akt): a multifunctional mediator of phosphatidylinositol 3-kinase activation. Biochem J 335 (Pt 1):1-13.

Cooper-Kuhn CM, Winkler J & Kuhn HG. (2004). Decreased neurogenesis after cholinergic forebrain lesion in the adult rat. J Neurosci Res 77(2):155-65.

Corotto FS, Henegar JA & Maruniak JA. (1993). Neurogenesis persists in the subependymal layer of the adult mouse brain. Neurosci Lett 149(2):111-4.

Dagyte G, Trentani A, Postema F, Luiten PG, Den Boer JA, Gabriel C, Mocaer E, Meerlo P & Van der Zee EA. (2010). The novel antidepressant agomelatine normalizes hippocampal neuronal activity and promotes neurogenesis in chronically stressed rats. CNS Neurosci Ther 16(4):195-207.

Dal-Pra S, Furthauer M, Van-Celst J, Thisse B & Thisse C. (2006). Noggin1 and Follistatin-like2 function redundantly to Chordin to antagonize BMP activity. Dev Biol 298(2):514-26.

Dash PK, Mach SA & Moore AN. (2001). Enhanced neurogenesis in the rodent hippocampus following traumatic brain injury. J Neurosci Res 63(4):313-9.

Datta SR, Brunet A & Greenberg ME. (1999). Cellular survival: a play in three Akts. Genes Dev 13(22):2905-27.

Dawirs RR, Hildebrandt K & Teuchert-Noodt G. (1998). Adult treatment with haloperidol increases dentate granule cell proliferation in the gerbil hippocampus. J Neural Transm 105(2-3):317-27.

de Lima MN, Presti-Torres J, Vedana G, Alcalde LA, Stertz L, Fries GR, Roesler R, Andersen ML, Quevedo J, Kapczinski F and others. (2011). Early life stress decreases hippocampal BDNF content and exacerbates recognition memory deficits induced by repeated d-amphetamine exposure. Behav Brain Res 224(1):100-6.

Deng W, Saxe MD, Gallina IS & Gage FH. (2009). Adult-born hippocampal dentate granule cells undergoing maturation modulate learning and memory in the brain. J Neurosci 29(43):13532-42.

Ding Q, Ying Z & Gomez-Pinilla F. (2011). Exercise influences hippocampal plasticity by modulating brain-derived neurotrophic factor processing. Neuroscience.

Doetsch F, Garcia-Verdugo JM & Alvarez-Buylla A. (1997). Cellular composition and three-dimensional organization of the subventricular germinal zone in the adult mammalian brain. J Neurosci 17(13):5046-61.

Duman RS, Heninger GR & Nestler EJ. (1997). A molecular and cellular theory of depression. Arch Gen Psychiatry 54(7):597-606.

Duman RS, Malberg J, Nakagawa S & D'Sa C. (2000). Neuronal plasticity and survival in mood disorders. Biol Psychiatry 48(8):732-9.

Ebara S & Nakayama K. (2002). Mechanism for the action of bone morphogenetic proteins and regulation of their activity. Spine (Phila Pa 1976) 27(16 Suppl 1):S10-5.

Eckenstein FP, Baughman RW & Quinn J. (1988). An anatomical study of cholinergic innervation in rat cerebral cortex. Neuroscience 25(2):457-74.

Ehm O, Goritz C, Covic M, Schaffner I, Schwarz TJ, Karaca E, Kempkes B, Kremmer E, Pfrieger FW, Espinosa L and others. (2010). RBPJkappa-dependent signaling is essential for long-term maintenance of neural stem cells in the adult hippocampus. J Neurosci 30(41):13794-807.

Ehninger D & Kempermann G. (2003). Regional effects of wheel running and environmental enrichment on cell genesis and microglia proliferation in the adult murine neocortex. Cereb Cortex 13(8):845-51.

Eisch AJ, Barrot M, Schad CA, Self DW & Nestler EJ. (2000). Opiates inhibit neurogenesis in the adult rat hippocampus. Proc Natl Acad Sci U S A 97(13):7579-84.

Englund C, Fink A, Lau C, Pham D, Daza RA, Bulfone A, Kowalczyk T & Hevner RF. (2005). Pax6, Tbr2, and Tbr1 are expressed sequentially by radial glia, intermediate progenitor cells, and postmitotic neurons in developing neocortex. J Neurosci 25(1):247-51.

Fallon J, Reid S, Kinyamu R, Opole I, Opole R, Baratta J, Korc M, Endo TL, Duong A, Nguyen G and others. (2000). In vivo induction of massive proliferation, directed migration, and differentiation of neural cells in the adult mammalian brain. Proc Natl Acad Sci U S A 97(26):14686-91.

Fan XT, Xu HW, Cai WQ, Yang H & Liu S. (2004). Antisense Noggin oligodeoxynucleotide administration decreases cell proliferation in the dentate gyrus of adult rats. Neurosci Lett 366(1):107-11.

Ferrer I, Boada Rovira M, Sanchez Guerra ML, Rey MJ & Costa-Jussa F. (2004). Neuropathology and pathogenesis of encephalitis following amyloid-beta immunization in Alzheimer's disease. Brain Pathol 14(1):11-20.

Filippov V, Kronenberg G, Pivneva T, Reuter K, Steiner B, Wang LP, Yamaguchi M, Kettenmann H & Kempermann G. (2003). Subpopulation of nestin-expressing progenitor cells in the adult murine hippocampus shows electrophysiological and morphological characteristics of astrocytes. Mol Cell Neurosci 23(3):373-82.

Fiorentini A, Rosi MC, Grossi C, Luccarini I & Casamenti F. (2010). Lithium improves hippocampal neurogenesis, neuropathology and cognitive functions in APP mutant mice. PLoS One 5(12):e14382.

Fukuda S, Kato F, Tozuka Y, Yamaguchi M, Miyamoto Y & Hisatsune T. (2003). Two distinct subpopulations of nestin-positive cells in adult mouse dentate gyrus. J Neurosci 23(28):9357-66.

Gage FH. (2000). Mammalian neural stem cells. Science 287(5457):1433-8.

Gage SL, Keim SR, Simon JR & Low WC. (1994). Cholinergic innervation of the retrosplenial cortex via the fornix pathway as determined by high affinity choline uptake, choline acetyltransferase activity, and muscarinic receptor binding in the rat. Neurochem Res 19(11):1379-86.

Gao X, Arlotta P, Macklis JD & Chen J. (2007). Conditional knock-out of beta-catenin in postnatal-born dentate gyrus granule neurons results in dendritic malformation. J Neurosci 27(52):14317-25.

Gaura V, Bachoud-Levi AC, Ribeiro MJ, Nguyen JP, Frouin V, Baudic S, Brugieres P, Mangin JF, Boisse MF, Palfi S and others. (2004). Striatal neural grafting improves cortical metabolism in Huntington's disease patients. Brain 127(Pt 1):65-72.

Gauthier S. (2002). Advances in the pharmacotherapy of Alzheimer's disease. Cmaj 166(5):616-23.

Geula C & Mesulam MM. (1989). Cortical cholinergic fibers in aging and Alzheimer's disease: a morphometric study. Neuroscience 33(3):469-81.

Gnjec A, Fonte JA, Atwood C & Martins RN. (2002). Transition metal chelator therapy--a potential treatment for Alzheimer's disease? Front Biosci 7:d1016-23.

Gobeske KT, Das S, Bonaguidi MA, Weiss C, Radulovic J, Disterhoft JF & Kessler JA. (2009). BMP signaling mediates effects of exercise on hippocampal neurogenesis and cognition in mice. PLoS One 4(10):e7506.

Gomez-Pinilla F, Dao L & So V. (1997). Physical exercise induces FGF-2 and its mRNA in the hippocampus. Brain Res 764(1-2):1-8.

Gotz M, Stoykova A & Gruss P. (1998). Pax6 controls radial glia differentiation in the cerebral cortex. Neuron 21(5):1031-44.

Gould E, Tanapat P, McEwen BS, Flugge G & Fuchs E. (1998). Proliferation of granule cell precursors in the dentate gyrus of adult monkeys is diminished by stress. Proc Natl Acad Sci U S A 95(6):3168-71.

Griffin EW, Mulally S, Foley C, Warmington SA, O'Mara SM & Kelly AM. (2011). Aerobic exercise improves hippocampal function and increases BDNF in the serum of young adult males. Physiol Behav.

Han YG, Spassky N, Romaguera-Ros M, Garcia-Verdugo JM, Aguilar A, Schneider-Maunoury S & Alvarez-Buylla A. (2008). Hedgehog signaling and primary cilia are required for the formation of adult neural stem cells. Nat Neurosci 11(3):277-84.

Hanson ND, Nemeroff CB & Owens MJ. (2011). Lithium, but not fluoxetine or the corticotropin-releasing factor receptor 1 receptor antagonist R121919, increases cell proliferation in the adult dentate gyrus. J Pharmacol Exp Ther 337(1):180-6.

Hastings NB & Gould E. (1999). Rapid extension of axons into the CA3 region by adult-generated granule cells. J Comp Neurol 413(1):146-54.

Hattiangady B, Rao MS & Shetty AK. (2004). Chronic temporal lobe epilepsy is associated with severely declined dentate neurogenesis in the adult hippocampus. Neurobiol Dis 17(3):473-90.

Heinrich C, Nitta N, Flubacher A, Müller M, Fahrner A, Kirsch M, Freiman T, Suzuki F, Depaulis A, Frotscher M, Haas CA. (2006). Reelin deficiency and displacement of mature neurons, but not neurogenesis, underlie the formation of granule cell dispersion in the epileptic hippocampus. J Neurosci 26(17):4701-13.

Heins N, Malatesta P, Cecconi F, Nakafuku M, Tucker KL, Hack MA, Chapouton P, Barde YA & Gotz M. (2002). Glial cells generate neurons: the role of the transcription factor Pax6. Nat Neurosci 5(4):308-15.

Heppner FL, Gandy S & McLaurin J. (2004). Current concepts and future prospects for Alzheimer disease vaccines. Alzheimer Dis Assoc Disord 18(1):38-43.

Hong JH & Jang SH. (2010). Neural pathway from nucleus basalis of Meynert passing through the cingulum in the human brain. Brain Res:(Epub ahead of print).

Huangfu D, Liu A, Rakeman AS, Murcia NS, Niswander L & Anderson KV. (2003). Hedgehog signalling in the mouse requires intraflagellar transport proteins. Nature 426(6962):83-7.

Huangfu D & Anderson KV. (2005). Cilia and Hedgehog responsiveness in the mouse. Proc Natl Acad Sci U S A 102(32):11325-30.

Hwang IK, Chung JY, Yoo DY, Yi SS, Youn HY, Seong JK & Yoon YS. (2010). Comparing the effects of acupuncture and electroacupuncture at Zusanli and Baihui on cell proliferation and neuroblast differentiation in the rat hippocampus. J Vet Med Sci 72(3):279-84.

Imayoshi I, Sakamoto M, Yamaguchi M, Mori K & Kageyama R. (2010). Essential roles of Notch signaling in maintenance of neural stem cells in developing and adult brains. J Neurosci 30(9):3489-98.

Jessberger S & Kempermann G. (2003). Adult-born hippocampal neurons mature into activity-dependent responsiveness. Eur J Neurosci 18(10):2707-12.

Jessberger S, Romer B, Babu H & Kempermann G. (2005). Seizures induce proliferation and dispersion of doublecortin-positive hippocampal progenitor cells. Exp Neurol 196(2):342-51.

Jin K, Minami M, Lan JQ, Mao XO, Batteur S, Simon RP & Greenberg DA. (2001). Neurogenesis in dentate subgranular zone and rostral subventricular zone after focal cerebral ischemia in the rat. Proc Natl Acad Sci U S A 98(8):4710-5.

Johansson CB, Momma S, Clarke DL, Risling M, Lendahl U & Frisen J. (1999). Identification of a neural stem cell in the adult mammalian central nervous system. Cell 96(1):25-34.

Kaneko Y, Sakakibara S, Imai T, Suzuki A, Nakamura Y, Sawamoto K, Ogawa Y, Toyama Y, Miyata T & Okano H. (2000). Musashi1: an evolutionally conserved marker for CNS progenitor cells including neural stem cells. Dev Neurosci 22(1-2):139-53.

Kawai T, Takagi N, Miyake-Takagi K, Okuyama N, Mochizuki N & Takeo S. (2004). Characterization of BrdU-positive neurons induced by transient global ischemia in adult hippocampus. J Cereb Blood Flow Metab 24(5):548-55.

Kempermann G, Kuhn HG & Gage FH. (1997). More hippocampal neurons in adult mice living in an enriched environment. Nature 386(6624):493-5.

Kempermann G & Gage FH. (2000). Neurogenesis in the adult hippocampus. Novartis Found Symp 231:220-35; discussion 235-41, 302-6.

Kempermann G, Gast D & Gage FH. (2002). Neuroplasticity in old age: sustained fivefold induction of hippocampal neurogenesis by long-term environmental enrichment. Ann Neurol 52(2):135-43.

Kempermann G, Gast D, Kronenberg G, Yamaguchi M & Gage FH. (2003). Early determination and long-term persistence of adult-generated new neurons in the hippocampus of mice. Development 130(2):391-9.

Kempermann G, Jessberger S, Steiner B & Kronenberg G. (2004). Milestones of neuronal development in the adult hippocampus. Trends Neurosci 27(8):447-52.

Kim EH, Kim YJ, Lee HJ, Huh Y, Chung JH, Seo JC, Kang JE, Lee HJ, Yim SV & Kim CJ. (2001). Acupuncture increases cell proliferation in dentate gyrus after transient global ischemia in gerbils. Neurosci Lett 297(1):21-4.

Kim EH, Jang MH, Shin MC, Lim BV, Kim HB, Kim YJ, Chung JH & Kim CJ. (2002). Acupuncture increases cell proliferation and neuropeptide Y expression in dentate gyrus of streptozotocin-induced diabetic rats. Neurosci Lett 327(1):33-6.

Kim EJ, Battiste J, Nakagawa Y & Johnson JE. (2008). Ascl1 (Mash1) lineage cells contribute to discrete cell populations in CNS architecture. Mol Cell Neurosci 38(4):595-606.

Kim JS, Chang MY, Yu IT, Kim JH, Lee SH, Lee YS & Son H. (2004). Lithium selectively increases neuronal differentiation of hippocampal neural progenitor cells both in vitro and in vivo. J Neurochem 89(2):324-36.

Kitamura Y, Doi M, Kuwatsuka K, Onoue Y, Miyazaki I, Shinomiya K, Koyama T, Sendo T, Kawasaki H, Asanuma M and others. (2011). Chronic treatment with imipramine and lithium increases cell proliferation in the hippocampus in adrenocorticotropic hormone-treated rats. Biol Pharm Bull 34(1):77-81.

Kobayashi K, Ikeda Y, Sakai A, Yamasaki N, Haneda E, Miyakawa T & Suzuki H. (2010). Reversal of hippocampal neuronal maturation by serotonergic antidepressants. Proc Natl Acad Sci U S A 107(18):8434-9.

Kohl Z, Kuhn HG, Cooper-Kuhn CM, Winkler J, Aigner L & Kempermann G. (2002). Preweaning enrichment has no lasting effects on adult hippocampal neurogenesis in four-month-old mice. Genes Brain Behav 1(1):46-54.

Kralic JE, Ledergerber DA & Fritschy JM. (2005). Disruption of the neurogenic potential of the dentate gyrus in a mouse model of temporal lobe epilepsy with focal seizures. Eur J Neurosci 22(8):1916-27.

Kronenberg G, Reuter K, Steiner B, Brandt MD, Jessberger S, Yamaguchi M & Kempermann G. (2003). Subpopulations of proliferating cells of the adult hippocampus respond differently to physiologic neurogenic stimuli. J Comp Neurol 467(4):455-63.

Lai K, Kaspar BK, Gage FH & Schaffer DV. (2003). Sonic hedgehog regulates adult neural progenitor proliferation in vitro and in vivo. Nat Neurosci 6(1):21-7.

Lanz TA, Himes CS, Pallante G, Adams L, Yamazaki S, Amore B & Merchant KM. (2003). The gamma-secretase inhibitor N-[N-(3,5-difluorophenacetyl)-L-alanyl]-S-phenylglycine t-butyl ester reduces A beta levels in vivo in plasma and cerebrospinal fluid in young (plaque-free) and aged (plaque-bearing) Tg2576 mice. J Pharmacol Exp Ther 305(3):864-71.

Leanza G. (1998). Chronic elevation of amyloid precursor protein expression in the neocortex and hippocampus of rats with selective cholinergic lesions. Neurosci Lett 257(1):53-6.

Lee BH & Kim YK. (2010). The roles of BDNF in the pathophysiology of major depression and in antidepressant treatment. Psychiatry Investig 7(4):231-5.

Lee JE, Hollenberg SM, Snider L, Turner DL, Lipnick N & Weintraub H. (1995). Conversion of Xenopus ectoderm into neurons by NeuroD, a basic helix-loop-helix protein. Science 268(5212):836-44.

Leuner B, Glasper ER & Gould E. (2010). Sexual experience promotes adult neurogenesis in the hippocampus despite an initial elevation in stress hormones. PLoS One 5(7):e11597.

Li Z, Gao C, Huang H, Sun W, Yi H, Fan X & Xu H. (2010). Neurotransmitter phenotype differentiation and synapse formation of neural precursors engrafting in amyloid-beta injured rat hippocampus. J Alzheimers Dis 21(4):1233-47.

Lie DC, Colamarino SA, Song HJ, Desire L, Mira H, Consiglio A, Lein ES, Jessberger S, Lansford H, Dearie AR and others. (2005). Wnt signalling regulates adult hippocampal neurogenesis. Nature 437(7063):1370-5.

Lin L, LeBlanc CJ, Deacon TW & Isacson O. (1998). Chronic cognitive deficits and amyloid precursor protein elevation after selective immunotoxin lesions of the basal forebrain cholinergic system. Neuroreport 9(3):547-52.

Liu CZ, Yu JC, Zhang XZ, Fu WW, Wang T & Han JX. (2006). Acupuncture prevents cognitive deficits and oxidative stress in cerebral multi-infarction rats. Neurosci Lett 393(1):45-50.

Liu L, Ikonen S, Heikkinen T, Heikkila M, Puolivali J, van Groen T & Tanila H. (2002). Effects of fimbria-fornix lesion and amyloid pathology on spatial learning and memory in transgenic APP+PS1 mice. Behav Brain Res 134(1-2):433-45.

Liu M, Pleasure SJ, Collins AE, Noebels JL, Naya FJ, Tsai MJ & Lowenstein DH. (2000). Loss of BETA2/NeuroD leads to malformation of the dentate gyrus and epilepsy. Proc Natl Acad Sci U S A 97(2):865-70.

Liu Q, Yu J, Mi WL, Mao-Ying QL, Yang R, Wang YQ & Wu GC. (2007). Electroacupuncture attenuates the decrease of hippocampal progenitor cell proliferation in the adult rats exposed to chronic unpredictable stress. Life Sci 81(21-22):1489-95.

Liu Y, Namba T, Liu J, Suzuki R, Shioda S & Seki T. (2010). Glial fibrillary acidic protein-expressing neural progenitors give rise to immature neurons via early intermediate progenitors expressing both glial fibrillary acidic protein and neuronal markers in the adult hippocampus. Neuroscience 166(1):241-51.

Llorens-Martin M, Tejeda GS & Trejo JL. (2010). Differential regulation of the variations induced by environmental richness in adult neurogenesis as a function of time: a dual birthdating analysis. PLoS One 5(8):e12188.

Lois C & Alvarez-Buylla A. (1994). Long-distance neuronal migration in the adult mammalian brain. Science 264(5162):1145-8.

Lois C, Garcia-Verdugo JM & Alvarez-Buylla A. (1996). Chain migration of neuronal precursors. Science 271(5251):978-81.

Lugert S, Basak O, Knuckles P, Haussler U, Fabel K, Gotz M, Haas CA, Kempermann G, Taylor V & Giachino C. (2010). Quiescent and active hippocampal neural stem cells with distinct morphologies respond selectively to physiological and pathological stimuli and aging. Cell Stem Cell 6(5):445-56.

Machold R, Hayashi S, Rutlin M, Muzumdar MD, Nery S, Corbin JG, Gritli-Linde A, Dellovade T, Porter JA, Rubin LL and others. (2003). Sonic hedgehog is required for progenitor cell maintenance in telencephalic stem cell niches. Neuron 39(6):937-50.

Maekawa M, Takashima N, Arai Y, Nomura T, Inokuchi K, Yuasa S & Osumi N. (2005). Pax6 is required for production and maintenance of progenitor cells in postnatal hippocampal neurogenesis. Genes Cells 10(10):1001-14.

Magavi SS & Macklis JD. (2002). Induction of neuronal type-specific neurogenesis in the cerebral cortex of adult mice: manipulation of neural precursors in situ. Brain Res Dev Brain Res 134(1-2):57-76.

Malberg JE, Eisch AJ, Nestler EJ & Duman RS. (2000). Chronic antidepressant treatment increases neurogenesis in adult rat hippocampus. J Neurosci 20(24):9104-10.

Markakis EA & Gage FH. (1999). Adult-generated neurons in the dentate gyrus send axonal projections to field CA3 and are surrounded by synaptic vesicles. J Comp Neurol 406(4):449-60.

Martino G & Pluchino S. (2006). The therapeutic potential of neural stem cells. Nat Rev Neurosci 7(5):395-406.

Mason HA, Rakowiecki SM, Gridley T & Fishell G. (2006). Loss of notch activity in the developing central nervous system leads to increased cell death. Dev Neurosci 28(1-2):49-57.

McDonald HY & Wojtowicz JM. (2005). Dynamics of neurogenesis in the dentate gyrus of adult rats. Neurosci Lett 385(1):70-5.

Mehler MF, Mabie PC, Zhu G, Gokhan S & Kessler JA. (2000). Developmental changes in progenitor cell responsiveness to bone morphogenetic proteins differentially modulate progressive CNS lineage fate. Dev Neurosci 22(1-2):74-85.
Miles DK & Kernie SG. (2008). Hypoxic-ischemic brain injury activates early hippocampal stem/progenitor cells to replace vulnerable neuroblasts. Hippocampus 18(8):793-806.
Ming GL & Song H. (2005). Adult neurogenesis in the mammalian central nervous system. Annu Rev Neurosci 28:223-50.
Miyata T, Maeda T & Lee JE. (1999). NeuroD is required for differentiation of the granule cells in the cerebellum and hippocampus. Genes Dev 13(13):1647-52.
Murakami S, Imbe H, Morikawa Y, Kubo C & Senba E. (2005). Chronic stress, as well as acute stress, reduces BDNF mRNA expression in the rat hippocampus but less robustly. Neurosci Res 53(2):129-39.
Muroyama Y, Fujihara M, Ikeya M, Kondoh H & Takada S. (2002). Wnt signaling plays an essential role in neuronal specification of the dorsal spinal cord. Genes Dev 16(5):548-53.
Nacher J, Varea E, Blasco-Ibanez JM, Castillo-Gomez E, Crespo C, Martinez-Guijarro FJ & McEwen BS. (2005). Expression of the transcription factor Pax 6 in the adult rat dentate gyrus. J Neurosci Res 81(6):753-61.
Nakatomi H, Kuriu T, Okabe S, Yamamoto S, Hatano O, Kawahara N, Tamura A, Kirino T & Nakafuku M. (2002). Regeneration of hippocampal pyramidal neurons after ischemic brain injury by recruitment of endogenous neural progenitors. Cell 110(4):429-41.
Nasrallah HA, Hopkins T & Pixley SK. (2011). Differential effects of antipsychotic and antidepressant drugs on neurogenic regions in rats. Brain Res 1354:23-9.
Olson AK, Eadie BD, Ernst C & Christie BR. (2006). Environmental enrichment and voluntary exercise massively increase neurogenesis in the adult hippocampus via dissociable pathways. Hippocampus 16(3):250-60.
Palma V, Lim DA, Dahmane N, Sanchez P, Brionne TC, Herzberg CD, Gitton Y, Carleton A, Alvarez-Buylla A & Ruiz i Altaba A. (2005). Sonic hedgehog controls stem cell behavior in the postnatal and adult brain. Development 132(2):335-44.
Palmer TD, Willhoite AR & Gage FH. (2000). Vascular niche for adult hippocampal neurogenesis. J Comp Neurol 425(4):479-94.
Parent JM, Yu TW, Leibowitz RT, Geschwind DH, Sloviter RS & Lowenstein DH. (1997). Dentate granule cell neurogenesis is increased by seizures and contributes to aberrant network reorganization in the adult rat hippocampus. J Neurosci 17(10):3727-38.
Parent JM, Vexler ZS, Gong C, Derugin N & Ferriero DM. (2002). Rat forebrain neurogenesis and striatal neuron replacement after focal stroke. Ann Neurol 52(6):802-13.
Parent JM & Murphy GG. (2008). Mechanisms and functional significance of aberrant seizure-induced hippocampal neurogenesis. Epilepsia 49 Suppl 5:19-25.
Pilar-Cuellar F, Vidal R & Pazos A. (2011). Subchronic treatment with fluoxetine and the 5-HT(2A) antagonist ketanserin upregulates hippocampal BDNF and beta-catenin in parallel with antidepressant-like effect. Br J Pharmacol.
Pluchino S, Zanotti L, Rossi B, Brambilla E, Ottoboni L, Salani G, Martinello M, Cattalini A, Bergami A, Furlan R and others. (2005). Neurosphere-derived multipotent

precursors promote neuroprotection by an immunomodulatory mechanism. Nature 436(7048):266-71.

Reus GZ, Stringari RB, Ribeiro KF, Ferraro AK, Vitto MF, Cesconetto P, Souza CT & Quevedo J. (2011). Ketamine plus imipramine treatment induces antidepressant-like behavior and increases CREB and BDNF protein levels and PKA and PKC phosphorylation in rat brain. Behav Brain Res 221(1):166-71.

Rogoz Z, Skuza G & Legutko B. (2008). Repeated co-treatment with fluoxetine and amantadine induces brain-derived neurotrophic factor gene expression in rats. Pharmacol Rep 60(6):817-26.

Russo-Neustadt A, Beard RC & Cotman CW. (1999). Exercise, antidepressant medications, and enhanced brain derived neurotrophic factor expression. Neuropsychopharmacology 21(5):679-82.

Sakakibara S, Imai T, Hamaguchi K, Okabe M, Aruga J, Nakajima K, Yasutomi D, Nagata T, Kurihara Y, Uesugi S and others. (1996). Mouse-Musashi-1, a neural RNA-binding protein highly enriched in the mammalian CNS stem cell. Dev Biol 176(2):230-42.

Sakurai K & Osumi N. (2008). The neurogenesis-controlling factor, Pax6, inhibits proliferation and promotes maturation in murine astrocytes. J Neurosci 28(18):4604-12.

Santarelli L, Saxe M, Gross C, Surget A, Battaglia F, Dulawa S, Weisstaub N, Lee J, Duman R, Arancio O and others. (2003). Requirement of hippocampal neurogenesis for the behavioral effects of antidepressants. Science 301(5634):805-9.

Saper CB. (1984). Organization of cerebral cortical afferent systems in the rat. II. Magnocellular basal nucleus. J Comp Neurol 222(3):313-42.

Schliebs R & Arendt T. (2006). The significance of the cholinergic system in the brain during aging and in Alzheimer's disease. J Neural Transm 113(11):1625-44.

Schmidt RH, Scholten KJ & Maughan PH. (1999). Time course for recovery of water maze performance and central cholinergic innervation after fluid percussion injury. J Neurotrauma 16(12):1139-47.

Seri B, Garcia-Verdugo JM, McEwen BS & Alvarez-Buylla A. (2001). Astrocytes give rise to new neurons in the adult mammalian hippocampus. J Neurosci 21(18):7153-60.

Seri B, Garcia-Verdugo JM, Collado-Morente L, McEwen BS & Alvarez-Buylla A. (2004). Cell types, lineage, and architecture of the germinal zone in the adult dentate gyrus. J Comp Neurol 478(4):359-78.

Silva R, Mesquita AR, Bessa J, Sousa JC, Sotiropoulos I, Leao P, Almeida OF & Sousa N. (2008). Lithium blocks stress-induced changes in depressive-like behavior and hippocampal cell fate: the role of glycogen-synthase-kinase-3beta. Neuroscience 152(3):656-69.

Son H, Yu IT, Hwang SJ, Kim JS, Lee SH, Lee YS & Kaang BK. (2003). Lithium enhances long-term potentiation independently of hippocampal neurogenesis in the rat dentate gyrus. J Neurochem 85(4):872-81.

Song H, Kempermann G, Overstreet Wadiche L, Zhao C, Schinder AF & Bischofberger J. (2005). New neurons in the adult mammalian brain: synaptogenesis and functional integration. J Neurosci 25(45):10366-8.

Stanfield BB & Trice JE. (1988). Evidence that granule cells generated in the dentate gyrus of adult rats extend axonal projections. Exp Brain Res 72(2):399-406.

Steiner B, Kronenberg G, Jessberger S, Brandt MD, Reuter K & Kempermann G. (2004). Differential regulation of gliogenesis in the context of adult hippocampal neurogenesis in mice. Glia 46(1):41-52.

Steiner B, Klempin F, Wang L, Kott M, Kettenmann H & Kempermann G. (2006). Type-2 cells as link between glial and neuronal lineage in adult hippocampal neurogenesis. Glia 54(8):805-14.

Steiner B, Zurborg S, Horster H, Fabel K & Kempermann G. (2008). Differential 24 h responsiveness of Prox1-expressing precursor cells in adult hippocampal neurogenesis to physical activity, environmental enrichment, and kainic acid-induced seizures. Neuroscience 154(2):521-9.

Su XW, Li XY, Banasr M & Duman RS. (2009). Eszopiclone and fluoxetine enhance the survival of newborn neurons in the adult rat hippocampus. Int J Neuropsychopharmacol 12(10):1421-8.

Suh H, Consiglio A, Ray J, Sawai T, D'Amour KA & Gage FH. (2007). In vivo fate analysis reveals the multipotent and self-renewal capacities of Sox2+ neural stem cells in the adult hippocampus. Cell Stem Cell 1(5):515-28.

Sui Y, Zhang Z, Guo Y, Sun Y, Zhang X, Xie C, Li Y & Xi G. (2009). The function of Notch1 signaling was increased in parallel with neurogenesis in rat hippocampus after chronic fluoxetine administration. Biol Pharm Bull 32(10):1776-82.

Szenborn M. (1993). Neuropathological study on the nucleus basalis of Meynert in mature and old age. Patol Pol 44(4):211-6.

Traiffort E, Charytoniuk DA, Faure H & Ruat M. (1998). Regional distribution of Sonic Hedgehog, patched, and smoothened mRNA in the adult rat brain. J Neurochem 70(3):1327-30.

Urrea C, Castellanos DA, Sagen J, Tsoulfas P, Bramlett HM & Dietrich WD. (2007). Widespread cellular proliferation and focal neurogenesis after traumatic brain injury in the rat. Restor Neurol Neurosci 25(1):65-76.

van Praag H, Christie BR, Sejnowski TJ & Gage FH. (1999a). Running enhances neurogenesis, learning, and long-term potentiation in mice. Proc Natl Acad Sci U S A 96(23):13427-31.

van Praag H, Kempermann G & Gage FH. (1999b). Running increases cell proliferation and neurogenesis in the adult mouse dentate gyrus. Nat Neurosci 2(3):266-70.

van Praag H, Schinder AF, Christie BR, Toni N, Palmer TD & Gage FH. (2002). Functional neurogenesis in the adult hippocampus. Nature 415(6875):1030-4.

Wang C, Zhang M, Sun C, Cai Y, You Y, Huang L & Liu F. (2011). Sustained increase in adult neurogenesis in the rat hippocampal dentate gyrus after transient brain ischemia. Neurosci Lett 488(1):70-5.

Wang JW, David DJ, Monckton JE, Battaglia F & Hen R. (2008). Chronic fluoxetine stimulates maturation and synaptic plasticity of adult-born hippocampal granule cells. J Neurosci 28(6):1374-84.

Wexler EM, Geschwind DH & Palmer TD. (2008). Lithium regulates adult hippocampal progenitor development through canonical Wnt pathway activation. Mol Psychiatry 13(3):285-92.

Willert K, Brown JD, Danenberg E, Duncan AW, Weissman IL, Reya T, Yates JR, 3rd & Nusse R. (2003). Wnt proteins are lipid-modified and can act as stem cell growth factors. Nature 423(6938):448-52.

Xuan AG, Luo M, Ji WD & Long DH. (2009). Effects of engrafted neural stem cells in Alzheimer's disease rats. Neurosci Lett 450(2):167-71.

Yagita Y, Kitagawa K, Ohtsuki T, Takasawa K, Miyata T, Okano H, Hori M & Matsumoto M. (2001). Neurogenesis by progenitor cells in the ischemic adult rat hippocampus. Stroke 32(8):1890-6.

Yamamoto T, Hirayama A, Hosoe N, Furube M & Hirano S. (2009). Soft-diet feeding inhibits adult neurogenesis in hippocampus of mice. Bull Tokyo Dent Coll 50(3):117-24.

Yap JJ, Takase LF, Kochman LJ, Fornal CA, Miczek KA & Jacobs BL. (2006). Repeated brief social defeat episodes in mice: effects on cell proliferation in the dentate gyrus. Behav Brain Res 172(2):344-50.

Yoshimura S, Takagi Y, Harada J, Teramoto T, Thomas SS, Waeber C, Bakowska JC, Breakefield XO & Moskowitz MA. (2001). FGF-2 regulation of neurogenesis in adult hippocampus after brain injury. Proc Natl Acad Sci U S A 98(10):5874-9.

Yu J, Liu C, Zhang X & Han J. (2005). Acupuncture improved cognitive impairment caused by multi-infarct dementia in rats. Physiol Behav 86(4):434-41.

Yu TS, Zhang G, Liebl DJ & Kernie SG. (2008). Traumatic brain injury-induced hippocampal neurogenesis requires activation of early nestin-expressing progenitors. J Neurosci 28(48):12901-12.

Zhang X, Jin G, Tian M, Qin J & Huang Z. (2007). The denervated hippocampus provides proper microenvironment for the survival and differentiation of neural progenitors. Neurosci Lett 414(2):115-20.

Zhao C, Teng EM, Summers RG, Jr., Ming GL & Gage FH. (2006). Distinct morphological stages of dentate granule neuron maturation in the adult mouse hippocampus. J Neurosci 26(1):3-11.

Zigova T, Pencea V, Wiegand SJ & Luskin MB. (1998). Intraventricular administration of BDNF increases the number of newly generated neurons in the adult olfactory bulb. Mol Cell Neurosci 11(4):234-45.

3

Cellular Organization of the Subventricular Zone in the Adult Human Brain: A Niche of Neural Stem Cells

Oscar Gonzalez-Perez
Laboratory of Neuroscience, Facultad de Psicología, Universidad de Colima,
Department of Neuroscience,
Centro Universitario de Ciencias de la Salud,
Universidad de Guadalajara
Mexico

1. Introduction

The dogma that the brain is a quiescent organ incapable of postnatal neuron generation was first challenged in the sixties by Joseph Altman (Altman, 1962). He described the presence of thymidine-labeled cells in the subependymal zone located along the ventricular walls, which suggested the presence of dividing neurons in this brain region (Altman and Gopal, 1965; Altman and Das, 1967). A decade after, these findings were confirmed by other group using electron microscopy analyses (Kaplan and Hinds, 1977). Later, further studies described ongoing neurogenesis in female canaries (Goldman and Nottebohm, 1983), lizards (Pérez-Cañellas and García-Verdugo, 1996) and the adult mammalian brain (McDermott and Lantos, 1990; McDermott and Lantos, 1991; Lois and Alvarez-Buylla, 1993; Kornack and Rakic, 1995; Huang et al., 1998; Garcia-Verdugo et al., 2002). This process is mainly confined to the subventricular zone (SVZ) of the forebrain and the subgranular zone (SGZ) of the dentate gyrus in the hippocampus (Reznikov, 1991; Luskin, 1993; Lois and Alvarez-Buylla, 1994). The SVZ is the largest neurogenic niche in the adult brain (Luskin, 1993; Alvarez-Buylla and Garcia-Verdugo, 2002). Within this region resides a subpopulation of astrocytes with stem-cell-like features (Doetsch et al., 1999; Laywell et al., 2000; Imura et al., 2003; Morshead et al., 2003; Garcia et al., 2004). Recently, it has been suggested that the SVZ may be not only a source of neural precursor for brain repair, but also a source of brain tumors (Ignatova et al., 2002; Galli et al., 2004; Sanai et al., 2005; Vescovi et al., 2006). These hypotheses highlight the importance of studying and understanding the organization and regulation of the SVZ precursors. This chapter discusses and analyzes the cytoarchitecture and cellular composition of the human SVZ, as well as, its potential implications on the clinical treatment of neurodegenerative diseases and brain tumors.

2. Human neural stem cells

The gold standard for determining the presence of neural stem cells is the neurosphere assay (Reynolds and Rietze, 2005). This assay consists in plating a suspension of cells under serum-free, growth-factor-supplemented, non-adherent conditions in-vitro; thus, stem-like cells are able to divide and form multipotent undifferentiated clones called neurospheres (Reynolds and Weiss, 1992). The neurospheres can be serially dissociated and their single-cell clones are able to generate further spheres, while cells not capable of self-renewal eventually die (Reynolds and Rietze, 2005). These neurospheres are multipotent and can generate neurons, astrocytes and/or oligodendrocytes after the removal of mitogens and transfer to adherent plates (Reynolds and Weiss, 1992; Doetsch et al., 2002).

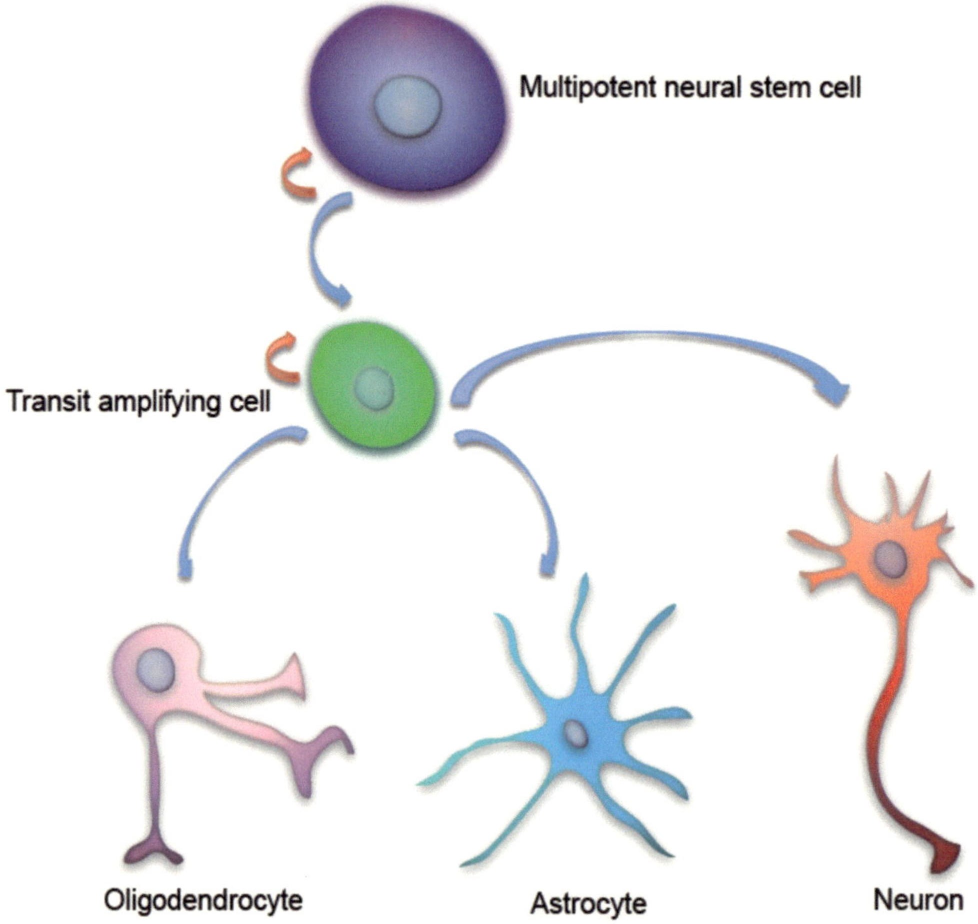

Fig. 1. Neural stem cells and their progeny in the adult SVZ. When a multipotent type B cell (on the top) divides, it generates to a type C cell, also known as transit-amplifying precursors, which can give rise to neurons and glial cells. The petite curved arrows represent the self-renewal capacity of type B and type C cells. *Figure reproduced with permission from: Alvarez-Palazuelos et al. Current Signal Transduction Therapy 2011;6(3) (Alvarez-Palazuelos et al., 2011). Copyright 2011 Bentham Science Publishers.*

Using neurosphere assays, neural stem cells have been isolated in human fetal cells (Chalmers-Redman et al., 1997). These multipotent human cells are also capable of self-renewal when maintained under serum-free conditions (Nunes et al., 2003). In the adult human brain, neural stem cells can be isolated from the SVZ and SGZ and give rise to neurons, oligodendrocytes and astrocytes in vitro (Figure 1) (Kukekov et al., 1999). Further evidence indicates that SVZ explants isolated from temporal lobectomies in patients with refractory epilepsy are capable of producing neurons in vitro (Kirschenbaum et al., 1994; Pincus et al., 1997). It has been suggested that new neurons are generated in the SGZ of the human hippocampus in vivo (Eriksson et al., 1998). This evidence has been obtained from postmortem brain tissue derived from patients with lung squamous cell carcinomas, who were diagnostically infused with bromodeoxyuridine to label mitotic cells. Nevertheless, despite this promising advances, none of these studies can demonstrate that the adult human brain possess neural stem cells per se, namely with self-renewal and multipotency properties (Vescovi et al., 2006).

2.1 The subventricular zone in the adult mammalian brain

The SVZ is the largest source of new neurons in the adult brain. This neurogenic region is located adjacent to the ependyma at the lateral wall of the lateral ventricles (Figure 2). The epithelial layer is composed by multiciliated non-mitotic ependymal cells, which contribute to the flow of cerebrospinal fluid and appear to play a role in the modulation of the stem cell niche (Lim et al., 2000; Spassky et al., 2005; Sawamoto et al., 2006; Mirzadeh et al., 2008). The SVZ contains a slowly dividing primary progeny (type B cells) and rapidly dividing cell precursors (type C cells) (Figure 2). Type B cells have been identified as the primary neural progenitors i.e., neural stem cells in the adult brain (Doetsch et al., 1999). Interestingly, based on differences in their location and morphology, type B progenitors are a subpopulation of astrocytes that can be categorized into two types: B1 and B2 astrocytes (Doetsch et al., 1997). At the ependymal side of the SVZ, type B1 astocytes are usually closely associated with the ependymal layer through adherens and gap junctions, and frequently extend a short apical process that reaches the ventricle (Mirzadeh et al., 2008). At the parenchymal side of the SVZ, type B1 astrocytes contact the basal lamina and blood vessels that underlie the SVZ (Shen et al., 2004; Mirzadeh et al., 2008). The ventricular end of the apical process of type B1 cells contains a non-motile primary cilium that contacts the cerebrospinal fluid (Mirzadeh et al., 2008). In contrast, type B2 astrocytes are located close to the brain parenchyma (Mirzadeh et al., 2008). It has been suggested that SVZ astrocytes play a dual role in neurogenesis, serving as both neural stem cells per se and supporting cells that promote neurogenesis (Lim and Alvarez-Buylla, 1999; Song et al., 2002).

The immediate progeny of type B1 astrocytes is known as transit amplifying progenitors or type C cells, which give rise to migrating neuroblasts (type A cells) (Figure 2)(Kriegstein and Alvarez-Buylla, 2009). These young neurons are surrounded by a glial sheath and migrate anteriorly toward the olfactory bulb (Jankovski and Sotelo, 1996; Lois et al., 1996; Doetsch et al., 1997). The adult SVZ also generates oligodendrocytes, although in much lower numbers than neuroblasts (Menn et al., 2006; Gonzalez-Perez et al., 2009; Gonzalez-Perez and Quinones-Hinojosa, 2010; Gonzalez-Perez et al., 2010b; Gonzalez-Perez and Alvarez-Buylla, 2011). The mechanisms that control the cell proliferation and renewal in the SVZ are not well-known, but increasing evidence indicates that neural stem cells are instructed via cell-cell contacts and extracellular signals from ependymal cells, immunological cells, the extracellular matrix, microglia, the local vasculature, neuronal inputs and the cerebrospinal fluid (Gonzalez-Perez et al., 2010a; Gonzalez-Perez and Alvarez-Buylla, 2011; Ihrie and Alvarez-Buylla, 2011).

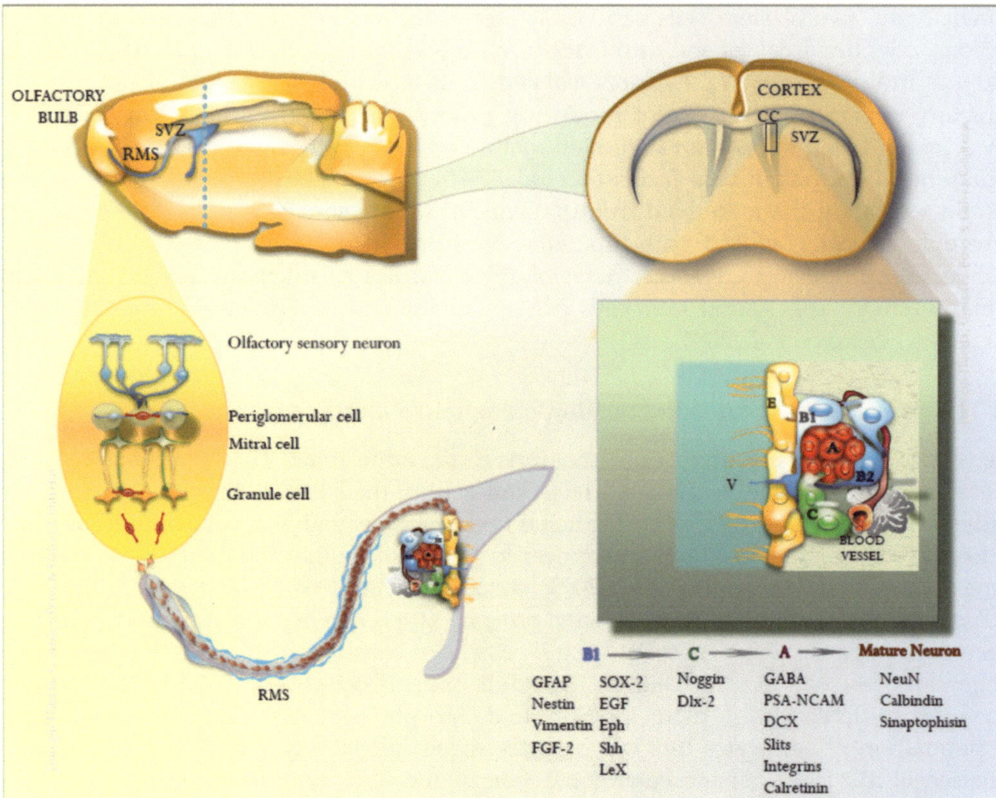

Fig. 2. Schematic representation of the localization and cellular composition of the adult subventricular zone (SVZ) in the rodent brain. Neuroblasts generated in the SVZ niche migrate to the olfactory bulb and, then, differentiate into granular and periglomerular GABAergic interneurons. Cell markers expressed by type B, type C, type A and mature neurons are listed under each cell label. V: Ventricle; E: Ependymal cell; CC: Corpus callosum; RMS: Rostral migratory stream. *Figure reproduced with permission from: Gonzalez-Perez et al. Current Immunology Reviews 2010;6(3):167 (Gonzalez-Perez et al., 2010b). Copyright 2010 Bentham Science Publishers.*

2.2 Cell type markers of the SVZ progenitors

As mentioned above, type B1 cells have astrocytic morphology and ultrastructure and express molecular markers that have been usually associated with astroglia, such as: the glial fibrillary acidic protein (GFAP), nestin, vimentin, connexin 30, the astrocyte-specific glutamate transporter (GLAST) and the brain-lipid-binding protein (BLBP) (Doetsch et al., 1999; Hartfuss et al., 2001; Kriegstein and Alvarez-Buylla, 2009). Type B1 astrocytes also express the cell surface carbohydrate Lewis X (LeX)/CD15/SSEA-, which has been proposed as a marker of neural stem cells in the SVZ (Capela and Temple, 2002). In addition, type B1 cells express prominin-1, also known as CD133, a protein commonly used as a stem-cell marker (Coskun et al., 2008; Shmelkov et al., 2008; Beckervordersandforth et al., 2010). However, prominin-1 expression at the apical endings of type B1 cells appears to be dynamically regulated (Mirzadeh et al., 2008). Therefore, given that Type B1 cells have many astroglial characteristics, finding potential markers to distinguish the B1 cell progeny from other non-multipotent astrocytes would be very useful in future studies. Some markers generally used to identify type-C cells are the epidermal growth factor receptor (EGFR), Dlx2 and Ascl1 (also known as Mash1) transcription factors (Doetsch et al., 2002; Parras et al., 2004), while doublecortin and the polysialylated neural cell adhesion molecule are useful to identify A-cell progeny (SVZ neuroblasts) (Lois and Alvarez-Buylla, 1994; Rousselot et al., 1995; Francis et al., 1999). Ependymal cells express S100beta and CD24 (Raponi et al., 2007; Mirzadeh et al., 2008).

Longitudinal analysis of molecular markers within the SVZ progenitor cells indicates that many of these proteins are expressed at particular points along the cell differentiation of neural stem cells. For instance, while GFAP expression is restricted to B cell progeny, GLAST and the orphan nuclear receptor Tlx is also present in a subpopulation of type C cells (Pastrana et al., 2009). Similarly, EGFR and Mash1are expressed in a limited number of type B cells, and they possibly may be useful to label ''activated'' type B cells (Doetsch et al., 2002; Gonzalez-Perez et al., 2010a; Gonzalez-Perez and Alvarez-Buylla, 2011). In addition, nestin expression that was thought to be exclusive to adult neural stem cells has been found broadly expressed within the brain (Hendrickson et al., 2011). Taken together, this evidence indicates that marker for stem and/or progenitor cells are likely to identify overlapping, but not identical subpopulations of SVZ cells. Therefore, researchers should be cautious when assigning biological characteristics to a subset of SVZ cells (Chojnacki et al., 2009).

2.3 The cell composition and architecture of the human subventricular zone

The human SVZ is located within the lateral wall of the lateral ventricles and consist of four layers with very particular cell compositions (Figure 3) (Quinones-Hinojosa et al., 2006). The layer adjacent to the lateral ventricle (Layer I) is formed by a monolayer of multiciliated ependymal cells with basal cytoplasm expansions that are either tangential or perpendicular to the ventricular surface. The Layer II or hypocellular layer is comprised of some of ependymal cytoplasm expansions interconnected with a number of astrocyte processes and very rare astrocytic and neuronal cell bodies (Figure 3) (Quinones-Hinojosa et al., 2006). The biological relevance of this hypocellular gap, is unknown, but it may be a remnant of the brain development at embryonic stages, because from this region a number of new neurons born and migrate radially and tangentially toward cortical and subcortical structures (Guerrero-Cazares et al., 2011). Other hypotheses suggest that the astrocytic and ependymal interconnections within this layer regulate neuronal functions or preserve metabolic homeostasis in the SVZ (Ihrie and Alvarez-Buylla, 2011; Ihrie et al., 2011). Abutting the hypocellular layer is a ribbon of astrocyte somata (Layer III) (Figure 3), which shows some proliferative activity as indicated by postmortem Ki67 expression (Sanai et al., 2004; Quinones-Hinojosa et al., 2006). It is believed that a subpopulation of astrocytes within this ribbon can proliferate in vivo, as well as form multipotent neurospheres (Sanai et al., 2004; Quinones-Hinojosa et al., 2007). Based on differences in their location and morphology by electron microscopy, the SVZ astrocytes can be subdivided into three types (Quinones-Hinojosa et al., 2006): The small astrocytes that are predominantly found in the hypocellular layer, and possess long, tangential cytoplasm processes. These astrocytes contain scarce cytoplasm, very dense bundles of intermediate filaments and sparse organelles. The second type of astroglia is the large astrocyte that has large cytoplasm expansions, abundant organelles and is found at the interface between Layer II and III and within the ribbon itself. This type of astrocyte is primarily found in the medial wall at the level of the body of the lateral ventricle. The third type of astrocyte is also large, but it possesses few organelles and is primarily found in the ventral temporal horn overlying the hippocampus. To date, the physiological relevance n of these three types of astrocytes is unknown differences, but in vitro evidence sugests that neural stem cells may belong to one of these astrocytic subtypes (Sanai et al., 2004). On the other hand, small clusters of displaced ependymal ells can be occasionally found embedded within this ribbon. This type of cells has abundant cilia, junctional complexes and microvilli (Figure 3). Finally, a few oligodendrocytes that do not appear to be myelinating axons are also seen in the Layer III (Figure 3). The deepest layer, the Layer IV is comprised of a number of myelin tracts and is considered a transition zone between the astrocytic ribbon and the brain parenchyma (Quinones-Hinojosa et al., 2007).

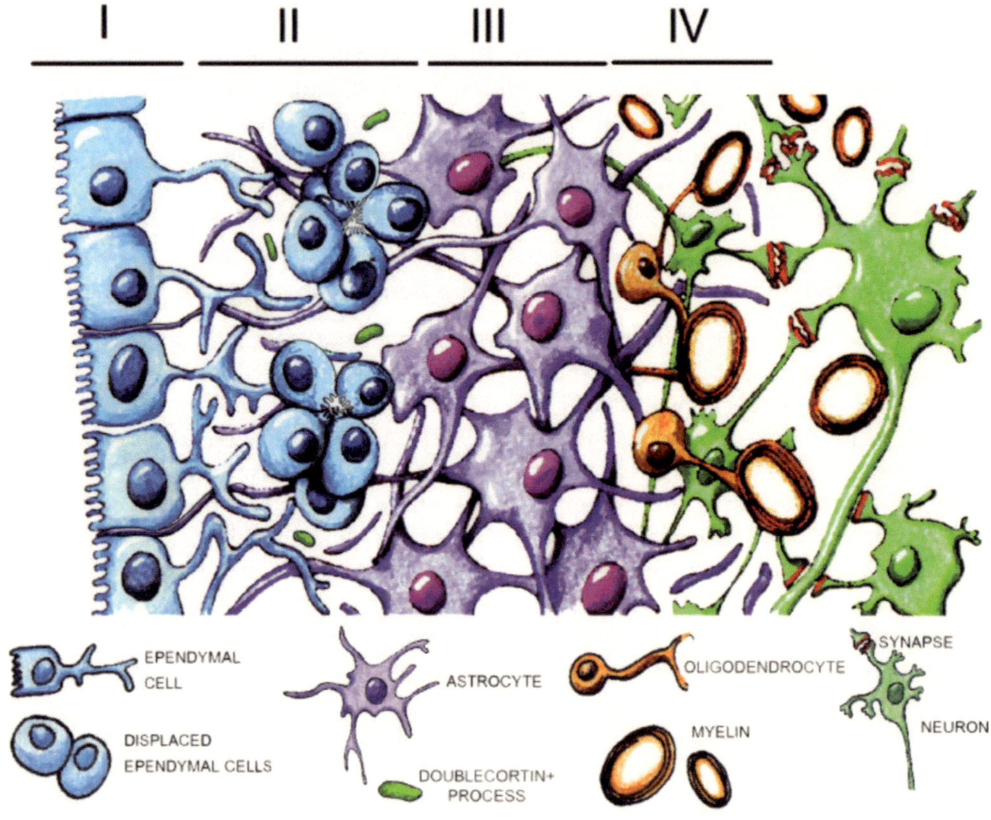

Fig. 3. Cellular organization within the human SVZ. The human SVZ displays unique characteristics as compared to the rodent or primate SVZ. Briefly, layer II devoid of cell bodies, type B cells (astrocytes) are organized as a ribbon of GFAP+ cells, which is not in close contact with the ependymal layer, no chains of migrating neuroblasts are found along the ventricular wall, and very few neuronal cell bodies as well as proliferating cells can be found within the human SVZ. *Figure reproduced with permission from: Alvarez-Palazuelos et al. Current Signal Transduction Therapy 2011;6(3) (Alvarez-Palazuelos et al., 2011). Copyright 2011 Bentham Science Publishers.*

As described above, many features of the human SVZ (Figure 3) are dissimilar to the well-studied rodent SVZ (Figure 2). Some of these fundamental differences are: First, the presence of a layer devoid of cell bodies (Layer II), which contrast with findings reported in the lizard, rodent, feline, canine or primate SVZ that show that all of them have type B cells in close contact to ependymal cells (Doetsch et al., 1997). The second dissimilarity is that the human SVZ lacks chains of migrating neuroblasts (Sanai et al., 2004; Quinones-Hinojosa et al., 2006; Sanai et al., 2007; Wang et al., 2011). Although some authors have suggested that other regions might have migrating cells in the human brain (Bernier et al., 2000; Curtis et al., 2007). Third, the number of proliferating cells (Ki-67 or PCNA expressing cells) in the human SVZ is significantly less than that reported in the rodent SVZ (Sanai et al., 2004; Quinones-Hinojosa et al., 2006; Sanai et al., 2007). Finally, the human SVZ has also very few neuronal cell bodies as compared to other species (Doetsch et al., 1997; Sanai et al., 2004; Quinones-Hinojosa et al., 2006; Sanai et al., 2007). In summary, all these obvious differences between the cell compositions of the human versus the rodent SVZ may also indicate functional differences that need to be studied in detail.

3. Conclusion

Until the end of the twenty century, the brain was perceived as a quiescent organ, with only glia able to have postnatal mitosis. This view was challenged with the isolation of neural stem cells within the adult brain. These multipotent and self-renewing cells are primary located within two germinal niches, the SVZ and SGZ of the hippocampus. The SVZ is the largest source of new cells in adult mammals; thus, a detailed understanding of this neurogenic region may have fundamental medical implications. Nevertheless, a number of questions remain to be elucidated, including the understanding of the role of SVZ neurogenesis in physiological processes such as learning, memory and cell migration. Moreover, SVZ neural stem cells might have some medical uses for a number of neurological disorders including Alzheimer's disease, multiple sclerosis, ischemia, Parkinson's disease, schizophrenia, depression and others. In contrast, since genetic alterations can be acquired through our life time, some groups have proposed that the SVZ may also represent a source of cells for the development of malignant brain tumors but, so far, there is no concluding evidence to support this hypothesis. In summary, the study of neural stem cells in the human SVZ, which is distinct region from those of other animal species, is a vital step with potential medical implications. Therefore new research on the human brain tissue is very important to elucidate these questions.

4. Acknowledgment

This work was supported by grants from the Consejo Nacional de Ciencia y Tecnologia (CONACyT; CB-2008-101476) and The National Institute of Health and the National Institute of Neurological Disorders and Stroke (NIH/NINDS; R01 NS070024-02).

5. References

Altman J (1962) Are new neurons formed in the brains of adult mammals? Science 135:1127-1128.

Altman J, Gopal DD (1965) Post-natal origin of microneurones in the rat brain. Nature 207:953-956.

Altman J, Das GD (1967) Autoradiographic and Histological Studies of Postnatal Neurogenesis. I. A Longitudinal Investigation of the Kinetics, Migration and Transformation of Cells Incorporating Tritiated Thymidine in Neonate Rats, with Special Reference to Postnatal Neurogenesis in Some Brain Regions. JCompNeurol 126:337-390.

Alvarez-Buylla A, Garcia-Verdugo JM (2002) Neurogenesis in adult subventricular zone. J Neurosci 22:629-634.

Alvarez-Palazuelos LE, Robles-Cervantes MS, Castillo-Velázquez G, Rivas-Souza M, Guzman-Muniz J, Moy-Lopez N, González-Castañeda RE, Luquín S, González-Pérez O (2011) Regulation of neural stem cell in the SVZ by thopic and morphogenic factors. Current Signal Transduction Therapy 6:in press.

Beckervordersandforth R, Tripathi P, Ninkovic J, Bayam E, Lepier A, Stempfhuber B, Kirchhoff F, Hirrlinger J, Haslinger A, Lie DC, Beckers J, Yoder B, Irmler M, Gotz M (2010) In vivo fate mapping and expression analysis reveals molecular hallmarks of prospectively isolated adult neural stem cells. Cell Stem Cell 7:744-758.

Bernier PJ, Vinet J, Cossette M, Parent A (2000) Characterization of the subventricular zone of the adult human brain: evidence for the involvement of Bcl-2. Neurosci Res 37:67-78.

Capela A, Temple S (2002) LeX/ssea-1 is expressed by adult mouse CNS stem cells, identifying them as nonependymal. Neuron 35:865-875.

Coskun V, Wu H, Blanchi B, Tsao S, Kim K, Zhao J, Biancotti JC, Hutnick L, Krueger RC, Jr., Fan G, de Vellis J, Sun YE (2008) CD133+ neural stem cells in the ependyma of mammalian postnatal forebrain. Proc Natl Acad Sci U S A 105:1026-1031.

Curtis MA, Kam M, Nannmark U, Anderson MF, Axell MZ, Wikkelso C, Holtas S, van Roon-Mom WM, Bjork-Eriksson T, Nordborg C, Frisen J, Dragunow M, Faull RL, Eriksson PS (2007) Human neuroblasts migrate to the olfactory bulb via a lateral ventricular extension. Science 315:1243-1249.

Chalmers-Redman RM, Priestley T, Kemp JA, Fine A (1997) In vitro propagation and inducible differentiation of multipotential progenitor cells from human fetal brain. Neuroscience 76:1121-1128.

Chojnacki AK, Mak GK, Weiss S (2009) Identity crisis for adult periventricular neural stem cells: subventricular zone astrocytes, ependymal cells or both? Nat Rev Neurosci 10:153-163.

Doetsch F, Garcia-Verdugo JM, Alvarez-Buylla A (1997) Cellular composition and three-dimensional organization of the subventricular germinal zone in the adult mammalian brain. JNeurosci 17:5046-5061.

Doetsch F, Caille I, Lim DA, Garcia-Verdugo JM, Alvarez-Buylla A (1999) Subventricular zone astrocytes are neural stem cells in the adult mammalian brain. Cell 97:703-716.

Doetsch F, Petreanu L, Caille I, Garcia-Verdugo JM, Alvarez-Buylla A (2002) EGF converts transit-amplifying neurogenic precursors in the adult brain into multipotent stem cells. Neuron 36:1021-1034.

Eriksson PS, Perfilieva E, Bjork-Eriksson T, Alborn A, Nordborg C, Peterson DA, Gage FH (1998) Neurogenesis in the adult human hippocampus. Nature Medicine 4:1313-1317.

Francis F, Koulakoff A, Boucher D, Chafey P, Schaar B, Vinet M-C, G., McDonnell N, Reiner O, Kahn A, McConnell SK, Berwald-Netter Y, Denoulet P, Chelly J (1999)

Doublecortin Is a Developmentally Regulated, Microtubule-Associated Protein Expressed in Migrating and Differentiating Neurons. Neuron 23:247-256.

Galli R, Binda E, Orfanelli U, Cipelletti B, Gritti A, De Vitis S, Fiocco R, Foroni C, Dimeco F, Vescovi A (2004) Isolation and characterization of tumorigenic, stem-like neural precursors from human glioblastoma. Cancer Res 64:7011-7021.

Garcia-Verdugo JM, Ferron S, Flames N, Collado L, Desfilis E, Font E (2002) The proliferative ventricular zone in adult vertebrates: a comparative study using reptiles, birds, and mammals. Brain Res Bull 57:765-775.

Garcia AD, Doan NB, Imura T, Bush TG, Sofroniew MV (2004) GFAP-expressing progenitors are the principal source of constitutive neurogenesis in adult mouse forebrain. Nat Neurosci 7:1233-1241.

Goldman SA, Nottebohm F (1983) Neuronal production, migration, and differentiation in a vocal control nucleus of the adult female canary brain. ProcNatlAcadSciUSA 80:2390-2394.

Gonzalez-Perez O, Quinones-Hinojosa A (2010) Dose-dependent effect of EGF on migration and differentiation of adult subventricular zone astrocytes. Glia 58:975-983.

Gonzalez-Perez O, Alvarez-Buylla A (2011) Oligodendrogenesis in the subventricular zone and the role of epidermal growth factor. Brain Res Rev 67:147-156.

Gonzalez-Perez O, Quinones-Hinojosa A, Garcia-Verdugo JM (2010a) Immunological control of adult neural stem cells. J Stem Cells 5:23-31.

Gonzalez-Perez O, Jauregui-Huerta F, Galvez-Contreras AY (2010b) Immune system modulates the function of adult neural stem cells. Curr Immunol Rev 6:167-173.

Gonzalez-Perez O, Romero-Rodriguez R, Soriano-Navarro M, Garcia-Verdugo JM, Alvarez-Buylla A (2009) Epidermal growth factor induces the progeny of subventricular zone type B cells to migrate and differentiate into oligodendrocytes. Stem Cells 27:2032-2043.

Guerrero-Cazares H, Gonzalez-Perez O, Soriano-Navarro M, Zamora-Berridi G, Garcia-Verdugo JM, Quinones-Hinojosa A (2011) Cytoarchitecture of the lateral ganglionic eminence and rostral extension of the lateral ventricle in the human fetal brain. J Comp Neurol 519:1165-1180.

Hartfuss E, Galli R, Heins N, Gotz M (2001) Characterization of CNS precursor subtypes and radial glia. Dev Biol 229:15-30.

Hendrickson ML, Rao AJ, Demerdash ON, Kalil RE (2011) Expression of nestin by neural cells in the adult rat and human brain. PLoS One 6:e18535.

Huang L, DeVries GJ, Bittman EL (1998) Photoperiod Regulates Neuronal Bromodeoxyuridine Labeling in the Brain of a Seasonally Breeding Mammal. JNeurobiol 36:410-420.

Ignatova TN, Kukekov VG, Laywell ED, Suslov ON, Vrionis FD, Steindler DA (2002) Human cortical glial tumors contain neural stem-like cells expressing astroglial and neuronal markers in vitro. Glia 39:193-206.

Ihrie RA, Alvarez-Buylla A (2011) Lake-front property: a unique germinal niche by the lateral ventricles of the adult brain. Neuron 70:674-686.

Ihrie RA, Shah JK, Harwell CC, Levine JH, Guinto CD, Lezameta M, Kriegstein AR, Alvarez-Buylla A (2011) Persistent sonic hedgehog signaling in adult brain determines neural stem cell positional identity. Neuron 71:250-262.

Imura T, Kornblum HI, Sofroniew MV (2003) The Predominant Neural Stem Cell Isolated from Postnatal and Adult Forebrain But Not Early Embryonic Forebrain Expresses GFAP. J Neurosci 23:2824-2832.

Jankovski A, Sotelo C (1996) Subventricular zone-olfactory bulb migratory pathway in the adult mouse: cellular composition and specificity as determined by heterochronic and heterotopic transplantation. J Comp Neurol 371:376-396.

Kaplan MS, Hinds JW (1977) Neurogenesis in the adult rat: Electron microscopic analysis of light radioautographs. Science 197:1092-1094.

Kirschenbaum B, Nedergaard M, Preuss A, Barami K, Fraser RAR, Goldman SA (1994) In vitro Neuronal Production and Differentiation by Precursor Cells Derived from the Adult Human Forebrain. CerebCortex 6:576-589.

Kornack DR, Rakic P (1995) Radial and horizontal deployment of clonally related cells in the primate neocortex: Relationship to distinct mitotic lineages. Neuron 15:311-321.

Kriegstein A, Alvarez-Buylla A (2009) The glial nature of embryonic and adult neural stem cells. Annu Rev Neurosci 32:149-184.

Kukekov VG, Laywell ED, Suslov O, Davies K, Scheffler B, Thomas LB, O'Brien TF, Kusakabe M, Steindler DA (1999) Multipotent stem/progenitor cells with similar properties arise from two neurogenic regions of adult human brain. Exp Neurol 156:333-344.

Laywell ED, Rakic P, Kukekov VG, Holland EC, Steindler DA (2000) Identification of a multipotent astrocytic stem cell in the immature and adult mouse brain. Proc Natl Acad Sci U S A 97:13883-13888.

Lim DA, Alvarez-Buylla A (1999) Interaction between astrocytes and adult subventricular zone precursors stimulates neurogenesis. Proc Natl Acad Sci U S A 96. 96:7526-7531.

Lim DA, D. TA, Trevejo JM, Herrera DG, García-Verdugo JM, Alvarez-Buylla A (2000) Noggin Antagonizes BMP Signaling to Create a Niche for Adult Neurogenesis. Neuron 28:713-726.

Lois C, Alvarez-Buylla A (1993) Proliferating subventricular zone cells in the adult mammalian forebrain can differentiate into neurons and glia. ProcNatlAcadSciUSA 90:2074-2077.

Lois C, Alvarez-Buylla A (1994) Long-distance neuronal migration in the adult mammalian brain. Science 264:1145-1148.

Lois C, Garcia-Verdugo JM, Alvarez-Buylla A (1996) Chain migration of neuronal precursors. Science 271:978-981.

Luskin MB (1993) Restricted proliferation and migration of postnatally generated neurons derived from the forebrain subventricular zone. Neuron 11:173-189.

McDermott KW, Lantos PL (1991) Distribution and fine structural analysis of undifferentiated cells in the primate subependymal layer. JAnat 178:45-63.

McDermott KWG, Lantos PL (1990) Cell proliferation in the subependymal layer of the postnatal marmoset, Callithrix jacchus. DevBrain Res 57:269-277.

Menn B, Garcia-Verdugo JM, Yaschine C, Gonzalez-Perez O, Rowitch D, Alvarez-Buylla A (2006) Origin of oligodendrocytes in the subventricular zone of the adult brain. J Neurosci 26:7907-7918.

Mirzadeh Z, Merkle FT, Soriano-Navarro M, Garcia-Verdugo JM, Alvarez-Buylla A (2008) Neural stem cells confer unique pinwheel architecture to the ventricular surface in neurogenic regions of the adult brain. Cell Stem Cell 3:265-278.

Morshead CM, Garcia AD, Sofroniew MV, van Der Kooy D (2003) The ablation of glial fibrillary acidic protein-positive cells from the adult central nervous system results in the loss of forebrain neural stem cells but not retinal stem cells. Eur J Neurosci 18:76-84.

Nunes MC, Roy NS, Keyoung HM, Goodman RR, McKhann G, 2nd, Jiang L, Kang J, Nedergaard M, Goldman SA (2003) Identification and isolation of multipotential neural progenitor cells from the subcortical white matter of the adult human brain. Nat Med 9:439-447.

Parras CM, Galli R, Britz O, Soares S, Galichet C, Battiste J, Johnson JE, Nakafuku M, Vescovi A, Guillemot F (2004) Mash1 specifies neurons and oligodendrocytes in the postnatal brain. Embo J 23:4495-4505.

Pastrana E, Cheng LC, Doetsch F (2009) Simultaneous prospective purification of adult subventricular zone neural stem cells and their progeny. Proc Natl Acad Sci U S A 106:6387-6392.

Pérez-Cañellas MR, García-Verdugo JM (1996) Adult neurogenesis in the telencephalon of a lizard: A [3H]thymidine autoradiographic and bromodeoxyuridine immunocytochemical study. DevBrain Res 93:49-61.

Pincus DW, Harrison-Restelli C, Barry J, Goodman RR, Fraser RA, Nedergaard M, Goldman SA (1997) In vitro neurogenesis by adult human epileptic temporal neocortex. Clin Neurosurg 44:17-25.

Quinones-Hinojosa A, Sanai N, Gonzalez-Perez O, Garcia-Verdugo JM (2007) The human brain subventricular zone: stem cells in this niche and its organization. Neurosurg Clin N Am 18:15-20, vii.

Quinones-Hinojosa A, Sanai N, Soriano-Navarro M, Gonzalez-Perez O, Mirzadeh Z, Gil-Perotin S, Romero-Rodriguez R, Berger MS, Garcia-Verdugo JM, Alvarez-Buylla A (2006) Cellular composition and cytoarchitecture of the adult human subventricular zone: a niche of neural stem cells. J Comp Neurol 494:415-434.

Raponi E, Agenes F, Delphin C, Assard N, Baudier J, Legraverend C, Deloulme JC (2007) S100B expression defines a state in which GFAP-expressing cells lose their neural stem cell potential and acquire a more mature developmental stage. Glia 55:165-177.

Reynolds B, Weiss S (1992) Generation of neurons and astrocytes from isolated cells of the adult mammalian central nervous system. Science 255:1707-1710.

Reynolds BA, Rietze RL (2005) Neural stem cells and neurospheres--re-evaluating the relationship. Nat Methods 2:333-336.

Reznikov KY (1991) Cell proliferation and cytogenesis in the mouse hippocampus. Adv Anat Embryol Cell Biol 122:1-74.

Rousselot P, Lois C, Alvarez-Buylla A (1995) Embryonic (PSA) N-CAM reveals chains of migrating neuroblasts between the lateral ventricle and the olfactory bulb of adult mice. JCompNeurol 351:51-61.

Sanai N, Alvarez-Buylla A, Berger MS (2005) Neural stem cells and the origin of gliomas. N Engl J Med 353:811-822.

Sanai N, Berger MS, Garcia-Verdugo JM, Alvarez-Buylla A (2007) Comment on "Human neuroblasts migrate to the olfactory bulb via a lateral ventricular extension". Science 318:393; author reply 393.

Sanai N, Tramontin AD, Quinones-Hinojosa A, Barbaro NM, Gupta N, Kunwar S, Lawton MT, McDermott MW, Parsa AT, Manuel-Garcia Verdugo J, Berger MS, Alvarez-Buylla A (2004) Unique astrocyte ribbon in adult human brain contains neural stem cells but lacks chain migration. Nature 427:740-744.

Sawamoto K, Wichterle H, Gonzalez-Perez O, Cholfin JA, Yamada M, Spassky N, Murcia NS, Garcia-Verdugo JM, Marin O, Rubenstein JL, Tessier-Lavigne M, Okano H, Alvarez-Buylla A (2006) New neurons follow the flow of cerebrospinal fluid in the adult brain. Science 311:629-632.

Shen Q, Goderie SK, Jin L, Karanth N, Sun Y, Abramova N, Vincent P, Pumiglia K, Temple S (2004) Endothelial cells stimulate self-renewal and expand neurogenesis of neural stem cells. Science 304:1338-1340.

Shmelkov SV, Butler JM, Hooper AT, Hormigo A, Kushner J, Milde T, St Clair R, Baljevic M, White I, Jin DK, Chadburn A, Murphy AJ, Valenzuela DM, Gale NW, Thurston G, Yancopoulos GD, D'Angelica M, Kemeny N, Lyden D, Rafii S (2008) CD133 expression is not restricted to stem cells, and both CD133+ and CD133- metastatic colon cancer cells initiate tumors. J Clin Invest 118:2111-2120.

Song H, Stevens CF, Gage FH (2002) Astroglia induce neurogenesis from adult neural stem cells. Nature 417:39-44.

Spassky N, Merkle FT, Flames N, Tramontin AD, Garcia-Verdugo JM, Alvarez-Buylla A (2005) Adult ependymal cells are postmitotic and are derived from radial glial cells during embryogenesis. J Neurosci 25:10-18.

Vescovi AL, Galli R, Reynolds BA (2006) Brain tumour stem cells. Nat Rev Cancer 6:425-436.

Wang C, Liu F, Liu YY, Zhao CH, You Y, Wang L, Zhang J, Wei B, Ma T, Zhang Q, Zhang Y, Chen R, Song H, Yang Z (2011) Identification and characterization of neuroblasts in the subventricular zone and rostral migratory stream of the adult human brain. Cell Res.

4

The Spinal Cord Neural Stem Cell Niche

Jean-Philippe Hugnot
University of Montpellier 2, INSERM U1051,
Institute for Neurosciences of Montpellier
France

1. Introduction

The spinal cord is the caudal portion of the central nervous system (CNS) that extends from the lower part of the brain stem (the medulla) to the cauda equina. It receives several types of sensory information from the joints, muscles, organs and skin and contains the motoneurons responsible for voluntary/reflex movements and for the function of the autonomic nervous system. The spinal cord is divided into i) gray matter, which notably contains motoneurons and interneurons that form the spinal cord circuitry; ii) white matter, which surrounds the gray matter and is made up of ascending and descending longitudinal tracts; and iii) the central canal or ependymal region, which is organized as an oval or round-shaped epithelium whose apical pole abuts the cerebral spinal fluid. The spinal cord is not simply a relay that carries information between the brain and body, but it also contains a complex circuitry that is implicated in the generation and coordination of reflexive responses to sensory inputs. Furthermore, the spinal cord is involved in the formation of rhythmic movements, such as locomotion and swimming in animals. One emerging field of research concerns spinal cord plasticity, as this structure should not be considered a static and hard-wired system. Instead, the spinal cord displays considerable activity-dependent adaptation and, similar to other CNS regions, can learn and remember throughout life (Guertin 2008; Wolpaw 2010; Wolpaw and Tennissen 2001). Plasticity plays an important role in the acquisition and maintenance of motor skills. In pathology, it could be manipulated to alleviate spinal cord lesions that originate from traumas or degenerative diseases.

In parallel with spinal cord plasticity, one field of research that is rapidly growing concerns the presence of neural stem cells and progenitor cells in the adult spinal cord. In this review, I will describe recent findings regarding stem cells and attempt to formulate hypotheses concerning their role in spinal cord physiology and plasticity. The presence of stem cells in the spinal of lower vertebrates, such as salamanders and newts, has been reported for decades. These stem cells are at the basis of the phenomenal regeneration capacity of these animals that is observed when the spinal cord is transected. There are excellent reviews on this topic (for instance, see (Tanaka 2003)), and I will thus focus on the adult spinal cord stem cells in mammals.

2. Discovery and properties of mouse spinal cord neural stem cells

2.1 Spinal cord neurospheres

Definite proof of the presence of neural stem cells in the adult CNS using in vitro assays dates back to the early nineties (Gage, Ray, and Fisher 1995; Reynolds, W., and Weiss 1992). Since then, much attention has been given to stem and progenitor cells in the brain, whereas little is known about these cells in the spinal cord. The persistence of stem cells in this caudal region of the CNS was reported using adherent and non-adherent culture conditions in the late nineties (Shihabuddin, Ray, and Gage 1997; Weiss et al. 1996). The neurosphere assay (Deleyrolle and Reynolds 2009) was instrumental in their discovery, as this assay is particularly suited to demonstrate, at the clonal level, the cardinal properties of stem cells, i.e., multipotentiality, self-renewal and extended proliferation capabilities. Indeed, in 1996 Weiss et al. reported that in mice, 0.1 and 0.6% of isolated thoracic and lumbar spinal cord cells, respectively, that were grown in the presence of FGF2 and EGF, were able to form multipotent and passageable neurospheres (Weiss et al. 1996). Using microdissection and cytometric analysis, these cells were located primarily in the central canal region (Martens, Seaberg, and van der Kooy 2002; Meletis et al. 2008; Sabourin et al. 2009). Progenitor cells with a more limited proliferation potential are also present in the parenchyma (Horner et al. 2000; Kulbatski et al. 2007; Sabourin et al. 2009; Martens, Seaberg, and van der Kooy 2002; Yamamoto et al. 2001). More recently, we were able to show that the dorsal part of the central canal region is enriched in neurosphere-forming cells (Sabourin et al. 2009). Even when clonally-expanded, these neurospheres appear to be heterogeneous entities that are composed of different types of Nestin$^+$ cells, which express various levels of stem cell (CD133), astrocytic (GFAP, Adhl1l1), radial glial cell (CD15, Blbp, Glast, RC2) and oligodendrocytic-lineage (NG2, A2B5, PDGFRα) markers (Fig. 1).

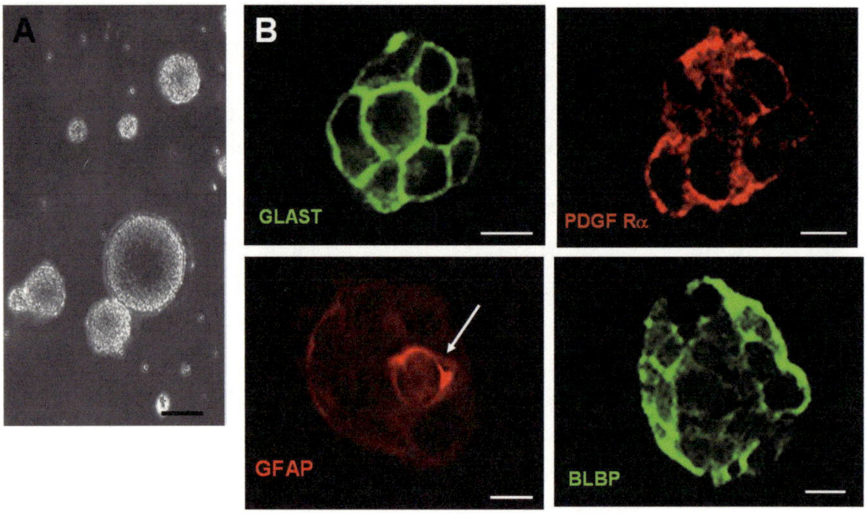

Fig. 1. A) Examples of adult spinal cord-derived neurospheres that were clonally expanded. Note the differences in the sizes of the neurospheres. Scale bar=500 μm. B) Examples of markers that were detected by immunofluorescence in the neurospheres. The white arrow shows a unique GFAP+ cell in a neurosphere. Scale bars=10 μm.

The widespread expression of the latter coincides with the higher propensity of the neurospheres to differentiate into oligodendrocytes vs. neurons in vitro and in vivo after spinal cord injury (SCI) (Kulbatski et al. 2007; Meletis et al. 2008). A few cells that express so-called neuronal markers, such as Map2 and Dcx, are also observed in some neurospheres (Fig. 2).

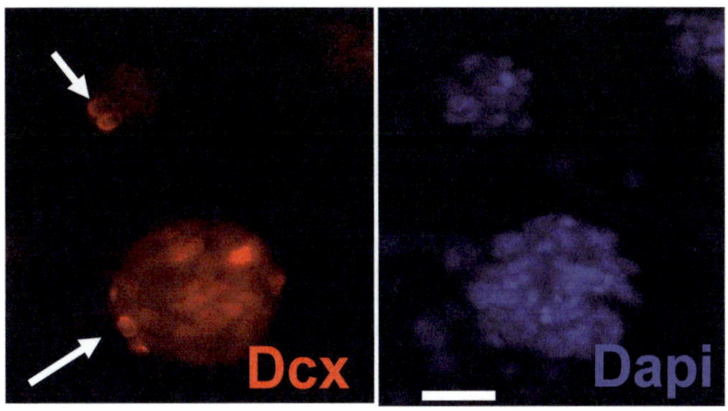

Fig. 2. Example of Dcx+ cells detected in spinal cord-derived neurospheres. Scale bar= 100 μm.

Often, several of these markers are coexpressed in the same cells, which may reflect the presence of intermediates and uncommitted states of differentiation. $GFAP^+$ cells in these spheres (Fig. 1) should not be considered to be astrocytic-differentiated cells but rather, as they frequently co-express immature markers (CD133, CD15), these cells likely represent $GFAP^+$ neural stem cells. Within neurospheres, only a fraction of the cells, typically between 1 to 10%, are able to generate new neurospheres and are considered to be bona fide neural stem cells. Other cells are considered to be progenitors because they give rise to small neurospheres with limited proliferative and self-renewal capabilities (Louis et al. 2008). The prospective isolation of stem vs. progenitor cells in neurospheres remains challenging, as no definitive cell surface marker has been clearly identified to distinguish these two types of cells.

Notably, as observed for neurospheres that were derived from different brain regions (Armando et al. 2007; Conti and Cattaneo 2010), neurospheres derived from different parts of the spinal cord have different growth and differentiation properties (Kulbatski and Tator 2009; Sabourin et al. 2009). We showed that neurospheres that were derived from the cervical, thoracic and lumbar regions maintained their expression of specific and different rostro-caudal combinations of developmental homeogenes of the Hox family. These data indicate that even after several passages in vitro, these cells maintain molecular cues from their original position. This phenomenon might affect considerations regarding cellular therapy with adult neural stem cells because not all cells might have equivalent capacities to replenish cell loss and to integrate into the adult host tissue. In addition to Hox genes, the adult spinal cord neurospheres express high levels of a set of transcription factors, including Dlx2, Nkx2.2, Nkx6.1, Olig2, Pax6, Sox2, Sox4 and Sox9 (Moreno-Manzano et al. 2009; Sabourin et al. 2009; Yamamoto et al. 2001) which are involved in spinal cord embryonic development. This expression likely reflects the maintenance, in adult stem cells, of embryonic transcriptional programs and active signaling pathways. Yet, these neurospheres appear to express only a limited range of developmental gene networks, as the transcription factors that are involved in motoneuron development, such as Islet1, Lim1 and HB9, are not expressed (Yamamoto et al. 2001), which illustrates a somewhat restricted fate for these cells. Indeed, upon differentiation in vitro, these cells primarily generate GABAergic neurons, oligodendrocytes and astrocytes (Moreno-Manzano et al. 2009; Sabourin et al. 2009). Importantly, these cells appear to remain competent to respond to morphogens to redirect their differentiation into other neuronal cell subtypes. Indeed, treatment with embryonic morphogens that are involved in spinal cord caudal regionalization and motoneuron development, i.e., retinoic acid and sonic hedgehog, was able to drive their differentiation toward electrophysiological active motoneurons (Moreno-Manzano et al. 2009).

2.2 Neurosphere differentiation

The differentiation of adult spinal cord neurospheres into neuronal and glial cells is generally achieved by plating them onto an adhesive substrate and by removing growth factors. This differentiation occurs even without the addition of serum, suggesting that endogenous cytokines are implicated in the differentiation process. Indeed, we found that undifferentiated and differentiated neurospheres expressed at a high level, a wide range of endogenous cytokines (Deleyrolle et al. 2005). The expression of several of them is controlled by the FGF2 and EGF which are present in the medium. Notably, upon differentiation, a striking upregulation of astrocytic differentiating factors, such as BMP4, BMP6 and CNTF, are observed. The active role of these endogenous cytokines is illustrated by inhibition experiments that impair neurosphere differentiation (Deleyrolle et al. 2005). This production of endogenous cytokines by adult spinal cord stem cells could be considered to be beneficial in the context of cellular therapy because these cytokines have been shown to actively participate in the therapeutic effects observed following neural precursor cell transplantation (Pluchino et al. 2003; Redmond et al. 2007). In contrast, these endogenous factors, together with host factors, might also contribute to the absence or the low rate of neuronal formation, which is frequently observed in grafting experiments. Further investigations of the neurosphere differentiation process would yield important insights on how to direct the fate of endogenous and exogenous neural stem cells into the most appropriate cell type to achieve rational and effective spinal cord cellular therapy.

3. The central canal niche and identity of adult spinal cord stem cells

Many adult organs harbor a pool of stem cells in specialized structures called niches. These act as a nest and a barrier to protect, nourish and regulate the fate of stem cells. They do so by providing, in highly organized structures, cellular and molecular cues suitable for the strict control of stem cell properties (e.g., self-renewal, differentiation, quiescence). Typically, these niches contain a high level of canonical developmental signaling pathways, notably, BMP, SHH, Wnt and Notch (Li and Clevers 2010). These signaling pathways precisely regulate the proliferation/quiescence, differentiation/self-renewal and migratory/stationary balances of the stem cell pool. In addition, their particular architecture favors interactions between stem cells and specific cells, such as vascular cells. Historically, the best-characterized niches in mammals are the hematopoietic and intestinal niches, and more recently, the CNS niches have been studied. In the brain, the hippocampus, the recently discovered sub-callosal zone and the subcortical white matter contain neural progenitors, whereas bona fide stem cells that are capable of sustained proliferation are preferentially found in the subventricular zone (SVZ) (Seaberg, Smukler, and van der Kooy

2005). In the latter, contacts with the cerebral spinal fluid (CSF) located in the ventricles appear to be essential for the maintenance of SVZ stem cells (Lehtinen et al. 2010). Stem and progenitor cells have also been identified in the peripheral nervous system, i.e., in the carotid body, the enteric nervous system and the adult dorsal root ganglia (Pardal et al. 2007; Schafer, Van Ginneken, and Copray 2009; Singh et al. 2009). Whereas there have been a tremendous number of publications on the brain niches, few have addressed this issue in the spinal cord.

3.1 Cellular diversity in the central canal region

The central canal region is composed of several cell types, which are located either in direct contact with the lumen or in a subependymal position, evoking a pseudo-stratified epithelium (Fig. 3). However, a distinct subependymal layer, as observed in the SVZ, is not present. Ependymocytes are the primary cell type found around the central canal. A second frequently observed cell type is the tanycyte (also referred to as radial ependymocytes) (Seitz, Lohler, and Schwendemann 1981), which is mostly observed on the lateral sides of the central canal region. This ependymal cell type sends a long basal process that terminates at the blood vessels (Horstmann 1954). Tanycytes are in contact with the lumen, but their soma can either be subependymally or ependymally located (Meletis et al. 2008; Rafols and Goshgarian 1985). As in the brain, they bridge the CSF to the capillaries, thereby providing a potential link between the CSF and the blood.

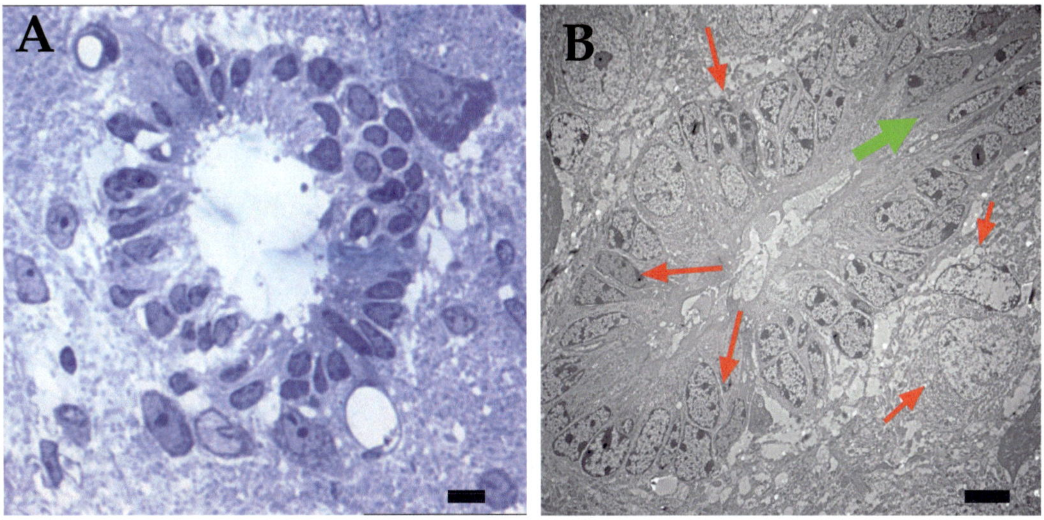

Fig. 3. A) Semi-thin section in the adult mouse thoracic spinal cord ependymal region (toluidine blue staining). Note the diversity of the cell types in contact with the lumen or in a subependymal position. B) An electron micrograph showing different types of cells around the central canal (red arrows). The green arrow indicates the dorsal region of the canal, which contains densely packed cells.

The central canal region contains also neuronal-like cells which contact the CSF. These are very common and well-described in several lower vertebrates especially fishes and amphibians (see for review (Vigh et al. 2004)). In mammals, their presence has been reported in several species including primates and rodents (Hugnot and Franzen 2010). These cells are sporadically distributed around the canal with a soma in an ependymal or subependymal position and they send a single thick dendritic-like process terminated by a large bulge in the lumen (Sabourin et al. 2009). Even in adults, these cells continue to express PSA-NCAM (Marichal et al. 2009; Seki and Arai 1993), Dcx (Marichal et al. 2009; Sabourin et al. 2009) and GAP43 (Stoeckel et al. 2003), three proteins that are involved in plasticity and migration, suggesting that they are endowed with some degree of immaturity. In rodents, these neuronal cells appear to be mainly GABAergic, and their function remains elusive. The expression of acidic pH-activated channels suggests that these cells might be implicated in CSF homeostasis (Huang et al. 2006; Marichal et al. 2009). They could also act as mechanoreceptors, which are sensitive to CSF pressure or flow or to spinal cord flexion. Importantly, these cells are not produced from continuous adult spinal cord neurogenesis, and a study performed in rats demonstrated that they are in fact produced during embryogenesis (Marichal et al. 2009).

The dorsal and ventral regions of the central canal display a divergent organization of a higher density of cells with a radial morphology, which are situated in ependymal and subependymal positions (Hamilton et al. 2009; Meletis et al. 2008; Sabourin et al. 2009). GFAP+ cells are frequently observed in these regions. Particularly in the dorsal region, some of these GFAP+ cells have long basal processes that extend along the dorsal midline up to the dorsal column white matter or pial surface (Bodega et al. 1994; Hamilton et al. 2009; Sabourin et al. 2009). GFAP+ cells can also be observed in the lateral region of the canal. These GFAP+ cells can lie in the ependymal layer adjacent to the canal lumen but are often located in a subependymal position, where they send a process toward the canal. Transgenic mice expressing a hGFAP promoter-GFP construct are particularly suited to visualize these cells (Fig. 4). Interestingly, some cells that express radial glia markers, such as CD15 or BLBP (some of which are GFAP+), are also occasionally detected in the dorsal region (Fig. 5).

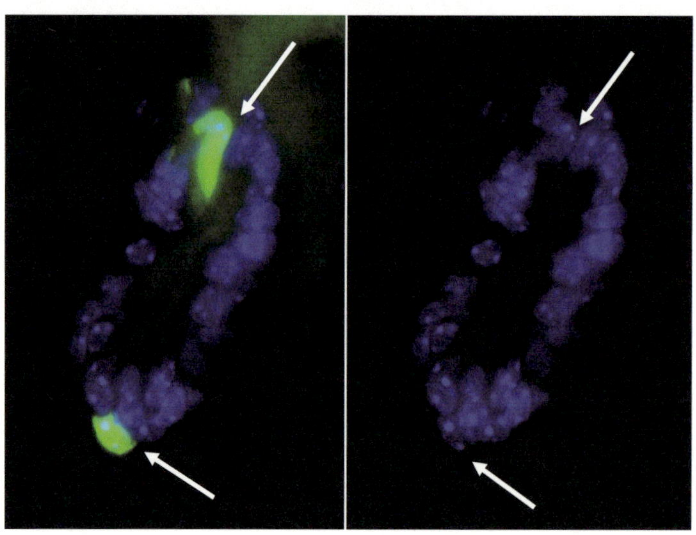

Fig. 4. Examples of GFP+ cells detected in the dorsal and ventral regions of the central canal region of hGFAP-GFP mice (Nolte et al. 2001). The dorsal GFP+ cell is located in the ependymal layer, whereas the ventral cell is located in the subependymal layer.

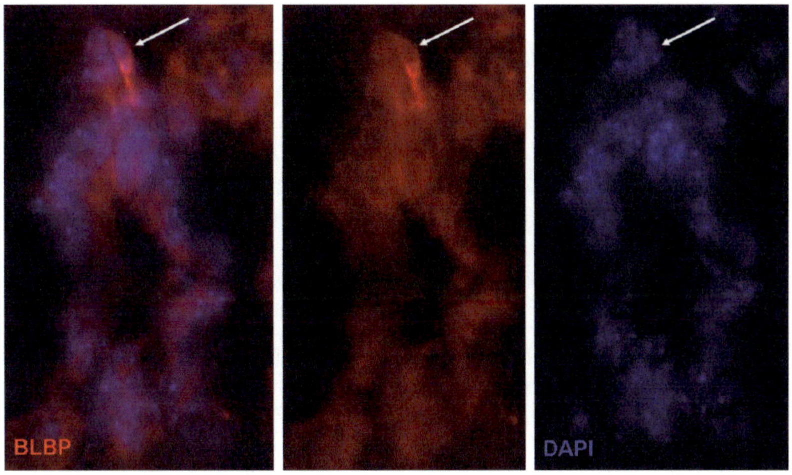

Fig. 5. Example of a subependymal BLBP+ cell detected in the dorsal part of the central canal of the mouse lumbar spinal cord (arrow).

The central canal region is surrounded by an abundant vasculature, and cellular proliferation within the niche occurs in close proximity to the vessels (Hamilton et al. 2009). This observation is consistent with the recently well-described interactions between neural stem/progenitor cells and endothelial cells (the so-called neurovascular niche (Palmer, Willhoite, and Gage 2000)).

Collectively, it appears that the central canal region is composed of several cell types that are localized at specific locations and express characteristic markers with different morphologies and potentially different functions. A schematic drawing of the lumbar central canal region is presented in Fig. 6.

3.2 Signaling within the niche

The persistence of stem cells within the central canal region implies that specific pathways are active or readily activated in the niche. These pathways will maintain the proliferation potential and multipotency of stem cells. Equally, the niche may contain molecules that act in a passive mode to protect stem cells from local or circulating growth and differentiation factors. Our lab used two approaches to identify important cues in the mouse spinal cord niche. First, we used immunofluorescence to screen for the presence of receptors, ligands and transcription factors that are associated with the Notch, SHH, Wnt and epithelial-mesenchymal transition (EMT) pathways. Second, we extensively screened online gene expression databases (notably, Allen Brain and the Gensat Atlas) for genes that are specifically expressed in the spinal cord central canal region. These approaches allowed us to identify the expression, at the transcript and/or the protein level, of several molecules involved in the Notch (Jagged, Hes1), Wnt (Wnt7a, Fzd3), BMP (DAN, BMP6) and Hedgehog (SHH) pathways (Hugnot and Franzen 2010; Sabourin et al. 2009). These genes are expressed by most of the cells in the central canal region or by restricted subpopulations. Unexpectedly, we also found that cells in this region highly expressed Zeb1 (Sabourin et al. 2009) (also known as δ-EF1, TCF8, AREB6), a zinc finger-homeodomain transcription factor, which has been described as an important regulator of EMT (Liu et al. 2008). Zeb1 protein is detected in the majority of cells surrounding the lumen but is present at higher levels in the cells that are located in the medial dorsal region, notably, the previously mentioned GFAP+ cells. Zeb1 is also detected in subpopulations of cells in the white and grey matter. Zeb1 is involved in the regulation of several cellular processes, such as migration, senescence and apoptosis. It exerts control by acting as a repressor for a number of genes, such as P15Ink4b, P21 Cdkn1, E-cadherin, CRB3 and myogenic transcription factors (Browne, Sayan, and Tulchinsky 2010). Conversely, it also acts as an activator for a group of genes that are typically expressed in mesenchymal cells, such as collagens, vimentin and smooth muscle actin (acta2) (Nishimura et al. 2006). As the two latter proteins are expressed in the central canal and SVZ cells (Sabourin et al, 2009), their expression might be under the control of Zeb1. Consistent with a role in adult precursor cells, Zeb1 and 2 are expressed by neurosphere cells derived from the adult spinal cord. These transcription factors are required for neurosphere formation and expansion because we demonstrated that the transfection of a dominant-negative form of Zeb1 and 2 induced massive apoptosis in vitro. In Drosophila, the Zeb1 orthologous protein Zfh-1 was recently shown to have a critical role in the maintenance of the somatic stem cell compartment in the testis stem cell niche (Leatherman and Dinardo 2008). These data suggest that the role of this family of transcription factors in the maintenance of immature properties has been conserved throughout evolution.

3.3 Identity of stem cells in the niche

Considering the diversity of cell types in the central canal region, the precise identity of the cells that are able to generate passageable neurospheres needed to be addressed by methods based on cell purification. A common and powerful technique is based on the cytometry of cells isolated from GFP transgenic animals, where a specific cell type is tagged using a cell-specific promoter. Alternatively, specific membranous markers and antibodies can be exploited for purification; however, this method can be challenging for studies of the adult spinal cord, as enzymes required for cellular dissociation could damage membrane-bound markers and lead to erroneous conclusions. To explore whether the GFAP+ cells we observed in the central canal region were endowed with stem cell properties, we used the hGFAP-GFP transgenic line established by Dr Kettenmann's group (Nolte et al. 2001). Sorting GFP+ cells by cytometry revealed that compared with GFP$^-$, the vast majority of neurospheres (>80%) are derived from GFP+ cells (Sabourin et al. 2009). In total, 0.2% of GFP+ cells were able to generate neurospheres. This frequency might appear low, but one must consider that the purification of GFP+ cells from these animals cannot discriminate GFP$^+$ cells located in the central canal region from those of the parenchyma, which are much more numerous. Consistent with our hypothesis that central canal GFAP+ cells are endowed with stem cells properties, we found that most primary neurospheres derived from these transgenic animals contained one or several GFAP+/GFP+ cells. This result supports the notion that as observed in the SVZ, the central canal cells with astrocytic features have neural stem cell properties. Another team conducted a second transgenic mouse approach with FoxJ1-GFP animals (Meletis et al. 2008). FoxJ1 expression is restricted to the central canal region, and it was assumed that there was no expression in GFAP+ cells. Using this line, Meletis et al. reported that the majority of spheres are derived from the GFP$^+$ fraction with an approximate frequency of 0.2% of GFP+ cells giving rise to neurospheres. As no GFAP+ cell was observed around the canal in this study, it was concluded that adult spinal cord neural stem cells are

GFAP- ependymal cells. Yet, in contrast with the GFP+ population purified from hGFAP-GFP animals that contain GFP+ cells from both the central canal region and the parenchyma, the cells obtained from the FoxJ1-GFP mice appear to be exclusively derived from the cells in the central canal region. Thus, the obtained frequency of 0.2% for neurosphere formation in FoxJ1+ cells signifies that only a small subpopulation of undefined ependymal cells (1/5000) would be endowed with neural stem cell properties. Moreover, a recent elegant transcriptome analysis from Beckervordersandforth, et al. clearly indicated that in the SVZ, GFAP+ neural stem cells highly expressed FoxJ1 (Beckervordersandforth et al. 2010).

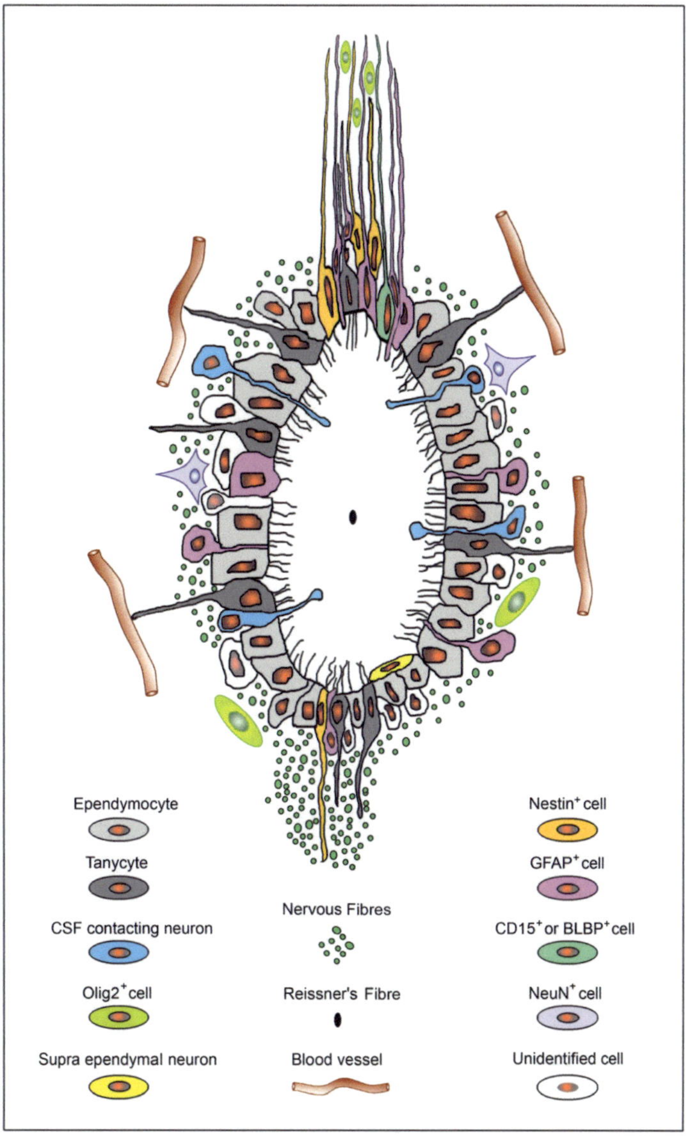

Fig. 6. Schematic drawing of the adult mouse ependymal region. Reproduced from (Sabourin et al. 2009)

Using immunofluorescence for FoxJ1 on hGFAP-GFP sections, we could readily observe double-labeled cells (Fig. 7, unpublished data). This indicated that in addition to the ependymocytes, the FoxJ1 transcription factor is also present in neural stem cells as also suggested by the study reported by Jacquet, BV et al (Jacquet et al. 2009).

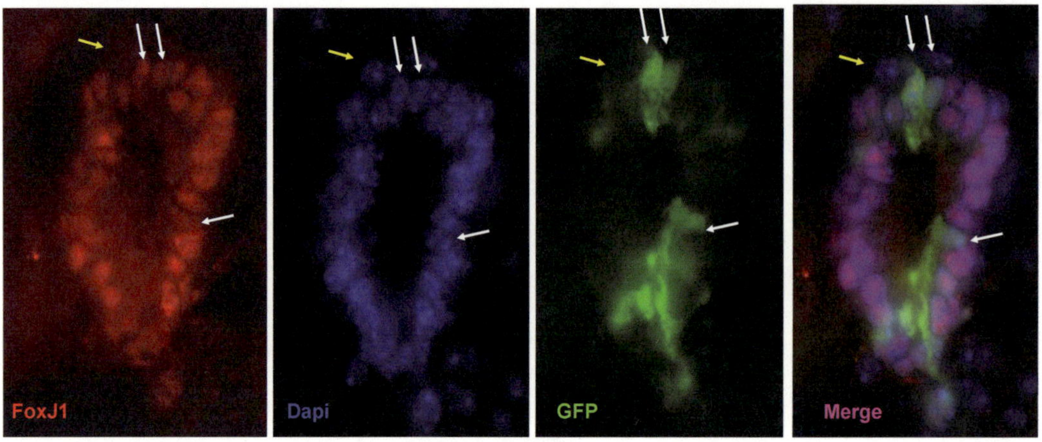

Fig. 7. Immunodetection of FoxJ1 in the cells around the central canal of the hGFAP-GFP adult mouse. The white arrows point to two GFP+ cells expressing FoxJ1. The yellow arrow indicates a subependymal FoxJ1-negative cell.

In the SVZ, neural stem cells are currently primarily considered to be CD133+ GFAP+ cells, whereas ependymocytes are endowed with a more restricted proliferative potential. In the spinal cord, most of the cells around the canal are $CD133^{+}$ (Sabourin et al. 2009), and in primary neurospheres, CD133+ GFAP+ cells are frequently observed. Thus, it is likely that GFAP+ CD133+ cells in the adult spinal cord represent all or at least a substantial fraction of neural stem cells. Considering that in the SVZ, $GFAP^{-}$ transit amplifying cells can be converted into stem cells with EGF (Doetsch et al. 2002), the possibility still exists that a fraction of neurospheres could be derived from central canal $GFAP^{-}$ cells.

3.4 Developmental origin of the central canal region

Neurospheres derived from the mouse adult lumbar spinal cord express the Nkx6.1 transcription factor (Sabourin et al. 2009). In vivo, most cells around the central lumen also express this factor; however, there was a variable level of expression and the CSF-contacting neurons displayed the highest staining in our studies. During spinal cord development, Nkx6.1 is expressed in the ventral neural tube, which contains progenitor cells of three main neuronal classes: V2, MN and V3 interneurons (Briscoe et al. 2000). Using marker analysis, Fu et al. concluded that the central canal region in mice and chicks is derived from the $Nkx6.1^{+}$ $Nkx2.2^{-}$ domain, which expresses the Olig2 transcription factor (Fu et al. 2003). The potential role of Nkx6.1 in the formation of the central canal niche might be to maintain adult stem cells, as suggested by its highly conserved expression in the chick, rodent and human (Fig. 9).

4. Neural precursor cells in the adult human spinal cord

Much of our knowledge of adult neural stem cells is derived from studies performed in rodents, but much less is known about these stem cells in humans. Adult neural stem and progenitor cells have been isolated from different parts of the human brain, i.e., the olfactory bulb, the SVZ, the hippocampus and the cortex, using the neurosphere assay (Arsenijevic et al. 2001; Roy et al. 1999; Roy et al. 2000; Nunes et al. 2003; Kukekov et al. 1999; Pagano et al. 2000; Scolding, Rayner, and Compston 1999; Sanai et al. 2004; Akiyama et al. 2001). There are major differences between the rodent and primate brain, not only concerning size and organization but also for the cell diversity and SVZ organization. Consequently, we set out to explore the organization of the central canal region and the presence of neural stem cells in the adult human spinal cord (Dromard et al. 2008). For this purpose, we used lumbar spinal cords from brain-dead organ-donor patients (24–70 years old) (Fig. 8) and cervical spinal cords from autopsy tissues.

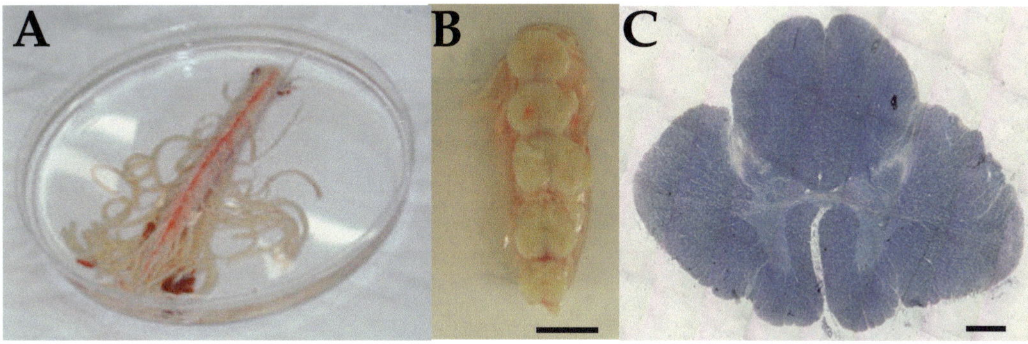

Fig. 8. A) Dissected human spinal cord (T9 thoracic region to caudal end). B) Thoracic, lumbar and sacral spinal cord fragments. Scale bar= 10 mm. C) Luxol staining of a thoraco-lumbar section. Scale bar=1 mm.

Histology and immunohistology studies revealed both similarities and dissimilarities between rodents and man. First, in contrast to mice, the central canal of the human spinal cord is often occluded (Dromard et al. 2008), as previously reported by (Fuller and Burger 1997; Milhorat, Kotzen, and Anzil 1994), and the ependymal region appeared disorganized, with the frequent presence of rosettes or microcanals. Reminiscent of the reported difference between the SVZ in rodents and humans (Quinones-Hinojosa et al. 2006), the human central canal is surrounded by a hypocellular region containing a high density of GFAP filaments and nerve fibers (Dromard et al. 2008). Equally contrasting with rodents, a cluster of Nestin$^+$ subependymal cells was repeatedly observed in the ventral regions of cervical and lumbar spinal cords. Immunolabeling for CD15, Nestin, Nkx6.1, PSA-NCAM and Sox2 revealed, as observed in rodents, that this region retains immature features and is composed of several cell types in contact with the canal or in a subependymal position (Fig. 9).

Whereas most central canal cells expressed Nkx6.1 and Sox2, Nestin, GFAP and CD15 expression was restricted to cell subpopulations (Fig. 9). Nestin is only expressed by a fraction of cells in direct contact with the lumen or in a subependymal position. In contrast, to rodents, GFAP+ cells are numerous and more frequently located in direct contact with the lumen. Strikingly, CD15 expression is most often observed in cells in the dorsal region of the canal. In one lumbar spinal cord, we observed the expression of PSA-NCAM by a subpopulation of cells primarily localized on the ventral side of the central canal region. Taken together, these results represent the first analysis and reveal the complexity of this region in humans; however, further characterization, notably with new markers for neural stem cells, such as Bmi1, is required.

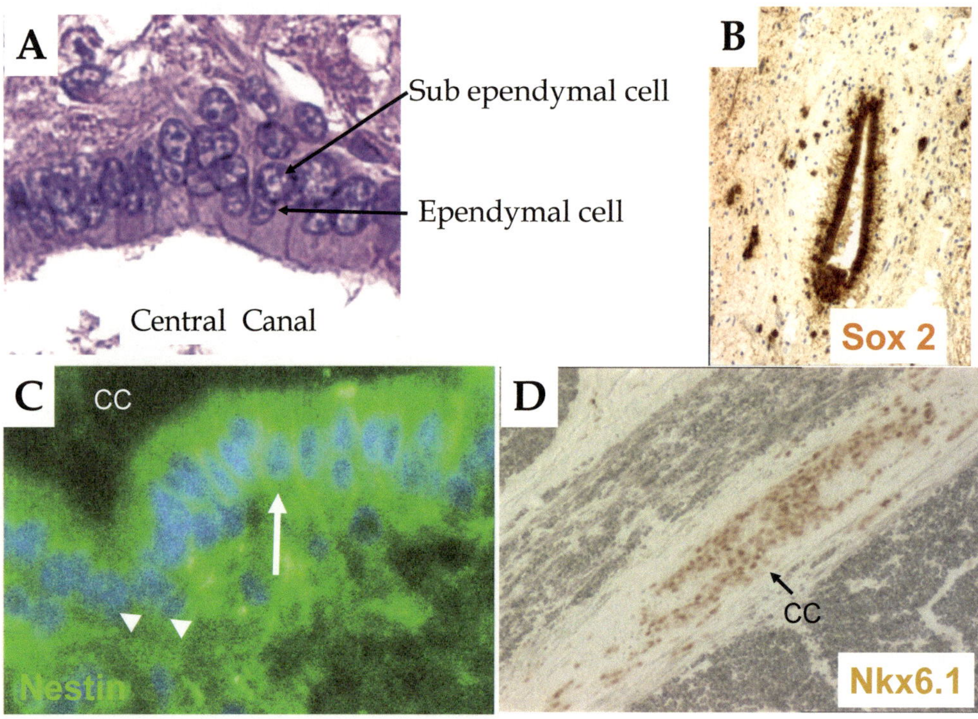

Fig. 9. A) Hematoxylin and eosin staining of a lumbar spinal cord section demonstrating the presence of several layers of cells in the central canal region. B, C, D) Immunodetection of the indicated protein. Note the intense Sox2 staining around the lumen (B), the presence of Nestin+ (arrows) and Nestin- (arrowheads) cells (C) and the expression of Nkx6.1 by central canal cells (D). In D, the central canal region is disorganized without a well-delimited lumen. These data may represent the actual organization or an artifact that occurred as a consequence of tissue processing for histology.

To explore the presence of neural stem or progenitor cells in the human spinal cord, we used the classical neurosphere assay. As compared with rodents, the frequency of neurosphere formation was at least 10 times lower (0.01-0.03 % of isolated cells). These neurospheres were Nestin+ Sox2+ and contained proliferative cells, as evidenced by Ki67 labeling and BrdU incorporation. In one sample, we were able to separate the central region from the surrounding tissue and observed that most of the spheres were derived from the central region. Upon differentiation (Fig. 10), these spheres generated glial (predominantly GFAP+ cells) and neuronal cells. These cells were mostly GABAergic, but some 5-HT neurons could be observed.

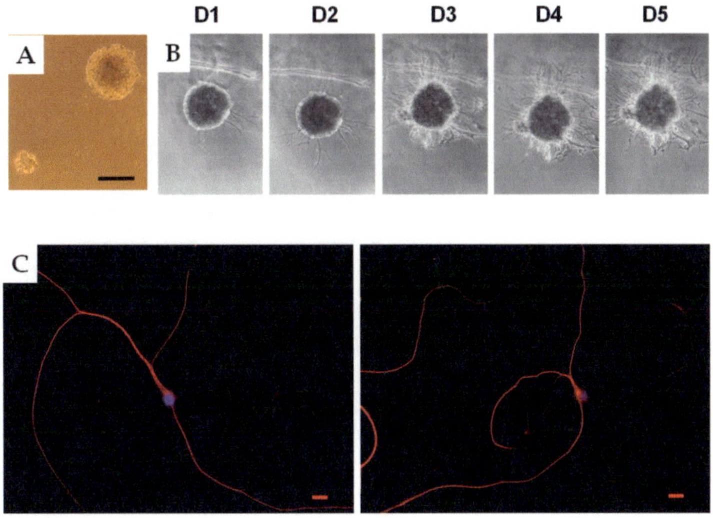

Fig. 10. A) Example of primary neurospheres derived from culturing dissociated human spinal cord cells (two weeks of culture). Scale bar= 500 µm. B) Neurosphere differentiation obtained by plating on an adhesive substrate for 5 days (D1-D5). C) Example of long Map2+ cells observed in differentiated neurospheres. Scale bars= 20 µm.

Importantly, despite intense attempts, we were not able to passage these neurospheres. This suggests that either these cells were progenitors with a limited proliferation potential or that the culture conditions were not adequate to sustain human neural stem cells self-renewal. As previously reported the long-term propagation of human neural stem cell culture is more challenging than that of rodents and might require specific techniques, such the absence of complete dissociation during the passaging process (Svendsen et al. 1998).

In addition to the neurosphere method, adherent conditions and a specific media containing serum could be used to isolate and propagate neural stem cells from the human temporal cortex and the hippocampus (Walton et al. 2006). We recently used these conditions as an attempt to isolate stem cells from the adult human spinal cord (Mamaeva et al. 2011). This attempt resulted in the isolation of a proliferating Nkx6.1+ Nestin+ cell population, which could be maintained for up to 9 passages. However, an in depth cellular characterization showed that these cells were acta2+ caldesmon+ calponin+ smooth muscle cells. These cells also displayed mesenchymal cell features, as evidenced by the high level of expression of the two transcription factors Snai2 and Twist1. We also observed that in vivo, Nkx6.1 was expressed by a subset of spinal cord vascular cells in addition to the central canal cells. Attempts to differentiate these cells into glial cells, neurons, chondrocytes and adipocytes were unproductive; however, these cells could readily become mineralized (formation of $CaPO_4$) when placed under osteogenic conditions. The calcification of CNS muscular cell vessels has been observed since 1884 (Compston 2007; Obersteiner 1884) in pathological situations or as part of the aging process in the brain (Makariou and Patsalides 2009). These cells constitute a useful model for studying CNS vessel calcification in vitro.

Finally, preliminary work from our lab suggests that in addition to the central canal niche, neural precursor cells might also reside in the parenchyma. For instance, Dcx^+ cells can be observed in the white matter, whereas scattered Sox2+ are notably present in the gray matter (Fig. 11). These results indicate that immature neural cells might represent a significant population of spinal cord cells with a completely unknown function.

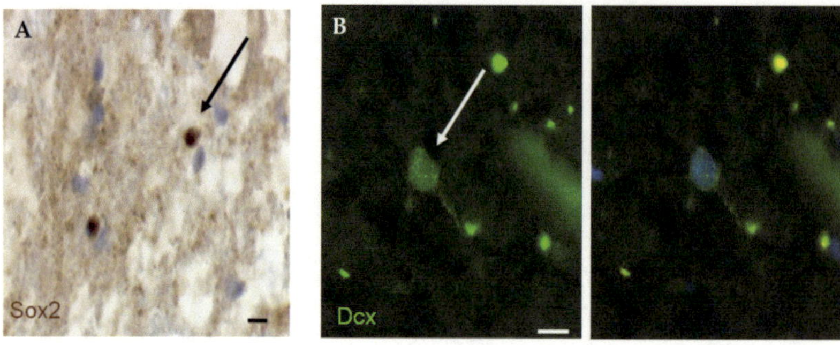

Fig. 11. A) Example of $Sox2^{+}$ cells detected in the gray matter of the adult human spinal cord. B) Example of one Dcx^{+} cell detected in the white matter of the adult human spinal cord. Scale bars= 10 µm.

5. Role of the niche in spinal cord physiology and plasticity

5.1 Proliferation in the niche

The spinal cord elongates extensively during the post-natal period, and its size increases 2.5-fold in mice from birth to 13 weeks (Sabourin et al. 2009). In parallel with this extension, active proliferation is detected postnatally in the central canal region, notably in the ependymocytes, and subsequently declines at 12-13 weeks, which corresponds to the end of spinal cord elongation (Fu et al. 2003; Sabourin et al. 2009). This production of new cells is likely to be necessary for extending the length of the central canal. In addition, as observed postnatally in the brain SVZ (Suzuki and Goldman 2003), it is likely that cells are produced from the central region for myelination or for integration into the developing spinal cord circuitries. In contrast to young animals, in adult rodents and humans, proliferation is low or absent in the ependymal region. Thus, it appears that in adults, most central canal cells are quiescent, meaning that they are in an arrested but reversible state. Consistent with the proliferative decline observed in the postnatal period, we also noted that several markers, such as nestin, cadherin-13, and Sox4, are more highly expressed in young animals than in adults (Sabourin et al. 2009).

The central canal cells could, however, readily re-initiate proliferation in spinal cord traumas (Vaquero et al. 1981) and in several models for neurodegenerative diseases, such as in multiple sclerosis (Danilov et al. 2006) and amyotrophic lateral sclerosis (Chi et al. 2006). In these situations, cells migrate from the central canal region toward the lesion sites by a mechanism that may involve the SDF1/CXCR4 pathway which is present in these cells (Hugnot and Franzen 2010; Shechter, Ziv, and Schwartz 2007). Their fate appear to become mostly macroglial cells and not neurons, despite an increased expression of the transcription factor Pax-6 (Yamamoto et al. 2001; Johansson et al. 1999; Coksaygan et al. 2006). Indeed, they preferentially differentiate into GFAP+ astrocytes contributing to scar formation (Meletis et al. 2008; Johansson et al. 1999; Takahashi et al. 2003; Mothe and Tator 2005) and to some extent into myelinating Olig2 oligodendrocytes (Meletis et al. 2008) making them interesting candidates for myelin degeneration diseases like multiple sclerosis.

Collectively, these data emphasize an important postnatal role for the cells of the central canal region. In adults, these cells enter a dormant state, which could be reactivated in specific physiological situations or under pathological conditions.

5.2 Spinal cord central canal niche, physical exercise and spinal plasticity

Whereas the role of forebrain stem cell niches in memory and learning is being elucidated, the functions of the spinal cord niche, if there are any, remain elusive in adults. One important observation reported by (Cizkova et al. 2009; Foret et al. 2009; Krityakiarana et al. 2010) is that physical exercise (treadmill training and wheel running) can reactivate proliferation within the niche and increase nestin expression. The identity and fate of newly formed cells needs to be fully addressed using reliable labeling techniques, such as permanent genetic lineage tracing with Cre recombinase. One possibility is that the cells in the niche generate new CSF-contacting neurons upon training. Work by Marichal et al. showed that the latter are at different stages of maturation, which raises the possibility that they are in "standby mode" and, under some circumstances (e.g., injury or training), might complete maturation to integrate spinal circuits (Marichal et al. 2009). Another exciting possibility would be that newly formed cells contribute to spinal cord activity-dependent plasticity. For instance, in addition to the hippocampus, long-term potentiation (LTP) has recently been detected in the spinal cord dorsal horn following natural noxious stimulation (Randic, Jiang, and Cerne 1993; Rygh et al. 1999). In the brain, the experimental modification of neurogenesis alters hippocampal LTP (Snyder, Kee, and Wojtowicz 2001; Saxe et al. 2006). Similarly, one could consider the implication of cells generated in the central canal niche for dorsal horn LTP. Along this line, Shechter et al. reported the presence of newly formed GABAergic Dcx+ in the dorsal spinal cord region (Shechter, Ziv, and Schwartz 2007). Further investigations are needed to understand the origin and role of these cells.

The capacity of the spinal cord to learn and memorize is now acknowledged (Guertin 2008; Wolpaw 2010; Wolpaw and Tennissen 2001). For instance, basic spinal cord functions, such as stretch reflexes, can be conditioned even after complete spine transection. Locomotion is another flexible function. It is largely generated by spinal cord circuits localized, notably, in the lumbar region and called central pattern generators (Guertin and Steuer 2009). These circuits generate basic signals for walking. After spinal cord transection, these networks do not die and can be modified and reactivated by several days of specific training sessions. This learning allows rats and cats to display involuntary but structured walking on a treadmill (Rossignol et al. 2009). Whereas these examples of plasticity are largely mediated by synaptic modulations, one could also envision that long-term adaptation would involve the recruitment of new glial and potentially neuronal cells. By harboring cells with a high proliferation and differentiation potential, the central canal niche would be a particularly suited source for providing such new cells.

6. Conclusion

In addition to SVZ and hippocampus, the spinal cord central canal region constitutes a-third adult neural stem cell niche (Hugnot and Franzen 2010). In contrast with the brain, a sustained cell production is not observed in adults in which the niche appears to be largely dormant. In several spinal cord pathologies, the niche reactivates promptly to generate new astrocytes and oligodendrocytes, which participate in repopulating the damaged tissue in addition to forming the glial scar. It appears, therefore, that cells in the niche are in a stand-by but ready-to-go mode. This outstanding capacity is associated with the presence of fetal features in the niche that is well illustrated by the maintenance of activated developmental pathways and genes. These pathways will maintain cells at a high potential for proliferation, differentiation, delamination and migration. Far from being a simple layer of homogenous cells, both in man and in rodents, this region is highly organized and is composed of several cell types. Further work is required to exactly determine the different properties of these cells, which is particularly true for the Dcx^+ CSF-contacting neuronal cells whose function is completely unknown. New cell-specific GFP transgenic mice together with fate mapping using Cre-Lox techniques will be valuable tools to address these issues.

Regarding spinal cord lesions, one important goal is to accurately control and redirect stem cell fate to endogenously repopulate the lesioned tissue with the appropriate cell type. This control and redirection would be an invaluable step for designing new therapeutic strategies against pathologies, such as amyotrophic lateral sclerosis or multiple sclerosis. It would require a detailed understanding of the molecular mechanisms involved in controlling spinal cord neural stem cell fate together with a broad knowledge of how stem cells generate specific neuronal cell subtypes during development. Thanks to intense developmental studies over the last 3 decades (Alaynick, Jessell, and Pfaff 2010), spinal cord development is particularly well described. These data are beginning to be exploited for redirecting adult stem cells into the appropriate cell types.

Investigating the presence and properties of neural precursor cells in humans is a prerequisite for designing regenerative strategies based on the endogenous cellular pool. However, work on adult human tissues is far more complicated than in rodents and raises important ethical, security, reproducibility, organizational and accessibility issues. These issues become exacerbated when research concerning non-pathological, non-fixed and well-preserved CNS is developed. The spinal cord is accessible in organ-donor patients, where thoracic and abdominal organs are removed for transplantation purposes (Dromard et al. 2008). Neurosurgeons could have access to this region of the CNS within hours after the patient is declared brain-dead. This access offers a unique opportunity to characterize the human spinal cord niche and to isolate different neural precursor cell types.

Finally, the spinal cord is an old (oldest ?) component of the CNS but is not just a simple bundle of fibers that connect the brain to the organs and the environment. Recent work has clearly revealed that this primitive CNS region harbors highly-organized internal circuits that are able to display short and long-term plasticity for adapting and memorizing. The spine cord possesses a well-described development and a relatively simple organization with a neural stem cell niche at the center. In addition, its activity can be easily and accurately manipulated by modulating different sensory components (e.g., nociception, temperature,…) and locomotor activity. Therefore, the spinal cord appears as a particularly interesting and simple model for studying the role of adult neural stem cells in CNS physiology. Further characterization of the niche in pathological situations will provide interesting clues on how to utilize this endogenous cell pool to treat spinal cord damage.

7. Acknowledgment

We would like to thank the Planiol foundation, the association Francaise contre les myopathies (AFM), the Rescue FP6 EU program, the Princess Grace de Monaco foundation and the Vertical Association for funding our work on spinal cord stem cells.

8. References

Akiyama, Y., O. Honmou, T. Kato, T. Uede, K. Hashi, and J. D. Kocsis. 2001. Transplantation of clonal neural precursor cells derived from adult human brain establishes functional peripheral myelin in the rat spinal cord. *Exp Neurol* 167 (1):27-39.

Alaynick, W. A., T. M. Jessell, and S. L. Pfaff. 2010. SnapShot: spinal cord development. *Cell* 146 (1):178-178 e1.

Armando, S., A. Lebrun, J. P. Hugnot, C. Ripoll, M. Saunier, and L. Simonneau. 2007. Neurosphere-derived neural cells show region-specific behaviour in vitro. *Neuroreport* 18 (15):1539-42.

Arsenijevic, Y., J. G. Villemure, J. F. Brunet, J. J. Bloch, N. Deglon, C. Kostic, A. Zurn, and P. Aebischer. 2001. Isolation of multipotent neural precursors residing in the cortex of the adult human brain. *Exp Neurol* 170 (1):48-62.

Beckervordersandforth, R., P. Tripathi, J. Ninkovic, E. Bayam, A. Lepier, B. Stempfhuber, F. Kirchhoff, J. Hirrlinger, A. Haslinger, D. C. Lie, J. Beckers, B. Yoder, M. Irmler, and M. Gotz. 2010. In vivo fate mapping and expression analysis reveals molecular hallmarks of prospectively isolated adult neural stem cells. *Cell Stem Cell* 7 (6):744-58.

Bodega, G., I. Suarez, M. Rubio, and B. Fernandez. 1994. Ependyma: phylogenetic evolution of glial fibrillary acidic protein (GFAP) and vimentin expression in vertebrate spinal cord. *Histochemistry* 102 (2):113-22.

Briscoe, J., A. Pierani, T. M. Jessell, and J. Ericson. 2000. A homeodomain protein code specifies progenitor cell identity and neuronal fate in the ventral neural tube. *Cell* 101 (4):435-45.

Browne, G., A. E. Sayan, and E. Tulchinsky. 2010. ZEB proteins link cell motility with cell cycle control and cell survival in cancer. *Cell Cycle* 9 (5):886-91.

Chi, L., Y. Ke, C. Luo, B. Li, D. Gozal, B. Kalyanaraman, and R. Liu. 2006. Motor neuron degeneration promotes neural progenitor cell proliferation, migration, and neurogenesis in the spinal cords of amyotrophic lateral sclerosis mice. *Stem Cells* 24 (1):34-43.

Cizkova, D., M. Nagyova, L. Slovinska, I. Novotna, J. Radonak, M. Cizek, E. Mechirova, Z. Tomori, J. Hlucilova, J. Motlik, I. Sulla, Jr., and I. Vanicky. 2009. Response of ependymal progenitors to spinal cord injury or enhanced physical activity in adult rat. *Cell Mol Neurobiol* 29 (6-7):999-1013.

Coksaygan, T., T. Magnus, J. Cai, M. Mughal, A. Lepore, H. Xue, I. Fischer, and M. S. Rao. 2006. Neurogenesis in Talpha-1 tubulin transgenic mice during development and after injury. *Exp Neurol* 197 (2):475-85.

Compston, A. 2007. From the archive "The cerebral blood-vessels in health and disease" by Pr H. Obersteiner. *Brain* 130:3057-3059.

Conti, L., and E. Cattaneo. 2010. Neural stem cell systems: physiological players or in vitro entities? *Nat Rev Neurosci* 11 (3):176-87.

Danilov, A. I., R. Covacu, M. C. Moe, I. A. Langmoen, C. B. Johansson, T. Olsson, and L. Brundin. 2006. Neurogenesis in the adult spinal cord in an experimental model of multiple sclerosis. *Eur J Neurosci* 23 (2):394-400.

Deleyrolle, L., S. Marchal-Victorion, C. Dromard, V. Fritz, M. Saunier, J. C. Sabourin, C. Tran Van Ba, A. Privat, and J. P. Hugnot. 2005. Exogenous and FGF2/EGF-regulated endogenous cytokines regulate neural precursor cell growth and differentiation. *Stem Cells.*

Deleyrolle, L. P., and B. A. Reynolds. 2009. Isolation, expansion, and differentiation of adult Mammalian neural stem and progenitor cells using the neurosphere assay. *Methods Mol Biol* 549:91-101.

Doetsch, F., L. Petreanu, I. Caille, J. M. Garcia-Verdugo, and A. Alvarez-Buylla. 2002. EGF converts transit-amplifying neurogenic precursors in the adult brain into multipotent stem cells. *Neuron* 36 (6):1021-34.

Dromard, C., H. Guillon, V. Rigau, C. Ripoll, J. C. Sabourin, F. E. Perrin, F. Scamps, S. Bozza, P. Sabatier, N. Lonjon, H. Duffau, F. Vachiery-Lahaye, M. Prieto, C. Tran Van Ba, L. Deleyrolle, A. Boularan, K. Langley, M. Gaviria, A. Privat, J. P. Hugnot, and L. Bauchet. 2008. Adult human spinal cord harbors neural precursor cells that generate neurons and glial cells in vitro. *J Neurosci Res* 86 (9):1916-26.

Foret, A., R. Quertainmont, O. Botman, D. Bouhy, P. Amabili, G. Brook, J. Schoenen, and R. Franzen. 2009. Stem cells in the adult rat spinal cord: plasticity after injury and treadmill training exercise. *J Neurochem* 112 (3):762-72.

Fu, H., Y. Qi, M. Tan, J. Cai, X. Hu, Z. Liu, J. Jensen, and M. Qiu. 2003. Molecular mapping of the origin of postnatal spinal cord ependymal cells: evidence that adult ependymal cells are derived from Nkx6.1+ ventral neural progenitor cells. *J Comp Neurol* 456 (3):237-44.

Fuller, G.N., and P.C. Burger. 1997. Central Nervous System. *Histology for Pathologists* Chapter 6:145-168.

Gage, F. H., J. Ray, and L. J. Fisher. 1995. Isolation, characterization, and use of stem cells from the CNS. *Annu Rev Neurosci* 18:159-92.

Guertin, P. 2008. Can the spinal cord learn and remember? *ScientificWorldJournal* 8:757-61.

Guertin, P. A., and I. Steuer. 2009. Key central pattern generators of the spinal cord. *J Neurosci Res* 87 (11):2399-405.

Hamilton, L. K., M. K. Truong, M. R. Bednarczyk, A. Aumont, and K. J. Fernandes. 2009. Cellular organization of the central canal ependymal zone, a niche of latent neural stem cells in the adult mammalian spinal cord. *Neuroscience* 164 (3):1044-56.

Horner, P. J., A. E. Power, G. Kempermann, H. G. Kuhn, T. D. Palmer, J. Winkler, L. J. Thal, and F. H. Gage. 2000. Proliferation and differentiation of progenitor cells throughout the intact adult rat spinal cord. *J Neurosci* 20 (6):2218-28.

Horstmann, E. 1954. [The fiber glia of selacean brain.]. *Z Zellforsch Mikrosk Anat* 39 (6):588-617.

Huang, A. L., X. Chen, M. A. Hoon, J. Chandrashekar, W. Guo, D. Trankner, N. J. Ryba, and C. S. Zuker. 2006. The cells and logic for mammalian sour taste detection. *Nature* 442 (7105):934-8.

Hugnot, J.P., and R. Franzen. 2010. The spinal cord ependymal region: A stem cell niche in the caudal central nervous system. *Frontiers in Bioscience* 16:1044-59.

Jacquet, B. V., R. Salinas-Mondragon, H. Liang, B. Therit, J. D. Buie, M. Dykstra, K. Campbell, L. E. Ostrowski, S. L. Brody, and H. T. Ghashghaei. 2009. FoxJ1-dependent gene expression is required for differentiation of radial glia into ependymal cells and a subset of astrocytes in the postnatal brain. *Development* 136 (23):4021-31.

Johansson, C. B., S. Momma, D. L. Clarke, M. Risling, U. Lendahl, and J. Frisen. 1999. Identification of a neural stem cell in the adult mammalian central nervous system. *Cell* 96 (1):25-34.

Krityakiarana, W., A. Espinosa-Jeffrey, C. A. Ghiani, P. M. Zhao, N. Topaldjikian, F. Gomez-Pinilla, M. Yamaguchi, N. Kotchabhakdi, and J. de Vellis. 2010. Voluntary exercise increases oligodendrogenesis in spinal cord. *Int J Neurosci* 120 (4):280-90.

Kukekov, V. G., E. D. Laywell, O. Suslov, K. Davies, B. Scheffler, L. B. Thomas, T. F. O'Brien, M. Kusakabe, and D. A. Steindler. 1999. Multipotent stem/progenitor cells with similar properties arise from two neurogenic regions of adult human brain. *Exp Neurol* 156 (2):333-44.

Kulbatski, I., A. J. Mothe, A. Keating, Y. Hakamata, E. Kobayashi, and C. H. Tator. 2007. Oligodendrocytes and radial glia derived from adult rat spinal cord progenitors: morphological and immunocytochemical characterization. *J Histochem Cytochem* 55 (3):209-22.

Kulbatski, I., and C. H. Tator. 2009. Region-specific differentiation potential of adult rat spinal cord neural stem/precursors and their plasticity in response to in vitro manipulation. *J Histochem Cytochem* 57 (5):405-23.

Leatherman, J. L., and S. Dinardo. 2008. Zfh-1 controls somatic stem cell self-renewal in the Drosophila testis and nonautonomously influences germline stem cell self-renewal. *Cell Stem Cell* 3 (1):44-54.

Lehtinen, M. K., M. W. Zappaterra, X. Chen, Y. J. Yang, A. D. Hill, M. Lun, T. Maynard, D. Gonzalez, S. Kim, P. Ye, A. J. D'Ercole, E. T. Wong, A. S. LaMantia, and C. A. Walsh. 2010. The cerebrospinal fluid provides a proliferative niche for neural progenitor cells. *Neuron* 69 (5):893-905.

Li, L., and H. Clevers. 2010. Coexistence of quiescent and active adult stem cells in mammals. *Science* 327 (5965):542-5.

Liu, Y., S. El-Naggar, D. S. Darling, Y. Higashi, and D. C. Dean. 2008. Zeb1 links epithelial-mesenchymal transition and cellular senescence. *Development* 135 (3):579-88.

Louis, S. A., R. L. Rietze, L. Deleyrolle, R. E. Wagey, T. E. Thomas, A. C. Eaves, and B. A. Reynolds. 2008. Enumeration of neural stem and progenitor cells in the neural colony-forming cell assay. *Stem Cells* 26 (4):988-96.

Makariou, E., and A.D. Patsalides. 2009. Intracranial calcifications. *Appl Radiol* 38 (11):48-50.

Mamaeva, D., C. Ripoll, C. Bony, M. Teigell, F. E. Perrin, B. Rothhut, I. Bieche, R. Lidereau, A. Privat, V. Rigau, H. Guillon, F. Vachiery-Lahaye, D. Noel, L. Bauchet, and J. P. Hugnot. Isolation of mineralizing Nestin+ Nkx6.1+ vascular muscular cells from the adult human spinal cord. *BMC Neurosci* 12 (1):99.

Marichal, N., G. Garcia, M. Radmilovich, O. Trujillo-Cenoz, and R. E. Russo. 2009. Enigmatic central canal contacting cells: immature neurons in "standby mode"? *J Neurosci* 29 (32):10010-24.

Martens, D. J., R. M. Seaberg, and D. van der Kooy. 2002. In vivo infusions of exogenous growth factors into the fourth ventricle of the adult mouse brain increase the proliferation of neural progenitors around the fourth ventricle and the central canal of the spinal cord. *Eur J Neurosci* 16 (6):1045-57.

Meletis, K., F. Barnabe-Heider, M. Carlen, E. Evergren, N. Tomilin, O. Shupliakov, and J. Frisen. 2008. Spinal cord injury reveals multilineage differentiation of ependymal cells. *PLoS Biol* 6 (7):e182.

Milhorat, T. H., R. M. Kotzen, and A. P. Anzil. 1994. Stenosis of central canal of spinal cord in man: incidence and pathological findings in 232 autopsy cases. *J Neurosurg* 80 (4):716-22.

Moreno-Manzano, V., F. J. Rodriguez-Jimenez, M. Garcia-Rosello, S. Lainez, S. Erceg, M. T. Calvo, M. Ronaghi, M. Lloret, R. Planells-Cases, J. M. Sanchez-Puelles, and M. Stojkovic. 2009. Activated spinal cord ependymal stem cells rescue neurological function. *Stem Cells* 27 (3):733-43.

Mothe, A. J., and C. H. Tator. 2005. Proliferation, migration, and differentiation of endogenous ependymal region stem/progenitor cells following minimal spinal cord injury in the adult rat. *Neuroscience* 131 (1):177-87.

Nishimura, G., I. Manabe, K. Tsushima, K. Fujiu, Y. Oishi, Y. Imai, K. Maemura, M. Miyagishi, Y. Higashi, H. Kondoh, and R. Nagai. 2006. DeltaEF1 mediates TGF-beta signaling in vascular smooth muscle cell differentiation. *Dev Cell* 11 (1):93-104.

Nolte, C., M. Matyash, T. Pivneva, C. G. Schipke, C. Ohlemeyer, U. K. Hanisch, F. Kirchhoff, and H. Kettenmann. 2001. GFAP promoter-controlled EGFP-expressing transgenic mice: a tool to visualize astrocytes and astrogliosis in living brain tissue. *Glia* 33 (1):72-86.

Nunes, M. C., N. S. Roy, H. M. Keyoung, R. R. Goodman, G. McKhann, 2nd, L. Jiang, J. Kang, M. Nedergaard, and S. A. Goldman. 2003. Identification and isolation of multipotential neural progenitor cells from the subcortical white matter of the adult human brain. *Nat Med* 9 (4):439-47.

Obersteiner, H. 1884. The cerebral blood-vessels in health and disease. *Brain* 7:289-389.

Pagano, S. F., F. Impagnatiello, M. Girelli, L. Cova, E. Grioni, M. Onofri, M. Cavallaro, S. Etteri, F. Vitello, S. Giombini, C. L. Solero, and E. A. Parati. 2000. Isolation and characterization of neural stem cells from the adult human olfactory bulb. *Stem Cells* 18 (4):295-300.

Palmer, T. D., A. R. Willhoite, and F. H. Gage. 2000. Vascular niche for adult hippocampal neurogenesis. *J Comp Neurol* 425 (4):479-94.

Pardal, R., P. , P. Ortega-Saenz, R. Duran, and J. Lopez-Barneo. 2007. Glia-like stem cells sustain physiologic neurogenesis in the adult mammalian carotid body. *Cell* 131 (2):364-77

Pluchino, S., A. Quattrini, E. Brambilla, A. Gritti, G. Salani, G. Dina, R. Galli, U. Del Carro, S. Amadio, A. Bergami, R. Furlan, G. Comi, A. L. Vescovi, and G. Martino. 2003. Injection of adult neurospheres induces recovery in a chronic model of multiple sclerosis. *Nature* 422 (6933):688-94.

Quinones-Hinojosa, A., N. Sanai, M. Soriano-Navarro, O. Gonzalez-Perez, Z. Mirzadeh, S. Gil-Perotin, R. Romero-Rodriguez, M. S. Berger, J. M. Garcia-Verdugo, and A. Alvarez-Buylla. 2006. Cellular composition and cytoarchitecture of the adult human subventricular zone: a niche of neural stem cells. *J Comp Neurol* 494 (3):415-34.

Rafols, J. A., and H. G. Goshgarian. 1985. Spinal tanycytes in the adult rat: a correlative Golgi gold-toning study. *Anat Rec* 211 (1):75-86.

Randic, M., M. C. Jiang, and R. Cerne. 1993. Long-term potentiation and long-term depression of primary afferent neurotransmission in the rat spinal cord. *J Neurosci* 13 (12):5228-41.

Redmond, D. E., Jr., K. B. Bjugstad, Y. D. Teng, V. Ourednik, J. Ourednik, D. R. Wakeman, X. H. Parsons, R. Gonzalez, B. C. Blanchard, S. U. Kim, Z. Gu, S. A. Lipton, E. A. Markakis, R. H. Roth, J. D. Elsworth, J. R. Sladek, Jr., R. L. Sidman, and E. Y. Snyder. 2007. Behavioral improvement in a primate Parkinson's model is associated with multiple homeostatic effects of human neural stem cells. *Proc Natl Acad Sci U S A* 104 (29):12175-80.

Reynolds, B. A., Tetzlaff W., and S. Weiss. 1992. A multipotent EGF-responsive striatal embryonic progenitor cell produces neurons and astrocytes. *J. Neurosci.* 12:4564-4574.

Rossignol, S., G. Barriere, O. Alluin, and A. Frigon. 2009. Re-expression of locomotor function after partial spinal cord injury. *Physiology (Bethesda)* 24:127-39.

Roy, N. S., S. Wang, C. Harrison-Restelli, A. Benraiss, R. A. Fraser, M. Gravel, P. E. Braun, and S. A. Goldman. 1999. Identification, isolation, and promoter-defined separation of mitotic oligodendrocyte progenitor cells from the adult human subcortical white matter. *J Neurosci* 19 (22):9986-95.

Roy, N. S., S. Wang, L. Jiang, J. Kang, A. Benraiss, C. Harrison-Restelli, R. A. Fraser, W. T. Couldwell, A. Kawaguchi, H. Okano, M. Nedergaard, and S. A. Goldman. 2000. In vitro neurogenesis by progenitor cells isolated from the adult human hippocampus. *Nat Med* 6 (3):271-7.

Rygh, L. J., F. Svendsen, K. Hole, and A. Tjolsen. 1999. Natural noxious stimulation can induce long-term increase of spinal nociceptive responses. *Pain* 82 (3):305-10.

Sabourin, J. C., K. B. Ackema, D. Ohayon, P. O. Guichet, F. E. Perrin, A. Garces, C. Ripoll, J. Charite, L. Simonneau, H. Kettenmann, A. Zine, A. Privat, J. Valmier, A. Pattyn, and J. P. Hugnot. 2009. A mesenchymal-like ZEB1(+) niche harbors dorsal radial glial fibrillary acidic protein-positive stem cells in the spinal cord. *Stem Cells* 27 (11):2722-33.

Sanai, N., A. D. Tramontin, A. Quinones-Hinojosa, N. M. Barbaro, N. Gupta, S. Kunwar, M. T. Lawton, M. W. McDermott, A. T. Parsa, J. Manuel-Garcia Verdugo, M. S. Berger, and A. Alvarez-Buylla. 2004. Unique astrocyte ribbon in adult human brain contains neural stem cells but lacks chain migration. *Nature* 427 (6976):740-4.

Saxe, M. D., F. Battaglia, J. W. Wang, G. Malleret, D. J. David, J. E. Monckton, A. D. Garcia, M. V. Sofroniew, E. R. Kandel, L. Santarelli, R. Hen, and M. R. Drew. 2006. Ablation of hippocampal neurogenesis impairs contextual fear conditioning and synaptic plasticity in the dentate gyrus. *Proc Natl Acad Sci U S A* 103 (46):17501-6.

Schafer, K. H., C. Van Ginneken, and S. Copray. 2009. Plasticity and neural stem cells in the enteric nervous system. *Anat Rec (Hoboken)* 292 (12):1940-52.

Scolding, N. J., P. J. Rayner, and D. A. Compston. 1999. Identification of A2B5-positive putative oligodendrocyte progenitor cells and A2B5-positive astrocytes in adult human white matter. *Neuroscience* 89 (1):1-4.

Seaberg, R. M., S. R. Smukler, and D. van der Kooy. 2005. Intrinsic differences distinguish transiently neurogenic progenitors from neural stem cells in the early postnatal brain. *Dev Biol* 278 (1):71-85.

Seitz, R., J. Lohler, and G. Schwendemann. 1981. Ependyma and meninges of the spinal cord of the mouse. A light-and electron-microscopic study. *Cell Tissue Res* 220 (1):61-72.

Seki, T., and Y. Arai. 1993. Highly polysialylated NCAM expression in the developing and adult rat spinal cord. *Brain Res Dev Brain Res* 73 (1):141-5.

Shechter, R., Y. Ziv, and M. Schwartz. 2007. New GABAergic interneurons supported by myelin-specific T cells are formed in intact adult spinal cord. *Stem Cells* 25 (9):2277-82.

Shihabuddin, L. S., J. Ray, and F. H. Gage. 1997. FGF-2 is sufficient to isolate progenitors found in the adult mammalian spinal cord. *Exp Neurol* 148 (2):577-86.

Singh, R. P. , Y. H. Cheng, P. Nelson, and F. C. Zhou. 2009. Retentive multipotency of adult dorsal root ganglia stem cells. *Cell Transplant* 18 (1):55-68

Snyder, J. S., N. Kee, and J. M. Wojtowicz. 2001. Effects of adult neurogenesis on synaptic plasticity in the rat dentate gyrus. *J Neurophysiol* 85 (6):2423-31.

Stoeckel, M. E., S. Uhl-Bronner, S. Hugel, P. Veinante, M. J. Klein, J. Mutterer, M. J. Freund-Mercier, and R. Schlichter. 2003. Cerebrospinal fluid-contacting neurons in the rat spinal cord, a gamma-aminobutyric acidergic system expressing the P2X2 subunit of purinergic receptors, PSA-NCAM, and GAP-43 immunoreactivities: light and electron microscopic study. *J Comp Neurol* 457 (2):159-74.

Suzuki, S. O., and J. E. Goldman. 2003. Multiple cell populations in the early postnatal subventricular zone take distinct migratory pathways: a dynamic study of glial and neuronal progenitor migration. *J Neurosci* 23 (10):4240-50.

Svendsen, C. N., M. G. ter Borg, R. J. Armstrong, A. E. Rosser, S. Chandran, T. Ostenfeld, and M. A. Caldwell. 1998. A new method for the rapid and long term growth of human neural precursor cells. *J Neurosci Methods* 85 (2):141-52.

Takahashi, M., Y. Arai, H. Kurosawa, N. Sueyoshi, and S. Shirai. 2003. Ependymal cell reactions in spinal cord segments after compression injury in adult rat. *J Neuropathol Exp Neurol* 62 (2):185-94.

Tanaka, E. M. 2003. Cell differentiation and cell fate during urodele tail and limb regeneration. *Curr Opin Genet Dev* 13 (5):497-501.

Vaquero, J., M. J. Ramiro, S. Oya, and J. M. Cabezudo. 1981. Ependymal reaction after experimental spinal cord injury. *Acta Neurochir (Wien)* 55 (3-4):295-302.

Vigh, B., M. J. Manzano e Silva, C. L. Frank, C. Vincze, S. J. Czirok, A. Szabo, A. Lukats, and A. Szel. 2004. The system of cerebrospinal fluid-contacting neurons. Its supposed role in the nonsynaptic signal transmission of the brain. *Histol Histopathol* 19 (2):607-28.

Walton, N. M., B. M. Sutter, H. X. Chen, L. J. Chang, S. N. Roper, B. Scheffler, and D. A. Steindler. 2006. Derivation and large-scale expansion of multipotent astroglial neural progenitors from adult human brain. *Development* 133 (18):3671-81.

Weiss, S., C. Dunne, J. Hewson, C. Wohl, M. Wheatley, A. C. Peterson, and B. A. Reynolds. 1996. Multipotent CNS stem cells are present in the adult mammalian spinal cord and ventricular neuroaxis. *J Neurosci* 16 (23):7599-609.

Wolpaw, J. R. 2010. What can the spinal cord teach us about learning and memory? *Neuroscientist* 16 (5):532-49.

Wolpaw, J. R., and A. M. Tennissen. 2001. Activity-dependent spinal cord plasticity in health and disease. *Annu Rev Neurosci* 24:807-43.

Yamamoto, S., M. Nagao, M. Sugimori, H. Kosako, H. Nakatomi, N. Yamamoto, H. Takebayashi, Y. Nabeshima, T. Kitamura, G. Weinmaster, K. Nakamura, and M. Nakafuku. 2001. Transcription factor expression and Notch-dependent regulation of neural progenitors in the adult rat spinal cord. *J Neurosci* 21 (24):9814-23.

Development of New Monoclonal Antibodies for Immunocytochemical Characterization of Neural Stem and Differentiated Cells

Aavo-Valdur Mikelsaar[1], Alar Sünter[2], Peeter Toomik[3], Kalmer Karpson[4] and Erkki Juronen[5]

[1]*Institute of General and Molecular Pathology, University of Tartu and LabAs Ltd., Tartu*
[2]*Institute of General and Molecular Pathology, University of Tartu*
[3]*Department of Food Science and Hygiene, Estonian University of Life Sciences, Tartu*
[4]*LabAs Ltd., Tartu*
[5]*Institute of General and Molecular Pathology, University of Tartu*
Estonia

1. Introduction

Neural stem cells are present in both the developing and adult nervous system of all mammals, including humans. Due to their therapeutic promise, considerable attention has been focused on identifying the sources of stem cells, the signals that regulate their proliferation and the specification of neural stem cells towards more differentiated cell lineages (Bauer et al., 2006). Presently, neural stem cells are often identified based upon the presence of molecular markers that are correlated with the stem and/or progenitor state along with the absence of a more differentiated phenotype as assessed through marker analysis. Nevertheless, from the very beginning of NSC research the frame was set by the search for such markers and some have been identified, which, at least, allow for the identification of NSC clonal cells in cell culture. Still, the task proves to be difficult because of the changing identity that NSC can undergo and the demands that are imposed on markers. A reliable marker should identify NSC not only in the embryonic brain but also in the adult brain. Generally, markers may either be selected for cell function or for some phenotypic differences. There exist several commonly used immunomarkers for the identification of cells of neural lineages. CD 15 (SSEA1), CD133 (Prominin-1), CD184 (CXCR4) , CD271 (p75-NTR), CDw338 (ABCG2), Ki67, Musashi-1, Musashi-2, nestin, Notch-1, PAX-6, SOX-1, SOX-2 are known as neural stem cell markers; PSA-NCAM and CD271(p75-NTR) as neuronal restricted progenitor markers; CD56 (N-CAM), MAP2, DCX, β-III- A2B5 and NG2. GFAP, FGFR3, Ran-2, S100B and CD44 (H-CAM) are known as Type 1 astrocyte, and GFAP, A2B5, CD44 and S100B as Type 2 astrocyte markers. The tubulin and neurofilament NF-H are used as neuron markers. Glial restricted progenitor markers are differentiation of glial restricted progenitors to oligodendrocyte progenitors has been marked by A2B5, NG2, Olig2 and CD140a (PDGFRα). GalC, MBP, CD140a (PDGFRα), O1, O4, Olig1, Olig2 and Olig3 have been used as mature oligodendrocyte markers. In addition CD57, CD271 (p75-NTR), MASH1, Neurogenin 3 and Notch-1 have been used as neural crest stem cell markers (Kennea & Mehmet, 2002; Schwartz et al., 2008; Yuan et al., 2011). The expression overlap of markers requires a use of different immunomarkers for the identification of specific cells in neural lineages. The most problematic point, however, is the potential pitfall in identifying the phenotype of any newborn cell by a single marker. There exists a great need for more and specific monoclonal antibodies as immunomarkers for the characterization of both normal and malignant neural cells.

2. Aim of the study

The main aim of the study is the development of new monoclonal antibodies (MAbs) against neuronal tissue cells to investigate the differentiation and malignization of human nerve cells. The antibody producing hybridomas were obtained by immunizing Balb/c mice with the native fragments of human glioblastoma and foetal neural stem/progenitor cells (see Fig. 1).

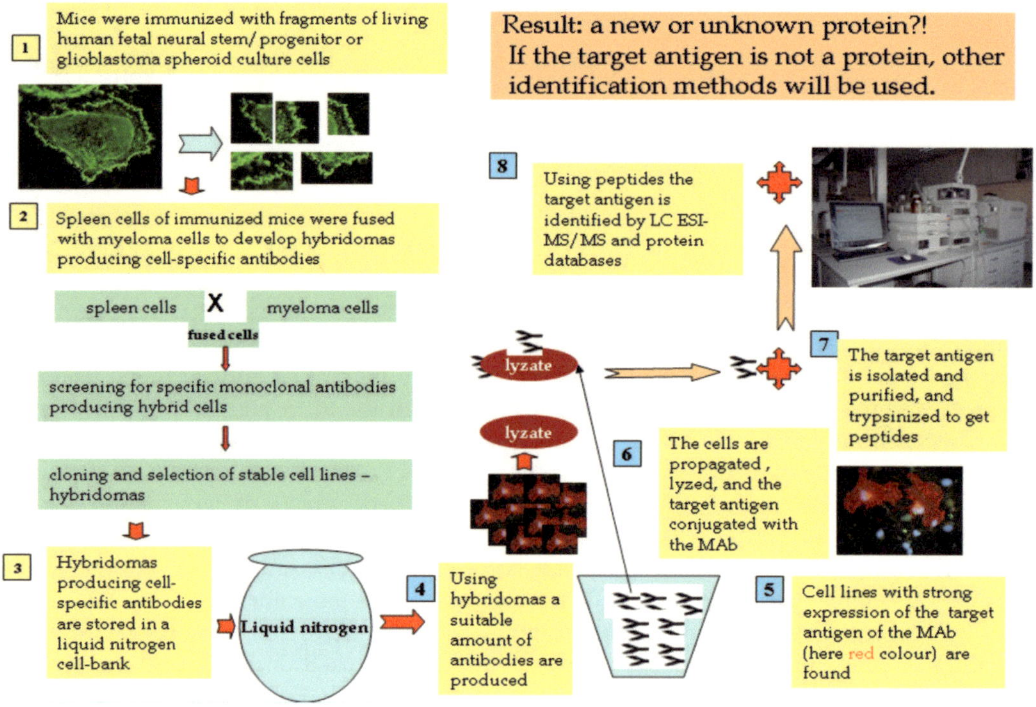

Fig. 1. Flow chart of developing hybridomas and monoclonal antibodies.

In this way it is possible to obtain MAbs against all kinds of antigenic determinants (proteins, glycans and lipids and conformational complex epitopes) that are co-expressed in the living cell, also including those determinants that are expressed on the cell surface. Used immunization scheme allows to obtain MAb panels which characterise the given cell type, including tumour cells. It is possible to obtain MAbs, differentiating between normal and tumour cells. Such specific antibodies against cell surface antigens may be the primary candidates for therapeutic antibodies and/or diagnostic purposes.

3. Methods

3.1 The isolation of human fetal neural stem/progenitor and glioblastoma spheroid cells

Human fetal neural stem/progenitor and glioblastoma spheroid cell lines were developed as described earlier (Kalev et al., 2006). In a greater detail, stem cells were isolated from the brains of 18-21-weeks old fetuses aborted due to medical indications (pregnancy problems). The study was approved by the Ethics Review Committe on Human Research of the University of Tartu. The tissue was mechanically dispersed into a cell suspension in the DMEM/F12 medium, containing gentamicin as antibiotic (Gibco BRL, Gaithersburg, USA). The cell suspension was centrifuged and washed once with the same medium and seeded into 6-well tissue culture plates with a density of 5000-10000 living cells per ml in the medium composed of DMEM/F12, B27 supplement (Gibco BRL) and growth factors bFGF (20 ng/ml; Peprotech), EGF (20 ng/ml; Peprotech), LIF (20 ng/ml, Chemicon) and gentamicin (Gibco BRL). The stem/progenitor cells were grown as neurospheres, the medium was changed every three days, the spheres were dissociated by mechanical trituration after every 12-15 days. Glioblastoma biopsy materials (obtained from Dr E.Jõeste, Department of Pathology, North-Estonian Regional Hospital, Tallinn) were from patients who had signed a consent form. Materials were manipulated by the same method as in the isolation of CNS stem/progenitor cells. The isolation of both CNS stem/progenitor cells and neurosphere-like growing cells from glioblastoma biopsies were performed earlier in LabAs Ltd. In this study five neural stem/progenitor cell lines (hBrSc003, hBrSc004, hBrSc005, hBrSc006 and hBrSc009) and two glioblastoma spheric cell lines (glioblastoma TiVi M-cells and glioblastoma OtAi M-cells) were used for the development and characterization of new monoclonal antibodies (Fig. 2). During differentiation all these cell lines produced three main neural cell types: neurons, astrocytes and oligodendrocytes as it was shown by staining cells with β-III-tubulin, GFAP and GalC (Fig. 3) although in glioblastoma spheroid cultures the cellular and nuclear heterogeneity was significantly higher than in fetal neural stem cell cultures (unpublished). Differentiation of fetal neurospheres and neurosphere-like growing glioblastoma cells (spheroids) was initiated by plating cells onto laminin-coated cover-glasses in the growth media containing all-trans retinoic acid (RA; 10-6 M) and dibutyryl cyclic AMP (dBcAMP; 1mM) and the cells were fixed on different days after the initiation of differentiation.

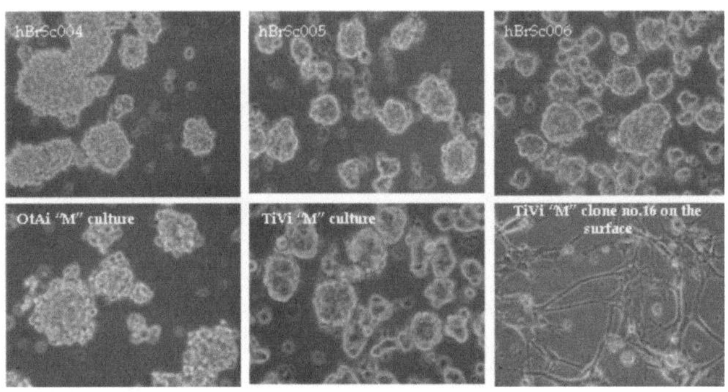

Fig. 2. Neurospheres in fetal neural stem/progenitor cell lines (hBrSc004, hBrSc005 and hBrSc006) and spheroids in glioblastoma OtAi "M" and TiVi "M" cell lines. The last picture shows a clone of TiVi M culture, adapted to grow on surface (Inverted microscope, obj. LWD A20PL 0.40 160/1.2).

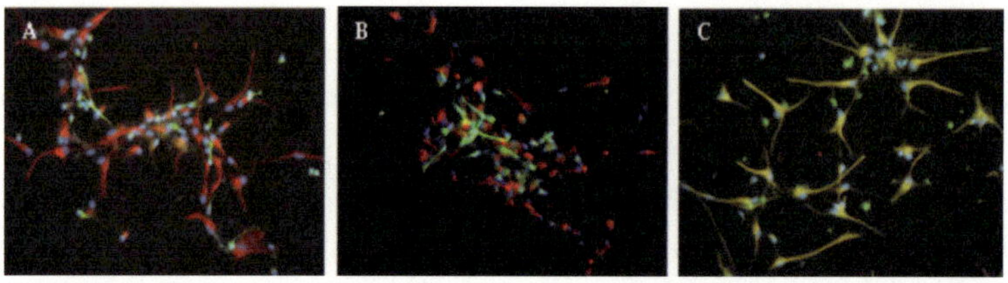

Fig. 3. Expression of different neural immunomarkers in fetal neural stem/progenitor cell line hBrSc006. Neurospheres were differentiated for 7 days. Cells were fixed with 4% PFA and permeabilized with Tritone X-100. (A) Double staining for β-III-tubulin (green - Alexa 488) and GFAP (red - Alexa 594); (B) Double staining for β-III-tubulin (green - Alexa 488) and GalC (red - Alexa 594); (C) Double staining for GFAP (green - Alexa 488) and GalC (red - Alexa 594), note co-location of GFAP and GalC - yellow staining. Obj. 40x. DAPI is used for nuclei staining (blue).

3.2 Development of new mouse monoclonal antibodies

3.2.1 Preparation of immunogen

The three neural stem/progenitor cell lines (hBrSc003, hBrSc004 and hBrSc005) and one glioblastoma spheric cell line (glioblastoma TiVi M-cells) were used as immunogens in this study. The stem/progenitor cells were grown as neurospheres, the DMEM/F12 medium with B27 supplement (Gibco BRL) and growth factors bFGF (20 ng/ml; Pepro Tech, Princeton, USA), EGF (20 ng/ml; Pepro Tech, LIF (20 ng/ml, Chemicon, Temecula, USA) and gentamicin (Gibco BRL) was changed every 3 days. Two-three-week old neurospheres from one 25cm^2 cell culture flask were mechanically dispersed into cell suspension and thereafter repeatedly frozen and thawn to get cell fragments.

3.2.2 Development of mouse hybridoma cells

About 100 µl of the disrupted cell suspension were injected intraperitoneally into 6-week-old normal female Balb/c mice. The injections were repeated 4 times at 4-week intervals.

The cells from the spleens of immunized Balb/c mice were fused with the Sp2/0 myeloma cells by standard procedure (Mikelsaar et al., 2009). Hybridomas were grown in the RPMI 1640/HAT medium containing 10% FCS (Gibco BRL) and gentamicin. Primary screening of supernatants from hybridomas was performed starting from the 10th day of growth. Both unfixed and/or non-permeabilized paraformaldehyde (PFA) fixed human fetal stem/progenitor cells were tested immunocytochemically for the reaction with MAbs. The unfixed cells were incubated with hybridoma supernatants for 1 h at 37°C or overnight at 4°C, washed three times with PBS and a specific reaction of the MAb was revealed by incubating cells with fluorochrome ALEXA 594® - conjugated secondary goat anti-mouse IgG antibody (Molecular Probes, Eugene, USA) for 1 h. The cell fixation without permeabilization was performed with 8% PFA in PBS for 15 min at room temperature (RT), washed three times with PBS, the excess of aldehyde being quenched with 50mM NH_4Cl in PBS (10 min) and blocked (0.3% casein, 0.01% Tween 20 in PBS) for 1 h at RT or overnight at 4°C. Ice-cold 100% methanol producing permeabilization was used for a further characterization of MAbs. The cells were incubated for 15 min at RT, washed three times with PBS and blocked. From plenty of MAb-producing clones, only those that revealed heterogeneity of reaction with stem/progenitor cells, and glioblastoma spheroid cultures, were further investigated. The expression of antigens of selected monoclonal antibodies was further characterized besides neural stem/progenitor cells also on other living and/or fixed cells of different origin. It is important to find the cells where the target antigen of MAbs has the strongest expression in order to use these cells for molecular characterization of the antigen.

3.2.3 MAb cloning and isotyping

Selected hybridomas were cloned by limiting dilution and isotypes of the MAbs were determined by using goat anti-mouse Ig isotype specific antibodies developed in LabAs Ltd Cloned hybridomas are stored in a liquid nitrogen cell bank.

3.3 Immunocytochemical characterization of monoclonal antibodies

3.3.1 Cells and cell lines used to characterize monoclonal antibodies

Besides the cells used as immunogens (fetal neural stem/progenitor and glioblastoma spheroid cells) other types of cells and cell lines were used to characterize monoclonal antibodies. This was necessary to study the specificity of antibodies and is also useful to find out the positive cell lines with shorter duplication time. These cell lines were often used for getting more cellular material in shorter time for the identification of target antigens of MAbs (see also 3.4.2). The following additional cells and cell lines were used for the characterization of monoclonal antibodies: normal human cells - human blood thrombocytes, human sperms, normal fetal and adult skin fibroblast cell lines NL011 (LabAs Ltd) and SA-54, respectively, human normal amniotic epithelial cell line KM (LabAs Ltd.); malignant cell lines - human glioblastoma TiViMNBFCS10 cell line, glioblastoma TiVi M clone 16, glioblastoma OjArMNBFCS and OjFeMNBFCS cell lines (all developed in LabAs Ltd), Bowes melanoma cell line; cells of other species - COS-1 cell line (simian origin), rat granulare cell culture (kindly provided by Prof. A. Žarkovsky).

3.3.2 Fixation methods

Paraformaldehyde or methanol fixation methods were used. In PFA fixation the coverslips were transferred without any previous washing into dishes containing prewarmed 4% PFA in PBS and left for 5 min at RT. Then the coverslips were washed for 3x 5 min with PBS and the excess of aldehyde was quenched with 50mM of NH_4Cl in PBS (10 min). After washing twice with PBS, the cells were permeabilized for 10 min with 0.1% Tritone X-100 in PBS, washed with PBS and blocked. (In methanol fixation the coverslips were treated for 5 min with ice cold methanol and washed with PBS). The coverslips were then transferred into a blocking solution (0.3% casein, 0.01% Tween 20 in PBS) for 1 h at RT or overnight at 4°C.

3.3.3 Staining methods

The blocking solution was removed by aspiration and the cells were stained as follows: they were incubated for 1 hour at RT with the MAb supernatant. The coverslips were washed at least for 3x5 min with PBS and immunolabeling was visualized by incubating the cells with the goat anti-mouse IgG antibody conjugated with fluorochrome Alexa 594 (Molecular Probes) for 1 h at RT. In all cases, the coverslips were washed at least for 3x5 min with PBS, 10 µl of DAPI solution 1µg/ml was added into the last PBS and then the coverslips were incubated for 5 minutes at RT. After quick rinsing in distilled water, the coverslips were mounted in the anti-fading mounting medium Prolong Gold Antifade (Molecular Probes). The cells were checked by a visual microscoping system (Olympus BX, using objectives UplanFI 20x/0.50, 40x/0.75, or 100x/1.30 Oil Iris and the Olympus DP50-CU Photographing System).

3.4 Characterization and identification of target antigens of new monoclonal antibodies

3.4.1 Strategy for target antigen identification

Our strategy for the identification of target antigens of monoclonal antibodies was the following: (1) lysing , electrophoresis and the immunoblotting of suitable cellular material were performed to identify where the bands reacting with specific monoclonal antibodies are located; (2) for immunoprecipitation antibodies were caught to protein-G-conjugated Sepharose beads from the culture medium or used DVS-activated beads for purified antibodies; (3) the immunoprecipitated antigen was separated from the antibody and nonspecifically associated material by using electrophoresis in the SDS-PAGE gel; (4) the bands containing antigen were identified by using immunoblotting from the same gel; (5) the bands containing the antigen were cut out from the gel and trypsinized; (6) tryptic peptides and target antigens were identified with aid of mass-spectroscopy (LC ESI-MS-MS) and protein databases.

3.4.2 Cells and cell lines used for identification of target antigenes

For molecular identification of the target antigens of MAbs different cells and cell lines were used (see also 3.3.1). However, in this study mainly the Bowes melanoma cell line (neural crest origin) and glioblastoma cell line TiViMNBFCS10 (glioblastoma TiVi M spheroid culture, but adapted to growing on the surface) were used. Both the cell lines were propagated in medium DMEM/F12 with 10% FCS and gentamicin (Gibco BRL).

3.4.3 Purification of antigens

3.4.3.1 Purification of antigens from cells growing on surface

The cells of the human Bowes melanoma or TiViMNBFCS10 cell lines were lysed in 8M urea, 3% SDS, 50mM Tris-HCL, pH 6.8 and diluted with 30 volumes of TBS. About 50 µl of Sepharose-bound MAbs were added to 30 ml of the diluted sample solution and incubated overnight at 4°C. The beads were washed, incubated in an electrophoresis sample buffer for 10 min at 95°C and loaded onto the top of the 10% SDS-PAGE gel. After electrophoresis immunopositive bands were located by immunoblotting an one part of the gel, whereas the rest of the gel was stained with colloidal Coomassie G-250 for the confirmation of protein location. This electrophoresis step was absolutely necessary to avoid interferences with non-specifically bound proteins. The immunopositive band was cut out, minced, washed and dried (about 15 min at RT) by CentriVac. The dried pieces of the gel were rehydrated and the proteins trypsinized overnight at 37°C with sequencing grade trypsin (Promega, Madison WI, USA).

3.4.3.2 Purification of human blood thrombocytes

Purification of platelets was performed essentially according to the method of P.J. Canvar and co-workers (Canvar et al., 2007). Shortly, human platelet-rich plasma from the blood center of Tartu University Clinicum was further purified by centrifuging at 1400g for 15 min to pellet any remaining white or red blood cells and the platelet-rich plasma was decanted. Protease inhibitors (Roche Complete, Roche Diagnostics GmbH, Mannheim, Germany) were added as recommended by manufacturer and incubated 10 min at RT. Platelets were pelleted by centrifuging in at 4°C and washed once in Tyrode's buffer by centrifuging at 2400g for 15 min at 4°C. The pellets were stored in Tyrode's buffer at -80°C until used.

3.4.3.3 Cell lysis

Radioimmunoprecipitation assay buffer (RIPA) was used containing 50 mM of Tris/HCl pH 7.4 (ultrapure, AppliChem, Darmstadt, Germany); 0.1% SDS (ultrapure, AppliChem, Darmstadt, Germany); 1% NP40 (Octylphenoxy polyethoxy ethanol, reagent grade, AMRESCO, Solon Ind. Ohio, USA); 1% Tritone X-100 (Schuchardt, München, Germany); 0.5% DOC (Natriumdeoxycholat, AppliChem, Darmstadt, Germany); 500 mM NaCl (AppliChem, Darmstadt, Germany); protease inhibitor tablets (Roche Complete, Roche Diagnostics GmbH, Mannheim, Germany). For more gently lysis of the cells simply 1%NP40 solution in PBS (AppliChem, Darmstadt, Germany) was used.

3.4.3.4 SDS-polyacrylamid gel electrophoresis (SDS-PAGE)

Cells for electrophoresis were lysed in the RIPA buffer or with 1% NP-40 in PBS and centrifuged at maximum speed in Eppendorf centrifuge 5415C for 10 min at 4°C. The supernatant was diluted 1:1 in the SDS-electrophoresis sample buffer with DTT (AppliChem, Darmstadt, Germany) and heated for 10 min at 95°C or at 60°C for sensitive antigenes. Before electrophoresis 8M of urea was added to the sample solution until 2M of the end concentration. Electrophoresis was run in a mighty small Hoefer electrophoresis system (Hoefer Scientific Instruments, CA, USA) with the glasses of 10 x 12 cm and the spacers of 0.75 mm. The end concentration of the gel buffer was 0.43 M and pH 8.4. Usually the samples were separated in the gradient (8-24%) SDS-PAGE gel for achieving sharper bands, yet the concentrating gel was not used. The parameters for prerunning electrophoresis were 60 V and 30 min. Electrophoresis was run for 4 hours at max 210V and 16mA.

3.4.3.5 Colloid Coomassie G250 staining

Colloid Coomassie G250 staining was performed essentially according to the D.Kang and co-workers (Kang et al., 2002). Shortly, the SDS-PAGE gel was fixed in a mixture of 30% ethanol and 2% phosphoric acid for 15 min and washed in distilled water 3x for 15 minutes. Staining was performed in a solution containing 5% w/v aluminium sulphate (Applichem, Darmstadt, Germany), 0.02% Coomassie G-250 (Serva Electrophoresis GmbH, Heidelberg, Germany), 2% phosphoric acid and 10% ethanol for 1 hour. Destaining was performed in distilled water until the bands became clearly visible.

3.4.3.6 Western blot

The method described in the Millipore manual for immunoblotting to the PVDF membrane was used. Buffer components and the stain were from AppliChem, Darmstadt, Germany. The procedure was as follows: the membrane was wetted in methanol for 1 to 3 seconds.

Methanol was eluted incubating the membrane for 5 min in distilled water on a shaker. Thereafter the membrane was equilibrated in the transfer buffer for 2 to 3 min. SDS-PAGE gel was equilibrated in the cathode buffer for 15 min on the shaker. GEHealthcare electrophoresis unit Multiphor II and a semidry immunoblotting kit for electroblotting were used. The anode buffers were Tris 0.3 M (no.1) and Tris 15 mM (no.2); the cathode buffer contained 40 mM of glycine and 25 mM of Tris. Two sheets of chromatography papers (Whatman CHR17, Mainstone, England) were soaked in anode buffer no.1 and one sheet in the anode buffer no.2. The wetted PVDF membrane was placed on the top of the soaked anode sheets and then equilibrated SDS-PAGE gel was placed on the top of them. After that three chromatography papers with the cathode buffer were located on the top of the gel and immunoblotting was performed at 0.8 mA /cm^2 for 2 hours. After electroblotting nonspecific binding of antibodies was blocked by incubating the membrane in TBS with 0.1% Tween 20 (Sigma, MO, USA). For vimentin-specific antibody GB26 10G3 a mixture of 1% horse serum and 0.1% Tween 20 in TBS was used as a blocking solution. The membrane was incubated with monoclonal antibody in the blocking buffer for overnight at 4°C. After washing 3 x for 10 min in PBS 0.1% Tween 20 the membrane was incubated with a secondary antibody (goat anti-mouse IgG polyclonal antibody conjugated with horseradish peroxidase) in PBS 0.1% Tween 20 for one hour at RT. After that the membrane was washed with 0.1% Tween 20 in PBS 3 times for 10 min and then 1 x for 10 min in PBS. The membrane was incubated for 10 minutes in the staining solution containing 50 mg of DAB (diaminobenzidin trihydrochloride, AppliChem, Darmstadt, Germany), 6 ml chloronaphtole (Sigma, MO,USA) solution (3 mg/ml) in ethanol and 20 µl of 30% H_2O_2 in PBS and washed 3 times for 5 minutes in distilled water. After drying the membrane can be saved in the archive.

3.4.3.7 Conjugation of antibody with horseradish peroxidase

Antibodies were conjugated to HRP as described previously (Tjissen, 1985). Shortly, 1 mg of horseradish peroxidase (Boehringer Mannheim, Germany) was solubilized in the Eppendorf tube in 0.1 ml of the freshly prepared 0.1M $NaHCO_3$ solution and 0.1 ml of 8-16 mM $NaIO_4$ (Merck, Darmstadt, Germany) was added and incubated for 2 hours in the dark room at RT. 3 mg of antibody was solubilized in 1 ml of the sodium carbonate buffer pH 9.2 and dry Sephadex G-25 (GE Healthcare, Sweden) was added of about 1/6 from the total amount of the solution and the mix was incubated for 3 hours in the dark room at RT. The conjugate was eluted from Sephadex and mixed (1/20 from the total volume) with the freshly prepared $NaBH_4$ (Sigma, MO, USA) solution in 0.1M of NaOH (5 mg/ml) on a shaker. After 30 min an additional amount of the $NaBH_4$ solution (1/10 from the total volume) was added and the mix was incubated for 1 hour at 4°C. Then 50% of glycerol (AppliChem, Darmstadt, Germany) was added and the conjugate was stored at -20°C. For immunoblotting HRP-conjugated antibodies were diluted from 1:500 to 1:1500.

3.4.3.8 Activating of Sepharose granules with divinylsulphone (DVS) and binding of antibodies to activated granules

Sepharose CL-4B granules (Pharmacia, Uppsala, Sweden) were washed with distilled water of 4 to 5 gel volumes. The granules were equilibrated with 4 to 5 gel volumes of 0.5 M carbonate buffer (pH 11), 10 ml of the 0.5 M carbonate buffer (pH 11) containing 1 ml of divinylsulphone (Sigma, MO, USA) were added to 10 ml of the gel and incubated on a shaker for 1.5 hr. The activated granules were washed once with the carbonate buffer, then 2 x with distilled water and stored in PBS with 0.01% NaN_3 at 4°C. Activated granules are useable during about 1 year. For the binding of antibodies the granules were washed with 4 to 5 volumes of distilled water and then equilibrated in 0.1 M carbonate buffer pH 9.0. The buffer for antibodies was changed to the carbonate buffer pH 9.0 using the PD 10 column (GE Healthcare, Sweden). DVS-activated Sepharose CL-4B granules (200 µl) were added to antibody solution (1-2 mg in 3.5 ml) in the carbonate buffer and incubated for 10 min at RT. Then 5% PEG 20,000 was added and the granules were incubated overnight at RT on a shaker. Afterwards the granules with conjugated antibodies were washed and blocked in TBS buffer and stored in TBS at 4°C with 0.01% NaN_3 until used.

3.4.3.9 Immunoprecipitation

For immunoprecipitation the lysates in RIPA buffer, PBS with 1%NP40 or PBS/TBS diluted 1:30 electrophoresis sample buffer were used. Before immunoprecipitation MAbs from 1.5 ml of the hybridoma medium (sometimes more) were conjugated to 50 µl of Protein-G Sepharose 4 Fast Flow granules (GE Healthcare, Sweden). The granules were washed 3 times in the immunoprecipitation buffer and incubated with cell lysate at 4°C overnight. Then the granules were washed in one of the immunoprecipitation buffers 4 times and once with distilled water. The washed granules (50 µl) were diluted in 150 µl of the SDS-PAGE sample buffer and incubated for 5 min at 95°C or in case of sensitive antigenes for 10 min at 60°C. Then 50 µl of 8M urea were added and the granules were incubated for 15 min at RT on a shaker and thereafter put on the top of the electrophoresis gel. When purified monoclonal antibodies not hybridoma supernatants were used for immunoprecipitation, MAbs were directly conjugated to the divinyl-sulphone-activated granules.

3.4.3.10 Trypsinization of isolated antigenes

After electrophoresis, immunopositive bands were located by immunoblotting a part of the gel, whereas the rest of the gel was stained with colloidal Coomassie G-250 for the confirmation of protein location. This electrophoresis step is necessary to avoid interferences with non-specifically bound proteins. The immunopositive band was cut out from the gel and minced to pieces of about 1 x 1 mm using a scalpel. Pieces of the gel were washed once with methanol on a shaker for 1-2 minutes, then with 10 mM $(NH_4)_2CO_3$ 2 times for 5 min and thereafter with 50% acetonitrile in 10 mM $(NH_4)_2CO_3$. The gel pieces were dried with the CentriVac for 15 min and treated for 10 min with the solution containing 2µg of trypsine in 1 ml of 50 mM $(NH_4)_2CO_3$ with 0.05% ProteasMax surfactant (Promega, USA). 10 mM of $(NH_4)_2CO_3$ was added to fully coated gel pieces and incubated at 37°C overnight. The solution from the top of the gel and the washing solutions were collected into one tube. The gel pieces were washed two times with 50% acetonitrile in 10 mM $(NH_4)_2CO_3$ and once in pure acetonitrile. The solution containing peptides was concentrated until the amount of 100 µl and analyzed using mass-spectroscopy. Sometimes for confirming the exact band location in the gel direct staining of proteins on the membrane is needed. To that end colloid Coomassie stain diluted 1:1 in distilled water was used. After that we destained the membrane washing with distilled water about ten times 10 min or until protein bands became visible. On the dry membrane we saw strongly stained black bands containing the antigen or antibody components and other blue-stained bands containing different co-immunoprecipitated nonantigenic proteins.

3.4.3.11 Peptide analysis by LC ESI-MS/MS and protein identification with Mascot and the Global Proteome Machine

The Agilent 1100 Series chromatograph with LC/MSD Trap XCT (Agilent, Santa Cruz, USA) was used for LC/MS experiments applying the 2.1 x 150 mm Agilent 300Extend C18 column of 3.5 µm particle size. 50 µl of the peptide mixture was injected into the column and eluted with a gradient from 0.1% HCOOH/5% acetonitrile to 75% acetonitrile during 120 minutes at 0.3 ml/min. The column was thermostated at 35°C. Positive ions were detected in a "smart mode" with the target mass set to 1000 m/z, whereas doubly charged ions were preferred. The data were analyzed with both Mascot www.matrixscience.com and GPM www.thegpm.org search engines (1000 most abundant ions). For the positive identification of the antigen the consent of both search engines was necessary.

4. Results

During our study we have developed several hundreds of new monoclonal antibodies. The characterization and identification of the target antigens of them are in progress. Some results of this work have been recently published (Mikelsaar et al., 2009). Here we present data on the five new monoclonal antibodies (see Table 1), which we have developed according to the strategy described in this chapter.

Name of monoclonal antibodies	**Ig sub-class**	**Immunogen**	**Identified target antigen**	**Neural cell types identified by antibody**
E14G2	IgG1	Mix of fetal neural cell lines hBrSc003/ hBrSc004/ hBrSc005	Annexin A1	Annexin A1$^-$ and $^+$ cells among differentiated fetal neural stem cells, glioblastoma spheroid cells and in glioblastomas; microglia; annexin A1$^+$ glial cells and Purkinje (?) cells in rat granule cell culture
E15F10.B9	IgG1	Mix of fetal neural cell lines hBrSc003/ hBrSc004/ hBrSc005	Calnexin	Calnexin$^-$ and $^+$ cells among differentiated fetal neural stem cells and glioblastoma spheroid cells; shows cellular heterogeneity in glioblastomas
W4A8.F4	Ig2a	hBrSc003	14-3-3 ζ/δ	Both human and rat glial and neuronal cells
GB26 10G3	IgG1	Mix of different glioblas-toma cells	Vimetin	The precursors cells of both neuronal and glial lineages; shows cellular heterogeneity in glioblastomas
A3G2.B4	IgG1	hBrSc003	Lupus La protein	La protein$^-$ and $^+$ cells among differentiated fetal neural stem cells and glioblastoma spheroid cells; shows cellular heterogeneity in glioblastomas

Table 1. Summary of new monoclonal antibodies described in this chapter. All these antibodies may be useful for detailed analysis of the expression of target antigens in all types of neural cells .

4.1 New monoclonal antibodies

4.1.1 Monoclonal antibody E14G2 (MAb E14G2)

Here we describe the results of the immunocytochemical analysis of MAb E14G2 (Table 1) with different cell types and the biochemical/molecular analysis for identifying the target antigen of the antibody. In Fig. 4 we can see a distinct heterogenous expression of MAb E14G2 antigen in differentiated fetal neural stem/progenitor (A,B,C,D), in glioblastoma spheroid cells (E) and also in the rat granule primary culture (H). However, the expression of the target antigen of MAb E14G2 was strong and homogenous in glioblastoma OjAr secondary culture and in Bowes melanoma cell lines (F and G, respectively).

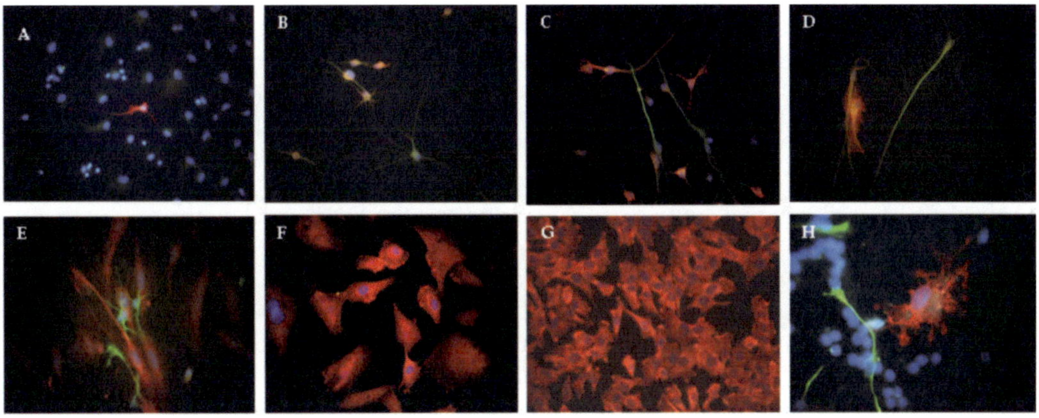

Fig. 4. Presence of MAb E14G2 target antigen (Annexin A1) in different cells. (A) Human fetal hBrSc003 neurosphere cells differentiated for 11 days. Double immunofluorescent staining, cells were fixed with 4% PFA and permeabilized. Annexin A1 expression was revealed by red staining - Alexa 594, β-III-tubulin by green staining - Alexa 488, obj. 40x. (B-D). Human fetal hBrSc005 neurosphere cells differentiated for 3 days (B,C) and 7 days (D). Double immunofluorescent staining, cells were fixed with 4% PFA and permeabilized. Annexin A1 expression was revealed by red staining - Alexa 594, β-III-tubulin by green staining - Alexa 488, obj. 40x. (E) Human glioblastoma TiVi spheroid cell line differentiated for 7 days. Double immunofluorescent staining, cells were fixed with 4% PFA and permeabilized. Annexin A1 expression was revealed by red staining - Alexa 594, β-III-tubulin by green staining - Alexa 488, obj. 40x. (F) Human glioblastoma OjAr secondary cell culture on the 6th day of cultivation. Cells were fixed with 4% PFA and permeabilized. Annexin A1 expression was revealed by red staining - Alexa 594, obj. 40x. (G) Bowes melanoma cell line on the 2nd day of cultivation, fixed with 4%PFA, no permeabilization. Specific Annexin A1 staining was revealed by red staining - Alexa 594, obj. 40x. (H) Rat granulare cell culture. Double immunofluorescent staining, cells were fixed with 4% PFA and permeabilized. Annexin A1 expression was revealed by red staining - Alexa 594, GFAP by green staining - Alexa 488, obj. 100x. DAPI is used for nuclei staining (blue).

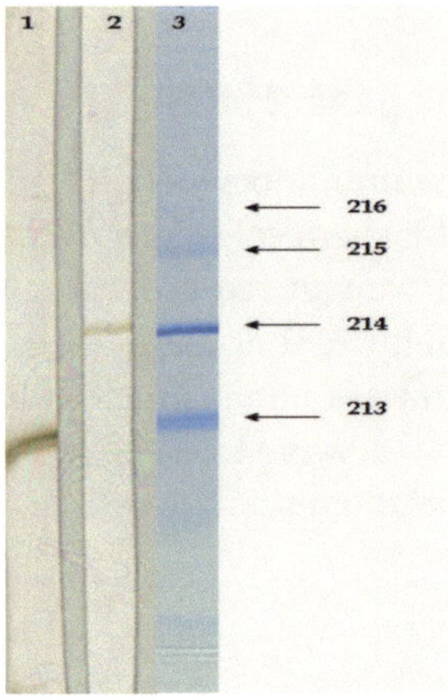

Fig. 5. Electrophoresis and Western blot of MAb E14G2 target antigen. (1) negative control, MAb Y1C7 directly conjugated with peroxidase; (2) MAb E14G2 directly conjugated with peroxidase; (3) gel stained with colloid Coomassie G-250. Arrows with numbers indicate bands analyzed by LC ESI-MS/MS.

In Fig. 5 we present the data on electrophoresis and immunoblot of the target antigen of MAb 14G2. The antigen was immunoprecipitated from the Bowes melanoma cell lysate and the precipitated cells (100 µl) were lysed in 10 ml of RIPA buffer. MAb E14G2 was conjugated directly to the divinylsulphone-activated Sepharose granules. Then 50 µl of the activated granules were incubated in 10 ml of the cleared Bowes melanoma cell lysate overnight at 4°C, washed 4x with the buffer and 1x with distilled water. Then 150 µl of SDS-PAGE sample buffer was added and the sample heated 5 min at 95°C before electrophoresis.

The bands, which were positive according to the reaction with specific monoclonal antibody MAb E14G2 (Fig. 4, arrows 213-216) were analyzed by LC ESI-MS/MS and the proteins were identified using Mascot and the GPM software databases (see Table 2).

Band	**Identified proteins**	**Mass**	**Mascot score**	**Sequence coverage %**
213	ANXA1 - Annexin A1	38690	41	8
	KV401 - Ig kappa chain V-IV region	13372	37	7
214	ANXA1 - Annexin A1	38690	2274	71
215	ANXA1 - Annexin A1	38690	198	34
	TBB5 - Tubulin beta chain	49639	113	24
	K2C1- Keratin, type II cytoskeletal 1	65999	84	6
216	DHSA- Succinate dehydrogenase [ubiquinone] flavoprotein subunit, mit	72645	103	26
	K2C1- Keratin, type II cytoskeletal 1	65999	82	7
	ANXA1 - Annexin A1	38690	65	18

Table 2. LC ESI-MS/MS and Mascot software analysis of immunoprecipitated with MAb 14G2 proteins in bands 213, 214, 215 and 216 (Fig. 4, arrows 213-216).

```
  1 MAMVSEFLKQ AWFIENEEQE YVQTVKSSKG GPGSAVSPYP TFNPSSDVAA
 51 LHKAIMVKGV DEATIIDILT KRNNAQRQQI KAAYLQETGK PLDETLKKAL
101 TGHLEEVVLA LLKTPAQFDA DELRAAMKGL GTDEDTLIEI LASRTNKEIR
151 DINRVYREEL KRDLAKDITS DTSGDFRNAL LSLAKGDRSE DFGVNEDLAD
201 SDARALYEAG ERRKGTDVNV FNTILTTRSY PQLRRVFQKY TKYSKHDMNK
251 VLDLELKGDI EKCLTAIVKC ATSKPAFFAE KLHQAMKGVG TRHKALIRIM
301 VSRSEIDMND IKAFYQKMYG ISLCQAILDE TKGDYEKILV ALCGGN
```

Fig. 6. Amino acid sequences of Annexin A1, ANXA1. In red are marked amino acids sequences identified by mass-spectroscopy (band 214 in Fig. 5).

Conclusion

On the basis of data presented above we conclude that the target antigen of MAb E14G2 is protein Annexin A1. According to UniProtKB/Swiss-Prot database Annexin A1 is a calcium/phospholipid-binding protein which promotes membrane fusion and is involved in exocytosis. This protein regulates phospholipase A2 activity. (http://www.uniprot.org/uniprot/P04083#section_comments).

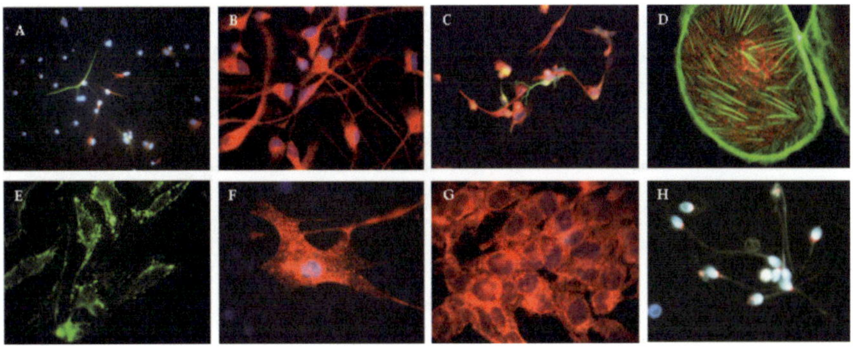

Fig. 7. Presence of MAb E15F10.B9 target antigen (calnexin) in different cells. (A) Human fetal hBrSc003 neurosphere cells differentiated for 11 days. Double immunofluorescent staining, cells were fixed with 4% PFA and permeabilized. Calnexin expression was revealed by red staining - Alexa 594, β-III-tubulin by green staining - Alexa 488, obj. 40x. (B) Human fetal hBrSc003 neurosphere cells differentiated for 7 days, cells were fixed with 4% PFA and permeabilized. Calnexin expression was revealed by red staining - Alexa 594, obj. 100x. (C) Human glioblastoma TiVi spheroid cell line differentiated for 7 days. Double immunofluorescent staining, cells were fixed with 4% PFA and permeabilized. Calnexin expression was revealed by red staining - Alexa 594, β-III-tubulin by green staining - Alexa 488, obj. 40x. (D) Human normal amniotic epithelial cell line KM on the fifth days of cultivation. Double immunofluorescent staining, cells were fixed with 4% PFA and permeabilized. Calnexin expression was revealed by red staining - Alexa 594, actin by green staining with Alexa 488 conjugated Phalloidin, obj. 100x. (E) Unfixed cells of adult skin fibroblast cell line SA-54 on the 3rd day of cultivation. Specific calnexin staining revealed by green staining - Alexa 488, obj. 20x. (F) Human glioblastoma TiViMNBFCS10 cells on the 3rd day of cultivation, fixed with 4%PFA, no permeabilization. Specific calnexin staining revealed by red staining - Alexa 594, obj. 100x. (G) Bowes melanoma cell line on the 2nd day of cultivation, fixed with 4%PFA, no permeabilization. Specific calnexin staining revealed by red staining - Alexa 594, obj.100x. (H) Unfixed sperms of a normal human male. Specific calnexin staining revealed by red staining - Alexa 488, obj. 100x (note a red staining between the sperm head and tail). DAPI is used for nuclei staining (blue).

4.1.2 Monoclonal antibody E15F10.B9 (MAb E15F10.B9)

In Fig. 7 we can see different expression of MAb E15F10.B9 (Table 1) detected antigen in differentiated fetal neural stem/progenitor (A) and glioblastoma spheroid cells (C). In some regions of hBrSc003 cell line differentiated for 7 days there was a very strong expression of MAb E15F10.B9 target antigen (B). Strong expression of the antigen one can see also in unfixed adult skin fibroblasts and in fixed, but not permeabilized glioblastoma and Bowes melanoma cells. Interestly, in unfixed human sperms the staining of MAb E15F10.B9 was seen as two separate points between the head and neck of the sperm (H). Double staining with actin in normal amniotic epithelial cells (D) shows an independent staining of actin and target antigen of MAb E15F10.B9, whereas the last one shows nuclear membrane and punctate cytoplasma staining.

The LC ESI-MS/MS and Mascot database analysis of peptides from the immunoprecipitated band 208 (Fig. 8) identified two possible candidates for the target antigen of MAb E15F10.B9 - calnexin and Keratin, type II cytoskeletal 1. However, as immunocytochemical analysis on different cell types including normal epithelial amniotic cells (Fig. 7) showed no typical cytoskeletal staining for cytokeratin we consider the keratin to be in immunoprecipitate as a contaminant substance. In Fig. 8 and 9 and Table 3 the results of Western blot and molecular identification of the target antigen of MAb E15F10.B9 are shown.

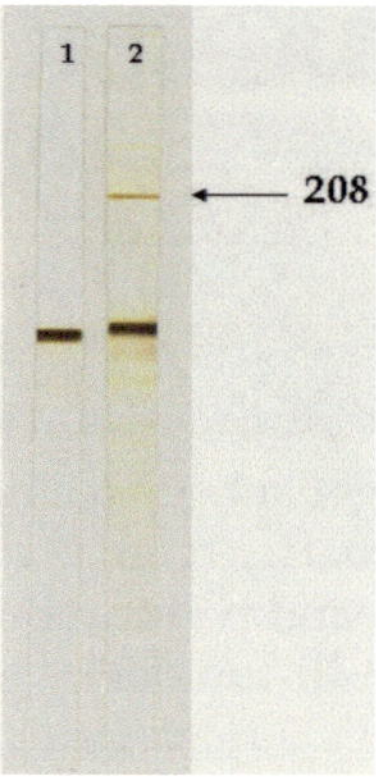

Fig. 8. Western blot of MAb E15F10.B9 target antigen (calnexin). MAb E15F10.B9 from hybridoma supernatant was conjugated to Protein G granules and thereafter target antigen immunoprecipitated from Bowes melanoma cell line lysed with RIPA buffer. Proteins were separated in SDS-PAGE 4-12% gradient gel. (1) negative control with HRP-conjugated goat anti-mouse IgG secondary antibody; (2) incubation with MAb E15F10.B9 and secondary antibody. Arrow indicates the band analyzed by ESI-MS/MS.

Band	Identified proteins	Mass	Mascot score	Sequence coverage %
208	CALX - Calnexin K2C1- Keratin, type II cytoskeletal 1	67526 65999	384 313	22 20

Table 3. LC ESI-MS/MS and Mascot software analysis of the proteins in the band which was immunoprecipitated with MAb E15F10.B9 proteins (Fig. 8).

1 MEGKWLLCML LVLGTAIVEA HDGHDDDVID IEDDLDDVIE EVEDSKPDTT
51 APPSSPKVTY K**APVPTGEVY FADSFDRGTL SGWILSK**AK**K DDTDDEIAK**Y
101 DGKWEVEEMK ESKLPGDKGL VLMSRAKHHA ISAKLNKPFL FDTKPLIVQY
151 EVNFQNGIEC GGAYVKLLSK **TPELNLDQFH DKTPYTIMFG PDK**CGEDYKL
201 HFIFRHKNPK **TGIYEEK**HAK **RPDADLKTYF TDKK**THLYTL ILNPDNSFEI
251 LVDQSVVNSG NLLNDMTPPV NPSR**EIEDPE DR**KPEDWDER PKIPDPEAVK
301 PDDWDEDAPA KIPDEEATKP EGWLDDEPEY VPDPDAEKPE DWDEDMDGEW
351 EAPQIANPRC ESAPGCGVWQ RPVIDNPNYK GKWKPPMIDN PSYQGIWKPR
401 **KIPNPDFFED LEPFR**MTPFS AIGLELWSMT SDIFFDNFII CADRR**IVDDW**
451 **ANDGWGLKK**A ADGAAEPGVV GQMIEAAEER PWLWVVYILT VALPVFLVIL
501 FCCSGKKQTS GMEYK**KTDAP QPDVK**EEEEE KEEEKDKGDE EEGEEKLEE
551 KQKSDAEEDG GTVSQEEEDR KPK**AEEDEIL NR**SPRNRKPR RE

Fig. 9. Amino acid sequences of Calnexin. With red colour are marked amino acids sequences identified by mass-spectroscopy (band 208 in Fig. 8).

Conclusion

On the basis of data presented above we conclude that the target antigen of MAb E15F10.B9 is protein Calnexin. According to UniProtKB/Swiss-Prot database Calnexin is a calcium-binding protein that interacts with newly synthesized glycoproteins in the endoplasmic reticulum. It may act in assisting protein assembly and/or in the retention within the ER of unassembled protein subunits. It seems to play a major role in the quality control apparatus of the ER by the retention of incorrectly folded proteins (http://www.uniprot.org/uniprot/P27824).

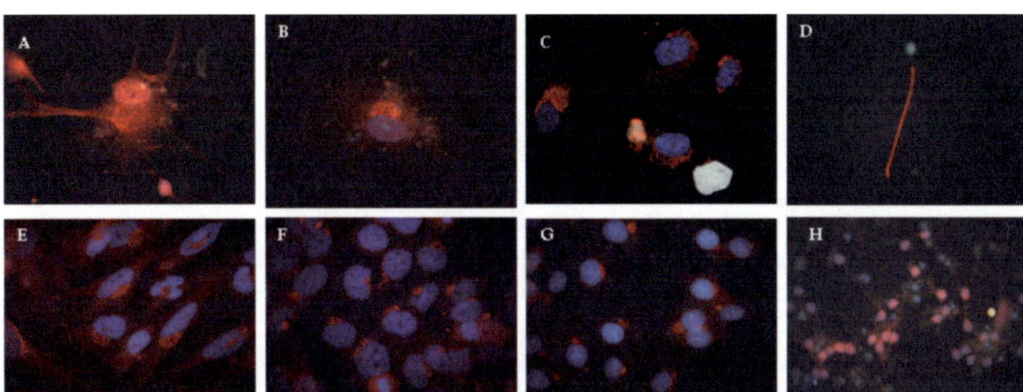

Fig. 10. Presence of MAb W4A8.F4 target antigen (protein 14-3-3) in different cells. Specifc staining of MAb W4A8.F4 was revealed by red staining –Alexa 594. (A,B) Human fetal hBrSc006 (A) and hBrSc009 (B) neurosphere cells differentiated for 3 days. Cells were fixed with 4% PFA and permeabilized, obj. 100x. (C) Human fetal skin fibroblast culture on the 3rd day of cultivation. Cells were fixed with 4% PFA and permeabilized, obj. 100x. (D) Unfixed sperms of a normal human male, obj. 100x. (E-F) Human glioblastoma cell line TiViMNBFCS10 (E), Bowes melanoma cell line (F), COS-1 cell line and rat granulare cell culture (H) on the 3rd day of cultivation. Cells were fixed with 4% PFA and permeabilized, obj. 100x. DAPI is used for nuclei staining (blue).

4.1.3 Monoclonal antibody W4A8.F4 (MAb W4A8.F4)

In Fig. 10 we can see the expression of MAb W4A8.F4 (Table 1) in the fetal neural stem/progenitor cell lines hBrSc006 (A) and hBrSc009 (B). The strong staining was seen in the cytoplasm in Golgi region but also in cell projections and nuclei. In the fetal skin fibroblasts, glioblastoma, Bowes melanoma and COS-1 cell lines the cytoplasm, especially Golgi region was strongly stained (C,E,F and G respectively). In the rat granulare cell culture there was a heterogeneous staining of the cytoplasm and nuclei (H). Interestly, in the unfixed human sperms the entire tail, except the neck, was strongly and homogeneously stained (D).

In Fig. 11 and 12 and Table 4 the results of Western blot and molecular identification of the target antigen of MAb W4A8.F4 are shown.

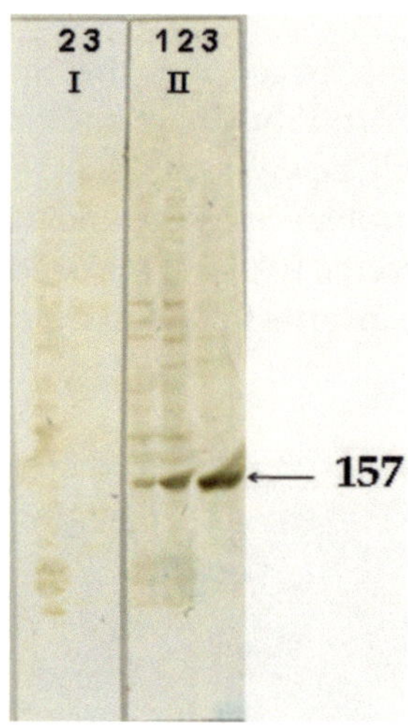

Fig. 11. Western blot with MAb W4A8.F4. Proteins in RIPA lysate were separated electrophoretically in SDS-PAGE gradient (8-25%) gel and transferred to PVDF membrane: (1) Bowes cell line, (2) glioblastoma TiViMNBFCS10 cell line and (3) human thromocytes. I - negative control with HRP-conjugated goat anti-mouse IgG secondary antibody; II - reaction with MAb W4A8.F4 and secondary antibody. Arrow indicates the band 157, analyzed by ESI-MS/MS.

Band	Identified proteins	Mass	Mascot score	Sequence coverage %
157	14-3-3 protein zeta/delta	27728	139	**40**
	14-3-3 protein gamma	28295	64	**18**
	14-3-3 protein eta	28201	64	**19**
	14-3-3 protein sigma	27757	64	**12**
	Chloride intracellular channel protein 1	26906	62	**19**

Table 4. LC ESI-MS/MS and Mascot software analysis of immunoprecipitated with MAb W4A8.F4 proteins in band 157 (Fig. 11).

```
  1 MDKNELVQKA KLAEQAERYD DMAACMKSVT EQGAELSNEE RNLLSVAYKN
 51 VVGARRSSWR VVSSIEQKTE GAEKKQQMAR EYREKIETEL RDICNDVLSL
101 LEKFLIPNAS QAESKVFYLK MKGDYYRYLA EVAAGDDKKG IVDQSQQAYQ
151 EAFEISKKEM QPTHPIRLGL ALNFSVFYYE ILNSPEKACS LAKTAFDEAI
201 AELDTLSEES YKDSTLIMQL LRDNLTLWTS DTQGDEAEAG EGGEN
```

Fig. 12. Amino acid sequences of protein 14-3-3 protein zeta/delta. With red colour are marked amino acids sequences identified by mass-spectroscopy (band 157 in Fig. 11).

Conclusion

On the basis of data presented above we conclude that most probably the target antigen of MAb W4A8.F4 is protein 14-3-3 zeta/delta. According to UniProtKB/ Swiss-Prot database protein 14-3-3 zeta/delta is an adapter protein implicated in the regulation of a large spectrum of both general and specialized signaling pathways. Binds to a large number of partners, usually by recognition of a phosphoserine or phosphothreonine motif. Binding generally results in the modulation of the activity of the binding partner (http://www.uniprot.org/uniprot/P63104).

4.1.4 Monoclonal antibody GB26 10G3 (MAb GB26 10G3)

In Fig. 13 we can see the filamentous staining of MAb GB26 10G3 (Table 1) antigen in human adult skin fibroblasts (A) and in glioblastoma cell line TiViMNBFCS10 (B). In glioblastoma cell lines the staining pattern was very heterogeneous, some cells did not show any signs of staining. Double staining for β-III-tubulin and the MAb GB26 10G3 (C) showed an independent staining pattern, some cells were double stained, some cells showed only β-III-tubulin staining and the majority of cells was only MAb GB26 10G3 positive. In D the double staining of target antigen of MAb GB26 10G3 and glial fibrillar acid protein (GFAP) is shown. Note that the both antigens are located in the same cells, but show an independent staining pattern.

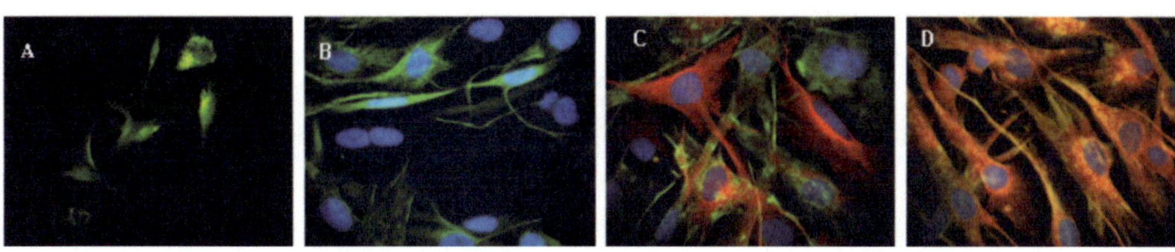

Fig. 13. Presence of MAb GB26 10G3 target antigen in different cells. (A,B) Human adult skin fibroblast cell line SA-54 (A) and human glioblastoma cell line TiViMNBFCS10 (B) on the 3rd day of cultivation. Cells were fixed with 4% PFA and permeabilized. The target antigen of MAb GB26 10G3 expression was revealed by green staining - Alexa 488, obj. 100x. (C) Human glioblastoma cell line TiViMNBFCS10 on the 3rd day of cultivation. Double immunofluorescent staining, cells were fixed with 4% PFA and permeabilized. The target antigen of MAb GB26 10G3 expression was revealed by green staining - Alexa 488, β-III-tubulin by red staining - Alexa 488, obj. 100x. (D) Human glioblastoma cell line TiViMNBFCS10 on the 3rd day of cultivation. Double immunofluorescent staining, cells were fixed with 4% PFA and permeabilized. The target antigen of MAb GB26 10G3 expression was revealed by green staining - Alexa 488, GFAP by red staining - Alexa 488, obj. 100x. DAPI is used for nuclei staining (blue).

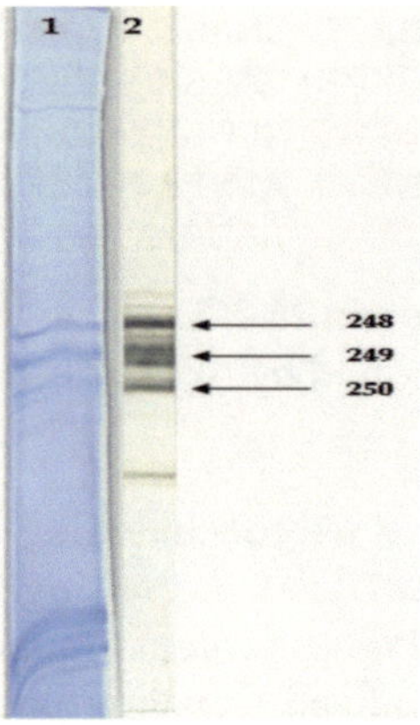

Fig. 14. SDS-PAGE gradient gel (8-24%) (1) and immunoblot (2) with MAb GB26 10G3. The sample was in RIPA buffer insoluble fraction of glioblastoma TiViNBFCS10 cells that was heated with SDS-PAGE sample buffer with DTT at 60°C for 10 min and 8M urea was added to 2M final concentration. Arrows with numbers indicate bands analyzed by ESI-MS/MS.

As molecular masses of antibody light chain and vimentin are similar and vimentin binds nonspecifically to the secondary antibody, we could not purify vimentin by immunoprecipitation as we did for other MAbs. Only gradient electrophoresis of lysate and immunoblotting were used to separate the target antigen of MAb GB26 10G3.

In Fig. 14 and 15 and Table 5 the results of electrophoresis, Western blot and molecular identification of the target antigen of MAb GB26 10G3 are shown.

Band	**Identified proteins**	**Mass**	**Mascot score**	**Sequence coverage %**
248	Vimentin	53619	198	39
249	Glial fibrillar acid protein - GFAP Vimentin	49850 53619	135 123	32 28
250	Actin, cytoplasmic 1	41710	43	14

Table 5. LC ESI-MS/MS and Mascot software analysis of proteins in bands 248- 250 (Fig. 14).

```
  1 MSTRSVSSSS YRRMFGGPGT ASRPSSSRSY VTTSTRTYSL GSALRPSTSR
 51 SLYASSPGGV YATRSSAVRL RSSVPGVRLL QDSVDFSLAD AINTEFKNTR
101 TNEKVELQEL NDRFANYIDK VRFLEQQNKI LLAELEQLKG QGKSRLGDLY
151 EEEMRELRRQ VDQLTNDKAR VEVERDNLAE DIMRLREKLQ EEMLQREEAE
201 NTLQSFRQDV DNASLARLDL ERKVESLQEE IAFLKKLHEE EIQELQAQIQ
251 EQHVQIDVDV SKPDLTAALR DVRQQYESVA AKNLQEAEEW YKSKFADLSE
301 AANRNNDALR QAKQESTEYR RQVQSLTCEV DALKGTNESL ERQMREMEEN
351 FAVEAANYQD TIGRLQDEIQ NMKEEMARHL REYQDLLNVK MALDIEIATY
401 RKLLEGEESR ISLPLPNFSS LNLRETNLDS LPLVDTHSKR TLLIKTVETR
451 DGQVINETSQ HHDDLE
```

Fig. 15. Amino acid sequences of Vimentin. With red colour are marked amino acids sequences identified by mass-spectroscopy (band 248 in Fig. 14.).

Conclusion

On the basis of data presented above we conclude that the target antigen of MAb GB26 10G3 is protein Vimentin. According to UniProtKB/Swiss-Prot database vimentins are class-III intermediate filaments found in various non-epithelial cells, especially mesenchymal cells (http://www.uniprot.org/uniprot/P08670).

4.1.5 Monoclonal antibody A3G2.B4 (MAb A3G2.B4)

In Fig. 16 we can see the expression of MAb A3G2.B4 antigen (Table 1) in differentiated fetal neural stem/progenitor cell lines hBrSc003 (A,B) and hBrSc005 (B). Note a clear heterogeneous staining of cell nuclei. In D, E,F and G the staining of glioblastoma spheroid cell line differentiated for 7 days, normal epithelial amniotic cells, adult skin fibroblasts and glioblastoma OjFe secondary cell culture are shown, respectively. Note the heterogeneous nuclear and nucleolar staining and a weak cytoplasma staining. However, in Bowes melanoma cell line both the nuclear and strong granular cytoplasmic staining was observed (H).

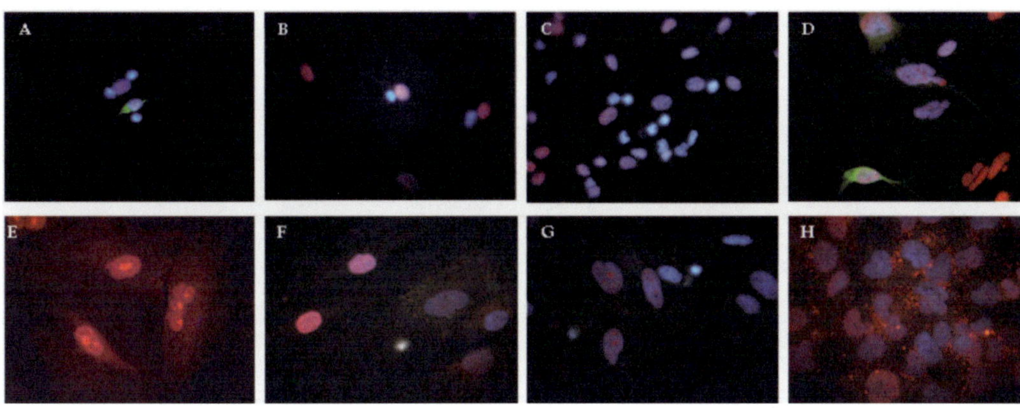

Fig. 16. Presence of MAb A3G2.B4 antigen in different cells. In all cases the cells were fixed with 4% PFA, permeabilized with Tritone X-100 and specifc staining of MAb A3G2.B4 revealed by red staining -Alexa 594. (A,B) Human fetal hBrSc003 neurosphere cells differentiated for 11 days. Double immunofluorescent staining. β-III-tubulin is revealed by green staining - Alexa 488, obj. 40x. (C) Human fetal hBrSc005 neurosphere cells differentiated for 7 days, obj. 40x. (D) Human glioblastoma TiVi spheroid cell line differentiated for 7 days. Double immunofluorescent staining. β-III-tubulin is revealed by green staining - Alexa 488obj. 100x. (E-H) Human normal amniotic epithelial cell line KM on the 2nd days of cultivation, human adult skin fibroblast cell line SA-54 on the 3rd day of cultivation, human glioblastoma OjFe secondary cell culture on the 3rd day of cultivation , and Bowes melanoma cell line on the 2nd day of cultivation, respectively, obj. 100x. DAPI is used for nuclei staining (blue).

In Fig. 17 and 18 the results of Western blot and molecular identification of the target antigen of MAb A3G2.B4 are shown.

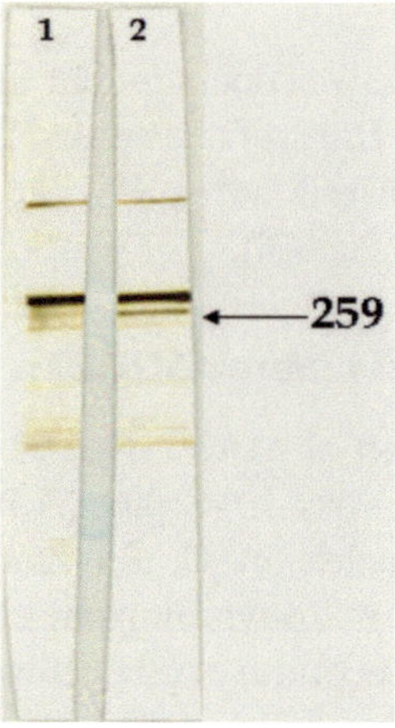

Fig. 17. Western blot with MAb A3G2.B4. Target antigen Lupus La antigen was first immunoprecipitated with MAb A3G4.B4 in 1% NP-40 PBS solution from the concentrated lysate of Bowes melanoma cells. Precipitate was heated at 60°C 15 min, added urea up to end concentration of 2M and proteins separated in SDS-PAGE gradient (8-24%) gel. 1 - negative control with HRP-conjugated goat anti-mouse IgG secondary antibody; 2 - reaction with MAb A3G2.B4 and secondary antibody.

1 MAENGDNEKM AALEAKICHQ IEYYFGDFNL PRDKFLKEQI KLDEGWVPLE
51 IMIKFNRLNR **LTTDFNVIVE ALSK**SKAELM EISEDKTKIR RSPSKPLPEV
101 TDEYKNDVKN RSVYIKGFPT DATLDDIKEW LEDKGQVLNI QMRRTLHKAF
151 K**GSIFVVFDS IESAK**KFVET PGQKYKETDL LILFKDDYFA KKNEERKQNK
201 VEAKLRAKQE QEAKQKLEED AEMKSLEEKI GCLLKFSGDL DDQTCREDLH
251 ILFSNHGEIK WIDFVRGAKE GIILFKEKAK EALGKAKDAN NGNLQLRNKE
301 VTWEVLEGEV EKEALKKIIE DQQESLNKWK SKGRRFKGKG KGNKAAQPGS
351 GKGKVQFQGK KTKFASDDEH DEHDENGATG PVKRAREETD KEEPASKQQK
401 TENGAGDQ

Fig. 18. Amino acid sequences of Lupus La protein. With red colour are marked amino acids sequences identified by mass-spectroscopy (band 259 in Fig. 17).

Conclusion

On the basis of data presented above we conclude that the target antigen of MAb A3G2.B4 is protein Lupus La protein. According to UniProtKB/Swiss-Prot database Lupus La protein binds to the 3' poly(U) terminii of nascent RNA polymerase III transcripts, protecting them from exonuclease digestion and facilitating their folding and maturation (http://www.uniprot.org/uniprot/P05455).

5. Discussion and conclusions

The main aim of present study is the development of new monoclonal antibodies against neuronal tissue cells to investigate the differentiation and malignization of human nerve cells. The antibody producing hybridomas were obtained by immunizing mice with the native fragments of human glioblastoma and foetal neural stem/progenitor cells to obtain MAbs against all kinds of antigenic determinants, that are expressed in the living cells, including those determinants, that are expressed on the cell surface. From plenty of MAb-producing clones only those that revealed heterogeneity of reaction with stem/progenitor cells and glioblastoma spheroid cultures were further characterize and the target antigens identified. In this way there is possible to obtain panels of MAbs, that characterizes the given cell types, including tumour cells. The spectrum of the monoclonal antibodies obtained using our method is quite large, including antibodies against proteins as well as against their different modifications. Previously, using described approach we have developed MAb F10H2.B3 specific to Ku80 (ATP-dependent DNA helicase 2 subunit 2). We suggest this antibody could be used in certain conditions as a proliferation marker for cells of different origin (Mikelsaar et al., 2009). In this chapter we present the data about development the five new monoclonal antibodies against neural antigens

Annexin A1 has been reported to take part in different functions as both inhibition of phospholipase A2, acute inflammation, pituitary hormone regulation, fever, neutrophil migration, cell proliferation, and stimulation of cell proliferation, differentiation, apoptosis, membrane repair, macrophage phagocytosis and neuroprotection (Solito et al., 2008). Less information is about annexin A1 expression in the developing brain. It seems to have limited neuronal distribution, but is strongly expressed in glia and ependymocytes (Fava et al., 1989). The studies have shown that annexin A1 (LC-1) positive cells carry other microglial markers and are quite distinct from astrocytes identified by S100*B* immunoreactivity (McKanna, 1993). It has been also proposed that LC1 can be a comprehensive and reliable marker for microglia (McKanna & Zhang, 1997). Annexins are generally cytosolic proteins, soluble or reversibly associated with components of the cytoskeleton or proteins that mediate interactions between the cell and the extracellular matrix (matricellular proteins) (Moss & Morgan, 2004). In certain cases, annexins may be expressed at the cell surface, despite the absence of any secretory signal peptide (Solito et al., 1994). In differentiated fetal neural stem/progenitor (Fig. 4, A,B,C,D) and glioblastoma spheroid cells (Fig.4, E) MAb E14G2 antigen (annexin A1) is expressed mainly in glial cells but not in β-III-tubulin positive neuronal cells.This observation is in accordance with data on the developing brain obtained by R.A.Fava and coworkers (Fava et al.,1989). Similar picture of expressing of annexin A1 only in limited cell types we have seen also in rat granule primary culture (Fig. 4, H). We propose that the cell expressing annexin A1 might be a Purkinje cell, which have been shown to be annexin A1 positive in adult rat cerebellum (Solito et al., 2008). In rat granule primary culture we have also seen a small amount of $GFAP^+$/annexin $A1^+$ double positive glial cells (data not shown here). The majority of glial cells were only GFAP positive. The expression of annexin A1 was strong and homogenous in glioblastoma OjAr secondary culture and in Bowes melanoma cell lines (Fig. 4, F and G respectively). We propose that MAb E14G2 may be a perspective marker for some distinct neural cell types..

Calnexin is a calcium-binding protein that interacts with newly synthesized glycoproteins in the endoplasmic reticulum (Ellgaard & Helenius, 2003; Ou et al, 1995). Like calreticulin, calnexin is predominantly located in the ER but it has also been identified at the cell surface of a number of cells. Okazaki Y. and co-workers (Okazaki et al., 2000) reported that a small fraction of calnexin is normally expressed on the surface of various cells. The results of these authors suggest that there is continuous exocytosis and endocytosis of calnexin, and the amount of calnexin on the plasma membrane results from the balance of the rates of these two events. The findings suggest that the surface expression of calnexin depends on the association with glycoproteins and that calnexin may play a certain role as a chaperone on the plasma membrane as well (Okazaki et al., 2000). Our observations are in good accordance with previous data. We observed the expression of of MAb E15F10.B9 detected calnexin on the surface of unfixed human adult skin fibroblasts and on fixed, but not permeabilized cells of glioblastoma TiViMNBFCS10 and Bowes melanoma cell line, respectively (Fig. 7, E-G). In Fig. 7 we can see a different expression of MAb E15F10.B9 detected calnexin in differentiated fetal neural stem/progenitor (A) and glioblastoma spheroid cells (C) in which β-III-tubulin-positive cells are negative or very weakly calnexin-positive. In the human fetal hBrSc003 neurosphere culture differentiated for 11 days we see at least three populations of cells, namely the calnexin$^+$/β-III-tubulin$^-$cells, calnexin-/β-III-tubulin+ cells and also cells which are negative for both antigens. It shows that MAb E15F10.B9 may be used for detection of some distinct population of neural cells (see also a strongly calnexin$^+$ cell among other totally calnexin$^-$ cells in TiViMNBFCS10 glioblastoma cell culture, Fig.7, F). Interestingly, in the unfixed human sperms we see two separate points between the head and neck of the sperm (H). This phenomenon needs further investigation.

14-3-3 proteins. There are seven genes that encode 14-3-3s in most mammals (Takashi, 2003). 14-3-3 proteins are abundantly expressed in the brain and have been detected in the cerebrospinal fluid of patients with different neurological disorders. By their interaction with more than 100 bindingpartners, 14-3-3 proteins modulate the action of proteins that are involved in cell cycle and transcriptional control, signal transduction, intracellular trafficking and regulation of ion channels. The study of some of these interactions is sheding light on the role of 14-3-3 proteins in processes such as apoptosis and neurodegeneration (Berg et al., 2003). The immunohistological and subcellular location of the 14-3-3 proteins was studied using different isoform-specific antisera (Martin et al., 1994). The immunohistochemical examination using the specific antibody showed significant staining of the cytoplasm, including neuronal axons and dendrites. This result was confirmed by the ultracentrifugal cellular fractionation method, indicating that 14-3-3 is mainly localized in the neuronal cytoplasm and a portion of 14-3-3 may be bound to the plasma membrane, endoplasmic reticulum, and Golgi membrane. This is in good accordance with our *in vitro* study on different cell lines. In Fig. 10 we see the expression of the protein 14-3-3 in the fetal neural stem/progenitor cell lines hBrSc006 (A) and hBrSc009 (B). The strong staining was seen in the cytoplasm in Golgi region, but also in cell projections and nuclei in many cells. In the rat granulare cell culture there are two different 14-3-3 stained cell populations, detected by MAb W4A8.F4: negatively stained cells and cells with positively stained cytoplasm and nuclei (Fig.10, H). It shows that the MAb W4A8.F4 may work in certain conditions as a marker for some types of neural cells. In the fetal skin fibroblasts, glioblastoma, Bowes melanoma and COS-1 cell lines the cytoplasm, especially Golgi region was strongly stained (Fig.10, C,E,F and G respectively).

Interestingly, in the unfixed human sperms the entire tail, except the neck, was strongly and homogeneously stained (Fig.10, D). This is a very interesting fact and needs further investigation.

Vimentins are class-III intermediate filaments found in various non-epithelial cells, especially mesenchymal cells. During the development of the nervous system, vimentin is transiently expressed in virtually all the precursors cells of both neuronal and glial lineages. In the astroglial cell lineage, vimentin is the only IF protein expressed in radial glia and immature astrocytes in the embryonic nervous system (Alonso, 2001; Colluci-Guyon et al., 1999; Schnitzer et al., 1981). The expression of glial fibrillary acidic protein (GFAP), vimentin and fibronectin (Fn) was studied in cells cultured from human glioma and fetal brain by indirect immunofluorescence (IIF) microscopy and multiple labelling experiments (Paetau 1988). The results of the study demonstrate a general coexpression of GFAP and vimentin in cultured astroglial cells, in addition to cells expressing only vimentin. This is in good accordance with our data. In Fig. 13 we see the filamentous staining of the vimentin with MAb GB26 10G3 in human adult skin fibroblasts (A) and in glioblastoma cell line TiViMNBFCS10 (B). In glioblastoma cell lines the staining pattern was very heterogeneous, some cells did not show any signs of staining. Double staining for β-III-tubulin and vimentin (Fig. 13,C) showed an independent staining pattern, some cells were double stained (vimentin^{+}/β-III-tubulin^{+}), some cells showed only β-III-tubulin staining (vimentin^{-}/β-III-tubulin^{+}) and the majority of cells were only vimentin positive. The double staining of vimentin and glial fibrillar acid protein (GFAP) is shown in Fig. 13, D. Note that all the cells are vimentin^{+}/GFAP^{+}, but show an independent staining pattern. We propose that our MAb GB26 10G3 may be a good additional tool for detecting and characterization of expression of vimentin in different types of neural cells, including glioblastomas.

Lupus La protein (known also as Sjogren syndrome antigen B and autoantigen La) is ubiquitous in eukaryotic cells and associates with the 3′ termini of many newly synthesized small RNAs. The La protein protects the 3′ ends of these RNAs from exonucleases (Wolin & Cedervall, 2002). The immunohistochemical location of La antigen was shown to be the nucleus but an intense staining of the nucleolus was seen in human cerebral cortical neurons as well as a subset of neurons of rat brain (Graus et al., 1985). Further it was shown that La ribonucleoproteins (RNP) exist in distinct states that differ in subcellular localization (Intine et al., 2003). This is in good accordance with our results of the immunocytochemical study. In Fig. 16 we see the staining of La protein with MAb A3G2.B4 in nuclei of all cell lines studied. However, the staining of nuclei was very heterogeneous in differentiated fetal neural stem cells (A-C), clearly showing the existence of two cell population for MabA3G2.B4 detected La protein – La protein^{+} and La protein^{-} cells. This may be a sign of real existence of two different cell populations and needs further investigation. A clear nucleolar staining was also seen in many cells, especially in normal amniocytes and glioblastoma cells (E-G). In the Bowes melanoma cell line both the nuclear and strong granular cytoplasmic staining was observed (H).

Further perspectives. The characterization of the target antigenes and epitopes of all other monoclonal antibodies obtained during our main project is in progress.

6. Acknowledgment

This work was partly supported by target financing SF018809s08 of the Estonian Ministry of Science and Education.

7. References

Alonso, G. (2001). Proliferation of Progenitor Cells in the Adult Rat Brain Correlates With the Presence of Vimentin-Expressing Astrocytes. *Glia*, 34, No.4 , (May), pp. 253-266, ISSN 1098-1136

Bauer, H.C.; Tempfer, H.; Bernroider, G. & Bauer, H. (2006). Neuronal Stem Cells in Adults. *Experimental Gerontology* , Vol.41, No.2, (February), pp. 111-116, ISSN 0531-5565

Bedard, K.; Szabo, E.; Michalak, M. & Opas, M. (2005). Cellular Functions of Endoplasmic Reticulum Chaperones Calreticulin, Calnexin and Erp57. *International Review of Cytology*, Vol.245, pp. 91-121, ISSN 0074-7696

Berg, D.; Holzmann, C. & Riess, O. (2003). 14-3-3 Proteins in the Nervous System. *Nature Reviews Neuroscience*, Vol.4, No.9, (September), pp. 752-762, ISSN 1471-0048

Bez, A.; Corsini, E.; Curti, D.; Biggiogera, M.; Colombo, A.; Nicosia, R.F.; Pagano, S.F. & Parati, E.A. (2003). Neurosphere and neurosphere-forming cells: morphological and ultrastructural characterization. *Brain Research*, Vol.993, No.1-2., (December), pp. 18-29, ISSN 0006-8993

Canvar, P.J.; Olenych, S.G.& Keller III, T.C.S. (2007). Molecular Identification and Localization of Cellular Titin, A Novel Titin Isoform in the Fibroblast Stress Fiber. *Cell Motility and the Cytosceleton*, Vol 64, No. 6, (June), pp. 418-433, ISSN 1097-0169

Colluci-Guyon, E.; Ribotta, M.G.Y.; Maurice, T.; Babinet, C. & Privat, A. (1999). Cerebellar Defect and Impaired Motor Coordination in Mica Lacking Vimentin. *Glia*, Vol.25, No.1, (January), pp. 33-43,ISSN 1098-1136

Ellgaard, L. & Helenius, A.(2003). Quality control in the endoplasmic reticulum. *Nature Reviews Molecular Cell Biology*, Vol. 4, No.3, (March), pp. 181–191, ISSN 1471-0072

Fava, R.A.; McKanna, J. & Cohen, S. (1989). Lipocortin I (p35) is abundant in a restricted number of differentiated cell types in adult organs. Journal of Cellular Physiology, Vol.141, No.2, (November), pp. 284-293, ISSN 0021-9541

Graus, F.; Cordon-Cardo, C.; Bonfa, E. & Elkon, K.B. (1985). Immunohistochemical Localization of La Nuclear Antigen in Brain. Selective Concentration of the La Protein in Neuronal Nucleoli. *Journal of Neuroimmunolgy*, Vol.9, No.5, (September), pp. 307-319, ISSN 0165-5728

Intine, R.V., Tenenbaum, S.A.; Sakulich, A.L,; Keene, J.D. & Maraia, R.J. (2003). Differential phosphorylation and subcellular localization of La RNPs associated with precursor tRNAs and translation-related mRNAs. *Molecular Cell*, Vol.12, No.5, (November), pp. 1301-1307, ISSN 1097-2765

Kalev, I.; Kaasik, A.; Žarkovski, A.& Mikelsaar, A.-V. (2006). Chemokine Receptor CCR5 Expression in *in vitro* Differentiating Human Fetal Neural Stem/Progenitor and Glioblastoma Cells. *Neuroscience Letters*, Vol.394, No.1, pp. 22-27, ISSN 0304-3940

Kang, D.; Gho, Y.S.; Suh, M.& Kang, C.(2002). Highly Sensitive and Fast Protein Detection with Coomassie Brilliant Blue in Sodium Dodecyl Sulfate-Polyacrylamide Gel Electrophoresis. *Bulletin of the Korean Chemical Society*, Vol.23, No.11, pp. 1511-1512, ISSN 0253-2964

Kennea, N.L. & Mehmet, H. (2002). Neural Stem Cells. *Journal of Pathology*, Vol.197, No.4, (July), pp. 536–550, ISSN 0022-3417

Kraus, A.; Groenendyk, J.; Bedard, K.; Baldwin, T.A., Krause, K.-H.; Dubois-Dauphin, M.; Dyck, J.; Rosenbaum, E.E.; Korngut, L.; Colley, N.J.; Gosgnach, S.; Zochodne, D.; Todd, K.; Agellon, L.B. & Michalak, M. (2010). Calnexin Deficiency Leads to Desmyelination. *Journal of Biological Chemistry*, Vol.285, No.24, (June), pp. 18928-18938, ISSN 0021-9258

Martin, H.; Rostas, J.; Patel, Y. & Aitken, A. (1994). Subcellular localization of 14-3-3 isoforms in rat brain using specific antibodies. *Journal of Neurochemistry*, Vol. 63, No. 6, (December), pp. 2259–2265, ISSN 0022-3042

McArthur, S.; Cristante, E.; Paterno, M.; Christian, H; Roncaroli, F.; Gillies, G.E. & Solito, E. (2010). Annexin A1: A Central Player in the Anti-Inflammatory and Neuroprotective Role of Microglia. *Journal of Immunology*, Vol.185, No.10, (November), pp. 6317-6328, ISSN 0022-1767

McKanna, J.A. (1993). Primitive glial compartments in the floor plate of mammalian embryos: distinct progenitors of adult astrocytes and microglia support the notoplate hypothesis. *Perspectives on Developmental Neurobiology*, Vol.1, No.4, pp. 245–255, ISSN 1064-0517

McKanna, J.A. & Zhang, M.-Z. (1997). Immunohistochemical Localization of Lipocortin 1 in Rat Brain is Sensitive to pH, Freezing and Dehydratation. *Journal of Histochemistry & Cytochemistry*, Vol.45, No.4, (April), pp. 527-538, ISSN 0022-1554

Mikelsaar, A.-V.; Sünter, A.; Toomik, P.; Karpson, K.& Juronen, E. (2009). New Anti-Ku80 Monoclonal Antibody F10H2.B3 As a Useful Marker for Dividing Cells in Culture. *Hybridoma*, Vol.28, No.2, (April), pp. 107-111, ISSN 1554-0014

Moss, S.E. & Morgan, R.O. (2004). The annexins. *Genome Biology*, Vol.5, No.4, (March), pp. 219-219.8, ISSN 1465-6906

Ou, W. J.; Cameron, P. H.; Thomas, D. Y. & Bergeron, J. J. (1993). Association of folding intermediates of glycoproteins with calnexin during protein maturation. *Nature*, Vol. 364, No.6440, (August), pp. 771–776, ISSN 0028-0836

Okazaki, Y.; Ohno, H.; Takase, K.; Ochiai, T. & Takashi, S. (2000). Cell Surface Expression of Calnexin, a Molecular Chaperone in the Endoplasmic Reticulum. *The Journal of Biological Chemistry*, Vol.275, No.46, (November), pp. 35751-35758, ISSN 0021-9258

Paetau, A. (1988). Glial fibrillary acidic protein, vimentin and fibronectin in primary cultures of human glioma and fetal brain. *Acta Neuropathologica (Berl.)*, Vol.75, 5, (May), pp. 448-455, ISSN 0001-6322

Perretti, M. & D'Acquisto, F. (2009). Annexin A1 and Glucocorticoid as Effectors of the Resolution of Inflammation. *Nature Reviews Immunology*, Vol.9, No.1, (January), pp. 62-70, ISSN 1474-1733

Riemekasten, G. & Hahn, B.H. (2005). Key Autoantigens in SLE. *Rheumatology*, Vol.44, No.8 , (August), pp. 975-982, ISSN 0315-162X

Schnitzer, J.; Franke, W.W. & Schachner, M. (1981). Immunocytochemical Demonstration of Vimentin in Astrocytes and Ependymal Cells of Developing and Adult Mouse Nervous System. *The Journal of Cell Biology*, Vol. 90, No.2, (August), pp. 435-447, ISSN 0021-9525

Schwartz, P.H.; Brick, D.J.; Alexander E. Stover, A.E.; Loring, J.F. & Müller, F.-J. (2008). Differentiation of Neural Lineage Cells from Human Pluripotent Stem Cells. *Methods*, Vol.45, No.2, (June), pp. 142–158, ISSN 1046-2023

Solito, E.; Nuti, S. & Parente, L. (1994). Dexamethasone-induced translocation of lipocortin (annexin) 1 to the cell membrane of U-937 cells. *British Journal of Pharmacology*, Vol.112, No.2, (June), pp. 347-348, ISSN 0007-1188

Solito, E.; McArthur, S.; Christian, H., Gavins, F.; Buckingham, J.C. & Gillies, G.E. Annexin A1 in the Brain – Undiscovered Roles? *Trends in Pharmacological Sciences*, Vol.29, No.3, (March), pp. 135-142, ISSN 0165-6147

Takahashi, Y. (2003). The 14-3-3 Proteins: Gene, Gene Expression and Function. Neurochemical Research. Vol. 28, No.8, (August), pp. 1265-1273, ISSN 0364-3190

Young, K.A.; Hirst, W.D.; Solito, E. & Wilkin, G.P. (1999). De Novo Expression of Lipocortin-1 in Reactive Microglia and Astrocytes in Kainic Acid Lesioned Rat Cerebellum. *Glia*, Vol.26, No.4, (June), pp. 333-343, ISSN 1098-1136

Tijssen P.,(1985) *Practice and theory of enzyme immunoassays*, in: Laboratory techniques in biochemistry and molecular biology, R.H.Burdon and P.H. van Knippenberg Elsevier pp. 221-278, Elsevier, ISBN 0-444-80633-4, Netherlands

Yuan, S.H.; Martin, J.; Elia, J.; Flippin, J.; Paramban, R.I.; Hefferan, M.P.; Vidal, J.G.; Mu, Y.; Killian, R.L.; Mason A. Israel, M.A.; Emre, N.; Marsala, S.; Marsala, M.; Gage, F.H.; Goldstein, L.S.B. & Carson, Ch.T. (2011). Cell-Surface Marker Signatures for the Isolation of Neural Stem Cells, Glia and Neurons Derived from Human Pluripotent Stem Cells. *PLoS ONE*, Vol.6, No.3, (March), pp. 1-16, ISSN 1932-6203

Zhang, X.; Szabo, E.; Michalak, M. & Opas, M. (2007). Endoplasmic Reticulum Stress During the Embryonic Development of the Central Nervous System in the Mouse. Intenational *Journal of Developmental Neuroscience*, Vol. 25, No.7 , (November), pp. 455-463, ISSN 0736-5748

Witusik-Perkowska, M.; Rieske, P.; Hulas-Bigoszewska, K.; Zakrzewska, M.; Stawski, R.; Kulczycka-Wojdala, D.; Bienkowski, M.; Stoczynska-Fidelus, E.; Grešner, S.M.; Piaskowski, S.; Jaskolski, D.J.; Papierz, W.; Zakrzewski, K.; Kolasa, M.; Ironside, J.W. & Liberski, P.P. (2011). Glioblastoma-Derived Spheroid Cultures as an Experimental Model for Analysis of EGFR Anomalies. *Journal of Neuro-Oncology*, Vol.102, No.3, (May), pp. 395-407, ISSN 0167-594X

Wolin, S.L. & Cedervall, T. (2002). The La Protein. *Annual Review of Biochemistry*, Vol. 71, No.1, pp. 375-403, ISSN 15454509, 0066-4154

Part 2

Neural Stem Cells in Invertebrates

6

Formation of Nervous Systems and Neural Stem Cells in Ascidians

Kiyoshi Terakado
Saitama University
Japan

1. Introduction

Phylum Chordata comprises three subphyla, Cephalochordata, Urochordata and Vertebrata. Urochordates (tunicates) are morphologically very diverse, but recent phylogenetic analyses revealed that urochordates and not chephalochordates are the closest living relatives of vertebrates (Blair and Hedges, 2005; Delsuc et al., 2006, 2008; Putnam et al., 2008).

The central nervous system of the ascidian embryo is formed from the neural plate by its rolling into a hollow tube on the dorsal surface. This feature is unique to and common in chordates (synapomorphy). Cell number of mature tadpole remains nearly constant until metamorphosis begins. From earlier, two histological features are noticed in the tissues of the swimming larva; the first is the cessation of cell division in all the larval tissues and also in rudiments of adult organs, and the second is that functional cells are restricted to the larval organs. These cells differentiate almost synchronously in the larva. Larvae swim first for distribution and then for search the settlement places for metamorphosis without feeding. After onset of metamorphosis, the cells of larval organ disintegrate and/or rearrangement and those of adult rudiments begin to divide and differentiate into functional organs. Body wall muscle begins to contract intermittently, concomitant with the beginning of the feeding. That is, the cells of adult organs inhibited by some factors during swimming stage. The adult ascidian neural complex comprises the cerebral ganglion and the neural gland/its derived organ. The former is formed from the primodium of the cerebral ganglion constructed by rearrangement of the larval central nervous system during metamorphosis, and the latter is formed from a thin tube called the neurohypophyseal duct, respectively. The larval central nervous system contains functional neurons and glial cells, called the ependymal cells. Most of functional neurons and the glial cells in the tail region are lost during metamorphosis. Neurohypophyseal duct cells, located in the anterior left side of the sensory vesicle of swimming larvae, are derived from the anterior embryonic neural plate, which expresses common transcription factors in vertebrates and urochordates. After metamorphosis begins, the duct elongates anteriorly and fuses with the stomodeal ectoderm, where the dorsal tubercle, a large ciliated structure that opens into the upper part of the pharynx, later develops. The rudiment of the cerebral ganglion and the duct elongate posteriorly. The duct also differentiates into the neural gland. The dorsal wall of the neural gland in more developed ascidians has a thick epithelium (placode), the central part of which forms the dorsal strand by repeated invaginations along the visceral nerve. Both gonadotropin-releasing hormone (GnRH) neurons and prolactin-like (non-GnRH) neurons are generated in the dorsal strand and migrate to the cerebral ganglion along the visceral nerve throughout adulthood. Thus, the epithelium of the dorsal strand derived from the neurohypophyseal duct possesses neurogenic potential for such neurons during life (neural stem cells). The GnRH and prolactin-like neurons are mutually in close contact in the dorsal strand and their concurrent seasonal changes also occur in relation to the reproduction. The generation of GnRH and prolactin-like neurons and their migration into the brain suggest that the ascidian dorsal strand is homologous to the craniate olfactory placode, and provide unequivocal support for the clade Olfactores.

2. Formation of central nervous systems

Recently, the notion about the formation of adult central nervous system in ascidians is greatly changed; earlier view is that the neural complex comprising the cerebral ganglion and the neural gland is formed from the neurohypophyseal duct positioned at left side of the cerebral ganglion (Wiley, 1893; Mackie and Burighel, 2005; Manni et al., 2004; Manni et al., 2005). However, Takamura (2002) using neuron-specific antibody and Horie et al. (2011) using light-labeled fluorescent protein clearly revealed that the larval central nervous system contributes to the formation of the adult central nervous system. The contributed cells, called the ependymal cells, remain unchanged in number and state of differentiation until metamorphosis. Based on the formation of the adult central nervous system from the larval ependymal cells, these cells are claimed to act as neural stem cell-like cells (Horie et al., 2011). The anterior-posterior axis of the larval central nervous system is also inherited to form the adult central nervous system, indicating that the anterior-posterior axis of the central nervous system is already determined by developmental regulatory genes (Wada et al., 1998; Horie et al., 2011). Experimental ablation of the cerebral ganglion in *C. intestinalis*, it regenerates in its entirety within a few weeks (Bollner et al., 1997). This indicates that the regeneration of the adult central nervous system does not require preexisting cells (neurons and glial cells) from the central nervous system; i.e., it accomplishes entirely by cell supply from other tissues and organs than those of the central nervous system. The regenerated central nervous system may also acquire the anterior-posterior axis identical to that of normal development. In the regenerating brain, no mitotic figures were detected, indicating that migration of post-mitotic cells to the site of ganglion regeneration (Bollner et al., 1995, 1997). Initial concentration of GnRH-like cells along the ventral surface of the regenerating brain in *C. intestinalis* (Bollner et al., 1997) suggests that these cells originate in the dorsal strand and migrate to the surface of regenerating brain as those in normal development. Because the adult brain possesses various types of neurons (cholinergic, GABAergic, glycinergic, and glutamatergic ones), trans-differentiation of the cells from existing tissue must be involved in brain regeneration. The regeneration of the neural gland epithelium and then the neural gland luminal cells after neural complex ablation suggests to be caused by extensive cell divisions of the remaining epithelial cells of the dorsal strand by a heavy BrdU labeling (Bollner et al., 1995).

In the larval central nervous system, non-functional cells are generated in excess as compared with the functional neurons; i.e., larval functional neurons are approximately 100 and non-functional cells, called the ependymal cells, are 245 in *Ciona* central nervous system

(Horie et al., 2011). With regard to the ependymal cells, another explanation may be possible that they are neural progenitor cells that remain quiescent during larval stages and differentiate into adult neurons (and provably also to glial cells) after metamorphosis. They might not be neural stem cells because they occupy large numbers of cells (over 70%) of the larval central nervous system. It has not also been ascertained whether these cells have the ability of self-renew or persistent cell division to yield new neurons and/or glial cells, that the ability is the primary characteristic of neural stem cells.

2.1 Formation of peripheral nervous system

There are two common routs of origin of neurons in vertebrates and urochordates, i.e., the origin in central nervous system (most neurons) and that in peripheral nervous system (very restricted neurons including GnRH neurons). The presence of the olfactory placode in vertebrates and that of its homologous organ in ascidians which both generate the GnRH neurons is a very conspicuous phenomenon in animal kingdom and makes an important morphological characteristic of the clade Olfactores (vertebrates + tunicates).

GnRH is the hypothalamic neurohormone that activates the release of gonadotropin from gonadotropes of the pituitary in vertebrates. These GnRH neurons originate from the olfactory placode and migrate into the brain (GnRH1 or GnRH3 in some teleosts). The other GnRH neurons that originate in the midbrain tegmentum (GnRH2) seem to have a co-transmitter or a neuromodulater function. In ascidians, GnRH that is so called even in invertebrates devoid of the pituitary, because of the composition of the identical number of amino acid residues, conserved sequences, and may involved in reproduction (Powell et al., 1996; Terakado 2001; Adams et al., 2003) or in neuromodulation (Tsutsui et al., 1998). GnRH fibers distribute widely through the body, such as innerside of gonoduct, surface of gonads, branchial basket, surface of muscle bands, ciliated epithelium of pharynx, tentacles, etc.,. Synaptic button was not observed in fiber tips, suggesting that one of the functions of ascidian GnRHs is a neuromodulator or a paracrine secretion. Collectively, there are, at least, four common characteristics between vertebrate and ascidian GnRHs, such as decapeptide, conserved sequence, relation to reproduction, origination in peripheral organ and persistent neurogenesis throughout the adult life.

The evolutionary origin of neurogenic placodes remains controversial because of morphological divergence in chordates. Despite the importance of neurogenic placodes for understanding real functions and phylogenetic relationships among chordates, morphological and developmental data remain scarce. In craniates, peripheral GnRH neurons arise from the olfactory placode, whose cells are derived from the anterior region of the embryonic neural plate (Okubo et al., 2006; Cariboni et al., 2007; Schwarting et al., 2007; Bhattacharyya and Bronner-Fraser, 2008; Chen et al., 2009; Kanaho et al., 2009). Generation of peripheral neurons is a unique phenomenon that does not conform to the central nervous system origin of most neurons. The possible presence of a neurogenic placodal structure in invertebrate chordates has long been debated (Manni et al., 2004, 2005, 2006; Mackie and Burighel, 2005; Mazet et al., 2005, 2006; Schlosser, 2005). Recently, it was directly shown that, using one of the biggest solitary ascidian *Halocynthia roretzi* (Fig.1), GnRH neurons are generated in the dorsal strand and migrate into the cerebral ganglion (Terakado, 2009). It was then hypothesized that the dorsal strand is homologous to the craniate olfactory placode, an idea that are based on the topological relations and generation of peripheral GnRH neurons, which are commonly derived from the anterior region of the embryonic neural plate (Elwyn, 1937; Satoh, 1994; Cole and Meinertzhagen, 2001, 2004). The anterior region of the embryonic neural plate has been suggested to be the territory of the olfactory/adenohypophyseal placodes of vertebrates (Mazet et al., 2005) and expresses certain transcription factors common in craniates. Similarly, prolactin (PRL)-like neurons generated in close contact with GnRH neurons in the dorsal strand, which is formed from the dorsal epithelium of the neural gland by repeated invaginations (Fig. 2A).

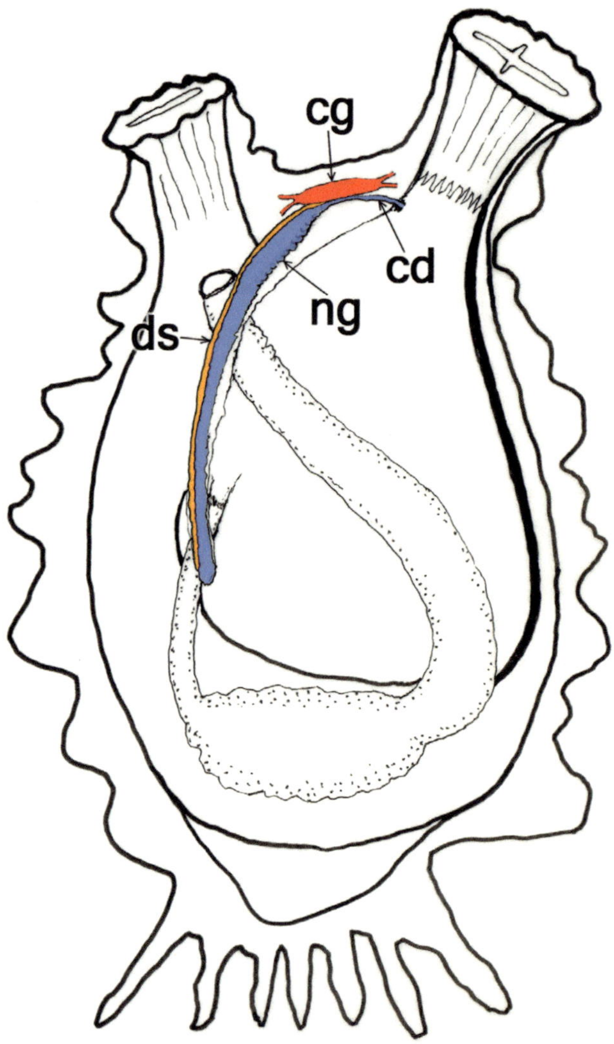

Fig. 1. Schematic drawing of the neural complex in a 3-year-old Halocynthia roretzi. The cerebral ganglion (cg) lies between the atrial and branchial siphons. The neural gland (ng) is located just beneath the ganglion, and its lumen opens anteriorly to upper part of the pharynx through the ciliated duct (cd). The neural gland extends posteriorly with the dorsal strand (ds).

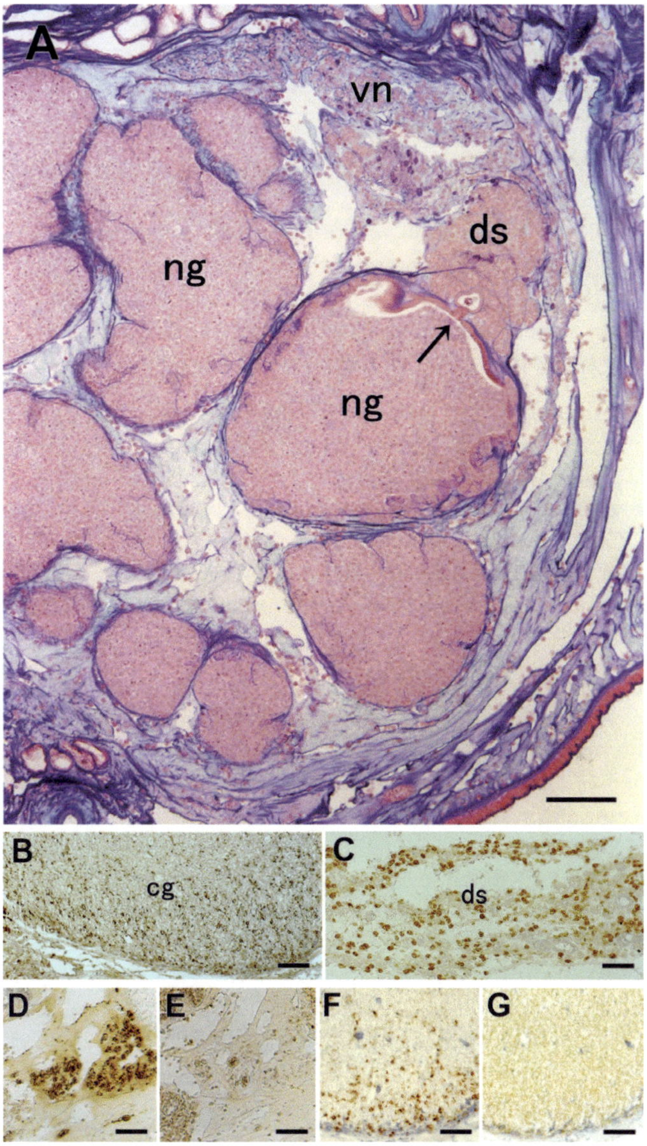

Fig. 2. Formation of the dorsal strand, localization of GnRH and PRL-like neurons and controls.

A. Formation of the dorsal strand (ds) from the placode (arrow) in the dorsal epithelium of the neural gland (ng) by invagination facing the visceral nerve (vn). Aldehyde fuchsin stain. B and C. Occurrence of PRL-like neurons in the cerebral ganglion (cg) and the dorsal strand (ds). PRL-like neurons in the cerebral ganglion are located mostly in the cortical region and possess long neurites, while neurons in the dorsal strand possess very short or lack neurites. Anti–bullfrog PRL stain. D and E. Specificity of GnRH immunoreactivity. Anti-human GnRH reactivity (D) is completely abolished by treatment with antiserum (E) that had been preabsorbed with the same antigen (2 ug/ml). F and G. Specificity of PRL-like immunoreactivity. Anti–bullfrog PRL reactivity (F) is completely abolished by treatment with antiserum (G) that had been preabsorbed with the same antigen (10 ug/ml). Bars: (A) 100 um; (B-G) 50 um.

Urochordates are now thought to be the sister group of vertebrates. Therefore, it is likely that developmental novelties of chordates (neural crest and placode) arose during evolution of the common ancestor of urochordates and vertebrates.

In cephalochordates (amphioxus), the similarity to vertebrates during embryonic development and in the expression of certain common transcription factors in the anterior region suggests the possible presence of a placode(s) that is homologous with those in craniates (Gorbman, 1995; Yasui et al., 2000; Boorman and Shimeld, 2002). However, the proposed placode generates neither neuroendocrine cells that give rise to GnRH neurons nor endocrine cells that give rise to adenohypophyseal endocrine cells. Thus, the first chordate may have lacked placodes (Meulemans and Bronner-Fraser, 2007). These observations provide support the clade Olfactores.

Several neuronal populations are generated from the olfactory placode of vertebrates. Recent results have demonstrated certain morphological, developmental (Burighel et al., 1998; Manni et al., 2004; Terakado, 2009), and molecular (such as *Six, Pitx, Eya, Pax, Coe, Dach, POUIV* gene families) commonalities between urochordates and vertebrates (Bassham and Postlethwait, 2005; Boorman and Shimeld, 2002; Christiaen et al., 2002; Mazet et al., 2005; Mazet and Shimeld, 2005; Schlosser, 2005), but information on the localization of these transcription factors and on the sites of emergence of neuroendocrine/endocrine cells remains controversial.

Because the novel structure, termed the dorsal strand placode, is for the first time histologically observed in invertebrate chordates, definition of the structure is seriously evaluated based the previous criteria (Northcut and Gans, 1983; Schlosser, 2005). The dorsal strand "placode" has the following characteristics; (1) generation from thickened epithelium. (2) invagination at the center of thickened epithelium. (3) generation of many types of cells including neurons. (4) delamination from the invaginated epithelium. (5) migration into the brain and to some other regions. All criteria are fit for the dorsal strand "placode", indicating that generation of placode really occurred in ascidians as an important developmental novelty in a common ancestor of vertebrates and urochordates. Additionally, expression of some transcription factors suggest the establishment of gene network to yield neuroendocrine/endocrine cells, but are often contradict between expression site and real occurrence of cells (see review by Schlosser, 2005) by provably an incomplete establishment of network or by its regression.

The aim of this review is to describe the recent findings on the origin of adult nervous systems, the persistent proliferation of GnRH and PRL-like neurons in the dorsal strand through lifetime, the morphological features of GnRH and PRL-like non-GnRH neurons, and to provide the morphological bases for further cellular, molecular and neural stem cell studies.

2.1.1 Animals

Following species are used in most ascidian studies.

Ciona intestinalis (Enterogona) is a cosmopolitan solitary ascidian that has become the model species of urochordates. Generation time is about 3 months.

Halocynthia roretzi (Pleurogona) is one of the biggest solitary ascidian and produce large eggs. Generation time is about 3-4 years (Fig.1).

Phallusia mammillata is a large solitary ascidian and found in the Atlantic and Mediterranean Sea. Generation time is about 8 months.

Botryllus schlosseri is a cosmopolitan colonial ascidian that has become the model species for the study of blastogenesis (asexual reproduction).

Description of conventional histological and immunohistlogical methods is omitted. Some sections were immunostained with a swine ACTH primary antiserum (Tanaka and Kurosumi, 1986) to illustrate the intimate topological relationships between the neural gland, the dorsal strand, and the visceral nerve.

2.1.2 Immunoelectron microscopy

The neural complexes were cut into small pieces and fixed for 4 hrs at 4°C in 0.2 M sodium phosphate buffer, pH 7.4, containing 4% paraformaldehyde (Merck) and 0.5% glutaraldehyde (TAAB). Tissue was washed overnight at 4°C in phosphate-buffered saline, post-fixed for 1 hr at 4°C with 1% osmium tetroxide in 0.15 M phosphate buffer, dehydrated, and embedded in Epon-Araldite. Ultrathin sections (approximately 8 nm thick) were collected and mounted on nickel grids (Nisshin EM, Tokyo) coated with Formvar (TAAB) which had been stored in a refrigerator at –20°C in order to ensure the adhesion of ultrathin sections, and treated with 1% meta-periodic acid for 10-30 min. As the primany antibody, an antiserum to bullfrog PRL (1:120) or to salmon GnRH (1:200) was used. The sections were then treated with gold-labeled (10 nm) secondary antibody (1:20; British BioCell International, Cardiff, UK) for 1 hr, washed, double-stained lightly with aqueous uranyl acetate and lead citrate, and examined with an electron microscope.

2.1.3 Specificity of immunoreactivity for PRL-like or GnRH neurons

Immunocytochemistry using anti–bullfrog PRL revealed numerous PRL-like neurons in the dorsal strand and the cerebral ganglion (Terakado et al., 1997; Figs. 2B, C). In control sections using preabsorbed antibody, immunoreactivity was undetectable (compare Figs. 2F and G). Using immunoelectron microscopy in control sections with the same preabsorbed antibody, no gold particles were observed in any neurons or endocrine cells. Because GnRH and PRL-like neurons are often localized side by side, the specificity of the GnRH antiserum for the dorsal strand or the cerebral ganglion was also examined. We observed no GnRH immunoreactivity when sections were stained with preabsorbed GnRH antiserum (compare Figs. 2D and E).

2.2 Development of the neurohypophyseal duct

The neurohypophyseal duct (ND) is a thin duct located on the anterior left side of the sensory vesicle of larvae, and remains a duct-like structure until the onset of metamorphosis (Satoh, 1994; Cole and Meinertzhagen, 2001, 2004; Manni et al., 2005). In metamorphosed juveniles, the rudiment of the cerebral ganglion appears as a small mass of cells on the dorsal side of the preexisting neurohypophyseal duct (Figs. 3A, B, C). Brain development proceeds slowly in *H. roretzi*, and thickening of the epidermis (placode formation) and delamination/migration of pioneer cells into the brain is not discernible with light microscopy (Figs. 3A, B, C). The ND elongates anteriorly and fuses with the stomodeal ectoderm (3B), later forming the dorsal tubercle. The cerebral ganglion elongates posteriorly to form a thin, long ganglion in young juveniles (Fig. 3D). Formation and development of the neural gland are more delayed than those of the adult central nervous sytem (Fig. 3D).

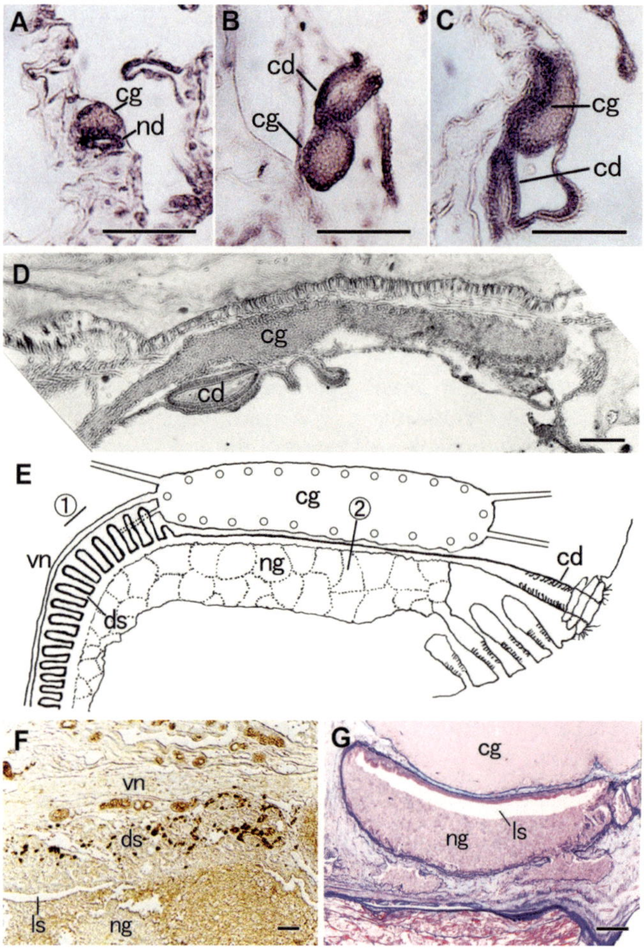

Fig. 3. Development of the neural complex and the topological relationships of the dorsal strand.

A, B, C. Early stages of neural complex formation. The rudiment of the cerebral ganglion (cg) appears on the dorsal side of the neurohypophyseal duct (nd), attaching closely to it (A, anterior is bottom). The neurohypophyseal duct elongates to fuse with the stomodial epithelium (B, anterior is top. cd; ciliated duct). The cerebral ganglion begins to elongate posteriorly (C, anterior is bottom). Hematoxylin/eosin stain. D. The cerebral ganglion in a 5-mm juvenile. The cerebral ganglion (cg) elongates further posteriorly to form a thin, long structure. The nerve fibers run along the long axis. The ciliated duct (cd) is seen in the anterior ventral side (anterior to the left). Hematoxylin stain. E. Schematic representation of the adult neural complex. The epithelium of the dorsal strand (ds) is continuous to the ciliated duct (cd) through the dorsal region of the neural gland (ng, anterior to the right. vn; visceral nerve). F. Longitudinal section along the neural gland (ng)–dorsal strand (ds)–visceral nerve (vn) axis at the position indicated by ① in panel E. The dorsal strand (ds) is closely associated with the visceral nerve (vn). The luminal space of the neural gland lies just under the dorsal epithelium. Dense cells in the dorsal strand are ACTH-positive cells. Anti–swine ACTH stain. G. Transverse section at the position indicated by ② in panel E. The cerebral ganglion (cg) is separated by the dorsal epithelium of the neural gland (ng). The luminal space (ls) lies in the most dorsal region of the neural gland. Aldehyde fuchsin stain. Bars: 50 um.

2.3 Formation of neural complex

The adult neural complex of *H. roretzi* is schematically described, with special reference to the continuation of the epithelium of dorsal strand (ds, Fig. 3E). The most anterior part is the dorsal tubercle, which opens into the upper part of the pharyngeal cavity, where it presents as a screw-like structure in which cilia on the outer surface move towards the pharyngeal cavity. In contrast, cilia on the inner surface move towards the interior of the body. Next to the dorsal tubercle towards the inside is the ciliated duct, in which cells are elongated. The ciliated duct contains both young cells, as judged by their small size and high nucleus/cytoplasm ratio, and degenerating cells, as judged by degenerating cells, as judged by disintegrating organelles. Cilia are embedded in prominent microvilli, and are arranged obliquely to point towards the interior. The region between the ciliated duct and the anterior part of the neural gland body is non-ciliated and has an exocrine function; some large granules (0.3-0.8 um in diameter) lie on the apical side and are often exposed to the lumen (Terakado et al., 1997). In the neural gland region, the epithelial structure is apparent in the dorsal side that faces the cerebral ganglion or the visceral nerve, but is indistinct in other regions. The luminal spaces that are continuous through the ciliated duct, the neural gland, and the dorsal strand terminate at the tips of the tubular structures of the dorsal strand (Figs. 3E, F, G).

Other than the dorsal epithelial cells, the cells of the neural gland are mostly, if not entirely, binucleate and loosely connected, and have no secretory granules. The neural gland elongates posteriorly along the visceral nerve and forms the dorsal strand from the dorsal epithelium by invagination towards the visceral nerve (Fig. 2A). There is an intimate topological relationship between the neural gland, the dorsal strand, and the visceral nerve (Figs. 3E, F). Why is the dorsal strand formed along the visceral nerve? One possibility may be that induction phenomenon might exist between them. Neuroendocrine cells containing GnRH and non-GnRH (PRL-like) neurons are localized in the dorsal strand and the cerebral ganglion. Adenohypophyseal-like cells such as adrenocorticotropic hormone-, growth hormone-, prolactin-, and gonadotropic hormone-immunoreactive cells were also present in the dorsal strand of *H. roretzi* (unpublished observation), which are compatible with the close proximal development of the olfactory and hypophyseal placodes in vertebrates (Mazet et al., 2005), though no such immunoreactivities are obtained in *C. intestinalis* which are compatible with the genome (Holland et al., 2008) and peptidomic (Kawada et al., 2011) analyses. Those differences between two species might be caused by loss of hypophyseal hormone genes in *C. intestinalis.* However, its exact understanding requires further studies.

2.4 Formation of the neural gland and the dorsal strand

The origin of the neural gland is unclear because of the absence of a specific marker (Takamura, 2002; Horie et al., 2011) and least studies on wide range of adult development. Our observations clearly revealed that the epithelium of the ciliated duct is continuous with the dorsal epithelium of the neural gland and further with that of the dorsal strand. Therefore, it is evident that the ciliated funnel (ciliated duct)–neural gland–dorsal strand system is a single entity (Fig. 3E) that is derived from the neurohypophyseal duct. The rudiment of the cerebral ganglion in *H. roretzi* appears immediately on the dorsal side of the neurohypophyseal duct of metamorphosed larvae (Fig. 3A). Delamination/migration of cells from the neurohypophyseal duct to the rudiment of the cerebral ganglion was not ascertained in our studies. The above results suggest that the adult neural complex is formed from dual origins and that its components may develop separately in an early phase (before migration of neurons to the cerebral ganglion) in normal development, although the rudiments of the cerebral ganglion and the neural gland are closely associated. This hypothesis is compatible with the proposition that the larval central nervous system contributes to the formation of the adult central nervous system (Takamura, 2002; Horie et al., 2011), rather than the hypothesis that the entire adult neural complex is generated from the neurohypophyseal duct (Willey, 1893). Peripheral GnRH and PRL-like neurons are generated in the dorsal strand, the epithelium of which is derived from the dorsal wall of the neurohypophyseal duct. In colonial ascidians, neurogenesis may also occur in the dorsal strand (if present), which remains fused with the cerebral ganglion for a long time, supplying neural cells (Manni et al., 1999; Koyama, 2002). The vertebrate olfactory epithelium (derived from the olfactory placode) is capable of prolonged neurogenesis that continues throughout adulthood (Beites et al., 2005; Murdoch and Roskams, 2007). This phenomenon also occurs during regeneration of neurons in the olfactory epithelium (Beites et al., 2005). This observation shows that the neurohypophyseal duct (and its derivatives) is homologous to the olfactory placode of vertebrates and can generate neural cells (neural stem cells) throughout the life of the organism, and that this phenomenon has been maintained throughout evolution from urochordates to mammals. Urochordate species that lack a dorsal strand, such as thaliaceans, appendicularians, and some colonial ascidians, seem not to generate peripheral neurons such as the GnRH and PRL-like neurons. These species may have been regressed the peripheral neurogenesis as an adaptation to asexual reproduction and/or perhaps because of a gross downsizing of body size.

It is well known that in appendicularians, cell division does not occur throughout the body after metamorphosis, and adults of chephalochordate (amphioxus) do not regenerate when injured, suggesting devoid of multipotent stem cells including neural ones.

2.5 Neurogenesis in the neural complex

Although ascidian neural complex contains several organs and distinct regions (cerebral ganglion, neural gland, dorsal strand, ciliated duct, and non-ciliated duct), neurogenesis occurs in the cerebral ganglion and the dorsal strand. The latter is probably the exclusive site of peripheral neurogenesis under normal conditions. Using immunostaining with anti–bullfrog PRL, we observed PRL-like cells along the dorsal strand and in the cerebral ganglion (Terakado et al., 1997). Using electron microscopy, two types of morphologically distinct neurons that occur side by side were discernible (Fig. 4A). In contrast to the generation of GnRH neurons, which occurs both within and adjacent to the epithelium (Terakado, 2009), PRL-like neurons were primarily generated adjacent to the epithelium (Figs. 4A, B, 5A). GnRH neurons contained a single kind of moderately dense secretory granules. On the other hand, PRL-like neurons contained both very dense and moderately dense granules of similar diameter (Figs. 4A, B). Young PRL-like neurons, as judged by a high nucleus/cytoplasm ratio and the presence of a few secretory granules, were frequently found within cell masses lying beside the epithelium (Fig. 5B). Using immunoelectron microscopy, PRL-immunoreactive material was detected in dense granules but not in moderately dense ones (compare Figs. 6A, B). Granules of one cell were often immunopositive for GnRH, whereas those of a neighboring cell were immunonegative (Fig.

6C). Similarly, granules of one cell were frequently immunopositive for PRL, whereas those of a neighboring cell were immunonegative (Fig. 6D). Most notably, GnRH and PRL immunoreactivities were mutually exclusive, suggesting that these neurons are distinct.

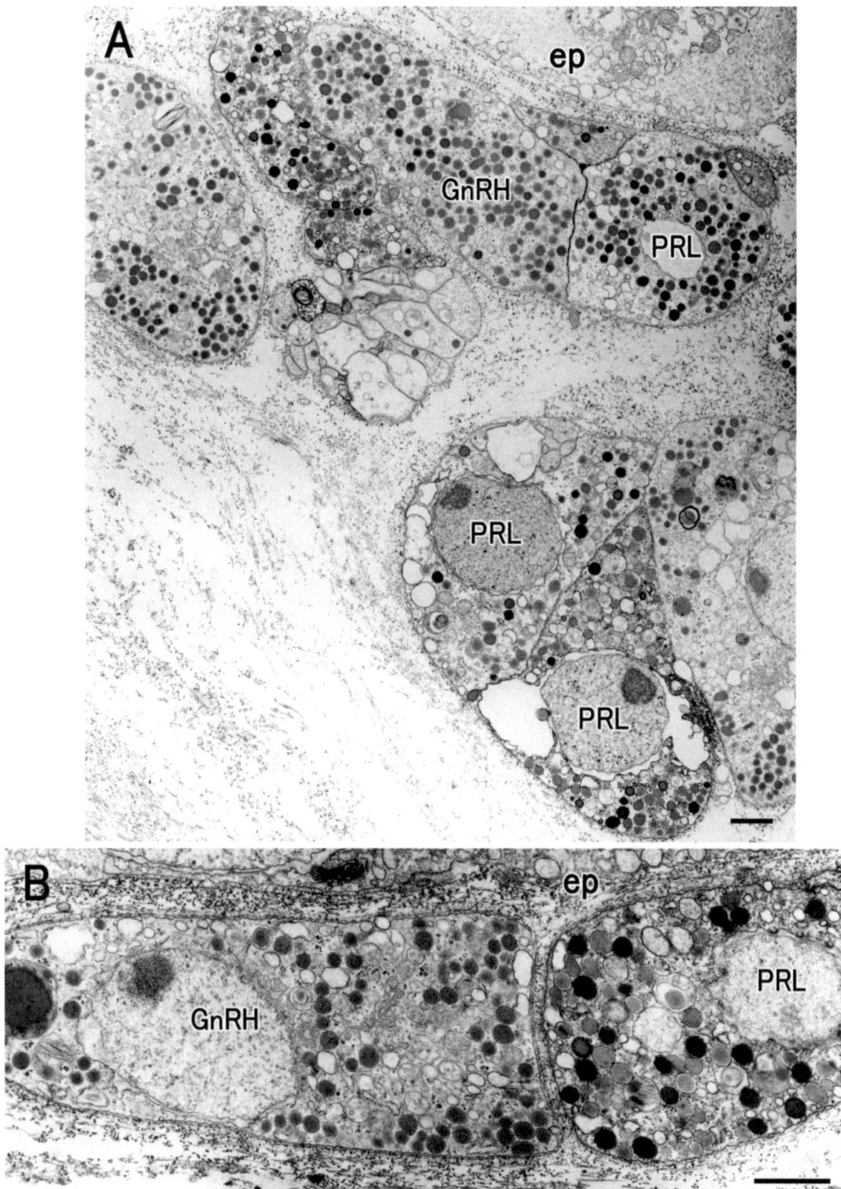

Fig. 4. Electron micrographs of parts of the dorsal strand.

A. Many young and developing GnRH (GnRH) and PRL-like (PRL) neurons are localized beside the epithelium (ep) of the dorsal strand. B. Two types of neurons in the dorsal strand. PRL-like neurons possess dense and moderately dense secretory granules, while GnRH neurons contain similar, moderately dense granules. Uranyl acetate-lead citrate double stain. Bars: 1 um.

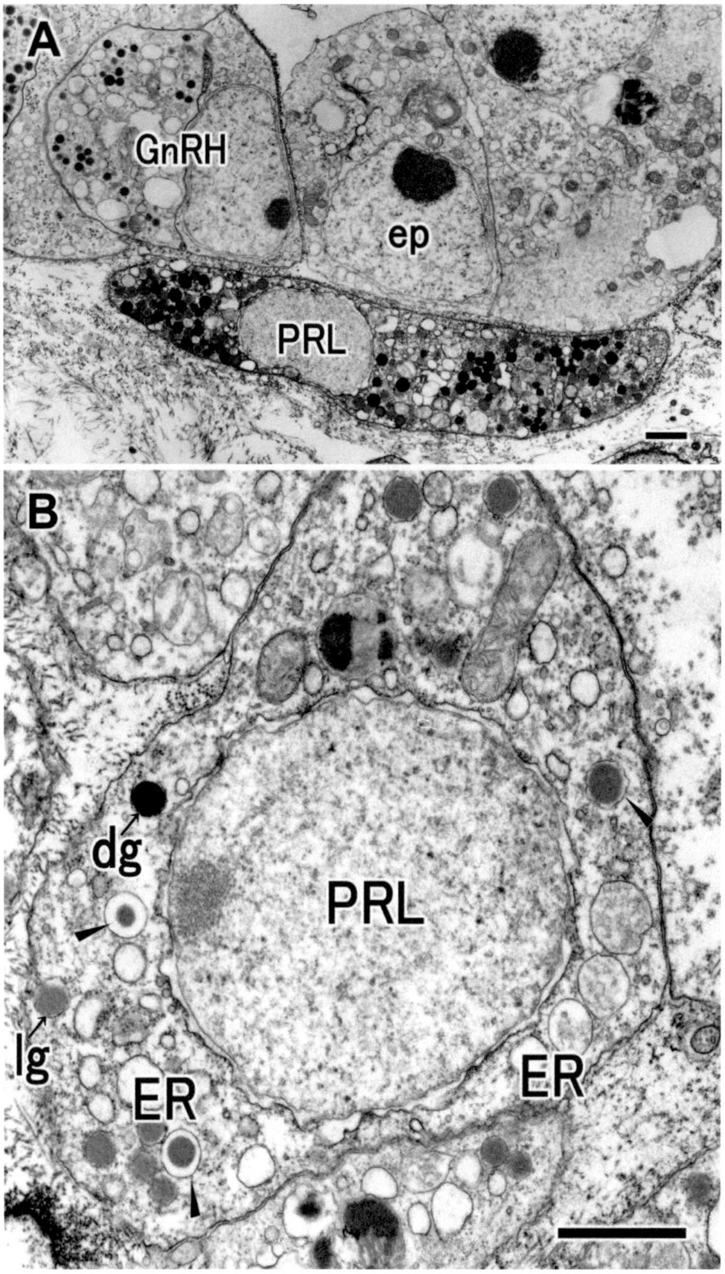

Fig. 5. Localization and ultrastructural features of young PRL-like neurons.

A. PRL-like neurons (PRL) with dense and moderately dense granules are often located beside the epithelium (ep). Growing GnRH neurons (GnRH) are seen in the epithelium of the dorsal strand. B. A young PRL-like neuron is shown possessing a few dense (dg) and moderately dense (lg) granules (arrows) that are membrane-bound. Developing granules (arrowheads), which are centrally condensed, were often observed. Rough endoplasmic reticulum (ER) is distended in several places, which is indicative of extensive protein synthesis. Uranyl acetate-lead citrate double stain. Bars: 1 um.

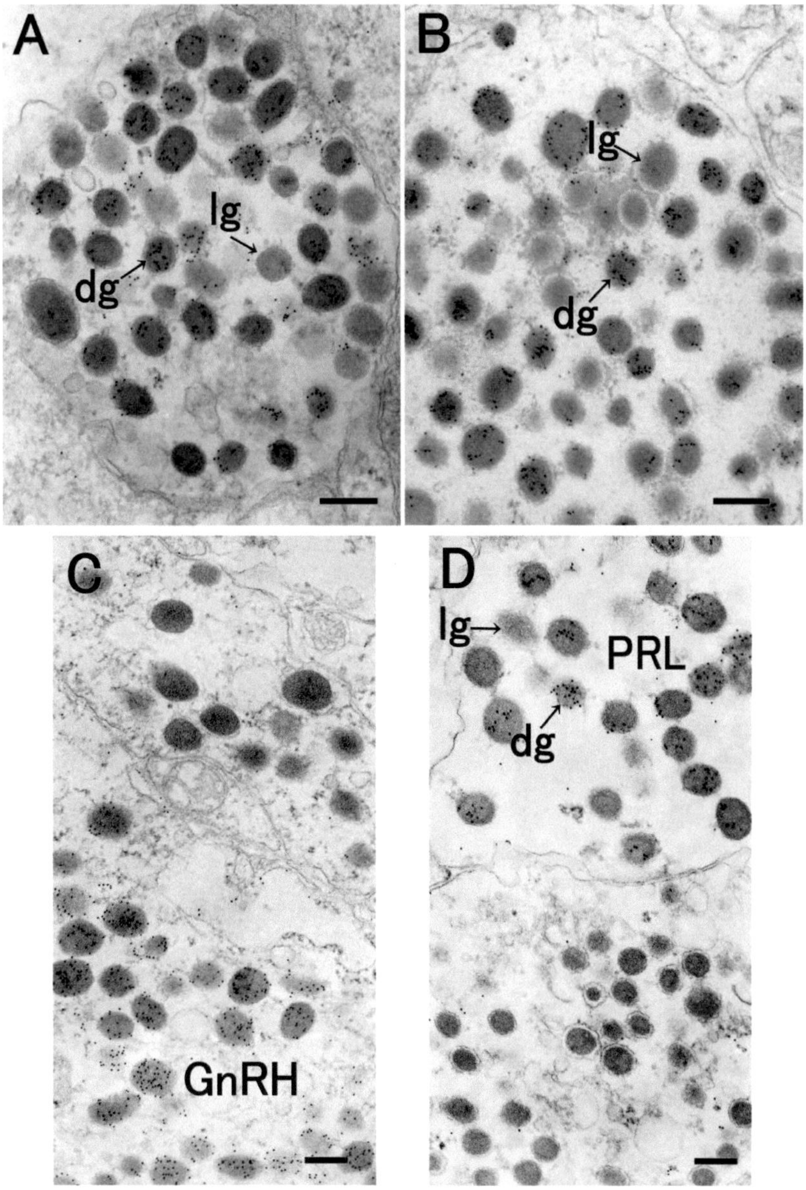

Fig. 6. Distinction of PRL-like neurons from GnRH neurons.

A and B. Immunoelectron micrographs of PRL-like neurons in the cerebral ganglion (A) and the dorsal strand (B). Gold particles are localized on the dense granules (dg) in both neurons. Gold particles are not localized in the moderately dense granules (lg). Anti–bullfrog PRL labeling. C and D. Comparison of anti-GnRH and anti-PRL immunoreactivity between GnRH neuron and probable PRL-like neuron, and between PRL-like neuron and probable GnRH neuron. C. GnRH immunoreactivity in the cerebral ganglion reveals that granules in one cell type are GnRH immunopositive (bottom), whereas those in the other cell type are GnRH immunonegative (top). D. PRL-like immunoreactivity in the cerebral ganglion reveals that granules in the cell on the top are PRL-like immunopositive, whereas those of the cell on the bottom are immunonegative. Bars: (A, B) 250 nm; (C, D) 500 nm.

2.6 Homology of the dorsal strand to the olfactory placode

The dorsal strand of *H. roretzi* is generated by repeated invaginations of the dorsal strand placode, and it produces GnRH and PRL-like neurons, some of which migrate into the cerebral ganglion through the visceral nerve (Terakado, 2009, 2010). Cells derived from the anterior region of the embryonic neural plate and their topological relationships indicate striking similarities between the dorsal strand and the olfactory placode, which suggests that the dorsal strand of urochordate ascidians is homologous to the olfactory placode of vertebrates (Terakado, 2009). Because the ascidian dorsal strand is a single organ, this notion is compatible with the formation of a single olfactory placode in agnathans (Uchida et al., 2003). The dorsal strand also generates many PRL-like (non-GnRH) neurons (Figs. 2C, 4A, B), a finding that has been reported in other species. Anti–salmon PRL also stains some cells of the dorsal strand in *H. roretzi* (unpublished observation). The presence of PRL-like cells has been reported in the cerebral ganglion of *Ciona* (Fritsch et al., 1982) and *Styela* (Pestalino, 1983), and in the brain of vertebrates (Fuxe et al., 1977; Krieger and Liotta, 1979; Toubeau et al., 1979; Hansen and Hansen, 1982). We previously suggested that these PRL-immunoreactive neurons in the vertebrate brain may be homologous to those in the cerebral ganglion of ascidians (Terakado et al., 1997). It is well known that molecular features of prolactin in the adenohypophysis resemble those of vertebrate growth hormones (Kawauchi and Sower, 2006). Immunoreactivity of some neurons of the cerebral ganglion and some cells of the dorsal strand to anti-PRL and anti–growth hormone antisera (unpublished observation) suggests the presence of ancestral molecule(s) of the growth hormone family in ascidians. Even the presence of a single molecule in ascidians raises the possibility that multiple antisera raised against molecules belonging to growth hormone family members react to ascidian prolactin due to the presence of common epitopes. Additional informations are needed to elucidate this problem.

2.7 Significance of neurogenesis in the peripheral organ

Most neurons are generated in the central nervous system in vertebrates and invertebrates. However, the vertebrate olfactory placode (peripheral organ) commonly generates GnRH neurons as well as other non-GnRH neurons (Murakami and Arai, 1994; Hilal et al., 1996; Yamamoto et al., 1996). Prior to or during the breeding season, the number of secretory granules greatly increases in the visceral nerve (Fig. 8). Generation of PRL-like neurons in the peripheral organ of chordates has not been reported other than the present species. The reason(s) why the GnRH neurons originate peripherally in vertebrates and ascidians is unknown; however, it is evident that GnRH neurons generated in peripheral organs are crucial for reproduction. PRL-like neurons in *H. roretzi* also concomitantly originate with the GnRH neurons side by side and migrate into the brain. Quantity of both neurons reveals a year-round change on a large scale which provably accompanies neuron loss after breeding season or winter in mature individuals and recovery thereafter by cell supply to the cerebral ganglion from the dorsal strand.

2.8 Migration of GnRH and PRL-like neurons to the brain

The observation that granulated PRL-like neurons are often found among the fibers of the visceral nerve (Fig. 7) suggests that they migrate towards the cerebral ganglion, similar to the GnRH neurons (Terakado, 2009). The PRL-like neurons in the dorsal strand mostly oval, whereas those in the cerebral ganglion have long neurites (compare Figs. 2B and C). This suggests that PRL-like neurons elongate after entering the brain. Unattached GnRH and PRL-like neurons were numerous and were located also near the visceral nerve (Fig.7 left). The intimate morphological relationship between the dorsal strand and the visceral nerve (Figs. 3E, F) has long been emphasized (see review by Goodbody, 1974; Chiba et al., 2004). In *H. roretzi*, this relationship may now be explained by the possibility that GnRH and PRL-like neurons generated in the dorsal strand invade the bundles of visceral nerve fibers and migrate towards the brain. This close topological relationship may be very important for invasion and migration from the site of generation (the dorsal strand) to the cerebral ganglion. This morphological relationship may partly explain the observation that the brain of *H. roretzi* easily reverts from a thin, cord-like structure (after breeding season, in winter) to the normal brain shape via migration of neurons from the dorsal strand. Similar phenomena are observed during regeneration of other brain structures; for example, after extirpation of the brain in *C. intestinalis*, the regenerating brain contains GnRH neurons, suggesting that GnRH neurons originate in the dorsal strand and subsequently migrate into the regenerating brain (Bollner et al., 1997). Together with the previous demonstration which GnRH neurons are generated in the dorsal strand and migrate into the brain (Terakado, 2009), the current results reveal that PRL-like neurons are generated in the dorsal strand and migrate into the brain during normal development. They both maintain the brain function throughout the life of the organism with other neurons. Even in colonial ascidians, dorsal strand cells divide frequently and become incorporated directly into the cerebral ganglion (Koyama, 2002) and/or enter into the circulation and participate in the formation of blastozooids as neural stem cells.

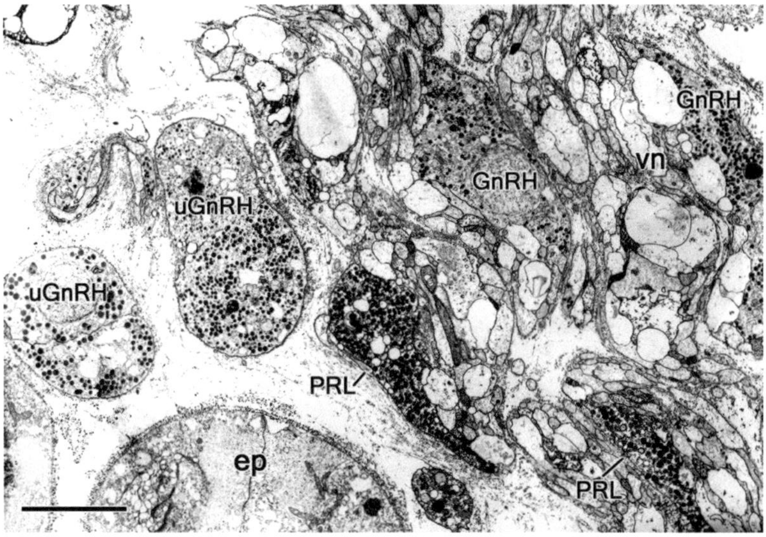

Fig. 7. Contact region between the dorsal strand and the visceral nerve (vn).

GnRH neurons and PRL-like neurons are often attached to the nerve, elongated along the nerve fibers, and distributed continuously towards the cerebral ganglion. Unattached GnRH neurons (uGnRH) are seen in the left. Uranyl acetate-lead citrate double stain. Bar: 5 um.

The neurohypophyseal duct and its derived tissues and cells generate a number of cell types including the ciliated duct cells, epithelial cells of the neural gland, luminal cells of the neural gland, epithelial cells of the dorsal strand, and neuroendocrine/endocrine cells in the dorsal strand. Similarly, the rudiment of the adult central nervous system may generate various kinds of neural cells in the developing brain. Of these cell types, those that are generated in the dorsal strand and migrate into the brain via the visceral nerve may be exclusively GnRH neurons in *Ciona* or GnRH and PRL-like neurons in *Halocynthia*. Abundance of both neurons in the cerebral ganglion and the dorsal strand in *H. roretzi* may correspond to the gigantism seen in this species and the corresponding necessity for a large neural network (provably relating to gametogenesis at expense of the body-wall muscle) to adjust to external/internal changes in the environment.

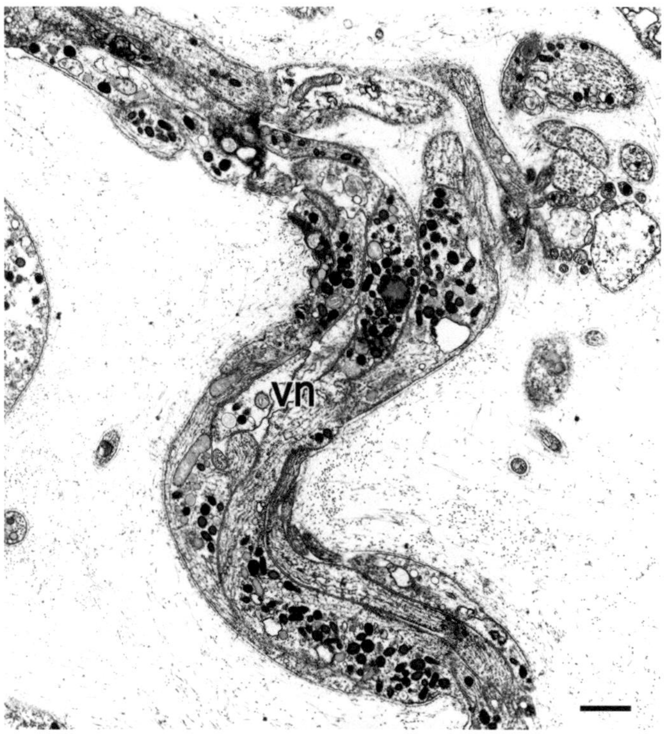

Fig. 8. Electron micrograph of the visceral nerve fibers at breeding season. The secretory granules in the fibers (vn) increase greatly prior to or during the breeding season. Bar: 1 um

2.9 Role of the neurohypophyseal duct

What is the significance of the neurohypophyseal duct in ascidians, in spite of solitary or colonial ones? From the neurohypophyseal duct, the neural gland and then the dorsal strand are formed in all the solitary ascidians and in some colonial ascidians. In the vegetative reproduction (the latter), migratory cells are produced from the dorsal strand (Manni et al, 1999; Koyama, 2002) and may participate in the formation of adult central nervous system of blastozooids as neural stem cells. Above results lead to the hypothesis that the neurohypophyseal duct is a cell reservoir that sets the undifferentiated cells aside for the post-metamorphic formation of the neural gland and the dorsal strand that contains neural stem cells. Peripheral organ origin of the GnRH neurons and their migration into the brain shares in vertebrates and urochordates, and may have evolved in the common ancestor of vertebrates and urochordates. In colonial ascidians that lack or reduced immunological staining to GnRH may be the result of regression due to the exceeding of asexual reproduction.

3. Conclusion

Urochordate ascidians share some morphological and developmental characteristics with those of vertebrates. About one thirds of the larval central nervous system is functional neurons and the rest is glial cells, called the ependymal cells. At metamorphosis, most of functional neurons and glial cells in the tail region disappear. Adult central nervous system is generated from the rearranged ependymal cells (undifferentiated neural cells) of the larval central nervous system after onset of metamorphosis. On the other hand, the peripheral nervous system is later generated from the dorsal strand which is formed by repeated invaginations of the thickened dorsal epithelium of the neural gland. The olfactory placode in vertebrates and the dorsal strand in ascidians are both derived from the anterior region of embryonic neural plate and generate GnRH and some other neurons. In some solitary ascidians, GnRH and PRL-like neurons are continuously generated in the dorsal strand throughout life. Therefore, ascidians are very useful for neural stem cell studies in providing important informations about fundamental processes of neural stem cell formation.

4. Acknowledgment

I am grateful to Emer. Prof. Hideshi Kobayashi of The University of Tokyo for continuous encouragement during studies and to the members of the Department of Regulation Biology, Saitama University for facilitation of the study. The author is also grateful to Emer. Prof. Sakae Kikuyama of Waseda University for gift of the bullfrog prolactin PRL antiserum.

5. References

Adams BA, Tello JA, Erchegyi J, Warby C, Hong DJ, Akinsanya KO, Mackie GO, Vale W, Rivier JE, Sherwood NM (2003) Six novel gonadotropin-releasing hormones are encoded as triplets on each of two genes in the protochordate, *Ciona intestinalis.* Endocrinology 144: 1907-1919

Bassham S, Postlethwait JH (2005) The evolutionary history of placodes: a molecular genetic investigation of the larvacean urochordate Oikopleura dioica. Development 132: 4259–4272

Beites CL, Kawauchi S, Crocker CE, Calof AL (2005) Identification and molecular regulation of neural stem cells in the olfactory epithelium. Exp Cell Res 306: 309–316

Bhattacharyya S, Bronner-Fraser M (2008) Conpetence, specification and commitment to an olfactory placode fate. Development 135: 4165–4177

Blair JE, Hedges SB (2005) Molecular phylogeny and divergence times of deuterostome animals. Mol Biol Evol 22: 2275–2284

Bollner T, Beesley PW, Thorndyke MC (1997) Investigation of the contribution from peripheral GnRH-like immunoreactive 'neuroblast' to the regenerating central nervous system in the protochordate *Ciona intestinalis.* Proc Roy Soc Lond B264: 1117–1123

Boorman CJ, Shimeld SM (2002) *Pitx* homeobox genes in *Ciona* and amphioxus show left-right asymmetry is a conserved chordate character and defines the ascidian adenohypophysis. Evol Dev 4: 354–365

Burighel P, Lane NJ, Zaniolo G, Manni L (1998) Neurogenic role of the neural gland in the development of the ascidian, *Botryllus schlosseri* (Tunicata, Urochordata). J Comp Neurol 394: 230-241

Cariboni A, Maggi R, Parnavelas JG (2007) From nose to fertility: the long migratory journey of gonadotropin-releasing hormone neurons. Trends Neurosci 30: 638-644

Chen B, Kim EH, XU PX (2009) Initiation of olfactory placode development and neurogenesis is blocked in mice lacking both Six1and Six4. Bev Biol 326: 75-85

Chiba S, Sasaki A, Nakayama A, Takamura K, Satoh N (2004) Development of *Ciona intestinalis* juveniles (through 2nd ascidian stage). Zool Sci 21: 285-298

Christiaen L, Burighel P, Smith WC, Vernier P, Bourrat F, Joly J-S (2002) *Pitx* genes in Tunicates provide new molecular insight into the evolutionary origin of pituitary. Gene 287: 107-113

Cole AG, Meinertzhagen IA (2001) Tailbud embryogenesis and the development of the neurohypophysis in the ascidian *Ciona intestinalis.* In "Biology of ascidians" Ed by H Sawada, H Yokosawa, CC Lambert, Tokyo, Springer-Verlag pp 137-141

Cole AG, Meinertzhagen IA (2004) The central nervous system of the ascidian larva: mitotic history of cells forming the neural tube in late embryonic *Ciona intestinalis*. Dev Biol 271: 239-262

Delsuc F, Brinkmann H, Chourrout D, Philippe H (2006) Tunicate and not cephalochordates are the closest living relatives of vertebrates. Nature 439: 965-968

Delsuc F, Tsagkogeorga G, Lartillot N, Phillippe H (2008) Additional Molecular support for the new chordate phylogeny. Genesis 46: 592-604

Elwyn A (1937) Some stages in the development of the neural complex in Ecteinascida turbinate. Bull Neurol Inst NY6: 163-177

Fritsch HA, Van Noorden S, Pearse AG (1982) Gastro-intestinal and neurohormonal peptides in the alimentary tract and cerebral complex of *Ciona intestinalis* (Ascidiaceae). Cell Tissue Res 223: 369-402

Fuxe K, Haekfelt T, Eneroth P, Gustafsson JA, Skett P (1977) Prolactin-like immunoreactivity: localization in nerve terminals of rat hypothalamus. Science 196: 899-900

Goodbody I (1974) The physiology of ascidians. Adv Mar Biol 12: 1-140

Gorbman A (1995) Olfactory origins and evolution of the brain-pituitary endocrine system: facts and speculation. Gen Comp Endocrinol 97: 171-178

Hansen BL, Hansen GN (1982) Immunocytochemical demonstration of somatotropin-like and prolactin-like activity in the brain of *Calamoichtes calabaricus* (Actinopterygii). Cell Tissue Res 222: 615-6627

Hilal EM, Chen JH, Silverman AJ (1996) Joint migration of gonadotropin-releasing hormone (GnRH) and neuropeptide Y (NPY) neurons from olfactory placode to central nervous system. J Neurobiol 31: 487-502

Holland LZ, Albalat R, Azumi K, Benito-Gutierrez E, Blow MJ, Bronner-Fraser M, Brunet F, Butts T, Candiani S, Dishaw LJ, Ferrir DEK, Garcia-Fernandez J, Gibson-Brown JJ, Gissi C, Godzik A, Hallbook F, Hirose D, Hosomichi K, Ikuta T, Inoko H, Kasahara M, Kasamatsu L, Kawashima T, Kimura A, Kobayashi M, Kozmik Z, Kubokawa K, Laudet V, Litman GW, McHardy AC, Meulemans D, Nonaka M, Olinski RP, Pancer Z, Pennacchio LA, Pestarino M, Rast JP, Rigoutsos I, Robinson-Rechavi M, Roch G, Saiga H, Sasakura Y, Satake M, Satou Y, Schubert M, Sherwood N, Shiina T, Takatori N, Tello J, Vopalensky P, Wada S, Xu A, Ye Y, Yoshida K, Yoshizaki F, Yu Jr-K, Zhang Q, Zmasek CM, Putnam NH, Rokhsar DS, Satoh N, Holland PWH (2008) Genome Res 18: 1100-1111

Horie T, Shinki R, Ogura Y, Kusakabe TG, Satoh N, Sasakura Y (2011) Ependymal cells of chordate larvae are stem-like cells that form the adult nervous system. Nature 469: 525-528

Kanaho Y, Enomoto M, Endo D, Maehiro S, Park MK, Murakami S (2009) Neurotrophic effect of gonadotropin-releasing hormone of neurite extension and neuronal migration of embryonic gonadotropin-releasing hormone neurons in chick olfactory nerve bundle culture. J Neurosci Res 87: 2237-2244

Kawada K, Ogasawara M, Sekiguchi T, Aoyama M, Hotta K, Oka K, Satake H (2011) Peptidomic analysis of the central nervous system of the protochordate, *Ciona intestinalis*: Homologs and prototypes of vertebrate peptides and novel peptides. Endocrinology 152: 2416-2427

Kawauchi H, Sower SA (2006) The down and evolution of hormones in the adenohypophysis. Gen Comp Endocrinol 148: 3-14

Koyama H (2002) The dorsal strand of *Polyandrocarpa misakiensis* (Protochordata: Ascidiacea): a light and electron microscope study. Acta Zool (Stockh), 83: 231-243

Krieger DT, Liotta AS (1979) Pituitary hormones in brain: where, how, and why? Science 205: 366-372

Mackie GO, Burighel P (2005) The nervous system in adult tunicates: current research directions. Can J Zool. 83: 151-220

Manni L, Lane NJ, Sorrentino M, Zaniolo G, Burighel P (1999) Mechanism of neurogenesis during the embryonic development of a tunicate. J Comp Neurol 412: 527-541

Manni L, Lane NJ, Joly JS, Gasparini F, Tiozzo S, Caicci F, Zaniolo G, Burighel P (2004) Neurogenic and non-neurogenic placodes in ascidians. J Exp Zool (Mol Dev Evol) 302: 483-504

Manni L, Agnelletto A, Zaniolo G., Burighel P (2005) Stomodial and neurohypophysial placodes in *Ciona intestinalis*: Insights into the origin of the pituitary gland. J Exp Zool (Mol Dev Evol) 300B: 324-339

Manni L, Mackie GO, Caicci F, Zaniolo G, Burighel P (2006) Coronal organ of ascidians and the evolutionary significance of secondary sensory cells in chordates. J Comp Neurol 495: 363-37

Mazet F (2006) The evolution of sensory placodes. Sci World J 6: 1841-1850

Mazet F, Hutt JAA, Milloz J, Millard J, Graham A, Shimeld SM (2005) Molecular evidence from Ciona intestinalis for the evolutionary origin of vertebrate sensoiry placodes. Dev Biol 282: 494-508

Mazet F, Shimeld SM (2005) Molecular evidence from ascidians for the evolutionary origin of vertebrate cranial sensory placodes. J Exp Zool (Mol Dev Evol) 304B:340-346

Meulemans D, Bronner-Fraser M (2007) The amphioxus Sox family: implications for the evolution of vertebrate placodes. Int J Biol Sci 3: 356-364

Murakami S, Arai Y (1994) Transient expression of somatostatin immunoreactivity in the olfactory-forebrain region in the chick embryo. Brain Res Dev Brain Res 82: 277-285

Murdoch B, Roskams AJ (2007) Olfactory epithelium progenitors: Insights from transgenic mice and in vitro biology. J Mol Histol 38: 581-899

Northcutt RG, Gans C (1983) The genesis of neural crest and epidermal placode: a reinterpretation of vertebrate origins. Q Rev Biol 58: 1-28

Okubo K, Sakai F, Lau EL, Yoshizaki G, Takeuchi Y, Naruse K, Aida K, Nagahama Y (2006) Forebrain gonadotropin-releasing hormone neuronal development: insights from transgenic medaka and the relevance to X-linked Kallman syndrome. Endocrinology 147: 1076-1084

Pestalino M (1983) Prolactinergic neurons in a protochordate. Cell Tissue Res 233: 471-474

Powell JF, Reska-Skinner SM, Prakash MO, Fischer WH, Park M, Rivier JE, Craig AG, Mackie GO, Sherwood NM (1996) Two new forms gonadotropin-releasing hormone in a protochordate and the evolutionary implications. Proc Natl Acad Sci USA 93: 10461-10464

Putnam NH, Butts T, Ferrier DEK, Furlong RF, Hellsten U, Kawashima T, Robinson-Rechavi M, Shoguchi E, Terry A, Yu J-K, Benito-Gutierrez E, Dubchak I, Garcia-Fernandez J, Gibson-Brown JJ, Grigoriev IV, Horton AC, de Jong PJ, Jurka J, Kapitonov VV, Kohara Y, Kuroki Y, Lindquist E, Lucas S, Osoegawa K, Pennacchio LA, Salamov AA, Satou Y, Sauka-Spengler T, Schmutz J, Shin-I T, Toyoda A, Bronner-Fraser M, Fujiyama A, Holland LZ, Holland PWH, Satoh N, Rokhsar DS (2008) The amphioxus genome and the evolution of the chordate karyotype. Nature 453: 1064-1071

Satoh N (1994) Developmental biology of ascidians. Cambridge University Press, Cambridge, UK

Schlosser G (2005) Evolutionary origins of vertebrate placodes: insights from developmental studies and from comparisons with other deuterostomes. J Exp Zool (Mol Dev Evol) 304B: 347-399

Schwarting GA, Wierman ME, Tobet SA (2007) Gonadotropin-releasing hormone neuronal migration. Semin Reprod Med 25: 305-312

Takamura K (2002) Changes in the nervous systems from larva to juvenile in *Ciona intestinalis*. Rep Res Inst Mar Biores Fukuyama Univ 12: 27-35

Tanaka S, Kurosumi K (1986) Differential subcellular localization of ACTH and α-MSH in corticotropes of the rat anterior pituitary. Cell Tissue Res 243: 229-238

Terakado K (2010) Generation of prolactin-like neurons in the dorsal strand of ascidians. Zool Sci 27: 581-588

Terakado K (2009) Placode formation and generation of gonadotropin-releasing hormone (GnRH) neurons in ascidians. Zool Sci 26: 389-405

Terakado K (2001) Induction of gamete release by gonadotropin-releasing hormone in a protochordate, *Ciona intestinalis*. Gen Comp Endocrinol 124:277-284

Terakado K, Ogawa M, Inoue K, Yamamoto K, Kikuyama S (1997) Prolactin-like immunoreactivity in the granules of neural complex cells in the ascidian *Halocynthia roretzi*. Cell Tissue Res 289: 63-71

Toubeau G, Desclin J, Parmentier M, Pasteels JL (1979) Compared localization of prolactin-like and adrenocorticotropin immunoreactivity within the brain of the rat. Neuroendocrinology 29: 374-384

Tsutsui H, Yamamoto N, Ito H, Oka Y (1998) GnRH-immunoeactive neuronal system in the presumptive ancestral chordate, Ciona intestinalis (Ascidian). Gen Comp Endocrinol 112: 426-432

Uchida K, Murakami Y, Kuraku S, Hirano S, Kuratani S (2003) Development of the adenohypophysis in the lamprey: evolution of epigenetic patterning programs in organogenesis. J Exp Zool (Mol Dev Evol) 300B: 32-47

Wada H, Saiga H, Satoh N, Holland PW (1998) Tripartite organization of the ancestral chordate brain and the antiquity of placodes: insight from ascidian *Pax-2/5/8, Hox* and *Otx* genes. Development 125: 1113-1122

Willey A (1893) Studies on the Protochordata. II The development of neuro-hypophyseal system in *Ciona intestinalis* and *Clavelina lepadiformis*, with an account of the origin of sense organs in Ascidia mentula. Q J Microsc Sci 35: 295-334

Yamamoto N, Uchiyama H, Ohki-Hamazaki H, Tanaka H, Ito H (1996) Migration of GnRH-immunoreactive neurons from the olfactory placode to the brain: a study using avian embryonic chimeras. Brain Res Dev Brain Res 95: 234-244

Yasui K, Zhang S, Uemura M, Saiga H (2000) Left-right asymmetric expression of BbPtx-related gene, in a lancelet species and the developmental left-sidedness in deuterostomes. Development (Camb.) 127: 187-195

7

Regeneration of Brain and Dopaminergic Neurons Utilizing Pluripotent Stem Cells: Lessons from Planarians

Kaneyasu Nishimura[1,3], Yoshihisa Kitamura[2] and Kiyokazu Agata[1]
[1]*Department of Biophysics, Graduate School of Science, Kyoto University*
[2]*Department of Neurobiology, Kyoto Pharmaceutical University*
[3]*Department of Cell Growth and Differentiation, Center for iPS Cell Research and Application (CiRA), Kyoto University*
Japan

1. Introduction

Cell-transplantation therapy for Parkinson's disease is close to becoming a reality thanks to the recent development of methods for the differentiation of dopaminergic neurons and/or dopaminergic progenitor cells from embryonic stem cells (ESCs) and induced pluripotent stem cells (iPSCs) under *in vitro* conditions (Kawasaki et al., 2000, Perrier et al., 2004). There have been several reports concerning pre-clinical trial research for cell-transplantation therapy for Parkinson's disease with dopaminergic progenitor cells derived from either ESCs or iPSCs using rodent and non-human primate disease models before clinical trial (Björklund et al., 2002; Takagi et al., 2005; Wernig et al., 2008). Many researchers have contributed to improve the technology to create more efficient differentiation methods of donor cells for clinical applications (Chambers et al., 2009; Morizane et al., 2011). However, we still need to overcome many problems before such technology can be used in clinical settings. Even if we succeed in obtaining an optimized donor cell population for cell-transplantation, the rate of success of the transplantation may depend not only on the quality of donor cells but also on the host brain environment. One important issue is how to integrate dopaminergic neurons or dopaminergic progenitor cells into target regions after transplantation. However, we do not know what kind of donor cells will be efficiently integrated into the neural networks of the host brain. Also, we do not know whether fully differentiated neurons will really survive in the host brain. In addition, we need to know what state of the host brain environment will allow the participation of donor cells in the neural networks of the host brain. In order to solve such problems, planarians provide unique opportunities because they show robust regenerative ability based on their pluripotent stem cell system.

Planarians can regenerate lost tissues, including the nervous system, via their pluripotent stem cells (neoblasts) that are distributed throughout their body. In contrast, it is difficult for higher vertebrates to achieve the regeneration of the nervous system, in spite of their possession of neural stem cells. The success of tissue regeneration requires not only the presence of proliferating stem cells as a source but also the presence of the regulatory system for stem cells. Knowledge gained about the planarian stem cell system can provide hints about how to conduct cell-transplantation therapy for regenerative medicine in the future.

In this chapter, we focus on two different regenerative phenomena utilizing the stem cell system in planarians. The first one is brain regeneration after decapitation. The second is brain neurogenesis after selective neuronal degeneration (without decapitation). Both of them are achieved by regulation of the pluripotent stem cells distributed throughout the body. We address the following questions: (1) what type(s) of cells recognize the loss of the organs or cells? (2) What signal(s) initiate the regeneration or neurogenesis? (3) What signal(s) are necessary for recruitment of stem cells to defined type(s) of cells and the replacement in the proper positions.

2. Pluripotent stem cells of planarians

The flatworm *Dugesia japonica* is a common species of freshwater planarian in Japan, and has been extensively used as an experimental animal for regeneration and neuroscience studies. When planarians are artificially amputated, they can regenerate their whole body from even very small fragments (is the smallest competent fragment reported was 1/279th of the body; Morgan, 1898). This strong regenerative ability is supported by pluripotent stem cells called neoblasts. The neoblasts are the only mitotic cell population, and are distributed in the mesenchymal space throughout the body except for the region around the brain and the pharynx of *D. japonica* (Shibata et al., 1999; 2010) (Fig. 1). The neoblasts can differentiate in all types of cells and self-renew under both homeostatic and injured conditions. X-ray-irradiation induces selective elimination of proliferating stem cells in planarians, resulting in the loss of regenerative ability (Shibata et al., 1999; Hayashi et al., 2006). Therefore, X-ray irradiation is a powerful experimental tool for analyzing the stem cell system. We identified a *vasa*-like gene (*Djvlg*) as the first reported gene specifically expressed in neoblasts (Shibata et al., 1999). Recently, many reliable molecular markers for neoblasts, such as *piwi* homologue genes, have been identified (Fig. 1) (Salvetti et al., 2000, 2005, Orii et al., 2005; Reddien et al., 2005; Eisenfoffer et al., 2007; Shibata et al., 2010). Since pluripotent stem cells are the only proliferating and mitotic cell population, experimental methods using 5-bromo-2'-deoxyuridine (BrdU) (Newmark & Sánchez Alvarado, 1999), and immunostaining using anti-phosphohistone H3 (pH3) antibody (Hendzel et al., 1997; Newmark & Sánchez Alvarado, 1999) are also useful tools for staining neoblasts.Recently, the pluripotency of these cells was demonstrated by single Icell-transplantation experiments (Wagner et al., 2011). In addition, we found that pluripotent stem cells can be categorized into several cell populations by electromicroscopy analysis, suggesting that pluripotent stem cells are not a homogenous population, but may have heterogeneity like stem cell systems in higher animals (Higuchi et al., 2007). In addition, we recently developed single-cell PCR technology that is able to analyze the gene expression profile in individual cells at the single cell level (Hayashi et al., 2010). This method is a powerful tool for determining gene characteristics of not only pluripotent stem cells but also of tissues such as the nervous system.

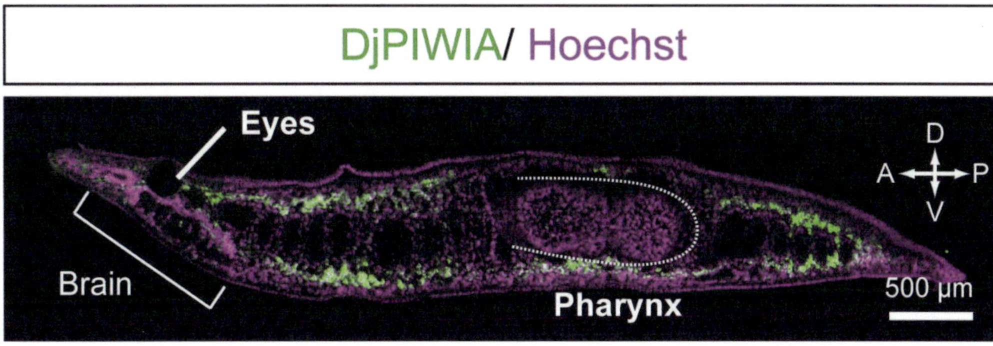

Fig. 1. Distribution of pluripotent stem cells of planarian *D. japonica*. Immunostaining using anti-DJPIWIA antibody (a marker of pluripotent stem cells) (Shibata et al., 2010) in a transverse section. Planarian stem cells are distributed in the mesenchymal space throughout the body.

3. Fundamental brain structure and function

Planarians have a simple body shape with cephalization, a dorso-ventral axis and bilateral symmetry, and are thought to be primitive animals, that acquired a central nervous system (CNS) at an early stage of evolution. The planarian CNS composed of a bilobed brain and a pair of ventral nerve cords (VNCs) (Agata et al., 1998; Tazaki et al., 1999). The brain is located in the anterior region of the body, and forms an inverted U-shaped structure (Fig. 2A). A pair of VNCs are located more ventral by relative to the brain, extending along the anterior-posterior (A-P) axis. The VNCs are a structure independent of the brain, although they are directly connected to it (Okamoto et al., 2005). The brain can be divided into several functional domains (Cebrià et al., 2002a; Nakazawa et al., 2003). The nine pairs of lateral branches of the brain project to the head margin, and function as the sensory system (Okamoto et al., 2005). A pair of eyes is located on the dorsal side of the brain, and the optic nerves forms the optic chiasm, and project to the dorso-medial position of the brain, which functions as the photosensory center (Sakai et al., 2000). The two main lobes of the brain consist of a mass of interneurons that function in the integration of multiple stimuli.

When planarians are exposed to some stimuli such as light-, chemo-, thermo- and mechano-stimulations, they can integrate different stimuli in the brain and decide on a response to these multiple stimuli. Planarians show light avoidance behavior known as negative phototaxis. We established a quantitative analytical method for this behavior that involves measuring the distance, direction, and speed of movement (Inoue et al., 2004). By using this method and RNA interference (RNAi), we showed that several molecules such as a planarian synaptosome-associated protein of 25 kDa (*Djsnap-25*) and a planarian glutamic acid decarboxylase (*DjGAD*) play important roles in photorecognition (Takano et al., 2007; Nishimura et al., 2008a). These results indicate that planarian behavior is regulated the molecular level via brain functions that are similar to mammalian brain functions.

3.1 Functional domain structure

We found that functional domains in the brain were defined by three *orthodenticle* and *orthopedia* homeobox genes (*DjotxA, DjotxB* and *Djotp*) that are exclusively expressed in specific regions of the brain (Umesono et al., 1997; 1999). *DjotxA* is expressed in the optic nerves and medial region of the brain, which form a photosensory domain. *DjotxB* is expressed in the main lobes of the brain, which form a signal processing domain containing a variety of interneurons. *Djotp* is expressed in the lateral branches, which form chemosensory domains. The lateral side of the head region, where *Otx/otp* expression is not detected, contains mechanosensory neurons. In addition, A-P patterning of the brain was shown to be regulated by the expression of *wnt*-family genes (*DjwntA* and *DjfzA*) (Kobayashi et al., 2007). Whereas *DjotxA, DjotxB* and *Djotp* genes were shown to be expressed medio-laterally, *DjwntA* and *DjfzA* genes were expressed antero-posteriorly in the brain. *Wnt* family genes and *Otx/otp* family genes play important roles in domain formation in planarians, as in mammals.

DNA microarray analysis comparing the head region versus the body region of planarians identified many genes that are specifically expressed in the head region (Nakazawa et al., 2003; Mineta et al., 2003). Expression analysis based on whole-mount *in situ* hybridization revealed that many neural genes that are conserved in the vertebrate brain are also expressed in several distinct domains of the planarian CNS (Cebrià et al., 2002a; Mineta et al., 2003). These results indicate that the planarian CNS is functionally regionalized by discrete expression of neural-specific genes.

3.2 Variations of neurotransmitters

Recently, we revealed that planarians have various neural populations defined by neurotransmitters, such as dopamine (DA), serotonin (5-HT), γ-aminobutyric acid (GABA), octopamine (OA; a counterpart of noradrenaline of vertebrates) and acetylcholine (ACh) (Nishimura et al., 2007a, 2007b, 2008a, 2008b, 2008c, 2010; Takeda et al., 2008) (Fig. 2). Immunostaining with specific antibodies against these neurons enables us to visualize their cell morphology and localizations at the single-cell level (Fig. 2). These neurons are distributed in restricted regions in the planarian CNS. In addition, each neuron exclusively uses one neurotransmitter, and forms distinct neural networks in the planarian CNS.

These neurons have also distinct functions, such as locomotion activity and photorecognition. Combined RNAi and pharmacological approaches revealed that dopaminergic neurons positively regulate muscule-mediated behavior. Upregulation of the DA level induced by methamphetamine (DA releaser) caused hyperkinetic conditions such as screw-like hyperkinesia and C-like hyperkinesia, and treatment with DA receptor antagonists (sulpride and reserpine) and reduction of the DA level by RNAi suppressed these hyperkinetic conditions (Nishimura et al., 2007a). Moreover, although an increase of the ACh level by physostigmine (acetylcholinesterase inhibitor) treatment induced sudden muscular contraction, treatment with ACh receptor antagonists (tubocrarine and atropine) or reduction of the ACh level by RNAi extended these behavioral changes (Nishimura et al., 2010). Our histological analysis indicated that cholinergic neurons elongated at neighboring positions of the body-wall musculature (DjMHC-B-positive cells), but dopaminergic neurons did not elongate to the body-wall musculature. These results suggest that although both dopaminergic and cholinergic neurons regulate motor functions, cholinergic neurons act as motor neurons whilst dopaminergic neurons act as interneurons in planarians. These results also indicate that similar gene sets function in both the planarian CNS and the vertebrate CNS.

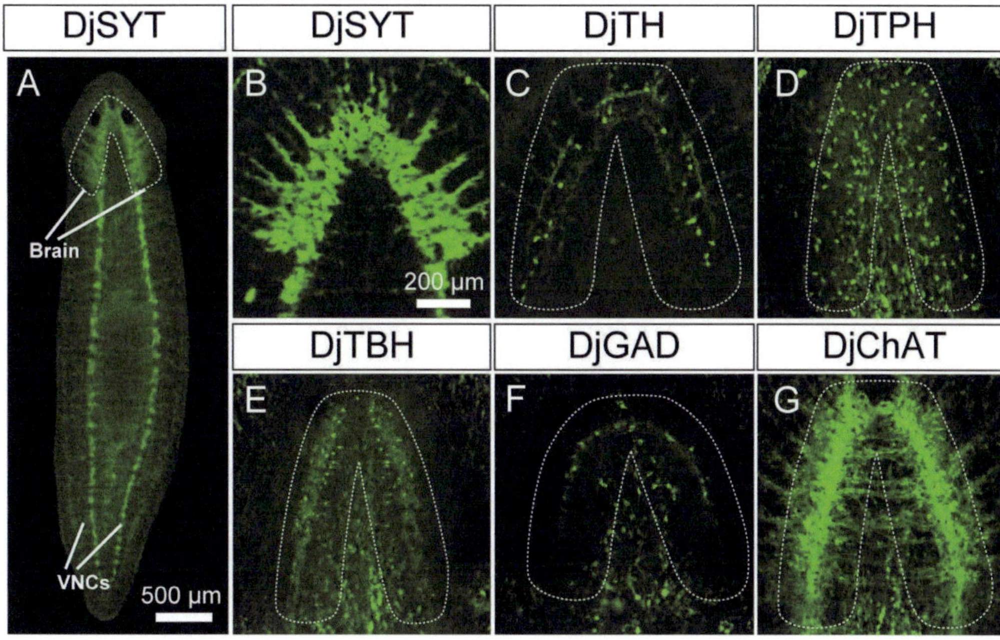

Fig. 2. The neural networks of neurotransmitter-synthesizing neurons. Distribution of pan-neural networks (DjSYT-positive neurons) of the whole body (A) and head (B). Distribution of dopaminergic neurons (DjTH-positive neurons) (C), serotonergic neurons (DjTPH-positive neurons) (D), octopaminergic neurons (DjTBH-positive neurons) (E), GABAergic neurons (DjGAD-positive neurons) (F), and cholinergic neurons (DjChAT-positive neurons) (G) in intact planarian head. White broken line indicates the outline of the brain (B-G).

4. Whole brain regeneration after head amputation

One of most interesting regeneration phenomena in planarians is that they can regenerate a functional brain from any portion of the body within 7-10 days after amputation, utilizing the pluripotent stem cell system. Although non-brain fragments just after decapitation show very little response external stimulation, they can restore normal behaviors such as feeding and negative phototaxis within one week. How can planarians regenerate their CNS not only morphologically but also functionally in one week? This regenerative process can be divided into at least five steps as defined by sequential gene expression alterations, which are similar to those in mammalian brain development (Agata & Umesono et al., 2008). That is, (1) anterior blastema formation, (2) brain rudiment formation, (3) pattern formation, (4) neural network formation, and (5) functional recovery (Fig. 3).

4.1 The stem cell system for brain regeneration

The first step of head regeneration after decapitation involves wound healing and subsequently the formation of the blastema, which is defined by a mass of morphologically undifferentiated cells at the edge of the amputated site. Dorso-ventral attachment induces initiation of the expression of *noggin-like gene A* (*DjnlgA*) at the edge of the amputated site after wound healing, and this expression leads to blastema formation in the first step of planarian regeneration (Ogawa et al., 2002). Mitotic cells are never observed in the blastema, in spite of the increasing mass of the blastema during regeneration (Wenemoser & Reddien, 2010; Tasaki et al, 2001a, 2001b). Recently, it was shown that the blastema cells are supplied from the postblastema region via mitosis from G2 phase-pluripotent stem cells, and that c-Jun-N-terminal kinase (JNK) is involved in this G2/M transition, and that extracellular signal-related kinase (ERK) is required for exit from the proliferative undifferentiated state during blastema formation (Tasaki et al., 2011a, 2011b). It is thought that BMP/noggin signal might be involved in activation of the ERK signal in cooperation with the JNK signal to form the blastema after wound closure.

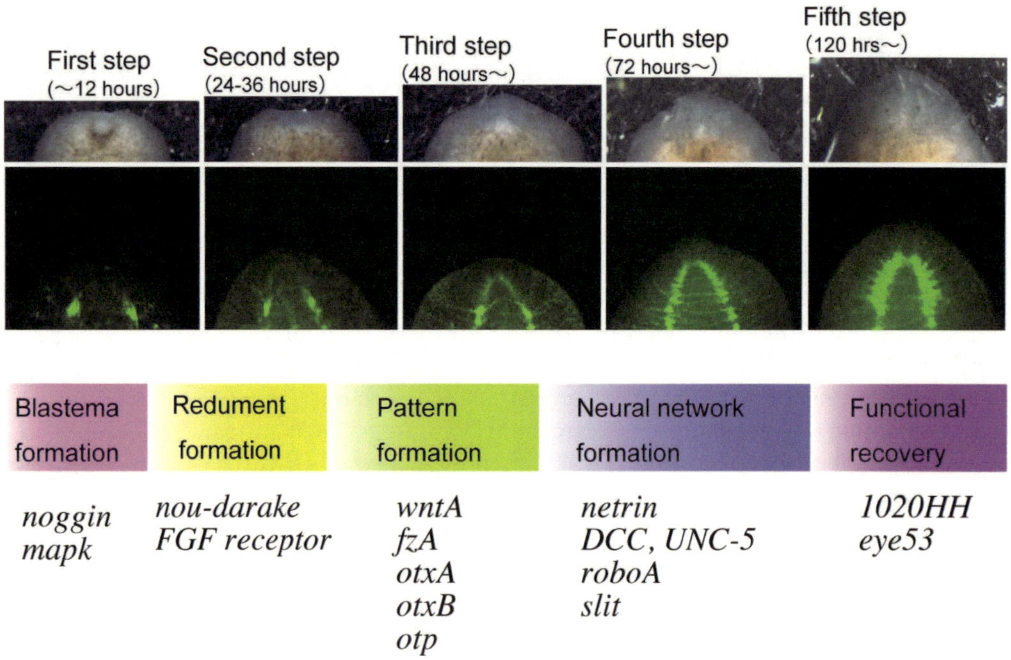

Fig. 3. Brain regeneration process after decapitation. This process can be divided into at least five steps according to sequential gene expression alterations. Abbreviations used; mapk, mitogen-activated protein kinase; FGF, fibroblast growth factor; DCC, deleted in colorectal cancer; UNC-5, uncoordinated-5; robo, roundabout.

After the formation of blastemas, the ERK signal is suppressed in the posterior blastema, but enhanced in the anterior blastema. Recently, we found that the hedgehog (Hh) signal has an important role in causing the difference between the anterior and posterior blastemas. In planarians, Hh is produced in the nervous system and Hh-containing vesicles might be transported from anterior to posterior along microtubules inside of the neurites (Yazawa et al., 2009). After amputation of the planarian body, Hh may be secreted from the posterior end of the amputated neurites, and then the Hh signal activates the Wnt signal in the posterior blastema to suppress the ERK signal and activate posterior-specific genes. In contrast, in the anterior blastema, the ERK signal forms a positive feedback loop to activate brain rudiment formation. A fibroblast growth factor receptor (FGFR)-like molecule, *nou-*

darake (*ndk*; meaning "brains everywhere" in Japanese), may have an important role in defining the region forming the positive feedback loop of the ERK signal in the anterior blastema (Cebrià et al., 2002b). The *ndk* gene was identified in *D. japonica* as a gene expressed in the brain rudiment at an early stage of brain regeneration. Interestingly, silencing of the *ndk* gene by RNAi induces the ectopic brain formation in all regions of the body. Thus, *ndk* is essential for defining the region where the brain rudiment is formed.

After formation of the brain rudiment, the Wnt and bone morphogenic protein (BMP) signaling pathways may regulate pattern formation of the brain along the A-P (Kobayashi et al., 2007; Gurley et al., 2008; Petersen & Reddien, 2008) and D-V (Molina et al., 2011; Gavino & Reddien, 2011) polarity, respectively. In conclusion, stem cells may be regulated by various signals in spatial- and temporal manners to form a functional brain.

4.2 Axon guidance and neural network formation during brain regeneration

New brain neurons have to project toward appropriate target sites to reconstruct their neural networks during regeneration. Recently, several axon guidance molecules, including netrin, uncoordinated-5 (UNC-5), deleted in colorectal cancer (DCC), slit, and roundabout (robo) were identified as key molecules regulating axon guidance during eye and brain regeneration in planarians (Cebrià & Newmark 2005, 2007; Cebrià et al., 2007; Yamamoto & Agata, 2011). It is known that netrin is a secreted protein that regulates the direction of axon growth by chemo-attractive and repulsive responses mediated by two types of receptor, UNC-5 and DCC (Hong et al., 1999). Slit is also a secreted protein, and acts as a chemo-repulsive factor for commisure axons by binding to robo in various animals (Brose et al., 1999). RNAi-mediated functional analysis revealed that the silencing of these guidance molecules caused abnormal neural network formation in the CNS and optic nerves during regeneration.

4.3 Functional recovery after completion of whole brain regeneration

In order to analyze the brain function during brain regeneration, we focused on negative phototaxis behavior. We found that there is a time gap between morphological and functional recovery. Although the optic nerves were reconstructed within 4 days after decapitation, negative phototaxis behavior began to recover from 5 days after decapitation (Inoue et al., 2004). Interestingly, two genes, *1020HH* and *eye53* genes, were activated just after completion of the morphological recovery (Cebrià et al., 2002c). Silencing of either *1020HH* or *eye53* caused a defect of the complete recovery of negative phototaxis. These findings suggest that these genes might be involved in the functional recovery, and morphological regeneration and functional regeneration can be distinguished according to their respective gene expression alterations (Inoue et al., 2004).

5. Neurogenesis after selective neuronal lesioning

Recently, we established an experimental model system for selective neuronal elimination to analyze the neurogenesis after selective neuronal lesioning without amputation. For this, we employed 6-hydroxydopamine (6-OHDA)-induced lesioning. 6-OHDA is a cytotoxic substance that induces dopaminergic neuronal cell death, and is widely used for killing dopaminergic neurons and creating parkinsonian animal models (Ungerstedt & Arbuthnott,

1970; Schwarting & Huston, 1996; Nass et al., 2002; Parish et al., 2007). In rodents, the nigro-striatal dopaminergic system is acutely and selectively degenerated by 6-OHDA-microinjection into the substantia nigra, and never recovers the missing neurons (Ungerstedt & Arbuthnott, 1970; Schwarting & Huston, 1996). We succeeded in selective degeneration of dopaminergic neurons in planarians, like that in higher animals. Interestingly, we found that planarians can regenerate only the dopaminergic neurons within 14 days after 6-OHDA-incuded selective dopaminergic neural degeneration (Fig. 4A). Although it has been reported that dopaminergic neurons are also regenerated during the head regeneration process after decapitation (Nishimura et al., 2007a; Takeda et al., 2009), our findings with 6-OHDA are the first showing that planarians are able to regenerate dopaminergic neurons after the selective degeneration of only dopaminergic neurons in the brains of non-amputated animals (Nishimura et al., 2011). According to our observations, dopaminergic neurons were completely degenerated and this degeneration was accompanied by reductions of DA content and locomotion activity within 24 hours after

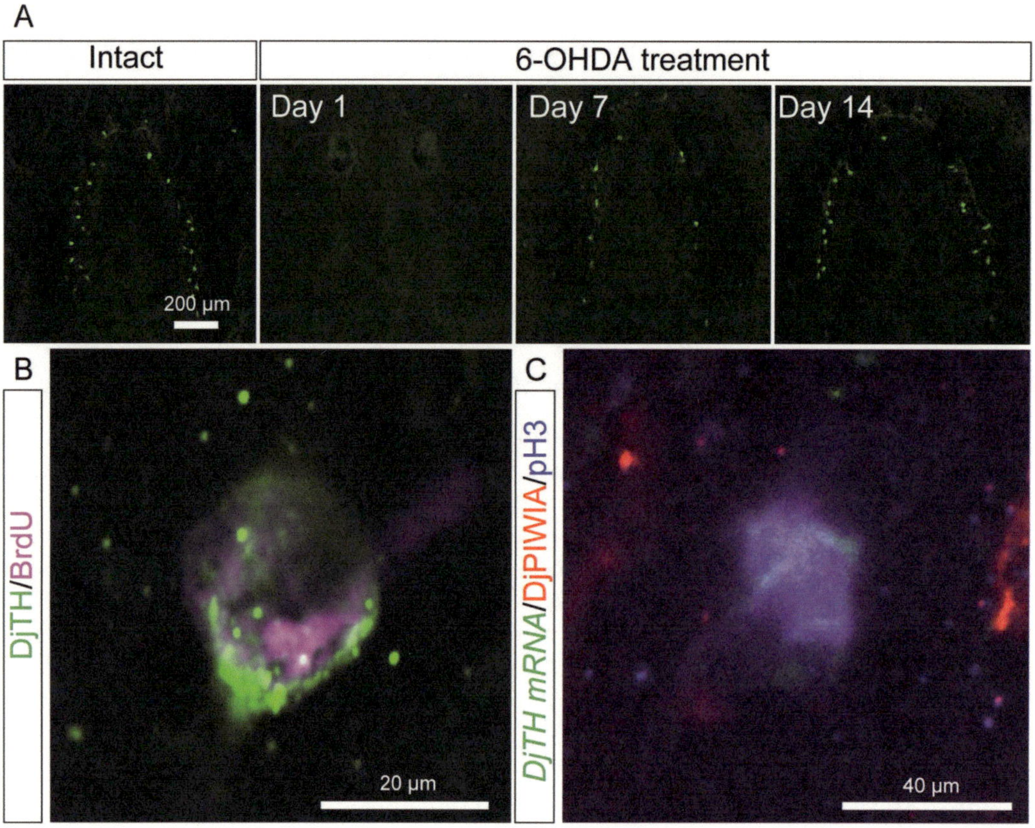

Fig. 4. Process of dopaminergic neurogenesis in the brain after 6-OHDA-induced-lesioning. Immunostaining of brain dopaminergic neurons in intact planarian and 1 day, 7 days, and 14 days after 6-OHDA-administration (A). BrdU-signal can be detected in newly generated dopaminergic neurons 10 days after 6-OHDA-administration (B). Newly generated dopaminergic neurons are produced from stem cells via cell division (C).

6-OHDA-administration. Then, newly generated dopaminergic neurons began to be detected in the brain 4 days after the 6-OHDA-induced lesion. Thereafter, the numbers and axons of dopaminergic neurons gradually recovered over a period of several days. Finally, dopaminergic neurons were completely recovered within 14 days after the 6-OHDA-induced lesion. We confirmed that in this process (1) X-ray-irradiated planarians never regenerate dopaminergic neurons after the 6-OHDA-induced lesion, (2) newly generated dopaminergic neurons are derived from pluripotent stem cells, as demonstrated by long-term trace experiments using BrdU. The dopaminergic neurogenesis after selective degeneration can be divided into three steps: (i) selective dopaminergic neurodegeneration (~24 hr after 6-OHDA-induced lesion), (ii) a transition period (24~72 hr), (iii) dopaminergic neurogenesis and dopaminergic neural network regeneration (96 hr~).

5.1 Recruitment of new dopaminergic neurons from pluripotent stem cells

Long-term chase experiments after BrdU-labeling clearly demonstrated that newly generated dopaminergic neurons are derived from proliferative stem cells. However, a BrdU-pulse chase analysis revealed that BrdU-incorporating cells were detected only in the trunk region but not around the brain region at all. In addition, immunohistochemical analysis using anti-proliferating cell nuclear antigen (PCNA) antibody revealed that PCNA-positive cells were never observed around the brain region (Orii et al., 2005). These results support the notion that essentially no proliferating stem cells that enter S-phase exist around the brain region. Thus, BrdU-positive cells detected in the brain by long term-chase experiments may migrate from the trunk region after proliferation (Newmark & Sánchez Alvarado 1999) (Fig. 4B). Therefore, we carefully investigated when proliferating stem cells are committed to differentiate into dopaminergic neurons during regeneration. Finally, we found that G2 phase stem cells are committed around the brain area to differentiate into dopaminergic neurons after lesioning. The most critical result was obtained by triple staining experiments immunostaining with anti-DjPIWIA antibody and anti-pH3 antibody and *in situ* hybridization using *a planarian tyrosine hydroxylase homologue (DjTH)* riboprobe. We detected *DjTH* mRNA/DjPIWIA protein/pH3-triple positive cells around the brain (Fig. 4C), suggesting that G2 phase stem cells may be accumulated in the head region and that these cells may participate in both regeneration and homeostatic events of the brain. It has already been suggested that the pluripotent stem cells may be committed at G2 phase into appropriate cell types (Hayashi et al., 2010), consistent with dividing stem cells immediately starting to differentiate to dopaminergic neurons. Based on these observations, we speculate that after proliferating in the trunk region, stem cells may migrate into the head region at G2 phase and then some of them might become committed to producing dopaminergic neurons (Nishimura et al., 2011).

5.2 System for recognition of the ablation of dopaminergic neurons

In planarians, it is known that older differentiated cells are constantly eliminated by apoptosis, and are then replaced by new cells by proliferation of stem cells under physiological conditions in planarians (Inoue et al., 2007; Pellettieri & Sánchez Alvarado, 2007). In our observation, a few BrdU-positive dopaminergic neurons were detected in vehicle-control-injected planarians, indicating that dopaminergic neurons could be replaced by stem cell proliferation in physiological conditions via homeostasis. Importantly, 6-

OHDA-induced lesioning accelerated the number and rate of the brain dopaminergic neurogenesis compared to that under physiological conditions in planarians. These results suggest that the number of dopaminergic neurons might be monitored by their surrounding environment. In the case of newts, a lower vertebrate, neurogenic potential for the repair of lost dopaminergic neurons is maintained even in adults (Parish et al., 2007), and this potential may work under conditions of injury-responsive cell-replacement that are induced by dopaminergic signals mediated by the DA receptor, but not under homeostatic conditions (Berg et al., 2010, 2011). In contrast, rodents have neural stem cells in restrict regions. It is known that the activity (proliferation and migration) of endogenous neural stem cells is enhanced in response to acute brain lesions caused by insults such as stroke and neurotoxin-exposure in the adult state (Arvidsson et al., 2002, Höglinger et al., 2004), suggesting that neural stem cells present in the adult brain can be responsive to alterations of the surrounding environment. In the future, it will be possible to identify the cellular and molecular systems that contribute to the recognition of dopaminergic ablation and the recruitment of new dopaminergic neurons, and it will become possible to use RNAi-mediated gene-knockdown and pharmacological drugs to further clarify the regulatory system of dopaminergic neurogenesis/regeneration.

6. Characterization of stem cell participation in brain regeneration

In both types of regeneration processes (*i.e.*, dopaminergic neurogenesis during brain regeneration and after selective degeneration of dopaminergic neurons), we have never observed the neural stem cell-like cells in planarians. Although commitment occurs at G2 phase, one committed stem cell produces only two differentiated cells. Committed stem cells can never enter into S phase after mitosis. Thus, we speculate that planarians have not yet invented a neural stem cell system. Histological analysis during regeneration supported the notion that pluripotent stem cells may directly give rise to fully differentiated neurons. First, we never observed proliferating cells in the brain rudiment during brain regeneration or in the intact brain. Second, the expression of the planarian *musashi* family genes supports the above hypothesis. *Musashi*, an RNA binding protein, is expressed in neural stem cells and/or progenitor cells in various animals (Okano et al., 2002). We isolated three *musashi*-like genes (*DjmlgA*, *DjmlgB* and *DjmlgC*) from planarians (Higuchi et al., 2008). Although they were expressed in the planarian CNS, their expression was not eliminated by X-ray irradiation, indicating that these genes were expressed after cells entered the differentiated state, not in the proliferative stem cells. Based on these observations, we hypothesized that the neural stem cell system probably evolved at a later stage of evolution independently in higher animals such as insects and vertebrates (Agata *et al.* 2006).

In the case of brain regeneration after decapitation, the brain rudiment is formed inside of the anterior blastema. The cells participating into blastema formation have already existed the proliferative state (Tasaki et al., 2011a, 2011b). A part of these cells then start to form the brain rudiment. Thus, commitment of dopaminergic neurons may occur after pattern formation of the brain. And then the neurons forming the primary brain might start to recruit G2 phase stem cells into brain neurons during enlargement of the brain and homeostasis (Takeda et al., 2009). In the case of dopaminergic neurogenesis after 6-OHDA-induced lesioning, G2 phase stem cells located around the brain may be recruited into dopaminergic neurons. The remaining neurons in the brain after 6-OHDA-induced lesioning may have an important role for sensing loss of dopaminergic neurons and recruiting G2 phase stem cells into dopaminergic neurons. Planarians thus have two different ways to regenerate dopaminergic neurons, although pluripotent stem cells become the source of regeneration in both cases. The latter case may provide a unique system for considering how to recruit dopaminergic neuron-committed cells into the lesioned regions (Nishimura et al., 2011). One of the important findings is that commitment occurs at the G2 phase of stem cells. We should consider to what extent committed cells can be incorporated into the lesioned regions, and whether the location of commitment is an important factor for future incorporation of committed stem cells into appropriate positions. As our future work, we will make an attempt to answer several important questions. "How do the remaining cells recognize the loss of dopaminergic neurons?" "How are surrounding stem cells recruited into dopaminergic neurons?" "What kind of signaling pathway(s) are activated in the G2 phase stem cells to differentiate dopaminergic neurons" "How do the committed cells find the pathways to the lesion points?" Answers to the above questions may provide hints about how to realize cell-transplantation therapy in the future.

7. Conclusion

It is difficult to analyze whether dopaminergic neurogenesis/neuroregeneration occurs in the adult mammalian midbrain (Zhao et al., 2003; Frielingsdorf et al., 2004), although it has been demonstrated that neurogenesis occurs in the restricted regions of the adult mammalian brain (Doetsch et al., 1997; Eriksson et al., 1998). However, it is still controversial whether dopaminergic neurogenesis/neuroregeneration potential is "lost" or "quiescent" in the adult mammalian midbrain. In any case, the potential for dopaminergic neurogenesis/neuroregeneration is not sufficient to recover the missing dopaminergic neurons in mammals. Our findings in planarians provide unique opportunities to consider how pluripotent stem cells respond to their surrounding environment, and how new dopaminergic neurons are recruited after the degeneration of dopaminergic neurons.

Cell-transplantation therapy is one possible way to compensate the missing dopaminergic neurons in Parkinson's disease patients. One of the important issues for cell-transplantation therapy is what state of dopaminergic neural precursor cells can be accepted into the host brain environment. For clinical application, non-regulated proliferative ability of donor cells may cause abnormal conditions such as tumor formation after grafting, and therefore, proliferative cells, including undifferentiated cells, should be eliminated as donor cells (Fukuda et al., 2006). Another approach would be to block proliferative ability artificially before grafting. Recently, it was demonstrated that N-[N-(3,5-difluorophenacetyl)-L-alanyl]-S-phenylglycine t-butyl ester (DAPT)-mediated Notch inhibition delays G1/S-phase transition of human ESC-derived neural stem cells, and promotes the onset of neuronal differentiation. However, the outcome of striatal transplantation of DAPT-treated neural stem cells was not different from that of non-DAPT-treated neural stem cells at a late period after grafting (Borghese et al., 2010). Consequently, inhibition of the G1/S-phase transition of donor cells to block proliferation may not enhance the efficiency of transplantation. Our findings from planarian studies suggest that G2-phase stem cells may be in a suitable cell state for harmonization with the host brain environment. Planarians are suitable model animals for analyzing the system that recognizes the ablation of dopaminergic signals and the system for recruitment of new dopaminergic neurons. Thus, our findings give useful suggestions about which state and type(s) of cells would be suitable for cell-replacement therapy with integration into the host brain environment using ESCs and/or iPSC-derived neural precursor cells to treat diseases such as Parkinson's disease.

8. Acknowledgments

We thank Dr. Elizabeth Nakajima (gCOE program A06, Kyoto University) for critical reading of the manuscript. This study was supported in part by a Grant-in-Aid for Global COE Program A06, Scientific Research on Innovative Areas to K. A. (22124001), Creative Scientific Research to K. A. (17GS0318) from the Ministry of Education, Culture, Sports, Science and Technology of Japan, and the Naito Foundation.

9. References

Agata, K.; Soejima, Y.; Kato, K.; Kobayashi, C.; Umesono, Y. & Watanabe K. (1998) Structure of the planarian central nervous system (CNS) revealed by neuronal cell markers. *Zoological Science,* Vol.15, No.3, (Jun 1998), pp. 433-440, ISSN 0289-0003

Agata, K.; Nakajima, E.; Funayama, N; Shibata, N; Saito, Y. & Umesono Y. (2006) Two different evolutionary origins of stem cell systems and their molecular basis. *Seminars in Cell & Developmental Biology,* Vol.17, No.4, (August 2006), pp. 503-509, ISSN 1084-9521

Agata, K. & Umesono, Y. (2008) Brain regeneration from pluripotent stem cells in planarian. *Philosophical Transactions The Royal Society B Biological Sciences,* Vol.363, No.1500, (Jun 2008), pp. 2071-2078, ISSN 1471-2970

Arvidsson, A.; Collin, T.; Kirik, D.; Kokaia, Z. & Lindvall, O. (2002) Neuronal replacement from endogenous precursors in the adult brain after stroke. *Nature Medicine,* Vol.8, No.9, (September 2002), pp. 963-970, ISSN 1078-8956

Borghese, L.; Dolezalova, D.; Opitz, T.; Haupt, S.; Leinhaas, A.; Steinfarz, B.; Koch, P.; Edenhofer, F.; Hampl, A. & Brüstle, O. (2010) Inhibition of notch signaling in human embryonic stem cell-derived neural stem cells delays G1/S phase transition and accelerates neuronal differentiation *in vitro* and *in vivo*. *Stem Cells,* Vol.28, No.5, (May 2010), pp. 955-964, ISSN 1066-5099

Berg, D.A.; Kirkham, M.; Beljajeva, A.; Knapp, D.; Habermann, B.; Ryge, J.; Tanaka, E.M. & Simon, A. (2010) Efficient regeneration by activation of neurogenesis in homeostatically quiescent regions of the adult vertebrate brain. *Development,* Vol.137, No.24, (December 2010), pp. 4127-4134, ISSN 0950-1991

Berg, D.A.; Kirkham, M.; Wang, H.; Frisén, J. & Simon, A. (2011) Dopamine controls neurogenesis in the adult salamander midbrain in homeostasis and during regeneration of dopamine neurons. *Cell Stem Cell,* Vol.8, No.4, (April 2011), pp. 426-433, ISSN 1934-5909

Björklund, L.M.; Sánchez-Pernaute, R.; Chung, S.; Andersson, T.; Chen, I.Y.; McNaught, K.S.; Brownell, A.L.; Jenkins, B.G.; Wahlestedt, C.; Kim, K.S. & Isacson, O. (2002) Embryonic stem cells develop into functional dopaminergic neurons after transplantation in a Parkinson rat model. *National Academy of Sciences of the United States of America,* Vol.99, No.4, (February 2002), pp. 2344-2349. ISSN 0027-8424

Brose, K.; Bland, K.S.; Wang, K.H.; Arnott, D.; Henzel, W.; Goodman, C.S.; Tessier-Lavigne, M. & Kidd, T. (1999) Slit proteins bind Robo receptors and have an evolutionarily conserved role in repulsive axon guidance. *Cell,* Vol.96, No.6, (March 1999), pp. 795-806, ISSN 0092-8674

Chambers, S.M.; Fasano, C.A.; Papapetrou, E.P.; Tomishima, M.; Sadelain, M. & Studer, L. (2009) Highly efficient neural conversion of human ES and iPS cells by dual inhibition of SMAD signaling. *Nature Biotechnology,* Vol.27, No.3, (May 2009), pp. 275-280, ISSN 1087-0156

Cebrià, F.; Kudome, T.; Nakazawa, M.; Mineta, K.; Ikeo, K.; Gojobori, T. & Agata, K. (2002a) The expression of neural-specific genes reveals the structural and molecular complexity of the planarian central nervous system. *Mechanisms of Development,* Vol.116, No.1-2, (August 2002), pp. 199-204, ISSN 0925-4773

Cebrià, F.; Kobayashi, C.; Umesono, Y.; Nakazawa, M.; Mineta, K.; Ikeo, K.; Gojobori, T.; Itoh, M.; Taira, M.; Sánchez Alvarado, A. & Agata, K. (2002b) FGFR-related gene *nou-darake* restricts brain tissues to the head region of planarians. *Nature,* Vol.419, No.6907, (October 2002), pp. 620-624, ISSN 0028-0836

Cebrià, F.; Nakazawa, M.; Mineta, K.; Ikeo, K.; Gojobori, T. & Agata, K. (2002c) Dissecting planarian central nervous system regeneration by the expression of neural-specific genes. *Development Growth & Differentiation,* Vol.44, No.2, (April 2002), pp. 135-146, ISSN 0012-1592

Cebrià, F. & Newmark, P.A. (2005) Planarian homologs of *netrin* and *netrin receptor* are required for proper regeneration of the central nervous system and the maintenance of nervous system architecture. *Development,* Vol.132, No.16, (August 2005), pp. 3691-3703, ISSN 0950-1991

Cebrià, F, Guo, T, Jopek, J. & Newmark, P.A. (2007) Regeneration and maintenance of the planarian midline is regulated by a *slit* orthologue. *Developmental Biology,* Vol.307, No.2, (July 2007), pp. 394-406, ISSN 0012-1606

Cebrià, F. & Newmark, P.A. (2007) Morphogenesis defects are associated with abnormal nervous system regeneration following *roboA* RNAi in planarians. *Development,* Vol.134, No.5, (March 2007), pp. 833-837, ISSN 0950-1991

Cebrià, F. (2008) Organization of the nervous system in the model planarian *Schmidtea mediterranea*: an immunocytochemical study. *Neuroscience Resersch,* Vol.61, No.4, (August 2008), pp. 375-384, ISSN 0168-0102

Couch, J.A.; Chen, J.; Rieff, H.I.; Uri, E.M. & Condron, B.G. (2004) *robo2* and *robo3* interact with eagle to regulate serotonergic neuron differentiation. *Development,* Vol.131, No.5, (March 2004) pp. 997-1006, ISSN 0950-1991

Doetsch, F.; García-Verdugo, J.M. & Alvarez-Buylla, A. (1997) Cellular composition and three-dimensional organization of the subventricular germinal zone in the adult mammalian brain. *The Journal of Neuroscience,* Vol.17, No.13, (July 1997), pp. 5046-5061, ISSN 270-6474

Eisenhoffer, G.T.; Kang, H. & Sánchez Alvarado, A. (2008) Molecular analysis of stem cells and their descendants during cell turnover and regeneration in the planarian *Schmidtea mediterranea. Cell Stem Cell,* Vol.3, No.3, (September 2008) pp. 327-339, ISSN 1934-5909

Eriksson, P.S.; Perfilieva, E.; Björk-Eriksson, T.; Alborn, A.M.; Nordborg, C.; Peterson, D.A. & Gage, F.H. (1998) Neurogenesis in the adult human hippocampus. *Nature Medicine,* Vol.4, No.11, (November 1998), pp. 1313-1317, ISSN 1078-8956

Fukuda, H.; Takahashi, J.; Watanabe, K.; Hayashi, H.; Morizane, A.; Koyanagi, M.; Sasai, Y. & Hashimoto, N. (2006) Fluorescence-activated cell sorting-based purification of embryonic stem cell-derived neural precursors averts tumor formation after transplantation. *Stem Cells,* Vol.24, No.3, (March 2006), pp. 763-771, ISSN 1066-5099.

Gaviño, M.A. & Reddien, P.W. (2011) A Bmp/Admp regulatory circuit controls maintenance and regeneration of dorsal-ventral polarity in planarians. *Current Biology,* Vol.21, No.4, (February 2011), pp.294-299, ISSN 0960-9822

Gurley, K.A.; Rink, J.C. & Sánchez Alvarado, A. (2008) *ß-catenin* defines head versus tail identity during planarian regeneration and homeostasis. *Science,* Vol.319, No.5861, (January 2008), pp. 323-327, ISSN 0036-8075

Hayashi, T.; Asami, M.; Higuchi, S.; Shibata, N. & Agata K. (2006) Isolation of planarian X-ray-sensitive stem cells by fluorescence-activated cell sorting. *Development Growth & Differentiation,* Vol.48, No.6, (August 2006), pp. 371-380. ISSN 0012-1592

Hayashi, T.; Shibata, N.; Okumura, R.; Kudome, T.; Nishimura, O.; Tarui, H. & Agata, K. (2010) Single-cell gene profiling of planarian stem cells using fluorescent activated cell sorting and its "index sorting" function for stem cell research. *Development Growth & Differentiation,* Vol.52, No.1, (January 2010), pp. 131-144, ISSN 0012-1592

Hendzel, M.J.; Wei, Y.; Mancini, M.A.; Van Hooser, A.; Ranalli, T.; Brinkley, B.R.; Bazett-Jones, D.P. & Allis, C.D. (1997) Mitosis-specific phosphorylation of histone H3 initiates primarily within pericentromeric heterochromatin during G_2 and spreads in an ordered fashion coincident with mitotic chromosome condensation. *Chromosoma,* Vol.106, No.6, (August 2006), pp. 348-360, ISSN 0009-5915

Heikkila, R.E. & Nicklas, W.J. (1971) Inhibition of biogenic amine uptake by hydrogen peroxide: mechanism for toxic effects of 6-hydroxydopamine. *Science,* Vol.172, No.989, (Jun 1971), pp. 1257-1258, ISSN 0036-8075

Higuchi, S.; Hayashi, T.; Hori, I.; Shibata, N.; Sakamoto, H. & Agata K. (2007) Characterization and categorization of fluorescence activated cell sorted planarian stem cells by ultrastructural analysis. *Development Growth & Differentiation,* Vol.49, No.7, (September 2007), pp. 571-581, ISSN 0012-1592

Higuchi, S.; Hayashi, T.; Tarui, H.; Nishimura, O.; Nishimura, K.; Shibata, N.; Sakamoto, H. & Agata, K. (2008) Expression and functional analysis of *musashi*-like genes in planarian CNS regeneration. *Mechanisms of Development,* Vol.125, No.7, (July 2008), pp. 631-645, ISSN 0925-4773

Höglinger, G.U.; Rizk, P.; Muriel, M.P.; Duyckaerts, C.; Oertel, W.H.; Caille, I. & Hirsch, E.C. (2004) Dopamine depletion impairs precursor cell proliferation in Parkinson disease. *Nature Neuroscience,* Vol.7, No.7, (July 2004), pp. 726-735, ISSN 1097-6256

Hong, K.; Hinck, L.; Nishiyama, M.; Poo, M.M.; Tessier-Lavigne, M. & Stein, E. (1999) A ligand-gated association between cytoplasmic domains of UNC5 and DCC family receptors converts netrin-induced growth cone attraction to repulsion. *Cell,* Vol.97, No.7, (Jun 1999), pp. 927-941, ISSN 0092-8674

Inoue, T.; Kumamoto, H.; Okamoto, K.; Umesono, Y.; Sakai, M., Sánchez Alvarado, A. & Agata, K. (2004) Morphological and functional recovery of the planarian photosensing system during head regeneration. *Zoological Science,* Vol.21, No.3, (March 2004), pp. 275-283, ISSN 0289-0003

Inoue, T.; Hayashi, T.; Takechi, K. & Agata, K. (2007) Clathrin-mediated endocytic signals are required for the regeneration of, as well as homeostasis in, the planarian CNS. *Development,* Vol.134, No.9, (May 2007), pp. 1679-1689, ISSN 0950-1991

Kato, K.; Orii, H.; Watanabe, K. & Agata, K. (1999) The role of dorsoventral interaction in the onset of planarian regeneration. *Development,* Vol.126, No.5, (February 1999), pp. 1031-1040, ISSN 0950-1991

Kawasaki, H.; Mizuseki, K.; Nishikawa, S.; Kaneko, S.; Kuwana, Y.; Nakanishi, S.; Nishikawa, S.I. & Sasai, Y. (2000) Induction of midbrain dopaminergic neurons from ES cells by stromal cell-derived inducing activity. *Neuron,* Vol.28, No.1, (October, 2000), pp. 31-40, ISSN 0896-6273

Kobayashi, C.; Saito, Y.; Ogawa K. & Agata K. (2007) *Wnt* signaling is required for antero-posterior patterning of the planarian brain. *Developmental Biology,* Vol.306, No.2, (Jun 2007), pp. 714-724, ISSN 0012-1606

Mineta, K.; Nakazawa, M.; Cebrià, F.; Ikeo, K.; Agata, K. & Gojobori, T. (2003) Origin and evolutionary process of the CNS elucidated by comparative genomics analysis of planarian ESTs. *National Academy of Sciences of the United States of America,* Vol.100, No.13, (Jun 2003), pp. 7666-7671, ISSN 0027-8424

Molina, M.D.; Neto, A.; Maeso, I.; Gómez-Skarmeta, J.L.; Saló, E. & Cebrià, F. (2011) Noggin and noggin-like genes control dorsoventral axis regeneration in planarians. *Current Biology,* Vol.21, No.4, (February 2011), pp. 300-305, ISSN 0960-9822

Morgan, T.H. (1989) Experimental Studies of the regeneration of *Planaria maculata. Arch Entwicklungsmech Org.*, Vol.7, pp. 364-397.

Morizane, A.; Doi, D.; Kikuchi, T.; Nishimura, K. & Takahashi, J. (2011) Small-molecule inhibitors of bone morphogenic protein and activin/nodal signals promote highly efficient neural induction from human pluripotent stem cells. *Journal of Neuroscience Research,* Vol.89, No.2, (February 2011), pp. 117-126, ISSN 0360-4012

Nakazawa, M.; Cebrià, F.; Mineta, K.; Ikeo, K.; Agata, K. & Gojobori T. (2003) Search for the evolutionary origin of a brain: planarian brain characterized by microarray. *Molecular Biology & Evolution,* Vol.20, No.5, (May 2003), pp. 784-791, ISSN 0737-4038

Nass, R.; Hall, D.H.; Miller, D.M. 3rd; & Blakely, R.D. (2002) Neurotoxin-induced degeneration of dopamine neurons in *Caenorhabditis elegans. National Academy of Sciences of the United States of America,* Vol.99, No.5, (March 2002), pp. 3264-3269, ISSN 0027-8424

Newmark, P.A. & Sánchez Alvarado, A. (2000) Bromodeoxyuridine specifically labels the regenerative stem cells of planarians. *Developmental Biology,* Vol.220, No.2, (April 2000), pp. 142-153, ISSN 0012-1606

Nishimura, K.; Kitamura, Y.; Inoue, T.; Umesono, Y.; Sano, S.; Yoshimoto, K.; Inden, M.; Takata, K.; Taniguchi, T.; Shimohama, S. & Agata, K. (2007a) Reconstruction of dopaminergic neural network and locomotion function in planarian regenerates. *Developmental Neurobiology,* Vol.67, No.8, (July 2007), pp. 1059-1078, ISSN 1932-8451

Nishimura, K.; Kitamura, Y.; Inoue, T.; Umesono, Y.; Yoshimoto, K.; Takeuchi, K.; Taniguchi, T. & Agata, K. (2007b) Identification and distribution of tryptophan hydroxylase (TPH)-positive neurons in the planarian *Dugesia japonica. Neuroscience Research,* Vol.59, No.1, (September 2007), pp. 101-106, ISSN 0168-0102

Nishimura, K.; Kitamura, Y.; Umesono, Y.; Takeuchi, K.; Takata, K.; Taniguchi, T. & Agata, K. (2008a) Identification of glutamic acid decarboxylase gene and distribution of GABAergic nervous system in the planarian, *Dugesia japonica. Neuroscience,* Vol.153, No.4, (Jun, 2008), pp. 1103-1114, ISSN 0306-4522

Nishimura, K.; Kitamura, Y.; Inoue, T.; Umesono, Y.; Taniguchi, T. & Agata, K. (2008b) Characterization of tyramine ß-hydroxylase in planarian *Dugesia japonica*: Cloning and expression. *Neurochemistry International,* Vol.53, No.6-8, (December 2008), pp. 184-192, ISSN 0197-0186

Nishimura, K.; Yamamoto, H.; Kitamura, Y.; & Agata, K. (2008c) Brain and Neural Network, In: *Planaria: A Model for Drug Action and Abuse*, (Raffa, R.B. & Rawls, S.M., (Ed.), 4-12 Landes Bioscience, ISBN: 978-1-58706-332-9, Texas, USA

Nishimura, K.; Kitamura, Y.; Taniguchi, T. & Agata, K. (2010) Analysis of motor function modulated by cholinergic neurons in planarian *Dugesia japonica. Neuroscience,* Vol.168, No.1, (Jun 2010), pp. 18-30, ISSN 0306-4522

Nishimura, K.; Inoue, T.; Yoshimoto, K.; Taniguchi, T.; Kitamura, Y. & Agata K. Regeneration of dopaminergic neurons after 6-hydroxydopamine-induced lesion in planarian brain. *Journal of Neurochemistry,* Vol.119, No.6, (December 2011), pp. 1219-1231 ISSN 0022-3042

Ogawa, K.; Ishihara, S.; Saito, Y.; Mineta, K.; Nakazawa, M.; Ikeo, K.; Gojobori, T.; Watanabe, K. & Agata K. (2002) Induction of a *noggin*-like gene by ectopic DV interaction during planarian regeneration. *Developmental Biology, Vol.*250, No.1, (October 2002), pp. 59-70, ISSN 0012-1606

Okamoto, K.; Takeuchi, K. & Agata, K. (2005) Neural projections in planarian brain revealed by fluorescent dye tracing. *Zoological* Science, Vol.22, No.5, (May 2005), pp. 535-546. ISSN 0289-0003

Okano, H.; Imai, T. & Okabe, M. (2002) *Musashi*: a translational regulator of cell fate. *Journal of Cell Science,* 2002 Apr Vol.115, No.Pt 7, (April 2002), pp. 1355-1359, ISSN 0021-9533

Orii, H., Sakurai, T. & Watanabe, K. (2005) Distribution of the stem cells (neoblasts) in the planarian *Dugesia japonica. Development Genes & Evolution,* Vol.215, No.3, (March 2005), pp. 143-157. ISSN 0949-944X

Parish, C.L.; Beljajeva, A.; Arenas, E. and Simon, A. (2007) Midbrain dopaminergic neurogenesis and behavioural recovery in a salamander lesion-induced regeneration model. *Development,* Vol.134, No.15, (August 2007), pp. 2881-2887, ISSN 0950-1991

Pellettieri, J. & Sánchez Alvarado, A. (2007) Cell turnover and adult tissue homeostasis: from humans to planarians. *Annual Review of Genetics,* Vol.41, pp. 83-105, ISSN 0066-4197

Perrier, A.L.; Tabar, V.; Barberi, T.; Rubio, M.E.; Bruses, J.; Topf, N.; Harrison, N.L. & Studer, L. (2004) Derivation of midbrain dopamine neurons from human embryonic stem cells. *National Academy of Sciences of the United States of America,* Vol.101, No.34, (August 2004), pp. 12543-12548, ISSN 0027-8424

Petersen, C.P. & Reddien, P.W. (2008) Smed-*ßcatenin-1* is required for anteroposterior blastema polarity in planarian regeneration. *Science,* Vol.319, No.5861, (January 2008), pp. 327-330, ISSN 0036-8075

Petersen, C.P. & Reddien, P.W. (2009) Wnt signaling and the polarity of the primary body axis. *Cell,* Vol.139, No.6, (December 2009), pp. 1056-1068, ISSN 0092-8674

Reddien, P.W. & Sánchez Alvarado, A. (2004) Fundamentals of planarian regeneration. *Annual Review of Cell Developmental Biology,* Vol.20, pp. 725-757, ISSN 1081-0706

Reddien, P.W.; Oviedo, N.J.; Jenning, J.R.; Jenkin, J.C. & Sánchez Alvarado, A. (2005) SMEDWI-2 is a PIWI-like protein that regulates planarian stem cells. *Science,* Vol.310, No.5752, (November 2005), pp. 1327-1330, ISSN 0036-8075

Reddien, P.W. (2011) Constitutive gene expression and the specification of tissue identity in adult planarian biology. *Trends in Genetics,* Vol.27, No.7, (July 2011), pp. 277-285, ISSN 0168-9525

Rink, J.C.; Gurley, K.A.; Elliott, S.A. & Sánchez Alvarado, A. (2009) Planarian Hh signaling regulates regeneration polarity and links Hh pathway evolution to cilia. *Science,* Vol.326, No.5958, (December 2009), pp. 1406-1410, ISSN 0036-8075

Sachs, C. & Jonsson, G. (1975) Mechanisms of action of 6-hydroxydopamine. *Biochemical Pharmacology,* Vo.24, No.1, (January 1975), pp. 1-8, ISSN 0006-2952

Saito, Y.; Koinuma, S.; Watanabe, K. & Agata, K. (2003) Mediolateral intercalation in planarians revealed by grafting experiments. *Developmental Dynamics,* Vol.226, No.2, (February 2003), pp. 334-340, ISSN 1058-8388

Salvetti, A.; Rossi, L.; Deri, P. & Batistoni, R. (2000) An MCM2-related gene is expressed in proliferating cells of intact and regenerating planarians. *Developmental Dynamics,* Vol.218, No.4, (August 2000), pp. 603-614, ISSN 1058-8388

Salvetti, A.; Rossi, L.; Lena, A.; Batistoni, R.; Deri, P.; Rainaldi, G.; Locci.; M.T.; Evangelista, M. & Gremigni, V. (2005) *DjPum,* a homologue of *Drosophila Pumilio,* is essential to planarian stem cell maintenance. *Development,* Vol.132, No.8, (April 2005), pp. 1863-1874, ISSN 0950-1991

Sakai, F.; Agata, K.; Orii, H. & Watanabe, K. (2000) Organization and regeneration ability of spontaneous supernumerary eyes in planarians -eye regeneration field and pathway selection by optic nerves-. *Zoological Science,* Vol.17, No.3, (April 2000), pp. 375-81, ISSN 0289-0003

Sanchez-Pernaute, R.; Lee, H.; Patterson, M.; Reske-Nielsen, C.; Yoshizaki, T.; Sonntag, K.C.; Studer, L. & Isacson, O. (2008) Parthenogenetic dopamine neurons from primate embryonic stem cells restore function in experimental Parkinson's disease. *Brain,* Vol.131, No. Pt.8, (August 2008), pp. 2127-2139, ISSN 0006-8950

Schwarting, R.K. & Huston, J.P. (1996) The unilateral 6-hydroxydopamine lesion model in behavioral brain research. Analysis of functional deficits, recovery and treatments. *Progress in Neurobiolgy,* Vol.50, No.2-3, (October 1996), pp. 275-331, ISSN 0301-0082

Shibata, N.; Umesono, Y.; Orii, H.; Sakurai, T.; Watanabe, K. & Agata, K. (1999) Expression of *vasa(vas)*-related genes in germline cells and totipotent somatic stem cells of planarians. *Developmental Biology,* Vol.206, No.1, (February 1999), pp. 73-87, ISSN 0012-1606

Shibata, N.; Rouhana, L. & Agata K. (2010) Cellular and molecular dissection of pluripotent adult somatic stem cells in planarians. *Development Growth & Differentiation,* Vol.52, No.1, (January 2010), pp. 27-41, ISSN 0012-1592

Tasaki, J.; Shibata, N.; Nishimura, O.; Itomi, K.; Tabata, Y.; Son, F.; Suzuki, N.; Araki, R.; Abe, M.; Agata, K. & Umesono, Y. (2011a) ERK signaling controls blastemacell differentiation during planarian regeneration. *Development,* Vol.138, No. 12, (Jun 2011), pp. 2417-2427, ISSN 0950-1991

Tasaki., J.; Shibata, N.; Sakurai, T.; Agata, K. & Umesono, Y. (2011b) Role of c-JunN-terminal kinase activity in blastema formation during planarian regeneration. *Development Growth & Differentiation,* Vol.53, No.3, (April 2011), 389-400, ISSN 0012-1592

Takagi, Y.; Takahashi, J.; Saiki, H.; Morizane, A.; Hayashi, T.; Kishi, Y.; Fukuda, H.; Okamoto, Y.; Koyanagi, M.; Ideguchi, M.; Hayashi, H.; Imazato, T.; Kawasaki, H.; Suemori, H.; Omachi, S.; Iida, H.; Itoh, N.; Nakatsuji, N.; Sasai, Y. & Hashimoto, N. (2005) Dopaminergic neurons generated from monkey embryonic stem cells function in a Parkinson primate model. *The Journal of Clinical Investigation,* Vol.115, No.1, (January 2005), pp. 102-109. ISSN 0021-9738

Takano, T.; Pulvers, J.N.; Inoue, T.; Tarui, H.; Sakamoto, H.; Agata, K. & Umesono, Y. (2007) Regeneration-dependent conditional gene knockdown (Readyknock) in planarian: demonstration of requirement for Djsnap-25 expression in the brain for negative phototactic behavior. *Development Growth & Differentiation,* Vol.49, No.5, (Jun 2007), pp. 383-394, ISSN 0012-1592

Takeda, H.; Nishimura, K. & Agata K. (Jun 2008) Planaria nervous system, In: *Scholarpedia*, 3, 5558, 2008, Available from http://www.scholarpedia.org/article/Planaria_nervous_system

Takeda, H.; Nishimura, K. & Agata K. (2009) Planarians maintain a constant ratio of different cell types during changes in body size by using the stem cell system. *Zoological Science*, Vol.26, No.12, (December 2009), pp. 805-813, ISSN 0289-0003

Tazaki, A.; Gaudieri, S.; Ikeo, K.; Gojobori, T.; Watanabe, K. & Agata, K. (1999) Neural network in planarian revealed by an antibody against planarian synaptotagmin homologue. *Biochemical and Biophysical Research Communications* Vol.260, No.2, (July 1999), pp. 426-432. ISSN 0006-291X

Umesono, Y.; Watanabe, K. & Agata, K. (1997) A planarian *orthopedia* homolog is specifically expressed in the branch region of both the mature and regenerating brain. *Development Growth & Differentiation*, Vol.39, No.6, (December 1997), pp. 723-727, ISSN 0012-1592

Umesono, Y.; Watanabe, K. & Agata, K. (1999) Distinct structural domains in the planarian brain defined by the expression of evolutionarily conserved homeobox genes. *Development Genes & Evolution*, Vol.209, No.1, (January 1999), pp. 31-39, ISSN 0949-944X

Umesono, Y.; & Agata, K. (2009) Evolution and regeneration of the planarian central nervous system. *Development Growth & Differentiation*, Vol.51, No.3, (April 2009), pp. 185-195, ISSN 0012-1592

Umesono, Y.; Tasaki, J.; Nishimura, K.; Inoue, T. & Agata, K. Regeneration in an evolutionarily primitive brain: the planarian *Dugesia japonica* model. *European Journal of Neuroscience.*,Vol.34, No.6, (September 2011), pp. 863-839. ISSN 0953-816X

Ungerstedt, U. & Arbuthnott, G.W. (1970) Quantitative recording of rotational behavior in rats after 6-hydroxy-dopamine lesions of the nigrostriatal dopamine system. *Brain Research*, Vol.24, No.3, (Decmber 1970), pp. 485-493, ISSN 0006-8993

Yamamoto, H. & Agata, K. (2011) Optic chiasm formation in planarian I: Cooperative *netrin*- and *robo*-mediated signals are required for the early stage of optic chiasm formation. *Development Growth & Differentiation*, Vol.53, No.3, (April 2011), pp. 300-311, ISSN 0012-1592

Yazawa, S.; Umesono, Y.; Hayashi, T.; Tarui, H. & Agata, K. (2009) Planarian Hedgehog/Patched establishes anterior-posterior polarity by regulating Wnt signaling. *National Academy of Sciences of the United States of America*, Vol.106, No.52, (December 2009), pp. 22329-22334, ISSN 0027-8424

Wagner, D.E.; Wang, I.E., & Reddien, P.W. (2011) Clonogenic neoblasts are pluripotent adult stem cells that underlie planarian regeneration. *Science*, Vol.332, No.6031, (May 2011), pp. 811-816, ISSN 0036-8075

Wenemoser, D. & Reddien, P.W. (2010) Planarian regeneration involves distinct stem cell responses to wounds and tissue absence. *Developmental Biology*, Vol.344, No.2, (August 2010), pp. 979-991, ISSN 0012-1606

Wernig, M.; Zhao, J.P.; Pruszak, J.; Hedlund, E.; Fu, D.; Soldner, F.; Broccoli, V.; Constantine-Paton, M.; Isacson, O. & Jaenisch, R. (2008) Neurons derived from reprogrammed fibroblasts functionally integrate into the fetal brain and improve symptoms of rats with Parkinson's disease. *National Academy of Sciences of the United States of America*, Vol.105, No.15, (April 2008), pp. 5856-5861, ISSN 0027-8424

Part 3

Regulation of Neural Stem Cell Development

8

γ-Secretase-Regulated Signaling Mechanisms: Notch and Amyloid Precursor Protein

Kohzo Nakayama[1], Hisashi Nagase[2], Chang-Sung Koh[3] and Takeshi Ohkawara[1]
[1]*Department of Anatomy,*
[2]*Department of Immunology and Infectious Diseases,*
Shinshu University, School of Medicine,
[3]*Department of Biomedical Sciences,*
Shinshu University, School of Health Sciences, Matsumoto
Japan

1. Introduction

In *Drosophila*, Notch mutations lost a lateral signaling ability and produced a neurogenic phenotype, where cells destined to become epidermis switch fate and give rise to neural tissue (Artavanis-Tsakonas *et al.* 1995; Lewis 1998). Therefore, when Notch signaling was disrupted, too many neurons were generated. Notch attracted further interest because sel-12, which appears to facilitate the reception of signaling mediated by lin-12 (*C. elegans* Notch), was identified by screening for a suppressor of lin-12 gain-of-function mutation (Levitan and Greenwald 1995). Since sel-12 is thought to be a counterpart of human presenilin (PS), which is a catalytic component of γ-secretase and has been implicated in Alzheimer's disease (AD), it was thought that the Notch signaling pathway might have a close relation with AD. Thus, many scientists have investigated the relationship between Notch signaling and AD. As we focused below, it has become clear that the Notch signaling pathway is controlled by γ-secretase-mediated proteolysis.

Both Notch receptors and their ligands are evolutionally conserved single transmembrane-spanning proteins (type 1 transmembrane protein; amino terminus is extracellular and carboxyl terminus is cytoplasmic.) that control the fates of numerous cells in both invertebrates and vertebrates (Artavanis-Tsakonas *et al.* 1995; Artavanis-Tsakonas *et al.* 1999; Justice and Jan 2002). For example, Delta, a major Notch ligand, expressing cells inhibit the neural determination of neighboring Notch-expressing neural stem cells (NSCs) during neurogenesis (Nakayama *et al.* 2008a). In addition, it is well known that isoforms of Notch mediate somitogenesis, differentiation of lymphoid cells as well as differentiation of NSCs, and that dysregulation of Notch signaling causes developmental defects or cancer in mammals (Bolos *et al.* 2007).

The molecular mechanism of Notch signaling is quite unique in that it is controlled by proteolytic cleavage reactions (Artavanis-Tsakonas *et al.* 1999; Justice and Jan 2002). In the canonical Notch signaling pathway, ligands bind to the extracellular domain of Notch on neighboring cells, and trigger sequential proteolytic cleavage. Finally, the intracellular domain (ICD) of Notch (NICD) is released from the cell membrane by γ-secretase and translocates to the nucleus to modulate gene expression through binding to transcription factors. Therefore, γ-secretase plays a central regulatory role in Notch signaling. First, we give a detailed interpretation of Notch itself and Notch signaling as well as its role in differentiation of NSCs.

The Notch signaling pathway has long been believed to be mono-directional because ligands for Notch were generally considered unable to transmit signals into the cells expressing them (Fitzgerald and Greenwald 1995; Henderson *et al.* 1997). However, several groups have shown that Delta is cleaved sequentially by proteases, probably including ADAM and γ-secretase (Ikeuchi and Sisodia 2003; LaVoie and Selkoe 2003; Six *et al.* 2003), and ICD of Delta is released from the cell membrane and translocates to the nucleus (LaVoie and Selkoe 2003; Six *et al.* 2003). We have also shown that ICD of mouse Delta binds to Smads, which are transcription factors for TGF-β/Activin signaling pathway, and enhances transcription of specific genes required for neuronal differentiation (Hiratochi *et al.* 2007). These results suggest that Delta also has a signaling mechanism similar to Notch signaling. Thus, we also review this issue that the Notch-Delta signaling pathway is bi-directional and similar mechanisms regulated by γ-secretase are involved in both directions of the Notch-Delta signaling pathway in developing NSCs.

γ-Secretase was first identified as a protease that cleaves amyloid precursor protein (APP) within the transmembrane (TM) domain and produces Aβ peptides (Haass and Selkoe 1993), which are thought to be pathogenic in AD (Hardy 1997; Selkoe 2001). However, the physiological functions of γ-secretase have not been clarified (Kopan and Ilagan 2004; Selkoe and Wolfe 2007). Recently, it was demonstrated that more than 50 type 1 transmembrane proteins, including APP, Notch and Delta, are substrates for γ-secretase (McCarthy *et al.* 2009) and their ICDs are also released from the cell membrane, similar to Notch. These observations that the common enzyme, γ-secretase, modulates proteolysis and the turnover of putative signaling molecules have led to the attractive hypothesis that mechanisms similar to the Notch signaling pathway may contribute widely to γ-secretase-regulated signaling pathways (Koo and Kopan 2004; Nakayama *et al.* 2008a; Nakayama *et al.* 2011).

Interestingly, it has also been reported that ICD of APP (AICD), which is released from the cell membrane by γ-secretase, translocates to the nucleus (Cupers *et al.* 2001; Gao and Pimplikar 2001; Kimberly *et al.* 2001) and may function as a transcriptional regulator (Cao and Sudhof 2001; Guenette 2002). As the apoptotic potential of AICD has been demonstrated, it is likely that APP signaling induces cell death, which leads to AD.

To explore APP signaling, we established embryonic carcinoma P19 cell lines overexpressing AICD (Nakayama *et al.* 2008b). Although neurons were differentiated from these cell lines with all-*trans*-retinoic acid (RA) treatment, AICD expression induced neuron-specific apoptosis. The effects of AICD were restricted to neurons, with no effects observed on non-neural cells. Furthermore, we evaluated changes in gene expression induced by AICD during this process of neuron-specific cell death using DNA microarrays (Ohkawara *et al.* 2011). The results of microarray analysis indicated that AICD induces dynamic changes in the gene expression profile. Therefore, it is likely that APP also has a signaling mechanism and that AICD may play a role in APP signaling, which leads to AD.

Here, we focus on molecular mechanisms of the Notch-Delta signaling pathway in a bi-directional manner and discuss the possibility that γ-secretase-regulated mechanisms similar to the Notch-Delta signaling pathway may play a potential role in signaling events involving type 1 transmembrane proteins. In addition, we introduce the current topics of γ-secretase. We also discuss the possibility that APP signaling induces dynamic changes in gene expression, which may be closely correlated with AICD-induced neuron-specific apoptosis, leading to AD.

2. Notch

2.1 Notch and its ligands

The typical Notch gene encodes a 300-kD type 1 transmembrane protein with the large extracellular domain which contains about 36 tandem epidermal growth factor (EGF)-like repeats (Wharton *et al.* 1985). The 11th and 12th EGF-like repeats are necessary and sufficient for binding to its ligands in *Drosophila* (Rebay *et al.* 1991). NICD is also large and has six tandem ankyrin-like (CDC10) repeats (Wharton *et al.* 1985). The fundamental structures are well conserved throughout evolution, although the numbers of EGF-like repeats vary from 10 in *C. elegans* (Glp-1) (Yochem and Greenwald 1989) to 36 in *Drosophila* and some vertebrate Notch .

While *Drosophila* has only one Notch gene, four Notch isoforms (Notch1 to 4) have been found in mammals. TAN1 (Notch1), which is a first identified mammalian homolog of Notch, was cloned as a gene responsible for human T cell acute lymphoblastic leukemia (T-ALL) (Ellisen *et al.* 1991). Notch2 was also cloned as an oncogene of cat thymic lymphoma (Rohn *et al.* 1996). A mutation of the Notch3 gene causes cerebral autosomal dominant arteriopathy with subcortical infarcts and leukoencephalopathy (CADASIL) (Joutel *et al.* 1996), in which the main symptom is cerebral vascular disorder. Interestingly, Notch4 is a cellular counterpart of the oncogene of mouse mammary tumor virus (int3) and expresses in vascular endothelial cells (Sarkar *et al.* 1994).

While *Drosophila* has two different ligands Delta (Kopczynski *et al.* 1988) and Serrate (Fleming *et al.* 1990), two families of ligands, Delta family (Delta-like protein: Dll1, 3 and 4) (Bettenhausen *et al.* 1995; Dunwoodie *et al.* 1997; Shutter *et al.* 2000) and Jagged family (Jagged1 and 2) (Lindsell *et al.* 1995; Shawber *et al.* 1996), have been identified in mammals to date. *C. elegans* has two ligands, Lag-2 (Tax *et al.* 1994) and Apx-1 (Mello *et al.* 1994). The extracellular domains of all these ligands also contain variable numbers of EGF-like repeats; for example *Drosophila* Delta has nine, most vertebrate Deltas have eight, and *C. elegans* Lag-2 has two repeats. All of these ligands also share a single copy of a second cysteine-rich conservative motif called the DSL (Delta: Serrate: Lag-2) domain (Tax *et al.* 1994), which is essential for binding to Notch (Henderson *et al.* 1997). In addition, spondylocostal dysostosis (SCD), which is characterized by abnormal vertebral segmentation, is caused by mutations of Dll3 gene (Sparrow *et al.* 2002). Alagille syndrome, which is a multi-system disorder characterized by paucity of bile ducts and congenital heart disease, is associated with a Jagged1 mutation (Oda *et al.* 1997) as well as a Notch2 mutation (McCright *et al.* 2002).

ICDs of all ligands are relatively short compared to those of Notch and it was thought that none of ICDs of Notch ligands display any significant sequence similarity throughout evolution (Henderson *et al.* 1994). As described below, structural evidence supports the idea that ICDs of these ligands are non-functional. However, we have revealed that Delta homologues display significant sequence similarity, which is restricted to vertebrates, in their ICDs (Hiratochi *et al.* 2007). There is no homology between these vertebrate Deltas and *Drosophila* Delta. In addition, Dll3, a divergent type of Delta, does not show any homology to other Delta in ICD.

2.2 The molecular mechanism of the Notch signaling pathway

Fig. 1 shows a diagram of the Notch signaling pathway. In the canonical Notch signaling pathway, ligands bind to the extracellular domain of Notch on neighboring cells. Both Notch (Gupta-Rossi *et al.* 2004) and its ligands (Itoh *et al.* 2003) undergo ubiquitin-regulated internalization. Mind bomb (Mib) is essential for efficient activation of Notch signaling in this step. Mib is a RING-type E3 ubiquitin ligase that ubiquitylates ICDs of Notch ligands and promotes internalization of these ligands in a ubiquitination-dependent manner (Itoh *et al.* 2003). As a result of these reactions, conformations of Notch and its ligands may be changed by pulling each other to trigger sequential proteolytic cleavages called the regulated intramembrane proteolysis (RIP) mechanism (Brown *et al.* 2000). The RIP mechanism requires sequential cleavage steps to occur within the juxtamembrane (JM) and TM domains, and these steps are carried out by metalloproteases and γ-secretase, respectively (Selkoe and Kopan 2003). Since precise steps of Notch processing are recently made clear and those steps are very similar to that of APP, we mentioned about details of these processes in "3.2 Processing mechanisms of several γ-secretase substrates, such as APP, are very similar to that of Notch."

Finally, γ-secretase serves to release NICD from the cell membrane to the cytoplasm and released NICD translocates to the nucleus. Thus, γ-secretase plays a central role in the regulation of Notch signaling. Although NICD has a nuclear localization signal and is accumulated in the nucleus as an activated form of Notch, mechanisms of the transport of NICD from cytoplasm to nucleus have not yet been clarified.

In the nucleus, NICD binds to transcription factors and controls expressions of certain genes. Members of the CSL family (CBF1/RBP-jκ in mammals, Su(H) in *Drosophila* and Lag-1 in *C. elegans*) are major downstream transcription factors of Notch signaling (Artavanis-Tsakonas *et al.* 1995; Kimble and Simpson 1997; Artavanis-Tsakonas *et al.* 1999). NICD binds to CSL transcription factors; six tandem ankyrin-like repeats lying in NICD are essential for binding to CSL transcriptional factors (Roehl *et al.* 1996). As NICD also binds to Mastermind-like proteins (MAML family in mammals) (Wu *et al.* 2000), the CSL-NICD-MAML complex is formed. As a result of forming these complexes, co-repressors are dispersed from CSL and co-activators such as P/CAF and P300 are recruited by these complexes (Wallberg *et al.* 2002). Therefore, the function of CSL complexes is converted from a transcriptional repressor to an activator. Finally, activated CSL complexes bind to the *cis*-acting DNA sequences of target genes and enhance the transcriptional activity of these genes.

The most established target genes for Notch signaling are Hes (Hairy/Enhancer of split in *Drosophila*) genes, which code for the basic helix-loop-helix (bHLH) transcriptional repressor for tissue-specific genes (Kageyama *et al.* 2007). Seven mammalian Hes, designated Hes1 to Hes7, have been identified to date, although the mouse does not have Hes4. Hes1 and Hes5 bind to their target DNA sequences called N box (CACNAG) by forming homodimers or heterodimers with Hey (Hes-related with YRPW motif) 1 or Hey2, and to recruit histone deacetylase (HDAC) activity by associating with Groucho, resulting in transcriptional repression (Akazawa *et al.* 1992; Leimeister *et al.* 1999; Iso *et al.* 2001). Moreover, they associate with E proteins which are ubiquitously expressed bHLH factors and prevent proneural bHLH factors, such as Neurogenin, from forming functional complexes with E protein (Kageyama *et al.* 2007). In this manner, Notch represses the differentiation of cells to specific lineages. In addition, Delta expression is induced by proneural genes that code for bHLH transcriptional factor, although multiple POU-binding factors are also important for Delta expression in mammalian NSCs (Nakayama *et al.* 2004). Thus, Notch signaling strongly inhibits Delta expression.

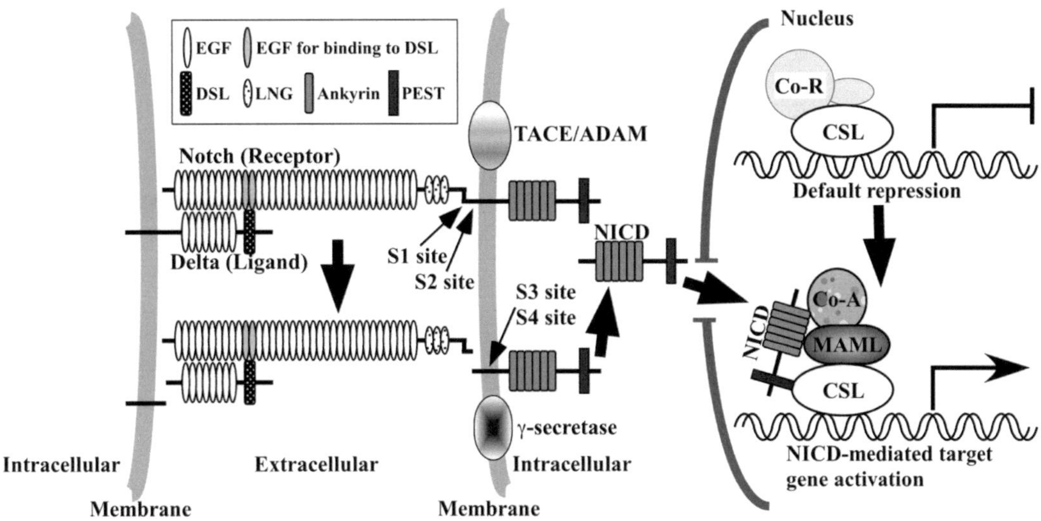

Notch proteins are expressed on the cell surface as heterodimers after cleavage at the S1 site by furin. The binding of Notch to the ligand triggers sequential proteolytic cleavage of RIP. When Notch binds to the ligand, Notch is cleaved at the S2 site in the juxtamembrane region by TACE or ADAM protease. Next, the remaining protein stub is further cleaved by γ-secretase at the S3 and S4 sites within the transmembrane domain and NICD is released from the membrane. Then, NICD translocates into the nucleus and binds to the CSL together with MAML. The resultant CSL-NICD-MAML complex removes co-repressors from CSL transcription factor and recruits a co-activator, resulting in conversion from repressor to activator. Finally, the complexes of CSL-NICD-MAML-co-activators promote transcription of the target genes

Fig. 1. Notch signaling pathway.

2.3 Notch signaling in the differentiation of NSCs

The definition of NSCs is that cells can self-renew and are capable of differentiating into main phenotypes of the nervous system, such as neurons, astrocytes and oligodendrocytes. In mammals, such cells have been isolated from the developing neural tube and more recently from the adult brain. Although there are some data showing that Notch signaling plays a role in the adult NSCs, a large majority of evidence for Notch signaling in controlling NSCs differentiation comes from analysis of embryonic neurogenesis.

In the developing mammalian central nervous system, NSCs repeat self-renewal by symmetric cell division to increase the total number of NSCs as a first step (NSCs/progenitor cells expansion phase). In this phase, Notch signaling is thought to maintain those NSCs in the proliferating and undifferentiating state (Fortini 2009; Kopan and Ilagan 2009; Pierfelice *et al.* 2011). Recently, it has been shown that expression of Hes1, the target genes for Notch signaling, oscillates with a period of about 2 hours in this phase (Kageyama *et al.* 2007). Hes1 expression may induce the oscillatory expression of Dll1 gene and the proneural Neurogenin2 (Ngn2) gene by periodic repression. Thus, concentrations of Dll1 mRNA and both Ngn2 mRNA and protein also oscillate with an inverse correlation with Hes1 (Kageyama *et al.* 2008; Shimojo *et al.* 2011). It is thought that Ngn2 cannot induce differentiation of neuron when Ngn2 expression oscillated and Dll1 leads to the activation of Notch signaling to maintain NSCs in proliferating state.

In next phase (neurogenic phase), NSCs undergo asymmetric cell division, where each NSC divides into two distinct types, NSC and neuron. In this phase, oscillatory expressions of Hes1 disappear (Kageyama *et al.* 2008; Shimojo *et al.* 2011). Since Hes1 expression is repressed, Dll1 and Ngn2 are constitutively expressed in a sustained manner and Ngn2 induces neuronal differentiation. Although the role of Notch signaling is not well understood, numb, which is an antagonist of Notch signaling (Frise *et al.* 1996; Guo *et al.* 1996; Spana and Doe 1996), is thought to be a critical component of NSCs asymmetrical division. During NSC divisions in this phase, numb appeared to be asymmetrically distributed to the neuronal daughter cells and was absent in undifferentiated NSCs (Zhong *et al.* 1996; Zhong *et al.* 1997; Chenn 2005). Thus, these observations suggest that numb inhibits Notch signaling and promotes differentiation to neuron in the neuronal daughter cells. After the generation of neurons, NSCs differentiate into oligodendrocytes and ependymal cells, followed by differentiation into astrocytes (gliogenic phase).

Recently, it has been shown that Notch signaling may also play an essential role in maintenance and differentiation of adult NSCs (Imayoshi and Kageyama 2011). Usually, adult NSCs are in the dormant state (quiescent state) and Notch signaling may maintain this state of adult NSCs. It is thought that NSCs turn from dormant state into dividing state, when the activity of Notch signaling falls down. Thus, Notch signaling controls the balance between dormant state and differentiation state of adult NSCs.

2.4 Delta signaling may be involved in neuronal differentiation

The Notch signaling pathway has long been thought to be mono-directional because ligands for Notch were generally considered unable to transmit signals into cells expressing these ligands (Henderson *et al.* 1994; Fitzgerald and Greenwald 1995). Indeed, it was thought that none of ICDs of putative Notch ligands display any significant sequence similarity throughout evolution (Henderson *et al.* 1994). Moreover, replacement of most of ICD of LAG-2, a *C. elegans* lin-12 (Notch) ligand, with a β-galactosidase fusion protein has no discernible effect on LAG-2 function (Henderson *et al.* 1994). In contrast, however, it has been reported that the extracellular domain of Notch expressed in the mesoderm provided a positive signal to the overlaying ectoderm in *Drosophila* as mentioned below (Baker and Schubiger 1996). Since these observations suggest that signaling in the opposite direction also exists, the important and critical question is whether signaling events occur not only from ligand-expressing cells to Notch-expressing cells but also vice versa, i.e., in a bi-directional manner.

Recently, evidence has been accumulating in support of a functional role of ICD of Notch ligands, which implies the existence of bi-directional signaling mechanisms. For example, Delta has been shown to release ICD from the cell membrane when cleaved by ADAM protease and γ-secretase (Qi *et al.* 1999; Ikeuchi and Sisodia 2003; LaVoie and Selkoe 2003; Six *et al.* 2003). Several groups have reported evidences supporting the nuclear localization of Delta ICD (Bland *et al.* 2003; LaVoie and Selkoe 2003; Six *et al.* 2003). These observations suggest that Delta ICD is released from the cell membrane by RIP. Indeed, we have shown that Delta homologues display significant sequence similarity, which is restricted to vertebrates, in their ICDs (Hiratochi *et al.* 2007). It is likely that conservation of these amino acid sequences reflect the functional importance of Delta ICD.

To clarify the question of whether the Notch-Delta signaling pathway is bi-directional, we investigated the effect of Notch on differentiation of NSCs isolated from mouse embryos (Hiratochi *et al.* 2007). When NSCs were co-cultured on a monolayer of mouse Dll1-expressing COS7 cells, the rate at which neurons emerged was lower than that in controls. As mentioned above, Notch signaling maintains the proliferating and undifferentiating state of NSCs and inhibits the differentiation into neurons (Fortini 2009; Kopan and Ilagan 2009; Pierfelice *et al.* 2011). Therefore, these observations indicate that Dll1 on COS7 cells generates signals to neighboring NSCs that express Notch and thus activates Notch signaling. Conversely, when NSCs were co-cultured on a monolayer of mouse Notch1-expressing COS7 cells, the rate of neurons developing from NSCs was significantly higher than that in control cultures. These results suggest that Notch1 on COS7 cells may also generate signals to neighboring NSCs and these ligands, probably Delta, may then transmit signals into cells expressing them to promote neuronal differentiation. Thus, signaling events may occur not only from Delta-expressing cells to Notch-expressing cells but also vice versa, that is, in a bi-directional manner, during differentiation of NSCs. Indeed, Baker and Schubiger published results from a mosaic experiment in *Drosophila*, which showed that expression of Notch in the mesoderm of Notch mutant suppressed the ectodermal defects of this mutant (Baker and Schubiger 1996). This effect was inferred to be due to the extracellular domain of the protein and not its signaling function, since activated Notch failed to produce non-autonomous suppression (Baker and Schubiger 1996). These results indicate that the extracellular domain of Notch expressed in the mesoderm sent a positive signal to the overlying ectoderm. Thus, these observations further support the hypothesis that Notch-expressing cells also send a signal to Delta-expressing cells.

2.5 Delta ICD may modify expression of certain genes

The nuclear localization of Delta ICD suggests that this domain may have effects on the transcription of a specific target gene similar to NICD. To examine this possibility, we searched for transcriptional factors capable of binding to ICD of Dll1 (Dll1IC) using a new method and identified Smads as a Dll1IC binding transcription factor through the differentiation process of mouse NSCs (Hiratochi *et al.* 2007).

Smads are transcription factors and have been shown to act as mediators of signaling by the TGF-β superfamily. Eight Smads, designated Smad1 to Smad8, have been identified to date in mammal (Miyazawa *et al.* 2002; Derynck and Zhang 2003). Smad2 and Smad3 are activated by TGF-β and activin (Eppert *et al.* 1996; Zhang *et al.* 1996; Nakao *et al.* 1997), while

Smad1 and Smad5 are major components that are activated by bone morphogenic proteins (BMPs) (Hoodless *et al.* 1996; Kretzschmar *et al.* 1997; Suzuki *et al.* 1997). Although Smad1 and Smad5 did not bind to Dll1IC, Smad2 and Smad3 showed strong binding (Hiratochi *et al.* 2007). These observations indicate that Dll1IC can modify TGF-β/Activin signaling through binding to Smad2 and/or Smad3. However, BMP signaling, which is known to inhibit neurogenesis and to enhance the appearance of astrocytes, may not be affected by Dll1IC, because Dll1IC did not bind to Smad1 or Smad5.

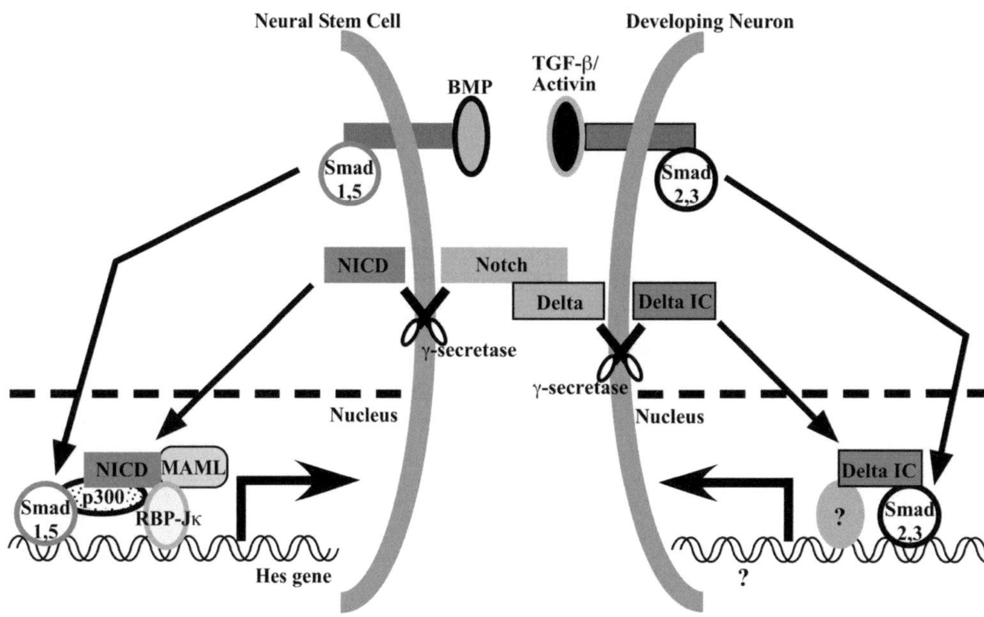

Notch receptor also generates signals to Delta expressed on the surface of neighboring NSCs. Delta is cleaved sequentially by proteases, probably including ADAM and γ-secretase, and finally the intracellular domain of Delta (DeltaIC) is released from the cell membrane and translocates to the nucleus, where it mediates TGF-β/Activin signaling through binding to Smad2/3 and enhances transcription of specific genes leading to neuronal differentiation. It is well known that NICD is also released from the cell membrane by proteases similar to the ones involved in the cleaving of Delta, then translocates to the nucleus to modulate gene expression through binding to the transcription factor, that is, Suppressor of Hairless (Su(H), RBP-jκ in mammals) together with MAML. This means that similar mechanisms are involved in both directions of the bi-directional Notch-Delta signaling pathway. BMPs, another group belonging to the TGF-β superfamily, have recently been shown to inhibit neurogenesis and to enhance the generation of astrocytes from NSCs. It has also been demonstrated that NICD and activated Smad1/5 form a complex with p300 in the specific promoter sequence, which contains both the RBP-jκ and Smad binding sequences. It is therefore possible that the TGF-β superfamily mediates both neurogenesis and gliogenesis from NSCs coupled with the bi-directional Notch-Delta signaling pathway.

Fig. 2. Schematic of the bi-directional model of Notch-Delta signaling pathway in the process of NSC differentiation.

Although we have yet to determine the actual target genes for the Dll1IC-Smad complex, we showed that binding of Dll1IC to Smad enhanced its transcriptional activity using the 9XCAGA-Luc promoter-reporter system that responds specifically to Smad3 (Dennler *et al.* 1998; Jonk *et al.* 1998), as a model system. These results strongly suggest that Dll1IC mediates transcription of certain genes, which are targets of TGF-β/Activin signaling, through binding to Smad2 and/or Smad3.

As mentioned above, it is likely that Delta transmits signals into NSCs expressing them to promote neuronal differentiation. To test this possibility, we established embryonic carcinoma P19 cells stably overexpressing Dll1IC (Hiratochi *et al.* 2007). Although control P19 cells have been shown to be induced to differentiate into neurons, RA stimulation is essential for the induction of neurons from these P19 cells. However, neurons could be induced from P19 cells stably overexpressing Dll1IC without RA stimulation and this induction was strongly inhibited by SB431542, a specific inhibitor of TGF-β type1 receptor (Laping *et al.* 2002) that activates Smad2 and Smad3. These results suggest that overexpression of Dll1IC in P19 cells induced neurons through binding to Smad2 and/or Smad3. Therefore, it is highly possible that Delta signaling also plays an important role in neuronal differentiation. Recently, it has been reported that TGF-β inhibits proliferation and accelerates differentiation of the hippocampal granule neuron (Lu *et al.* 2005). This observation also supports our hypothesis.

A schematic model of Notch-Delta signaling pathway in the process of NSC differentiation is shown in Fig. 2.

3. γ-Secretase

3.1 Overview of γ-secretase

γ-Secretase was first identified as a protease that cleaves APP within the TM domain and produces Aβ peptides (Haass and Selkoe 1993), which are thought to have pathogenic roles in AD. However, the physiological functions of this enzyme have not been clarified (Kopan and Ilagan 2004; Selkoe and Wolfe 2007). The γ-secretase is a complex composed of PS, nicastrin (NCT), anterior pharynx defective-1 (Aph-1), and PS enhancer-2 protein (Pen-2) (Iwatsubo 2004; Kopan and Ilagan 2004; Selkoe and Wolfe 2007). PS is a catalytic component of the γ-secretase complex, and the two PS genes, PS1 gene (*PSEN1*) (Sherrington *et al.* 1995) located on chromosome 14 and PS2 gene (*PSEN2*) (Levy-Lahad *et al.* 1995; Rogaev *et al.* 1995) located on chromosome 1, were identified by genetic linkage analyses as the genes responsible for several forms of early-onset familial AD (FAD). *PSEN1* and *PSEN2* encode polytopic transmembrane proteins of 467 and 448 amino acids, respectively, which show about 65% sequence identity between the two proteins. While PS1 expression level is higher than that of PS2, both proteins are expressed ubiquitously in the brain and peripheral tissues of adult mammals (Lee *et al.* 1996). The model for PS with eight or nine transmembrane domains is generally accepted and PS has a hydrophilic loop domain between the putative 6th and 7th transmembrane domains facing the cytoplasm (Doan *et al.* 1996) and is cleaved by an unidentified protease within this loop resulting into two fragments, N- and C-terminal fragment (NTF and CTF), that remain associated as a heterodimer (Thinakaran *et al.* 1996). This proteolytic cleavage is thought to occur when nascent PS assembles with NCT, Aph-1, and Pen-2 as a γ-secretase complex and activates PS as the catalytic component of aspartyl protease (Iwatsubo 2004; Kopan and Ilagan 2004; Selkoe and Wolfe 2007).

The single-pass membrane protein NCT may recognize the substrate proteins of γ-secretase (Yu *et al.* 2000; Shah *et al.* 2005). The extracellular domain of NCT resembles an aminopeptidase, but lacks catalytic residues, and can interact with the N-terminal stubs of γ-secretase substrates after ectodomain shedding (Shah *et al.* 2005). Thus, shedding of membrane proteins may be essential for the production of free N-termini of these proteins retained in the membrane, which can then be recognized by NCT. Aph-1 is thought to act as a scaffold during the process of γ-secretase complex assembly, and Pen-2 was suggested to act as a trigger for the proteolytic cleavage of PS to regulate PS activity (Kopan and Ilagan 2004; Selkoe and Wolfe 2007).

3.2 Processing mechanisms of several γ-secretase substrates, such as APP, are very similar to that of Notch

Precise steps of Notch processing are recently made clear (Fig.3). After translation, Notch is cleaved by furin-like covertase at the S1 site in the *trans*-Golgi network, and the two resulting fragments remain associated to form a functional heterodimer that is expressed on the cell surface (Logeat *et al.* 1998).

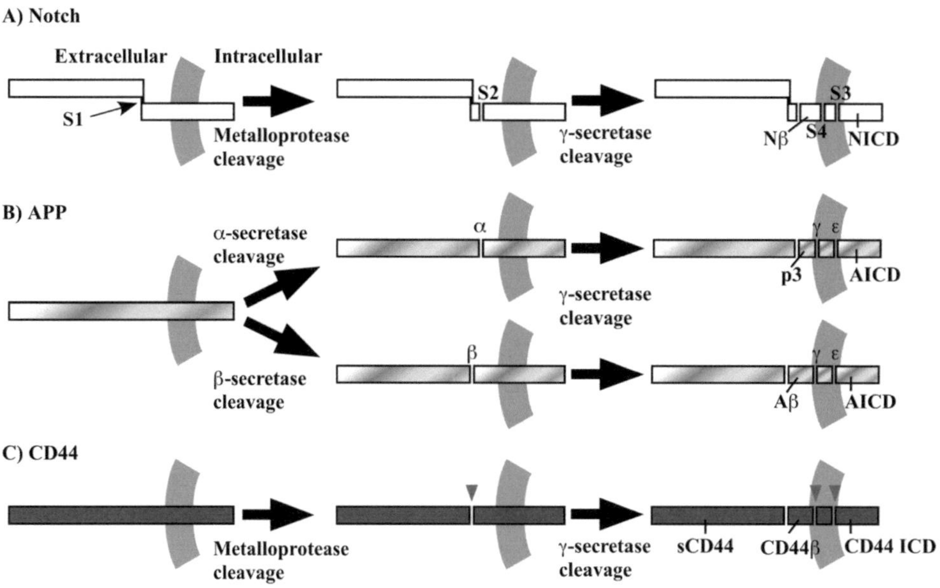

(A) In response to ligand binding, Notch undergoes shedding due to metalloprotease cleavage at the S2 site within the JM domain. After shedding the extracellular domain, the remaining Notch stub is further cleaved by γ-secretase at S3 and S4 sites within the TM domain. This sequential proteolysis produces NICD and Nβ fragment. (B) Cleavage of APP by α-secretase or β-secretase at the α-site or β-site, respectively, within the JM domain results in shedding of almost the entire extracellular domain and generates membrane-tethered α- or β-carboxy terminal fragments (CTFs). Several zinc metalloproteinases and BACE2 can cleave APP at the α-site, while BACE1 cleaves APP at the β-site. After shedding the extracellular domain, the remaining stub is further cleaved at least twice within the TM domain at γ- and ε-sites by γ-secretase, producing either p3 peptide (in combination with α-secretase) or Aβ (in combination with BACE1), respectively, and AICD. (C) Several stimuli, such as PKC activation and Ca^{2+} influx, trigger ectodomain cleavage of CD44 by a metalloprotease at the site within the JM domain, resulting in the secretion of soluble CD44 (sCD44). After shedding the extracellular domain, the remaining CD44 stub is further cleaved by γ-secretase at two sites within the TM domain. This sequential proteolysis produces the CD44 ICD and CD44β, an Aβ-like peptide.

Fig. 3. Similarities in the proteolytic processes among Notch, APP, and CD44.

As mentioned above, the sequential proteolytic cleavage called RIP mechanism is initiated by ligand binding and shedding at the S2 site by TACE or ADAM protease making the truncated Notch (Pan and Rubin 1997; Brou *et al.* 2000). Truncated Notch is further cleaved by γ-secretase in at least two sites within the TM domain, *i.e.*, at the S3 site to release NICD and at the S4 site to release the remaining small peptide (Nβ) (Kopan *et al.* 1996; Schroeter *et al.* 1998; Okochi *et al.* 2002), which resembles Aβ.

The proteolytic process of APP resembles that of Notch and also follows the RIP mechanism (Fig.3). Cleavage of APP by α-secretase (Esch *et al.* 1990) or β-secretase (Vassar *et al.* 1999) at the α-site or β-site, respectively, within the JM region results in shedding of almost the entire extracellular domain and generates membrane-tethered α- or β- CTFs. Several zinc metalloproteinases (Buxbaum *et al.* 1998; Lammich *et al.* 1999) and the aspartyl protease BACE2 can cleave APP at the α-site (Farzan *et al.* 2000), while BACE1 (β-site APP cleaving enzyme) cleaves APP at the β-site (Vassar *et al.* 1999). After shedding, the remaining stub is further cleaved at least twice by γ-secretase within the TM domain at γ- and ε-sites resulting in production of either non-amyloidogenic p3 peptide (in combination with α-secretase) or amyloidogenic Aβ (in combination with BACE1), respectively, and AICD (Kopan and Ilagan 2004; Selkoe and Wolfe 2007). In addition, AICD was shown to be a substrate of caspase and to be cleaved at the group III caspase consensus sequence 16 amino acids from the membrane border within the AICD.

It has been reported that several γ-secretase substrates also follow the RIP mechanism with release of their ICDs from the cell membrane. As shown in Fig. 3, the process of sequential proteolytic cleavage of CD44, which is important for immune system function, is very similar to those of Notch and APP and follows the RIP mechanism (Nagase *et al.* 2011). In addition, the ICD of this protein (CD44ICD) is also translocated to the nucleus, suggesting that CD44ICD may also mediate the gene expression.

3.3 Does γ-secretase mediate signaling events of type 1 transmembrane proteins?

γ-Secretase seems to cleave a diverse set of type1 transmembrane proteins, which have been shed their extracellular domains, in a sequence-independent manner (Struhl and Adachi 2000). As reflected by the flexible sequence specificity of γ-secretase activity, more than 50 type 1 transmembrane proteins have been reported as substrates of γ-secretase (McCarthy *et al.* 2009). As shown in Table 1, these substrates also have a wide range of functions, including roles in cell differentiation (Notch, Delta, and Jagged), cell adhesion (N-cadherin, E-cadherin, and CD44), synaptic adhesion (Nectin-1α), ion conductance regulation (voltage-gated sodium channel β2 subunit), axon guidance and tumor suppression (DCC), neurotrophin receptor (P75NTR), and its homolog (NRADD), lipoprotein receptor (ApoER2), and growth factor-dependent receptor tyrosine kinase (ERBB4).

As mentioned above, proteolytic cleavages of several γ-secretase substrates, such as APP and CD44, follow the RIP mechanism. The ICDs of these substrates are released from the cell membrane to cytoplasm by γ-secretase, and finally these ICDs translocate to the nucleus. These processes are very similar to those involved in Notch signaling. Thus, the observations that the common enzyme, γ-secretase, modulates proteolysis and the turnover of possible signaling molecules led to the signaling hypothesis suggesting that mechanisms similar to those occurring in the Notch signaling pathway may contribute widely to γ-secretase-regulated signaling pathways (Koo and Kopan 2004; Nakayama *et al.* 2008a; Nakayama *et al.* 2011).

Indeed, as mentioned above, Dll1 is cleaved sequentially by proteases, probably including ADAM and γ-secretase, and Dll1IC is released from the cell membrane and undergoes translocation to the nucleus (Hiratochi *et al.* 2007). In the nucleus, Dll1IC enhances transcription of specific genes through binding to Smads. These observations suggest that Dll1 may also have a signaling mechanism similar to Notch signaling.

Substrate	Function	PS or ICD function
ApoER2	Lipoprotein receptor, neuronal migration	Activate nuclear reporter
APP	Precursor to Aβ, adhesion, trophic properties, axonal transport?	Aβ generation, release of ICD, complex with Fe65/Tip60, Cell death?
APLP1/2	Cell adhesion?	Form complex with Fe65 and Tip60
E-cadherin	Cell adhesion	Promote disassembly of adhesion complex
N-cadherin	Cell adhesion	Promote CBP degradation
β-catenin	Transduce Wnt signals stabilize adherens junctions	Facilitate phosphorylation
CD43	Signal transduction	Signaling molecule?
CD44	Cell adhesion	Activate TRE-mediated nuclear transcription
CSF1-R	Protein tyrosine kinase	Unknown
CXCL16 & CX3CL1	Membrane chemokine ligands	Unknown
DCC	Axon guidance, tumor suppressor	Activate nuclear reporter
Delta	Notch ligand	Transcription regulation
ERBB4	Receptor tyrosine kinase	Regulate heregulin-induced growth inhibition
HLA-A2	MHC class I molecule	Unknown
IGIF-R	Receptor tyrosine kinase	Unknown
IFN-αR2	Subunit of type I IFN-α receptor	Transcriptional regulation
IL-1RI	Cytokine receptor	Unknown
IL-1RII	Cytokine receptor	Unknown
Jagged	Notch ligand	Modulate AP-1 mediated transcription
LDLR	Lipoprotein receptor	Unkown
LRP	Scavenger and signaling receptor	Activate nuclear reporter
Na channel β-subunit	Cell adhesion, an auxiliary subunit of voltage-gated Na channel	Alter cell adhesion and migration
Nectin-1α	Adherens junction, synapse receptor	Remodeling of cell junctions?

Substrate	Function	PS or ICD function
Notch1-4	Signaling receptor	Transcription regulation
NRADD	Apoptosis in neuronal cells	Modulate glycosylation/matutaion of NRADD
P75NTR	Neurotrophin co-receptor, dependence receptor	Modulate p75-TrkA complex? Nuclear singaling?
γ-protocadherin	Cell adhesion, neuronal differentiation	Regulation of gene transcription?
Syndecan-3	Cell surface proteoglycan co-receptor	Regulation of membrane-targeting of CASK
Telencephalin	Cell adhesion	Turnover of telencephalin
Tyrosinase, Tyrosinase-related protein 1/2	Pigment synthesis	Intracellular transport of Post-Golgi Tyr-containing vesicles

PS, presenilin; ICD, intracellular domain; APLP, APP like protein; CBP, CREB (cAMP-responsive element binding protein)-binding protein; TRE, TPA (12-o-tetradecanoylphorbol 13-acetate)-responsive element; AP-1, activator protein-1; CASK, calmodulin-dependent serine kinase; Tyr, Tyrosinase.

Table 1. Substrates for γ-secretase

3.4 Is γ-secretase a proteasome of the membrane?

As mentioned above, more than 50 type 1 membrane proteins have been reported as substrates of γ-secretase. This observation raises the simple question of why so many membrane proteins can transmit signals to the nucleus. In contrast to the signaling hypothesis, Kopan and Ilagan proposed another possibility that γ-secretase may act as a proteasome for membrane proteins (Kopan and Ilagan 2004). They pointed out that generally the ICDs of these substrates including AICD, which are released by γ-secretase, are rapidly degraded. Moreover, ectodomain shedding seems to be constitutive for some substrates, and ligand binding has been reported to enhance only intramembrane cleavage of Notch (Schroeter *et al.* 1998), Delta (Hiratochi *et al.* 2007), Syndecan-3 (Schulz *et al.* 2003), and ERBB4 (Ni *et al.* 2001). In addition, they also pointed out that the most evidence supporting the signaling hypothesis was obtained in overexpression experiments that differ somewhat from physiological conditions. Based on these observations, they proposed the proteasome hypothesis that the primary function of γ-secretase is to facilitate the selective disposal of type 1 membrane proteins (Kopan and Ilagan 2004).

While the proteasome hypothesis of γ-secretase is reasonable, there is no doubt that γ-secretase regulates signaling pathways of some substrates, such as Notch (Artavanis-Tsakonas *et al.* 1999; Selkoe and Kopan 2003; Koo and Kopan 2004). Although further studies are required to elucidate this issue, it is likely that γ-secretases are not uniform complexes but that different γ-secretase complexes may exist in different combinations with components such as Aph-1, Pen2, and/or PS isoforms, with different cellular functions, such as roles in signaling or degradation (Kopan and Ilagan 2004). Since γ-secretase substrates such as APP are generally more abundant than transcription factors, which are usually rare molecules, it is uncertain whether the majority of the ICDs of these substrates released by γ-secretase are required for the signaling mechanisms (Nakayama *et al.* 2008a; Nakayama *et al.*

2011). Although a large proportion of ICDs of these substrates are rapidly degraded, it is likely that a small amount of the remaining ICDs may be suitable for their functions with a small quantity of transcription factors. Thus, the greater part of ICDs of these substrates may be degraded and only a small proportion may play a role in signaling.

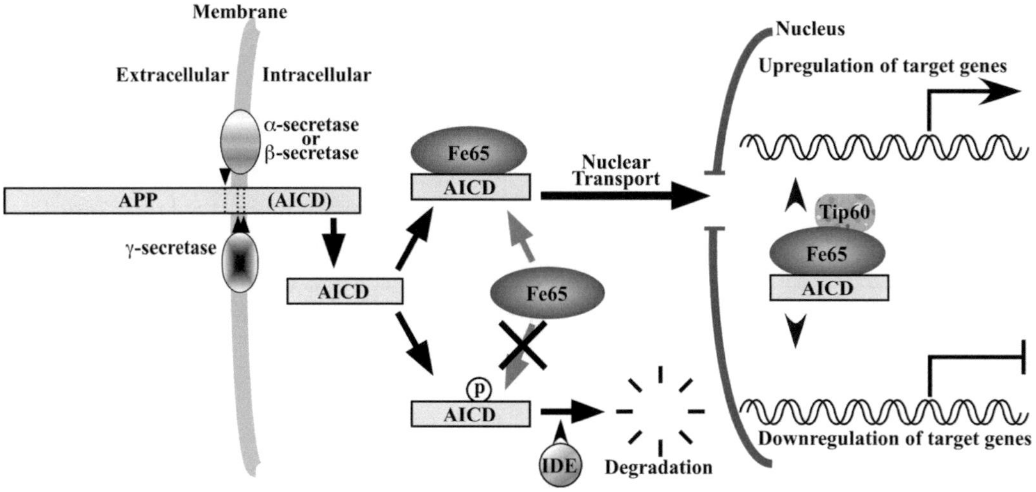

After cleavege of JM domain by α- or β-secretase, AICD is released from the membrane by γ-secretase. Non-phosphorylated AICD binds to the nuclear adaptor protein Fe65, which is thought to be essential for translocation of AICD to the nucleus, and forms complexes, alone or with the histone acetyltransferase Tip60. These complexes can immediately translocate to the nucleus, where they meidate up- and downregulation of certain target genes in association with Tip60. On the other hand, phospholylated AICD cannot translocate to the nucleus due to the inhibition of binding to Fe65, leading to rapid degradation by the proteasome and/or insulin-degrading enzyme (IDE).

Fig. 4. Putative APP signaling pathway.

In relation to this issue, an attractive model has been proposed (Fig.4) (Buoso *et al.* 2011). Binding to nuclear adaptor protein Fe65 is thought to be essential for translocation of AICD to the nucleus. In this model, since non-phosphorylated AICD binds to Fe65 and forms complexes, these complexes can immediately translocate to the nucleus, where they control the expression of certain genes in association with the histone acetyltransferase Tip60. On the other hand, as other stimuli induce phosphorylation of AICD, which strongly inhibits binding to Fe65, AICD without Fe65 cannot translocate to the nucleus. Phosphorylated AICD left in the cytosol is rapidly degraded, most likely by the proteasome and/or insulin-degrading enzyme (IDE) (Edbauer *et al.* 2002). Indeed, it has been reported that when phosphorylated at Thr^{668} in the APP-695 isoform, AICD cannot bind to Fe65 (Kimberly *et al.* 2005).

4. APP signaling?

4.1 Overview of APP

APP was first identified as a cDNA cloned using a partial amino acid sequence of Aβ fragment from the amyloid plaque of AD brains (Kang *et al.* 1987). APP is a type 1 membrane protein expressed in many tissues, especially concentrated in the synapses of neurons. In humans, the APP gene contains at least 18 exons in a total length of 240 kb (Yoshikai *et al.* 1990), and several alternative splicing isoforms of APP have been observed, differing mainly in the absence (APP-695 which is predominately expressed in neurons) or presence (APP-751 and APP-770) of a Kunitz protease inhibitor (KPI) domain located toward the N-terminus of the protein (Sisodia *et al.* 1993). As mentioned above, APP undergoes sequential proteolytic cleavage reactions to yield the extracellular fragment, intracellular fragment (AICD), and Aβ fragment located in the membrane-spanning domain, which is thought to be the main cause of the onset of AD.

While APP has central roles in AD (Hardy 1997; Selkoe 2001), the physiological functions of this protein also remain to be clarified (Zheng and Koo 2006). It has been reported that APP acts as a cell adhesion molecule for cell-cell interaction (Soba *et al.* 2005), and as a neurotrophic and/or synaptogenic factor (Hung *et al.* 1992; Bibel *et al.* 2004; Leyssen *et al.* 2005). In addition, the possibility that APP is a cell-surface receptor is interesting from the signaling perspective. Several evidences support this idea; *e.g.*, Aβ can bind to APP and thus may be a candidate ligand for APP (Lorenzo *et al.* 2000). It has also been reported that F-spondin (Ho and Sudhof 2004) and Nogo-66 receptor (Park *et al.* 2006) could bind to the extracellular domain of APP and regulate Aβ production. Furthermore, the extracellular domain of APP may potentially interact in *trans* suggesting that APP molecules can bind to each other (Wang and Ha 2004).

APP homologs show significant evolutionary sequence conservation in ICD (Nakayama *et al.* 2008b), which may reflect the functional importance of AICD. However, the Aβ region of this protein is not well conserved across species. As mentioned, AICD is thought to form complexes with Fe65 and these complexes translocate to the nucleus. In the nucleus, these complexes may associate with Tip60 and may bind to the *cis*-acting DNA sequence of the tetraspanin protein KAI1 gene to control transcriptional activity (Baek *et al.* 2002).

4.2 AICD induces neuron-specific apoptosis

There is accumulating evidence in support of the idea that APP signaling exists and contributes to the onset of AD. For example, transgenic mice overexpressing both AICD and Fe65 showed abnormal high activity of glycogen synthase kinase 3 beta (*Gsk3b* protein) (Ryan and Pimplikar 2005), leading to hyperphosphorylation and aggregation of TAU, resulting in microtubule destabilization, and reduction of nuclear β-catenin levels causing a loss of cell-cell contact mechanisms that may give rise to neurodegeneration in AD brain. In addition, it was also shown that c-Abl modulates AICD-dependent transcriptional induction, as well as apoptotic responses (Vazquez *et al.* 2009). Interestingly, elevated AICD levels have also been observed in AD brains (Ghosal *et al.* 2009). Therefore, it is highly possible that APP signaling changes expression patterns of certain genes and induces cell death, which may lead to AD pathology.

To explore APP signaling, we established several AICD-overexpressing embryonic carcinoma P19 cell lines (Nakayama *et al.* 2008b). Although neurons were differentiated from these cell lines after aggregation culture with RA treatment, AICD expression induced neuron-specific cell death. Indeed, while neurons from control cells which carried vector alone were healthy, almost all neurons from AICD-overexpressing P19 cells showed severe degeneration four days after induction of differentiation (Fig. 5). Moreover, DNA fragmentation was detected, and all of terminal deoxynucleotidyl transferase (TdT)-mediated deoxyuridine triphosphate (dUTP)-biotin nick end-labeling (TUNEL)-positive cells were also Tuj1-positive neurons. Based on these observations, we concluded that AICD can induce neuron-specific apoptosis (Nakayama *et al.* 2008b). The effects of AICD were restricted to neurons, with no effects observed on non-neural cells. Thus, although further studies are required, these results strongly suggest that AICD plays a role in APP signaling, which leads to the onset of AD.

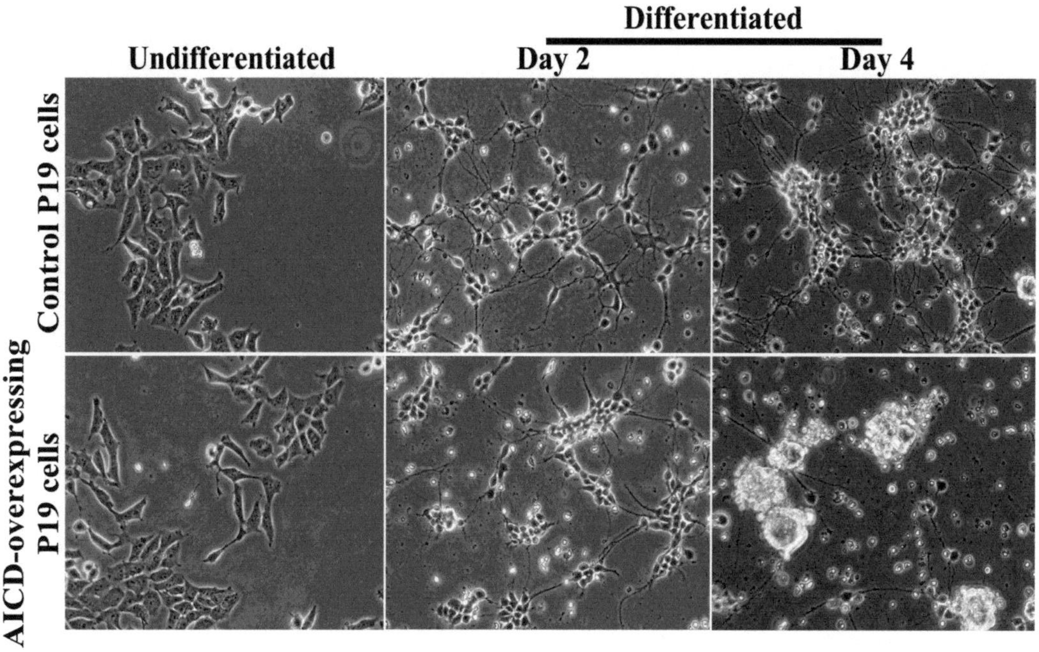

After aggregation culture with RA, AICD-overexpressing P19 and control P19 cells carrying vector alone were replated and cultured for the indicated periods on dishes and allowed to differentiate. Undifferentiated AICD-overexpressing P19 cells retained epithelial cell-like morphology similar to control cells, while the differentiated cells became round and showed a bipolar morphology with neurite extension. Two days after replating (Day 2), all cell lines grew well and neurons with long neurites appeared. Four days after replating (Day 4), control cells still grew well as clusters and many neurons had differentiated from these cells. However, many AICD-overexpressing P19 cells showed severe degeneration, becoming spherical with numerous vacuoles and detached from the culture dishes.

Fig. 5. Overexpression of AICD in P19 cells induces neuronal cell death.

4.3 AICD induces dynamic changes in the gene expression profile

If APP signaling exists, AICD should change expression of certain genes. To test this hypothesis and identify the genes involved in this process of neuron-specific apoptosis, we employed both AICD-overexpressing P19 cells and control P19 cells again and monitored AICD-induced changes in expressions of more than 20,000 independent genes by DNA microarray analysis at 3 time points during culture: the undifferentiated state, after 4 days of aggregation with RA (aggregated state), and 2 days after replating (differentiated state) (Ohkawara *et al.* 2011). Surprisingly, AICD can change expressions of a great many genes:

the expression levels of 277 genes were upregulated by more than 10-fold in the presence of AICD, while 341 genes showed downregulation of expression to less than 10% of the original level (Fig.6).

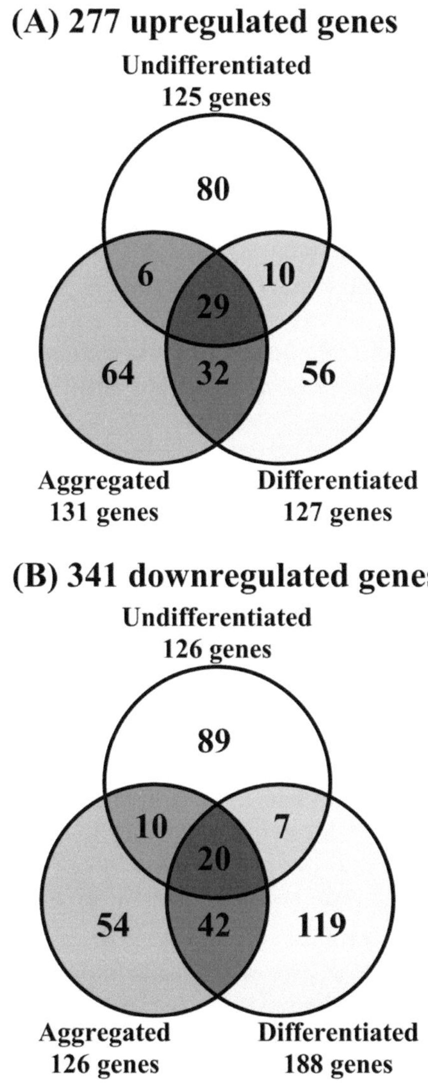

Venn diagrams showing the total numbers of genes upregulated by more than 10-fold in the presence of AICD (A) and genes downregulated to less than 10% of their original level (B) at three states of neural differentiation in P19 cells: undifferentiated, aggregated, and differentiated.

Fig. 6. Upregulated and downregulated genes by AICD.

AICD strongly induced expressions of several genes. For example, AICD-overexpressing P19 cells showed strong expression of protein tyrosine phosphatase receptor T (*Ptprt*) gene at all sampling points: 906-fold, 204-fold, and 116-fold upregulation, in undifferentiated, aggregated, and differentiated states, respectively, estimated from the intensity of hybridization signals. In contrast to these upregulated genes, AICD also strongly inhibited the expression of several genes (Ohkawara *et al.* 2011). For example, *Hes5* was markedly increased through the process of neural differentiation: an increase of almost 300-fold in control P19 cells. However, this extreme induction in control P19 cells could not be detected in AICD-overexpressing P19 cells, indicating that AICD inhibits this induction. These results show that AICD induces both upregulation and downregulation of many genes, suggesting that AICD plays an important role in APP signaling.

We performed Gene Ontology (GO) analysis and classified these upregulated and downregulated genes according to GO terms (Ohkawara *et al.* 2011). While a few genes were classified into GO terms related to cell death, many genes were classified into GO terms unrelated to cell death. Furthermore, we evaluated AICD-induced changes in expression of genes thought to be involved in cell death in AD; however, we found no significant changes in expression of these genes. Therefore, it is likely that AICD does not directly induce the expression of genes involved in cell death, but the extreme dynamic changes in gene expression disrupt the homeostasis of certain neurons and thus give rise to neuron-specific cell death.

4.4 Amyloid hypothesis

Genetic studies indicate that both APP itself and its proteolytic processing are responsible for the onset of AD (Nakayama *et al.* 2011). The amyloid hypothesis is generally accepted as the mechanism of the onset of AD. The traditional amyloid hypothesis is that overproduced Aβ forms insoluble amyloid plaques, which are commonly observed in the AD brain and are believed to be the toxic form of APP responsible for neurodegeneration (Hardy and Selkoe 2002).

However, several issues have been raised regarding these hypotheses. One of the most significant arguments against the amyloid hypothesis is the presence of high levels of Aβ deposition in many non-demented elderly people (Terry RD 1999). This observation implies that Aβ amyloid plaques are not toxic. Based on these observations, the interesting possibility has been proposed that AD may be caused by an APP-derived protein, other than Aβ (Schnabel 2009). As both extracellular fragment and AICD are generated at the same time as Aβ, acceleration of proteolytic processing leads to overproduction of not only Aβ but also of both extracellular fragment and AICD. Therefore, it is likely that neuron-specific apoptosis induced by AICD may also be involved in the onset of AD.

5. Conclusion

Although γ-secretase plays central roles in AD, the physiological functions of this enzyme have yet to be fully elucidated. As reviewed here, Notch signaling is controlled by γ-secretase: intramembrane cleavage of Notch by γ-secretase serves to release ICD that has activity in the nucleus through binding to transcription factors. Recently, it was reported that many type 1 transmembrane proteins are substrates for γ-secretase, and ICDs of these substrates are released from the cell membrane by γ-secretase. These observations that the common enzyme, γ-secretase, modulates proteolysis and the turnover of possible signaling molecules have led to the attractive hypothesis that mechanisms similar to the Notch signaling pathway may widely contribute to γ-secretase-regulated signaling pathways. Indeed, APP signaling induces dynamic changes in gene expression, which may be closely correlated with AICD-induced neuron-specific apoptosis and the onset of AD. Thus, it is likely that γ-secretase controls Notch signaling in NSCs and APP signaling in neurons that may lead to the onset of AD.

6. Abbreviations

AD	Alzheimer's disease
APP	amyloid precursor protein
AICD	the intracellular domain of APP
Aph-1	anterior pharynx defective-1
bHLH	basic helix-loop-helix
BMP	bone morphogenic protein
CTF	C-terminal fragment
Dll	Delta-like protein
Dll1IC	the intracellular domain of Dll1
EGF	epidermal growth factor
FAD	familial AD
GO	gene ontology
Hes	Hairy/Enhancer of split
ICD	intracellular domain
IDE	insulin-degrading enzyme
JM	juxtamembrane
Mib	Mind bomb
NCT	nicastrin
Ngn2	Neurogenin2
NICD	the intracellular domain of Notch
NSC	neural stem cell
NTF	N-terminal fragment
Pen-2	PS enhancer-2
PS	presenilin
RA	all-*trans*-retinoic acid
RIP	the regulated intramembrane proteolysis
TM	transmembrane

7. References

Akazawa, C., Y. Sasai, S. Nakanishi and R. Kageyama (1992) Molecular characterization of a rat negative regulator with a basic helix-loop-helix structure predominantly expressed in the developing nervous system. *J Biol Chem*. 267, 21879-85.

Artavanis-Tsakonas, S., K. Matsuno and M. E. Fortini (1995) Notch signaling. *Science*. 268, 225-32.

Artavanis-Tsakonas, S., M. D. Rand and R. J. Lake (1999) Notch signaling: cell fate control and signal integration in development. *Science*. 284, 770-6.

Baek, S. H., K. A. Ohgi, D. W. Rose, E. H. Koo, C. K. Glass and M. G. Rosenfeld (2002) Exchange of N-CoR corepressor and Tip60 coactivator complexes links gene expression by NF-kappaB and beta-amyloid precursor protein. *Cell*. 110, 55-67.

Baker, R. and G. Schubiger (1996) Autonomous and nonautonomous Notch functions for embryonic muscle and epidermis development in Drosophila. *Development*. 122, 617-26.

Bettenhausen, B., M. Hrabe de Angelis, D. Simon, J. L. Guenet and A. Gossler (1995) Transient and restricted expression during mouse embryogenesis of Dll1, a murine gene closely related to Drosophila Delta. *Development*. 121, 2407-18.

Bibel, M., J. Richter, K. Schrenk, K. L. Tucker, V. Staiger, M. Korte, M. Goetz and Y. A. Barde (2004) Differentiation of mouse embryonic stem cells into a defined neuronal lineage. *Nat Neurosci*. 7, 1003-9.

Bland, C. E., P. Kimberly and M. D. Rand (2003) Notch-induced proteolysis and nuclear localization of the Delta ligand. *J Biol Chem*. 278, 13607-10.

Bolos, V., J. Grego-Bessa and J. L. de la Pompa (2007) Notch signaling in development and cancer. *Endocr Rev*. 28, 339-63.

Brou, C., F. Logeat, N. Gupta, C. Bessia, O. LeBail, J. R. Doedens, A. Cumano, P. Roux, R. A. Black and A. Israel (2000) A novel proteolytic cleavage involved in Notch signaling: the role of the disintegrin-metalloprotease TACE. *Mol Cell*. 5, 207-16.

Brown, M. S., J. Ye, R. B. Rawson and J. L. Goldstein (2000) Regulated intramembrane proteolysis: a control mechanism conserved from bacteria to humans. *Cell*. 100, 391-8.

Buoso, E., C. Lanni, G. Schettini, S. Govoni and M. Racchi (2011) beta-Amyloid precursor protein metabolism: focus on the functions and degradation of its intracellular domain. *Pharmacol Res*. 62, 308-17.

Buxbaum, J. D., K. N. Liu, Y. Luo, J. L. Slack, K. L. Stocking, J. J. Peschon, R. S. Johnson, B. J. Castner, D. P. Cerretti and R. A. Black (1998) Evidence that tumor necrosis factor alpha converting enzyme is involved in regulated alpha-secretase cleavage of the Alzheimer amyloid protein precursor. *J Biol Chem*. 273, 27765-7.

Cao, X. and T. C. Sudhof (2001) A transcriptionally [correction of transcriptively] active complex of APP with Fe65 and histone acetyltransferase Tip60. *Science*. 293, 115-20.

Chenn, A. (2005) The simple life (of cortical progenitors). *Neuron*. 45, 817-9.

Cupers, P., I. Orlans, K. Craessaerts, W. Annaert and B. De Strooper (2001) The amyloid precursor protein (APP)-cytoplasmic fragment generated by gamma-secretase is rapidly degraded but distributes partially in a nuclear fraction of neurones in culture. *J Neurochem*. 78, 1168-78.

Dennler, S., S. Itoh, D. Vivien, P. ten Dijke, S. Huet and J. M. Gauthier (1998) Direct binding of Smad3 and Smad4 to critical TGF beta-inducible elements in the promoter of human plasminogen activator inhibitor-type 1 gene. *EMBO J*. 17, 3091-100.

Derynck, R. and Y. E. Zhang (2003) Smad-dependent and Smad-independent pathways in TGF-beta family signalling. *Nature*. 425, 577-84.

Doan, A., G. Thinakaran, D. R. Borchelt, H. H. Slunt, T. Ratovitsky, M. Podlisny, D. J. Selkoe, M. Seeger, S. E. Gandy, D. L. Price and S. S. Sisodia (1996) Protein topology of presenilin 1. *Neuron*. 17, 1023-30.

Dunwoodie, S. L., D. Henrique, S. M. Harrison and R. S. Beddington (1997) Mouse Dll3: a novel divergent Delta gene which may complement the function of other Delta homologues during early pattern formation in the mouse embryo. *Development*. 124, 3065-76.

Edbauer, D., M. Willem, S. Lammich, H. Steiner and C. Haass (2002) Insulin-degrading enzyme rapidly removes the beta-amyloid precursor protein intracellular domain (AICD). *J Biol Chem*. 277, 13389-93.

Ellisen, L. W., J. Bird, D. C. West, A. L. Soreng, T. C. Reynolds, S. D. Smith and J. Sklar (1991) TAN-1, the human homolog of the Drosophila notch gene, is broken by chromosomal translocations in T lymphoblastic neoplasms. *Cell*. 66, 649-61.

Eppert, K., S. W. Scherer, H. Ozcelik, R. Pirone, P. Hoodless, H. Kim, L. C. Tsui, B. Bapat, S. Gallinger, I. L. Andrulis, G. H. Thomsen, J. L. Wrana and L. Attisano (1996) MADR2 maps to 18q21 and encodes a TGFbeta-regulated MAD-related protein that is functionally mutated in colorectal carcinoma. *Cell*. 86, 543-52.

Esch, F. S., P. S. Keim, E. C. Beattie, R. W. Blacher, A. R. Culwell, T. Oltersdorf, D. McClure and P. J. Ward (1990) Cleavage of amyloid beta peptide during constitutive processing of its precursor. *Science*. 248, 1122-4.

Farzan, M., C. E. Schnitzler, N. Vasilieva, D. Leung and H. Choe (2000) BACE2, a beta - secretase homolog, cleaves at the beta site and within the amyloid-beta region of the amyloid-beta precursor protein. *Proc Natl Acad Sci U S A*. 97, 9712-7.

Fitzgerald, K. and I. Greenwald (1995) Interchangeability of Caenorhabditis elegans DSL proteins and intrinsic signalling activity of their extracellular domains in vivo. *Development*. 121, 4275-82.

Fleming, R. J., T. N. Scottgale, R. J. Diederich and S. Artavanis-Tsakonas (1990) The gene Serrate encodes a putative EGF-like transmembrane protein essential for proper ectodermal development in Drosophila melanogaster. *Genes Dev*. 4, 2188-201.

Fortini, M. E. (2009) Notch signaling: the core pathway and its posttranslational regulation. *Dev Cell*. 16, 633-47.

Frise, E., J. A. Knoblich, S. Younger-Shepherd, L. Y. Jan and Y. N. Jan (1996) The Drosophila Numb protein inhibits signaling of the Notch receptor during cell-cell interaction in sensory organ lineage. *Proc Natl Acad Sci U S A*. 93, 11925-32.

Gao, Y. and S. W. Pimplikar (2001) The gamma -secretase-cleaved C-terminal fragment of amyloid precursor protein mediates signaling to the nucleus. *Proc Natl Acad Sci U S A*. 98, 14979-84.

Ghosal, K., D. L. Vogt, M. Liang, Y. Shen, B. T. Lamb and S. W. Pimplikar (2009) Alzheimer's disease-like pathological features in transgenic mice expressing the APP intracellular domain. *Proc Natl Acad Sci U S A*. 106, 18367-72.

Guenette, S. Y. (2002) A role for APP in motility and transcription? *Trends Pharmacol Sci*. 23, 203-5; discussion 205-6.

Guo, M., L. Y. Jan and Y. N. Jan (1996) Control of daughter cell fates during asymmetric division: interaction of Numb and Notch. *Neuron*. 17, 27-41.

Gupta-Rossi, N., E. Six, O. LeBail, F. Logeat, P. Chastagner, A. Olry, A. Israel and C. Brou (2004) Monoubiquitination and endocytosis direct gamma-secretase cleavage of activated Notch receptor. *J Cell Biol*. 166, 73-83.

Haass, C. and D. J. Selkoe (1993) Cellular processing of beta-amyloid precursor protein and the genesis of amyloid beta-peptide. *Cell*. 75, 1039-42.

Hardy, J. (1997) Amyloid, the presenilins and Alzheimer's disease. *Trends Neurosci*. 20, 154-9.

Hardy, J. and D. J. Selkoe (2002) The amyloid hypothesis of Alzheimer's disease: progress and problems on the road to therapeutics. *Science*. 297, 353-6.

Henderson, S. T., D. Gao, E. J. Lambie and J. Kimble (1994) lag-2 may encode a signaling ligand for the GLP-1 and LIN-12 receptors of C. elegans. *Development*. 120, 2913-24.

Henderson, S. T., D. Gao, S. Christensen and J. Kimble (1997) Functional domains of LAG-2, a putative signaling ligand for LIN-12 and GLP-1 receptors in Caenorhabditis elegans. *Mol Biol Cell*. 8, 1751-62.

Hiratochi, M., H. Nagase, Y. Kuramochi, C. S. Koh, T. Ohkawara and K. Nakayama (2007) The Delta intracellular domain mediates TGF-beta/Activin signaling through binding to Smads and has an important bi-directional function in the Notch-Delta signaling pathway. *Nucleic Acids Res*. 35, 912-22.

Ho, A. and T. C. Sudhof (2004) Binding of F-spondin to amyloid-beta precursor protein: a candidate amyloid-beta precursor protein ligand that modulates amyloid-beta precursor protein cleavage. *Proc Natl Acad Sci U S A*. 101, 2548-53.

Hoodless, P. A., T. Haerry, S. Abdollah, M. Stapleton, M. B. O'Connor, L. Attisano and J. L. Wrana (1996) MADR1, a MAD-related protein that functions in BMP2 signaling pathways. *Cell*. 85, 489-500.

Hung, A. Y., E. H. Koo, C. Haass and D. J. Selkoe (1992) Increased expression of beta-amyloid precursor protein during neuronal differentiation is not accompanied by secretory cleavage. *Proc Natl Acad Sci U S A*. 89, 9439-43.

Ikeuchi, T. and S. S. Sisodia (2003) The Notch ligands, Delta1 and Jagged2, are substrates for presenilin-dependent "gamma-secretase" cleavage. *J Biol Chem*. 278, 7751-4.

Imayoshi, I. and R. Kageyama (2011) The role of notch signaling in adult neurogenesis. *Mol Neurobiol*. 44, 7-12.

Iso, T., V. Sartorelli, C. Poizat, S. Iezzi, H. Y. Wu, G. Chung, L. Kedes and Y. Hamamori (2001) HERP, a novel heterodimer partner of HES/E(spl) in Notch signaling. *Mol Cell Biol*. 21, 6080-9.

Itoh, M., C. H. Kim, G. Palardy, T. Oda, Y. J. Jiang, D. Maust, S. Y. Yeo, K. Lorick, G. J. Wright, L. Ariza-McNaughton, A. M. Weissman, J. Lewis, S. C. Chandrasekharappa and A. B. Chitnis (2003) Mind bomb is a ubiquitin ligase that is essential for efficient activation of Notch signaling by Delta. *Dev Cell*. 4, 67-82.

Iwatsubo, T. (2004) The gamma-secretase complex: machinery for intramembrane proteolysis. *Curr Opin Neurobiol*. 14, 379-83.

Jonk, L. J., S. Itoh, C. H. Heldin, P. ten Dijke and W. Kruijer (1998) Identification and functional characterization of a Smad binding element (SBE) in the JunB promoter that acts as a transforming growth factor-beta, activin, and bone morphogenetic protein-inducible enhancer. *J Biol Chem*. 273, 21145-52.

Joutel, A., C. Corpechot, A. Ducros, K. Vahedi, H. Chabriat, P. Mouton, S. Alamowitch, V. Domenga, M. Cecillion, E. Marechal, J. Maciazek, C. Vayssiere, C. Cruaud, E. A. Cabanis, M. M. Ruchoux, J. Weissenbach, J. F. Bach, M. G. Bousser and E. Tournier-Lasserve (1996) Notch3 mutations in CADASIL, a hereditary adult-onset condition causing stroke and dementia. *Nature*. 383, 707-10.

Justice, N. J. and Y. N. Jan (2002) Variations on the Notch pathway in neural development. *Curr Opin Neurobiol*. 12, 64-70.

Kageyama, R., T. Ohtsuka and T. Kobayashi (2007) The Hes gene family: repressors and oscillators that orchestrate embryogenesis. *Development*. 134, 1243-51.

Kageyama, R., T. Ohtsuka, H. Shimojo and I. Imayoshi (2008) Dynamic Notch signaling in neural progenitor cells and a revised view of lateral inhibition. *Nat Neurosci.* 11, 1247-51.

Kang, J., H. G. Lemaire, A. Unterbeck, J. M. Salbaum, C. L. Masters, K. H. Grzeschik, G. Multhaup, K. Beyreuther and B. Muller-Hill (1987) The precursor of Alzheimer's disease amyloid A4 protein resembles a cell-surface receptor. *Nature.* 325, 733-6.

Kimberly, W. T., J. B. Zheng, S. Y. Guenette and D. J. Selkoe (2001) The intracellular domain of the beta-amyloid precursor protein is stabilized by Fe65 and translocates to the nucleus in a notch-like manner. *J Biol Chem.* 276, 40288-92.

Kimberly, W. T., J. B. Zheng, T. Town, R. A. Flavell and D. J. Selkoe (2005) Physiological regulation of the beta-amyloid precursor protein signaling domain by c-Jun N-terminal kinase JNK3 during neuronal differentiation. *J Neurosci.* 25, 5533-43.

Kimble, J. and P. Simpson (1997) The LIN-12/Notch signaling pathway and its regulation. *Annu Rev Cell Dev Biol.* 13, 333-61.

Koo, E. H. and R. Kopan (2004) Potential role of presenilin-regulated signaling pathways in sporadic neurodegeneration. *Nat Med.* 10 Suppl, S26-33.

Kopan, R., E. H. Schroeter, H. Weintraub and J. S. Nye (1996) Signal transduction by activated mNotch: importance of proteolytic processing and its regulation by the extracellular domain. *Proc Natl Acad Sci U S A.* 93, 1683-8.

Kopan, R. and M. X. Ilagan (2004) Gamma-secretase: proteasome of the membrane? *Nat Rev Mol Cell Biol.* 5, 499-504.

Kopan, R. and M. X. Ilagan (2009) The canonical Notch signaling pathway: unfolding the activation mechanism. *Cell.* 137, 216-33.

Kopczynski, C. C., A. K. Alton, K. Fechtel, P. J. Kooh and M. A. Muskavitch (1988) Delta, a Drosophila neurogenic gene, is transcriptionally complex and encodes a protein related to blood coagulation factors and epidermal growth factor of vertebrates. *Genes Dev.* 2, 1723-35.

Kretzschmar, M., F. Liu, A. Hata, J. Doody and J. Massague (1997) The TGF-beta family mediator Smad1 is phosphorylated directly and activated functionally by the BMP receptor kinase. *Genes Dev.* 11, 984-95.

Lammich, S., E. Kojro, R. Postina, S. Gilbert, R. Pfeiffer, M. Jasionowski, C. Haass and F. Fahrenholz (1999) Constitutive and regulated alpha-secretase cleavage of Alzheimer's amyloid precursor protein by a disintegrin metalloprotease. *Proc Natl Acad Sci U S A.* 96, 3922-7.

Laping, N. J., E. Grygielko, A. Mathur, S. Butter, J. Bomberger, C. Tweed, W. Martin, J. Fornwald, R. Lehr, J. Harling, L. Gaster, J. F. Callahan and B. A. Olson (2002) Inhibition of transforming growth factor (TGF)-beta1-induced extracellular matrix with a novel inhibitor of the TGF-beta type I receptor kinase activity: SB-431542. *Mol Pharmacol.* 62, 58-64.

LaVoie, M. J. and D. J. Selkoe (2003) The Notch ligands, Jagged and Delta, are sequentially processed by alpha-secretase and presenilin/gamma-secretase and release signaling fragments. *J Biol Chem.* 278, 34427-37.

Lee, M. K., H. H. Slunt, L. J. Martin, G. Thinakaran, G. Kim, S. E. Gandy, M. Seeger, E. Koo, D. L. Price and S. S. Sisodia (1996) Expression of presenilin 1 and 2 (PS1 and PS2) in human and murine tissues. *J Neurosci.* 16, 7513-25.

Leimeister, C., A. Externbrink, B. Klamt and M. Gessler (1999) Hey genes: a novel subfamily of hairy- and Enhancer of split related genes specifically expressed during mouse embryogenesis. *Mech Dev.* 85, 173-7.

Levitan, D. and I. Greenwald (1995) Facilitation of lin-12-mediated signalling by sel-12, a Caenorhabditis elegans S182 Alzheimer's disease gene. *Nature.* 377, 351-4.

Levy-Lahad, E., W. Wasco, P. Poorkaj, D. M. Romano, J. Oshima, W. H. Pettingell, C. E. Yu, P. D. Jondro, S. D. Schmidt, K. Wang and et al. (1995) Candidate gene for the chromosome 1 familial Alzheimer's disease locus. *Science.* 269, 973-7.

Lewis, J. (1998) Notch signalling and the control of cell fate choices in vertebrates. *Semin Cell Dev Biol.* 9, 583-9.

Leyssen, M., D. Ayaz, S. S. Hebert, S. Reeve, B. De Strooper and B. A. Hassan (2005) Amyloid precursor protein promotes post-developmental neurite arborization in the Drosophila brain. *EMBO J.* 24, 2944-55.

Lindsell, C. E., C. J. Shawber, J. Boulter and G. Weinmaster (1995) Jagged: a mammalian ligand that activates Notch1. *Cell.* 80, 909-17.

Logeat, F., C. Bessia, C. Brou, O. LeBail, S. Jarriault, N. G. Seidah and A. Israel (1998) The Notch1 receptor is cleaved constitutively by a furin-like convertase. *Proc Natl Acad Sci U S A.* 95, 8108-12.

Lorenzo, A., M. Yuan, Z. Zhang, P. A. Paganetti, C. Sturchler-Pierrat, M. Staufenbiel, J. Mautino, F. S. Vigo, B. Sommer and B. A. Yankner (2000) Amyloid beta interacts with the amyloid precursor protein: a potential toxic mechanism in Alzheimer's disease. *Nat Neurosci.* 3, 460-4.

Lu, J., Y. Wu, N. Sousa and O. F. Almeida (2005) SMAD pathway mediation of BDNF and TGF beta 2 regulation of proliferation and differentiation of hippocampal granule neurons. *Development.* 132, 3231-42.

McCarthy, J. V., C. Twomey and P. Wujek (2009) Presenilin-dependent regulated intramembrane proteolysis and gamma-secretase activity. *Cell Mol Life Sci.* 66, 1534-55.

McCright, B., J. Lozier and T. Gridley (2002) A mouse model of Alagille syndrome: Notch2 as a genetic modifier of Jag1 haploinsufficiency. *Development.* 129, 1075-82.

Mello, C. C., B. W. Draper and J. R. Priess (1994) The maternal genes apx-1 and glp-1 and establishment of dorsal-ventral polarity in the early C. elegans embryo. *Cell.* 77, 95-106.

Miyazawa, K., M. Shinozaki, T. Hara, T. Furuya and K. Miyazono (2002) Two major Smad pathways in TGF-beta superfamily signalling. *Genes Cells.* 7, 1191-204.

Nagase, H., C. S. Koh and K. Nakayama (2011) gamma-Secretase-regulated signaling pathways, such as notch signaling, mediate the differentiation of hematopoietic stem cells, development of the immune system, and peripheral immune responses. *Curr Stem Cell Res Ther.* 6, 131-41.

Nakao, A., T. Imamura, S. Souchelnytskyi, M. Kawabata, A. Ishisaki, E. Oeda, K. Tamaki, J. Hanai, C. H. Heldin, K. Miyazono and P. ten Dijke (1997) TGF-beta receptor-mediated signalling through Smad2, Smad3 and Smad4. *EMBO J.* 16, 5353-62.

Nakayama, K., K. Nagase, Y. Tokutake, C. S. Koh, M. Hiratochi, T. Ohkawara and N. Nakayama (2004) Multiple POU-binding motifs, recognized by tissue-specific nuclear factors, are important for Dll1 gene expression in neural stem cells. *Biochem Biophys Res Commun.* 325, 991-6.

Nakayama, K., H. Nagase, M. Hiratochi, C. S. Koh and T. Ohkawara (2008a) Similar mechanisms regulated by gamma-secretase are involved in both directions of the bi-directional Notch-Delta signaling pathway as well as play a potential role in signaling events involving type 1 transmembrane proteins. *Curr Stem Cell Res Ther.* 3, 288-302.

Nakayama, K., T. Ohkawara, M. Hiratochi, C. S. Koh and H. Nagase (2008b) The intracellular domain of amyloid precursor protein induces neuron-specific apoptosis. *Neurosci Lett.* 444, 127-31.

Nakayama, K., H. Nagase, C. S. Koh and T. Ohkawara (2011) gamma-Secretase-Regulated Mechanisms Similar to Notch Signaling May Play a Role in Signaling Events, Including APP Signaling, Which Leads to Alzheimer's Disease. *Cell Mol Neurobiol.* 31, 887-900.

Ni, C. Y., M. P. Murphy, T. E. Golde and G. Carpenter (2001) gamma -Secretase cleavage and nuclear localization of ErbB-4 receptor tyrosine kinase. *Science.* 294, 2179-81.

Oda, T., A. G. Elkahloun, B. L. Pike, K. Okajima, I. D. Krantz, A. Genin, D. A. Piccoli, P. S. Meltzer, N. B. Spinner, F. S. Collins and S. C. Chandrasekharappa (1997) Mutations in the human Jagged1 gene are responsible for Alagille syndrome. *Nat Genet.* 16, 235-42.

Ohkawara, T., H. Nagase, C. S. Koh and K. Nakayama (2011) The amyloid precursor protein intracellular domain alters gene expression and induces neuron-specific apoptosis. *Gene.* 475, 1-9.

Okochi, M., H. Steiner, A. Fukumori, H. Tanii, T. Tomita, T. Tanaka, T. Iwatsubo, T. Kudo, M. Takeda and C. Haass (2002) Presenilins mediate a dual intramembranous gamma-secretase cleavage of Notch-1. *EMBO J.* 21, 5408-16.

Pan, D. and G. M. Rubin (1997) Kuzbanian controls proteolytic processing of Notch and mediates lateral inhibition during Drosophila and vertebrate neurogenesis. *Cell.* 90, 271-80.

Park, J. H., D. A. Gimbel, T. GrandPre, J. K. Lee, J. E. Kim, W. Li, D. H. Lee and S. M. Strittmatter (2006) Alzheimer precursor protein interaction with the Nogo-66 receptor reduces amyloid-beta plaque deposition. *J Neurosci.* 26, 1386-95.

Pierfelice, T., L. Alberi and N. Gaiano (2011) Notch in the vertebrate nervous system: an old dog with new tricks. *Neuron.* 69, 840-55.

Qi, H., M. D. Rand, X. Wu, N. Sestan, W. Wang, P. Rakic, T. Xu and S. Artavanis-Tsakonas (1999) Processing of the notch ligand delta by the metalloprotease Kuzbanian. *Science.* 283, 91-4.

Rebay, I., R. J. Fleming, R. G. Fehon, L. Cherbas, P. Cherbas and S. Artavanis-Tsakonas (1991) Specific EGF repeats of Notch mediate interactions with Delta and Serrate: implications for Notch as a multifunctional receptor. *Cell.* 67, 687-99.

Roehl, H., M. Bosenberg, R. Blelloch and J. Kimble (1996) Roles of the RAM and ANK domains in signaling by the C. elegans GLP-1 receptor. *EMBO J.* 15, 7002-12.

Rogaev, E. I., R. Sherrington, E. A. Rogaeva, G. Levesque, M. Ikeda, Y. Liang, H. Chi, C. Lin, K. Holman, T. Tsuda and et al. (1995) Familial Alzheimer's disease in kindreds with missense mutations in a gene on chromosome 1 related to the Alzheimer's disease type 3 gene. *Nature.* 376, 775-8.

Rohn, J. L., A. S. Lauring, M. L. Linenberger and J. Overbaugh (1996) Transduction of Notch2 in feline leukemia virus-induced thymic lymphoma. *J Virol.* 70, 8071-80.

Ryan, K. A. and S. W. Pimplikar (2005) Activation of GSK-3 and phosphorylation of CRMP2 in transgenic mice expressing APP intracellular domain. *J Cell Biol*. 171, 327-35.

Sarkar, N. H., S. Haga, A. F. Lehner, W. Zhao, S. Imai and K. Moriwaki (1994) Insertional mutation of int protooncogenes in the mammary tumors of a new strain of mice derived from the wild in China: normal- and tumor-tissue-specific expression of int-3 transcripts. *Virology*. 203, 52-62.

Schnabel, J. (2009) Alzheimer's theory makes a splash. *Nature*. 459, 310.

Schroeter, E. H., J. A. Kisslinger and R. Kopan (1998) Notch-1 signalling requires ligand-induced proteolytic release of intracellular domain. *Nature*. 393, 382-6.

Schulz, J. G., W. Annaert, J. Vandekerckhove, P. Zimmermann, B. De Strooper and G. David (2003) Syndecan 3 intramembrane proteolysis is presenilin/gamma-secretase-dependent and modulates cytosolic signaling. *J Biol Chem*. 278, 48651-7.

Selkoe, D. J. (2001) Alzheimer's disease: genes, proteins, and therapy. *Physiol Rev*. 81, 741-66.

Selkoe, D. and R. Kopan (2003) Notch and Presenilin: regulated intramembrane proteolysis links development and degeneration. *Annu Rev Neurosci*. 26, 565-97.

Selkoe, D. J. and M. S. Wolfe (2007) Presenilin: running with scissors in the membrane. *Cell*. 131, 215-21.

Shah, S., S. F. Lee, K. Tabuchi, Y. H. Hao, C. Yu, Q. LaPlant, H. Ball, C. E. Dann, 3rd, T. Sudhof and G. Yu (2005) Nicastrin functions as a gamma-secretase-substrate receptor. *Cell*. 122, 435-47.

Shawber, C., J. Boulter, C. E. Lindsell and G. Weinmaster (1996) Jagged2: a serrate-like gene expressed during rat embryogenesis. *Dev Biol*. 180, 370-6.

Sherrington, R., E. I. Rogaev, Y. Liang, E. A. Rogaeva, G. Levesque, M. Ikeda, H. Chi, C. Lin, G. Li, K. Holman, T. Tsuda, L. Mar, J. F. Foncin, A. C. Bruni, M. P. Montesi, S. Sorbi, I. Rainero, L. Pinessi, L. Nee, I. Chumakov, D. Pollen, A. Brookes, P. Sanseau, R. J. Polinsky, W. Wasco, H. A. Da Silva, J. L. Haines, M. A. Perkicak-Vance, R. E. Tanzi, A. D. Roses, P. E. Fraser, J. M. Rommens and P. H. St George-Hyslop (1995) Cloning of a gene bearing missense mutations in early-onset familial Alzheimer's disease. *Nature*. 375, 754-60.

Shimojo, H., T. Ohtsuka and R. Kageyama (2011) Dynamic expression of notch signaling genes in neural stem/progenitor cells. *Front Neurosci*. 5, 78-84.

Shutter, J. R., S. Scully, W. Fan, W. G. Richards, J. Kitajewski, G. A. Deblandre, C. R. Kintner and K. L. Stark (2000) Dll4, a novel Notch ligand expressed in arterial endothelium. *Genes Dev*. 14, 1313-8.

Sisodia, S. S., E. H. Koo, P. N. Hoffman, G. Perry and D. L. Price (1993) Identification and transport of full-length amyloid precursor proteins in rat peripheral nervous system. *J Neurosci*. 13, 3136-42.

Six, E., D. Ndiaye, Y. Laabi, C. Brou, N. Gupta-Rossi, A. Israel and F. Logeat (2003) The Notch ligand Delta1 is sequentially cleaved by an ADAM protease and gamma-secretase. *Proc Natl Acad Sci U S A*. 100, 7638-43.

Soba, P., S. Eggert, K. Wagner, H. Zentgraf, K. Siehl, S. Kreger, A. Lower, A. Langer, G. Merdes, R. Paro, C. L. Masters, U. Muller, S. Kins and K. Beyreuther (2005) Homo- and heterodimerization of APP family members promotes intercellular adhesion. *EMBO J*. 24, 3624-34.

Spana, E. P. and C. Q. Doe (1996) Numb antagonizes Notch signaling to specify sibling neuron cell fates. *Neuron*. 17, 21-6.

Sparrow, D. B., M. Clements, S. L. Withington, A. N. Scott, J. Novotny, D. Sillence, K. Kusumi, R. S. Beddington and S. L. Dunwoodie (2002) Diverse requirements for Notch signalling in mammals. *Int J Dev Biol*. 46, 365-74.

Struhl, G. and A. Adachi (2000) Requirements for presenilin-dependent cleavage of notch and other transmembrane proteins. *Mol Cell*. 6, 625-36.

Suzuki, A., C. Chang, J. M. Yingling, X. F. Wang and A. Hemmati-Brivanlou (1997) Smad5 induces ventral fates in Xenopus embryo. *Dev Biol*. 184, 402-5.

Tax, F. E., J. J. Yeargers and J. H. Thomas (1994) Sequence of C. elegans lag-2 reveals a cell-signalling domain shared with Delta and Serrate of Drosophila. *Nature*. 368, 150-4.

Terry RD, K. R., Bick KL, Sisodia SS (1999) *Alzheimer's disease.*, Lippincott, Williams & Wilkins, Philadelphia

Thinakaran, G., D. R. Borchelt, M. K. Lee, H. H. Slunt, L. Spitzer, G. Kim, T. Ratovitsky, F. Davenport, C. Nordstedt, M. Seeger, J. Hardy, A. I. Levey, S. E. Gandy, N. A. Jenkins, N. G. Copeland, D. L. Price and S. S. Sisodia (1996) Endoproteolysis of presenilin 1 and accumulation of processed derivatives in vivo. *Neuron*. 17, 181-90.

Vassar, R., B. D. Bennett, S. Babu-Khan, S. Kahn, E. A. Mendiaz, P. Denis, D. B. Teplow, S. Ross, P. Amarante, R. Loeloff, Y. Luo, S. Fisher, J. Fuller, S. Edenson, J. Lile, M. A. Jarosinski, A. L. Biere, E. Curran, T. Burgess, J. C. Louis, F. Collins, J. Treanor, G. Rogers and M. Citron (1999) Beta-secretase cleavage of Alzheimer's amyloid precursor protein by the transmembrane aspartic protease BACE. *Science*. 286, 735-41.

Vazquez, M. C., L. M. Vargas, N. C. Inestrosa and A. R. Alvarez (2009) c-Abl modulates AICD dependent cellular responses: transcriptional induction and apoptosis. *J Cell Physiol*. 220, 136-43.

Wallberg, A. E., K. Pedersen, U. Lendahl and R. G. Roeder (2002) p300 and PCAF act cooperatively to mediate transcriptional activation from chromatin templates by notch intracellular domains in vitro. *Mol Cell Biol*. 22, 7812-9.

Wang, Y. and Y. Ha (2004) The X-ray structure of an antiparallel dimer of the human amyloid precursor protein E2 domain. *Mol Cell*. 15, 343-53.

Wharton, K. A., K. M. Johansen, T. Xu and S. Artavanis-Tsakonas (1985) Nucleotide sequence from the neurogenic locus notch implies a gene product that shares homology with proteins containing EGF-like repeats. *Cell*. 43, 567-81.

Wu, L., J. C. Aster, S. C. Blacklow, R. Lake, S. Artavanis-Tsakonas and J. D. Griffin (2000) MAML1, a human homologue of Drosophila mastermind, is a transcriptional co-activator for NOTCH receptors. *Nat Genet*. 26, 484-9.

Yochem, J. and I. Greenwald (1989) glp-1 and lin-12, genes implicated in distinct cell-cell interactions in C. elegans, encode similar transmembrane proteins. *Cell*. 58, 553-63.

Yoshikai, S., H. Sasaki, K. Doh-ura, H. Furuya and Y. Sakaki (1990) Genomic organization of the human amyloid beta-protein precursor gene. *Gene*. 87, 257-63.

Yu, G., M. Nishimura, S. Arawaka, D. Levitan, L. Zhang, A. Tandon, Y. Q. Song, E. Rogaeva, F. Chen, T. Kawarai, A. Supala, L. Levesque, H. Yu, D. S. Yang, E. Holmes, P. Milman, Y. Liang, D. M. Zhang, D. H. Xu, C. Sato, E. Rogaev, M. Smith, C. Janus, Y. Zhang, R. Aebersold, L. S. Farrer, S. Sorbi, A. Bruni, P. Fraser and P. St George-Hyslop (2000) Nicastrin modulates presenilin-mediated notch/glp-1 signal transduction and betaAPP processing. *Nature*. 407, 48-54.

Zhang, Y., X. Feng, R. We and R. Derynck (1996) Receptor-associated Mad homologues synergize as effectors of the TGF-beta response. *Nature*. 383, 168-72.

Zheng, H. and E. H. Koo (2006) The amyloid precursor protein: beyond amyloid. *Mol Neurodegener*. 1, 5.
Zhong, W., J. N. Feder, M. M. Jiang, L. Y. Jan and Y. N. Jan (1996) Asymmetric localization of a mammalian numb homolog during mouse cortical neurogenesis. *Neuron*. 17, 43-53.
Zhong, W., M. M. Jiang, G. Weinmaster, L. Y. Jan and Y. N. Jan (1997) Differential expression of mammalian Numb, Numblike and Notch1 suggests distinct roles during mouse cortical neurogenesis. *Development*. 124, 1887-97.

9

Role of Growth Factor Receptors in Neural Stem Cells Differentiation and Dopaminergic Neurons Generation

Lucía Calatrava, Rafael Gonzalo-Gobernado, Antonio S. Herranz, Diana Reimers, Maria J. Asensio, Cristina Miranda and Eulalia Bazán
Department of Neurobiology, Ramón y Cajal Institute for Health Research, Madrid Spain

1. Introduction

Neural stem cells (NSCs) are defined as clonogenic cells with self-renewal capacity and multilineage potential (Bazán et al., 2004). Cells with these characteristics have been isolated from the embryonic and adult Central Nervous System (CNS) (Gil-Perotin et al., 2009; Merkle & Alvarez-Buylla, 2006; Weiss et al., 1996). Under specific conditions, these cells proliferate in culture as cell clusters, called neurospheres, and differentiate into neurons, glia, and non-neural cell types (Kennea & Mehmet, 2002; Lobo et al., 2003; Reynolds & Weiss, 1992; Vescovi et al., 2002; Arias-Carrión & Yuan, 2009). Moreover, these cultures represent a potential source for cell replacement therapies in neurological diseases such as Parkinson's disease (PD) (Bjugstad et al., 2008; Pluchino et al., 2005; Reimers et al., 2011; Zhu et al, 2009).

Both basic fibroblast growth factor (bFGF) and epidermal growth factor (EGF) promote the proliferation of NSCs (Ciccolini & Svendsen, 1998; Gritti et al., 1996, 1999; Reynolds et al., 1992; Palmer et al., 1999). Moreover, growth factors (GFs) and intracellular mechanisms have been reported to influence or even determine NSCs phenotypic choice "in vivo" and "in vitro" (Daadi & Weiss, 1999; Hagg, 2005, 2009; Ninkovic & Götz, 2007; Redondo et al., 2007; Reimers et al., 2001, 2008). Thus, bFGF in combination with agents that increase cAMP levels and/or protein kinase C (PKC) activators induced the expression of tyrosine hydroxylase (TH), which is the rate-limiting enzyme involved in the synthesis of catecholaminergic neurotransmitters (Lopez-Toledano et al., 2004).

Growth factors exert their action through their interaction with specific receptors. A subset of FGF receptors (FGFRs) have been described in NSCs and their progeny (Reimers et al., 2001, Lobo et al., 2003), but to present nothing is known regarding the role played by these different FGFRs subtypes in the TH-inductive effect of bFGF above described. NSCs also express a 170 kD protein corresponding to the EGF receptor (EGFR) (Lobo et al., 2003). Since bFGF modulates EGF responsiveness in striatal precursors (Ciccolini & Svendsen, 1998), we wonderer whether TH induction in NSCs progeny could be associated with changes in EGFR protein expression and/or cellular localization.

2. Material and methods

2.1 Isolation of neural stem cells from the embryonic rat striatum

Striatal primordia from E15 Sprague-Dawley rat embryos were dissected and mechanically dissociated. Cells were grown in suspension in a defined medium (DF12) composed of Dulbecco's modified Eagle's medium and Ham's F-12 (1:1), 2 mM L-glutamine, 1 mM sodium piruvate (all from Gibco BRL, Life Technologies Inc, Grand Island, NY), 0.6% glucose, 25 μg/ml insulin, 20 nM progesterone, 60 μM putrescine, and 30 nM sodium selenite (all from Sigma Chemical Co, St Louis, MO), 100 μg/ml human transferrin (Boehringer Mannheim GmbH, Germany) and 20 ng/ml human recombinant EGF (PreproTech EC Ltd., London, England). After 48-72 hr in vitro, the cells grew as free-floating neurospheres and were passaged by mechanical dissociation every 2-3 days (Lobo et al., 2003; Reimers et al., 2001).

After a minimum of 4 and a maximum of 5 passages, neurospheres were dissociated and plated at a density of 20,000–30,000 cells/cm2 on 15μg/ml poly l-ornithine (Sigma)-coated round glass cover slips (ø12 mm) or plastic dishes (ø 35 mm). Cultures were maintained in DF12 and 20 ng/ml EGF for 3 days and then switched to DF12 without EGF for longer culture periods (control group). At 7 days postplating (dpp), parallel cultures were treated with 10 ng/ml human recombinant bFGF (Boehringer Mannheim) and 1 mM dibutyryladenosine 3,5-cyclic monophosphate (bFGF + dbcAMP) in the absence (vehicle group), or presence of 20μM PD98058 or 10^{-7}M staurosporin. Cellular phenotypes were determined immunocytochemically 24 hr later using antibodies to β-tubulin isotype III (β-tubulin III) for neurons, glial fibrillary acidic protein (GFAP) for astrocytes, A2B5, which stains bipotential O2A glial progenitors (Raff et al. 1984; Schnitzer & Schachner 1982) as well as subsets of neurons (Schnitzer & Schachner 1982), O4 for immature oligodendrocytes, O1 for mature oligodendrocytes, and TH for catecholaminergic phenotypes.

2.2 Immunocytochemical staining

The polyclonal antibodies used in this study were anti-β-tubulin III (BabCO; Richmond, CA), anti-TH (Chemicon International;Temecula, CA), anti-DOPA-decarboxilase (Sigma, Missouri, USA), anti-dopamine transporter (DAT, Chemicon), anti-vesicular monoamine transporter 2 (VMAT2, Pel-Freez, Arkansas LLC, USA) and anti-GFAP (Dako; Glostrup, Denmark). The antibodies used for the detection of FGF receptors (FGFR1, FGFR2, FGFR3), and EGF receptor (EGFR) were from Santa Cruz Biotechnology Inc., Burlingame, CA, USA. Monoclonal antibodies against β-tubulin III were obtained from Sigma, anti-TH was obtained from Chemicon, and anti-nestin (clone Rat 401) was from the Developmental Studies Hybridoma Bank (University of Iowa, Iowa City, IO). Monoclonal antibodies against A2B5, O4, and O1 were obtained in our laboratory as hybridoma supernatants. The secondary antibodies used were: biotinylated goat anti-mouse IgG (Zymed Laboratories; South San Francisco, CA), streptavidin–biotin–peroxidase complex (DakoCytomation), diaminobenzidine (DAB) + substrate–chromogen system (both from DakoCytomation), Alexa Fluor-568 goat anti-mouse IgG, Alexa Fluor-488 donkey anti-rat IgG, and Alexa Fluor-488 goat anti-rabbit IgG (1:400; all from Molecular Probes; Eugene, OR), Fluorescein-conjugated goat anti-mouse IgG (1:25; Jackson ImmunoResearch Laboratories Inc, West Grove, PA), Cy3-conjugated donkey anti-guinea pig IgG (1:500, Jackson ImmunoResearch Laboratories Inc.), and Rhodamine-conjugated goat anti-rabbit IgG (1:100, Chemicon International Inc.).

For immunocytochemical studies, cells were fixed with 4% paraformaldehyde for 10 min and immunostained for A2B5 (1:10), O4 (1:10), O1 (1:10), FGFR1 (1:100), FGFR2 (1:100), FGFR3 (1:50) and EGFR (1:50) as previously described (Reimers et al., 2001). Permeabilization for GFAP (1:500), β-tubulin III (1:200 for monoclonal and 1:3000 for polyclonal anti-β-tubulin III, and TH (1:500) was achieved by treating cultures with 0.05% Triton X-100 at 4°C for 5min. Immunofluorescent procedures were applied for neural antigens and FGFR3 detection, and immunoperoxidase methods for FGFR1 and FGFR2 visualization. Cover slips were mounted in a medium containing p-phenylenediamine and bis-Benzimide (Hoechst 33342; Sigma).

2.3 Western blot protein analysis

NSCs progeny were treated for 24 hr with TH inducers, in the absence or presence of 20 μM PD98058 or 10^{-7}M staurosporin, and proteins were processed for Western blot analysis to determine the relative levels of growth factor receptors (GFRs) and neural antigens. Cells were lysed with 0.5 M Tris-HCl buffer (pH 7.4) containing 0.24% Triton X-100, 10 mg/ml leupeptin, and 0.5 mM PMSF, all from Sigma. After 1 hr at 4°C, samples were centrifuged at 12,000g for 30 min. Total protein content was quantified using a BCA kit (Pierce; Rockford, IL). Aliquots of 30 μg of protein were separated by electrophoresis on 10% SDS-polyacrylamide minigels and transferred to nitrocellulose filters. Membranes were soaked in blocking solution (0.2 M Tris-HCl, 137 mM NaCl, and 3–5% dry skimmed milk, pH 7.6) and incubatedwith primary antibodies diluted in the same blocking solution: anti-FGFR1 (1:200), anti-FGFR3 (1:200), anti EGFR (1:250), anti-TH (1:10,000), and anti-GFAP (1:1000), anti-β-tubulin III (1:10,000), and anti-CNPasa (1:1000). After extensive washing membranes were incubated with the peroxidase-conjugated secondary antibodies diluted 1:1000 in blocking solution. The filters were developed with enhanced chemiluminescence Western blotting analysis, following the procedure described by the manufacturer (Amersham, Buckinghamshire, England). Membranes were immunolabeled for control charge using mouse anti-β actine (1:5000; Sigma Aldrich). Autoradiograms were quantified by computer-assisted videodensitometry.

2.4 Data analysis and cell counting

For Western blot analysis, results are expressed as mean ± SEM from two to four independent experiments. Where indicated, data represent the mean ± SEM of several cover slips. For each cover slip, stereological sampling of 25 visual fields (magnification of 200x or 400x) was performed by fluorescence microscopy. The number of cells was corrected for cover slip area. Statistical analyses were performed using Student's t-test or one-way ANOVA followed by Newman-Keuls multiple comparison test, and differences were considered significant at $p \leq 0.05$.

3. Results and discussion

3.1 Acquisition of a dopaminergic phenotype in the progeny of neural stem cells

Previously, we demonstrated that bFGF in combination with the PKA activator dbcAMP induced TH immunoreactivity in a subset of neurons and A2B5-positive progenitors derived from striatal EGF-expanded NSCs (striatal EGF-NSCs) (Lopez-Toledano et al., 2004). However, to present nothing is known regarding the ability of bFGF + dbcAMP to promote the expression of other features of dopaminergic mature neurons in these cells. As shown in Fig. 1A, bFGF + dbcAMP treatment increased by 1.5-fold TH protein expression in the progeny derived from striatal EGF-NSCs.

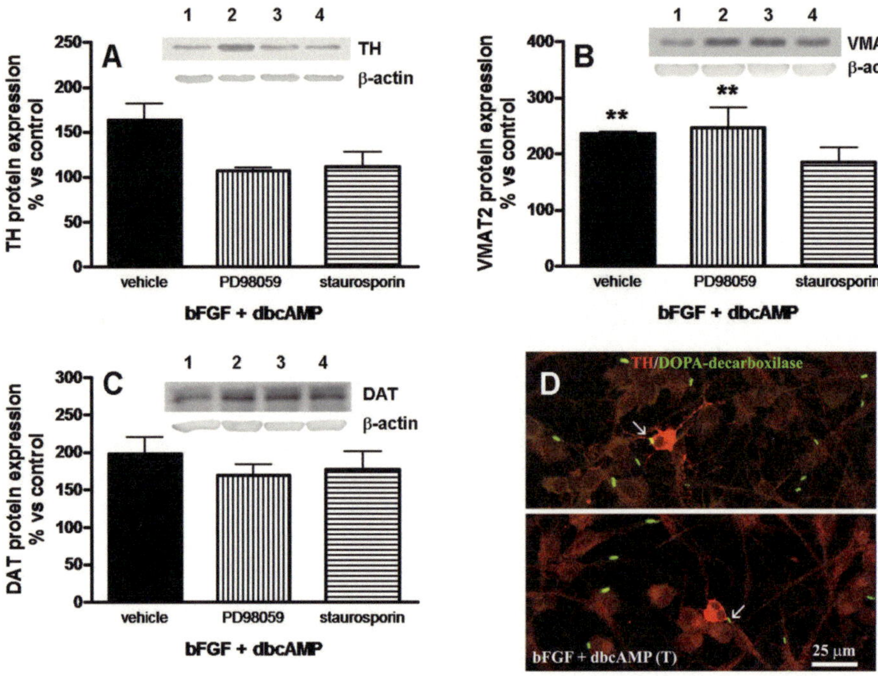

Fig. 1. Induction of a dopaminergic phenotype in the progeny derived from striatal EGF-NSCs. As shown in A-C, bFGF + dbcAMP treatment (black bars) increased the expression of TH (A), VMAT2 (B) and DAT (C). Note how 20 μM PD98059 (vertical line bars) and 10^{-7}M staurosporin (horizontal line bars) prevented TH protein overexpression (A). Line 1, control cultures, line 2, bFGF + dbcAMP treated cultures, line 3, bFGF + dbcAMP in the presence of PD98059, line 4, bFGF + dbcAMP in the presence of staurosporin. Results represent the mean ± SEM of 2 independent experiments. $^{**}p \leq 0.01$ vs control. D shows how in bFGF + dbcAMP treated cultures DOPA-decarboxilase immunoreactivity (D, green) associates with TH-positive cells (D, red, white arrows).

In the presence of 20 μM PD98059 or 10^{-7} M staurosporin, which are inhibitors of the mitogen activated protein kinase/extracellular signal-regulated kinase (MAPK/ERK1/2) and PKC, respectively, the raise in TH protein levels promoted by bFGF + dbcAMP treatment was prevented (Fig. 1A). These results are in agreement with our previous studies showing that the activation of the MAPK/ERK1/2 signaling pathway and PKC activity were required for the generation of TH-positive cells in striatal EGF-NSCs progeny (Lopez-Toledano et al., 2004). Besides TH, bFGF + dbcAMP treated cultures also showed DOPA-decarboxilase immunoreactivity that in some cases was associated with TH-positive cells (Fig. 1D). This enzyme catalyzes the conversion of levodopa (L-DOPA) in dopamine, so we may consider that under TH-inductive conditions striatal EGF-NSCs progeny is able to synthesize this neurotransmitter which deficit is involved in the progression of PD (Aron et al, 2011). As a matter of fact, our preliminary studies indicate that bFGF + dbcAMP treatment significantly increases L-DOPA levels by 1.7-fold ($p \leq 0.01$ vs control). Moreover, in the presence of the DOPA-decarboxilase inhibitor NSD-1015, L-DOPA levels were significantly higher than those observed under control conditions ($p \leq 0.001$), or bFGF + dbcAMP treatment ($p \leq 0.01$). Altogether, these results strongly suggest that under our experimental conditions TH and DOPA-decarboxilase are active enzymes.

We also analyzed the effect of the TH-inductive treatment in the expression of other dopaminergic markers such as VMAT2 and DAT. Thus, bFGF + dbcAMP increased by 2-fold VMAT2 and DAT protein levels (Fig. 1B, C). The expression of both proteins was not regulated by the MAPK/ERK1/2 signaling pathway or PKC activity because neither PD98059, nor staurosporin prevented VMAT2 or DAT upregulation (Fig. 1C, D). In our cultures, bFGF + dbcAMP treatment increased the phosphorylation of the cyclic AMP response element binding protein (CREB) probably through the activation of PKA by dbcAMP (Lopez-Toledano et al., 2004). Because this transcription factor regulates the expression of catecholamine biosynthetic enzymes and transporters (Lewis-Tuffin et al., 2004; Lim et al., 2000; Watson et al., 2001), its activation could be responsible for the increase in VMAT2 and DAT protein levels observed in this study. Besides, bFGF + dbcAMP could stimulate striatal EGF-NSCs to release neurotrophins that are able to increase the expression of TH and monoamine transporters in neural precursors (Maciaczyk et al., 2008; Sun et al., 2004), and in the damaged brain (Emborg et al., 2008).

3.2 Dopaminergic inductors modulate the expression and cellular localization of fibroblast growth factor receptors in the progeny of neural stem cells

Striatal EGF-NSCs and their progeny express a subset of FGFRs, so we were interested in to determine the role played by the different FGFR subtypes in the acquisition of dopaminergic features in these cells. FGFR1 is expressed in nestin-positive neural precursors (Lobo et al, 2003), and was down-regulated during their differentiation in neurons and glial cells (Reimers et al., 2001). Under dopaminergic-inductive conditions, FGFR1 immunoreactivity was higher than in controls (Fig.2A, C). Similarly, FGFR1 protein expression was significantly raised by more than 10-fold in bFGF + dbcAMP treated cultures (Fig. 2B). FGFR1 up-regulation was not affected in the presence of MAPK/ERK1/2 or PKC inhibitors (Fig. 2B, D), indicating that other signaling pathways are involved in FGFR1 over-expression. Growth factors are able to stimulate the phosphatidylinositol 3-kinase (PI3K)/Akt signaling transduction pathway in NSCs to promote their proliferation, survival and differentiation (Lim et al., 2007; Meng et al, 2011; Nguyen et al., 2009; Torroglosa et al., 2007). As mentioned above, FGFR1 is expressed in nestin-positive neural precursors. In our cultures, bFGF + dbcAMP treatment significantly increased nestin protein expression by 2-fold ($p \leq 0.05$ vs control), and neither PD98059 nor staurosporin were able to abolish this effect. Because similar results were observed under bFGF treatment (Reimers et al., 2001), our results strongly suggest that the GF promotes the survival and/or proliferation of resting FGFR1-/nestin-positive neural precursors probably through the stimulation of PI3K/Akt signaling pathway.

Dopaminergic-inductive conditions modulate FGFR3 protein expression and cellular localization. Under control conditions, FGFR3 was localized in the membrane of GFAP-positive astrocytes (Fig. 3A), in the cell bodies of β-tubulin III-positive neurons (Fig. 3B), and in the nuclei of O4-positive preoligodendrocytes (Fig. 3C). Western blot analysis revealed that FGFR3 levels were significantly up-regulated by bFGF + dbcAMP treatment (Fig. 3D). This experimental condition changed the morphology of GFAP-positive cells that showed longer and thinner processes where FGFR3 immunoreactivity was observed (Fig. 3E). Moreover, FGFR3 immunostaining was also detected in their nuclei (Fig. 3E). bFGF + dbcAMP treatment also affected to the morphology of O4-positive cells and increased FGFR3 immunoreactivity in their nuclei (Fig. 3G). Besides, some TH-positive cells showing a neuronal morphology co-expressed FGFR3 in their cell bodies (Fig. 3F). Other authors have shown the presence of two forms of FGFR3 in the nucleus of malignant and non-malignant epithelial cells (Johnston et al., 1995; Zammit et al., 2001). However, to our knowledge this is the first study reporting the nuclear localization of FGFR3 in neural cells. From our results, it is difficult to determine whether the higher levels of FGFR3 nuclear staining observed in GFAP-positive cells are due to the increase in protein expression, or to the translocation of FGFR3 to perinuclear localization.

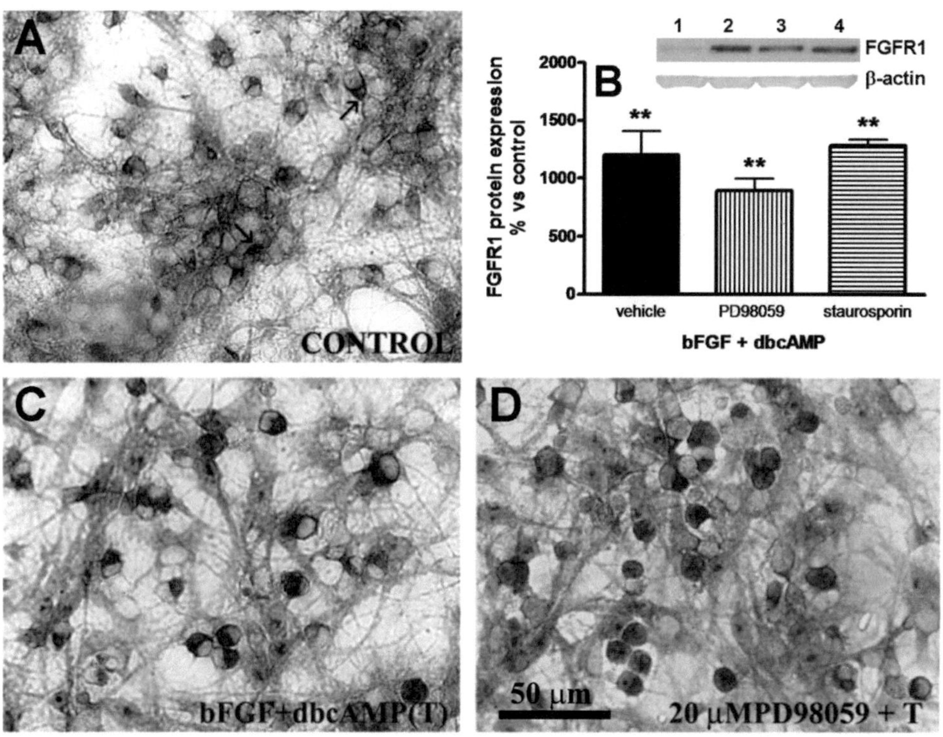

Fig. 2. Tyrosine hydroxylase inductors upregulate FGFR1 protein expression in the progeny derived from striatal EGF-NSCs. A, C an D show FGFR1 immunocytochemical staining in control conditions (A), and in cultures treated with bFGF + dbcAMP (T) in the absence (C) or presence of 20 μM PD98059 (D). Note how in the presence of TH-inductors FGFR1 immunoreactivity is increased (C). bFGF + dbcAMP treatment also upregulates FGFR1 protein expression (B, black bar), and this effect is not prevented by 20 μM PD98059 (B, vertical lines bar) or 10^{-7}M staurosporin (B, horizontal lines bar). Line 1, control cultures, line 2, bFGF + dbcAMP treated cultures, line 3, bFGF + dbcAMP in the presence of PD98059, line 4, bFGF + dbcAMP in the presence of staurosporin. Results represent the mean ± SEM of 2 independent experiments. **$p \leq 0.01$ vs control.

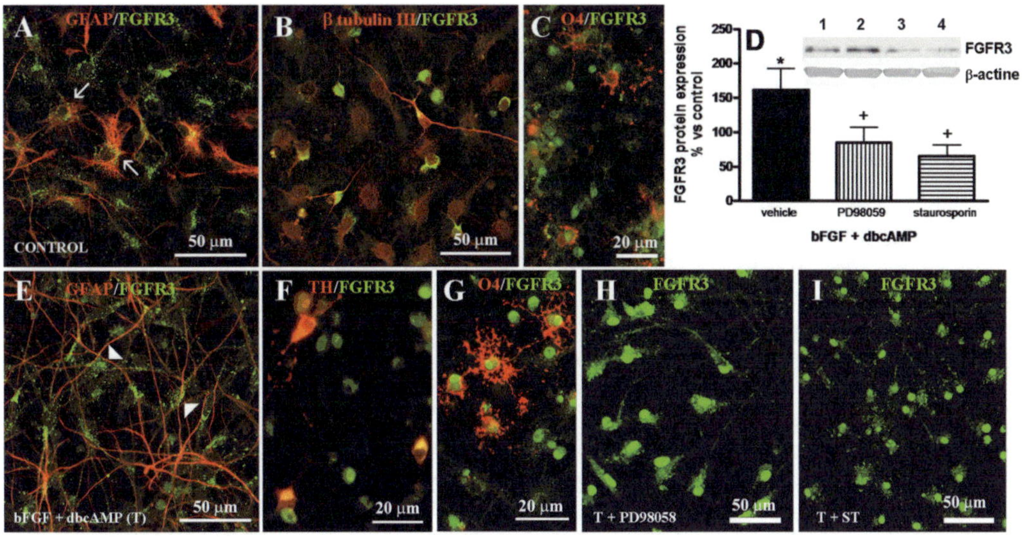

Fig. 3. Tyrosine hydroxylase inductors modulate FGFR3 protein expression and cellular localization in the progeny derived from striatal EGF-NSCs. A-C show how under control conditions FGFR3 immunoreactivity (green) is localized in the membranes of GFAP-positive astrocytes (A, white arrows), cell bodies of β-tubulin III-positive neurons (B, yellow), and nuclei of O4-positive preoligodendrocytes (C). E-I show FGFR3 immunostaining in basic FGF + dbcAMP treated cultures (T). Under TH-inductive conditions, FGFR3 is localized in the nuclei of GFAP-positive astrocytes (E, white arrowheads) and preoligodendrocytes (G), and in the cell bodies of TH-positive neurons (F, yellow). Note, how astrocytes (E, red) and O4-positive cells (G, red) show longer and thinner extensions in bFGF + dbcAMP treated cultures, and how FGFR3 immunoreactivity is decreased in the presence of 20 μM PD98059 (H, green) and 10^{-7}M staurosporin (I, green). TH inductors also upregulate FGFR3 protein expression (D, black bar), and this effect is prevented by 20 μM PD98059 (D, vertical lines bar) or 10^{-7}M staurosporin (D, horizontal lines bar). Line 1, control cultures, line 2, bFGF + dbcAMP treated cultures, line 3, bFGF + dbcAMP in the presence of PD98059, line 4, bFGF + dbcAMP in the presence of staurosporin. Results represent the mean ± SEM of 4 independent experiments. $*p \leq 0.05$ vs control, $+p \leq 0.05$ vs vehicle.

Interestingly, the nuclear translocation of FGFRs in response to bFGF has been reported in reactive astrocytes (Clarke et al., 2001) and Swiss 3T3 fibroblasts (Maher, 1996). Moreover, the antibody used in this study recognizes a 135 kDa form of FGFR3 that showed a mix of nuclear and cytoplasmic localization in epithelial cells that depends on its degree of activation by different members of the FGF family (Zammit et al., 2001).

Similarly to TH protein expression, FGFR3 up-regulation was not observed in the presence of 20 μM PD98059 or 10^{-7}M staurosporin (Fig. 3D). FGFR3 immunoreactivity was also reduced in both experimental conditions (Fig. 3H, I). Moreover, PKC inhibition prevented the morphological changes promoted by bFGF + dbcAMP treatment (Fig. 3I). These results suggest that the modulation of FGFR3 and TH induction are two events closely associated. As a matter of fact, FGFRs nuclear localization has been related with the differentiation of neural progenitor cells (Fang et al., 2005; Stachowiak et al., 2003), and TH gene expression

(Peng et al., 2002). We should comment that not all TH-positive cells were FGFR3 immunoreactive (our unpublished observations), so it seems that they are not direct targets for bFGF. Under our TH-inductive conditions, GFAP protein expression was significantly raised by more than 1.6-fold ($p \leq 0.05$ vs control). GFAP over-expression and morphological changes are features of reactive glia which is able to synthesize trophic factors involved in neuronal survival and differentiation (Barreto et al., 2011). Because glial derived factors are also involved in the differentiation of NSCs in TH-immunoreactive dopaminergic neurons (Anwar et al., 2008; Maciaczyk et al., 2008; Sun et al., 2004), from our results we propose that stimulation of FGFR3 localized in glial cells mediate the release of several unknown factors that in combination with PKA activators stimulate TH-induction in the target cells. In fact, glial conditioned medium in combination with PKA activators elicits the expression of TH in the progeny of striatal EGF-NSCs (Reimers et al., 2008).

We have also analyzed the effects of bFGF + dbcAMP treatment in FGFR2 expression. Under the experimental conditions presented in this study, FGFR2 nuclear localization and FGFR2 protein expression were not affected (data not shown), indicating that probably this FGFR is not involved in the acquisition of a dopaminergic phenotype in striatal EGF-NSCs progeny.

3.3 Tyrosine hydroxylase-inducing cues trigger nuclear epidermal growth factor receptor accumulation in the progeny of neural stem cells

EGFR stimulation is essential for the proliferation of striatal EGF-NSCs (Bazán et al., 1998, 2006; Reimers et al., 2001). EGFR protein expression has been detected during the differentiation of these cells in neurons and glia (Lobo et al., 2003); however, to present nothing is known regarding its cellular localization, and the role, if any, played by this receptor in the differentiation of striatal EGF-NSCs to TH-immunopositive cells. As shown in Fig. 4A, EGFR-positive cells were observed in 8 dpp control cultures. At this experimental time, EGFR imuunoreactivity was localized in the cell bodies of β-tubulin III- (Fig. 4B) and O4-positive cells (Fig. 4D). Moreover, EGFR immunostaining was also observed in the membrane of GFAP-positive astrocytes (Fig. 4C). TH-inductive conditions did not affect EGFR protein expression (Fig. 4K), but promoted the translocation of EGFR to the nuclei in many cells (Fig. 4E). Nuclear EGFR immunoreactivity was observed in small spots, suggesting that the EGFR could be localized in the nucleolar compartment. Functional nuclear EGFR have been described in normal and tumoral cells (Jaganathan et al., 2011; Lo & Hung, 2007; Xu et al., 2009), but to our knowledge this is the first study reporting the nuclear localization of EGFR in NSCs and its derived progeny. Interestingly, a recent report discusses the possibility that proliferation and differentiation of NSCs could be controlled by nuclear receptors (Katayama et al., 2005). Moreover, NSCs proliferation seems to be regulated by a nucleolar mechanism that involves the interaction of proteins located in their nucleoli (Tsai & McKay, 2002). Neither EGFR protein expression (Fig. 4K), nor nuclear localization (Fig. 4I) were affected in the presence of PD98059. However, both parameters were significantly reduced in the presence of the PKC inhibitor staurosporin (Fig. 4J, K). As a matter of fact, a recent report demonstrates that PKC activation triggers nuclear EGFR accumulation in a bronchial carcinoma cell line (Wanner et al., 2008). Besides, the nuclear translocalization of EGFR may require its phosphorylation at Ser-229 by Akt (Huang et al., 2011). Further experiments are warranted to determine whether the PI3K/Akt signaling pathway mediates bFGF + dbcAMP-induced EGFR nuclear translocalization in striatal EGF-NSCs.

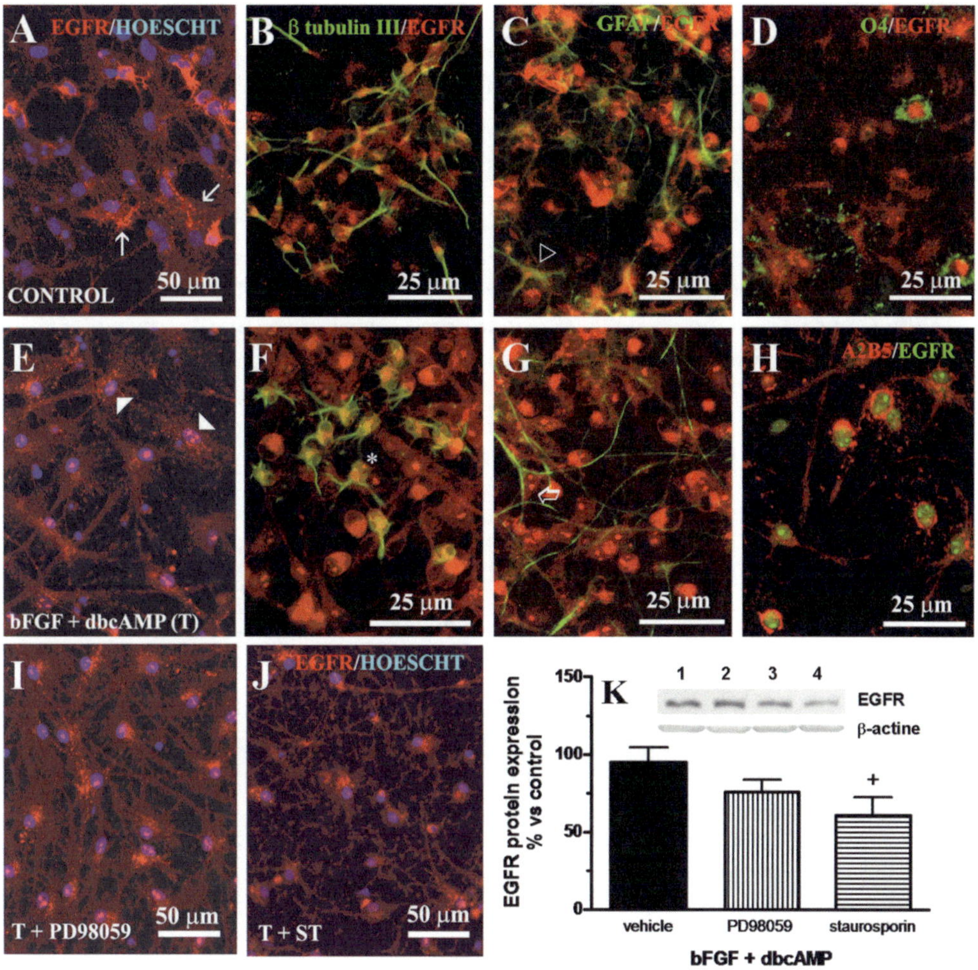

Fig. 4. Tyrosine hydroxylase inductors elicit nuclear translocalization of EGFR in the progeny derived from striatal EGF-NSCs. In control cultures EGFR immunoreactivity is localized the cell bodies (A, red), and in membranes (A, white arrows). Under this condition, EGFR is expressed in the cell bodies of neurons (B, yellow) and O1-positive oligodendrocytes (D), and in the membranes of GFAP-positive astrocytes (C, open triangle). Note how bFGF + dbcAMP (T) translocates EGFR to the nuclear compartment (E, white arrowheads), and how nuclear EGFR is observed in a few β-tubulin III-positive neurons (F, white star), a few astrocytes (G, open arrow), and in many A2B5-positive progenitors (H). Neither EGFR immunoreactivity nor nuclear EGFR localization are affected in the presence of 20 μM PD98059 (I, red), but 10^{-7} M staurosporine significantly decreases both parameters (J, red). bFGF + dbcAMP treatment in the absence (K, black bar) or presence of 20 μM PD98059 (K, vertical lines bar) does not affect EGFR protein expression; however, 10^{-7} M staurosporin significantly reduces EGFR levels (K, horizontal lines bar). Line 1, control cultures, line 2, bFGF + dbcAMP treated cultures, line 3, bFGF + dbcAMP in the presence of PD98059, line 4, bFGF + dbcAMP in the presence of staurosporin. Results represent the mean ± SEM of 2 independent experiments. $+p \leq 0.05$ vs vehicle.

Double immunostaining showed that some neurons (Fig. 4F) and a few GFAP-positive astrocytes (Fig.4G) presented nuclear EGFR immunoreactivity. Besides, a population of cells showing EGFR nuclear localization colabeled with A2B5-positive cells (Fig.4H). Emerging evidence suggest that EGFR nuclear translocalization regulates gene expression and mediate other cellular processes such as DNA repair (Chen & Nirodi, 2007; Dittmann et al., 2010; Lo, 2010). Its transcriptional activity depends on its C-terminal transactivation domain, and its physical and functional interaction with other transcription factors that lead the activation of genes (Lo & Hung, 2006). Under our experimental conditions TH is expressed in 4% of the total population of β-tubulin III-positive neurons, and 20% of the TH-positive cells colabel with A2B5 (Lopez-Toledano et al., 2004), so we may hypothesize that nuclear EGFR could regulate TH gene expression in both cell types.

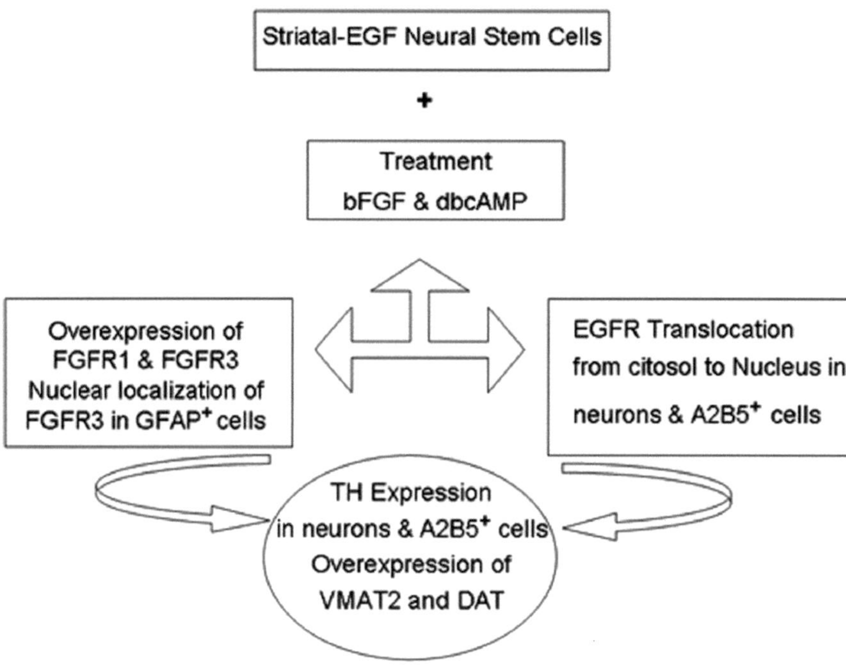

Fig. 5. Schematic diagram illustrating how bFGF + dbcAMP treatment induces the expression of specific features of dopaminergic neurons in the progeny derived from striatal EGF-NSCs through the modulation of GFRs expression and cellular localization.

4. Conclusions

1. In striatal EGF-NSCs, bFGF + dbcAMP treatment up-regulates the expression of the specific dopaminergic markers TH, DAT and VMAT2.
2. FGFR1 and nestin protein levels are significantly raised in bFGF + dbcAMP treated cultures, suggesting the survival and/or proliferation of undifferentiated neural precursors under this experimental condition.

3. bFGF + dbcAMP treatment increases FGFR3 protein expression and FGFR3 immunoreactivity in the glial progeny derived from striatal-EGF-NSCs. Moreover, this experimental condition up-regulates GFAP protein levels, and elicits the translocalization of FGFR3 to the nucleus in reactive GFAP-positive astrocytes.
4. In the presence of MAPK/ERK1/2 or PKC inhibitors, TH protein expression, FGFR3 up-regulation, and glial reactivity are partially prevented, suggesting that in bFGF + dbcAMP treated cultures, the modulation of FGFR3 in glial cells and TH induction are two events closely associated.
5. Under bFGF + dbcAMP treatment, EGFR immunostaining shows a nuclear localization in β-tubulin III-positive neurons and in A2B5-positive precursors derived from striatal EGF-NSCs. Because both cell types are able to express TH, we hypothesize that the nuclear translocalization of EGFR is necessary for the induction of TH in these cells.

As summarize in Fig. 5, our results demonstrate that in striatal EGF-NSCs, bFGF and dbcAMP treatment induces TH protein expression through the stimulation of FGFR3 expressed in glia, and nuclear translocalization of EGFR in the target cells. Under these TH-inductive conditions, striatal EGF-NSCs progeny acquire other specific features of mature dopaminergic neurons, so they can be considered as an excellent tool for stem cell-based replacement therapies in PD.

5. Acknowledgment

The authors wish to acknowledge the editorial for its invitation to contribute with a full chapter for the book under the working title "Neural Stem Cells", ISBN 978-953-307-795-6. This work was supported by the Fondo de Investigaciones Sanitarias (FISS PI060315) and Agencia Laín Entralgo (NDG7/09). RG-G and CM were the recipients of a FiBio HRyC fellowship, and LC was the recipient of an Agencia Laín Entralgo fellowship. We thank Macarena Rodríguez for technical help, and Dr. Maria José Casarejos for the determination of L-DOPA levels by HPLC.

6. References

Anwar MR, Andreasen CM, Lippert SK, Zimmer J, Martinez-Serrano A & Meyer M (2008). Dopaminergic differentiation of human neural stem cells mediated by co-cultured rat striatal brain slices. *J Neurochem.*,Vol. 105, No. 2, (April 2008), pp.(460-70), ISSN 0022-3042.

Arias-Carrión O & Yuan TF (2009). Autologous neural stem cell transplantation: A new treatment option for Parkinson's disease? *Med Hypotheses.*, Vol. 73, No. 5, (November 2009), pp. (757-759), ISSN 0306-9877.

Aron L & Klein R (2011). Repairing the parkinsonian brain with neurotrophic factors. *Trends Neurosci.*, Vol. 34, No. 2, (February 2011), pp. (88-100), ISSN 0166-2236.

Barreto GE, Gonzalez J, Torres Y & Morales L (2011). Astrocytic-neuronal crosstalk: Implications for neuroprotection from brain injury. *Neurosci Res.*, (June 2011) [Epub ahead of print]

Bazán E, López-Toledano MA, Redondo C, Alcázar A, Paíno CL & Herranz AS. (1998). Characterization of rat neural stem cells from embryonic striatum and mesencephalon during in vitro differentiation. In: *Understanding glial cells,*

Castellano B, González B, Nieto-Sampedro M, editors, pp. 133–147. Kluwer Academic Publishers, ISBN 0-7923-8140-8, Dordrecht, The Netherlands.

Bazán E, Alonso FJ, Redondo C, López-Toledano MA, Alfaro JM, Reimers D, Herranz AS, Paíno CL, Serrano AB, Cobacho N, Caso E & Lobo MV (2004). In vitro and in vivo characterization of neural stem cells. *Histol Histopathol,*Vol. 19, No. 4, (October 2004), pp. (1261-75), ISSN 0213-3911.

Bazán E, Herranz AS, Reimers D, Lobo MVT, Redondo C, López-Toledano MA, Gonzalo-Gobernado R, Asensio MJ & Alonso R. (2006). Neural stem cells and taurine. In: *Neural stem cell research,* Greer EV, editor, pp 99–114. Nova Science Publishers, Inc, ISBN 1-59454-846-3, New York.

Bjugstad KB, Teng YD, Redmond DE Jr, Elsworth JD, Roth RH, Cornelius SK, Snyder EY & Sladek JR Jr (2008). Human neural stem cells migrate along the nigrostriatal pathway in a primate model of Parkinson's disease. *Exp Neurol.,* Vol. 211, No. 2, (June 2008), pp. (362-369), ISSN 0014-4886.

Chen DJ & Nirodi CS (2007). The epidermal growth factor receptor: a role in repair of radiation-induced DNA damage. *Clin Cancer Res.,* Vol. 13, No. 22 Pt 1,(November 2007), pp. (6555-6560), ISSN 0008-5472.

Ciccolini F & Svendsen CN (1998). Fibroblast growth factor 2 (FGF-2) promotes acquisition of epidermal growth factor (EGF) responsiveness in mouse striatal precursor cells: Identification of neural precursors responding to both EGF and FGF-2. *J Neurosci.,* Vol. 18, No. 19, (October 1998), pp. (7869–7880), ISSN 0270-6474.

Clarke WE, Berry M, Smith C, Kent A & Logan A (2001).Coordination of fibroblast growth factor receptor 1 (FGFR1) and fibroblast growth factor-2 (FGF-2) trafficking to nuclei of reactive astrocytes around cerebral lesions in adult rats. *Mol Cell Neurosci,* Vol 17, No. 1, (January 2001), pp. (17-30), ISSN 1044-7431.

Daadi MM & Weiss S (1999). Generation of tyrosine hydroxylase-producing neurons from precursors of the embryonic and adult forebrain. *J Neurosci.,* Vol. 19, No. 11,(June 1999), pp. (4484-97), ISSN 0270-6474.

Dittmann K, Mayer C & Rodemann HP (2010). Nuclear EGFR as novel therapeutic target: insights into nuclear translocation and function. *Strahlenther Onkol.,* Vol. 186, No.1, (January 2010), pp.(1-6), ISSN 0179-7158.

Emborg ME, Ebert AD, Moirano J, Peng S, Suzuki M, Capowski E, Joers V, Roitberg BZ, Aebischer P & Svendsen CN (2008). GDNF-secreting human neural progenitor cells increase tyrosine hydroxylase and VMAT2 expression in MPTP-treated cynomolgus monkeys. *Cell Transplant.* Vol. 17, No. 4, pp, (383-95), E-ISSN 1555-3892.

Fang X, Stachowiak EK, Dunham-Ems SM, Klejbor I & Stachowiak MK (2005). Control of CREB-binding protein signaling by nuclear fibroblast growth factor receptor-1: a novel mechanism of gene regulation. *J Biol Chem.,* Vol. 280, No. 31, (August 2005), pp. (28451-28462), ISSN 0021-9258.

Gil-Perotín S, Alvarez-Buylla A & García-Verdugo JM (2009). Identification and characterization of neural progenitor cells in the adult mammalian brain. *Adv Anat Embryol Cell Biol.,* Vol. 203, pp.(1-101), ISSN 0301-5556.

Gritti A, Parati EA, Cova L, , Frölichsthal-Schoeller P, Galli R, Wanke E, Faravelli L, Morassutti DJ, Roisen F, Nickel DD & Vescovi AL (1996). Multipotent stem cells from the adult mouse brain proliferate and self-renewing response to basic

.broblast growth factor. *J Neurosci.*, Vol. 16, No. 3, (February 1996), pp.(1091–1100), ISSN 0270-6474.

Gritti A, Frölichsthal-Schoeller P, Galli R, Paraati EA, Cova L, Pagano SF, Bjornson CR & Vescovi AL (1999). Epidermal and fibroblast growth factors behave as mitogenic regulators for a single multipotent stem cell-like population from the subventricular region of the adult mouse forebrain. *J Neurosci.*, Vol. 19, No. 9,(May 1999), pp. (3287–3297), ISSN 0270-6474.

Hagg T (2005). Molecular regulation of adult CNS neurogenesis: an integrated view. *Trends Neurosci.*, Vol. 28, No. 11, (November 2005), pp.(589-595), ISSN 0166-2236.

Hagg T (2009). From neurotransmitters to neurotrophic factors to neurogenesis. *Neuroscientist.*, Vol. 15, No. 1, (February 2009), pp.(20-27), ISSN 1073-8584.

Huang WC, Chen YJ, Li LY, Wei YL, Hsu SC, Tsai SL, Chiu PC, Huang WP, Wang YN, Chen CH, Chang WC, Chang WC, Chen AJ, Tsai CH & Hung MC (2011). Nuclear translocation of epidermal growth factor receptor by Akt-dependent phosphorylation enhances breast cancer-resistant protein expression in gefitinib-resistant cells. *J Biol Chem*, Vol. 286, No. 23, (June 2011), pp. (20558-68), ISSN 0021-9258.

Jaganathan S, Yue P, Paladino DC, Bogdanovic J, Huo Q & Turkson J (2011). A functional nuclear epidermal growth factor receptor, SRC and Stat3 heteromeric complex in pancreatic cancer cells. *PLoS One.*, Vol. 6, No. 5, (May 2011), e19605, ISSN 1932-6203.

Johnston CL, Cox HC, Gomm JJ & Coombes RC (1995). Fibroblast growth factor receptors (FGFRs) localize in different cellular compartments. A splice variant of FGFR-3 localizes to the nucleus. *J Biol Chem*, Vol. 270, No. 51 (December 1995), pp. (30643-50), ISSN 0021-9258.

Katayama K, Wada K, Nakajima A, Kamisaki Y & Mayumi T (2005).Nuclear receptors as targets for drug development: the role of nuclear receptors during neural stem cell proliferation and differentiation. *J Pharmacol Sci*, Vol. 97, No. 2, (February 2005), pp. (171-6), ISSN 1347-8613.

Kennea NL & Mehmet H (2002). Neural stem cells. *J Pathol.*, Vol. 197 , No. 4, (July 2002), pp. (536-550), ISSN 1096-9896.

Lewis-Tuffin LJ, Quinn PG & Chikaraishi DM (2004). Tyrosine hydroxylase transcription depends primarily on cAMP response element activity, regardless of the type of inducing stimulus. *Mol Cell Neurosci.*,Vol. 25, No. 3, (March 2004), pp. (536-547), ISSN 1044-7431.

Lim J, Yang C, Hong SJ & Kim KS (2000). Regulation of tyrosine hydroxylase gene transcription by the cAMP-signaling pathway: involvement of multiple transcription factors. *Mol Cell Biochem* , Vol. 212, No. 1-2, (September 2000), pp. (51-60), ISSN 0300-8177.

Lim MS, Nam SH, Kim SJ, Kang SY, Lee YS & Kang KS (2007). Signaling pathways of the early differentiation of neural stem cells by neurotrophin-3. *Biochem Biophys Res Commun.*, Vol. 357, No. 4, (June 2007), pp. (903-909), ISSN 0006-291X.

Lo HW & Hung MC (2006). Nuclear EGFR signalling network in cancers: linking EGFR pathway to cell cycle progression, nitric oxide pathway and patient survival. *Br J Cancer*, Vol. 94, No. 2, (January 2011), pp. (184-8), ISSN 0007-0920

Lo HW & Hung MC (2007). Nuclear EGFR signalling network in cancers: linking EGFR pathway to cell cycle progression, nitric oxide pathway and patient survival. *Br J Cancer* ,Vol. 96 Suppl, pp. (16-20), ISSN 0007-0920.

Lo HW (2010). Nuclear mode of the EGFR signaling network: biology, prognostic value, and therapeutic implications. *Discov Med.*,Vol 10, No 50, (July 2010), pp. (44-51), ISSN 1539-6509.

Lobo MV, Alonso FJ, Redondo C, López-Toledano MA, Caso E, Herranz AS, Paíno CL, Reimers D & Bazán E (2003). Cellular characterization of epidermal growth factor-expanded free-floating neurospheres. *J Histochem Cytochem* ,Vol. 51, No. 1, (January 2003), pp. (89-103), ISSN 0022-1554.

López-Toledano MA, Redondo C, Lobo MV, Reimers D, Herranz AS, Paíno CL & Bazán E (2004). Tyrosine hydroxylase induction by basic fibroblast growth factor and cyclic AMP analogs in striatal neural stem cells: role of ERK1/ERK2 mitogen-activated protein kinase and protein kinase C. *J Histochem Cytochem.*, Vol. 52, No. 9, (September 2004), pp. (1177-1189), ISSN 0022-1554.

Maciaczyk J, Singec I, Maciaczyk D & Nikkhah G (2008). Combined use of BDNF, ascorbic acid, low oxygen, and prolonged differentiation time generates tyrosine hydroxylase-expressing neurons after long-term in vitro expansion of human fetal midbrain precursor cells. *Exp Neurol.*, Vol. 213, No. 2, (October 2008), pp. (354-362), ISSN 0014-4886.

Maher PA (1996). Nuclear Translocation of fibroblast growth factor (FGF) receptors in response to FGF-2. J Cell Biol, Vol. 134, No. 2, (July 1996), pp. (529-36), ISSN 0021-9525.

Meng X, Arocena M, Penninger J, Gage FH, Zhao M & Song B (2011). PI3K mediated electrotaxis of embryonic and adult neural progenitor cells in the presence of growth factors. *Exp Neurol.*, Vol. 227, No. 1, (January 2011), pp. (210-217), ISSN 0014-4886.

Merkle FT & Alvarez-Buylla A (2006). Neural stem cells in mammalian development. *Curr Opin Cell Biol.*, Vol 18, No. 6, (December 2006), pp. (704-709), ISSN 0955-0674.

Nguyen N, Lee SB, Lee YS, Lee KH & Ahn JY (2009). Neuroprotection by NGF and BDNF against neurotoxin-exerted apoptotic death in neural stem cells are mediated through Trk receptors, activating PI3-kinase and MAPK pathways. *Neurochem Res.*,Vol. 34, No. 5, (May 2009), pp. (942-951), ISSN 1573-6903.

Ninkovic J & Gotz M (2007). Signaling in adult neurogenesis: from stem cell niche to neuronal networks. *Curr Opin Neurobiol.*, Vol. 17, No. 3, (June 2007), pp. (338-344), ISSN 0959-4388.

Palmer TD, Markakis EA, Willhoite AR, Safar F & Gage FH (1999). Fibroblast growth factor-2 activates a latent neurogenic program in neural stem cells from diverse regions of the adult CNS. *J Neurosci.*, 19, No. 19, (October 1999), pp. (8487–8497), ISSN 0270-6474.

Peng H, Moffett J, Myers J, Fang X, Stachowiak EK, Maher P, Kratz E, Hines J, Fluharty SJ, Mizukoshi E, Bloom DC & Stachowiak MK (2001). Novel nuclear signaling pathway mediates activation of fibroblast growth factor-2 gene by type 1 and type 2 angiotensin II receptors. *Mol Biol Cell.*, Vol. 12, No. 2, (February 2001), pp. (449-462), ISSN 1059-1524.

Pluchino S, Zanotti L, Deleidi M & Martino G (2005). Neural stem cells and their use as therapeutic tool in neurological disorders. *Brain Res Brain Res Rev.*, Vol. 48, No. 2, (April 2005), pp. 211-219, ISSN 0165-0173.

Raff MC, Abney ER, Miller RH (1984) Two glial cell lineages diverge prenatally in rat optic nerve. *Dev Biol* .,Vol. 106, No. 1, (November 1984), pp.(53–60), ISSN 0012-1606.

Redondo C, López-Toledano MA, Lobo MV, Gonzalo-Gobernado R, Reimers D, Herranz AS, Paíno CL & Bazán E (2007). Kainic acid triggers oligodendrocyte precursor cell proliferation and neuronal differentiation from striatal neural stem cells. *J Neurosci Res.*,Vol. 85, No. 6, (May 2007), pp. (1170-1182), ISSN 0360-4012.

Reimers D, López-Toledano MA, Mason I, Cuevas P, Redondo C, Herranz AS, Lobo MV & Bazán E (2001). Developmental expression of fibroblast growth factor (FGF) receptors in neural stem cell progeny. Modulation of neuronal and glial lineages by basic FGF treatment. *Neurol Res.*, Vol. 23, No. 6, (September 2001), pp. (612-621), ISSN 0161-6412.

Reimers D, Gonzalo-Gobernado R, Herranz AS, Osuna C, Asensio MJ, Baena S, Rodríguez M & Bazán E (2008). Driving neural stem cells towards a desired phenotype. *Curr Stem Cell Res Ther.*,Vol. 3, No. 4, (December 2008), pp.(247-253), ISSN 1574-888X.

Reimers D, Osuna C, Gonzalo-Gobernado R, Herranz AS, Díaz-Gil JJ, Jiménez-Escrig A, Asensio MJ, Miranda C, Rodríguez-Serrano M & Bazán E (2010). Liver Growth Factor Promotes the Survival of Grafted Neural Stem Cells in a Rat Model of Parkinson's Disease. *Curr Stem Cell Res Ther.*, (December 2010), [Epub ahead of print], ISSN 1574-888X.

Reynolds BA & Weiss S (1992). Generation of neurons and astrocytes from isolated cells of the adult mammalian central nervous system. *Science.*, Vol. 255, No. 5052, (March 1992), pp. (1707-1710), ISSN 0036-8075.

Reynolds BA, Tetzlaff W & Weiss S (1992). A multipotent EGF responsive striatal embryonicprogenitor cell produces neurons and astrocytes. *J Neurosci.*, Vol. 12, No. 11, (November 1992), pp. (4565–4574), ISSN 0270-6474.

Schnitzer J & Schachner M (1982) Cell type specificity of a neural cell surface antigen recognized by the monoclonal antibody A2B5. *Cell Tissue Res.*, Vol. 224, No. 3, pp. (625–636), ISSN 0302-766X.

Stachowiak EK, Fang X, Myers J, Dunham S & Stachowiak MK (2003). cAMP-induced differentiation of human neuronal progenitor cells is mediated by nuclear fibroblast growth factor receptor-1 (FGFR1). *J Neurochem.*, Vol. 84, No. 6, (March 2003), pp. (1296-1312), ISSN 0022-3042.

Sun ZH, Lai YL, Li P, Zuo HC & Xie ZP (2004). GDNF augments survival and differentiation of TH-positive neurons in neural progenitor cells. *Cell Biol Int.*, Vol. 28, No. 4, pp. (323-325), ISSN 1065-6995.

Torroglosa A, Murillo-Carretero M, Romero-Grimaldi C, Matarredona ER, Campos-Caro A & Estrada C (2007). Nitric oxide decreases subventricular zone stem cell proliferation by inhibition of epidermal growth factor receptor and phosphoinositide-3-kinase/Akt pathway. *Stem Cells* Vol. 25, No. 1, (January 2007), pp. (88-97), ISSN 1066-5099.

Tsai RY & McKay RD (2002). A nucleolar mechanism controlling cell proliferation in stem cells and cancer cells. *Genes Dev*, Vol. 16, No. 23, (December 2002), pp. (2991-3003), ISSN 0890-9369.

Vescovi AL, Rietze R, Magli MC, Bjornson C (2002). Hematopoietic potential of neural stem cells. *Nat Med.,* Vol. 8, No. 6, (June 2002), pp. (535; author reply 536-537), ISSN 1078-8956.

Wanner G, Mayer C, Kehlbach R, Rodemann HP & Dittmann K (2008). Activation of protein kinase Cepsilon stimulates DNA-repair via epidermal growth factor receptor nuclear accumulation. *Radiother Oncol.,* Vol. 86, No. 3, (March 2008), pp. (383-390), ISSN 0167-8140.

Watson F, Kiernan RS, Deavall DG, Varro A & Dimaline R (2001). Transcriptional activation of the rat vesicular monoamine transporter 2 promoter in gastric epithelial cells: regulation by gastrin. *J Biol Chem.,*Vol. 276, No. 10, (March 2001), pp. (7661-7671), ISSN 0021-9258.

Weiss S, Reynolds BA, Vescovi AL, Morshead C, Craig CG & van der Kooy D (1996). Is there a neural stem cell in the mammalian forebrain? *Trends Neurosci.,* Vol. 19, No. 9, (September 1996), pp. (387-393), ISSN 0166-2236.

Xu Y, Shao Y, Zhou J, Voorhees JJ & Fisher GJ (2009). Ultraviolet irradiation-induces epidermal growth factor receptor (EGFR) nuclear translocation in human keratinocytes. *J Cell Biochem.,* Vol. 107, No. 5, (August 2009), pp. (873-880), ISSN 0730-2312.

Zammit C, Barnard R, Gomm J, Coope R, Shousha S, Coombes C & Johnston C (2001). Altered intracellular localization of fibroblast growth factor receptor 3 in human breast cancer. *J Pathol,* Vol. 194, No. 1, (May 2001), pp (27-34), ISSN 1096-9896.

Zhu Q, Ma J, Yu L & Yuan C (2009). Grafted neural stem cells migrate to substantia nigra and improve behavior in Parkinsonian rats. *Neurosci Lett.,* Vol. 462, No. 3, (October 2009), pp. (213-218), ISSN 0304-3940.

10

Musashi Proteins in Neural Stem/Progenitor Cells

Kenichi Horisawa and Hiroshi Yanagawa
Department of Bioscience and Informatics, Keio University
Japan

1. Introduction

Many RNA-binding proteins are encoded in the genomes of various organisms and play a critical role in several life systems. The human genome contains thousands of RNA-binding proteins (Glisovic *et al.*, 2008). The most important biological role for these proteins involves the post-transcriptional events in gene expression, *e.g.*, splicing, export, stabilization, localization, and translation. Recent studies have shown that these post-transcriptional events are of similar importance to transcriptional and post-translational events and are highly orchestrated (Keene *et al.*, 2007).

Musashi is an RNA-binding protein that contains typical RNA-recognition motifs (RRMs). The gene encoding the Musashi protein was originally identified in *Drosophila* and is responsible for the asymmetrical division of sensory organ precursor cells (Nakamura *et al.*, 1994). Later studies determined that Musashi proteins are RNA-binding proteins that bind to a sequence in the 3'untranslated region (UTR) of *tramtrack69* (*ttk69*) mRNA (Lai & Li, 1999). The binding of these proteins prevents the translation of *ttk69* mRNA, resulting in asymmetric cell division (Hirota *et al.*, 1999; Okabe *et al.*, 2001).

Subsequently, two highly conserved mammalian homolog proteins, Musashi1 (Msi1) and Musashi2 (Msi2), were discovered in mice (Sakakibara *et al.*, 1996, 2001). Over 90 Musashi and Musashi-like proteins have been discovered in various multicellular animals; however, these proteins have not been found in prokaryotes, plants, or monocellular organisms. The expression pattern and structure of these proteins are highly similar among various organisms (Good *et al.*, 1998; Yoda *et al.*, 2000; Kawashima *et al.*, 2000; Cuadrado *et al.*, 2002; Lowe *et al.*, 2003; Asai *et al.*, 2005; Higuchi *et al.*, 2008).

Musashi proteins are highly expressed in the vertebrate nervous system. (Kaneko *et al.*, 2000). In the mammalian central nervous system, Msi1 appears specifically in undifferentiated neural stem/precursor cells during both the embryonic and adult stages (Sakakibara *et al.*, 1996; Kaneko *et al.*, 2000; Sakakibara & Okano, 1997). Interestingly, Msi1 expression was also observed in many kinds of somatic stem cells in adult tissues, such as the eye (Raji *et al.*, 2007), intestine (Potten *et al.*, 2003), stomach (Akasaka *et al.*, 2005), mammary gland (Clarke *et al.*, 2005), and hair follicles (Sugiyama-Nakagiri *et al.*, 2006).

Later studies revealed that Msi1 maintains the stemness of the neural stem/precursor cells through the translational suppression of m-Numb, a regulator protein of the Notch signal pathway (Imai *et al.*, 2001; Kawahara *et al.*, 2008). Although other target mRNAs of Msi1 have been recently reported, the full function of Msi1 in maintaining stem/precursor cells remains to be elucidated (Battelli *et al.*, 2006; de Sousa Abreu *et al.*, 2009; Horisawa *et al.*, 2009).

Furthermore, the relationship between Musashi proteins and several disease states have recently been reported. Msi1 is reported to play a role in a variety of cancers (Toda *et al.*, 2001; Shu *et al.*, 2002; Sakatni *et al.*, 2005) and neural disorders (Lovell & Markesbery, 2005; Ziabreva *et al.*, 2006; O'Sullivan *et al.*, 2011; Oki *et al.*, 2010; Nakayama *et al.*, 2010; Crespel *et al.*, 2005).

In this chapter, we present an overview of Musashi proteins, especially mammalian Msi1, and consider possible directions for further research.

2. The discovery of Musashi proteins and their function

Musashi was originally identified in a *Drosophila* mutant with abnormal external sensory organs (Fig.1A&B). In the early '90s, Nakamura and co-workers showed that the *musashi* gene is responsible for the asymmetrical division of sensory organ precursor (SOP) cells in *Drosophila*, which are precursors for the ectodermal system common to both neural and non-neural cell lineages in loss-of-function experiments (Nakamura *et al.*, 1994). In wild-type animals, the SOP cell divides into a non-neural precursor cell (green in Fig.1C) and a neural precursor cell (white in Fig.1C), whereas in *musashi* mutants, two non-neural precursor cells (white in Fig.1D) are produced instead. The symmetrically divided non-neural precursor cells differentiate to hair-forming cells (Sf and So in Fig.1A&B), leading to a double-bristle phenotype instead of the single-hair wild-type phenotype (Fig.1 A&B). Based on this double-hair shape, the gene was named "Musashi" after a famous Japanese swordsman who fought with two swords, Musashi Miyamoto (A.D. 1584-1645).

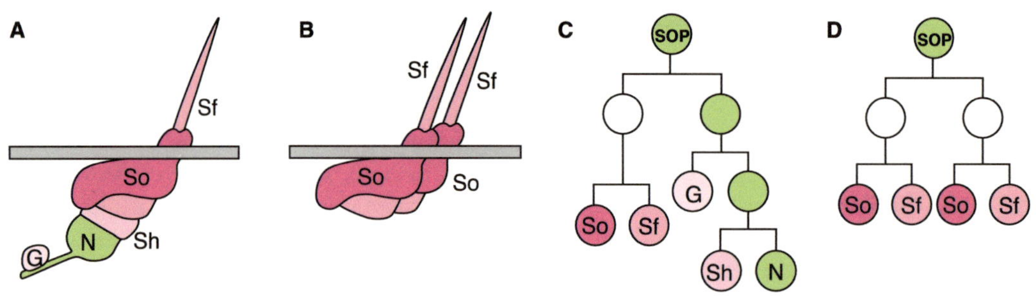

Fig. 1. Structures and cell lineages of the external sensory organs in *Drosophila*
(A & B) Structures of the adult external sensory organs (mechanosensory bristles) in wild-type (A) and musahi mutant (B) animals. The cell lineages contain neuron (N; green) and non-neuronal support cells (magenta), a shaft cell (Sf), a socket cell (So), a sheath cell (Sh) and glia (G); (C & D) Cell lineages of the mechanosensory bristle in wild-type (C) and *musashi* mutant (D) animals. Cells with neuronal potential are green, and non-neuronal cells are magenta and white.

Subsequent studies revealed that the Musashi protein, which has RNA-binding activity, introduces neural differentiation potential for one daughter cell of the SOP cell via selective translational repression of the mRNA of a neural differentiation inhibitory factor (a transcription repressor possessing a BTB domain and zinc-finger domain) called *ttk69*

(Okabe *et al.*, 2001). The ttk69 protein is located downstream in the Notch signaling pathway and acts as a determinant of non-neural identity.

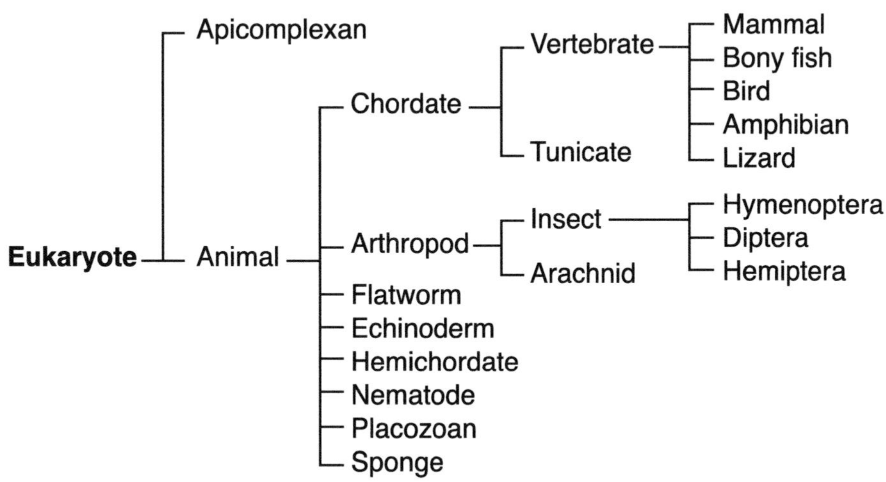

All listed organisms are multi-cellular eukaryotes.

Fig. 2. A dendrogram of the organisms bearing the *musashi* or *musashi*-related genes.

The Systematic Evolution of Ligands by EXponential enrichment (SELEX) assay (Tuerk & Gold, 1990), an *in vitro* selection method for RNA, was employed to identify the specific target RNA motifs of Musashi proteins from a synthesized random-sequence RNA library. Uridine-rich sequences containing two or three (GUU...UAG) or (GUU...UG) repeats were identified as Musashi binding targets (Okabe *et al.*, 2001). Indeed, *ttk69* mRNA contains 15 of these motif sequences in the 3′UTR, and it has been demonstrated that the Musashi protein binds to the 3′UTR of *ttk69* mRNA and inhibits the translation of a reporter gene linked to the 3′UTR *in vitro* (Okabe *et al.*, 2001).

Musashi and Musashi-like proteins have since been discovered in several multicellular organisms, but these genes have not been found in prokaryotes, plants, or monocellular organisms (Fig.2). This implies that the Musashi protein is specifically required for the development and evolution of multicellular animals.

3. The homologs of Musashi protein in mammals

Further studies in *Drosophila* have revealed that the Musashi protein is also expressed in the compound eye primordium (Hirota *et al.*, 1999), CNS (Nakamura *et al.*, 1994), and neural stem/precursor cells in the larval brain (Nakamura *et al.*, 1994), which have many characteristics in common with mammalian neural stem cells (NSCs) (Ito & Hotta, 1992). Thus, to elucidate the functions of the Musashi gene family in mammals, a homolog search and immunohistochemical studies were performed in mice.

Two highly conserved homolog genes, *musashi1* (*msi1*) (Sakakibara *et al.*, 1996) and *musashi2* (*msi2*) (Sakakibara *et al.*, 2001), were discovered in mice. While the length of these proteins are considerably shorter than that of *Drosophila* Musashi, the RNA-recognition motifs (RRMs) were highly conserved (~95% identical) between mammalian and insect systems (Fig.3).

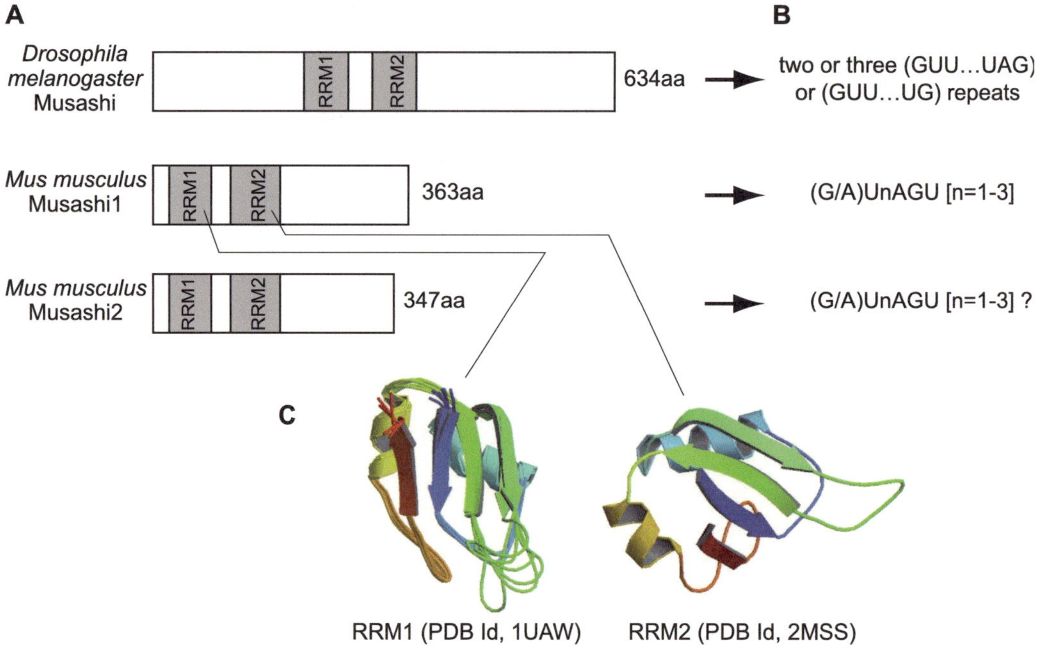

Fig. 3. The structures of Musashi proteins and their target RNA sequences
(A) Primary structures of Musashi proteins in Drosohila melanogaster and Mus musculus. (B) Musashi binding motifs in target mRNAs. (C) Partial 3D structures of RRMs of Msi1 protein.

A high level of expression of Msi1 in NSCs of the periventricular area and undifferentiated neural precursor cells (Sakakibara *et al.*, 1996; Kaneko *et al.*, 2000; Sakakibara & Okano, 1997) was observed. Therefore, Msi1 is now widely used as a marker of NSCs and progenitor cells in the CNS of a variety of vertebrates. These cells, which can form neurospheres, were identified in the adult human brain using this approach (Pincus *et al.*, 1998). Precise immunohistochemical analyses revealed that Msi1 is strongly expressed in the ventricular zone of the neural tube in embryos and in neurogenic sites within the postnatal brain, including the subventricular zone (SVZ), olfactory bulb, and rostral migratory stream (RMS) (Sakakibara & Okano, 1997). The Msi1 protein is expressed in neural stem/progenitor cells within these tissues and is rapidly down-regulated in post-mitotic neurons (Sakakibara *et al.*, 1996).

The Msi2 protein in mice is a paralog of Msi1, displaying more than 90% homology with the Msi1 protein in the RRMs (Sakakibara *et al.*, 2001) (Fig.3). Although the expression pattern in the CNS is very similar between the members of this family, Msi2 is also continuously expressed in a subset of neuronal lineage cells, such as parvalbumin-containing GABA neurons in the neocortex and neurons in several nuclei of the basal ganglia (Sakakibara *et al.*, 2001). Other reports have also shown a differential expression pattern of these genes in uroepithelial cells (Nikpour *et al.*, 2010). Although the functional properties (*e.g.*, RNA-binding specificity) of the two proteins are similar, these differences in expression might explain the functional assignation of these proteins.

Although the partial 3D structures of RRMs of Msi1 protein have been solved by NMR, the full-length 3D structures of Msi1 and Msi2 remain to be elucidated (Nagata *et al*, 1999; Miyanoiri *et al*, 2003).

The human genome also contains both *msi1* and *msi2* genes (Good *et al.*, 1998; Sakakibara *et al.*, 2001). The structure and expression pattern of these proteins in the CNS highly resemble those in mice. As described below, Musashi proteins are related to several diseases. Functional studies in mice will thus contribute to therapeutic developments for Musashi-related conditions.

Additionally, a small Msi2-like gene, LOC100504473, has been found near the Msi2 locus on the mouse genome, but its expression and function are yet to be defined.

4. Molecular and physiological functions of Musashi proteins in stem/progenitor cells

To identify the target RNAs of Msi1 in mammals, a SELEX analysis from a random-sequence RNA library was performed, similar to those done in *Drosophila*. The selected consensus sequence revealed that the mouse Msi1 protein binds specifically to RNAs that possess a (G/A)UnAGU [n=1-3] sequence (Imai *et al.*, 2001) (Fig.3).

A survey for the motif was performed in mRNAs expressed in the embryonic CNS. The 3'UTR region of *m-numb* mRNA (Zhong *et al.*, 1996) was highlighted as a candidate target. Subsequent experiments found that *m-munb* mRNA is a specific binding target of the Msi1 protein *in vitro* and *in vivo*. Its translation is repressed by the Msi1 protein (Imai *et al.*, 2001; Kawahara *et al.*, 2008).

The m-Numb protein binds to the intracellular domain of the Notch protein, which has nuclear translocation and transactivation activities, and inhibits the Notch signaling pathway (Berdnik *et al.*, 2002), which positively regulates neural stem cell self-renewal (Nakamura *et al.*, 2000; Hitoshi *et al.*, 2002; Tokunaga *et al.*, 2004) (Fig.4). In agreement with this hypothesis, oscillation in the expression of the *hes1* gene, a downstream target of Notch, controls the differentiation of embryonic stem cells to neural cells (Kobayashi *et al.*, 2009). Indeed, the Msi1 protein induces the expression of the *hes1* gene (Imai *et al.*, 2001; Yokota *et al.*, 2004).

Musashi proteins in both mammalian species and *Drosophila* contribute to maintaining the stem/progenitor cell status via translational repression of target mRNA. However, the target mRNAs are, interestingly, not orthologous. This result implies a highly conserved function of Musashi proteins in maintaining the stemness of progenitor/stem cells and the probable presence of other unknown target RNAs of the Musashi proteins.

Recently, mammalian Msi1 protein expression was identified not only in CNS, but also in other tissues and organs in embryonic or adult stages, including the eye (*e.g.*, corneal epithelium, corneal endothelium, stromal keratocyte, progenitor cell of the limbus, equatorial lens stem cell, differentiated lens fiber, and retinal pigment epithelium cells) (Raji *et al.*, 2007; Susaki *et al.*, 2009), intestine (small intestinal crypt, colon crypt, columnar cell, and epithelial cell) (Kayahara *et al.*, 2003; Nishimura *et al.*, 2003; Potten *et al.*, 2003; Asai *et al.*,

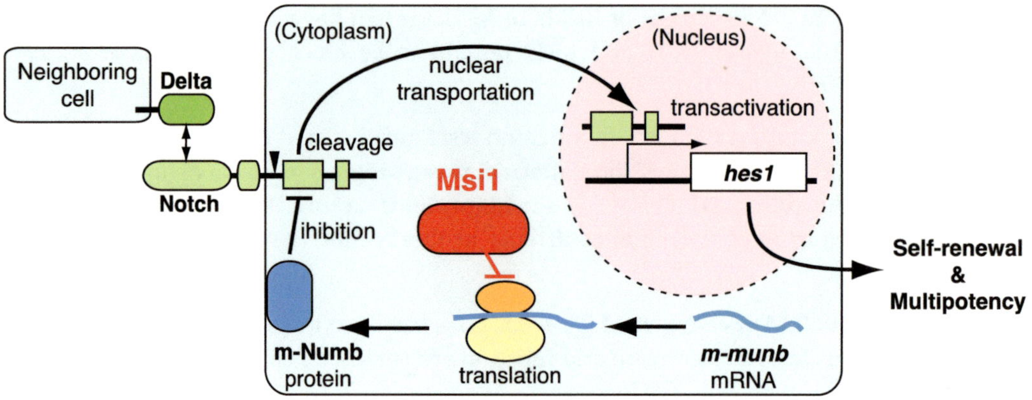

Fig. 4. A model for Msi1 function in the regulation of Notch signaling
Msi1 translationally regulates *m-numb* gene expression. Because m-Numb blocks the activation of the Notch signal induced by Delta on neighboring cells, translational repression of m-Numb by Msi1 stimulates Notch signaling and HES1 pathways.

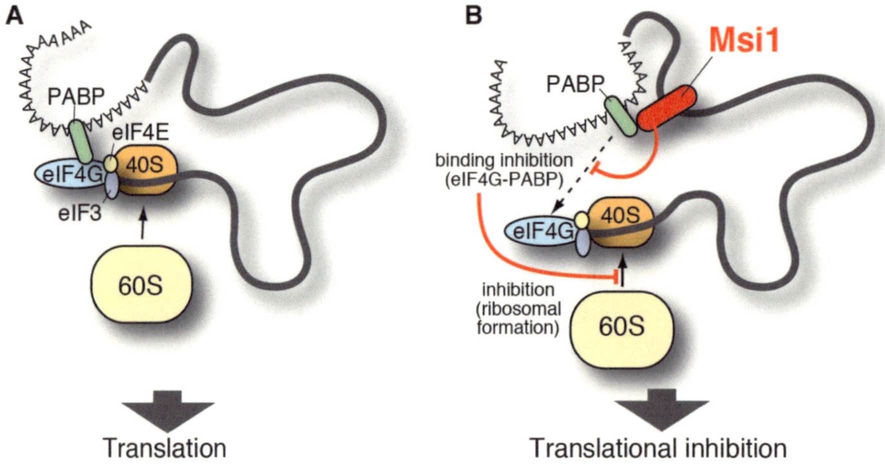

Fig. 5. Schematic representation of the molecular function of Msi1
(A) Translation of a non-target mRNA of Msi1; (B) Translational inhibition of a target mRNA of Msi1. The Msi1 protein interacts with the 3'UTR of its target mRNA and PABP and subsequently inhibits translation initiation by competing with eIF4G for PABP. These sequential events inhibit the formation of the 80S ribosome complex.

2005; Samuel *et al.*, 2008; Murata *et al.*, 2008), stomach (luminal compartment of the mucosa, isthmus/neck region, and fetal pyloric gland) (Nagata *et al.*, 2006; Akasaka *et al.*, 2005; Asai *et al.*, 2005; Murata *et al.*, 2008), breast (mammary gland epithelial cell) (Clarke *et al.*, 2005), and hair follicles (keratinocyte) (Sugiyama-Nakagiri *et al.*, 2006). These studies suggest that the Msi1 protein may be an effective marker for stem/progenitor cells in various tissues and acts as a regulator of the stem cell status of cells.

On the other hand, the function of the Msi2 protein in neural stem/progenitor cells is still unclear, though it is known that Msi1 and Msi2 have similar RNA-binding specificity

(Sakakibara *et al.*, 2001). The results of an Msi1 and Msi2 double knockout experiment suggested that these proteins have mutually complementary functions (Sakakibara *et al.*, 2002).

Recently, the molecular mechanism of translational repression by Msi1 has been uncovered. Kawahara *et al.* identified the poly(A)-binding protein (PABP) as an Msi1-binding protein and found that Msi1 competes with elF4G for PABP binding on its target mRNAs (Fig.5) (Kawahara *et al.*, 2008).

However, the molecular machinery of other functions of Musashi protein (described below) remains to be elucidated. A survey for the co-factors of Musashi proteins is an important future task in order to fully understand the functions of these proteins.

5. Musashi-related diseases - Cancers and neural disorders

Msi1 has been shown to play a role in a variety of cancers and neural disorders.

Several studies have reported high Msi1 protein expression in many types of tumors, including glioma (Toda *et al.*, 2001), hepatoma (Shu *et al.*, 2002), colorectal adenoma (Sakatani *et al.*, 2005; Schulenburg *et al.*, 2007), teratoid/rhabdoid tumors in eye (Fujita *et al.*, 2005), non-small cell lung cancer (Kanai *et al.*, 2006), retinoblastoma (Seigel *et al.*, 2007), medulloblastoma (Nakano *et al.*, 2007; Sanchez-Diaz *et al.*, 2008), ependymoma (Nakano *et al.*, 2007), endometrial carcinoma (Götte *et al.*, 2008), neurocytoma (Yano *et al.*, 2009), glioblastoma (Liu *et al.*, 2006), and cervical carcinoma (Ye *et al.*, 2008). This might because that many carcinoma cells are of epithelial stem cell lineage (Miller *et al.*, 2005) which express the Msi1 protein.

Although the exact function of Msi1 in these cancer cells remains unclear, the knockdown of Msi1 via siRNA resulted in arrested tumor growth in colon adenocarcinoma xenografts transplanted in athymic nude mice, reduced cancer cell proliferation, and increased apoptosis (Sureban *et al.*, 2008). These results suggest an important potential role for Msi1 in tumorigenesis and tumor proliferation.

It has also been shown that some tumors express the Msi2 protein in addition to Msi1 (Seigel *et al.*, 2007). This may indicate a complementary role of the two proteins in tumors.

Msi1 is also hypothesized to be a key player in neuronal disorders. Several reports indicate that Msi1 is relevant to neurodegenerative disorders, such as Alzheimer's disease (AD). Ectopic expression of Msi1 was observed in the hippocampus of AD patients (Lovell & Markesbery, 2005), while a significant decrease in Msi1-expressing cells was observed in the SVZ (Ziabreva *et al.*, 2006). Although it is difficult to explain these phenomena at present, the function of Msi1 in maintaining the stemness of NSCs might play a role in the pathogenesis of the disease.

Msi1 may also play a role in Parkinson's disease (PD), an another type of neurodegenerative disorder. A clinical experiment found that chronic treatment with an anti-PD drug increased Msi1-positive cells in the SVZ of PD patients. The authors suggested that impaired neurogenesis may contribute to the decline in this neurodegenerative disease (O'Sullivan *et al.*, 2011).

Oki *et al.* (2010) reported an up-regulation in Msi1 expression in collapsed nervous system tissue arising from a blood circulation defect. In the ischemic striatum induced by middle cerebral artery occlusion (MCAO), an increase in Msi1-immunoreactivity was observed in reactive astrocytes beginning at 2 days after MCAO and persisting until 14 days after MCAO. The proliferation of Msi1-positive cells was observed beginning at 4 days after MCAO and reached a peak at 7 days after MCAO (Oki *et al.*, 2010)

Nakayama *et al.* (2010) also observed an induction of Msi1-positive cells at the site of ischemic lesions beginning on day 1 after stroke in humans. This result indicates the presence of a regional regenerative response in the human cerebral cortex and the importance of Msi1 in this phenomenon (Nakayama *et al.*, 2010).

Interestingly, Msi1 protein-expressing cells are increased in the hippocampus of mesial temporal lobe epilepsy (MTLE) patients (Crespel *et al.*, 2005). Large numbers of Msi1-positive cells were also observed in the SVZ in these patients (Crespel *et al.*, 2005). Increased neurogenesis has been reported in animal models of MTLE (Crespel *et al.*, 2005). The abnormal proliferation of such Msi1-expressing neural progenitors in the hippocampus might cause epilepsy.

Unlike Msi1, the relevance of Msi2 to various diseases has not yet been elucidated. However, it was recently demonstrated that Msi2 triggers the acute transformation of chronic myelogenous leukaemia (CML) through translational control of the Numb protein in humans (Ito *et al.*, 2010; Kharas et al., 2010; Nishimoto & Okano, 2010). Byers et al. (2011) reported that Msi2 protein expression can be a clinical prognostic biomarker of human myeloid leukaemia (Byers *et al.*, 2011)

6. Novel finding for the functions of Musashi proteins

Although Musashi proteins are thought to act as translational suppressors, MacNicol and co-workers found a novel function of Musashi in *Xenopus* oocytes. In this system, it activates the translation of *mos* mRNA (Charlesworth *et al.*, 2006), a gene that is related to the meiotic cell cycle progression (Sagata *et al.*, 1988). This is an opposite result from previous findings on the translational effect of Msi1. Interestingly, in human oocytes, a parallel physiological phenomenon is controlled by factors other than the Musashi homolog proteins (Prasad *et al.*, 2008).

A similar translation-activating effect of Msi1 was also observed in mammals. Kuwako *et al.* (2010) found that the Msi1 protein up-regulates the translation of the Robo3 protein and controls midline crossing in precerebellar neurons. While previous studies reported that the Msi1-binding sites are in the 3'UTR of mRNAs, the Msi1-binding region in Robo3 mRNA is in the protein-coding region and does not bear the Msi1-binding consensus sequence (Kuwako *et al.*, 2010). This result implies that the discovery of novel Msi1 co-factors will be an important task for understanding the molecular mechanism of the Msi1 proteins in translational control.

Until recently, only a small number of Msi1-targeted mRNAs (Imai *et al.*, 2001; Battelli *et al.*, 2006) had been reported. De Sousa Abreu *et al.* (2009) performed an RNA immnoprecipitation (RIP)-Chip assay in HEK293T cells, a cell line derived from human embryonic kidney, to comprehensively identify the target mRNAs (de Sousa Abreu, 2009). They identified a group of 64 mRNAs whose genes belong to two main functional categories pertinent to tumorigenesis and protein modification. A subsequent proteomics study also revealed that

Msi1 can have not only negative but also positive effects on gene expression for some of the targets (de Sousa Abreu *et al.*, 2009). This result is consistent with other recent findings.

Our group also performed *in vitro* screening analysis to detect specific binding targets of Msi1 controlling the stem cell status of NSCs. We succeeded in identifying a novel target mRNA of Msi1, *doublecortin* (*dcx*), from an mRNA library of embryonic mouse brain tissue (Horisawa *et al.*, 2009).

The *dcx* is a gene related to the migration of newborn neurons and neural development. Mutations in this gene cause an X-linked dominant disorder characterized by classic lissencephaly with severe mental retardation and epilepsy in hemizygous males and subcortical laminar heterotopia, also known as double cortex syndrome, associated with milder mental retardation and epilepsy in heterozygous females (Gleeson *et al.*, 1998; des Portes *et al.*, 1998; Sossey-Alaoui *et al.*, 1998).

The Msi1 protein specifically bound to the 3'UTR region of the mRNA *in vitro*, which contains an Msi1 binding motif, and repressed translation of a reporter gene linked to the mRNA fragment (Horisawa *et al.*, 2009).

We hypothesize that the Msi1 protein prevents inappropriate migration of NSCs through translational inhibition of the *dcx* gene. Several findings support our hypothesis. First, the Dcx protein is expressed only in neuronal precursors just differentiated from NSCs (Couillard-Despres *et al.*, 2005). Secondly, mutually exclusive protein expression of Msi1 and Dcx in human brains was observed (Crespel *et al.*, 2005). Finally, a knock-out of the Musashi family genes reduced the number of neurospheres isolated from embryonic mouse brains, while the knock-down of *dcx* prevented migration of the cells from neurospheres, leaving their structure intact (Ocbina *et al.*, 2006).

Gene symbols	Functions of the encoded proteins	Spices	Effects	References
ttk69	Notch signaling	*Drosophila melanogaster*	Translational suppression	Okabe *et al*, 2001
m-numb	Notch signaling	*Mus musculus* (Msi1)	Translational suppression	Imai *et al*, 2001
p21^{WAF1}	Cell cycle control	*Mus musculus* (Msi1)	Translational suppression	Battelli *et al*, 2006
mos	Meiotic cell cycle progression	*Xenopus laevis* (Msi1)	Translational activation	Charlesworth *et al*, 2006
dcx	Neural migration	*Mus musculus* (Msi1)	Translational suppression?	Horisawa *et al*, 2009
robo3	Axonal guidance	*Mus musculus* (Msi1)	Translational activation	Kuwako *et al*, 2010
let-7	Non coding RNA	*Mus musculus* (Msi1)	Nuclear translocation of Lin28	Kawahara *et al*, 2010
numb	Notch signaling	*Homo sapiens* (Msi2)	Translational suppression	Ito *et al*, 2010

Table 1. Known target RNAs of Musashi proteins

In addition, we also identified another candidate Msi1-binding mRNA that is related to neuronal migration and axon outgrowth (unpublished data). Thus, Msi1 might repress the maturation of neural stem/progenitor cells to neurons through direct translational inhibition of the genes that influence neuronal maturation and migration.

All of the known target RNAs of Musashi proteins, which have been validated in previous studies, are listed in Table 1.

The precise mechanism through which the function of Msi1 is controlled remains unclear. Although Wang *et al.* (2008) proposed that the Msi1 protein is involved in both Notch and Wnt signaling pathways as a novel autocrine process (Nagata *et al.*, 2006; Glazer *et al.*, 2008), details of the mechanism remain unclear, and direct regulators of Msi1 have not been identified.

On the other hand, Ratti *et al.* (2006) reported post-transcriptional regulation of Msi1 mRNA by embryonic lethal abnormal vision (ELAV), an RNA-binding protein of *Drosophila* (Ratti *et al.*, 2006). This is an interesting result because it may imply that some kind of cascade of post-transcriptional regulation contributes to neurogenesis, in addition to other machinery, *i.e.*, signal transduction, transcriptional regulation, and post-translational modification.

Upstream mechanisms regulating Msi1 transcription have also been studied. Kawase *et al.* (2011) found that the sixth intron of the *msi1* gene has a regulatory element for *msi1* transcription in neural stem/progenitor cells. The identification of transcription factors for the *msi1* gene will help elucidate the role of the Msi1 protein in stem cells (Kawase *et al.*, 2011).

More recently, Kawahara *et al.* (2011) discovered a novel function of the Msi1 protein. They showed that Msi1 works in concert with Lin28 to regulate post-transcriptional microRNA (miRNA) biogenesis in the cropping step, which occurs in the nucleus. This indicates that Msi1 can influence stem cell maintenance and differentiation by controlling the subcellular localization of proteins involved in miRNA synthesis, as well as by regulating the translation of its target mRNA (Kawahara *et al.*, 2011).

7. Conclusion

Somatic stem/precursor cells, including neural stem cells, are promising targets for regenerative medicine. Because these cells are derived from the patients themselves, they are not subject to the ethical questions and possible immunological rejection that are common problems in regenerative therapies using embryonic stem (ES) cell.

The Musashi family proteins are key factors for the understanding and application of somatic stem cells. Musashi proteins control the stem cell state through the translational regulation of target mRNAs, and the Musashi family is a highly conserved RNA-binding protein group expressed in undifferentiated stem/precursor cells at both embryonic and adult stages.

Although several studies have revealed that a Notch signal inhibitor, *m-numb*, and a cell cycle regulator, $p21^{WAF1}$, are direct targets of Msi1, a mouse homolog of Musashi and that the machinery involved is located in the context of both the Notch and Wnt signaling pathways (Glazer *et al.*, 2008), the full picture of Msi1 function in neural stem/precursor cells remains to be uncovered. Recently, de Sousa Abreu *et al.* employed an RNA immunoprecipitation (RIP)-chip technique to comprehensively detect Msi1 targeted mRNAs, and they identified a group of 64 mRNAs from HEK293T cells, whose genes belong to two main functional categories pertinent to tumorigenesis: 1) cell cycle, proliferation, differentiation, and apoptosis and 2) protein modification (de Sousa Abreu *et al.*, 2009). However, a more specific survey is necessary to identify the key factors that regulate stemness. We speculate that Msi1 might have specific targets in each cell type or site, such as *dcx*, in addition to the previously discovered targets, which are related to the cell cycle, proliferation, and self-renewal.

While Musashi proteins were originally thought to act as translational suppressors, recent findings indicate that these proteins can activate the translation of specific targets (Charlesworth *et al.*, 2006; de Sousa Abreu *et al.*, 2009; Kuwako *et al.*, 2010). This result indicates that currently unknown molecular machinery may exist that differs from the translational suppression machinery (Kawahara *et al.*, 2008). Components of this machinery, *e.g.*, binding proteins of Msi1, need to be comprehensively clarified using high-throughput techniques (Rigaut *et al.*, 1999; Horisawa *et al.*, 2004, 2008).

Many researches indicate that Musashi proteins have strong associations with some diseases, such as cancers and neuronal disorders. A more complete understanding of the Musashi proteins will also contribute to the development of therapies for these diseases.

8. Acknowledgment

We thank Dr. Takao Imai and Prof. Hideyuki Okano (Department of Physiology, Keio University School of Medicine) for valuable advice and discussion.

9. References

Akasaka, Y.; Saikawa, Y.; Fujita, K.; Kubota, T.; Ishikawa, Y.; Fujimoto, A.; Ishii, T.; Okano, H. & Kitajima, M. (2005) Expression of a candidate marker for progenitor cells, Musashi-1, in the proliferative regions of human antrum and its decreased expression in intestinal metaplasia. *Histopathology*, Vol.47, No.4, (October 2005), pp.348-356, ISSN 1365-2559

Asai, R.; Okano, H. & Yasugi, S. (2005) Correlation between Musashi-1 and c-hairy-1 expression and cell proliferation activity in the developing intestine and stomach of both chicken and mouse. *Dev Growth Differ.*, Vol.47, No.8, (October 2005) pp.501-510, ISSN 0012-1592

Battelli, C.; Nikopoulos, G.N.; Mitchell, J.G. & Verdi, J.M. (2006) The RNA-binding protein Musashi-1 regulates neural development through the translational repression of $p21^{WAF-1}$. *Mol Cell Neurosci.*, Vol.31, No.1, (January 2006), pp.85-96, ISSN 1044-7431

Berdnik, D.; Török, T.; González-Gaitán, M. & Knoblich, J.A. (2002) The endocytic protein alpha-Adaptin is required for numb-mediated asymmetric cell division in *Drosophila*. *Dev Cell*. Vol.3, No.2, (August 2002), pp.221-231, ISSN 1534-5807

Byers, R.J.; Currie, T.; Tholouli, E.; Rodig, S.J. & Kutok, J.L. MSI2 protein expression predicts unfavorable outcome in acute myeloid leukemia. (2011) *Blood*, in press, (July 2011), ISSN 0006-4971

Charlesworth, A.; Wilczynska, A.; Thampi, P., Cox, L.L. & MacNicol, A.M. (2006) Musashi regulates the temporal order of mRNA translation during Xenopus oocyte maturation. *EMBO J.*, Vol.25, No.12, (June 2006), pp.2792-2801, ISSN 0261-4189

Clarke, R.B.; Spence, K.; Anderson, E.; Howell, A.; Okano, H. & Potten, C.S. (2005) A putative human breast stem cell population is enriched for steroid receptor-positive cells. *Dev Biol.*, Vol.277, No.2, (January 2005), pp.443-456, ISSN 0012-1606

Couillard-Despres, S.; Winner, B.; Schaubeck, S.; Aigner, R.; Vroemen, M.; Weidner, N.; Bogdahn, U.; Winkler, J.; Kuhn, H.G. & Aigner, L. (2005) Doublecortin expression levels in adult brain reflect neurogenesis. *Eur J Neurosci.*, Vol.21, No.1, (January 2005), pp.1-14, ISSN 1460-9568

Crespel, A.; Rigau, V.; Coubes, P.; Rousset, M.C.; de Bock, F.; Okano, H.; Baldy-Moulinier, M.; Bockaert, J. & Lerner-Natoli, M. (2005) Increased number of neural progenitors in human temporal lobe epilepsy. *Neurobiol Dis.*, Vol.19, No.3, (Augut 2005), pp.436-450, ISSN 0969-9961

Cuadrado, A.; García-Ferníndez, L.F.; Imai, T.; Okano, H. & Muñoz, A. (2002) Regulation of tau RNA maturation by thyroid hormone is mediated by the neural RNA-binding protein musashi-1. *Mol Cell Neurosci.*, Vol.20, No.2, (June 2002), pp.198-210, ISSN 1044-7431

de Sousa Abreu, R.; Sanchez-Diaz, P.C.; Vogel, C.; Burns, S.C.; Ko, D.; Burton, T.L.; Vo, D.T.; Chennasamudaram, S.; Le, S.Y.; Shapiro, B.A. & Penalva, L.O. (2009) Genomic analyses of musashi1 downstream targets show a strong association with cancer-related processes. *J Biol Chem.*, Vol.284, No.18, (May 2009), pp.12125-12135, ISSN 0021-9258

des Portes, V.; Pinard, J.M.; Billuart, P.; Vinet, M.C.; Koulakoff, A.; Carrié A.; Gelot, A.; Dupuis, E.; Motte, J.; Berwald-Netter, Y.; Catala, M.; Kahn, A.; Beldjord, C. & Chelly, J. (1998) A novel CNS gene required for neuronal migration and involved in X-linked subcortical laminar heterotopia and lissencephaly syndrome. *Cell*, Vol.92, No.1, (January 1998), pp.51-61, ISSN 0092-8674

Fujita, M.; Sato, M.; Nakamura, M.; Kudo, K.; Nagasaka, T.; Mizuno, M.; Amano, E.; Okamoto, Y.; Hotta, Y.; Hatano, H.; Nakahara, N.; Wakabayashi, T. & Yoshida J. (2005) Multicentric atypical teratoid/rhabdoid tumors occurring in the eye and fourth ventricle of an infant: case report. *J Neurosurg.*, Vol.102, No.3, (April 2005), pp.299-302, ISSN 0022-3085

Glazer, R.I.; Wang, X.Y.; Yuan, H. & Yin, Y. (2008) Musashi1: a stem cell marker no longer in search of a function. *Cell Cycle*, Vol.7, No.17, (September 2008) pp.2635-2639, ISSN 1538-4101

Gleeson, J.G.; Allen, K.M.; Fox, J.W.; Lamperti, E.D.; Berkovic, S.; Scheffer, I.; Cooper, E.C.; Dobyns, W.B.; Minnerath, S.R.; Ross, M.E. & Walsh, C.A. (1998) Doublecortin, a brain-specific gene mutated in human X-linked lissencephaly and double cortex syndrome, encodes a putative signaling protein. *Cell*, Vol.92, No.1, (January 1998), pp.63-72, ISSN 0092-8674

Glisovic, T.; Bachorik, J.L.; Yong, J. & Dreyfuss, G. RNA-binding proteins and post-transcriptional gene regulation. (2008) *FEBS Lett.*, Vol.582, No.14, (January 2008), pp.1977-1986, ISSN 0014-5793

Good, P.; Yoda, A.; Sakakibara, S.; Yamamoto, A.; Imai, T.; Sawa, H.; Ikeuchi, T.; Tsuji, S.; Satoh, H. & Okano, H. (1998) The human Musashi homolog 1 (MSI1) gene encoding the homologue of Musashi/Nrp-1, a neural RNA-binding protein putatively expressed in CNS stem cells and neural progenitor cells. *Genomics*, Vol.52, No.3, (September 1998), pp.382-384, ISSN 0888-7543

Götte, M.; Wolf, M.; Staebler, A.; Buchweitz, O.; Kelsch, R.; Schüring, A.N. Kiesel, L. (2008) Increased expression of the adult stem cell marker Musashi-1 in endometriosis and endometrial carcinoma. *J Pathol.*, Vol.215, No.3, (July 2008), pp.317-329, ISSN 1096-9896

Higuchi, S.; Hayashi, T.; Tarui, H.; Nishimura, O.; Nishimura, K.; Shibata, N.; Sakamoto, H. & Agata, K. (2008) Expression and functional analysis of musashi-like genes in planarian CNS regeneration. *Mech Dev.*, Vol.125, No.7, (July 2008), pp.631-645, ISSN 0047-6374

Hirota, Y.; Okabe, M.; Imai, T.; Kurusu, M.; Yamamoto, A.; Miyao, S.; Nakamura, M.; Sawamoto, K. & Okano, H. (1999) Musashi and seven in absentia downregulate Tramtrack through distinct mechanisms in *Drosophila* eye development. *Mech Dev.*, Vol.87, No.1-2, (September 1999), pp.93-101, ISSN 0047-6374

Hitoshi, S.; Alexson, T.; Tropepe, V.; Donoviel, D.; Elia, A.J.; Nye, J.S.; Conlon, R.A.; Mak, T.W.; Bernstein, A. & van der Kooy, D. (2002) Notch pathway molecules are essential for the maintenance, but not the generation, of mammalian neural stem cells. *Genes Dev.*, Vol.16, No.7, (April 2002), pp.846-858, ISSN 0890-9369

Horisawa, K.; Tateyama, S.; Ishizaka, M.; Matsumura, N.; Takashima, H.; Miyamoto-Sato, E.; Doi, N. & Yanagawa, H. (2004) *In vitro* selection of Jun-associated proteins using mRNA display. *Nucleic Acids Res.*, Vol.32, No.21, (December 2004), e169, ISSN 0305-1048

Horisawa, K.; Doi, N. & Yanagawa, H. (2008) Use of cDNA tiling arrays for identifying protein interactions selected by in vitro display technologies. *PLoS One*, Vol.3, No.2, (February 2008), e1646, ISSN 1932-6203

Horisawa, K.; Imai, T.; Okano, H. & Yanagawa, H. (2009) 3'-Untranslated region of doublecortin mRNA is a binding target of the Musashi1 RNA-binding protein. *FEBS Lett.*, Vol.583, No.14, (July 2009), pp.2429-2434, ISSN 0014-5793

Imai, T.; Tokunaga, A.; Yoshida, T.; Hashimoto, M.; Mikoshiba, K.; Weinmaster, G.; Nakafuku, M. & Okano, H. (2001) The neural RNA-binding protein Musashi1 translationally regulates mammalian numb gene expression by interacting with its mRNA. *Mol Cell Biol.*, Vol.21, No.12, (June 2001), pp.3888-3900, ISSN 0270-7306

Ito, K. & Hotta, Y. (1992) Proliferation pattern of postembryonic neuroblasts in the brain of *Drosophila melanogaster. Dev Biol.*, Vol.149, No.1 (January 1992), pp.134-148, ISSN 0012-1606

Ito, T.; Kwon, H.Y.; Zimdahl, B.; Congdon, K.L.; Blum, J.; Lento, W.E.; Zhao, C.; Lagoo, A.; Gerrard, G.; Foroni, L.; Goldman, J.; Goh, H.; Kim, S.H.; Kim, D.W.; Chuah, C.; Oehler, V.G.; Radich, J.P.; Jordan, C.T. & Reya, T. (2010) Regulation of myeloid leukaemia by the cell-fate determinant Musashi. *Nature*, Vol.466, No,7307, (August 2010), pp.765-768, ISSN 0028-0836

Kanai, R.; Eguchi, K.; Takahashi, M.; Goldman, S.; Okano, H.; Kawase, T. & Yazaki, T. (2006) Enhanced therapeutic efficacy of oncolytic herpes vector G207 against human non-small cell lung cancer--expression of an RNA-binding protein, Musashi1, as a marker for the tailored gene therapy. *J Gene Med.*, Vol.8, No.11, (November 2006), pp.1329-1340, ISSN 1099-498X

Kaneko, Y.; Sakakibara, S.; Imai, T.; Suzuki, A.; Nakamura, Y.; Sawamoto, K.; Ogawa, Y.; Toyama, Y.; Miyata, T. & Okano, H. (2000) Musashi1: an evolutionally conserved marker for CNS progenitor cells including neural stem cells. *Dev Neurosci.*, Vol.22, No.1-2, pp.139-153, ISSN 0378-5866

Kawahara, H.; Imai, T.; Imataka, H.; Tsujimoto, M.; Matsumoto, K. & Okano, H. (2008) Neural RNA-binding protein Musashi1 inhibits translation initiation by competing with eIF4G for PABP. *J Cell Biol.*, Vol.181, No.4, (May 2008), pp.639-653, ISSN 0021-9525

Kawahara, H.; Okada, Y.; Imai, T.; Iwanami, A.; Mischel, P.S. & Okano, H. (2011) Musashi1 cooperates in abnormal cell lineage protein 28 (Lin28)-mediated let-7 family microRNA biogenesis in early neural differentiation. *J Biol Chem.*, Vol.286, No.18, (May 2011), pp.16121-16130, ISSN 0021-9258

Kawase, S.; Imai, T.; Miyauchi-Hara, C.; Yaguchi, K.; Nishimoto, Y.; Fukami, S.; Matsuzaki, Y.; Miyawaki, A.; Itohara, S. & Okano H. (2011) Identification of a novel intronic enhancer responsible for the transcriptional regulation of musashi1 in neural stem/progenitor cells. *Mol Brain*, Vol.4, (April 2011), 14, ISSN 1756-6606

Kawashima, T.; Murakami, A.R.; Ogasawara, M.; Tanaka, K.; Isoda, R.; Sasakura, Y.; Nishikata, T.; Okano, H. & Makabe, K.W. (2000) Expression patterns of musashi homologs of the ascidians, *Halocynthia roretzi* and *Ciona intestinalis*. *Dev Genes Evol.*, Vol.210, No.3, (March 2000), pp.162-165, ISSN 0949-944X

Kayahara, T.; Sawada, M.; Takaishi, S.; Fukui, H.; Seno, H.; Fukuzawa, H.; Suzuki, K.; Hiai, H.; Kageyama, R.; Okano, H. & Chiba, T. (2003) Candidate markers for stem and early progenitor cells, Musashi-1 and Hes1, are expressed in crypt base columnar cells of mouse small intestine. *FEBS Lett.*, Vol.535, No.1-3, (January 2003), pp.131-135, ISSN 0014-5793

Keene, J.D. (2007) RNA regulons: coordination of post-transcriptional events. *Nat Rev Genet.*, Vol.8, No.7, (July 2007), pp.533-543, ISSN 1471-0056

Kharas, M.G.; Lengner, C.J.; Al-Shahrour, F.; Bullinger, L.; Ball, B.; Zaidi, S.; Morgan, K.; Tam, W.; Paktinat, M.; Okabe, R.; Gozo, M.; Einhorn, W.; Lane, S.W.; Scholl, C.; Fröhling, S.; Fleming, M.; Ebert, B.L.; Gilliland, D.G.; Jaenisch, R. & Daley, G.Q. Musashi-2 regulates normal hematopoiesis and promotes aggressive myeloid leukemia. (2010) *Nat. Med.*, Vol. 16, No.8, (August 2010), pp.903-908, ISSN 1078-8956

Kobayashi, T.; Mizuno. H.; Imayoshi, I.; Furusawa, C.; Shirahige, K. & Kageyama, R. (2009) The cyclic gene Hes1 contributes to diverse differentiation responses of embryonic stem cells. *Genes Dev.*, Vol.23, No.16, (August 2009), pp.1870-1875, ISSN 0890-9369

Lai, Z.C. & Li, Y. (1999) Tramtrack69 is positively and autonomously required for *Drosophila* photoreceptor development. *Genetics*, Vol.152, No.1, (May 1999), pp.299-305, ISSN 0016-6731

Liu, G.; Yuan, X.; Zeng, Z.; Tunici, P.; Ng, H.; Abdulkadir, I.R.; Lu, L.; Irvin, D.; Black, K.L. & Yu J.S. (2006) Analysis of gene expression and chemoresistance of $CD133^{+}$ cancer stem cells in glioblastoma. *Mol Cancer*, Vol.5, (December 2006), 67, ISSN 1476-4598

Lovell, M.A. & Markesbery, W.R. (2005) Ectopic expression of Musashi-1 in Alzheimer disease and Pick disease. *J Neuropathol Exp Neurol.*, Vol.64, No.8, (August 2005), pp.675-680, ISSN 0022-3069

Lowe, C.J.; Wu, M.; Salic, A.; Evans, L.; Lander, E.; Stange-Thomann, N.; Gruber, C.E.; Gerhart, J. & Kirschner, M. (2003) Anteroposterior patterning in hemichordates and the origins of the chordate nervous system. *Cell*, Vol.113, No.7, (June 2003), pp.853-865, ISSN 0092-8674

Miller, S.J.; Lavker, R.M. & Sun, T.T. (2005) Interpreting epithelial cancer biology in the context of stem cells: tumor properties and therapeutic implications. *Biochim Biophys Acta.*, Vol.1756, No.1, (September 2005), pp.25-52, ISSN 0006-3002

Miyanoiri, Y.; Kobayashi, H.; Imai, T.; Watanabe, M.; Nagata, T.; Uesugi, S.; Okano, H. and Katahira, M. (2003) Origin of higher affinity to RNA of the N-terminal RNA-binding domain than that of the C-terminal one of a mouse neural protein, musashi1, as revealed by comparison of their structures, modes of interaction, surface electrostatic potentials, and backbone dynamics. *J Biol Chem.*, Vol.278 No. 42, (October 2003), pp.41309-41315, ISSN 0021-9258

Murata, H.; Tsuji, S.; Tsujii, M.; Nakamura, T.; Fu, H.Y.; Eguchi, H.; Asahi, K.; Okano, H.; Kawano, S. & Hayashi, N. (2008) *Helicobacter pylori* infection induces candidate stem cell marker Musashi-1 in the human gastric epithelium. *Dig Dis Sci.*, Vol.53, No,2, (February 2008), pp.363-369, ISSN 0163-2116

Murayama, M.; Okamoto, R.; Tsuchiya, K.; Akiyama, J.; Nakamura, T.; Sakamoto, N.; Kanai, T. & Watanabe, M. (2009) Musashi-1 suppresses expression of Paneth cell-specific genes in human intestinal epithelial cells. *J Gastroenterol.*, Vol.44, No.3, (February 2009), pp.173-182, ISSN 0002-9270

Nagata, H.; Akiba, Y.; Suzuki, H.; Okano, H. and Hibi, T. (2006) Expression of Musashi-1 in the rat stomach and changes during mucosal injury and restitution. *FEBS Lett.*, Vol.580, No.1, (January 2006), pp.27-33, ISSN 0014-5793

Nagata, T.; Kanno, R.; Kurihara, Y.; Uesugi, S.; Imai, T.; Sakakibara, S.; Okano, H. and Katahira, M. (1999) Structure, backbone dynamics and interactions with RNA of the C-terminal RNA-binding domain of a mouse neural RNA-binding protein, Musashi1. *J Mol Biol.*, Vol.287, No.2, (March 1999), pp.315-330, ISSN 0022-2836

Nakamura, M.; Okano, H.; Blendy, J.A. and Montell, C. (1994) Musashi, a neural RNA-binding protein required for *Drosophila* adult external sensory organ development. *Neuron*,Vol.13, No.1, (July 1994), pp.67-81, ISSN 0896-6273

Nakamura, Y.; Sakakibara, S.; Miyata, T.; Ogawa, M.; Shimazaki, T.; Weiss, S.; Kageyama, R. & Okano, H. (2000) The bHLH gene hes1 as a repressor of the neuronal commitment of CNS stem cells. *J Neurosci.*, Vol.20, No.1, (January 2000), pp.283-293, ISSN 0270-6474

Nakano, A.; Kanemura, Y.; Mori, K.; Kodama, E.; Yamamoto, A.; Sakamoto, H.; Nakamura, Y.; Okano, H.; Yamasaki, M. and Arita, N. (2007) Expression of the neural RNA-binding protein Musashi1 in pediatric brain tumors. *Pediatr Neurosurg.*, Vol.43, No.4, pp.279-284, ISSN 8755-6863

Nakayama, D.; Matsuyama, T.; Ishibashi-Ueda, H.; Nakagomi, T.; Kasahara, Y.; Hirose, H; Kikuchi-Taura, A.; Stern, D.M.; Mori, H. & Taguchi, A. Injury-induced neural stem/progenitor cells in post-stroke human cerebral cortex. (2010) *Eur J Neurosci.*, Vol.31, No.1, (January 2010), pp.90-98, ISSN 0953-816X

Nikpour, P.; Baygi, M.E.; Steinhoff, C.; Hader, C.; Luca, A.C.; Mowla, S.J. & Schulz, W.A. (2011) The RNA binding protein Musashi1 regulates apoptosis, gene expression and stress granule formation in urothelial carcinoma cells. *J Cell Mol Med.*, Vol.15, No.5, (May 2011), pp.1210-1224, ISSN 1582-1838

Nishimoto, Y. and Okano, H. New insight into cancer therapeutics: induction of differentiation by regulating the Musashi/Numb/Notch pathway. (2010) *Cell Res.*, Vol.20, No.10, (October 2010), pp.1083-1085, ISSN 1001-0602

Nishimura, S.; Wakabayashi, N.; Toyoda, K.; Kashima, K. & Mitsufuji, S. (2003) Expression of Musashi-1 in human normal colon crypt cells: a possible stem cell marker of human colon epithelium. *Dig Dis Sci.*, Vol.48, No.8, (August 2003), pp.1523-1529, ISSN 0163-2116

Ocbina, P.J.; Dizon, M.L.; Shin, L. & Szele, F.G. (2006) Doublecortin is necessary for the migration of adult subventricular zone cells from neurospheres. *Mol. Cell Neurosci.*, Vol.33, No.2, (October 2006), pp.126-135, ISSN 1044-7431

Okabe, M.; Imai, T.; Kurusu, M.; Hiromi, Y. & Okano, H. (2001) Translational repression determines a neuronal potential in *Drosophila* asymmetric cell division. *Nature*, Vol.411, No.6833, (May 2001), pp.94-98, ISSN 0028-0836

Okano, H.; Imai, T. & Okabe, M. (2002) Musashi: a translational regulator of cell fate. *J Cell Sci.*, Vol.115, Pt7, (April 2002), pp.1355-1359, ISSN 0021-9533

Okano, H.; Kawahara, H.; Toriya, M.; Nakao, K.; Shibata, S. & Imai, T. (2005) Function of RNA-binding protein Musashi-1 in stem cells. *Exp Cell Res.*, Vol.306, No.2, (June 2005), pp.349-356, ISSN 0014-4827

Oki, K.; Kaneko, N.; Kanki, H.; Imai, T.; Suzuki, N.; Sawamoto, K. & Okano, H. (2010) Musashi1 as a marker of reactive astrocytes after transient focal brain ischemia. *Neurosci Res.*, Vol.66, No.4, (April 2010), pp.390-395, ISSN 0168-0102

O'Sullivan, S.S.; Johnson, M.; Williams, D.R.; Revesz, T.; Holton, J.L.; Lees, A.J. & Perry, E.K. (2011) The effect of drug treatment on neurogenesis in Parkinson's disease. *Mov Disord.*, Vol.26, No.1, (January 2011), pp.45-50, ISSN 0885-3185

Pincus, D.W.; Keyoung, H.M.; Harrison-Restelli, C.; Goodman, R.R.; Fraser, R.A.; Edgar, M.; Sakakibara, S.; Okano, H.; Nedergaard, M. & Goldman, S.A. (1998) Fibroblast growth factor-2/brain-derived neurotrophic factor-associated maturation of new neurons generated from adult human subependymal cells. *Ann Neurol.*, Vol.43, No.5, (May 1998), pp.576-585, ISSN 0364-5134

Potten, C.S.; Booth, C.; Tudor, G.L.; Booth, D.; Brady, G.; Hurley, P.; Ashton, G.; Clarke, R.; Sakakibara, S. & Okano, H. (2003) Identification of a putative intestinal stem cell and early lineage marker; musashi-1. *Differentiation*, Vol.71, No.1, (January 2003), pp.28-41, ISSN 0301-4681

Prasad, C.K.; Mahadevan, M.; MacNicol, M.C. & MacNicol, A.M. (2008) Mos 3' UTR regulatory differences underlie species-specific temporal patterns of Mos mRNA cytoplasmic polyadenylation and translational recruitment during oocyte maturation. *Mol Reprod Dev.*, Vol.75, No.8, (August 2008), pp.1258-1268, ISSN 1040-452X

Raji, B.; Dansault, A.; Leemput, J.; de la Houssaye, G.; Vieira, V.; Kobetz, A.; Arbogast, L.; Masson, C.; Menasche, M. & Abitbol, M. (2007) The RNA-binding protein Musashi-1 is produced in the developing and adult mouse eye. *Mol Vis.*, Vol.13, (August 2007), pp.1412-1427, ISSN 1090-0535

Ratti, A.; Fallini, C.; Cova, L.; Fantozzi, R.; Calzarossa, C.; Zennaro, E.; Pascale, A.; Quattrone, A. and Silani, V. (2006) A role for the ELAV RNA-binding proteins in neural stem cells: stabilization of Msi1 mRNA. *J Cell Sci.*, Vol.119, Pt7, (April 2006), pp.1442-1452, ISSN 0021-9533

Rigaut, G.; Shevchenko, A.; Rutz, B.; Wilm, M.; Mann, M. & Séraphin, B. (1999) A generic protein purification method for protein complex characterization and proteome exploration. *Nat Biotechnol.*, Vol.17, No.10, (October 1999), pp.1030-1032, ISSN 1087-0156

Sagata, N.; Oskarsson, M.; Copeland, T.; Brumbaugh, J. and Vande Woude, G.F. (1988) Function of c-mos proto-oncogene product in meiotic maturation in Xenopus oocytes. *Nature*, Vol.335, No.6190, (October 1988), pp.519-525, ISSN 0028-0836

Sakakibara, S.; Imai, T.; Hamaguchi, K.; Okabe, M.; Aruga, J.; Nakajima, K.; Yasutomi, D.; Nagata, T.; Kurihara, Y.; Uesugi, S.; Miyata, T.; Ogawa, M.; Mikoshiba, K. & Okano, H. (1996) Mouse-Musashi-1, a neural RNA-binding protein highly enriched in the mammalian CNS stem cell. *Dev Biol.*, Vol.176, No.2, (June 1996), pp.230-242, ISSN 0012-1606

Sakakibara, S. & Okano, H. (1997) Expression of neural RNA-binding proteins in the postnatal CNS: implications of their roles in neuronal and glial cell development. *J Neurosci.*, Vol.17, No.21, (November 1997), pp.8300-8312, ISSN 0270-6474

Sakakibara, S.; Nakamura, Y.; Satoh, H. & Okano, H. (2001) RNA-binding protein Musashi2: developmentally regulated expression in neural precursor cells and subpopulations of neurons in mammalian CNS. *J Neurosci.*, Vol.21, No.20, (October 2001), pp.8091-8107, ISSN 0270-6474

Sakakibara, S.; Nakamura, Y.; Yoshida, T.; Shibata, S.; Koike, M.; Takano, H.; Ueda, S.; Uchiyama, Y.; Noda, T. & Okano, H. (2002) RNA-binding protein Musashi family: roles for CNS stem cells and a subpopulation of ependymal cells revealed by targeted disruption and antisense ablation. *Proc Natl Acad Sci USA*, Vol.99, No.23, (November 2002), pp.15194-15199, ISSN 0027-8424

Sakatani, T.; Kaneda, A.; Iacobuzio-Donahue, C.A.; Carter, M.G.; de Boom Witzel, S.; Okano, H.; Ko, M.S.; Ohlsson, R.; Longo, D.L. & Feinberg, A.P. (2005) Loss of imprinting of Igf2 alters intestinal maturation and tumorigenesis in mice. *Science*, Vol.307, No.5717, (March 2005), pp.1976-1978, ISSN 0036-8075

Samuel, S.; Walsh, R.; Webb, J.; Robins, A.; Potten, C. & Mahida, Y.R. (2009) Characterization of putative stem cells in isolated human colonic crypt epithelial cells and their interactions with myofibroblasts. *Am J Physiol Cell Physiol.*, Vol.296, No.2, (February 2009), pp.C296-305, ISSN 0363-6143

Sanchez-Diaz, P.C.; Burton, T.L.; Burns, S.C.; Hung, J.Y. & Penalva, L.O. (2008) Musashi1 modulates cell proliferation genes in the medulloblastoma cell line Daoy. *BMC Cancer*, Vol.8, (September 2008), 280, ISSN 1471-2407

Schulenburg, A.; Cech, P.; Herbacek, I.; Marian, B.; Wrba, F.; Valent, P. Ulrich-Pur, H. (2007) CD44-positive colorectal adenoma cells express the potential stem cell markers musashi antigen (msi1) and ephrin B2 receptor (EphB2). *J Pathol.*, Vol.213, No.2, (October 2007), pp.152-160, ISSN 1096-9896

Seigel, G.M.; Hackam, A.S.; Ganguly, A.; Mandell, L.M. & Gonzalez-Fernandez, F. (2007) Human embryonic and neuronal stem cell markers in retinoblastoma. *Mol Vis.* Vol.13, (June 2007), pp.823-832, ISSN 1090-0535

Shu, H.J.; Saito, T.; Watanabe, H.; Ito, J.I.; Takeda, H.; Okano, H. & Kawata, S. (2002) Expression of the Musashi1 gene encoding the RNA-binding protein in human hepatoma cell lines. *Biochem Biophys Res Commun.*, Vol.293, No.1, (April 2002), pp.150-154, ISSN 0006-291X

Sossey-Alaoui, K.; Hartung, A.J.; Guerrini, R.; Manchester, D.K.; Posar, A.; Puche-Mira, A.; Andermann, E.; Dobyns, W.B. & Srivastava, A.K. (1998) Human doublecortin (DCX) and the homologous gene in mouse encode a putative $Ca2^{+}$-dependent signaling protein which is mutated in human X-linked neuronal migration defects. *Hum Mol Genet.*, Vol.7, No.8, (August 1998), pp.1327-1332, ISSN 0964-6906

Sugiyama-Nakagiri, Y.; Akiyama, M.; Shibata, S.; Okano, H. & Shimizu, H. (2006) Expression of RNA-binding protein Musashi in hair follicle development and hair cycle progression. *Am J Pathol.*, Vol.168, No.1, (January 2006), pp.80-92, ISSN 0887-8005

Sureban, S.M.; May, R.; George, R.J.; Dieckgraefe, B.K.; McLeod, H.L.; Ramalingam, S.; Bishnupuri, K.S.; Natarajan, G.; Anant, S. & Houchen, C.W. (2008) Knockdown of RNA binding protein musashi-1 leads to tumor regression *in vivo*. *Gastroenterology*, Vol.134, No.5, (May 2008), pp.1448-1458, ISSN 0016-5085

Susaki, K.; Kaneko, J.; Yamano, Y.; Nakamura, K.; Inami, W.; Yoshikawa, T.; Ozawa, Y.; Shibata, S.; Matsuzaki, O.; Okano, H. & Chiba, C. (2009) Musashi-1, an RNA-binding protein, is indispensable for survival of photoreceptors. *Exp Eye Res.*, Vol.88, No.3, (March 2009), pp.347-355, ISSN 0014-4835

Toda, M.; Iizuka, Y.; Yu, W.; Imai, T.; Ikeda, E.; Yoshida, K.; Kawase, T.; Kawakami, Y.; Okano, H. & Uyemura, K. (2001) Expression of the neural RNA-binding protein Musashi1 in human gliomas. *Glia*, Vol.34, No.1, (April 2001), pp.1-7, ISSN 1098-1136

Tokunaga, A.; Kohyama, J.; Yoshida, T.; Nakao, K.; Sawamoto, K. & Okano, H. (2004) Mapping spatio-temporal activation of Notch signaling during neurogenesis and gliogenesis in the developing mouse brain. *J Neurochem.*, Vol.90, No.1, (July 2004), pp.142-154, ISSN 0022-3042

Tuerk, C. & Gold, L. (1990) Systematic evolution of ligands by exponential enrichment: RNA ligands to bacteriophage T4 DNA polymerase. *Science*, Vol.249, No.4968, (August 1990), pp.505-510, ISSN 0036-8075

Wang, X.Y.; Yin, Y.; Yuan, H.; Sakamaki, T.; Okano, H. & Glazer, R.I. (2008) Musashi1 modulates mammary progenitor cell expansion through proliferin-mediated activation of the Wnt and Notch pathways. *Mol Cell Biol.*, Vol.28, No.11, (June 2008), pp.3589-3599, ISSN 0270-7306

Yano, H.; Ohe, N.; Shinoda, J.; Yoshimura, S. & Iwama, T. (2009) Immunohistochemical study concerning the origin of neurocytoma--a case report. *Pathol Oncol Res.*, Vol.15, No.2, (June 2009), pp.301-305, ISSN 1219-4956

Ye, F.; Zhou, C.; Cheng, Q.; Shen, J. and Chen, H. (2008) Stem-cell-abundant proteins Nanog, Nucleostemin and Musashi1 are highly expressed in malignant cervical epithelial cells. *BMC Cancer*, Vol.8, (April 2008), 108, ISSN 1471-2407

Yoda, A.; Sawa, H. & Okano, H. (2000) MSI-1, a neural RNA-binding protein, is involved in male mating behaviour in Caenorhabditis elegans. *Genes Cells*. Vol.5, No.11, (November 2000), pp.885-895, ISSN 1356-9597

Yokota, N.; Mainprize, T.G.; Taylor, M.D.; Kohata, T.; Loreto, M.; Ueda, S.; Dura, W.; Grajkowska, W.; Kuo, J.S. & Rutka, J.T. (2004) Identification of differentially expressed and developmentally regulated genes in medulloblastoma using suppression subtraction hybridization. *Oncogene*, Vol.23, No.19, (April 2004), pp.3444-3453, ISSN 0950-9232

Zhong, W.; Feder, J.N.; Jiang, M.M.; Jan, L.Y. & Jan, Y.N. (1996) Asymmetric localization of a mammalian numb homolog during mouse cortical neurogenesis. *Neuron*, Vol.17, No.1, (July 1996), pp.43-53, ISSN 0896-6273

Ziabreva, I.; Perry, E.; Perry, R.; Minger, S.L.; Ekonomou, A.; Przyborski, S. & Ballard, C. (2006) Altered neurogenesis in Alzheimer's disease. *J Psychosom Res.*, Vol.61, No.3, (September 2006), pp.311-316, ISSN 0022-3999

11

Active Expression of Retroelements in Neurons Differentiated from Adult Hippocampal Neural Stem Cells

Slawomir Antoszczyk, Kazuyuki Terashima, Masaki Warashina, Makoto Asashima and Tomoko Kuwabara
Research Center for Stem Cell Engineering, National Institute of Advanced Industrial Science and Technology (AIST) Japan

1. Introduction

In the mammalian brain, neurogenesis constitutively occurs in the subventricular zone (SVZ), the olfactory bulb, and the hippocampus throughout adulthood (Kuhn et al., 1996; Lois and Alvarez-Buylla, 1993; Gage, 2000; Pagano et al., 2000; Gritti et al., 2002), and adult hippocampal neurogenesis plays an important role in learning and memory (Deng *et al.*, 2010). In the hippocampus, multipotent neural stem cells (NSCs) reside in the inner layer of the subgranular zone (SGZ) of the dentate gyrus (Gage, 2000; Suh *et al.*, 2007). Undifferentiated NSCs express the high mobility group (HMG)-box transcription factor Sox2 (D'Amour and Gage, 2003; Suh *et al.*, 2007). Sox2 is an SRY-related transcription factor encoding an HMG DNA-binding motif, and is expressed in embryonic stem (ES) cells and neural epithelial cells during development (Avilion et al., 2003; Ferri et al., 2004). Sox2 is essential for the multipotency and self-renewal capacity of NSCs and also functions in pluripotent ES cells (Bylund et al., 2003; Graham et al., 2003; Ferri et al., 2004). Sox2 is known to prevent neurogenesis during development and is thought to be critical for maintaining NSC populations in the neonatal brain (Bylund et al., 2003; Graham et al., 2003; Bani-Yaghoub et al., 2006). In the dentate gyrus of the hippocampus, Sox2 expression is found in the undifferentiated stem cell population with self-renewal capacity, and these Sox2-positive stem cells are exclusive with TUJ1-positive early stage of neurons (Fig. 1).

Fate-tracing studies showed that Sox2-positive cells in the SGZ of the adult hippocampus have the potential to give rise to neurons and astrocytes, revealing their multipotency at both the cell population and single cell levels (Suh et al., 2007). Moreover, a subpopulation of Sox2-positive cells gives rise to cells that retain Sox2, highlighting the importance of Sox2-positive cells as a primary source of adult NSCs (Suh et al., 2007). In response to intracellular or extracellular signals, Sox2-positive NSCs undergo cell division, giving rise to more Sox2-positive NSCs as well as neuronal precursors (Avilion et al., 2003; Ferri et al., 2004). Adult hippocampal neural stem cells express receptors and signaling components for Wnt proteins, which are key regulators of NSCs (Lie et al., 2005).

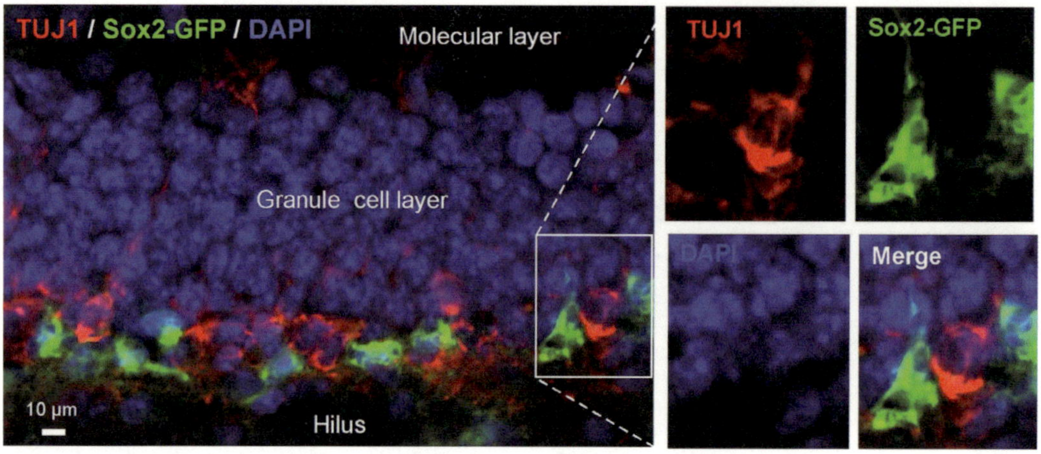

Hippocampal neural stem cells express multipotent marker Sox2. Sox2-positive cells (Sox2-GFP: green; immunohistochemistry analysis using Sox2 promoter-driven EGFP transgenic mice) are exclusive with early stage of neurons expressing β tubulin III (TUJ1: red). DAPI; blue.

Fig. 1. Adult neural stem cells in dentate gyrus of hippocampus

2. Wnt3 promotes the active expression of retroelements in adult hippocampal neurogenesis

Astrocytes are an essential cell population defining the hippocampal niche (Song *et al.*, 2002), and Wnt3 factors secreted from these cells are instructive in promoting adult neurogenesis (Lie *et al.*, 2005). The deletion of *Wnt3a* (*Wnt3a-/-* mice) prevents the formation of the dentate gyrus, which is the site of adult neurogenesis (Lee et al., 2000). Other Wnt proteins (Wnt1, Wnt2, Wnt5a, and Wnt7a/b) are detected and actively function in mature hippocampal neurons (Miyaoka T *et al.*, 1999; Gogolla et al., 2009; Cuitino et al., 2010; Okamoto *et al.*, 2010), suggesting that other Wnt proteins work in an autocrine manner to control neuronal functions, activities in the neuronal network, and synaptic connectivity in mature neurons. In contrast, astrocyte-secreted Wnt3/Wnt3a (Lie et al., 2005) acts in a paracrine manner to induce neurogenesis in NSCs by direct activation of the proneuronal gene neurogenic differentiation 1 (*NeuroD1)*, and to generate diversity in newborn neurons through retroelements (i.e., retrotransposons) (Kuwabara et al., 2009).

2.1 Astrocytes-secreting Wnt3 initiates the expression of retroelements in adult neural stem cells

We recently reported that Wnt3/Wnt3a released from underlying astrocytic layers in the dentate gyrus has an important role in triggering neuronal differentiation of hippocampal NSCs. *NeuroD1,* a target gene of Wnt signaling in the coding region of the mammalian genome, is a proneural basic helix-loop-helix (bHLH) transcription factor that is essential for the development of the CNS, particularly for the generation of granule cells in the hippocampus (Miyata et al., 1999; Liu et al., 2000; Deisseroth et al., 2004; Tozuka et al., 2005). The deletion of β-catenin leads to substantial loss of NeuroD1-positive cells, while the stem cell compartment remains intact in vivo, suggesting that Wnt/β-catenin-mediated neuronal differentiation is dependent on NeuroD1 (Kuwabara et al., 2009).

Importantly, paracrine Wnt factors secreted from astrocytes simultaneously target the genomic noncoding region through long interspersed nuclear elements (LINE-1, L1) (Kuwabara et al., 2009). L1 is a large family of mobile elements that constitutes up to 17% of the mammalian genome (Han and Boeke, 2005) and was recently found to be actively retrotransposed in the course of adult neurogenesis in rodents (Muotri et al., 2005) and humans (Coufal et al., 2009). The regulatory sequence recognized by both Sox2 and TCF/LEF/β-catenin is present in the promoter region of NeuroD1 and L1 (Kuwabara *et al.*, 2009). Sox2 can suppress L1 expression in adult NSCs. In contrast, canonical Wnt/β-catenin signaling triggers the active expression of NeuroD1 and L1, indicating the essential role of Sox2 and Wnt in balancing self-renewal of NSCs and neuronal differentiation in the adult hippocampal dentate gyrus (Kuwabara et al., 2009).

The balance of asymmetric lineage control of adult NSCs in maintaining the constant size of the neural stem cell pools and producing newly born neurons relies on the definitive molecular mechanism of Wnt target genes: the Sox/LEF overlapping regulatory sequence on NeuroD1 and L1 functions as a "molecular switch" between Sox2-mediated repression and Wnt signaling-mediated activation of target genes (Fig. 2).

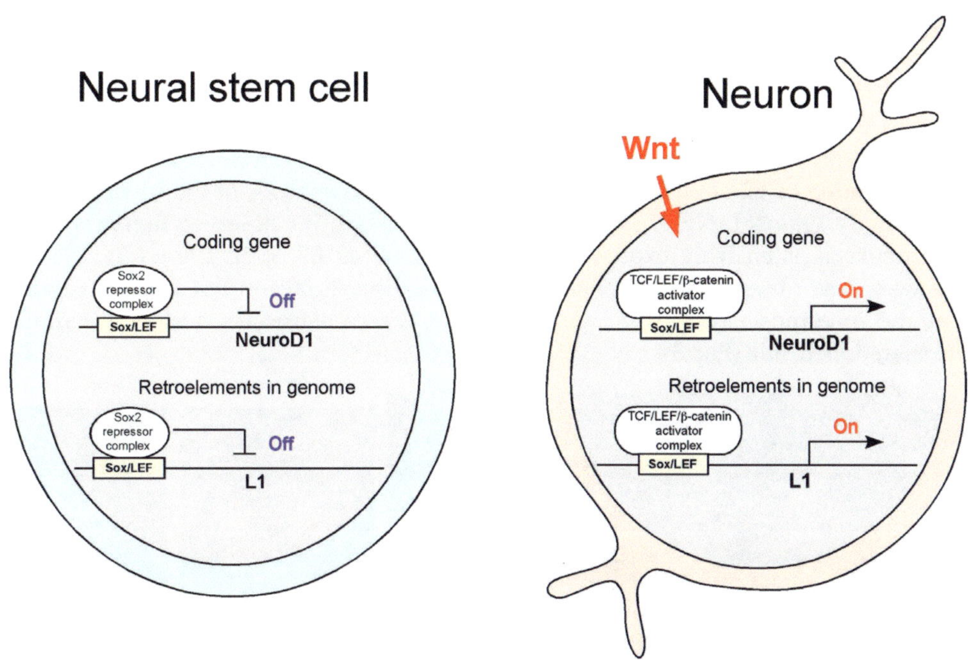

Overlapping Sox/LEF binding site is shown in yellow box. The overlapping DNA regulatory consensus sequence (A/T-A/T-C-A-A-G; yellow box) recognized by both Sox2 (A/T-A/T-C-A-A-A/T-G) and TCF/LEF (A/T-A/T-C-A-A-A). When Wnt3/Wnt3a stimulate NSCs, the Wnt-signaling activate NeuroD1 gene and L1 gene.

Fig. 2. Schematic representation of Wnt-mediated regulation in adult hippocampal NSCs.

The Sox/LEF regulatory elements reside within the 5′ UTR sequences of human, rat, and mouse L1. Several Sox/LEF-binding sites are also present throughout the entire L1 sequence, including several sites in the second open reading frame (ORF2). The discovery that L1 retroelements embedded in the mammalian genome can function as bidirectional promoters suggests that Sox/LEF regulatory sites may represent a general mechanism for transcriptional regulation. This led us to examine whether other retroelements have a similar ability, thereby expanding the role of retroelements in adult hippocampal neurogenesis.

2.2 Expression of B1 SINE RNA and B2 SINE RNAs in adult hippocampus

The human and rodent genomes harbor numerous non-autonomous retrotransposons, termed short interspersed elements (SINEs). SINEs are highly abundant components of mammalian genomes and are propagated via retrotransposition (Ferrigno et al., 2001). Non-autonomous SINEs recruit L1-encoded proteins for their own mobilization. Alu elements are the major SINEs in the human genome, whereas B1 and B2 elements are the major SINE families in the mouse genome. The B2 SINE family constitutes approximately 0.7% of total mouse genomic DNA (Bennett et al., 1984). These retroelements are widely distributed throughout the genome, although many are heavily truncated and only a few are thought to be active and able to retrotranspose.

We first investigated B2 SINE RNA expression in the adult hippocampus by in situ hybridization. Brains were dissected from freshly euthanized Fisher 344 rats and placed in ice-cold saline. The brains were then placed in plastic blocks in OCT compound (Tissue Tek) and frozen. Sections were cut at 15 μm thickness with a cryostat (LEICA CM1850, Leica). Brain sections on the slide glass were hybridized with labeled riboprobes. Following the in situ hybridization of B2 SINE RNA, immunohistochemical analysis of the L1 protein (1:300; rabbit antibody against LINE-1, SantaCruz) was carried out. We observed that hippocampal granule neurons extensively express L1 RNA, as well as B2 SINE RNA (Fig. 3). Strong signals were also observed in neuronal layers (granule cell layers) but were not found in cells at the innermost layer of the dentate gyrus where astrocytes and undifferentiated neural stem cells reside (Fig. 3).

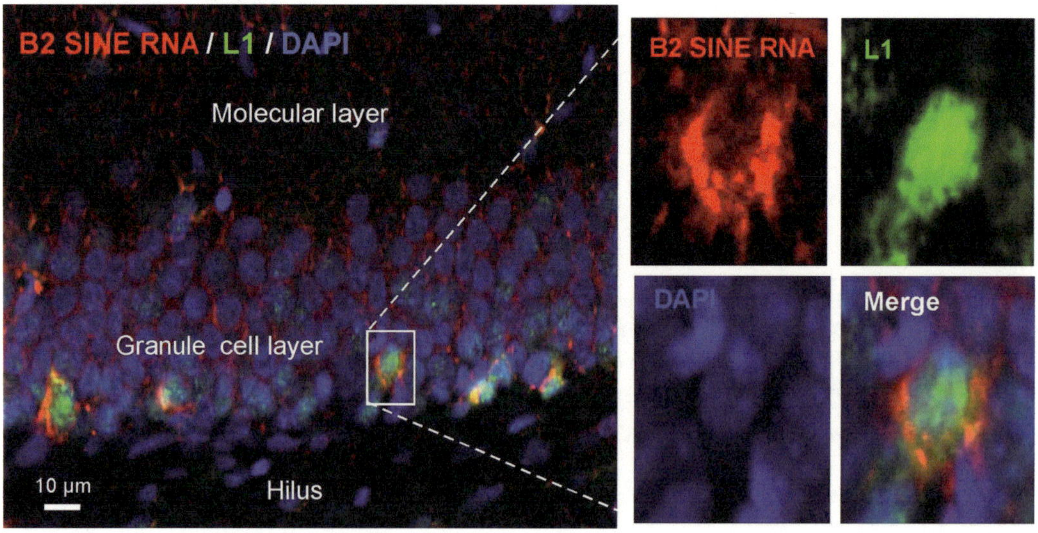

Fig. 3. Expression B2 SINE RNA and L1 in DG of adult hippocampus.

In situ hybridization for B2 SINE RNA (red) and immunohistochemistry of L1 protein (green) was carried out simultaneously on the DG of adult rat hippocampus. Hippocampal granule neurons extensively express both B2 SINE RNA and L1. The SINE RNA and L1 protein double-positive cells in the white square are magnified in the right panels. B2 SINE RNA: red, L1 protein: green, DAPI: blue.

2.3 Wnt-signaling regulatory sites on B1 SINE and B2 SINE DNA sequences

We identified that the 2 major classes of non-coding retroelements, B1 and B2 SINEs (B1 SINE, GenBank accession number X62249; B2 SINE, M31441) also carry Wnt-responsive elements (Fig. 4). SINEs originate from retrotransposition events of small RNAs (Batzer and Deininger, 2002; Hasler & Strub, 2006; Nishihara et al., 2006). Both Alu and B1 elements are derived from the 7SL RNA (Ullu & Tschudi; 1984), whereas B2 and most other SINEs are derived from tRNA genes. The eukaryotic RNA polymerase III (Pol III) system is responsible for synthesizing transfer RNA molecules and other transcripts, which in yeast include the U6 spliceosomal RNA, 7SL RNA, 5S ribosomal RNA, snr52 small nucleolar RNA, and the RNA component of RNaseP (Paule & White, 2000; Geiduschek & Kassavetis, 2001; Huang & Maraia, 2001). Transcription Factor for polymerase III C (TFIIIC) binds to 2 intragenic (lying within the transcribed DNA sequence) control sequences, the A-Box and B-Box (Fig. 4).

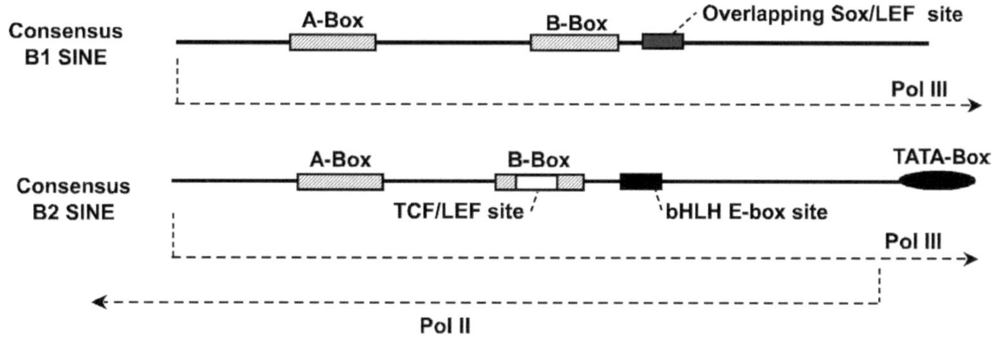

Fig. 4. Transcriptional regulation of B1 SINE RNA and B2 SINE RNA.

RNA polymerase III (also called Pol III) transcribes DNA by recognizing A-box and B-box (grey boxes). Overlapping Sox/LEF sequence (dark grey box) is contained in B1 SINE RNA. TCF/LEF regulatory sequence (white box) is involved in the B-box sequence of consensus B2 SINE RNA. The DNA of B2 SINE RNA contains also the bHLH transcription factor recognizing E-box sequence (black box). B2 SINE also carries an active RNA polymerase II (pol II) regulatory site that is located outside the tRNA region (Ferrigno et al., 2001). TATA box of the pol II promoter is indicated in black oval box.

As Figure 4 indicates, DNA sequences of both B1 SINE RNA and B2 SINE RNA include a Wnt signaling responsive element (Sox/LEF site in B1 SINE and TCF/LEF site in B2 SINE). Interestingly, the DNA sequence of B2 SINE RNA also contains an E-box sequence, which is recognized by a bHLH transcription factor, such as NeuroD1 (Fig. 4). The expression of NeuroD1 is triggered by Wnt-signaling, suggesting that B2 SINE RNA has additional active sites for NeuroD1 regulation in the adult neuronal lineage-differentiation. To explore the expression profile of these retroelement RNAs (B1 SINE RNA and B2 SINE RNA) in a lineage-specific manner in the adult hippocampus, we cultured adult hippocampal NSCs.

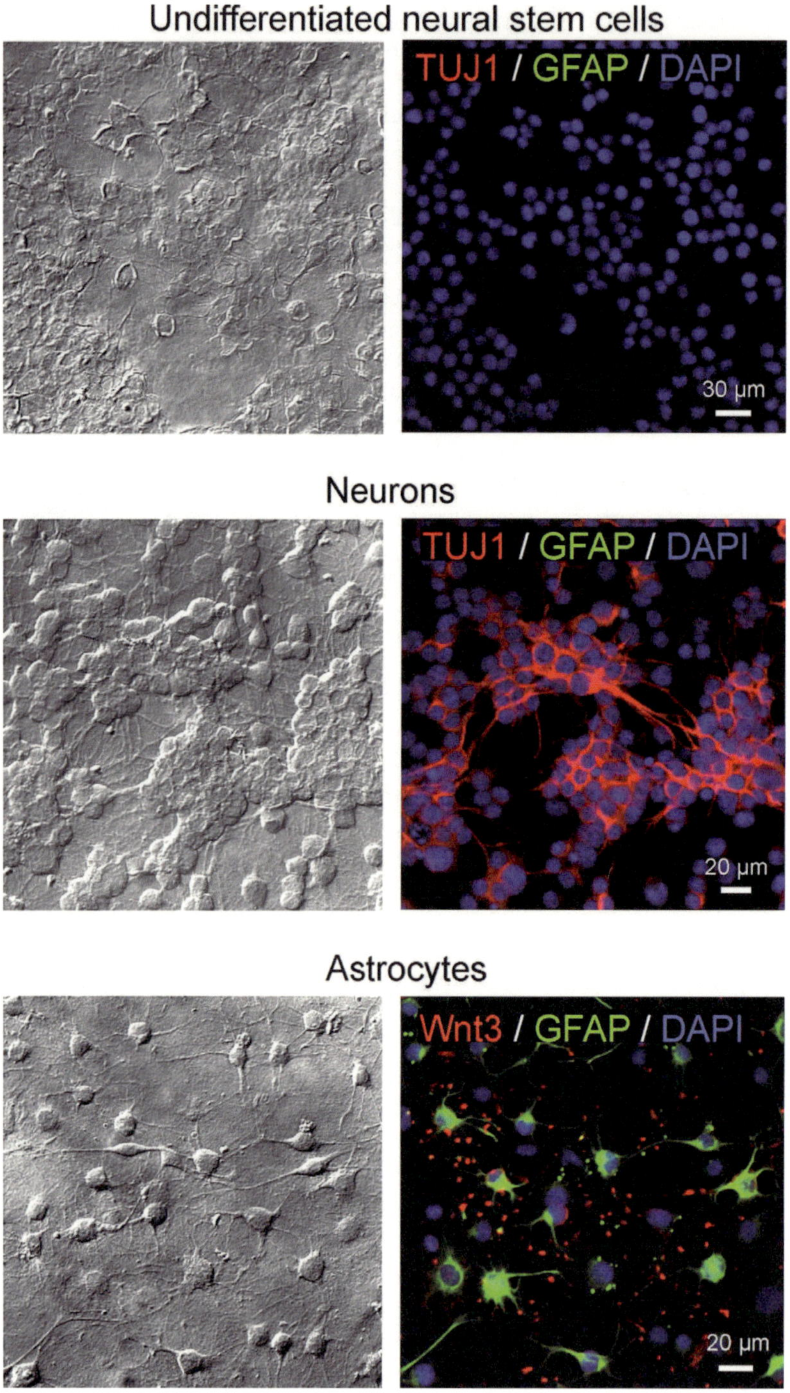

Neural stem cells (top), neurons (middle) and astrocyte cells (bottom) were examined by the immunohistochemistry analysis. TUJ1 (top and middle) and Wnt3 (bottom): red, GFAP: green, DAPI: blue.

Fig. 5. *In vitro* culture system of adult hippocampal neural stem cells

2.4 Expression of B1 SINE RNA and B2 SINE RNA in adult hippocampal neural stem cell culture

Adult hippocampal NSCs were round and retained their shape when expanded as a monolayer (top panels, Fig. 5). These neural progenitor cells have stem cell properties in vitro: (1) they undergo self-renewal in the presence of basic FGF-2; (2) single genetically marked clones can differentiate into neurons, oligodendrocytes, and astrocytes in vitro and when grafted back to the adult hippocampus in vivo; and (3) they express progenitor cell markers such as Sox2 and nestin (Gage et al., 1995; Palmer et al., 1997). Undifferentiated NSCs were negative for the neuronal marker β-tubulin III (TUJ1) and the astrocyte lineage marker glial fibrillary acidic protein (GFAP) (Fig. 5). Under neuronal differentiation conditions, we added 1 μM retinoic acid and 5 μM forskolin to the culture. The expression of β-tubulin III was remarkably up-regulated, as confirmed by the immunohistochemistry analysis. GFAP was found to be absent in the immunohistochemistry analysis, indicating that the adult NSCs in culture were committed to a neuronal lineage specifically (middle panels, Fig. 5). Differentiation into the astrocyte lineage was stimulated by the addition of 50 ng/mL leukemia inhibitory factor (LIF) and 50 ng/mL bone morphogenetic protein 2 (BMP2). Astrocytes prepared in vitro expressed Wnt3 factors, consistent with in vivo data (Lie et al., 2005; Kuwabara et al., 2009; Okamoto et al, 2011).

2.5 Neuronal-specific expression of B1 SINE RNA and B2 SINE RNA in the adult hippocampal neural stem cell culture

By using an *in vitro* culture system of adult hippocampal NSCs (Fig. 5), we investigated the relative expression levels of B1 SINE and B2 SINE RNA. Neurons differentiated from adult neural stem cells expressed high levels of both B1 RNA and B2 RNA, although the expression level of B2 SINE RNA was higher (Fig. 6).

Total RNAs were extracted cells and expression levels of B1 SINE RNA (black bars) and B2 SINE RNA (white bars) ware examined by quantitative real-time PCR (Q-PCR analysis). A) The level was normalized internal control gene GAPDH and platted in the graph. The expression level of B1 SINE RNA in undifferentiated neural stem cells (NSCs) was taken as 100% (asterisk; black bars). B) Time course analysis of B1 SINE RNAs and B2 SINE RNAs.

In the astrocyte lineage, the basal expression of B1 SINE RNA and B2 SINE RNA from undifferentiated cells was largely diminished (Fig. 6A). The neuronal specific expression of B2 SINE RNA by quantitative real-time PCR (QPCR) analysis in vitro was consistent with the results of in situ hybridization (Fig. 3).

Time course of B1 SINE RNA and B2 SINE RNA expression during the early stages of neurogenesis in cultured adult NSCs was assessed by QPCR analysis. Following neuronal differentiation in vitro, B1 SINE RNA and B2 SINE RNA expression peaked at 24 h after neuronal induction (20-fold increase in B1 SINE RNA, 50-fold increase in B2 SINE RNA). These expression levels gradually decreased until they reached a plateau by 48 h after neuronal induction (Fig. 6B). Both NeuroD1 and L1 expression was highest at 24 h after neuronal induction and gradually declined during neuronal differentiation (Kuwabara et al., 2009), similar to the expression profile of B1 SINE RNA and B2 SINE RNA, suggesting that these genes are under similar transcriptional regulation. Combined expression analysis of in situ hybridization and immunohistochemistry of B2 SINE RNA and L1 showed that early stage neuronal progenitors (i.e., neuroblast cells) contained high levels of L1 and B2 SINE RNAs (Fig. 3; inner layer of dentate gyrus, cells indicated by the white square). In the mature neuronal layer of the dentate gyrus (deeper neuronal layer of SGZ), the L1 and B2 SINE RNA signals were still clearly observed, although it was weaker than those in neuroblast cells. The in vivo expression profile obtained from in situ hybridization and the in vitro expression profile obtained from the time course analysis using adult NSC culture are consistent (Fig. 3 and Fig. 6).

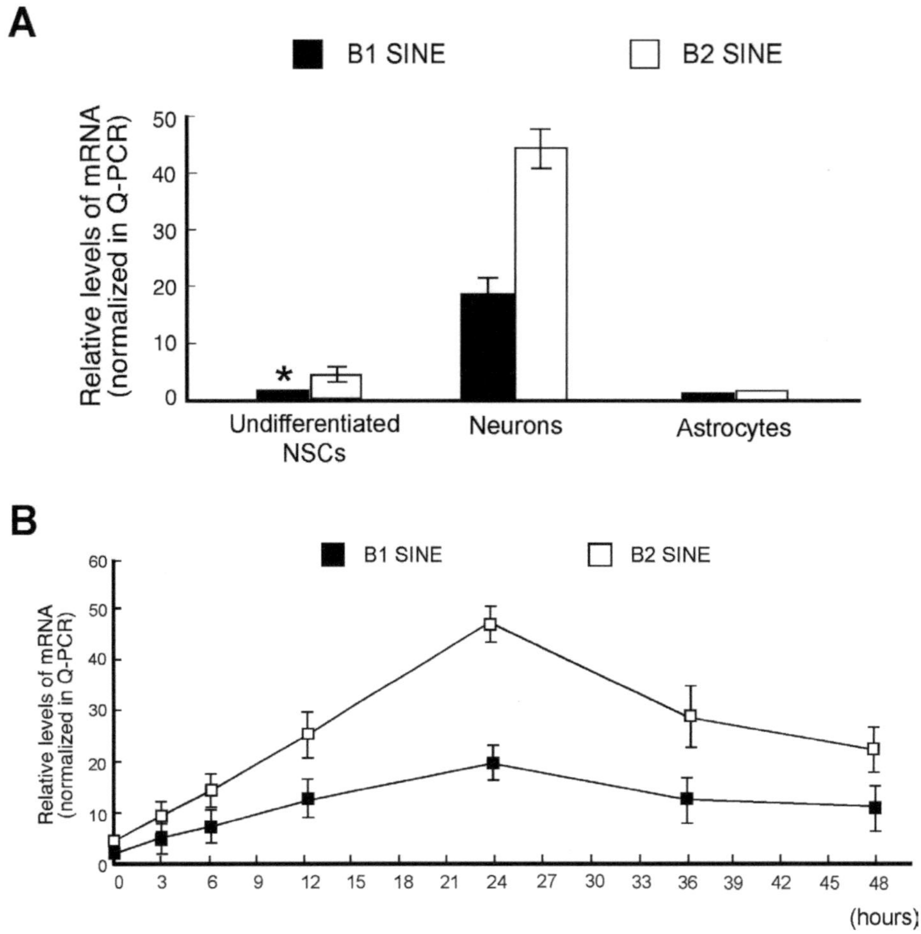

Fig. 6. Expression of B1 SINE RNA and B2 SINE RNA during adult neurogenesis.

2.6 Effect of Wnt3 on the expression of B1 SINE RNA and B2 SINE RNA

Next, we examined the effect of the Wnt3a ligand on the expression of B1 SINE RNA and B2 SINE RNA. Wnt3a ligands were added to the NSC culture at different concentrations (0 ng/mL, 10 ng/mL, 20 ng/mL, and 50 ng/mL of Wnt3a), and the expression levels of B1 SINE RNA and B2 SINE RNA were up-regulated in a dose-dependent manner (Fig. 7).

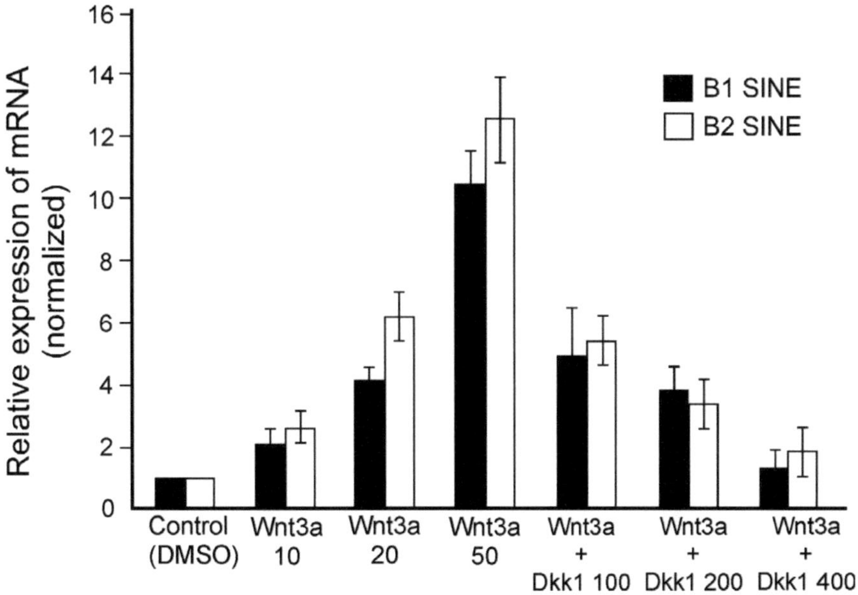

Fig. 7. Wnt signaling increases the expression of B1 SINE RNA and B2 SINE RNA in adult neural stem cell cultures. Ligand of Wnt3a and Dkk1 was added in the adult NSC cultures. Ligand concentration is indicated in the graph (ng/mL). The expression levels of B1 SINE RNA and B2 SINE RNA were normalized with internal control GAPDH. Cells treated with DMSO were taken as control and the relative value of the normalized expression of B1 SINE RNA and B2 SINE RNA was plotted in the graph. B1 SINE RNA: black bars, B2 SINE RNA: white bars.

To evaluate the observed positive effects of Wnt3a on the expression of B1 SINE RNA and B2 SINE RNA, the Wnt antagonist Dickkopf1 (Dkk1) was added into the adult hippocampal NSC culture. As the antagonist concentration was increased (0 ng/mL, 100 ng/mL, 200 ng/mL, and 400 ng/mL of Dkk1 with 50 ng/mL Wnt3a), Wnt3a-mediated activation of B1 SINE RNA and B2 SINE RNA was diminished (Fig. 7). From these data, we confirmed the contribution of Wnt signaling to the transcriptional activation of B1 SINE RNA and B2 SINE RNA in adult hippocampal NSCs.

2.7 Effect of Wnt-signaling on the chromatin regulation of B1 SINE RNA and B2 SINE RNA

To assess protein association on the regulatory region of B1 SINE RNA and B2 SINE RNA, we performed chromatin immunoprecipitation (ChIP). Association levels were quantitatively evaluated using real-time PCR (ChIP-QPCR). From in situ hybridization / immunohistochemistry in vivo data (Fig. 3) and the time course expression profile in vitro (Fig. 6), coordinated expression between SINE RNAs and L1 was observed. Therefore, we performed ChIP-QPCR analysis for SINE RNAs and L1 in parallel.

Addition of Wnt3a into the adult NSC culture promoted the enrichment of trimethyl histone H3 lysine 4 (triMetK4), a modified histone mark associated with gene activation, by more than 15-fold at the B1 SINE RNA locus and by more than 20-fold at the B2 SINE RNA locus, compared to the control cells treated with DMSO (Fig. 8).

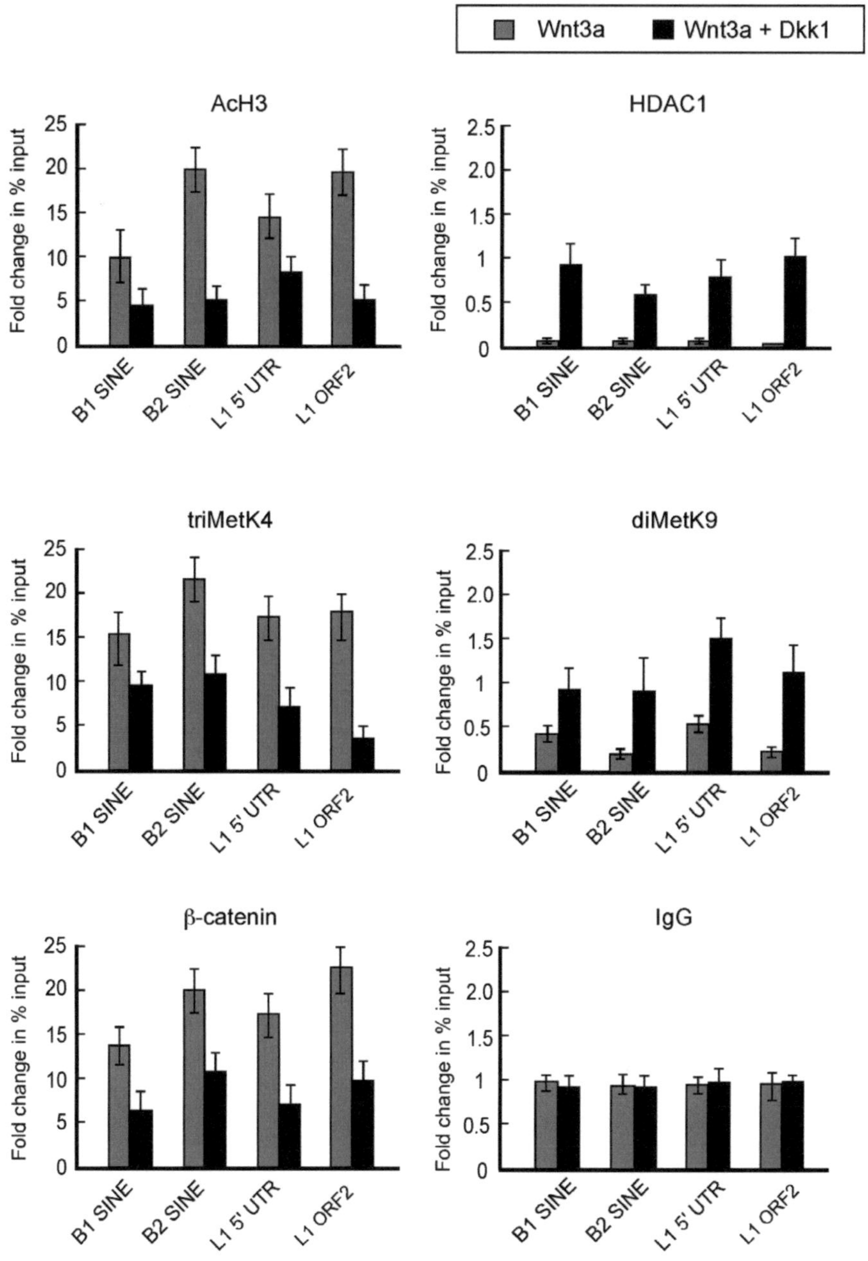

Fig. 8. Chromatin immunoprecipitation analysis (ChIP-QPCR) of retroelements in adult neural stem cells stimulated with Wnt3a and the antagonist Dkk1. PCR primers were designed to surround the Sox/LEF sequence on the L1 promoter. PCR primers for B1 SINE RNAs were designed to surround the Sox/LEF sequence on the B1 SINE DNA (Fig. 4). PCR primers for B2 SINE RNAs were designed to surround the TCF/LEF sequence and bHLH E-box sequence on the B2 SINE DNA (Fig. 4). From these results, we confirmed that similar molecular mechanism controlled SINE RNAs and L1 in the adult hippocampal neurogenesis and Wnt3 has essential role to activate the transcription of these retroelements.

The association of acetylated histone H3 (Ac-H3) and β-catenin increased with the Wnt3a treatment. Dimethylated H3 lysine 9 (diMetK9) and histone deacetylase 1 (HDAC1) were rarely associated, suggesting that the Wnt3a treatment promotes an active chromatin state (Fig. 6A). In contrast, the addition of Dkk1 inhibited the Wnt3a-mediated activation of chromatin at the B1 SINE RNA and B2 SINE RNA loci. We found that β-catenin, Ac-H3, and triMetK4 were associated at the L1 5′ UTR region and L1 ORF2 region, both of which include the Sox/LEF regulatory sequence (Kuwabara et al., 2009), similar to their association on the B1 SINE RNA and B2 SINE RNA loci. In contrast, Dkk1 diminished the Wnt3a-mediated activation process, as seen in the case of the B1 SINE RNA and B2 SINE RNA loci (Fig. 8).

2.8 Promoter activity of B2 SINE RNA in vivo

Although we determined the global expression of B2 SINE RNA in neuronal cells by in situ hybridization (Fig. 3), we further examined the detailed expression of B2 SINE RNA in the adult hippocampus. The transcriptional activity of B2 SINE RNA is controlled by regulatory sequences present on the internal promoter (Fig. 4). We prepared a lentivirus construct that carries the B2 SINE promoter sequence (not including termination sequences) and EGFP reporter cassette (B2 SINE promoter driven EGFP; LV SINE GFP).

SINE GFP lentivirus was stereotactically injected into the dentate gyrus of young adult rats. Three weeks later, mice were injected with BrdU daily for 10 days. Notably, we observed that SINE GFP expression was restricted to neurogenic areas, and that GFP-positive cells colocalized with NeuroD1-positive neuronal progenitors (Fig. 9). We also detected SINE GFP-positive cells that migrated further into the granule cell layer where NeuN-positive mature neurons reside.

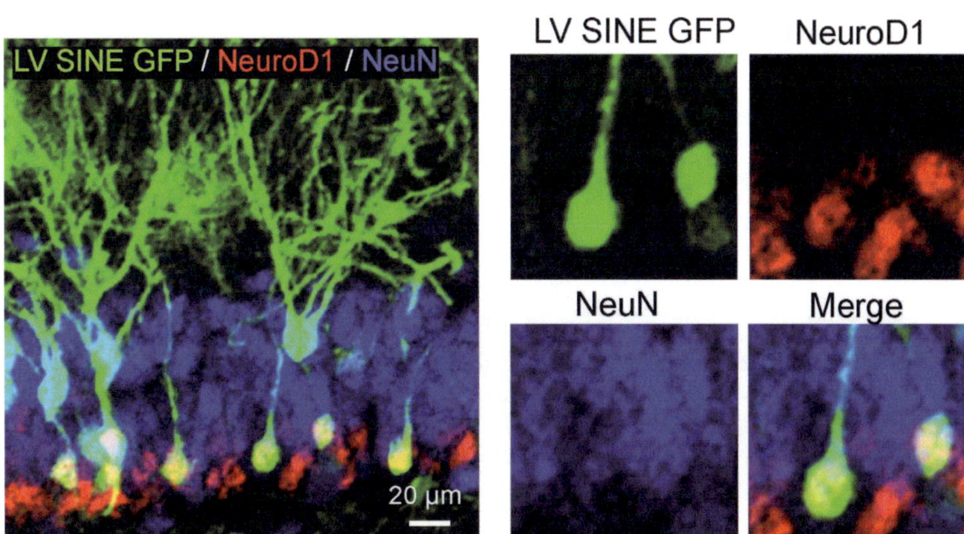

Fig. 9. Activity of B2 SINE as a promoter in adult rat hippocampus.

We examined the activity and specificity of the B2 SINE-based promoter in adult rat hippocampus. EGFP-expressing lentivirus, under the control of the B2 SINE-based promoter, was stereotactically microinjected into the dentate gyrus of adult rats and the population of GFP-positive cells (green) was analyzed by immunohistochemistry using antibodies to NeuroD1 (red) and NeuN (blue).

To identify the composition of cell types in which the B2 SINE-based promoter activity was turned on, we quantified the results of the immunohistochemical analysis. The number of SINE GFP and the lineage marker double-positive cells was counted and graphically plotted (Fig. 10).

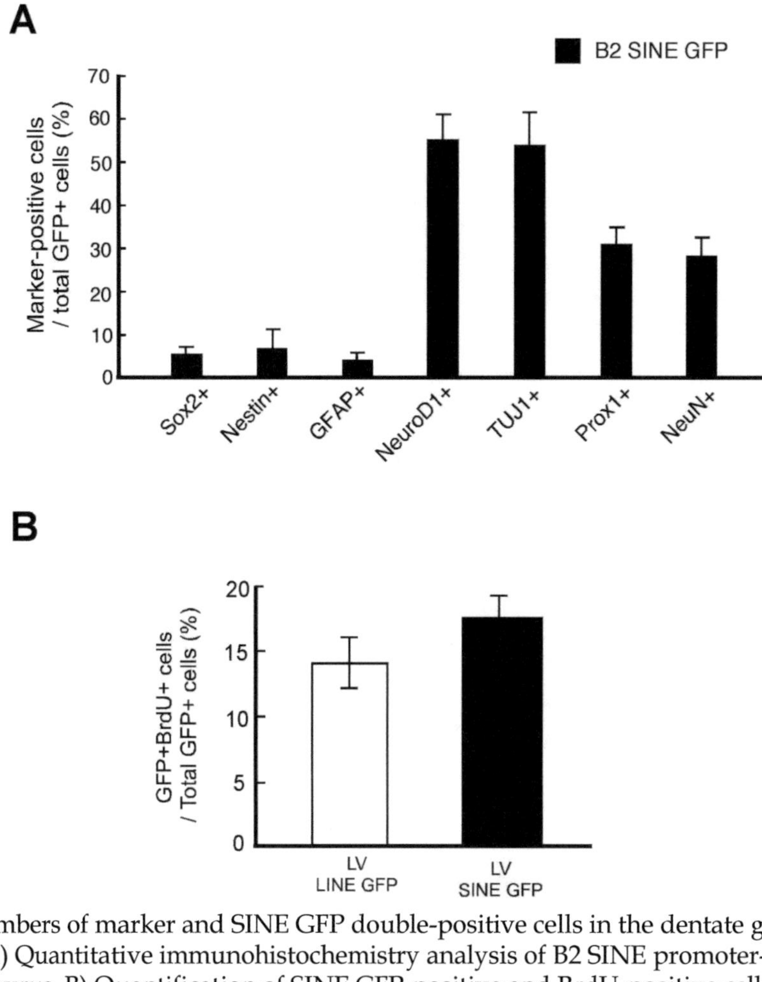

Fig. 10. Numbers of marker and SINE GFP double-positive cells in the dentate gyrus of adult rat. A) Quantitative immunohistochemistry analysis of B2 SINE promoter-active cells in dentate gyrus. B) Quantification of SINE GFP-positive and BrdU-positive cells in dentate gyrus of adult rat (black bar). LINE GFP-positive cells were also examined (white bar).

To examine SINE GFP-positive cells in the stem cell compartment, we stained the cells with Sox2 and nestin, a marker of radial stem-like cells. The proportion of Sox2 and SINE GFP double-positive cells was 6.4% of the total SINE GFP-positive cells. The proportion of nestin and SINE GFP double-positive cells was 7.4% of the total SINE GFP-positive cells. To examine SINE GFP-positive cells in the glial cell compartment, we stained the cells with GFAP. The proportion of GFAP and SINE GFP double-positive cells was 3.7% of total SINE GFP-positive cells.

In contrast, SINE GFP-positive cells in the neuronal cell compartment comprised a majority of the total GFP-positive cells (Fig. 10A). B2 SINE-based promoter activity was up-regulated in cells at the early stage of neuronal lineage, such as NeuroD1- and TUJ1-positive cells. The proportion of NeuroD1 and SINE GFP double-positive cells was 56% of the total SINE GFP-positive cells. The up-regulated activity of the SINE GFP promoter was retained in mature neurons (Prox-1- and NeuN-positive cells), but the proportion was almost half of neuroblast cells (NeuroD1-positive cells). Furthermore, SINE GFP-positive cells were labeled by BrdU, and the proportion of double-positive cells was found to be higher than that with LINE-GFP and BrdU staining (Fig. 10B), suggesting that the B2 SINE RNA-expressing cells actively proliferated in the adult hippocampus.

3. Conclusion

Our data indicate that both B1 SINE RNA and B2 SINE RNA are up-regulated during adult hippocampal neurogenesis, and that Wnt induction is required as the driving force. The specific response of these retroelements to paracrine Wnt factors is coupled with the differentiation progress of adult hippocampal NSCs into the neuronal pathway. Upon Wnt-mediated neuronal induction, the β-catenin activation complex up-regulated NeuroD1 and L1, as well as SINEs in adult hippocampal NSCs in vitro, while the Wnt antagonist Dkk1 down-regulated SINE RNA expression.

In ChIP-QPCR analysis, the association of β-catenin, triMetK4, and Ac-H3 in B1 and B2 SINEs was stimulated in cells treated with the Wnt3a ligand, suggesting that the genomic DNA of B1 and B2 SINEs was positively activated by canonical Wnt signaling. From these data, we propose that the activation of various genomic loci, where L1 and SINE fragments are present, may occur during adult hippocampal neurogenesis. Expression of retroelement RNAs are identified generally in neurogenic region in adult brain, although only hippocampal dentate gyrus is determined as the region that the retrotransposition (" jumping") occurs. Since various transcriptional activators are found in other neurogenic region in adult brain, not solely Wnt-signaling pathway but also other multiple molecules may control the expression of retroelement RNAs. We will actively extend current research to further determine potent regulators and the role of retroelement RNA during adult neurogenesis.

Since retroelement sequences are scattered throughout the genome and contain Wnt-responsive regulatory elements, retroelements retain the ability to act as promoters inducing the activity of nearby chromatin loci during adult neurogenesis. The functional relevance of adult hippocampal neurogenesis has been extensively studied, and the fact that physiological and behavioral events, such as aging, stress, diseases, seizures, learning, and exercise, can modulate neurogenesis is of particular interest. This suggests that the chromatin state at SINE/L1 loci potentially changes depending on paracrine stimuli. These epigenetic mechanisms may act as genomic "sensors" of environmental changes and function as fine modulators of adult hippocampal neurogenesis.

4. Acknowledgment

We thank Hideto Takimoto for providing assistance in the care of animals and in the experiments. SA, MW, KT, MA and TK were supported by various grants from AIST. TK was partly supported by the Grant-in-Aid for Young Scientists (B).

5. References

Avilion, . AA. Et al. (2003) Multipotent cell lineages in early mouse development depend on SOX2 function. *Genes & Development*, Vol. 17, No. 1, (January 1), 126-140, ISSN 0890-9369

Bani-Yaghoub, M. et al. (2006) Role of Sox2 in the development of the mouse neocortex. *Developmental Biology*, Vol. 295, No. 1, (July 1), pp. 52-66, ISSN 0012-1606

Batzer, M. A. & Deininger, P. L. (2002) Alu repeats and human genomic diversity. *Nature Reviews Genetics*, Vol. 3, No. 5, (May), pp. 370-379, ISSN 1471-0056

Bennett, K.L. *et al.* (1984) Most highly repeated dispersed DNA families in the mouse genome. *Molecular and Cellular Biology*. Vol. 4, No. 8, (August) 1561-1571, ISSN 0270-7306

Bylund, M. et al. (2003) Vertebrate neurogenesis is counteracted by Sox1-3 activity. *Nature Neuroscience*, Vol. 6, No. 11, (November), pp. 1162-1168, ISSN 1097-6256

Coufal, N. G. et al. (2009) L1 retrotransposition in human neural progenitor cells. *Nature*, Vol. 460, No. 7259, (August 27), pp. 1127-1131, ISSN 0028-0836

Cuitino, L. N. et al. (2010) Wnt5a is found to modulate recycling of functional GABAA receptors on hippocampal neurons. *The Journal of Neuroscience*, Vol. 30, No. 25, (June 23), pp. 8411-8420, ISSN 0270-6474

Deisseroth, K. et al. (2004) Excitation-neurogenesis coupling in adult neural stem/progenitor cells. *Neuron* Vol. 42, No. 4, (May 27), pp. 535-552, ISSN 0896-6273

Deng, W. et al. (2010) New neurons and new memories: how does adult hippocampal neurogenesis affect learning and memory? *Nature Reviews Neuroscience*, Vol.11, No.5, (May 11), pp. 339-350, ISSN 1471-0048

Ferri, A. L. et al. (2004) Sox2 deficiency causes neurodegeneration and impaired neurogenesis in the adult mouse brain. *Development*, Vol. 131, No. 15, (August), pp. 3805-3819, ISSN 1011-6370

Ferrigno, O. et al. (2001) Transposable B2 SINE elements can provide mobile RNA polymerase II promoters. *Nature Genetics*, Vol. 28, No. 1, (May 28), pp. 77-81, ISSN 1061-4036

Gage, F. H. (2000) Mammalian neural stem cells. *Science* Vol.287, No.5457, (February 25), pp.1433-1438, ISSN 1095-9203.

Gage, F. H. et al. (1995) Survival and differentiation of adult neuronal progenitor cells transplanted to the adult brain. *Proceedings of the National Academy of Sciences*, Vol. 92, No.25, (December 5), pp. 11879-11883, ISSN 1091-6490

Geiduschek, E. P. & Kassavetis, G. A. (2001) The RNA polymerase III transcription apparatus. Journal of Molecular Biology Vol. 310, No. 1, (June 29), pp. 1-26, ISSN 0022-2836

Gogolla, N. et al. (2009) Wnt signaling mediates experience-related regulation of synapse numbers and mossy fiber connectivities in the adult hippocampus. *Neuron* Vol. 62, No. 4, (May 28), pp. 510-525, ISSN 0896-6273

Graham, V. et al. (2003) SOX2 functions to maintain neural progenitor identity. *Neuron*, Vol. 39, No. 5, (August 28) pp. 749-765, ISSN 0896-6273

Gritti, A. et al. (2002) Multipotent neural stem cells reside into the rostral extension and olfactory bulb of adult rodents. *The Journal of Neuroscience*, Vol.22, No.2, (January 15), pp. 437-445, ISSN 0270-6474

Han, J. S. & Boeke, J. D. (2005) LINE-1 retrotransposons: modulators of quantity and quality of mammalian gene expression? *Bioessays,* Vo. 27, No. 8, (August), pp. 775-784, ISSN 1521-1878

Häsler, J. & Strub, K. (2006) Alu elements as regulators of gene expression. *Nucleic Acids Research,* Vol. 34, No. 19, pp. 5491-5497, ISSN 0305-1048

Huang, Y. & Maraia, RJ. (2001) Comparison of the RNA polymerase III transcription machinery in Schizosaccharomyces pombe, Saccharomyces cerevisiae and human. *Nucleic Acids Research,* Vol. 29, No. 13, (July 1), pp. 2675-2690, ISSN 0305-1048

Kuhn, H. G. et al. (1996) Neurogenesis in the dentate gyrus of the adult rat: age-related decrease of neuronal progenitor proliferation. *The Journal of Neuroscience,* Vol.16, No.6, (March 15), pp. 2027-2033, ISSN 0270-6474

Kuwabara, T. et al. (2009) Wnt-mediated activation of NeuroD1 and retro-elements during adult neurogenesis. *Nature Neuroscience,* Vol. 12, No. 9, (September), pp. 1097-1105, ISSN 1097-6256

Lee, S. M. et al. (2000) A local Wnt-3a signal is required for development of the mammalian hippocampus. *Development,* Vol. 127, No. 3, (February), pp. 457-467, ISSN 1011-6370

Lie, D. C. et al. (2005) Wnt signalling regulates adult hippocampal neurogenesis. *Nature,* Vol. 437, No. 7063, (October 27), pp. 1370-1375, ISSN 0028-0836

Liu, M. et al. (2000) Loss of BETA2/NeuroD leads to malformation of the dentate gyrus and epilepsy. *Proceedings of the National Academy of Sciences,* Vol.97, No2, (January 18), pp. 865-870, ISSN 1091-6490

Lois, &Alvarez-Buylla, A. (1993) Proliferating subventricular zone cells in the adult mammalian forebrain can differentiate into neurons and glia. *Proceedings of the National Academy of Sciences,* Vol.90, No.5, (March 1), pp. 2074-2077, ISSN 1091-6490

Miyaoka, T., Seno, H., and Ishino, H. (1999) Increased expression of Wnt-1 in schizophrenic brains. *Schizophrenia Research,* Vol. 38, No. 1, (July 27), pp. 1-6, ISSN 0920-9964

Miyata, T. et al. (1999) NeuroD is required for differentiation of the granule cells in the cerebellum and hippocampus. *Genes & Development,* Vol. 13, No. 13, (Jury 1), 1467-1652, ISSN 0890-9369

Muotri, A. R. et al. (2005) Somatic mosaicism in neuronal precursor cells mediated by L1 retrotransposition. *Nature,* Vol. 435, No. 7044, (June 16), pp. 903-910, ISSN 0028-0836

Nishihara, H. et al. (2006) Functional noncoding sequences derived from SINEs in the mammalian genome. *Genome Research,* Vol. 16, No. 7, (July), pp. 864-874, ISSN 1088-9051

Okamoto, H. et al. (2010) Wnt2 expression and signaling is increased by different classes of antidepressant treatments. *Biological Psychiatry,* Vol. 68, No. 6, (September 15), pp. 521-527, ISSN 0006-3223

Okamoto, M. et al. (2011) Reduction in paracrine Wnt3 factors during aging causes impaired adult neurogenesis. *The FASEB Journal,* Vol. 25, No. 8, (August), pp. xx-xx, ISSN 0892-6638

Pagano S. F. et al. (2000) Isolation and characterization of neural stem cells from the adult human olfactory bulb. *Stem Cells.* Vo.18, No.4, pp. 295-300, ISSN 1066-5099

Palmer, T. D. et al. (1995) FGF-2-responsive neuronal progenitors reside in proliferative and quiescent regions of the adult rodent brain. *Molecular and Cellular Neuroscience,* Vol. 6, No. 5, (October), pp. 474-486, ISSN 1044-7431

Paule, M. R. & White, R. J. (2000) Survey and summary: transcription by RNA polymerases I and III. *Nucleic Acids Research,* Vol. 28, No. 6, (March 15), pp. 1283-1298, ISSN 0305-1048

Song, H. et al. (2002) Astroglia induce neurogenesis from adult neural stem cells. *Nature,* Vol. 417, No. 6884, (May 2), pp. 39-44, ISSN 0028-0836

Suh, H., et al. (2007) In vivo fate analysis reveals the multipotent and self-renewal capacities of Sox2+ neural stem cells in the adult hippocampus. *Cell Stem Cell,* Vol. 1, No. 5, (November), pp.515-528, ISSN 1934-5909

Tozuka, Y. et al. (2005) GABAergic excitation promotes neuronal differentiation in adult hippocampal progenitor cells. *Neuron* Vol. 47, No. 6, (September 15), pp. 803-815, ISSN 0896-6273

Ullu, E. & Tschudi, C. (1984) Alu sequences are processed 7SL RNA genes. *Nature,* Vol. 312, No. 5990, November 8-14), pp. 171-172, ISSN 0028-0836

12

Noncoding RNAs in Neural Stem Cell Development

Shan Bian and Tao Sun
Department of Cell and Developmental Biology
Cornell University Weill Medical College
USA

1. Introduction

Neural stem cells and neural progenitors/precursors (NSCs/NPs) are identified in both embryonic and adult central nervous system (CNS). NSCs can self-renew and give rise to neurons and glia. The development of NSCs is controlled by precisely orchestrated gene expression regulation. Recently, emerging evidence has shown the importance of noncoding RNA regulation in NSC self-renewal, proliferation, survival and differentiation. In this chapter, we will present new research of noncoding RNA functions in NSC development. We will highlight the future directions of applying noncoding RNAs in stem cell-based therapy for neurological diseases.

2. Noncoding RNAs

Noncoding RNAs (ncRNAs) are functional RNA molecules that do not show protein translation capability. ncRNAs consist of ribosomal RNAs (rRNAs), transfer RNAs (tRNAs), piwi-interacting RNAs (piRNAs), microRNAs (miRNAs), and long noncoding RNAs (lncRNAs) and so on. ncRNAs have shown to play distinct but also conserved roles in normal development in invertebrates and vertebrates.

2.1 piRNAs

piRNAs are a group of small RNAs with size between 26-31 nucleotides (nt), which are only found in male and female germlines of invertebrates and within testes in mammalians (Houwing et al., 2007; 2009; Lander et al., 2001; Lau et al., 2006; Seto et al., 2007). piRNAs interact with piwi proteins to form RNA-protein complexes (Das et al., 2008; Houwing et al., 2007). The piRNA-protein complexes have been shown to silence transcription, specifically transposons (Brennecke et al., 2008; Das et al., 2008). Since piRNAs are mainly expressed during the germline stem cell development, they will not be discussed further in this chapter.

2.2 miRNAs

miRNAs are ~22 nt highly conserved small noncoding RNAs found in almost all eukaryotic cells (Khraiwesh et al., 2010) (Fig. 1). Like coding genes, miRNAs are mainly transcribed by the

RNA Polymerase II (Pol II) into long primary miRNA transcripts (pri-miRNAs). Pri-miRNAs are next cleaved by Drosha, a class 2 RNase III enzyme, to produce short hairpin stem-loop structures, known as precursor miRNAs (pre-miRNAs) (Ambros, 2008; Gregory et al., 2006). Pre-miRNAs are then processed into about 20-25 nt double-stranded mature miRNAs by RNase III enzyme Dicer (Lund and Dahlberg, 2006). The duplex undergoes unwinding and the strand with the weakest base pairing at its 5′ terminus, together with Dicer and many associated proteins including Argonaute proteins, form an active RNA-induced silencing complex (RISC) (Neilson and Sharp, 2008; Rana, 2007). A miRNA recognizes and binds to the 3′ untranslated region (3′ UTR) of target messenger RNAs (mRNAs) by imperfect complementary sequence recognition (Neilson and Sharp, 2008; Wang et al., 2004). Once bound, the RISC can negatively regulate miRNA targets by degrading the mRNA or repressing its translation (Khraiwesh et al., 2010; Pratt and MacRae, 2009) (Fig. 1).

miRNAs are classified into intergenic and intragenic miRNAs, depending on their genomic location (Lagos-Quintana et al., 2001; Lau et al., 2001; Lee and Ambros, 2001). Intergenic miRNAs are located at intergenic regions (between genes) in the genome. These miRNAs are usually transcribed using their own promoters. Intragenic miRNAs normally lie in intronic regions of coding genes with the same orientation. They are often transcribed together with the host genes.

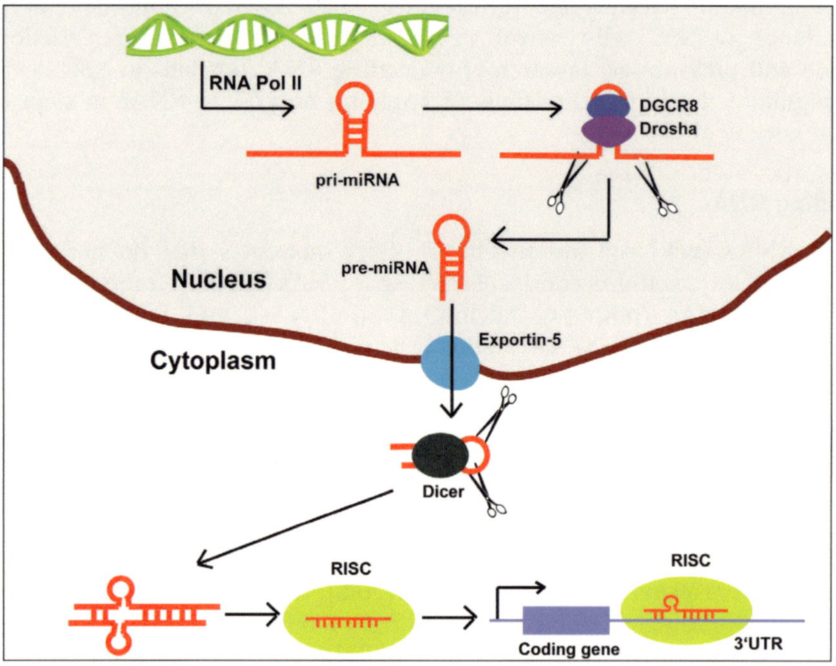

Fig. 1. A scheme of microRNA biogenesis. microRNAs silence target coding genes by binding to the 3′ untranslated region (3′ UTR).

2.3 lncRNAs

The majority of the human genome has previously been considered as "junk" DNA, since only about 1.5% of the human genome, which occupies over 3 billion DNA base pairs, consists of protein-coding genes (Lander et al., 2001). A recent study of a large-scale complementary DNA (cDNA) sequencing project has shown that four fifths of transcripts of the human genome are RNA transcripts that don't encode proteins (Kapranov et al., 2007). These RNA transcripts are normally longer than 200 nt, thus they are called long noncoding RNAs. Except no open reading frame (ORF) found within lncRNAs, they share many features with coding mRNAs such as 5′ capping. lncRNAs usually contain exons and introns (Carninci et al., 2005).

lncRNAs display different features of genomic location and orientation in the genome. Emerging evidence has demonstrated potential functions of lncRNAs in many biological events. lncRNAs have been shown to modulate gene transcription via different mechanisms. For example, some *cis*-antisense lncRNAs complementary to protein-coding transcripts regulate expression of the coding genes (Feng et al., 2006; Kotake et al., 2010; Pasmant et al., 2007; Tochitani and Hayashizaki, 2008; Yu et al., 2008); some lncRNAs modulate transcription factors by acting as co-regulators (Feng et al., 2006; Nguyen et al., 2001; Shamovsky et al., 2006; Wang et al., 2008; Willingham et al., 2005; Yang et al., 2001). Moreover, some studies have shown that lncRNAs are involved in epigenetic regulations such as chromatin and histone modification, and X-chromosome inactivation (Denisenko et al., 1998; Mancini-Dinardo et al., 2006; Mazo et al., 2007; Rinn et al., 2007; Sanchez-Elsner et al., 2006; Tsai et al., 2010; Wutz et al., 2002).

3. Noncoding RNAs and neural stem cell development

Noncoding RNAs such as miRNAs and lncRNAs participate gene expression regulation in many ways. The underlying mechanisms of noncoding RNA functions in normal development are beginning to be uncovered. In this book chapter, we will focus on reviewing functions of miRNAs and lncRNAs in neural stem cell development, such as NSC self-renewal, cell fate determination and survival.

3.1 miRNAs and self-renewal and proliferation of NSCs

The ability of self-renewal is essential for NSCs/NPs to perpetuate themselves to maintain an undifferentiated status during the embryonic stage and even in the adulthood (Gage, 2000; Shi et al., 2008; Temple, 2001). NSC proliferation and self-renewal are modulated by a complicated regulation network that consists of growth factors, epigenetic regulators, transcription factors and extrinsic signaling molecules from the NSC niche. Recent discovers have indicated that ncRNAs also play important roles in NSC self-renewal through a post-transcriptional regulation mechanism (Doe, 2008; Shi et al., 2008).

miRNAs have been shown to play essential roles in regulating NSC proliferation. Since Dicer is the key enzyme in miRNA processing, several studies have reported the global effects of miRNAs in NSC development by ablating *Dicer* and in turn blocking biogenesis of all miRNAs in the CNS using tissue specific Cre lines. Conditional deletion of *Dicer* from the mouse cerebral cortex using the *Emx1-Cre* line results in a significant reduction in cortical size and the cortical NP pool (De Pietri Tonelli et al., 2008; Kawase-Koga et al., 2010; Kawase-Koga et al., 2009). *Dicer* ablation from the mouse CNS using the *Emx1-Cre* and *Nestin-Cre* line causes a reduction of NSC numbers and abnormal differentiation (Andersson et al., 2010; Kawase-Koga et al., 2009) (Fig. 2A). *Dicer*-deficient NSCs display apoptosis when mitogens are withdrawn from the culture medium (Fig. 2A). Because Dicer is also involved in maintaining the heterochromatin assembly, the defects of NSCs in *Dicer* knockout mice need to be carefully interpreted (Fukagawa et al., 2004; Kanellopoulou et al., 2005). Examining functions of individual miRNAs will help reveal precise roles of miRNAs in the NSC self-renewal and proliferation (Fig. 3).

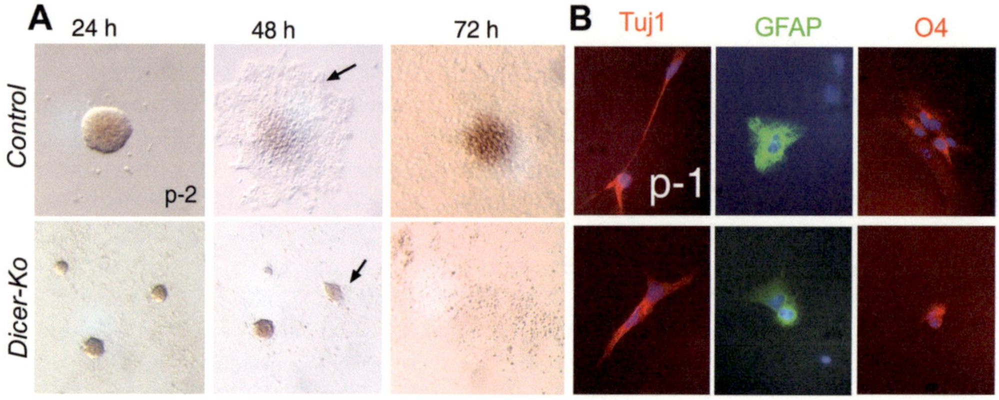

Fig. 2. A. *Dicer*-deficient (*Dicer-Ko*) neural stem cells (NSCs) did not survive well in a differentiation culture medium without mitogens. Most *Dicer-Ko* neurospheres died after 48 hours in culture. Many differentiated cells (arrows) migrated away from the control neurosphere but not from the *Dicer-Ko* neurospheres. B. Under the differentiation condition without mitogens, passaged (p-1) *Dicer-Ko* NSCs gave rise to cells expressing neuronal (Tuj1$^+$) and glial (GFAP$^+$ and O4$^+$) markers. However, their morphology was abnormal, as shown with shorter neurites and processes than controls (Kawase-Koga et al., 2010).

let-7, the first identified miRNA (Reinhart et al., 2000), has been shown to regulate NSC proliferation and differentiation by targeting the nuclear receptor TLX and the cell cycle regulator cyclin D1 (Zhao et al., 2010a) (Fig. 3). Overexpression of *let-7b* inhibits NSC proliferation and enhances differentiation, while knockdown of *let-7b* promotes NSC proliferation (Zhao et al., 2010a). It appears that the expression levels of let-7 in NSCs are controlled by a feedback regulation of Lin-28, a pluripotency factor that controls miRNA processing in NSCs (Rybak et al., 2008). Lin-28 binds to the *let-7* precursor and inhibits its processing by Dicer. On the other hand, the expression of Lin-28 is repressed by let-7 and miR-125, allowing the maturation of let-7 (Fig. 4A). This feedback loop reveals an autoregulation between miRNA let-7 and miR-125, and transcription factor Lin-28 during NSC development (Rybak et al., 2008).

miR-124 is identified as a CNS-enriched miRNA and its expression is upregulated during neuronal differentiation (Lagos-Quintana et al., 2002) (Fig. 3). In the adult brain, NSCs are identified in the subventricular zone (SVZ). In cultured adult NSCs derived from the SVZ and in the SVZ *in vivo*, knocking down *miR-124* results in an increase of NSC proliferation and a decrease of differentiation, while overexpressing *miR-124* reduces the number of dividing precursors and enhances neuronal differentiation (Cheng et al., 2009). Moreover, miR-124 modulates NSC proliferation and differentiation by suppressing Sox9 expression in adult NSCs (Cheng et al., 2009). A recent study has shown that miR-124 regulates neuronal differentiation through a mutual inhibition mechanism of Ephrin-B1 (Arvanitis et al., 2010). In addition, miR-124 promotes differentiation of NPs by modulating a network of nervous system-specific alternative splicing through suppressing expression of PTBP1, which encodes a global repressor of alternative pre-mRNA splicing (Makeyev et al., 2007). Together, miR-124 plays a general role in promoting differentiation of embryonic and adult NSCs and NPs. It appears that miR-124 executes its function through repressing various targets.

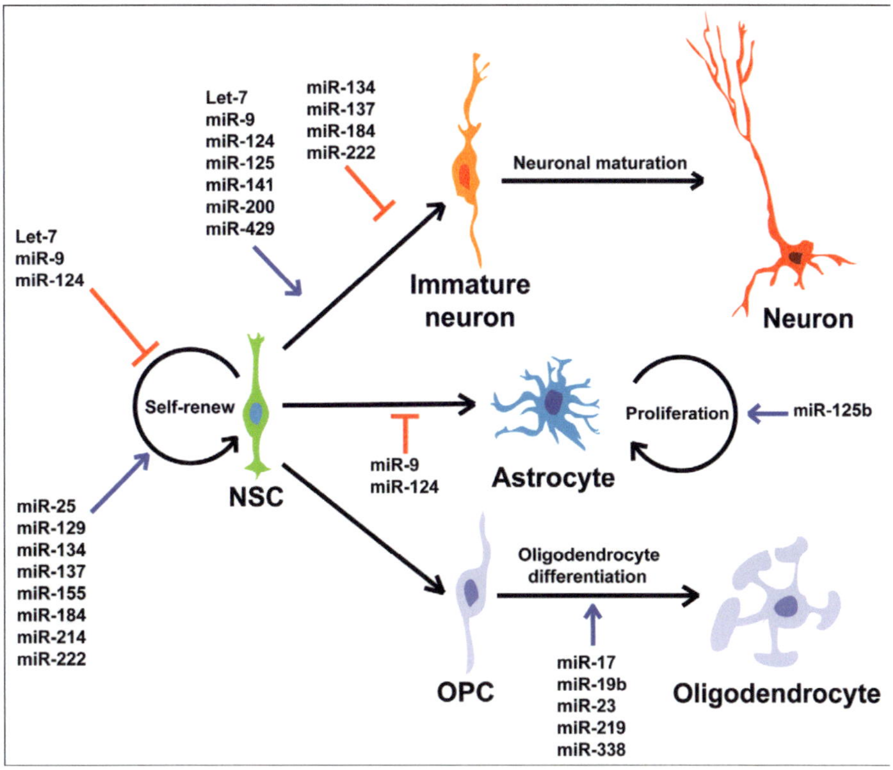

Fig. 3. Many miRNAs are involved in neural stem cell (NSC) self-renewal and differentiation into neurons, astrocytes and oligodendrocyte precursor cells (OPCs) and oligodendrocytes.

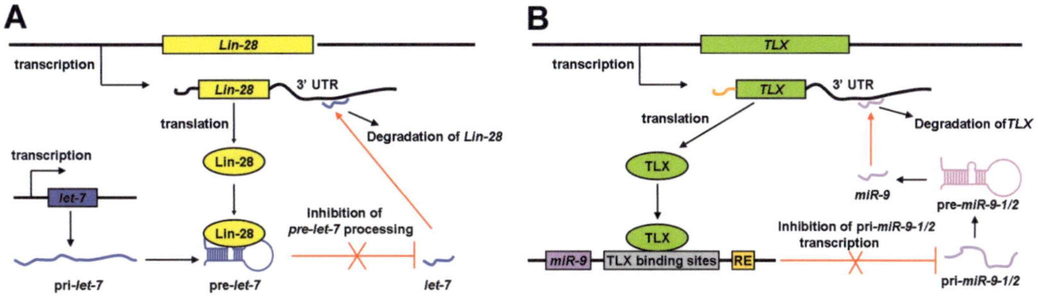

Fig. 4. Feedback loop regulation of miRNAs and their target genes. A. Let-7 processing is inhabited by Lin-28, and the 3′ untranslated region (3′ UTR) of Lin-28 has binding sites for Let-7. B. TLX inhibits miR-9 expression, while miR-9 displays silencing effects on TLX.

miR-9 is another CNS-enriched miRNA. miR-9 is shown to inhibit NSC proliferation but promote differentiation through a feedback regulation of a nuclear receptor TLX (Zhao et al., 2009) (Fig. 4B). In human embryonic stem cell (ESC) derived NPs, miR-9 is shown to have a positive effect on proliferation but a negative effect on migration by directly targeting Stmn1, which increases microtubule instability (Delaloy et al., 2010). The opposite effect of miR-9 on proliferation is perhaps caused by differential physical contacts of miR-9 with target genes and the different culture systems.

In the CNS of *Xenopus*, miR-9 knockdown promotes the proliferation of NPs in the hindbrain, leads to an increased expression of *cyclin D1* and a downregulation of *p27Xic1* (Bonev et al., 2011). miR-9 targets *Hairy1* and regulates proliferation of NPs (Bonev et al., 2011). In zebrafish, miR-9 promotes differentiation of NPs that give rise to neurons at the midbrain-hindbrain domain and controls the organization of the midbrain-hindbrain boundary by targeting several genes in the Fibroblast growth factor (Fgf) signaling, such as *fgf8-1* and *fgfr1* (Leucht et al., 2008). In the chick spinal cord, miR-9 specifies a subtype of motor neurons that project axons to the axial muscles from motor neuron progenitors by specifically targeting transcription factor FoxP1 (Otaegi et al., 2011).

In the mouse brain, miR-9 function is demonstrated by the generation of *miR-9-2* and *miR-9-3* double knockout mice. *miR-9* double mutants show reduced cortical layers, disordered migration of interneurons, and misrouted thalamocortical axons and cortical axon projections, suggesting an important role of miR-9 in NP proliferation, differentiation and migration during brain development (Shibata et al., 2011). Moreover, it appears that miR-9 regulates multiple target genes, including Foxg1, Pax6 and Gsh2, which have shown to be essential in cortical development (Shibata et al., 2011). Therefore, miR-9 plays an important role in controlling differentiation of NSCs/NPs in different regions in the CNS (Fig. 3).

The major role of let-7, miR-124 and miR-9 is to inhibit NSC/NP proliferation and to induce their differentiation into specific cell types. miRNAs that promote proliferation of NSCs and NPs have also been identified (Fig. 3). miR-134 plays a role in enhancing proliferation of cortical NPs by targeting doublecortin (Dcx) and/or Chordin-like 1 (Chrdl-1) (Gaughwin et al., 2011). miR-25 is shown to be a major player in the miR-106-25 cluster in neural development. Overexpression of *miR-25* but not *miR-106b* and *miR-93* promotes adult NP proliferation (Brett et al., 2011). Interestingly, the expression of the miR-106-25 cluster is regulated by FoxO3, a transcription factor maintaining the NSC population (Renault et al., 2009).

During the retina development, *otx2* and *vsx1* genes are shown to control the division of retinal precursors and differentiation into bipolar retina neurons. In early retinal precursors, the expression of *otx2* and *vsx1* is inhibited, accompanied with a rapid precursor division. miR-129, miR-155, miR-214, and miR-222, which are highly expressed in the embryonic retina, have been identified to target and repress translation of *otx2* and *vsx1*, by which they promote proliferation of retinal precursors (Decembrini et al., 2009).

miRNA expression is also controlled by epigenetic regulators in the NSC development. The expression of miR-137 is regulated by DNA methyl-CpG-binding protein (MeCP2) and transcription factor Sox2. miR-137 modulates adult NSC proliferation and cell fate determination by targeting Ezh2, a histone methyltransferase and polycomb group protein (Szulwach et al., 2010). Ectopic expression of *miR-137* in adult NSCs enhances proliferation, while knockdown of *miR-137* promotes differentiation of adult NSCs (Szulwach et al., 2010).

In addition, miR-184 expression is suppressed by methyl-CpG binding protein 1 (MBD1) and miR-184 promotes adult NSC proliferation by repressing the expression of Numb-like (Numbl) (Liu et al., 2010).

3.2 lncRNAs and proliferation of NSCs

The lncRNAs may also play a role in controlling NSC proliferation, even though studies of lncRNAs in NSC development are still sparse. Sox2 is a transcription factor and plays a key role in the maintenance of the undifferentiating state of embryonic and adult NSCs (Pevny and Placzek, 2005). *Sox2 overlapping transcript* (*Sox2OT*) is a lncRNA containing *Sox2* gene and shares the same transcriptional orientation with *Sox2* (Fig. 5A). Similar to *Sox2*, *Sox2OT* is stably expressed in mouse embryonic stem cells and down-regulated during differentiation. *Sox2OT* is expressed in the neurogenic regions of the adult mouse brain including olfactory bulb (OB), rostral migratory stream (RMS) and SVZ, and is dynamically regulated during vertebrate CNS development, implying its role in regulating NSC self-renewal and neurogenesis (Amaral et al., 2009; Mercer et al., 2008).

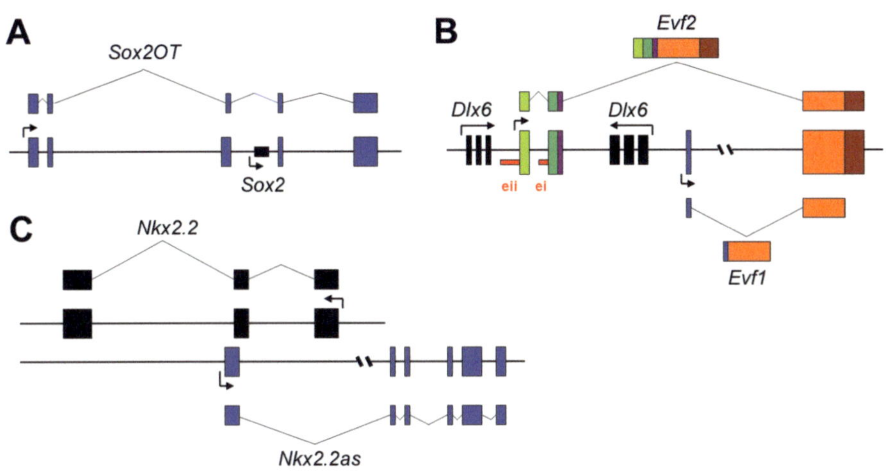

Fig. 5. Genomic location and potential functions of long noncoding RNAs (lncRNAs) (Bian and Sun, 2011). A. *Sox2 overlapping transcript Sox2OT* is a lncRNA containing *Sox2* gene and shares the same transcriptional orientation with *Sox2*. B. *Evf2* is transcribed from the intergenic region between the *Dlx-5* and *Dlx-6* loci, and is overlapped with *Dlx-5/6* enhancer i (ei) and enhancer ii (eii) sequences. *Evf2* acts as a transcriptional co-activator of Dlx-2 and activates the *Dlx5/6* enhancer. C. *Nkx2.2 antisense* (*Nkx2.2as*) is an antisense lncRNA to *Nkx2.2* gene and promotes *Nkx2*-2 expression.

3.3 Summary

Taken together, self-renewal and differentiation of NSCs and NPs are controlled by complex gene regulation networks that consist of both protein coding genes and noncoding miRNAs. During proliferation and differentiation of NSCs and NPs, one miRNA can have multiple target genes and features a feedback regulation with their targets (Figs. 3 and 4). The availability of physical contacts and the binding affinity of a miRNA and its targets perhaps determine interactions of the miRNA with the specific targets. The interactions of miRNAs and their target genes eventually produce proper protein output of key factors that directly control self-renewal, proliferation and differentiation of NSCs/NPs.

3.4 NSC survival controlled by noncoding RNAs

Several reports have shown that miRNAs play a general role in controlling cell survival. Conditional deletion of *Dicer* from neural crest cells using *Wnt1-Cre* mouse line results in an increased apoptosis of neural crest-derived cells (Zehir et al., 2010). Ablation of *Dicer* from postmitotic neurons in the cortex and the hippocampus using *calmoduln kinase II* (*CaMKII*) promoter-driven *Cre* transgenic mice results in smaller cortex, enhanced cortical cell death (Davis et al., 2008). *Emx1-Cre Dicer* conditional knockout mice have shown an increased apoptosis, especially in the ventricular zone (VZ) and SVZ (De Pietri Tonelli et al., 2008). Our own work of cortical NSCs of *Emx1-Cre Dicer* conditional knockout mice using proteomic analysis by mass spectrometry and bioinformatic assays has indicated that *Dicer* deletion results in an increase of pro-cell-death and a decrease of pro-survival proteins in *Dicer*-deficient NSCs (Kawase-Koga et al., 2010) (Fig. 2A). Interestingly, an upregulation of fragile X mental retardation protein (FMRP), a proven target for miR-124, and Caspase3, a key cell apoptosis molecule, are observed in *Dicer*-deficient NSCs. On the other hand, proteins such as transforming growth factor-beta receptor type II (TGFβR2) and SOD1 are downregulated in *Dicer*-deficient NSCs (Kawase-Koga et al., 2010). These observations suggest that miRNAs perhaps control survival of NSCs by modulating the balance of protein output of genes regulating apoptosis and survival.

Neurotrophins and their receptors play important roles in the NSC proliferation, survival and differentiation. miR-128 is shown to target the truncated non-catalytic form of the human neurotrophin-3 receptor (NTRK3), which affects membrane remodeling and cytoskeletal reorganization. Overexpression of *miR-128* in neuroblastoma cells leads to round cell body and shorter neurites, which is similar to knockdown of truncated NTRK3. miR-128 overexpression causes altered expression of genes involved in cell proliferation and apoptosis such as antiapoptotic factor Bcl-2, suggesting an important role of miR-128 on cell survival (Guidi et al., 2010). Moreover, miR-134 is shown to be required for inhibiting apoptosis initiated by Chrdl-1 in cortical progenitors (Gaughwin et al., 2011). Studies on an ethanol teratogenic culture model by exposing embryonic cortex-derived NPs in ethanol have revealed different roles of miRNAs during this pathological process (Sathyan et al., 2007). In NP cultures, miR-21 is suppressed by the ethanol exposure and the reduction of miR-21 causes cell apoptosis, suggesting an anti-apoptotic effect of miR-21 (Sathyan et al., 2007).

The BH3-only family is a group of pro-apoptotic regulators, including Bim, Hrk, Bmf, Puma and N-Bak, which induce cytochrome c release from mitochondria (Giam et al., 2008). Overexpression of miR-29b in neurons inhibits endogenous BH3-only proteins Bim, Puma and Bmf, and promotes neuronal survival (Giam et al., 2008; Kole et al., 2011). In the brain of the calorie-restricted mice, expression of three miRNAs, miR-181a-1*, miR-30e and miR-34a, is significantly downregulated with a corresponding upregulation of their target gene Bcl-2, a decrease of pro-apoptotic factor Bax and cleavage of Caspases (Khanna et al., 2011). Overexpressing these three genes results in an increased cell apoptosis, accompanied with a decrease in Bcl-2 expression (Khanna et al., 2011).

miRNAs also play an important role in neural tissue growth and organ development by regulating cell survival. In *Drosophila*, the Hippo pathway together with Yorkie transcriptional activator contribute to the regulation of tissue growth by stimulating cell proliferation and inhibiting apoptosis (Saucedo and Edgar, 2007). Recent studies have shown that Yorkie not only activates cyclin E and apoptosis inhibitor DIAP1, but also triggers the expression of *bantam* miRNA to promote proliferation and cell survival (Huang et al., 2005; Thompson and Cohen, 2006). A downregulation of *bantam* miRNA is found in dying Rim cells at the eye margin, and restoration of *bantam* miRNA to higher levels prevents apoptosis of these cells, suggesting a role of *bantam* miRNA in enhancing cell survival in eye development (Thompson and Cohen, 2006). In addition, as the largest miRNA family in *Drosophila*, miR-2/6/11/13/308 are required for inhibiting embryonic apoptosis by suppressing pro-apoptotic factors hid, grim, reaper and sickle (Leaman et al., 2005).

In the forebrain of *Xenopus*, miR-9 deletion results in apoptosis of NPs due to increased expression of p53 (Bonev et al., 2011). In addition, miR-24a is expressed in the retina of *Xenopus* (Walker and Harland, 2009). Overexpression of *miR-24a* in retinal cells prevents cells from death, while knockdown of *miR-24a* causes a reduction in eye size due to an increased apoptosis (Walker and Harland, 2009). miR-24a controls cell survival by a negative regulation of pro-apoptotic factors caspase9 and apaf1.

In summary, miRNAs play critical roles in regulating survival of both NSCs/NPs and postmitotic neurons. miRNAs either promote cell survival or lead to apoptosis, depending on functions of their target genes.

3.5 NSC differentiation and cell fate determination mediated by noncoding RNAs

In the mammalian CNS, different neural cell types arise and migrate in a precise temporospatial manner. In the developing mouse brain, neurons arise first by embryonic day 12 (E12), neurogenesis peaks at E14 and ceases by E18. Astrocytes appear around E18, with their numbers peaking in the postnatal period. Oligodendrocytes are generated after birth when the neurogenesis is almost complete. Studies have shown that ncRNAs play an important role in regulating both neurogenesis and gliogenesis.

3.5.1 Cell fate determination controlled by miRNAs

miRNAs play essential roles in NSC differentiation and the cell fate switch between neurons and glia (Cuellar et al., 2008; Hebert et al., 2010; Zheng et al., 2010). We have found that *Dicer*-deficient NSCs display abnormal differentiation, with shorter neurites in neurons and fewer processes in glial cells (Kawase-Koga et al., 2010) (Fig. 2B). Conditional deletion of *Dicer* from the mouse forebrain neurons using *CamKII-Cre* line results in neuronal degeneration and an increase in glial fibrillary acidic protein (GFAP)-positive astrocytes (Hebert et al., 2010). *Dicer* ablation in the dopaminoceptive neurons in the basal ganglia using a *dopamine receptor-1* (*DR-1*)-*Cre* line leads to astrogliogenesis, but not neurodegeneration (Cuellar et al., 2008). Interestingly, in the mouse spinal cord, conditional deletion of *Dicer* using *Olig1-Cre* line disrupts production of both oligodendrocytes and astrocytes (Zheng et al., 2010). These observations suggest that global loss of miRNAs in specific precursor cells affects production of distinct cell types.

miRNA expression profiling studies have shown that some miRNAs are preferentially expressed in neurons or glia. For example, miR-124 and miR-128 are highly expressed in neurons, while miR-23 is restrictively expressed in astrocytes. miR-26 and miR-29 display higher expression in astrocytes than in neurons; and miR-9 and miR-125 are evenly expressed in neurons and astrocytes (Smirnova et al., 2005). Overexpressing *miR-124* in cultured NSCs and in embryonic cortical NPs using lenti-virus and *in utero* electroporation, respectively, promotes neurogenesis and stimulates cortical progenitor migration (Maiorano and Mallamaci, 2009). In cultured adult NSCs, overexpressing *miR-124* enhances neuronal differentiation (Cheng et al., 2009). Ectopic expression of miR-124a and miR-9 in the embryonic stem cell-derived NPs results in a great reduction of GFAP-positive astrocytes compared to the control groups, while knockdown of miR-9, but not miR-124a, switches differentiation of NPs from neurogenesis to astrogliogenesis (Krichevsky et al., 2006). miR-124 and miR-9 promotes neurogenesis by targeting phospholated signal transducer and activator of transcription 3 (STAT3), a transcription factor normally initiating astrogliogenesis (Bonni et al., 1997; Krichevsky et al., 2006).

miR-200 family members, including miR-200a, miR-200b, miR-200c, miR-141 and miR-429, are highly expressed in the developing olfactory bulb. Loss of function of the *miR-200* family prevents normal differentiation of olfactory precursors into mature neurons (Choi et al., 2008). Foxg1, Zfhx1 and Lfng have been identified as the targets of the miR-200 family that affect neurogenesis of the olfactory bulb.

Specific miRNAs that promote gliogenesis have also been identified. Brain-enriched miR-125b is up-regulated in cultured interleulin-6 (IL-6)-induced human astrocytes. Loss of function of *miR-125b* causes an impaired proliferation of astrocytes, accompanied by an upregulation of a miR-125b target cyclin-dependent kinase inhibitor 2A (CDKN2A), which is a negative modulator for cell proliferation (Pogue et al., 2010). The *miR-17-92* cluster displays enriched expression in cultured oligodendrocytes. Specific deletion of the *miR-17-92* cluster from oligodendrocyte precursor cells (OPCs) results in a decreased number of Olig2-positive oligodendrocytes in the mouse brain (Budde et al., 2010). Overexpression of *miR-17* and *miR-19b* in cultures increases the number of oligodendrocytes. The *miR-17-92* cluster regulates oligodendrocyte development by targeting tumor suppressor *Pten* and activating its downstream Akt signaling pathway.

Moreover, *miR-219* and *miR-338* are identified in the oligodendrocyte lineage in the mouse spinal cord and brain. Overexpression of *miR-219* and *miR-338* in cultured OPCs and in the embryonic chick neural tube promotes differentiation of oligodendrocytes, while knockdown of these two miRNAs in OPC cultures and knockdown of *miR-219* in zebrafish abolish oligodendrocyte maturation (Zhao et al., 2010b). Oligodendrocyte differentiation inhibitors Sox6 and Hes5 are identified as targets of miR-219 and miR-338 during oligodendrocyte development (Zhao et al., 2010b).

Lamin B1 (LMNB1) is reported to be associated with autosomal domination leukodystrophy disease (ADLD), a CNS demyelination disorder (Padiath et al., 2006). Overexpression of Lamin B1 represses expression of oligodendrocyte-specific genes such as myelin basic protein (MBP) and myelin oligodendrocyte glycoprotein (MOG), and leads to impaired oligodendrocyte maturation. A recent study has shown that Lamin B1 is post-transcriptionally regulated by miR-23, a glia-specific miRNA. Overexpression of *miR-23* results in significantly increased number of oligodendrocytes and rescues the defects of oligodendrocyte differentiation caused by Lamin B1 (Lin and Fu, 2009).

3.5.2 Cell fate determination regulated by lncRNAs

Studies of lncRNA functions on NSC differentiation are emerging. Dlx-6 is a homeobox containing transcription factor and plays an important role in forebrain neurogenesis (Wang et al., 2010). *Embryonic ventral forebrain-1 (Evf1)* is a 2.7 kb lncRNA transcribed upstream of the mouse *Dlx-6* gene (Kohtz and Fishell, 2004). As an alternatively spliced form of *Evf1*, *Evf2* is transcribed from the intergenic region between the *Dlx-5* and *Dlx-6* loci, and is overlapped with the conserved *Dlx-5/6* intergenic enhancer (Feng et al., 2006; Zerucha et al., 2000) (Fig. 5B). Induced by the Sonic hedgehog (Shh) signaling pathway, *Evf2* has been proven to function as a transcriptional co-activator of Dlx-2 and activates the *Dlx5/6* enhancer during forebrain development (Feng et al., 2006). Deletion of *Evf2* results in a reduction of GABAergic interneurons and impaired synaptic inhibition in the developing hippocampus (Bond et al., 2009).

Nkx2.2 antisense (*Nkx2.2as*) is an antisense lncRNA to *Nkx2.2* gene, which is expressed in the developing mammalian forebrain and is required for oligodendrocyte development (Price et al., 1992) (Fig. 5C). Ectopic expression of *Nkx2.2as* in cultured NSCs induces oligodendrocyte differentiation through an upregulation of the *Nkx2.2* mRNA level, suggesting that *Nkx2.2as* regulates NSC differentiation and promotes gliogenesis by modulating protein coding gene *Nkx2.2* expression (Tochitani and Hayashizaki, 2008).

Retinal noncoding RNA 2 (*RNCR2*), an intergenic lncRNA also known as *Gomafu* and *Miat*, is an abundant polyadenylated RNA in the developing retina (Blackshaw et al., 2004). *RNCR2* is highly expressed in both mitotic and postmitotic retinal progenitors. Knockdown of *RNCR2* leads to an increase of amacrine cells and Müller glial cells in postnatal retina. Mis-localization of *RNCR2* from nuclear to cytoplasm photocopies the effects caused by *RNCR2* knockdown, suggesting that *RNCR2* is required for retinal precursor cell specification (Rapicavoli et al., 2010).

4. Noncoding RNAs as a tool for stem cell-based therapy

Because of the features of self-renewal and the ability to differentiate into many cell types in the CNS, applying NSCs for the treatment of neurological disorders, especially neurodegeneration diseases and injuries in the CNS, has become promising. Directing NSCs into specific cell types and transplanting these cells to replace damaged cells in the CNS have been proven to be successful in some mouse models (Kim and de Vellis, 2009).

Transplantation of NSCs into aged triple transgenic Alzheimer's disease mouse model (3×Tg-AD) rescues the spatial learning and memory defects in these mice (Blurton-Jones et al., 2009). Parkinson's disease (PD) results from a loss of dopaminergic neurons in the substantia nigra. It involves abnormalities in movement variably accompanied by sensory, mood and cognitive changes. Transplantation of undifferentiated human NSCs into PD primate models causes a significant behavioral improvement (Redmond et al., 2007). Directed differentiation of mouse ventral midbrain NSCs in the presence of Shh, FGF8 and Wnt5a produce 10-fold more dopaminergic neurons *in vitro* (Parish et al., 2008).

Transplantation of these pre-differentiated dopaminergic neurons into the brain of PD mouse models results in functional recovery (Parish et al., 2008). Implantation of human NSCs in the rat model of Huntington's disease (HD) is shown improved motor function (McBride et al., 2004). Furthermore, delayed transplantation of adult mouse NSCs surrounding the lesion site of the spinal cord promotes remyelination and functional recovery after spinal cord injuries in rats (Karimi-Abdolrezaee et al., 2006).

Stem cell-based therapeutic applications for neurological disorders also face problems. First, the molecular mechanisms that control NSC proliferation and differentiation into distinct cell types are still unclear. Second, to succeed in clinical applications, transplanting sufficient numbers of NSCs and specific neuronal cell types is critical. Third, to achieve functional recovery from neurological disorders, transplanted cells need to acquire connections with neighbor neurons and restore neural circuitry. Although little studies of using ncRNAs for therapeutic treatment have been done, the emerging reports of ncRNA functions in NSC proliferation and cell fate determination have shown promising future directions. Moreover, due to the technical advances in ncRNA *in vitro* synthesis and delivery, particularly miRNAs, manipulating ncRNA expressions in NSCs will provide a new means for stem cell based therapies for neurological diseases.

5. Acknowledgment

We thank members of the Sun lab for providing thoughtful comments. Owing to space limitations, we apologize for being unable to cite many excellent papers in this field. This work was supported by the Ellison Medical Foundation (T. S.), an award from the Hirschl/Weill-Caulier Trust (T. S.) and an R01-MH083680 grant from the NIH/NIMH (T. S.).

6. References

Amaral PP, Neyt C, Wilkins SJ, Askarian-Amiri ME, Sunkin SM, Perkins AC, et al. Complex architecture and regulated expression of the Sox2ot locus during vertebrate development. *RNA* 2009; 15: 2013-27.

Ambros V. The evolution of our thinking about microRNAs. *Nat Med* 2008; 14: 1036-40.

Andersson T, Rahman S, Sansom SN, Alsio JM, Kaneda M, Smith J, et al. Reversible block of mouse neural stem cell differentiation in the absence of dicer and microRNAs. *PLoS One* 2010; 5: e13453.

Arvanitis DN, Jungas T, Behar A, Davy A. Ephrin-B1 reverse signaling controls a posttranscriptional feedback mechanism via miR-124. Mol Cell Biol 2010; 30: 2508-17.

Bian S, Sun T. Functions of Noncoding RNAs in Neural Development and Neurological Diseases. *Mol Neurobiol* 2011; 44: 359-73.

Blackshaw S, Harpavat S, Trimarchi J, Cai L, Huang H, Kuo WP, et al. Genomic analysis of mouse retinal development. *PLoS Biol* 2004; 2: E247.

Blurton-Jones M, Kitazawa M, Martinez-Coria H, Castello NA, Muller FJ, Loring JF, et al. Neural stem cells improve cognition via BDNF in a transgenic model of Alzheimer disease. *Proc Natl Acad Sci U S A* 2009; 106: 13594-9.

Bond AM, Vangompel MJ, Sametsky EA, Clark MF, Savage JC, Disterhoft JF, et al. Balanced gene regulation by an embryonic brain ncRNA is critical for adult hippocampal GABA circuitry. *Nat Neurosci* 2009; 12: 1020-7.

Bonev B, Pisco A, Papalopulu N. MicroRNA-9 reveals regional diversity of neural progenitors along the anterior-posterior axis. *Dev Cell* 2011; 20: 19-32.

Bonni A, Sun Y, Nadal-Vicens M, Bhatt A, Frank DA, Rozovsky I, et al. Regulation of gliogenesis in the central nervous system by the JAK-STAT signaling pathway. *Science* 1997; 278: 477-83.

Brennecke J, Malone CD, Aravin AA, Sachidanandam R, Stark A, Hannon GJ. An epigenetic role for maternally inherited piRNAs in transposon silencing. *Science* 2008; 322: 1387-92.

Brett JO, Renault VM, Rafalski VA, Webb AE, Brunet A. The microRNA cluster miR-106b~25 regulates adult neural stem/progenitor cell proliferation and neuronal differentiation. *Aging (Albany NY)* 2011; 3: 108-24.

Budde H, Schmitt S, Fitzner D, Opitz L, Salinas-Riester G, Simons M. Control of oligodendroglial cell number by the miR-17-92 cluster. *Development* 2010; 137: 2127-32.

Carninci P, Kasukawa T, Katayama S, Gough J, Frith MC, Maeda N, et al. The transcriptional landscape of the mammalian genome. *Science* 2005; 309: 1559-63.

Cheng LC, Pastrana E, Tavazoie M, Doetsch F. miR-124 regulates adult neurogenesis in the subventricular zone stem cell niche. *Nat Neurosci* 2009; 12: 399-408.

Choi PS, Zakhary L, Choi WY, Caron S, Alvarez-Saavedra E, Miska EA, et al. Members of the miRNA-200 family regulate olfactory neurogenesis. *Neuron* 2008; 57: 41-55.

Cuellar TL, Davis TH, Nelson PT, Loeb GB, Harfe BD, Ullian E, et al. Dicer loss in striatal neurons produces behavioral and neuroanatomical phenotypes in the absence of neurodegeneration. *Proc Natl Acad Sci U S A* 2008; 105: 5614-9.

Das PP, Bagijn MP, Goldstein LD, Woolford JR, Lehrbach NJ, Sapetschnig A, et al. Piwi and piRNAs act upstream of an endogenous siRNA pathway to suppress Tc3 transposon mobility in the Caenorhabditis elegans germline. *Mol Cell* 2008; 31: 79-90.

Davis TH, Cuellar TL, Koch SM, Barker AJ, Harfe BD, McManus MT, et al. Conditional loss of Dicer disrupts cellular and tissue morphogenesis in the cortex and hippocampus. J *Neurosci* 2008; 28: 4322-30.

De Pietri Tonelli D, Pulvers JN, Haffner C, Murchison EP, Hannon GJ, Huttner WB. miRNAs are essential for survival and differentiation of newborn neurons but not for expansion of neural progenitors during early neurogenesis in the mouse embryonic neocortex. *Development* 2008; 135: 3911-21.

Decembrini S, Bressan D, Vignali R, Pitto L, Mariotti S, Rainaldi G, et al. MicroRNAs couple cell fate and developmental timing in retina. *Proc Natl Acad Sci U S A* 2009; 106: 21179-84.

Delaloy C, Liu L, Lee JA, Su H, Shen F, Yang GY, et al. MicroRNA-9 coordinates proliferation and migration of human embryonic stem cell-derived neural progenitors. *Cell Stem Cell* 2010; 6: 323-35.

Denisenko O, Shnyreva M, Suzuki H, Bomsztyk K. Point mutations in the WD40 domain of Eed block its interaction with Ezh2. *Mol Cell Biol* 1998; 18: 5634-42.

Doe CQ. Neural stem cells: balancing self-renewal with differentiation. *Development* 2008; 135: 1575-87.

Feng J, Bi C, Clark BS, Mady R, Shah P, Kohtz JD. The Evf-2 noncoding RNA is transcribed from the Dlx-5/6 ultraconserved region and functions as a Dlx-2 transcriptional coactivator. *Genes Dev* 2006; 20: 1470-84.

Fukagawa T, Nogami M, Yoshikawa M, Ikeno M, Okazaki T, Takami Y, et al. Dicer is essential for formation of the heterochromatin structure in vertebrate cells. *Nat Cell Biol* 2004; 6: 784-91.

Gage FH. Mammalian neural stem cells. *Science* 2000; 287: 1433-8.

Gaughwin P, Ciesla M, Yang H, Lim B, Brundin P. Stage-Specific Modulation of Cortical Neuronal Development by Mmu-miR-134. *Cereb Cortex* 2011.

Giam M, Huang DC, Bouillet P. BH3-only proteins and their roles in programmed cell death. *Oncogene* 2008; 27 Suppl 1: S128-36.

Gregory RI, Chendrimada TP, Shiekhattar R. MicroRNA biogenesis: isolation and characterization of the microprocessor complex. *Methods Mol Biol* 2006; 342: 33-47.

Guidi M, Muinos-Gimeno M, Kagerbauer B, Marti E, Estivill X, Espinosa-Parrilla Y. Overexpression of miR-128 specifically inhibits the truncated isoform of NTRK3 and upregulates BCL2 in SH-SY5Y neuroblastoma cells. *BMC Mol Biol* 2010; 11: 95.

Hebert SS, Papadopoulou AS, Smith P, Galas MC, Planel E, Silahtaroglu AN, et al. Genetic ablation of Dicer in adult forebrain neurons results in abnormal tau hyperphosphorylation and neurodegeneration. *Hum Mol Genet* 2010; 19: 3959-69.

Houwing S, Kamminga LM, Berezikov E, Cronembold D, Girard A, van den Elst H, et al. A role for Piwi and piRNAs in germ cell maintenance and transposon silencing in Zebrafish. *Cell* 2007; 129: 69-82.

Huang J, Wu S, Barrera J, Matthews K, Pan D. The Hippo signaling pathway coordinately regulates cell proliferation and apoptosis by inactivating Yorkie, the Drosophila Homolog of YAP. *Cell* 2005; 122: 421-34.

Kanellopoulou C, Muljo SA, Kung AL, Ganesan S, Drapkin R, Jenuwein T, et al. Dicer-deficient mouse embryonic stem cells are defective in differentiation and centromeric silencing. *Genes Dev* 2005; 19: 489-501.

Kapranov P, Cheng J, Dike S, Nix DA, Duttagupta R, Willingham AT, et al. RNA maps reveal new RNA classes and a possible function for pervasive transcription. *Science* 2007; 316: 1484-8.

Karimi-Abdolrezaee S, Eftekharpour E, Wang J, Morshead CM, Fehlings MG. Delayed transplantation of adult neural precursor cells promotes remyelination and functional neurological recovery after spinal cord injury. *J Neurosci* 2006; 26: 3377-89.

Kawase-Koga Y, Low R, Otaegi G, Pollock A, Deng H, Eisenhaber F, et al. RNAase-III enzyme Dicer maintains signaling pathways for differentiation and survival in mouse cortical neural stem cells. *J Cell Sci* 2010; 123: 586-94.

Kawase-Koga Y, Otaegi G, Sun T. Different timings of Dicer deletion affect neurogenesis and gliogenesis in the developing mouse central nervous system. *Dev Dyn* 2009; 238: 2800-12.

Khanna A, Muthusamy S, Liang R, Sarojini H, Wang E. Gain of survival signaling by down-regulation of three key miRNAs in brain of calorie-restricted mice. *Aging (Albany NY)* 2011; 3: 223-36.

Khraiwesh B, Arif MA, Seumel GI, Ossowski S, Weigel D, Reski R, et al. Transcriptional control of gene expression by microRNAs. *Cell* 2010; 140: 111-22.

Kim SU, de Vellis J. Stem cell-based cell therapy in neurological diseases: a review. *J Neurosci Res* 2009; 87: 2183-200.

Kohtz JD, Fishell G. Developmental regulation of EVF-1, a novel non-coding RNA transcribed upstream of the mouse Dlx6 gene. *Gene Expr Patterns* 2004; 4: 407-12.

Kole AJ, Swahari V, Hammond SM, Deshmukh M. miR-29b is activated during neuronal maturation and targets BH3-only genes to restrict apoptosis. *Genes Dev* 2011; 25: 125-30.

Kotake Y, Nakagawa T, Kitagawa K, Suzuki S, Liu N, Kitagawa M, et al. Long non-coding RNA ANRIL is required for the PRC2 recruitment to and silencing of p15(INK4B) tumor suppressor gene. *Oncogene* 2010; 30: 1956-62.

Krichevsky AM, Sonntag KC, Isacson O, Kosik KS. Specific microRNAs modulate embryonic stem cell-derived neurogenesis. *Stem Cells* 2006; 24: 857-64.

Lagos-Quintana M, Rauhut R, Lendeckel W, Tuschl T. Identification of novel genes coding for small expressed RNAs. *Science* 2001; 294: 853-8.

Lagos-Quintana M, Rauhut R, Yalcin A, Meyer J, Lendeckel W, Tuschl T. Identification of tissue-specific microRNAs from mouse. *Curr Biol* 2002; 12: 735-9.

Lander ES, Linton LM, Birren B, Nusbaum C, Zody MC, Baldwin J, et al. Initial sequencing and analysis of the human genome. *Nature* 2001; 409: 860-921.

Lau NC, Lim LP, Weinstein EG, Bartel DP. An abundant class of tiny RNAs with probable regulatory roles in Caenorhabditis elegans. *Science* 2001; 294: 858-62.

Lau NC, Seto AG, Kim J, Kuramochi-Miyagawa S, Nakano T, Bartel DP, et al. Characterization of the piRNA complex from rat testes. *Science* 2006; 313: 363-7.

Leaman D, Chen PY, Fak J, Yalcin A, Pearce M, Unnerstall U, et al. Antisense-mediated depletion reveals essential and specific functions of microRNAs in Drosophila development. *Cell* 2005; 121: 1097-108.

Lee RC, Ambros V. An extensive class of small RNAs in Caenorhabditis elegans. *Science* 2001; 294: 862-4.

Leucht C, Stigloher C, Wizenmann A, Klafke R, Folchert A, Bally-Cuif L. MicroRNA-9 directs late organizer activity of the midbrain-hindbrain boundary. *Nat Neurosci* 2008; 11: 641-8.

Lin ST, Fu YH. miR-23 regulation of lamin B1 is crucial for oligodendrocyte development and myelination. *Dis Model Mech* 2009; 2: 178-88.

Liu C, Teng ZQ, Santistevan NJ, Szulwach KE, Guo W, Jin P, et al. Epigenetic regulation of miR-184 by MBD1 governs neural stem cell proliferation and differentiation. *Cell Stem Cell* 2010; 6: 433-44.

Lund E, Dahlberg JE. Substrate selectivity of exportin 5 and Dicer in the biogenesis of microRNAs. *Cold Spring Harb Symp Quant Biol* 2006; 71: 59-66.

Maiorano NA, Mallamaci A. Promotion of embryonic cortico-cerebral neuronogenesis by miR-124. *Neural Dev* 2009; 4: 40.

Makeyev EV, Zhang J, Carrasco MA, Maniatis T. The MicroRNA miR-124 promotes neuronal differentiation by triggering brain-specific alternative pre-mRNA splicing. *Mol Cell* 2007; 27: 435-48.

Mancini-Dinardo D, Steele SJ, Levorse JM, Ingram RS, Tilghman SM. Elongation of the Kcnq1ot1 transcript is required for genomic imprinting of neighboring genes. *Genes Dev* 2006; 20: 1268-82.

Mazo A, Hodgson JW, Petruk S, Sedkov Y, Brock HW. Transcriptional interference: an unexpected layer of complexity in gene regulation. *J Cell Sci* 2007; 120: 2755-61.

McBride JL, Behrstock SP, Chen EY, Jakel RJ, Siegel I, Svendsen CN, et al. Human neural stem cell transplants improve motor function in a rat model of Huntington's disease. *J Comp Neurol* 2004; 475: 211-9.

Mercer TR, Dinger ME, Sunkin SM, Mehler MF, Mattick JS. Specific expression of long noncoding RNAs in the mouse brain. *Proc Natl Acad Sci U S A* 2008; 105: 716-21.

Neilson JR, Sharp PA. Small RNA regulators of gene expression. *Cell* 2008; 134: 899-902.

Nguyen VT, Kiss T, Michels AA, Bensaude O. 7SK small nuclear RNA binds to and inhibits the activity of CDK9/cyclin T complexes. *Nature* 2001; 414: 322-5.

Otaegi G, Pollock A, Hong J, Sun T. MicroRNA miR-9 modifies motor neuron columns by a tuning regulation of FoxP1 levels in developing spinal cords. *J Neurosci* 2011; 31: 809-18.

Padiath QS, Saigoh K, Schiffmann R, Asahara H, Yamada T, Koeppen A, et al. Lamin B1 duplications cause autosomal dominant leukodystrophy. *Nat Genet* 2006; 38: 1114-23.

Parish CL, Castelo-Branco G, Rawal N, Tonnesen J, Sorensen AT, Salto C, et al. Wnt5a-treated midbrain neural stem cells improve dopamine cell replacement therapy in parkinsonian mice. *J Clin Invest* 2008; 118: 149-60.

Pasmant E, Laurendeau I, Heron D, Vidaud M, Vidaud D, Bieche I. Characterization of a germ-line deletion, including the entire INK4/ARF locus, in a melanoma-neural system tumor family: identification of ANRIL, an antisense noncoding RNA whose expression coclusters with ARF. *Cancer Res* 2007; 67: 3963-9.

Pevny L, Placzek M. SOX genes and neural progenitor identity. *Curr Opin Neurobiol* 2005; 15: 7-13.

Pogue AI, Cui JG, Li YY, Zhao Y, Culicchia F, Lukiw WJ. Micro RNA-125b (miRNA-125b) function in astrogliosis and glial cell proliferation. *Neurosci Lett* 2010; 476: 18-22.

Pratt AJ, MacRae IJ. The RNA-induced silencing complex: a versatile gene-silencing machine. *J Biol Chem* 2009; 284: 17897-901.

Price M, Lazzaro D, Pohl T, Mattei MG, Ruther U, Olivo JC, et al. Regional expression of the homeobox gene Nkx-2.2 in the developing mammalian forebrain. *Neuron* 1992; 8: 241-55.

Rana TM. Illuminating the silence: understanding the structure and function of small RNAs. *Nat Rev Mol Cell Biol* 2007; 8: 23-36.

Rapicavoli NA, Poth EM, Blackshaw S. The long noncoding RNA RNCR2 directs mouse retinal cell specification. *BMC Dev Biol* 2010; 10: 49.

Redmond DE, Jr., Bjugstad KB, Teng YD, Ourednik V, Ourednik J, Wakeman DR, et al. Behavioral improvement in a primate Parkinson's model is associated with multiple homeostatic effects of human neural stem cells. *Proc Natl Acad Sci U S A* 2007; 104: 12175-80.

Reinhart BJ, Slack FJ, Basson M, Pasquinelli AE, Bettinger JC, Rougvie AE, et al. The 21-nucleotide let-7 RNA regulates developmental timing in Caenorhabditis elegans. *Nature* 2000; 403: 901-6.

Renault VM, Rafalski VA, Morgan AA, Salih DA, Brett JO, Webb AE, et al. FoxO3 regulates neural stem cell homeostasis. *Cell Stem Cell* 2009; 5: 527-39.

Rinn JL, Kertesz M, Wang JK, Squazzo SL, Xu X, Brugmann SA, et al. Functional demarcation of active and silent chromatin domains in human HOX loci by noncoding RNAs. *Cell* 2007; 129: 1311-23.

Rybak A, Fuchs H, Smirnova L, Brandt C, Pohl EE, Nitsch R, et al. A feedback loop comprising lin-28 and let-7 controls pre-let-7 maturation during neural stem-cell commitment. *Nat Cell Biol* 2008; 10: 987-93.

Sanchez-Elsner T, Gou D, Kremmer E, Sauer F. Noncoding RNAs of trithorax response elements recruit Drosophila Ash1 to Ultrabithorax. *Science* 2006; 311: 1118-23.

Sathyan P, Golden HB, Miranda RC. Competing interactions between micro-RNAs determine neural progenitor survival and proliferation after ethanol exposure: evidence from an ex vivo model of the fetal cerebral cortical neuroepithelium. *J Neurosci* 2007; 27: 8546-57.

Saucedo LJ, Edgar BA. Filling out the Hippo pathway. *Nat Rev Mol Cell Biol* 2007; 8: 613-21.

Seto AG, Kingston RE, Lau NC. The coming of age for Piwi proteins. *Mol Cell* 2007; 26: 603-9.

Shamovsky I, Ivannikov M, Kandel ES, Gershon D, Nudler E. RNA-mediated response to heat shock in mammalian cells. *Nature* 2006; 440: 556-60.

Shi Y, Sun G, Zhao C, Stewart R. Neural stem cell self-renewal. *Crit Rev Oncol Hematol* 2008; 65: 43-53.

Shibata M, Nakao H, Kiyonari H, Abe T, Aizawa S. MicroRNA-9 regulates neurogenesis in mouse telencephalon by targeting multiple transcription factors. *J Neurosci* 2011; 31: 3407-22.

Smirnova L, Grafe A, Seiler A, Schumacher S, Nitsch R, Wulczyn FG. Regulation of miRNA expression during neural cell specification. *Eur J Neurosci* 2005; 21: 1469-77.

Szulwach KE, Li X, Smrt RD, Li Y, Luo Y, Lin L, et al. Cross talk between microRNA and epigenetic regulation in adult neurogenesis. *J Cell Biol* 2010; 189: 127-41.

Temple S. The development of neural stem cells. *Nature* 2001; 414: 112-7.

Thompson BJ, Cohen SM. The Hippo pathway regulates the bantam microRNA to control cell proliferation and apoptosis in Drosophila. *Cell* 2006; 126: 767-74.

Tochitani S, Hayashizaki Y. Nkx2.2 antisense RNA overexpression enhanced oligodendrocytic differentiation. *Biochem Biophys Res Commun* 2008; 372: 691-6.

Tsai MC, Manor O, Wan Y, Mosammaparast N, Wang JK, Lan F, et al. Long noncoding RNA as modular scaffold of histone modification complexes. *Science* 2010; 329: 689-93.

Walker JC, Harland RM. microRNA-24a is required to repress apoptosis in the developing neural retina. *Genes Dev* 2009; 23: 1046-51.

Wang X, Arai S, Song X, Reichart D, Du K, Pascual G, et al. Induced ncRNAs allosterically modify RNA-binding proteins in cis to inhibit transcription. *Nature* 2008; 454: 126-30.

Wang XJ, Reyes JL, Chua NH, Gaasterland T. Prediction and identification of Arabidopsis thaliana microRNAs and their mRNA targets. *Genome Biol* 2004; 5: R65.

Wang Y, Dye CA, Sohal V, Long JE, Estrada RC, Roztocil T, et al. Dlx5 and Dlx6 regulate the development of parvalbumin-expressing cortical interneurons. *J Neurosci* 2010; 30: 5334-45.

Willingham AT, Orth AP, Batalov S, Peters EC, Wen BG, Aza-Blanc P, et al. A strategy for probing the function of noncoding RNAs finds a repressor of NFAT. *Science* 2005; 309: 1570-3.

Wutz A, Rasmussen TP, Jaenisch R. Chromosomal silencing and localization are mediated by different domains of Xist RNA. *Nat Genet* 2002; 30: 167-74.

Yang Z, Zhu Q, Luo K, Zhou Q. The 7SK small nuclear RNA inhibits the CDK9/cyclin T1 kinase to control transcription. *Nature* 2001; 414: 317-22.

Yu W, Gius D, Onyango P, Muldoon-Jacobs K, Karp J, Feinberg AP, et al. Epigenetic silencing of tumour suppressor gene p15 by its antisense RNA. *Nature* 2008; 451: 202-6.

Zehir A, Hua LL, Maska EL, Morikawa Y, Cserjesi P. Dicer is required for survival of differentiating neural crest cells. *Dev Biol* 2010; 340: 459-67.

Zerucha T, Stuhmer T, Hatch G, Park BK, Long Q, Yu G, et al. A highly conserved enhancer in the Dlx5/Dlx6 intergenic region is the site of cross-regulatory interactions between Dlx genes in the embryonic forebrain. *J Neurosci* 2000; 20: 709-21.

Zhao C, Sun G, Li S, Lang MF, Yang S, Li W, et al. MicroRNA let-7b regulates neural stem cell proliferation and differentiation by targeting nuclear receptor TLX signaling. *Proc Natl Acad Sci U S A* 2010a; 107: 1876-81.

Zhao C, Sun G, Li S, Shi Y. A feedback regulatory loop involving microRNA-9 and nuclear receptor TLX in neural stem cell fate determination. *Nat Struct Mol Biol* 2009; 16: 365-71.

Zhao X, He X, Han X, Yu Y, Ye F, Chen Y, et al. MicroRNA-mediated control of oligodendrocyte differentiation. *Neuron* 2010b; 65: 612-26.

Zheng K, Li H, Zhu Y, Zhu Q, Qiu M. MicroRNAs are essential for the developmental switch from neurogenesis to gliogenesis in the developing spinal cord. *J Neurosci* 2010; 30: 8245-50.

Part 4

Neural Stem Cells and Therapy

13

Neural Stem/Progenitor Cell Clones as Models for Neural Development and Transplantation

Hedong Li, He Zhao, Xiaoqiong Shu and Mei Jiang
West China Developmental & Stem Cell Institute,
Department of Obstetric & Gynecologic/Pediatric,
Key Laboratory of Obstetric & Gynecologic,
Pediatric Diseases and Birth Defects of Ministry of Education,
West China Second University Hospital,
Sichuan University, Chengdu, Sichuan Province,
P.R. China

1. Introduction

Neural stem cells (NSCs) are somatic stem cells, capable of giving rise to all the mature cell types of the adult nervous system. NSCs are generally "simple" in their cell shape, and acquire much more complex morphology as they differentiate into mature cell types. Morphological changes are not the only phenomena, both genetic and epigenetic alterations also mark the NSC differentiation, allowing development of series of genetic and immunological markers to identify and isolate certain cell types from developing neural tissue. During early phase of the development of central nervous system (CNS), NSCs occupy the newly formed neural tube as a thin layer of cells (called neuroepithelium), which later become thicker and more complex in cellular architecture by the process of cell proliferation, differentiation, migration and maturation in a precisely organized fashion. The underlying mechanisms of these processes are always the attraction to neurobiologists.

NSC differentiation does not occur overnight; rather, it occurs in a progressive manner hypothesized by the lineage-restriction theory. As development proceeds, NSCs acquire additional features (demonstrated by markers) and become regionally diversified under the influence of gradients of growth factors and cytokines. These cells still proliferate, but their differentiation potential may be limited capable of giving rise to fewer cell types as they divide. This group of dividing cells at any CNS developmental stages is often referred as "neural progenitor pool" (Fig. 1). Lineage restriction hypothesis indicates that NSCs undergo progressive loss of differentiation potential (i.e. the number of cell types that they can give rise to) as neural development proceeds. In support of this notion, certain cell surface markers (e.g. A2B5 and 5A5) have been found to label and fractionate lineage-restricted progenitors of the developing spinal cord in a non-overlapping manner (Mayer-Proschel et al., 1997). However, lineage restriction is not the only mechanism of NSC development. For example, Noble examined the expression patterns of A2B5 and 5A5 in the developing forebrain and identified a population of neural progenitors expressing both markers suggesting that NSC differentiation is a much more complicated process in the brain than in the spinal cord (Noble et al., 2003). Therefore, the original lineage-restriction theory based on the expression of A2B5 and 5A5 may be an oversimplified one and can not be generalized to explain NSC differentiation in the entire developing CNS. Nevertheless, it initiated a logical prediction, which will be modified or corrected by more upcoming results from different area of the developing CNS.

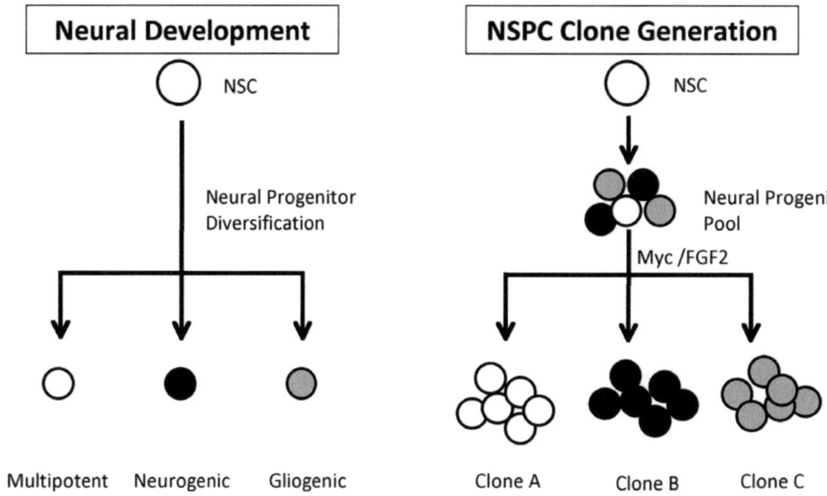

Fig. 1. Immortalized NSPC clones represent neural progenitor diversity during development.

NSCs differentiate at different rate and direction, which creates a diversity of neural progenitors at any given period of CNS development. One way to reveal neural progenitor diversity is to generate neural progenitor clones (Fig. 1). Over the past decades, numerous NSC and neural progenitor clones have been generated, and they have proven to be great models in delineating mechanisms of NSC differentiation. Many of these clones have also been transplanted into animal models of various neurological diseases and trauma, and demonstrated their great usefulness for obtaining mechanistic insights on how exogenous NSCs interact with host tissues and promising effects in animal's behavioural recovery. In our laboratory, we also took this approach and generated various neural stem/progenitor cell (NSPC) clones from rat embryonic forebrains. In this book chapter, we will review more recent discoveries with these "old" and more recently generated neural progenitor clones as in vitro models to unravel NSC differentiation mechanisms as well as cellular tools for testing their therapeutic potentials in paradigms of transplantation. In the era of translational research, we would like to stress on the need for neurobiologists to take on the mission seriously and format our thinking to materialize the transition from bench to bedside.

2. Generation of neural stem/progenitor cell clones

Neural cell lines have been generated by numerous approaches including isolation from spontaneous or induced neural tumors and somatic cell fusion with immortal cells. These cell lines have been widely used as in vitro models, some of which we are still experimenting on today such as Neuro-2a, PC12 cells. More recently, with the advance of our understanding on genetic networks controlling cell proliferation, the "purpose-driven" generation of NSPC clones has been systematically performed. Both growth factor stimulation (epigenetic) and viral introduction of immortalizing genes such as oncogenes (genetic) have proven to be efficient in generating NSPC clones. In this section, we will summarize genetic functions of commonly used immortalizing genes (Myc, neu, large T-antigen, adenoviral E1A, Tert and p53) in promoting cell proliferation and growth factor dependency of the resulting NSPC clones including human NSPC clones (summarized in Table 1 below).

2.1 Immortalizing genes

Myc has been most extensively studied in cancer research since its deregulation relates many different types of tumors (Grandori et al., 2000). Myc also plays a critical role in stem cell biology and development where its expression is tightly controlled (Laurenti et al., 2009). As such, many cancers have been suggested to derive from stem cells during development where Myc expression persists or is deregulated into the adulthood. Myc has various cellular functions ranging from proliferation, cell growth and differentiation to apoptosis. Thirty years after its discovery, the exact molecular mechanisms of Myc mediating these functions still remain elusive. As an immortalizing tool, however, Myc has been widely used to generate cell lines from different lineages, especially neural cell lines, some of which are still widely applied in neuroscience research.

Myc encodes a transcription factor that binds to E-box sequence CACGTG on genomic DNA. Myc protein dimerizes with another bHLH transcription factor Max through its bHLH/LZ domain, and its transregulatory domain can interact with numerous co-factors to execute its transcriptional activity on target genes including the transcription activator E2F-1 (Fig. 2). The critical function domains of Myc proteins have been mapped to their N- and C-terminus, which contain the transregulatory domain and the basic-helix-loop-helix-leucine zipper (bHLH/LZ) domain, respectively (Farina et al., 1992; Min et al., 1993; Min and Taparowsky, 1992).

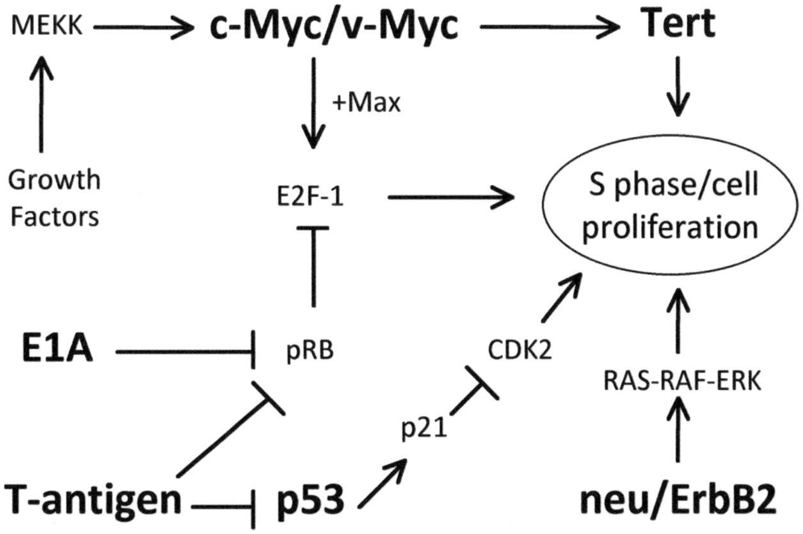

Fig. 2. Simplified cellular pathways involving immortalizing genes (labeled in bold).

V-Myc, the viral homologue of cellular Myc (c-Myc), was first discovered from a transforming retrovirus MC29 from chicken with spontaneous myelocytomatosis and subsequently cloned and sequenced (Alitalo et al., 1983; Reddy et al., 1983; Watson et al., 1983). V-Myc is expressed as a fusion protein Gag-Myc in MC29 and other related retroviruses. Although the Gag portion of the fusion protein seems to be dispensable in its transforming activity (Shaw et al., 1985), the first generation of v-Myc containing retrovirus for immortalization was created as fusion of v-Myc with part of Gag protein (Villa et al., 2000). Myc genes carry mutations, some of which potentiate their transforming activity. A frequently detected mutation Thr58 on c-Myc, which is equivalent to Thr61 on v-Myc, is often found in Burkitt's lyphomas where c-Myc was first discovered (Albert et al., 1994). A Thr58 to Ala substitution (T58A) in c-Myc has been found to promote its proliferative effect in NSCs, and a resulting NSC line immortalized by this mutant c-Myc exhibited enhanced cell proliferation compared with NSC line generated by the wild type gene (De Filippis et al., 2008).

Large T antigen is another popular immortalizing gene used to generate NSC lines. Numerous NSC lines have been derived by utilizing different forms of large T antigen (e.g. temperature-sensitive and N-terminally truncated mutants) from different regions of the developing CNS (Whittemore and Snyder, 1996). SV40 large T antigen is one of the early gene products during viral infection by polyomavirus SV40. SV40 is a double-stranded DNA virus first identified in 1960 and is responsible for formation of solid tumors upon infection. Large T antigen is involved in viral genome replication after infection and regulation of host cell cycle mainly through its perturbation of tumor suppressor protein p53 and the retinoblastoma protein (pRB) (Fig. 2), although several other cellular factors, including the transcriptional co-activators p300 and CBP, may also contribute to its transformation function (Ahuja et al., 2005).

p53 is a tumor suppressor protein and its function is involved in preventing cancer formation. P53 plays many cellular roles in its anti-cancer function, which include cell cycle regulation, apoptosis and genome stability. Upon activation, p53 can initiate a cell cycle arrest to allow DNA repair to take place, or signal cells to go on apoptosis if the DNA damage is unfixable. One mechanism of such function of p53 protein is that, upon activation, it can turn on the expression of p21, which forms complex with cyclin-dependent kinase 2 (CDK2) and inhibits its promoting activity in G1/S transition in the cell cycle (Wierod et al., 2008) (Fig. 2). Gene knockout of p53 or mutations that affect p53 binding to DNA results in unavailability of p21 to act as a stop signal in cell cycle. As such, cell will continue to divide and in some cases will form tumors. The inactivation of p53 can transform cells, and this also provides a way to immortalize cells. Indeed, NSPC clones have been reported to derive from p53 knockout mice (Tominaga et al., 2005; Yamada et al., 1999), but cautions have to be practiced in using these cells in transplantation since they are supposed to be sensitive to DNA mutations and therefore more likely to develop tumors in vivo.

E1A is one of the early gene products of adenoviruses and has also been reported to immortalize neural progenitor cells. The primary cellular target of E1A is pRB (Nevins, 1992). The tumor suppressor protein pRB prevents cell proliferation by inhibiting cell cycle progression. Dysfunction of pRB is often detected in many types of cancer. pRB is a member of pocket protein family and can bind and inhibit E2F-1 transcription activator, thereby preventing cell cycle from entering S phase. Upon adenoviral infection, E1A binds tumor suppressor protein pRB (Fig. 2) and transforms cells with the help of another early gene product E1B, which binds p53.

The oncogene neu was first discovered in a neural tumor, and later was found to share the same sequence with human epidermal growth factor receptor 2 (HER2) and avian erythroblastosis oncogene B2 (ErbB2). HER2/neu is an oncogene that is involved in many types of cancer (Dougall et al., 1994). HER2/neu is a transmembrane receptor tyrosine kinase, and it clusters with other members of the EGFR family and initiates signal transduction pathways that lead to cell proliferation and differentiation. The downstream signaling pathways include the RAS-RAF-MAP kinase, the phosphatidyl inositol 3-kinase (PI3K), and the Akt pathways, where RAS-RAF-MAP kinase pathway plays a more important role to cell cycle progression, thus neural progenitor immortalization when HER2/neu is overexpressed (Fig. 2). Immortalization with neu has been reported (Frederiksen et al., 1988; Sherman et al., 1999), but rare.

Telomerase (Tert) is a reverse transcriptase that adds DNA sequence repeats ("TTAGGG" in all vertebrates) to the 3'-end of DNA strands in the telomere regions of chromosome. In nearly all dividing mammalian cells, telomere shortening is prevented by the activity of Tert, which ensures the sufficient DNA replication during cell division. The decreased expression level of Tert accompanied by telomere shortening of chromosome is often observed in cells that lose the capacity of cell division and become senescence. Overexpression of Tert often leads to cancer formation and also provides a tool for immortalization of neural progenitors (Bai et al., 2004; Roy et al., 2004; Schwob et al., 2008) (Table 1).

hNSPC clone	Tissue origin	Immortalizing gene	Reference
B4, C2, C10 and others	13 wks, brain	Tet-off v-Myc	(Sah et al., 1997)
HNSC.100	10-10.5 wks, diencephalic and telencephalic regions	v-Myc	(Villa et al., 2000)
hSC11V-TERT	9-13 wks, spinal cord	hTERT	(Roy et al., 2004)
hNS2	10 wks, forebrain	v-Myc	(Villa et al., 2004)
hNPC-TERT	fetal SVZ (age?)	hTERT	(Bai et al., 2004)
CTX0E03	first trimester, brain	c-MycER, conditional	(Pollock et al., 2006)
hc-NSC-F7b and others	8-9 wks, cortex	v-Myc	(Cacci et al., 2007)
ReNcell VM	10 wks, midbrain	v-Myc	(Donato et al., 2007)
ReNcell CX	14 wks, cortex	c-Myc	(Donato et al., 2007)
T-IhNSC	10.5 wks, diencephalic and telencephalic tissue	c-Myc (T58A) mutant	(De Filippis et al., 2008)
hVM1	10 wks, ventral mesencephalon	v-Myc	(Villa et al., 2009)
HB2.G2	11–14 wks, telencephalic tissue	Tet-on v-Myc	(Kim et al., 2011)

Table 1. Summary of immortalized human NSPC clones.

2.2 Growth factor dependency

Genetic modification such as overexpressing an oncogene in a cell is expected to bypass growth factor stimulation and thus make possible the cell cycle progression. In reality, however, many neural progenitor cell lines generated by oncogenes mentioned above are heavily dependent on growth factors for their in vitro proliferation. Upon removal of these growth factors, progenitor cells cease proliferation and go on with differentiation towards mature cell types in correlation with the drastic down-regulation of the oncogene expression (such as in the case of v-Myc). Although the mechanism of this phenomenon (oncogene down-regulation upon growth factor withdrawal) is still unclear (perhaps some type of feedback loop system is controlling the exogenous oncogene expression), it provides an extremely important safety prerequisite in using these immortalized cells especially in a transplantation scenario, since the primary phenotypes of a tumorigenetic cell are growth factor independency and loss of contact-inhibition.

Although some of the early generated neural progenitor lines were propagated in mediums containing fetal calf serum (FCS) and sometimes in combination with growth factors, more defined serum-free medium has been widely applied in culturing these cells and FCS has been used as inducing reagent for differentiation into certain neural cell types such as astroglia. Growth factors such as fibroblast growth factor 2 (FGF2) and epidermal growth factor (EGF) have been shown to promote proliferation of immortalized neural progenitors (Kitchens et al., 1994) and are commonly used in serum-free medium as mitogens to promote neural progenitor cell division. In combination of growth factor stimulation and immortalizing genes, NSCs can be expanded in large quantity without obvious phenotypic abnormalities. Then, upon growth factor withdrawal, NSCs simultaneously differentiate along distinct cell lineages. Differentiation inducing factors such as cytokines and neurotrophic factors can be applied to facilitate differentiation and survival of certain cell types. For example, neurotrophin 3 (NT3) and brain-derived neurotrophic factor (BDNF) induce neuronal differentiation and promote neurite outgrowth, while bone morphogenic proteins (BMPs) and leukymia inhibitory factor (LIF) induce astroglial differentiation by upregulating expression of astrocytic marker, glial fibrillary acidic protein (GFAP). Sonic hedgehog homolog (SHH) and platelet-derived growth factor (PDGF) are mitogens to subpopulations of neural progenitors and favor differentiation towards oligodendrocyte lineage.

3. Neural stem/progenitor cell clones as in vitro models

As in vitro models, immortalized neural progenitor clones possess advantages over primary cells isolated from tissue. Firstly, one can produce large quantity of immortalized cells because of the activity of exogenously introduced oncogenes that keep cell proliferation almost infinitely in the presence of appropriate growth factors. Secondly, in vitro expanded immortalized progenitors have cellular homogeneity of higher degree than primary cells since most of these progenitor clones reported are clonal. Lastly, probably because of their promoted cell cycle properties, immortalized progenitor clones are more assessable to genetic modification and more likely to maintain gene expression over passages. There are excellent reviews in the past categorizing most of these clones and describing scientific insights resulted from them (Martinez-Serrano and Bjorklund, 1997; Vescovi and Snyder, 1999; Whittemore and Snyder, 1996). In this section, we will review more recent works on these clones and new clones generated from our lab and others, and focus on their value as in vitro models.

3.1 Models of neural progenitor development

Neural progenitors exhibit spatiotemporal diversity during CNS development (Li and Shi, 2010). To model neural progenitor development, immortalized clones have been isolated from different regions of the CNS at different developmental stages (Whittemore and Snyder, 1996).

Hippocampus plays pivotal roles in learning and memory, and is one of the few sites in adult CNS that retain the capacity of neurogenesis. Therefore, the proliferation and differentiation of hippocampal neural progenitor are of great interests to neurobiologists. HiB5 was isolated as a hippocampal neuronal progenitor from E16 rat (Renfranz et al., 1991), and has been used to model hippocampal neurogenesis. It was found that activation of RARgamma promotes proliferation of HiB5 cells in vitro (Chung et al., 2000). When HiB5 cells were treated with all-trans- or 9-cis-retinoic acid (RA), a significantly increased proportion of these cells were found in S-phase. This was accompanied by increased level of bcl-2 mRNA, while the level of bax mRNA was not affected, which suggests that retinoid treatment increases viable cells by enhancing proliferation rather than suppressing cell death. Furthermore, the proliferation promoting effect of retinoid on HiB5 cells can be mimicked by RARgamma-selective agonist and blocked by its antagonist (Chung et al., 2000). HiB5 cells can model differentiation of hippocampal neurons as well. To give an example, PKA signalling pathway was examined for its role in hippocampal neuronal differentiation process. Kim G et al. demonstrated that treatment of HiB5 cells by cyclic AMP (cAMP) in a serum-free medium containing N2 supplement induced drastic neurite outgrowth of these cells and inhibition of their proliferation (Kim et al., 2002). Phosphorylation of cAMP responsive element binding protein (CREB) was observed accompanied by increased expression of neuronal markers including neurofiliments and decreased expression of glial markers such as nestin and GFAP. Furthermore, overexpression of a GFP-fused protein kinase A (PKA) catalytic subunit alpha protein induced neurite outgrowth in HiB5 cells. Altogether, the authors demonstrated the critical involvement of PKA pathway in hippocampal neuronal differentiation using HiB5 clone as an in vitro model. More recently, another Korean group applied a traditional Korean medicine, Scutellaria baicalensis extract, in the culture of HiB5 cells and showed that it enhanced cell survival of HiB5 and increased their differentiation into choline acetyltransferase (ChAT) positive cholinergic neurons. Along with in vivo data, the authors suggested that Scutellaria baicalensis extract might be used as a neuroprotective medicine in cerebral ischemia (Heo et al., 2009).

Hippocampus maintains on-going neurogenesis in the adult CNS, and yet many of the newborn hippocampal neurons die shortly after birth especially when brain insults occur. Using HiB5 as a model cell, Cacci E et al. were able to show that increased cytokine release by activated microglia may contribute to the death of hippocampal neurons (Cacci et al., 2005). When TNFalpha, as well as conditioned medium from activated microglia, was added to the culture medium, HiB5 cells quickly ceased proliferation and underwent significant cell death. Both HiB5 and microglia express TNF receptors, TNF-R1 and TNF-R2, which may mediate this effect. Hippocampus is vulnerable to different types of insults that lead to neuronal cell death in this brain region. Glucocorticoid (GC), a steroid hormone, participates in normal glucose metabolism. However, when high hippocampal GC level is prolonged, for example in prolonged stressed condition, hippocampal neurons undergo apoptosis. Heat shock protein Hsp27 has been shown to antagonize GC-evoked apoptosis in HiB5 cells (Son et al., 2005). When HiB5 cells were treated with dexamethasone (DEX), a synthetic GC, apoptosis occurred. Interestingly, expression of Hsp27 was also induced upon this treatment. To evaluate possible function of Hsp27 in this process, Son GH et al. overexpressed several constructs in HiB5 cells before DEX treatment and demonstrated that Hsp27 protects hippocampal neurons from GC-induced apoptosis (Son et al., 2005). The same research group also demonstrated that another heat shock protein Hsp25 is involved in neuroprotective effect in hippocampal neurons as well. Furthermore, phosphorylation of Hsp25 mediated by MAPK and ERK signalling is important for its translocation from cytoplasma to nucleus, where it protects nuclear structure, thereby preventing neuronal cell death (Geum et al., 2002). P62, a ubiquitously expressed phosphoprotein, is implied to play a role in protecting hippocampal neuronal survival. When overexpressed in HiB5 cells, p62 not only reduces cell death, but also promotes neuronal differentiation of the cells (Joung et al., 2005). In addition, pre-treatment by vitamin D3 substantially reduced the degree of DEX-induced apoptosis in HiB5 and primary hippocampal neurons suggesting a cross-talk between vitamin D3 and GC pathways (Obradovic et al., 2006).

Isolated from rat E14 striatum, neural progenitor clone ST14A has been used as a model for striatum-derived neurons (Cattaneo and Conti, 1998). The availability of large number of cells made it possible and convenient to examine signalling pathways in these cells. Wnt signalling has been shown to involve in NSC differentiation by expressing necessary Wnt receptors (Lange et al., 2006). ST14A cells express JAK/STAT signalling components and are susceptible to cytokine stimulation leading to cell proliferation (Cattaneo et al., 1996). Ventrally born neurons such as cortical interneurons reach neocortex by tangential migration. The migratory property of ST14A has been realized and used as a model of neuronal migration in vitro. Hepatocyte growth factor /scatter factor (HGF/SF) has been involved in migration and proliferation in many types of epithelial cells. HGF/SF and its receptor Met are also present in the developing CNS as well as ST14A cells. When ST14A cells were exposed to HGF/SF in culture, the cells quickly changed morphology and increased cell motility, a process that involves PI3-K pathway as revealed by pharmacological blocking analysis (Cacci et al., 2003). The cytoskeletal rearrangement including actin network and dissociation of beta-catenin from N-cadherin were also observed in ST14A cells upon treatment of HGF/SF, but not nerve growth factor (NGF), BDNF, NT3 and ciliary neurotrophic factor (CNTF) (Soldati et al., 2008). ErbB family proteins play important function in neuronal migration. Gambarotta et al. demonstrated that ErbB4, but not ErbB1-3, is a crucial receptor of neuregulin1 (Nrg1) in activating migration of ST14A cells (Gambarotta et al., 2004). By gene expression profiling analysis, the same group subsequently identified the epidermal growth factor receptor pathway substrate 8 (Eps8), a multimodular regulator of actin dynamics, as a key mediator of Nrg1/ErbB4 induced neuronal migration (Fregnan et al., 2011).

Radial glia (RG) is a transient cell type during CNS development. It is known by their unique bipolar radial morphology, which is important for its function of supporting neuronal migration. Later, RG has been shown to give rise to neurons and probably the major neuronal precursors throughout the developing CNS (Noctor et al., 2002). We generated first RG clones by v-myc immortalization and demonstrated their properties in vitro including expression of RG specific markers and ability to support neuronal migration. Because of their transient nature, isolation and propagation of RG in vitro have been unsuccessful. The v-myc immortalized RG clones (RG3.6 and L2.3) showed high proliferation rate in the presence of FGF2 and maintained RG markers over many passages, which provides an in vitro model system to allow examination and manipulation on this cell type (Hasegawa et al., 2005; Li et al., 2004). We did observe a gradual change in expression of certain markers in cultured RG clones during passage. For example, brain lipid binding protein (BLBP), a specific RG marker, decreases while a marker for GRPs, A2B5, increases. This initial observation eventually led us to discover a transition from RG to restricted precursors during embryonic forebrain development (Li et al., 2004; Li and Grumet, 2007). To further stablize RG in culture, we introduced actived form of Notch1 into clone L2.3 by retroviral infection. We found that active Notch1 signalling inhibited GRP marker expression and enhanced RG morphology and gene expression (Li et al., 2008a).

Among series of neural progenitor clones generated in our lab, clone L2.2 showed neuronal restricted differentiation in vitro. Moreover, L2.2 gives rise to exclusively GABAergic neuronal subtype upon FGF2 withdrawal as evaluated by their expression of TuJ1, GADs, Dlxs and calretinin (Li et al., 2008b). Neurons derived from L2.2 fire action potential in culture, and this functional differentiation is accelerated when cocultured with RG clone RG3.6 probably through a cell-cell contact mechanism, because the conditioned medium from RG3.6 was not able to exert the same effect. The neuronal progeny of RG clone identified by TuJ1 positivity exhibits projection neuron phenotypes, e.g., bipolar simple morphology, and the majority ($87.4 \pm 1.5\%$) of which are glutamate immunoreactive. Based on our gene expression analysis, L2.3 differentiated culture also expressed T-brain-1 (Tbr-1) transcription factor, which, along with Pax-6 and Tbr-2, is an essential marker for projection neuron differentiation in vivo (Hevner, 2006). During cortical neurogenesis, most, if not all, of glutamatergic projection neurons come from RG, which later differentiate into glial cells. Therefore, clone L2.2 and L2.3 (or RG3.6) generate interneurons and projection neurons, respectively in culture, and they may serve as in vitro models to study interaction between these different neuronal subtypes during cortical development.

3.2 Gene profiling and high-throughput analysis

The availability of large quantity of cells by in vitro expansion of immortalized neural progenitors made large-scale analysis possible. For example, gene expression profiling experiment using ST14A cells overexpressing GDNF has revealed upregulated genes that are involved in neural differentiation and migration (Pahnke et al., 2004). ST14A cells overexpressing CNTF demonstrated increased proliferation, metabolic activity and resistance to stress during early differentiation (Weinelt et al., 2003). Similarly, gene expression profiling analysis confirmed this observation by showing upregulated genes that are involved in stress response pathway of this CNTF-ST14A cells (Bottcher et al., 2003). By overexpressing activated Notch1 gene in RG clone L2.3, we generated a new clone NL2.3 that exhibits enhanced RG marker expression and exaggerated RG morphology (Li et al., 2008a). To explore genes that are responsible for RG phenotype, we conducted gene expression comparison between NL2.3 and its parental clone L2.3. As expected, RG related genes such as BLBP, nestin, tenasin and vimentin were upregulated in NL2.3, and surprisingly we also found that cell adhesion molecules, especially nidogen1 (showing 50 fold increase comparing to L2.3), were upregulated, which may explain the better attachment of NL2.3 cells on laminin-coated substrate. We further confirmed the functional role of nidogen1 in mediating cell adhesion by antibody blocking experiments, and revealed a previously unrecognized link between Notch1 signalling and cell adhesion. In addition, we showed that primary RG cells also express nidogen1 in a secreted fashion demonstrating the physiological significance of this result (Li et al., 2008a).

Proteomics in NSCs is still in its early stage. Nevertheless, large quantity of cells from immortalized progenitors is well suited for this type of analysis. Clone ST14A and an immortalized human NSC clone ReNcell VM have been applied in 2-DE proteomic profiling leading to meaningful discoveries (Beyer et al., 2007; Hoffrogge et al., 2007). Another significant application of expandable neural progenitor cells is high throughput (HTP) screening. Beside drug screening in neural progenitors in pharmaceutical industry, there is also an increasing demand for HTP protocol for genetic analysis on a genome scale. Park JY et al. developed two HTP-optimized expression vector systems that allow generation of red fluorescent protein (RFP)-tagged target proteins (Park et al., 2007). Using these systems, the authors screened sixty representative human C2 domains for their neuronal promoting effect, and identified two C2 domains for their further study. This is another good example for taking advantages of large quantity of cells from immortalized NSC clones.

3.3 Models for neurological diseases

NSC clones are capable not only to facilitate research on NSC differentiation, but also provide model systems for neurological diseases. Here, we give an example on clone ST14A. Derived from embryonic striatum, ST14A cells have been used as in vitro model for Huntingtin's disease (HD). In fact, it has been shown that under serum-free condition, ST14A cells were able to differentiate into DARPP-32-positive medium spiny neurons spontaneously and displayed electrophysiological properties similar to those of medium spiny neurons (Ehrlich et al., 2001). Protein aggregation is the hallmark of neurodegenerative diseases including HD. Ossato G et al. developed a so-called number and brightness method to monitor aggregation of Huntingtin exon 1 protein directly in live ST14A cells and found that the mutant protein underwent a two-step aggregation process, an initial phase of monomer accumulation and oligomer formation followed by protein inclusion depleting monomers in the cytoplasma (Ossato et al., 2010). The pathology of HD has been well documented, and yet its underlying molecular mechanisms still remain elusive. Using the same clone, Sadri-Vakili G et al. demonstrated that epigenetic regulation plays a role in HD progression (Sadri-Vakili et al., 2007). It was found that despite no change in overall acetylated histone levels, histone H3 was hypo-acetylated at the promoter regions of certain down-regulated genes in ST14A cells as well as in R6/2 mice, an animal model for HD. Furthermore, histone deacetylase (HDAC) inhibitor treatment increased level of acetylated histones and seemed to correct expression of misregulated genes, suggesting a potential therapeutic application of HDAC inhibitors for HD (Sadri-Vakili et al., 2007). Altered cholesterol biosynthetic pathway has been reported to involve in HD. The expression level of several key genes in the cholesterol pathway is severely disrupted in brain tissues of HD mice and human patients (Valenza et al., 2005). Mutant Huntingtin was introduced into ST14A cells and it significantly reduced total cellular cholesterol mass. By adding cholesterol back to the cells, the authors were able to prevent cell death of mutant ST14A in a dose-dependent manner. This report uncovered the cholesterol pathway as a novel player in HD, which could be used as a potential target for HD treatments. Phosphorylation of Huntingtin protein appears to be protective in HD, which is mediated by phosphatase calcineurin and phosphokinase Akt. Regulator of calcineurin (RCAN1-1L) is suppressed in HD patient samples, and overexpression of RCAN1-1L in ST14A cells that contain mutant Huntingtin gene increases phophorylation of Huntingtin and reduces ST14A cell death (Ermak et al., 2009). Therefore, the authors claim that RCAN1-1L might be a mediator for HD progression and offer an alternative avenue for drug treatments.

3.4 Testing biomaterials

Biomaterial science has been a fast-growing field providing promising materials for tissue engineering. Biocompatibility of these materials is a big issue since they will be in contact with human cells and tissue. Large quantity of cells, especially cells derived from immortalized neural progenitors are well suited to test toxicity of biomaterials to cells in culture, because in many cases, transplantation of neural progenitor cells is accompanied by biomaterials in the hope that the latter can potentiate stem cell differentiation and migration. The biocompatibility of a variety of biomaterials including polymers and nanofibers were tested using immortalized progenitor cells as first step towards biological applications. Poly(lactic-co-glycolic acid) (PLGA) was compared with other types of polymers in culturing with clone HiB5, where it performed the best in terms of supporting cell viability and neurite outgrowth. This result provided evidence that PLGA could be used as a scaffold for NSC transplantation for nerve regeneration (Bhang et al., 2007). Polymers can also be created to have different patterns and dimensions by various means. Electrospun poly(l-lactide) (PLLA) fibers with different parameters were tested in culture with NSPC clone C17.2. The cells displayed significantly different growth and differentiation depending on fiber pattern and dimension they adhered (He et al., 2010). PLLA can also be modified by tethering laminin-deirved peptides through a cross-linking reagent, and the modified PLLA showed significant improvement in supporting cell survival and neurite outgrowth of C17.2 cells (He et al., 2009). A UV pre-irradiation followed by UV grafting technique can create gradients of carboxyl group on poly(acrylic acid) (PAA) substrates. It was shown that C17.2 cells adhered to these substrates and appeared to respond the carboxyl gradient by directionally sending out neurites (Li et al., 2005). Polymers modified to be electronic affected seeding density of C17.2 cells and may have effects on stem cell differentiation as well (Salto et al., 2008). This report provided an alternative electronic control over NSC differentiation.

Nanotechnology has revolutionized the biomaterial field. Polymers and scaffolds produced at nano-scale have proven to be superior than other material in biological applications. Biomaterials made by nanopolymers possess high surface-to-volume ratio and offer a variety of topographic features that may promote cellular behavior. Nanofibrous scaffolds fabricated with different ratios of poly(epsilon-caprolactone) (PCL) and gelatin were tested in their ability to promote neuronal differentiation of C17.2 cells, and the PCL/gelatin 70:30 ratio generated the best biomaterial suited for nerve regeneration (Ghasemi-Mobarakeh et al., 2008). Similar cell culture tests were also performed on electroconductive polymeric nanowire templates showing that polypyrrole coating improved their effects on cell adhesion and proliferation (Bechara et al., 2011).

4. Neural stem/progenitor cell clones for transplantation

The promise for stem cells including NSCs is someday they may be used as therapeutic cures for diseases. In animal models, there are numerous reports that support this promise and demonstrate great potential of these cells in tissue protection, replacing lost cells and restoring behavioural function when transplanted in a variety of diseases and trauma. Immortalized NSCs provide unlimited cell number and maintain stem cell characteristics such as differentiation potentials to certain cell lineages, and therefore are outstanding candidates for this purpose. Attempts have been made for NSCs to go into human patients hoping they behave similarly to transplanted cells in animals. However, the major concern is the safety of these cells. NSCs derived from either primary tissue or immortalized counterparts have tendency to form tumors upon transplantation since their cell cycle dynamics have been altered by stimulation of growth factors and/or exogenous oncogenes in the case of immortalized NSCs. Although researchers have claimed that some of the oncogenes are non-transforming (e.g. Myc) and their expression is drastically reduced after NSC differentiation and transplantation in vivo, precautionary measures have to be taken into practice to ensure the safety of these cells in clinical settings. In this section, we will review up-to-date reports from others' work on transplantation using immortalized neural progenitor clones in normal and diseased CNS. We will also describe our work using neural progenitor clones to treat spinal cord injury in rat. Finally, we will touch on tumor inhibition, an unexpected property of NSCs revealed by studies using immortalized NSPC clones.

4.1 Transplantation into normal CNS

In order for NSCs to fulfil neural reparative goal, a key property is to be able to differentiate into mature cell types in regions of transplantation. Neurogenesis and neuronal differentiation are already completed in most regions of the adult brain. Therefore, it is a bit of challenge for transplanted cells to differentiate under this "mature" environment. Surprisingly, however, permissive cues must be still present in neonatal and adult CNS allowing neuronal differentiation to occur. Immortalized NSCs have been transplanted first into normal animals and tested in their survival, migration and differentiation potentials under in vivo environments. For example, RN33B, a conditionally immortalized neural progenitor clone derived from E13 rat medullary raphe nucleus (Whittemore and White, 1993), was transplanted into various regions of neonatal and adult brains. In cerebral cortex and hippocampus formation, RN33B cells survived up to 24 weeks and differentiate with morphologies similar to pyramidal neurons, granule neurons and polymorphic neurons in a region-specific manner (Shihabuddin et al., 1995). Electron microscopy and immunohistochemistry demonstrated that differentiated RN33B cells received synapses from host neurons. In striatum, transplanted RN33B cells survived, integrated and differentiated into neurons and glia, some of which displayed morphological and phenotypic properties of medium-sized spiny neurons. These neurons were also found to form connections with primary striatal target, the globus pallidus, by retrograde tracing analysis (Lundberg et al., 1996). Furthermore, GFP-labelled RN33B cells transplanted into hippocampus exhibited remarkable neuronal morphology and were capable of firing action potentials and receiving synaptic inputs of both excitatory and inhibitory nature (Englund et al., 2002). In contrast to neuronal differentiation mentioned above, RN33B cells transplanted in the mesencephalon predominantly differentiated into astroglia (Lundberg et al., 2002). These observations suggest that regional cues are still present in the adult brain that can direct differentiation of transplanted neural progenitors into certain cell types, which can integrate into local tissue structures, although it has been shown that neonatal brains have greater capacity to encourage differentiation, especially neuronal differentiation, of transplanted cells than adult brains suggesting a decline of permissive cues during CNS development (Shihabuddin et al., 1995).

On the other hand, immortalized progenitors derived from one region of the brain can differentiate into cell types specific to the other indicating plasticity of these cells that may broaden their reparative application in different areas of the CNS. Indeed, when RN33B, along with another NSPC clone C17.2 derived from postnatal cerebellum, was transplanted into adult retina, they survived up to 4 weeks and differentiate into both neurons and glia in all major retinal cell layers including retinal pigment epithelium (Warfvinge et al., 2001). Intrinsic molecular natures of individual progenitor clones have to be considered when applying them into transplantation therapies. They all may be multipotent in culture, but have limited potentials to differentiate in vivo. For example, two weeks after transplantation into adult striatum, unlike RN33B cells to become both neurons and glia, clones HiB5 and ST14A differentiate mostly into glial cells (Lundberg et al., 1996), indicating their distinct intrinsic properties including possible distinct responsiveness to the local host environment. Worthy of noting is that these NSC-derived astroglia functionally integrated into host by showing reactive phenotype in response to brain damage (Lundberg and Bjorklund, 1996). By using immortalized NSC clones, the above mentioned results demonstrated the plasticity of NSCs to differentiate into distinct cell types by responding to local cues, and the fact that adult brains still retain, to a certain degree, the ability to direct differentiation of exogenous neural progenitor cells. C17.2 cells were also transplanted into lumbar region of normal spinal cord where they differentiated into nonmyelinating ensheathing cells. In addition, they appeared to induce host axons to form de novo tracts aiming at graft site probably through the action of their secreted neurotrophic factors, suggesting NSCs may also trigger regenerative potentials of host tissue needing for repair (Yan et al., 2004).

4.2 Neurodegenerative diseases

Neurodegenerative diseases are caused by region-specific, progressive cell death and include Parkinson's disease (PD), HD, Alzheimer's disease (AD) and amyotrophic lateral sclerosis (ALS). This is an area where cell replacement therapies are on high demand. We will take PD as an example to highlight the value of immortalized progenitors in testing effects of NSCs in PD animal models in terms of neuroprotection and differentiation into desired neuronal cell type. The common pathology of PD patients is the progressive loss of dopaminergic neurons in the substantial nigra that normally project axons onto striatum. As a result, dopamine level in the striatum greatly reduced, which is believed to induce PD symptoms such as dystonic cramps and dementia (Weidong et al., 2009). Attempts have been made to restore dopamine level by various means, which seem to alleviate symptoms to a certain degree, but they can not prevent PD progression and replenish the lost neurons.

Clone C17.2 has been transplanted into normal and lesioned striatum of the adult rat brain in testing their behaviour in local environment where dopaminergic neurons reside. Consistent with the notion mentioned above, these cells were able to turn on neuronal markers and tyrosine hydroxylase (TH), an enzyme critical to dopamine systhesis, suggesting that local cues are still present and powerful enough to direct multipotent immortalized NSCs to differentiate into local neuronal subtypes (Yang et al., 2002). Efforts have also been made to reinforce dopaminergic differentiation of transplanted progenitors to maximize their effects in PD animal models. For example, C17.2 cells have been genetically modified to overexpress enzymes (e.g. TH and GTP cyclohydrolase 1) (Ryu et al., 2005), Nuclear receptor related 1 protein (NURR1) (Li et al., 2007) and secreted factors (e.g. neurturin) (Liu et al., 2007) towards the goal of getting more dopaminergic neurons. Both TH and GTP cyclohydrolase 1 are required for efficient L-DOPA synthesis, since HPLC assays demonstrated that L-DOPA released from C17.2 cells that are transduced with both genes (C17.2-THGC) is 760-fold higher than that from C17.2 cells transduced with TH gene alone (C17.2-TH). Following transplantation into striatum of PD rats, C17.2-THGC was able to promote animals' behavioural improvement when compared with transplants with parental C17.2 cells (Ryu et al., 2005). NURR1 has been shown to induce downstream target genes such as TH and facilitate dopaminergic differentiation. When overexpressed in C17.2 cells, NURR1 enhanced cell differentiation into dopaminergic neurons and even promoted behavioural recovery to a certain extent (Li et al., 2007). Unlike in normal CNS, transplanted neural progenitors have to overcome unfavourable environment in diseased tissue to survive and differentiate. Therefore, manipulations that mitigate local environment may prove to facilitate beneficial outcomes. Neurturin is a secreted factor that belongs to glial cell-derived neurotrophic factor (GDNF) family. Like GDNF, neurturin exerted neuroprotective effects on host dopaminergic neurons after transplantation into lesioned striatum. In addition, neurturin also promoted dopaminergic differentiation of C17.2 cells when overexpressed, and animal behaviour as well (Liu et al., 2007). Similarly, interleukin-10 overexpression in C17.2 lowered immune response of host tissue to create a beneficial microenvironment for cell survival and differentiation (Wang et al., 2007). In addition, treatment of melatonin in combination with C17.2 showed neuroprotection in PD models (Sharma et al., 2007).

4.3 Brain and spinal cord injury

Traumatic brain injury (TBI), also called acquired brain injury or simply head injury, occurs when a sudden trauma causes damage to the brain. TBI can result when the head suddenly and violently hits an object, or when an object pierces the skull and enters brain tissue. Symptoms of a TBI can be mild, moderate, or severe, depending on the extent of the damage to the brain. Besides direct damage of brain tissue by trauma, secondary damage in TBI results from toxic effects of a variety of modulators that magnify the initial traumatic damage. Among these modulators are the excitatory transmitter glutamate, the intracellular messenger calcium, and the intercellular messenger nitric oxide. Glutamate-induced toxicity, also called excitotoxicity, occurs from excess glutamate release following trauma. Because of the very limited capacity of the brain for self-repair, cellular transplantation has been explored to improve repair. The goal of many studies has been to replace the lost neurons (Bjorklund, 2000; Gage, 2000; Lie et al., 2004; McKay, 1997). Although instructive cues for neuronal differentiation in normal adult brains seem still exist (albeit declining comparing to neonatal brains), the local environment around the injured brain region could be very different. Transplanted cells have to overcome extra hurdles such as glutamte excitotoxicity and high concentration of released cytokines to survive and differentiate. Nevertheless, studies in the past have shown that immortalized progenitors can survive and differentiate upon transplantation in TBI. C17.2 cells were injected into adult mouse brains at 3 days after lateral controlled cortical impact injury (Riess et al., 2002). The study demonstrated cell survival of C17.2 as long as 13 weeks post-transplantation and significant improvement in motor behavior of C17.2 transpalnted animals as compared with those received human embryonic kidney cells. It was also found in this study that C17.2 cells implanted contralateral to the impact side differentiated into mostly neurons, while the cells implanted ipsilateral to the impact side differentiated into astroglial cells as well as neurons consistent with the notion that injured environment favors glial differentiation of transplanted NSCs (Riess et al., 2002). In order to improve the efficacy of NSC mediated beneficial effects in TBI treatemnt, immortalized progenitors engineered to express neurotrophic factors have been utilized. As expected, these genetically modified cells showed neuroprotective effects and improved functional recovery of the animals. For example, C17.2 overexpressing GDNF (GDNF-C17.2) improved survival of transplanted cells, enhanced neuronal differentiation of these cells and promoted learning behavior of the TBI rats at 6 weeks after transplantation comparing to parental C17.2 cells (Bakshi et al., 2006). Similarly, HiB5 progenitor clone engineered to secrete NGF (NGF-HiB5), when transplanted peripheral to the TBI site, decreased apoptosis of the host hippocampal neurons and improved motor and cognitive function comparing to controls (Philips et al., 2001).

Spinal cord injury (SCI) is a severe CNS injury often resulting in long-term disability. Immediately after contusion there is limited histological evidence of damage followed by neuronal death (hours), and that is followed by macrophage infiltration, Wallerian degeneration and astrogliosis (days-weeks). Similarly to TBI, physical disruption of spinal cord causes membrane depolarization and results in massive glutamate release, which is not only excitotoxic to injured cells themselves but surrounding cells. Several weeks after contusion injury in humans as well as in rats (but not in mice), cystic cavities develop surrounded by gliotic scars associated with extracellular matrix including chondroitin sulfate proteoglycans (CSPG), which is not hospitable to axonal regeneration (Busch and Silver, 2007).

Advances in cell characterization and isolation are opening new opportunities for cell transplantation to repair tissue damage by replacing cells that restore lost function (Gage, 2000). Numerous immortalized NSCs have been implanted into SCI to test their differentiation potentials and efficacy on functional recovery. Embryonic raphe nucleus-derived progenitor clone RN33B morphologically differentiated into multipolar neurons resembling nearby endogenous ones when transplanted into gray matter of normal spinal cord. However, only relatively undifferentiated RN33B cells with bipolar morphology could be found at 2 weeks after transplantation into rat spinal cords with various types of injuries with depletion of endogenous neurons due to the lesions. The authors suggested that cell-cell contact mechanisms contribute to instructive local cues for permissive neuronal differentiation of transplanted progenitors, and that molecules released from the injury site may also have prevented these cells from becoming neurons (Onifer et al., 1997; Whittemore, 1999). On the other hand, implanted neural progenitors secrete neurotrophic factors themselves, which may alter the microenvironment they encounter inside the CNS tissue. Clone C17.2 cells have been shown to naturally express and secrete several trophic factors including NGF, BDNF and GDNF both in vitro and in vivo after transplantation (Lu et al., 2003). When implanted into adult rat spinal cords with cystic dorsal column lesion, C17.2 cells promoted extensive growth of endogenous axons. Elevated expression of one factor NT3 by genetic modification in C17.2 expanded this promoting effect (Lu et al., 2003), and improved C17.2 cell survival near the lesion site and functional recovery analyzed by Basso-Beattie-Bresnahan (BBB) scoring (Zhang et al., 2007). We reported that an immortalized neural progenitor clone sharing properties with NSPC and radial glia (RG3.6) migrated extensively in the injured rat spinal cord and improved open field walking when transplanted acutely following contusive SCI (Hasegawa et al., 2005). The transplanted RG3.6 cells partially protected the rat spinal cord against several aspects of secondary injury including loss of axons and myelin as well as accumulation of CSPG and macrophages (Hasegawa et al., 2005).

Patients with SCI not only lose motor function below the injury site, but often develop debilitating neuropathic pain (allodynia) over time (Siddall et al., 1999), which diminishes the quality of their lives. The mechanisms underlying allodynia may be very complex and many possible factors contribute to these symptoms (Hulsebosch, 2005). One direct factor is the loss or reduction of inhibitory tone in the spinal cord sensory processing due to injury. Therapeutic strategies that prevent induction of allodynia, such as cell transplants that release anti-nociceptive substances, can be used to enhance the endogenous descending inhibitory neurotransmitter systems, such as GABA and serotonin (5HT). Immortalized progenitor RN33B cells were therefore used as a vehicle to deliver GABA in SCI. RN33B cells overexpressing GAD67, an enzyme critical for GABA synthesis, were transplanted into lumbar subarachnoid space of the rat spinal cord with chronic constraint injury (CCI) (Eaton et al., 1999b). Seven weeks after transplantation, RN33B cells were found on the pial surface of the spinal cord, and the animals that received these cells showed significantly reduction of both tactile and temperature allodynia comparing to those that received control cells. In the same CCI injury model, RN33B cells overexpressing BDNF or galanin have also been shown to have beneficial effects in reducing allodynia (Cejas et al., 2000; Eaton et al., 1999a). When C17.2 NSCs were transplanted into injured spinal cord, they primarily differentiated into astrocytes, which may result in sprouting of dorsal horn nocioceptive neurons and in turn allodynia of the animals. GDNF, when overexpressed in transplanted C17.2 cells, reduced nocioceptive fiber sprouting and allodynia to a certain extent suggesting a protective or analgesic effect of GDNF on injury-induced neuropathic pain (Macias et al., 2006). We have isolated cortical GABAergic interneuron progenitor clones from rat embryonic forebrains and demonstrated their restricted interneuronal differentiation in culture. We are further testing these intrinsic GABAergic progenitor clones in vitro and after transplanation in SCI, hoping that the cells can differentiate into neurons with GABAergic phenotype in SCI, not only to release GABA to reduce allodynia, but also to integrate into local neuronal circuitry and permanently eliminate the pain-like syndrome. Towards that goal, we have demonstrated that one such clone exhibited spontaneous synaptic activity when cocultured with E17 hippocampal neurons (Li et al., 2011).

4.4 Tumor inhibition

One surprising and yet interesting feature of NSCs when implanted in vivo is their ability to target tumorous tissues and inhibit their growth. For example, immortalized neural progenitor clones isolated from different regions of the embryonic brain, HiB5 (hippocampus) and ST14A (striatum primordium), were transplanted into nucleus Caudatus of Fisher rats along with N29 glioma cells. Both progenitors exhibited anti-tumor activity and prolonged animal's survival. Clone HiB5 was also shown to inhibit an additional tumor type and even be effective when transplanted 1 week after tumor cells inoculation (Staflin et al., 2004). Clone C17.2 was also tested in their tumor-tropic capacities in a similar paradigm, and the results showed that these cells were able to inhibit tumors of both neural and nonneural origin (Brown et al., 2003). Furthermore, C17.2 cells were able to migrate into the tumor mass even when injected via peripheral vasculatures showing great homing capacity of these cells that can be used to deliver therapeutic agents. The mechanism of the NSC homing phenomena towards tumor is not very clear. C17.2 chemotactic migration was tested in vitro by conditioned medium prepared from glioma culture as well as 13 different tumor-associated growth factors (Heese et al., 2005). The results showed that scatter factor/hepatocyte growth factor (SF/HGF) was the most potent one in attracting C17.2 cells in culture. In addition, antibody against SF/HGF was able to block the migratory behaviour of these cells stimulated by glioma-conditioned medium. Furthermore, Allport JR et al. showed tumor-targeting acivity of C17.2 in vivo and identified two other factors that are involved in this NSC homing event (Allport et al., 2004). This study showed that C17.2 cells that were transduced to express luciferase (C17.2-luc) accumulated onto tumors in mice carrying Lewis lung carcinomas. In vitro analysis showed that accumulation of C17.2-luc cells on tumor-derived endothelium (TEC) can be inhibited by functional blocking antibodies against SDF-1alpha and CD49d suggesting the involvement of SDF-1alpha/CXCR4 receptor and alpha4-integrin in the recruitment of C17.2-luc cells (Allport et al., 2004).

Unlike C17.2, HiB5 cells have not been able to show homing capacity towards tumors, even though HiB5 cells exhibited growth-inhibitory effect when they were cotransplanted with tumor cells into animals. Honeth et al. introduced the chemokine receptor CXCR3 to HiB5 cells and demonstrated its functionality by responding to ligand stimulation and activating downstream signaling pathways such as ERK and SAPK/JNK (Honeth et al., 2006). Upon transplantation, these modified cells showed enhanced migration towards glioma that expressed CXCR3 ligands, IP-10 and I-TAC, in comparison with parental HiB5 cells. This study provided proof-of-concepts that immortalized progenitors can be genetically modified and acquire homing capacity towards tumor to either inhibit tumor growth on its own or deliver therapeutic agents for local treatments. Among good examples of this notion is the study that was carried out by Barresi V and colleagues, where they genetically engineered neural progenitor clone ST14A to express cytosine deaminase (CD), by which 5-fluorocytosine (5-FC) can be converted into 5-fluorouracil (5-FU) to suppress tumor growth. DiI prelabeled CD-expressing ST14A cells were cotransplanted into rat brains with C6 glioma. The data showed that ST14A cells survived inside C6 tumor mass for at least 10 days and significantly reduced the size of tumor comparing to controls presumably through the action of 5-FU (Barresi et al., 2003).

5. Conclusions and perspectives

NSCs hold enormous promises for treating neurological diseases. However, at the present time, we are still facing many hurdles, one of which is how to direct these multipotent cells to become the desired cell types for a certain disease. NSPC clones proliferate rapidly in culture and maintain certain important characteristic properties after passages, therefore provide invaluable models for in vitro studies and transplantable tools for testing hypothesis in treating neurological disorders. Immortalized neural progenitor clones are potentially tumorigenic since they are engineered to express oncogenes such as Myc. Even though many reports have demonstrated that Myc expression is growth factor-dependent and can be down-regulated to an undetectable level after FGF2 withdrawal both in vitro and after transplantation, cautions have been taken to ensure the shutdown of oncogene expression upon cell differentiation. For example, modified immortalizing oncogenes such as the temperature-sensitive mutant of large T antigen, tsA58, have been used to generate a series of neural progenitor clones from the developing CNS. The tsA58 gene product is stable at permissive temperature (33°C), but rapidly degraded when temperature goes higher such as body temperature in culture as well as after transplantation into animals. Generation of neural progenitor clones using controllable-Myc expression may provide another solution to the same problem. Human NSPC clones (summarized in Table 1) have been generated by tetracycline-controllable v-Myc (Kim et al., 2011; Sah et al., 1997) and c-MycER^{TAM} transgene (Pollock et al., 2006). On the other hand, functions of promoting proliferation and tumorigenesis can be uncoupled in an experimental setting (Johnson et al., 2008). Therefore, dissecting out functional domains of oncogenes that enable sufficient expansion of neural stem cells and limit (or eliminate) their tumorigenic activity will be advantageous to generate newer version of immortalizing reagents (Harvey et al., 2007; Truckenmiller et al., 2002). A recent report on induced pluripotent stem cells (iPSCs) indicated that L-Myc, a Myc gene isoform, promoted iPSC generation with little transformation activity (Nakagawa et al., 2010). These safety-oriented designs in generation of immortalized progenitors will prove to be crucial in minimizing tumorigenc potential of exogenous oncogenes, and more research should be performed to optimize NSC immortalizing techniques especially in regard to safety for future clinical applications.

Oncogenes that are frequently used for NSC "immortalizations" such as Myc, carry mutations and are prone to spontaneous mutation as well. Some of these mutations potentiate their ability of transformation and tumorigenesis. Therefore, the culturing conditions for in vitro expansion of immortalized NSC clones need to be optimized and controlled in the goal of less stress so that spontaneous mutations can be minimized. The low oxygen culture condition, which many researchers have applied in their human stem cell culture, would fit in this type of precaution in addition to the fact that NSCs grow and differentiate better under this condition. Furthermore, a "quality check" protocol needs to be developed to screen out and eliminate cells with such mutations and tumor-like growth patterns in the case where mutations do occur. Stem cell replacement therapy is at its "transforming" stage. We are very hopeful for its future, but a lot more work needs to be done before it can go on to clinic, especially in regard to safety. Nevertheless, immortalized NSPC clones will for certain be the milestones in the road towards the ultimate goal of stem cell replacement therapy and, in some instances, may very well become therapeutics themselves (Thomas et al., 2009).

6. Acknowledgments

This work was supported by grants from West China Second University Hospital Research Fund, National Natural Science Foundation of China (30971633) , Program for Changjiang Scholars and Innovative Research Team in University (PCSIRT) (IRT0935) and New Jersey Commission on Spinal Cord Research.

7. References

Ahuja, D., Saenz-Robles, M. T. and Pipas, J. M. (2005). SV40 large T antigen targets multiple cellular pathways to elicit cellular transformation. *Oncogene* 24, 7729-45.

Albert, T., Urlbauer, B., Kohlhuber, F., Hammersen, B. and Eick, D. (1994). Ongoing mutations in the N-terminal domain of c-Myc affect transactivation in Burkitt's lymphoma cell lines. *Oncogene* 9, 759-63.

Alitalo, K., Bishop, J. M., Smith, D. H., Chen, E. Y., Colby, W. W. and Levinson, A. D. (1983). Nucleotide sequence to the v-myc oncogene of avian retrovirus MC29. *Proc Natl Acad Sci U S A* 80, 100-4.

Allport, J. R., Shinde Patil, V. R. and Weissleder, R. (2004). Murine neuronal progenitor cells are preferentially recruited to tumor vasculature via alpha4-integrin and SDF-1alpha-dependent mechanisms. *Cancer Biol Ther* 3, 838-44.

Bai, Y., Hu, Q., Li, X., Wang, Y., Lin, C., Shen, L. and Li, L. (2004). Telomerase immortalization of human neural progenitor cells. *Neuroreport* 15, 245-9.

Bakshi, A., Shimizu, S., Keck, C. A., Cho, S., LeBold, D. G., Morales, D., Arenas, E., Snyder, E. Y., Watson, D. J. and McIntosh, T. K. (2006). Neural progenitor cells engineered to secrete GDNF show enhanced survival, neuronal differentiation and improve cognitive function following traumatic brain injury. *Eur J Neurosci* 23, 2119-34.

Barresi, V., Belluardo, N., Sipione, S., Mudo, G., Cattaneo, E. and Condorelli, D. F. (2003). Transplantation of prodrug-converting neural progenitor cells for brain tumor therapy. *Cancer Gene Ther* 10, 396-402.

Bechara, S., Wadman, L. and Popat, K. C. (2011). Electroconductive polymeric nanowire templates facilitates in vitro C17.2 neural stem cell line adhesion, proliferation and differentiation. *Acta Biomater* 7, 2892-901.

Beyer, S., Mix, E., Hoffrogge, R., Lunser, K., Volker, U. and Rolfs, A. (2007). Neuroproteomics in stem cell differentiation. *Proteomics Clin Appl* 1, 1513-23.

Bhang, S. H., Lim, J. S., Choi, C. Y., Kwon, Y. K. and Kim, B. S. (2007). The behavior of neural stem cells on biodegradable synthetic polymers. *J Biomater Sci Polym Ed* 18, 223-39.

Bjorklund, A. (2000). Cell replacement strategies for neurodegenerative disorders. *Novartis Found Symp* 231, 7-15.

Bottcher, T., Mix, E., Koczan, D., Bauer, P., Pahnke, J., Peters, S., Weinelt, S., Knoblich, R., Strauss, U., Cattaneo, E. et al. (2003). Gene expression profiling of ciliary neurotrophic factor-overexpressing rat striatal progenitor cells (ST14A) indicates improved stress response during the early stage of differentiation. *J Neurosci Res* 73, 42-53.

Brown, A. B., Yang, W., Schmidt, N. O., Carroll, R., Leishear, K. K., Rainov, N. G., Black, P. M., Breakefield, X. O. and Aboody, K. S. (2003). Intravascular delivery of neural stem cell lines to target intracranial and extracranial tumors of neural and non-neural origin. *Hum Gene Ther* 14, 1777-85.

Busch, S. A. and Silver, J. (2007). The role of extracellular matrix in CNS regeneration. *Curr Opin Neurobiol* 17, 120-7.

Cacci, E., Claasen, J. H. and Kokaia, Z. (2005). Microglia-derived tumor necrosis factor-alpha exaggerates death of newborn hippocampal progenitor cells in vitro. *J Neurosci Res* 80, 789-97.

Cacci, E., Salani, M., Anastasi, S., Perroteau, I., Poiana, G., Biagioni, S. and Augusti-Tocco, G. (2003). Hepatocyte growth factor stimulates cell motility in cultures of the striatal progenitor cells ST14A. *J Neurosci Res* 74, 760-8.

Cacci, E., Villa, A., Parmar, M., Cavallaro, M., Mandahl, N., Lindvall, O., Martinez-Serrano, A. and Kokaia, Z. (2007). Generation of human cortical neurons from a new immortal fetal neural stem cell line. *Exp Cell Res* 313, 588-601.

Cattaneo, E. and Conti, L. (1998). Generation and characterization of embryonic striatal conditionally immortalized ST14A cells. *J Neurosci Res* 53, 223-34.

Cattaneo, E., De Fraja, C., Conti, L., Reinach, B., Bolis, L., Govoni, S. and Liboi, E. (1996). Activation of the JAK/STAT pathway leads to proliferation of ST14A central nervous system progenitor cells. *J Biol Chem* 271, 23374-9.

Cejas, P. J., Martinez, M., Karmally, S., McKillop, M., McKillop, J., Plunkett, J. A., Oudega, M. and Eaton, M. J. (2000). Lumbar transplant of neurons genetically modified to secrete brain-derived neurotrophic factor attenuates allodynia and hyperalgesia after sciatic nerve constriction. *Pain* 86, 195-210.

Chung, J. J., Cho, S., Kwon, Y. K., Kim, D. H. and Kim, K. (2000). Activation of retinoic acid receptor gamma induces proliferation of immortalized hippocampal progenitor cells. *Brain Res Mol Brain Res* 83, 52-62.

De Filippis, L., Ferrari, D., Rota Nodari, L., Amati, B., Snyder, E. and Vescovi, A. L. (2008). Immortalization of human neural stem cells with the c-myc mutant T58A. *PLoS One* 3, e3310.

Donato, R., Miljan, E. A., Hines, S. J., Aouabdi, S., Pollock, K., Patel, S., Edwards, F. A. and Sinden, J. D. (2007). Differential development of neuronal physiological responsiveness in two human neural stem cell lines. *BMC Neurosci* 8, 36.

Dougall, W. C., Qian, X., Peterson, N. C., Miller, M. J., Samanta, A. and Greene, M. I. (1994). The neu-oncogene: signal transduction pathways, transformation mechanisms and evolving therapies. *Oncogene* 9, 2109-23.

Eaton, M. J., Karmally, S., Martinez, M. A., Plunkett, J. A., Lopez, T. and Cejas, P. J. (1999a). Lumbar transplant of neurons genetically modified to secrete galanin reverse pain-like behaviors after partial sciatic nerve injury. *J Peripher Nerv Syst* 4, 245-57.

Eaton, M. J., Plunkett, J. A., Martinez, M. A., Lopez, T., Karmally, S., Cejas, P. and Whittemore, S. R. (1999b). Transplants of neuronal cells bioengineered to synthesize GABA alleviate chronic neuropathic pain. *Cell Transplant* 8, 87-101.

Ehrlich, M. E., Conti, L., Toselli, M., Taglietti, L., Fiorillo, E., Taglietti, V., Ivkovic, S., Guinea, B., Tranberg, A., Sipione, S. et al. (2001). ST14A cells have properties of a medium-size spiny neuron. *Exp Neurol* 167, 215-26.

Englund, U., Bjorklund, A., Wictorin, K., Lindvall, O. and Kokaia, M. (2002). Grafted neural stem cells develop into functional pyramidal neurons and integrate into host cortical circuitry. *Proc Natl Acad Sci U S A* 99, 17089-94.

Ermak, G., Hench, K. J., Chang, K. T., Sachdev, S. and Davies, K. J. (2009). Regulator of calcineurin (RCAN1-1L) is deficient in Huntington disease and protective against mutant huntingtin toxicity in vitro. *J Biol Chem* 284, 11845-53.

Farina, S. F., Huff, J. L. and Parsons, J. T. (1992). Mutations within the 5' half of the avian retrovirus MC29 v-myc gene alter or abolish transformation of chicken embryo fibroblasts and macrophages. *J Virol* 66, 2698-708.

Frederiksen, K., Jat, P. S., Valtz, N., Levy, D. and McKay, R. (1988). Immortalization of precursor cells from the mammalian CNS. *Neuron* 1, 439-48.

Fregnan, F., Petrov, V., Garzotto, D., De Marchis, S., Offenhauser, N., Grosso, E., Chiorino, G., Perroteau, I. and Gambarotta, G. (2011). Eps8 involvement in neuregulin1-ErbB4 mediated migration in the neuronal progenitor cell line ST14A. *Exp Cell Res* 317, 757-69.

Gage, F. H. (2000). Mammalian neural stem cells. *Science* 287, 1433-8.

Gambarotta, G., Garzotto, D., Destro, E., Mautino, B., Giampietro, C., Cutrupi, S., Dati, C., Cattaneo, E., Fasolo, A. and Perroteau, I. (2004). ErbB4 expression in neural progenitor cells (ST14A) is necessary to mediate neuregulin-1beta1-induced migration. *J Biol Chem* 279, 48808-16.

Geum, D., Son, G. H. and Kim, K. (2002). Phosphorylation-dependent cellular localization and thermoprotective role of heat shock protein 25 in hippocampal progenitor cells. *J Biol Chem* 277, 19913-21.

Ghasemi-Mobarakeh, L., Prabhakaran, M. P., Morshed, M., Nasr-Esfahani, M. H. and Ramakrishna, S. (2008). Electrospun poly(epsilon-caprolactone)/gelatin nanofibrous scaffolds for nerve tissue engineering. *Biomaterials* 29, 4532-9.

Grandori, C., Cowley, S. M., James, L. P. and Eisenman, R. N. (2000). The Myc/Max/Mad network and the transcriptional control of cell behavior. *Annu Rev Cell Dev Biol* 16, 653-99.

Harvey, B. K., Chen, G. J., Schoen, C. J., Lee, C. T., Howard, D. B., Dillon-Carter, O., Coggiano, M., Freed, W. J., Wang, Y., Hoffer, B. J. et al. (2007). An immortalized rat ventral mesencephalic cell line, RTC4, is protective in a rodent model of stroke. *Cell Transplant* 16, 483-91.

Hasegawa, K., Chang, Y. W., Li, H., Berlin, Y., Ikeda, O., Kane-Goldsmith, N. and Grumet, M. (2005). Embryonic radial glia bridge spinal cord lesions and promote functional recovery following spinal cord injury. *Exp Neurol* 193, 394-410.

He, L., Liao, S., Quan, D., Ma, K., Chan, C., Ramakrishna, S. and Lu, J. (2010). Synergistic effects of electrospun PLLA fiber dimension and pattern on neonatal mouse cerebellum C17.2 stem cells. *Acta Biomater* 6, 2960-9.

He, L., Liao, S., Quan, D., Ngiam, M., Chan, C. K., Ramakrishna, S. and Lu, J. (2009). The influence of laminin-derived peptides conjugated to Lys-capped PLLA on neonatal mouse cerebellum C17.2 stem cells. *Biomaterials* 30, 1578-86.

Heese, O., Disko, A., Zirkel, D., Westphal, M. and Lamszus, K. (2005). Neural stem cell migration toward gliomas in vitro. *Neuro Oncol* 7, 476-84.

Heo, H., Shin, Y., Cho, W., Choi, Y., Kim, H. and Kwon, Y. K. (2009). Memory improvement in ibotenic acid induced model rats by extracts of Scutellaria baicalensis. *J Ethnopharmacol* 122, 20-7.

Hevner, R. F. (2006). From radial glia to pyramidal-projection neuron: transcription factor cascades in cerebral cortex development. *Mol Neurobiol* 33, 33-50.

Hoffrogge, R., Beyer, S., Hubner, R., Mikkat, S., Mix, E., Scharf, C., Schmitz, U., Pauleweit, S., Berth, M., Zubrzycki, I. Z. et al. (2007). 2-DE profiling of GDNF overexpression-related proteome changes in differentiating ST14A rat progenitor cells. *Proteomics* 7, 33-46.

Honeth, G., Staflin, K., Kalliomaki, S., Lindvall, M. and Kjellman, C. (2006). Chemokine-directed migration of tumor-inhibitory neural progenitor cells towards an intracranially growing glioma. *Exp Cell Res* 312, 1265-76.

Hulsebosch, C. E. (2005). From discovery to clinical trials: treatment strategies for central neuropathic pain after spinal cord injury. *Curr Pharm Des* 11, 1411-20.

Johnson, S. A., Dubeau, L. and Johnson, D. L. (2008). Enhanced RNA polymerase III-dependent transcription is required for oncogenic transformation. *J Biol Chem* 283, 19184-91.

Joung, I., Kim, H. J. and Kwon, Y. K. (2005). p62 modulates Akt activity via association with PKCzeta in neuronal survival and differentiation. *Biochem Biophys Res Commun* 334, 654-60.

Kim, G., Choe, Y., Park, J., Cho, S. and Kim, K. (2002). Activation of protein kinase A induces neuronal differentiation of HiB5 hippocampal progenitor cells. *Brain Res Mol Brain Res* 109, 134-45.

Kim, K. S., Lee, H. J., Jeong, H. S., Li, J., Teng, Y. D., Sidman, R. L., Snyder, E. Y. and Kim, S. U. (2011). Self-renewal induced efficiently, safely, and effective therapeutically with one regulatable gene in a human somatic progenitor cell. *Proc Natl Acad Sci U S A* 108, 4876-81.

Kitchens, D. L., Snyder, E. Y. and Gottlieb, D. I. (1994). FGF and EGF are mitogens for immortalized neural progenitors. *J Neurobiol* 25, 797-807.

Lange, C., Mix, E., Rateitschak, K. and Rolfs, A. (2006). Wnt signal pathways and neural stem cell differentiation. *Neurodegener Dis* 3, 76-86.

Laurenti, E., Wilson, A. and Trumpp, A. (2009). Myc's other life: stem cells and beyond. *Curr Opin Cell Biol* 21, 844-54.

Li, B., Ma, Y., Wang, S. and Moran, P. M. (2005). Influence of carboxyl group density on neuron cell attachment and differentiation behavior: gradient-guided neurite outgrowth. *Biomaterials* 26, 4956-63.

Li, H., Babiarz, J., Woodbury, J., Kane-Goldsmith, N. and Grumet, M. (2004). Spatiotemporal heterogeneity of CNS radial glial cells and their transition to restricted precursors. *Dev Biol* 271, 225-38.

Li, H., Chang, Y. W., Mohan, K., Su, H. W., Ricupero, C. L., Baridi, A., Hart, R. P. and Grumet, M. (2008a). Activated Notch1 maintains the phenotype of radial glial cells and promotes their adhesion to laminin by upregulating nidogen. *Glia* 56, 646-58.

Li, H. and Grumet, M. (2007). BMP and LIF signaling coordinately regulate lineage restriction of radial glia in the developing forebrain. *Glia* 55, 24-35.

Li, H., Hader, A. T., Han, Y. R., Wong, J. A., Babiarz, J., Ricupero, C. L., Godfrey, S. B., Corradi, J. P., Fennell, M., Hart, R. P., Plummer, M. R. & Grumet, M. (2011). Isolation of a novel rat neural progenitor clone that expresses Dlx family transcription factors and gives rise to functional GABAergic neurons in culture. *Dev Neurobiol* (in press)

Li, H. and Shi, W. (2010). Neural Progenitor Diversity and their Therapeutic Potential for Spinal Cord Repair. *Front. Biol.* 5, 386-395.

Li, Q. J., Tang, Y. M., Liu, J., Zhou, D. Y., Li, X. P., Xiao, S. H., Jian, D. X. and Xing, Y. G. (2007). Treatment of Parkinson disease with C17.2 neural stem cells overexpressing NURR1 with a recombined republic-deficit adenovirus containing the NURR1 gene. *Synapse* 61, 971-7.

Lie, D. C., Song, H., Colamarino, S. A., Ming, G. L. and Gage, F. H. (2004). Neurogenesis in the adult brain: new strategies for central nervous system diseases. *Annu Rev Pharmacol Toxicol* 44, 399-421.

Liu, W. G., Lu, G. Q., Li, B. and Chen, S. D. (2007). Dopaminergic neuroprotection by neurturin-expressing c17.2 neural stem cells in a rat model of Parkinson's disease. *Parkinsonism Relat Disord* 13, 77-88.

Lu, P., Jones, L. L., Snyder, E. Y. and Tuszynski, M. H. (2003). Neural stem cells constitutively secrete neurotrophic factors and promote extensive host axonal growth after spinal cord injury. *Exp Neurol* 181, 115-29.

Lundberg, C. and Bjorklund, A. (1996). Host regulation of glial markers in intrastriatal grafts of conditionally immortalized neural stem cell lines. *Neuroreport* 7, 847-52.

Lundberg, C., Englund, U., Trono, D., Bjorklund, A. and Wictorin, K. (2002). Differentiation of the RN33B cell line into forebrain projection neurons after transplantation into the neonatal rat brain. *Exp Neurol* 175, 370-87.

Lundberg, C., Winkler, C., Whittemore, S. R. and Bjorklund, A. (1996). Conditionally immortalized neural progenitor cells grafted to the striatum exhibit site-specific neuronal differentiation and establish connections with the host globus pallidus. *Neurobiol Dis* 3, 33-50.

Macias, M. Y., Syring, M. B., Pizzi, M. A., Crowe, M. J., Alexanian, A. R. and Kurpad, S. N. (2006). Pain with no gain: allodynia following neural stem cell transplantation in spinal cord injury. *Exp Neurol* 201, 335-48.

Martinez-Serrano, A. and Bjorklund, A. (1997). Immortalized neural progenitor cells for CNS gene transfer and repair. *Trends Neurosci* 20, 530-8.

Mayer-Proschel, M., Kalyani, A. J., Mujtaba, T. and Rao, M. S. (1997). Isolation of lineage-restricted neuronal precursors from multipotent neuroepithelial stem cells. *Neuron* 19, 773-85.

McKay, R. (1997). Stem cells in the central nervous system. *Science* 276, 66-71.

Min, S., Mascarenhas, N. T. and Taparowsky, E. J. (1993). Functional analysis of the carboxy-terminal transforming region of v-Myc: binding to Max is necessary, but not sufficient, for cellular transformation. *Oncogene* 8, 2691-701.

Min, S. and Taparowsky, E. J. (1992). v-Myc, but not Max, possesses domains that function in both transcription activation and cellular transformation. *Oncogene* 7, 1531-40.

Nakagawa, M., Takizawa, N., Narita, M., Ichisaka, T. and Yamanaka, S. (2010). Promotion of direct reprogramming by transformation-deficient Myc. *Proc Natl Acad Sci U S A* 107, 14152-7.

Nevins, J. R. (1992). E2F: a link between the Rb tumor suppressor protein and viral oncoproteins. *Science* 258, 424-9.

Noble, M., Arhin, A., Gass, D. and Mayer-Proschel, M. (2003). The cortical ancestry of oligodendrocytes: common principles and novel features. *Dev Neurosci* 25, 217-33.

Noctor, S. C., Flint, A. C., Weissman, T. A., Wong, W. S., Clinton, B. K. and Kriegstein, A. R. (2002). Dividing precursor cells of the embryonic cortical ventricular zone have morphological and molecular characteristics of radial glia. *J Neurosci* 22, 3161-73.

Obradovic, D., Gronemeyer, H., Lutz, B. and Rein, T. (2006). Cross-talk of vitamin D and glucocorticoids in hippocampal cells. *J Neurochem* 96, 500-9.

Onifer, S. M., Cannon, A. B. and Whittemore, S. R. (1997). Altered differentiation of CNS neural progenitor cells after transplantation into the injured adult rat spinal cord. *Cell Transplant* 6, 327-38.

Ossato, G., Digman, M. A., Aiken, C., Lukacsovich, T., Marsh, J. L. and Gratton, E. (2010). A two-step path to inclusion formation of huntingtin peptides revealed by number and brightness analysis. *Biophys J* 98, 3078-85.

Pahnke, J., Mix, E., Knoblich, R., Muller, J., Zschiesche, M., Schubert, B., Koczan, D., Bauer, P., Bottcher, T., Thiesen, H. J. et al. (2004). Overexpression of glial cell line-derived neurotrophic factor induces genes regulating migration and differentiation of neuronal progenitor cells. *Exp Cell Res* 297, 484-94.

Park, J. Y., Hwang, E. M., Park, N., Kim, E., Kim, D. G., Kang, D., Han, J., Choi, W. S., Ryu, P. D. and Hong, S. G. (2007). Gateway RFP-fusion vectors for high throughput functional analysis of genes. *Mol Cells* 23, 357-62.

Philips, M. F., Mattiasson, G., Wieloch, T., Bjorklund, A., Johansson, B. B., Tomasevic, G., Martinez-Serrano, A., Lenzlinger, P. M., Sinson, G., Grady, M. S. et al. (2001). Neuroprotective and behavioral efficacy of nerve growth factor-transfected hippocampal progenitor cell transplants after experimental traumatic brain injury. *J Neurosurg* 94, 765-74.

Pollock, K., Stroemer, P., Patel, S., Stevanato, L., Hope, A., Miljan, E., Dong, Z., Hodges, H., Price, J. and Sinden, J. D. (2006). A conditionally immortal clonal stem cell line from human cortical neuroepithelium for the treatment of ischemic stroke. *Exp Neurol* 199, 143-55.

Reddy, E. P., Reynolds, R. K., Watson, D. K., Schultz, R. A., Lautenberger, J. and Papas, T. S. (1983). Nucleotide sequence analysis of the proviral genome of avian myelocytomatosis virus (MC29). *Proc Natl Acad Sci U S A* 80, 2500-4.

Renfranz, P. J., Cunningham, M. G. and McKay, R. D. (1991). Region-specific differentiation of the hippocampal stem cell line HiB5 upon implantation into the developing mammalian brain. *Cell* 66, 713-29.

Riess, P., Zhang, C., Saatman, K. E., Laurer, H. L., Longhi, L. G., Raghupathi, R., Lenzlinger, P. M., Lifshitz, J., Boockvar, J., Neugebauer, E. et al. (2002). Transplanted neural stem cells survive, differentiate, and improve neurological motor function after experimental traumatic brain injury. *Neurosurgery* 51, 1043-52; discussion 1052-4.

Roy, N. S., Nakano, T., Keyoung, H. M., Windrem, M., Rashbaum, W. K., Alonso, M. L., Kang, J., Peng, W., Carpenter, M. K., Lin, J. et al. (2004). Telomerase immortalization of neuronally restricted progenitor cells derived from the human fetal spinal cord. *Nat Biotechnol* 22, 297-305.

Ryu, M. Y., Lee, M. A., Ahn, Y. H., Kim, K. S., Yoon, S. H., Snyder, E. Y., Cho, K. G. and Kim, S. U. (2005). Brain transplantation of neural stem cells cotransduced with tyrosine hydroxylase and GTP cyclohydrolase 1 in Parkinsonian rats. *Cell Transplant* 14, 193-202.

Sadri-Vakili, G., Bouzou, B., Benn, C. L., Kim, M. O., Chawla, P., Overland, R. P., Glajch, K. E., Xia, E., Qiu, Z., Hersch, S. M. et al. (2007). Histones associated with downregulated genes are hypo-acetylated in Huntington's disease models. *Hum Mol Genet* 16, 1293-306.

Sah, D. W., Ray, J. and Gage, F. H. (1997). Bipotent progenitor cell lines from the human CNS. *Nat Biotechnol* 15, 574-80.

Salto, C., Saindon, E., Bolin, M., Kanciurzewska, A., Fahlman, M., Jager, E. W., Tengvall, P., Arenas, E. and Berggren, M. (2008). Control of neural stem cell adhesion and density by an electronic polymer surface switch. *Langmuir* 24, 14133-8.

Schwob, A. E., Nguyen, L. J. and Meiri, K. F. (2008). Immortalization of neural precursors when telomerase is overexpressed in embryonal carcinomas and stem cells. *Mol Biol Cell* 19, 1548-60.

Sharma, R., McMillan, C. R. and Niles, L. P. (2007). Neural stem cell transplantation and melatonin treatment in a 6-hydroxydopamine model of Parkinson's disease. *J Pineal Res* 43, 245-54.

Shaw, J., Hayman, M. J. and Enrietto, P. J. (1985). Analysis of a deleted MC29 provirus: gag sequences are not required for fibroblast transformation. *J Virol* 56, 943-50.

Sherman, L., Sleeman, J. P., Hennigan, R. F., Herrlich, P. and Ratner, N. (1999). Overexpression of activated neu/erbB2 initiates immortalization and malignant transformation of immature Schwann cells in vitro. *Oncogene* 18, 6692-9.

Shihabuddin, L. S., Hertz, J. A., Holets, V. R. and Whittemore, S. R. (1995). The adult CNS retains the potential to direct region-specific differentiation of a transplanted neuronal precursor cell line. *J Neurosci* 15, 6666-78.

Siddall, P. J., Taylor, D. A., McClelland, J. M., Rutkowski, S. B. and Cousins, M. J. (1999). Pain report and the relationship of pain to physical factors in the first 6 months following spinal cord injury. *Pain* 81, 187-97.

Soldati, C., Biagioni, S., Poiana, G. and Augusti-Tocco, G. (2008). beta-Catenin and actin reorganization in HGF/SF response of ST14A cells. *J Neurosci Res* 86, 1044-52.

Son, G. H., Geum, D., Chung, S., Park, E., Lee, K. H., Choi, S. and Kim, K. (2005). A protective role of 27-kDa heat shock protein in glucocorticoid-evoked apoptotic cell death of hippocampal progenitor cells. *Biochem Biophys Res Commun* 338, 1751-8.

Staflin, K., Honeth, G., Kalliomaki, S., Kjellman, C., Edvardsen, K. and Lindvall, M. (2004). Neural progenitor cell lines inhibit rat tumor growth in vivo. *Cancer Res* 64, 5347-54.

Thomas, R. J., Hope, A. D., Hourd, P., Baradez, M., Miljan, E. A., Sinden, J. D. and Williams, D. J. (2009). Automated, serum-free production of CTX0E03: a therapeutic clinical grade human neural stem cell line. *Biotechnol Lett* 31, 1167-72.

Tominaga, M., Honda, S., Okada, A., Ikeda, A., Kinoshita, S. and Tomooka, Y. (2005). A bipotent neural progenitor cell line cloned from a cerebellum of an adult p53-deficient mouse generates both neurons and oligodendrocytes. *Eur J Neurosci* 21, 2903-11.

Truckenmiller, M. E., Vawter, M. P., Zhang, P., Conejero-Goldberg, C., Dillon-Carter, O., Morales, N., Cheadle, C., Becker, K. G. and Freed, W. J. (2002). AF5, a CNS cell line immortalized with an N-terminal fragment of SV40 large T: growth, differentiation, genetic stability, and gene expression. *Exp Neurol* 175, 318-37.

Valenza, M., Rigamonti, D., Goffredo, D., Zuccato, C., Fenu, S., Jamot, L., Strand, A., Tarditi, A., Woodman, B., Racchi, M. et al. (2005). Dysfunction of the cholesterol biosynthetic pathway in Huntington's disease. *J Neurosci* 25, 9932-9.

Vescovi, A. L. and Snyder, E. Y. (1999). Establishment and properties of neural stem cell clones: plasticity in vitro and in vivo. *Brain Pathol* 9, 569-98.

Villa, A., Liste, I., Courtois, E. T., Seiz, E. G., Ramos, M., Meyer, M., Juliusson, B., Kusk, P. and Martinez-Serrano, A. (2009). Generation and properties of a new human ventral mesencephalic neural stem cell line. *Exp Cell Res* 315, 1860-74.

Villa, A., Navarro-Galve, B., Bueno, C., Franco, S., Blasco, M. A. and Martinez-Serrano, A. (2004). Long-term molecular and cellular stability of human neural stem cell lines. *Exp Cell Res* 294, 559-70.

Villa, A., Snyder, E. Y., Vescovi, A. and Martinez-Serrano, A. (2000). Establishment and properties of a growth factor-dependent, perpetual neural stem cell line from the human CNS. *Exp Neurol* 161, 67-84.

Wang, X. J., Liu, W. G., Zhang, Y. H., Lu, G. Q. and Chen, S. D. (2007). Effect of transplantation of c17.2 cells transfected with interleukin-10 gene on intracerebral immune response in rat model of Parkinson's disease. *Neurosci Lett* 423, 95-9.

Warfvinge, K., Kamme, C., Englund, U. and Wictorin, K. (2001). Retinal integration of grafts of brain-derived precursor cell lines implanted subretinally into adult, normal rats. *Exp Neurol* 169, 1-12.

Watson, D. K., Psallidopoulos, M. C., Samuel, K. P., Dalla-Favera, R. and Papas, T. S. (1983). Nucleotide sequence analysis of human c-myc locus, chicken homologue, and myelocytomatosis virus MC29 transforming gene reveals a highly conserved gene product. *Proc Natl Acad Sci U S A* 80, 3642-5.

Weidong, L., Shen, C. and Jankovic, J. (2009). Etiopathogenesis of Parkinson disease: a new beginning? *Neuroscientist* 15, 28-35.

Weinelt, S., Peters, S., Bauer, P., Mix, E., Haas, S. J., Dittmann, A., Petrov, S., Wree, A., Cattaneo, E., Knoblich, R. et al. (2003). Ciliary neurotrophic factor overexpression in neural progenitor cells (ST14A) increases proliferation, metabolic activity, and resistance to stress during differentiation. *J Neurosci Res* 71, 228-36.

Whittemore, S. R. (1999). Neuronal replacement strategies for spinal cord injury. *J Neurotrauma* 16, 667-73.

Whittemore, S. R. and Snyder, E. Y. (1996). Physiological relevance and functional potential of central nervous system-derived cell lines. *Mol Neurobiol* 12, 13-38.

Whittemore, S. R. and White, L. A. (1993). Target regulation of neuronal differentiation in a temperature-sensitive cell line derived from medullary raphe. *Brain Res* 615, 27-40.

Wierod, L., Rosseland, C. M., Lindeman, B., Oksvold, M. P., Grosvik, H., Skarpen, E. and Huitfeldt, H. S. (2008). Activation of the p53-p21(Cip1) pathway is required for CDK2 activation and S-phase entry in primary rat hepatocytes. *Oncogene* 27, 2763-71.

Yamada, K., Hisatsune, T., Uchino, S., Nakamura, T., Kudo, Y. and Kaminogawa, S. (1999). NMDA receptor mediated Ca2+ responses in neurons differentiated from p53-/- immortalized Murine neural stem cells. *Neurosci Lett* 264, 165-7.

Yan, J., Welsh, A. M., Bora, S. H., Snyder, E. Y. and Koliatsos, V. E. (2004). Differentiation and tropic/trophic effects of exogenous neural precursors in the adult spinal cord. *J Comp Neurol* 480, 101-14.

Yang, M., Stull, N. D., Berk, M. A., Snyder, E. Y. and Iacovitti, L. (2002). Neural stem cells spontaneously express dopaminergic traits after transplantation into the intact or 6-hydroxydopamine-lesioned rat. *Exp Neurol* 177, 50-60.

Zhang, L., Gu, S., Zhao, C. and Wen, T. (2007). Combined treatment of neurotrophin-3 gene and neural stem cells is propitious to functional recovery after spinal cord injury. *Cell Transplant* 16, 475-81.

14

Endogenous Neural Stem/Progenitor Cells and Regenerative Responses to Brain Injury

Maria Dizon
Northwestern University/Children's Memorial Hospital
USA

1. Introduction

Neural stem and/or progenitor cells (NSPCs) have generated much excitement because of their envisioned potential to treat a variety of central nervous system diseases that span the human lifetime. This chapter is meant as a general introduction to the topic of NSPCs and how they might be thought of in the context of disease, with a particular focus on endogenous NSPCs. My bias is to focus on endogenous NSPCs as an introduction because the transplantation of NSPCs entails its own complex set of problems including, but not limited to, the effects of in vitro culture conditions, problems of delivery to desired anatomical sites and rejection of foreign cells. Endogenous NSPC populations, on the other hand, change with regard to their distribution, rates of proliferation and fate restriction at different points in brain development. Thus, the cells available to respond to disease necessarily differ depending on whether fetal, neonatal, pediatric or adult brain is affected.

Stem cells are defined by their ability to self-renew indefinitely and by their ability to give rise to cells of many phenotypes. True neural stem cells are tripotential and give rise to neurons, oligodendrocytes and astrocytes. Although stem cells, they are restricted in their fate potential as compared, for example, to pluripotent embryonic stem cells that can give rise to an even wider range of cell phenotypes from endoderm, mesoderm and ectoderm. In turn, embryonic stem cells are restricted in their fate potential in comparison to the totipotent cells from a blastocyst. These cells are able to give rise to any cell of an entire organism; indeed they can give rise to the entire organism itself. In contrast, progenitor cells can self-renew, but only for a limited number of generations, and they give rise to cells of limited phenotypes. For example, glial restricted progenitors can give rise only to astrocytes or oligodendrocytes, neuroblasts can give rise only to neurons, and oligodendrocyte progenitor cells can give rise only to oligodendrocytes.

2. Postnatal neurogenesis and neural stem cells discovered

Prior to the second half of the twentieth century, the dogma was that neurogenesis did not occur postnatally, much less in adult animals. This idea was based largely on the lack of observed neurons with mitotic figures in adult birds and adult mammals. This stance, firmly taken by the father of neuroscience Dr. Ramon y Cajal, likely led to its endurance for almost a century (Ming and Song, 2005). Starting in the 1940s, evidence began to emerge in a number of different species that neurogenesis did indeed occur after development.

2.1 Earliest evidence is human

Some of the earliest evidence for the birth of new neurons after development was actually from human studies. It had been known that a mitotically active region from which the cerebral cortex developed, called the subependymal layer, existed embryonically. In studying ependymomas from autopsies, Drs. Globus and Kuhlenheck noted histologic connections and similarities of features between these neoplasms and the subependymal layer of children and adults, and proposed that the tumors had arisen from undifferentiated cells in the subependymal region. This suggested the persistence of what they termed a bipotential mother cell throughout postnatal life. (Globus and Kuhlenheck, 1944).

2.2 Lower vertebrate evidence

Starting in the 1950s, it was shown that lower vertebrates, including fish, amphibians and lizards (Zupanc, 2008) could regenerate spinal cord (Anderson and Waxman, 1983), optic nerve (Meyer et al., 1985) and even brain (Zupanc, 1999), (Zupanc and Zupanc, 1992).

2.3 Rodent evidence

A breakthrough was made in 1961 when Dr. Smart applied the new technique of labeling dividing cells with tritiated thymidine to 3 day old and adult mice and showed that the subependymal layer of the brain retains its ability to form new cells indefinitely. Standard histologic techniques suggested that these mitotic cells gave rise to neuroblasts and spongioblasts (glial precursors) (Smart, 1961). Shortly thereafter, Dr. Altman began to assemble an impressive body of evidence showing that a large number of interneurons were made postnatally in rat. Like Smart, he used tritiated thymidine, but in contrast he labeled mitotic granule cells within the hippocampus. He showed that these hippocampal granule cells declined from birth with a transient increase at 15 days. Importantly, the decline corresponded with an increase in differentiated granule cells. Later, he extended observations to the olfactory bulb. In both the hippocampus and the olfactory bulb, new interneurons were found continuously at a low rate and were likely born near the ventricles. In contrast, in the cerebellum newly born interneurons were limited to the first three weeks of life. Interneurons were identified both histologically and electrophysiologically. (Altman, 1963; Altman and Das, 1965a; Altman and Das, 1965b; Altman and Das, 1966).

2.4 Avian evidence

By the 1980s, Dr. Nottebohm and colleagues had also clearly demonstrated neurogenesis within the brain of adult birds. Again using tritiated thymidine, they showed that a forebrain nucleus of songbirds (the hyperstriatum ventralis pars caudalis (HVc)), varied greatly in size depending on sex and stage of song development, and that this change in size was related to the birth of new neurons. Cells were identified as neurons ultrastructurally and electrophysiologically. Because labeling was heaviest in the ventricular zone, neurons were presumed to have been born in this zone and to have migrated into the overlying HVc (Goldman and Nottebohm, 1983; Nottebohm, 1985) Interestingly it was Dr. Alvarez-Buylla, who had previously worked on songbirds with Dr. Nottebohm, who revisited the subependymal origin of new neurons in rodent olfactory bulb in the 1990s, and more precisely characterized this region (Lois and Alvarez-Buylla, 1993; Lois and Alvarez-Buylla, 1994).

2.5 Primate evidence

Unfortunately, despite the accumulating evidence, widely held beliefs predominated. As adult neurogenesis was conceded to vertebrates, and even mammals, scientists continued to hold the view that primates were different. Perhaps related to the sensitivity of tritiated thymidine studies, as late as 1985 Dr. Rakic asserted that neurogenesis was limited to development and early postnatal life in primates (Rakic, 1985). It was not until the late 1990s that this view was debunked by Dr. Gould using the thymidine analogue BrdU in adult macaques to show neurogenesis in the hippocampus (Gould et al., 1999a) and even in the neocortex (association cortex) (Gould et al., 1999b). Existence of adult neurogenesis in humans was finally widely embraced by the scientific community in the late twentieth century when Dr. Gage's group demonstrated neurogenesis in adult human dentate gyrus using tissue from cancer victims who had been treated with BrdU (Eriksson et al., 1998). They also formally studied human neocortex using similarly BrdU-treated cancer victims combined with carbon 14 exposure from Cold War above ground nuclear bomb tests and ruled out neocortical neurogenesis after development (Bhardwaj et al., 2006). Recently, a human rostral migratory stream of neuroblasts from subventricular zone to the olfactory bulb was demonstrated by Dr. Curtis (Curtis et al., 2007).

2.6 Neurogenesis indicates neural stem cells

Implicit in the concept of neurogenesis is the existence of a precursor cell capable of giving rise to a neuron. This precursor cell's fate might be limited to neurons, e.g. a restricted neuronal progenitor. Alternatively, the precursor cell might be capable of giving rise to neurons and another cell type such as an oligodendrocyte, e.g. a less restricted bipotential progenitor. Or it might even be capable of giving rise to neurons, oligodendrocytes or astrocytes, e.g. a multipotential neural stem cell. Thus, the firm establishment of postnatal neurogenesis indicates the existence of postnatal neural stem/progenitor cells (NSPCs).

3. Regional distribution of NSPCs

We will discuss the distribution of neural stem cells and more restricted progenitors and how their distribution changes over time. Rodent development is discussed as the bulk of experimental evidence to date derives from rodent studies. Rat brain development has been shown to correlate with mouse brain development. Furthermore, P7 rodent brain development has been shown to correlate with preterm human brain development (Craig et al., 2003). Much insight has been gained from in vitro culture experiments. However, data from in vivo studies including fate mapping experiments using Cre-lox transgenic mouse technology overcomes the uncertainty introduced by culture effects, so the focus here will be on data gleaned from in vivo work. The overarching theme is that earlier in development, precursors tend to be multipotential but become more restricted with time.

3.1 Embryonic period

3.1.1 Early embryonic period

During embryogenesis, the open neural plate folds to form the neural tube. As a result, the primary germinal matrix or neuroepithelium comes to line the lumen that will become the ventricles. Initially, the primary neuroepithelium is a simple columnar epithelium composed of so-called radial glial cells that span from the ventricle to the pial surface. Radial glial cells are more than just glia but are, in fact, "mother cells." They divide after interkinetic nuclear migration from basal surface to the pial surface and back again. Internuclear kinetic migration gives the appearance of a pseudo-stratified epithelium (Altman and Bayer, 1991; Altman and Bayer, 2011). Radial glia divide either symmetrically to form two identical cells, or asymmetrically to give rise to one stem cell and one neuroblast. The neuroblast, using the radial glia's process as a guide, migrates radially toward the pial surface and differentiates into a projection neuron after reaching its destination (Rakic, 1971; Malatesta et al., 2000; Noctor et al., 2001). At E13 the primary neuroepithelium can be subdivided into a more compact zone adjacent to the ventricle called the ventricular zone (VZ), a less cell-dense area more distant from the ventricle called the mantle, and a cell poor area most distant from the ventricle called the marginal zone (Globus and Kuhlenheck, 1944; Altman, 2011). Generally, later migrating neuroblasts move past previously migrated neuroblasts, thus forming the six cortical layers in an inside-out fashion (although Layer I lies closest to the pial surface) (Altman and Bayer, 1991; Altman and Bayer, 2011). During the early embryonic period, the VZ is the most important source of NSPCs and gives rise to projection neurons. The VZ also is the source of cells that create the subsequent germinal matrices.

3.1.2 Mid-late embryonic period

The secondary germinal matrix, termed the subventricular zone (SVZ), begins to evolve as the lateral ganglionic eminence and medial ganglionic eminence enlarge between E12 and E14 as a result of mitotic cells contributed by the VZ, adjacent to it but more distant from the ventricle. NSPCs continue to expand within the SVZ creating a truly stratified epithelium that differs from the VZ. Here the NSPCs do not undergo interkinetic nuclear migration. Nor are the cleavage planes of the mitotic cells oriented in any particular way with regard to the ventricle, so the SVZ does not fit neatly within the symmetrical v. asymmetrical explanation for expansion of NSPCs (Altman and Bayer, 1991; Altman and Bayer, 2011). SVZ NSPCs are tripotential.

During mid-gestation, the SVZ gives rise to neuroblasts that migrate radially to the cortex where they differentiate into interneurons (Altman and Bayer, 2011). The SVZ also gives rise to some OPCs that migrate to the striatum, subcortical white matter and neocortex (Levison and Goldman, 1997; Suzuki and Goldman, 2003) where they persist as oligodendroglial progenitors (OPCs) or differentiate into mature oligodendrocytes (Dawson et al., 2003). Simultaneously the VZ is also a source of OPCs. Distinct populations of VZ NSPCs contribute OPCs at different timepoints in development. Between E11.5 and E14.5, Nkx2.1-expressing NSPCs in the ventral VZ overlying the medial ganglionic eminence and the anterior entopeduncular area give rise to OPCs that migrate radially to the cortex. Later by E16.5, Gsh-2-expressing NSPCs from VZ overlying the lateral ganglionic eminence and caudal ganglionic eminence give rise to OPCs that migrate radially to the cortex (Kessaris et al., 2006).

Starting at approximately E14, radial glial cells start to gradually transform into astrocytes. Nonetheless, the VZ peaks in size at E17 and decreases in size between E18 to E21. Thus the VZ transiently coexists with the SVZ (Temple, 2001).

A tertiary germinal matrix can be found in the hippocampus at E16. Starting at E14.5, NSPCs proliferate at the dentate notch, then differentiate into neuroblasts that migrate along radial glia before differentiating further into granule cell neurons to form the blades of the dentate gyrus. Subsequently, more neuroblasts migrate along the same path but accumulate as mitotic progenitors to form the hilus of the dentate gyrus. Thus, the dentate gyrus is essentially a specialized subventricular zone. The true tertiary germinal matrix is the border between the granule cell layer and the hilus, also known as the subgranular zone (SGZ) (Li and Pleasure, 2005). NSPCs residing at the SGZ give rise to granule cell neurons (Ray et al., 1993).

The SVZ is fully formed by E15-16, and becomes larger than the VZ by E18. The peak of neurogenesis takes place between E14-E17 with the VZ and the SVZ both contributing neurons at this time. Thus, during mid gestation, both the VZ and SVZ are important sources of NSPCs that generate projection neurons and interneurons respectively. By late embryonic life NSPCs are restricted to the SVZ and the SGZ and their fate is further restricted to either a glial or a neuronal fate in the SVZ and a neuronal fate in the SGZ. By late gestation, the SVZ becomes the most important source of NSPCs capable of generating interneurons and oligodendrocytes.

3.2 Postnatal period

At the time of birth, cortical neurogenesis has ended. The VZ no longer exists and has been replaced by the ependyma. The SVZ and SGZ persists postnatally throughout childhood into adulthood. Evidence shows an SVZ containing NSPCs exists in adult humans as well.

3.2.1 Early postnatal period

Postnatally, neurogenesis is confined to the olfactory system and the hippocampus. SVZ neuroblasts no longer migrate radially, but have changed direction and now migrate tangentially through the rostral migratory stream to the olfactory bulb where they differentiate into interneurons (Marshall et al., 2003). By P10, the SGZ is well established and NSPCs here continue to give rise to hippocampal granule cell neurons (Li and Pleasure, 2005).

Oligodendrogenesis, on the other hand, continues to be widespread during the first postnatal week and their source is diverse. At birth, Nkx2.1-derived OPCs are abundant in the cortex. Postnatally, Emx1-expressing NSPCs located dorsally in the cortex give rise to OPCs that populate the corpus callosum and cortex. By P10, the majority of OPCs and oligodendrocytes derive from Emx-1- and Gsh-2-expressing NSPCs, but very few OPCs derive from Nkx2.1-expressing NSPCs remain (Kessaris et al., 2006). As well, OPCs migrate from the SVZ to the striatum, white matter and medial, dorsal and lateral regions of the cortex where they reside as progenitors or differentiate into mature oligodendrocytes. A small proportion the OPCs in the corpus callosum comes from the SVZ (Rivers, 2008). This wave of migration ends by P14.

Also around the time of birth a distinct population of cells migrates from the SVZ to striatum, white matter and medial, dorsal and lateral regions of the cortex to form astrocytes.

In summary, the bulk of neurogenesis occurs prenatally while the bulk of oligodendrogenesis occurs postnatally with a peak in mouse between P7-P14 (Wright et al., 2010). As in late gestation, in the postnatal period the SVZ remains the most important source of NSPCs for the generation of olfactory interneurons and oligodendrocytes throughout the brain. The SGZ remains an important source of NSPCs for the generation of granule cell neurons within the hippocampus.

3.2.2 Adult period

Neurogenesis during adulthood is confined to the olfactory system and the hippocampus. As mentioned above, some of the OPCs that migrated in the late embryonic period and early postnatal period persist within the striatum, white matter and cortex as progenitors (Dawson et al., 2003). Thus, OPCs are found throughout the adult brain. OPCs are characterized by expression of markers including Olig2, PDGFRA and NG2. It is controversial whether NG2+/PDGFRA+ cells scattered throughout the neocortex are truly restricted oligodendrocyte progenitors. Recently, NG2 cells were shown to have characteristics of NSPCs, giving rise to both gray and white matter. In fact, there is some evidence that they are tripotential. This issue has been explored in vivo using Cre conditional mutants and Cre conditional inducible mutants. Using the NG2CreBAC:Z/EG mouse, Zhu et al., showed that EGFP+ cells residing in white matter give rise to OL, whereas those residing in gray matter give rise to OL and some astrocytes (Zhu et al., 2008). As NG2+ cells in adult mice co-express PDGFRA and vice versa, Rivers et al., showed using the PDGFRACreERT2;Rosa26-YFP mouse, that cells that had expressed PDGFRA and their progeny give rise to myelinating oligodendrocytes in the corpus callosum and to projection neurons in the piriform cortex but never astrocytes (Rivers et al., 2008). As NG2+ cells also express Olig2 and vice versa, Dimou et al, showed using the Olig2CreER mouse, that cells that had expressed Olig2 and their progeny give rise to myelinating oligodendrocytes in white matter but not gray matter and became post-mitotic in gray matter, suggesting a non-progenitor function in these regions (Dimou et al., 2008). Interestingly, both ventrally derived (Gsh2-derived) and dorsally derived (Emx1-derived) precursors contribute equally to dividing and non-dividing subpopulations of NG2 cells (Psachoulia, 2009). The conclusion is that the overwhelming progeny of NG2 cells are oligodendrocytes but much less often are astrocytes and neurons. NG2 cells may be a class of cells with a unique function in their own right.

Developmental Period	Location	Differentiated Progeny Phenotype
Early Embryonic	VZ	Projection Neurons, OPCs
Mid-Late Embryonic	VZ, SVZ, SGZ	Projection Neurons, OPCs/NG2 cells, Cortical Interneurons, Hippocampal Granule Cell Neurons, Oligodendrocytes
Postnatal	SVZ, SGZ, striatum, corpus callosum, cortex	Olfactory Cortical Interneurons, Hippocampal Granule Cell Neurons, OPCs/NG2 cells, Oligodendrocytes
Adulthood	SVZ, SGZ, striatum, corpus callosum, cortex	Olfactory Cortical Interneurons, Hippocampal Granule Cell Neurons, OPCs/NG2 cells, Oligodendrocytes

Table 1. Distribution and progeny of NSPCs during development.

In adulthood, the SVZ remains an important source of NSPCs for the generation of olfactory interneurons and the SGZ remains an important source of NSPCs for the generation of granule cell neurons. In addition, widely spread throughout the brain are NG2 cells that are accepted as OPCs but may also be source NSPCs for projection neurons in the piriform cortex and for gray matter astrocytes.

Precursor Cell Phenotype	Markers
Neural stem cells	Nestin, GFAP, Sox2
Neuroblasts	Dcx
Oligodendrocyte Progenitor Cells	Olig2, NG2, PDGFRA, Sox10

Table 2. Markers of neural stem cells and restricted progenitors.

4. NSPCs in perinatal hypoxia-ischemia

Despite the restricted fate for progenitors in development, it seems that injury can provoke a relaxation of this fate restriction. We will review how SVZ cells respond to perinatal hypoxia-ischemia in terms of proliferation, fate commitment and migration. The response of NSPCs to stroke is covered elsewhere in this book.

In vitro data suggest that the SVZ responds to perinatal hypoxia-ischemia by attempting to regenerate lost cells through increased proliferation and also a shift in fate potential. Neurospheres generated from the SVZ of neonatal rats subjected to hypoxia-ischemia yield oligodendrocytes more often than neurospheres generated from non-lesioned rats (Felling et al., 2006; Yang and Levison, 2006). In vivo data also support a regenerative response by the SVZ that includes increased emigration as well. Plane et al. showed an increase in neuroblasts (BrdU+/Doublecortin+ cells) in SVZ and striatum 2 weeks after perinatal hypoxia-ischemia in the mouse, however no mature neurons (BrdU+/NeuN+ cells) 3 weeks after injury (Plane et al., 2004). Similarly in rat, Ong et al. showed an increase in SVZ neuroblasts (BrdU+/Doublecortin+ cells) 2-3 weeks after perinatal hypoxia-ischemia but no increase in mature neurons (BrdU+/NeuN+ cells) in striatum 4 weeks after injury (Ong et al., 2005). By contrast, Yang and Levison did show an increase in neuroblasts (BrdU+/Doublecortin+ cells) and mature neurons (BrdU+/NeuN+ cells) up to 5 months after perinatal hypoxia-ischemia in the rat. They also marked SVZ cells with Retroviral-AP and showed that these newly born neuroblasts and mature neurons originated from the SVZ (Yang and Levison, 2007). Several groups have shown that, 4 weeks after perinatal hypoxic-ischemic injury, newly born oligodendrocytes (BrdU+/MBP+ cells, BrdU+/carbonic anhydrase+ cells, and BrdU+/RIP+ cells) are found in the striatum, corpus callosum and infarcted cortex (Back et al., 2002; Zaidi et al., 2004; Ong et al, 2005).

By contrast our work, focusing on the oligodendroglial lineage, did not show an increase in emigration of OPCs from SVZ after perinatal hypoxia-ischemia. Like others, we showed an increase in neural progenitors in SVZ in response to perinatal HI; specifically, we showed an increase in OPCs in vivo (Figure 1) (Dizon et al., 2010).

Given that OPCs are actively migrating during the timing of perinatal hypoxic-ischemic injury, we expected to see an increase in OPC migration. Rather, using multi-photon microscopy to image OPCs in slice cultures derived from lesioned Olig1-EGFP mice, we showed a paucity of OPC emigration from SVZ. Nonetheless, we showed an increase of Olig1-EGFP+ cells within cortex and striatum in response to perinatal HI (Figure 2) (Dizon et al., 2010).

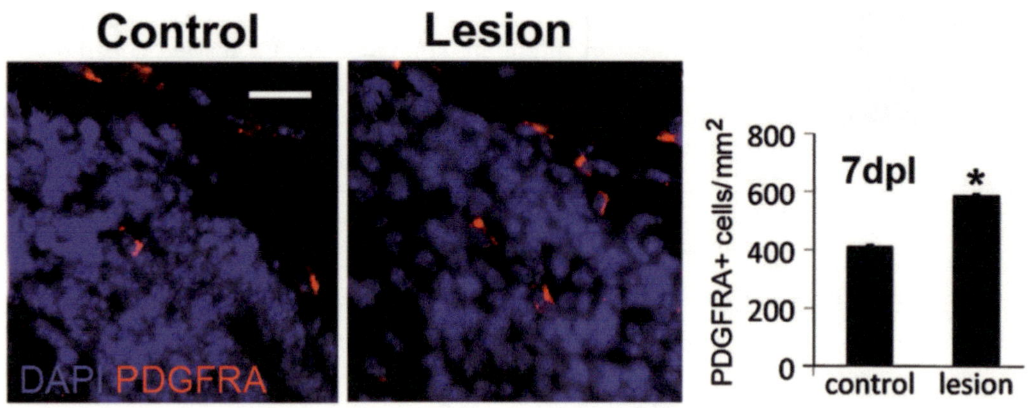

Fig. 1. Perinatal hypoxia-ischemia causes an increase in OPCs marked by PDGFRA within the SVZ by 7 days post lesion (dpl). Scale bar = 50 microns. *p=0.013.

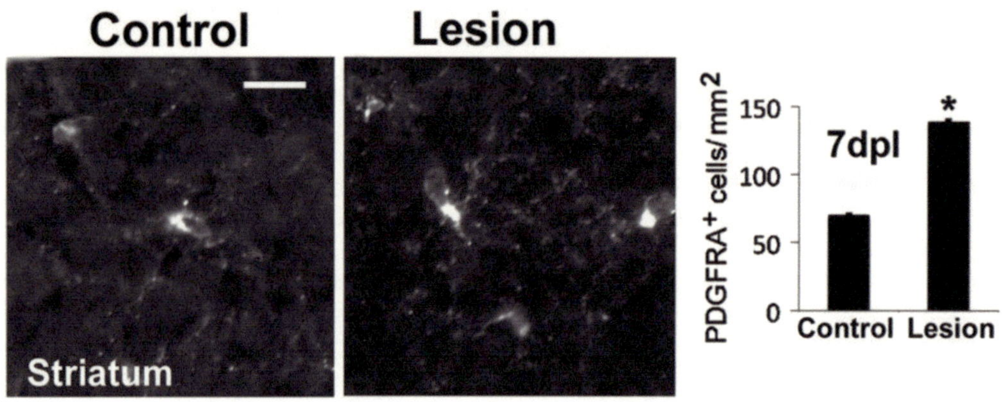

Fig. 2. Perinatal hypoxia-ischemia causes an increase in OPCs marked by PDGFRA within the striatum by 7 days post lesion (dpl). Scale bar = 50 microns. *p=0.003.

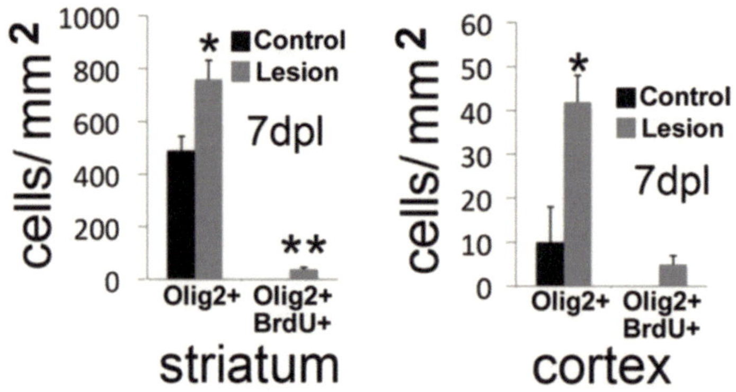

Fig. 3. Increases in OPCs marked by Olig2 result from increased proliferation of OPCs in the striatum but not in the cortex as evidenced by doublemarking with BrdU. Striatum: *p=0.044, **p=0.031. Cortex: *p=0.037.

We also found an increase in OPCs within the cortex. Interestingly, by marking newly born cells with BrdU, we showed that increased OPCs in the striatum arise through proliferation but this is not true for the cortex (Figure 3).

Therefore, we speculated that increased OPCs may arise via increased fate commitment of uncommitted neural progenitors. Thus, we have turned our attention to local neural progenitors and strategies to augment endogenous regenerative responses to white matter injury, specifically toward manipulation of regulators of NSPC fate, namely bone morphogenetic proteins (BMPs). BMPs negatively regulate an oligodendroglial fate choice by NSPCs. Recently, using a transgenic mouse that overexpresses the BMP antagonist noggin during the period of oligodendrogliogenesis, we were able to obtain increased OPCs and myelinating oligodendrocytes as well as improved motor function in lesioned noggin overexpressing mice compared to lesioned wildtype mice (Dizon et al., 2011). Subsequently, another group has independently shown improved outcomes in a rabbit model of perinatal hypoxia-ischemia when pups were treated with recombinant noggin protein after injury (Dummula et al., 2011). We are currently targeting BMP receptor subunits to more precisely downregulate signaling in in vivo experiments utilizing conditional inducible ablation of BMP receptor subunits BMPR1a, BMPR1b and BMPR2 following injury. We would anticipate that our strategies, if successful, could be applied to other diseases of white matter at other developmental timepoints including congenital dysmyelination and multiple sclerosis.

In conclusion, the abundance and fate restriction of available neural precursors to address disease states varies depending on the time during development. Nonetheless, manipulation of endogenous progenitors may be a more promising therapy than transplanted progenitors as there is no need to overcome problems with rejection.

5. Conclusions

In conclusion, the endogenous NSPC cell populations capable of replacing cells lost to disease are specific to the period of development during which injury is sustained as well as to the location of injury. Unlike NSPCs from SVZ or SGZ, NSPCs of the NG2 type are not restricted to these regions but reside in the many brain regions that may be injured throughout the lifetime. Thus, NG2 cells likely reside more proximal to injured regions and might more readily respond, so they may be the more appropriate target for research efforts rather than SVZ cells. In addition, more specific groups of NG2 cells might be targeted. For example, Emx1-derived NSPCs might give rise to OPCs that are capable of compensating for lost white matter in an adult animal, while Nkx2.1-derived NSPCs would more appropriately be targeted to regenerate white matter in newborns; these cells would not be present so could not be exploited in adult onset diseases. Rational therapy would target restricted population of NSPCs, thereby confining therapies to the cells best able to effect regeneration while also minimizing unwanted side effects.

6. References

Altman, J. (1963). Autoradiographic investigation of cell proliferation in the brains of rats and cats. *Anatomical Record*. 145, 573-91.

Altman, J., Das, G. D. (1965a). Autoradiographic and histological evidence of postnatal hippocampal neurogenesis in rats. *Journal of Comparative Neurolology*. 124, 319-35.

Altman, J., Das, G. D. (1965b). Post-natal origin of microneurones in the rat brain. *Nature*. 207, 953-6.

Altman, J., Das, G. D. (1966). Autoradiographic and histological studies of postnatal neurogenesis. I. A longitudinal investigation of the kinetics, migration and transformation of cells incorporating tritiated thymidine in neonate rats, with special reference to postnatal neurogenesis in some brain regions. *Journal of Comparative Neurology*. 126, 337-89.

Anderson, M. J., Waxman, S. G. (1983). Caudal spinal cord of the teleost Sternarchus albifrons resembles regenerating cord. *Anatomical Record*. 205, 85-92.

Bayer, S.A., and Altman, J. (1991). *Neocortical Development*, Raven Press, Retrieved from: <http://neurondevelopment.org>

Bayer, S.A., and Altman, J. (2011a) Laboratory Memoir, In: *Neuron Development*, Available from: <http://neurondevelopment.org>

Back, S.A., Han B.H., Luo, N.L., Chricton, C.A. Zanthoudakis, S., Tam, J. , Arvin, K.L., Holtzman, D.M. (2002). Selective vulnerabiliry of late oligodendrocyte progenitors to hypoxia-ischemia. *Journal of Neuroscience*. 22: 455-463.

Bhardwaj, R. D., Curtis, M.A., Spalding, K.L., Buchholz, B.A,, Fink, D., Bjork-Erikkson, T., Nordborg, C., Gage, F.H., Druid, H., Erikkson, P.S., Frisen, J.. (2006). Neocortical neurogenesis in humans is restricted to development. *Proceedings of the National Academy of Science U S A*. 103, 12564-8.

Craig, A., Ling, L.N., Beardsley, D.J., Wingate-Pearse, N., Walker, D.W., Hohimer, A.R., Back, S.A. (2003). Quantitative analysis of perinatal rodent oligodendrocyte lineage progression and its correlation with human. *Experimental Neurology*. 181, 231-40.

Curtis, M. A., Kam, M., Nannmark, U., Anderson, M.F., Axell, M.Z., Wikkelso, C., Holtas, S., van Roon-Mom, W.M., Bjork-Eriksson, T., Nordberg, C., Fresn, J., Dragunow, M., Faull, R.L., Eriksson, P.S. (2007). Human neuroblasts migrate to the olfactory bulb via a lateral ventricular extension. *Science*. 315, 1243-9.

Dawson., M.R., Polito, A., Levine, J.M., Reynolds, R. (2003). NG2-expressing glial progenitors cells: an abundant and widespread population of cycling cells in the adult rat CNS. *Molecular and Cellular Neuroscience*. 24:476-488.

Dimou, L., Simon, C., Kirchoff, F., Takebayashi, H., Gotz, M. (2008). Progeny of Olig2-expressing progenitors in gray and white matter of the adult mouse cerebral cortex. *Journal of Neuroscience*. 28, 10434-42.

Dizon, M., Szele, F., Kessler, J.A. (2010). Hypoxia-ischemia induces an endogenous reparative response by local neural progenitors in the postnatal mouse telencephalon. *Developmental Neuroscience*. 32, 173-83.

Dizon, M.L., Maa, T., Kessler, J.A. (2011).The bone morphogenetic protein antagonist noggin protects white matter after perinatal hypoxia-ischemia. *Neurobiology of Disease*. 42, 318.26.

Dummula, K., Vinukonda, G., Chu, P., Xing, Y., Hu, F., Mailk, S., Csiszar, A., Chua, C., Mouton, P., Kayton, R.J., Brumberg, J.C., Bansal, R., Ballabh, P. (2011). Bone morphogenetic protein inhibition promotes neurological recovery after intraventricular hemorrhage. *Journal of Neuroscience*. 31:12068-12082.

Eriksson, P. S., Perfilieva, E., Bjork-Eriksson, T., Alborn, A.M., Nordborg, C., Peterson, D.A., Gage, F.H. (1998). Neurogenesis in the adult human hippocampus. *Nature Medicine*. 4, 1313-7.

Felling, R.J., Snyder, M.J., Romanko, M.J., Rothstein R.P., Zeigler, A.N., Yang, Z. Givogri, M.I., Bongarzone, E.R., Levison, S.W. (2006). Neural stem/progenitor cells participate in the regenerative response to perinatal hypoxia/ischemia. *Journal of Neuroscience*. 26, 4359-4369.

Globus, J.H., Kuhlenbeck, H. (1944). The subependymal cell plate (matrix) and its relationship to brain tumors of the ependymal type. *Journal of Neuropathology and Experimental Neurology*. 3:1-35.

Goldman, S. A., Nottebohm, F. (1983). Neuronal production, migration, and differentiation in a vocal control nucleus of the adult female canary brain. *Proceedings of the National Academy of Sciences U S A*. 80, 2390-4.

Gould, E., Reeves, A.J., Fallah, M., Tanapat, P., Gross, C.G., Fuchs, E. (1999a). Hippocampal neurogenesis in adult Old World primates. *Proceedings of the National Academy of Sciences U S A*. 96, 5263-7.

Gould, E., Reeves, A.J., Graziano, M.S., Gross, C.G. (1999b). Neurogenesis in the neocortex of adult primates. *Science*. 286, 548-52.

Kessaris, N., Fogarty, M., Iannarelli, P., Grist, M., Wegner, M., Richardson, W.D. (2006). Competing waves of oligodendrocytes in the forebrain and postnatal elimination of an embryonic lineage. *Nature Neuroscience*. 9, 173-179.

Levison, S.W. and Goldman, J.E. (1997). Multipotential and lineage restricted precursors coexist in the mammalian perinatal subventricular zone. *Journal of Neuroscience Research*. 48, 83-94.

Li, G., and Pleasure, S.J. (2005). Morphogenesis of the dentate gyrus: what we are learning from mouse mutants. *Developmental Neuroscience*. 27:93-99.

Lois, C., Alvarez-Buylla, A. (1993). Proliferating subventricular zone cells in the adult mammalian forebrain can differentiate into neurons and glia. *Proceedings of the National Academy of Sciences U S A*. 90, 2074-7.

Lois, C., Alvarez-Buylla, A. (1994). Long-distance neuronal migration in the adult mammalian brain. *Science*. 264, 1145-8.

Malatesta, P., Hartfuss, E., Gotz, M. (2000) Isolation of radial glial cells by fluorescent-activated cell sorting reveals a neuronal lineage. *Development*. 127:5253-63.

Marshall, C. A., et al., (2003). Gliogenic and neurogenic progenitors of the subventricular zone: who are they, where did they come from, and where are they going? *Glia*. 43, 52-61.

Menn, B., Garcia-Verdugo, J.M., Yaschine, C., Gonzalez-Perez, O., Rowtich, D., Alvarez-Buylla, A. (2006). Origin of oligodendrocytes in the subventricular zone of the adult brain. *Journal of Neuroscience*. 26, 7907-7918.

Meyer, R. L., et al., (1985). Topography of regenerating optic fibers in goldfish traced with local wheat germ injections into retina: evidence for discontinuous microtopography in the retinotectal projection. *Journal of Comparative Neurology*. 239, 27-43.

Ming, G. L.and Song, H. (2005). Adult neurogenesis in the mammalian central nervous system. *Annual Reviews of Neuroscience*. 28, 223-50.

Noctor, S.C., Flint, A.C., Weissman, T.A., Dammerman, R.S., Kriegstein, A.R. (2001). Neurons derived from radial glial cells establish radial units in neocortex. *Nature*. 409, 714-720.

Nottebohm, F. (1985). Neuronal replacement in adulthood. Annals of the New York Academy of Sciences. 457, 143-61.

Ong, J., Plane, J.M., Parent, J.M., Silverstein, F.S. (2005). Hypoxic-ischemic injury stimulates subventricular zone proliferation and neurogenesis in the neonatal rat. *Pediatric Research*. 58:600-606.

Plane, J.M., Liu, R., Wang, T.W., Silverstein, F.S., Parent, J.M. (2004). Neonatal hypoxic-ischemic injury increases forebrain subventricular zone neurogenesis in the mouse. Neurobiology of Disease 16:585-595.

Psachoulia, K., Jamen, F., Young, K.M., Richardson, W.D. (2009). Cell cycle dynamics of NG2 cells in the postnatal and ageing brain. *Neuron Glia Biology*. 5, 57-67.

Ray, J., Peterson, D.A., Schinstine M., Gage, F.H.(1993). Proliferation, differentiation, and long term culture of primary hippocampal neurons. *Proceedings of the National Academy of Sciences*. 90:3602-6.

Rakic, P. (1971). Guidance of neurons migrating to the fetal monkey neocortex. *Brain Research*. 33, 471-476.

Rakic, P. (1985). Limits of neurogenesis in primates. *Science*. 227, 1054-6.

Rivers, L.E., Young, K.M., Rizzi, M., Jamen, F., Psachoulia, K., Wade, A., Kessaris, N., Richardson, W.D. (2008). PDGFRA/NG2 glia generate myleinating oligodendrocytes and piriform projection neurons in adult mice. Nature Neuroscience. 11: 1392-401.

Smart, I., (1961). The subependymal layer of the mouse brain and its cell production as shown by autography after [H^3}-thymidine injection. *Journal of Comparative Neurology*. 116:325-27.

Suzuki, S.O., and Goldman, J.E. (2003). Multiple cell populations in the early postnatal subventricular zone take distinct migratory pathways: a dynamic study of glial and neuronal progenitor migration. *Journal of Neuroscience*. 23. 4240-4250.

Wright, J. Zhang, G., Yu, T.S., Kernie, S.G. (2010). Age related changes in the oligodendrocyte progenitor pool influence brain remodeling after injury. *Developmental Neuroscience*. 32:499-509.

Yang, Z., and Levison, S.W. (2006). Hypoxia/ischemia expands the regenerative capacity of progenitors in the perinatal subventricular zone. Neuroscience. 139:555-64.

Yang, Z., and Levison, S.W. (2007). Perinatal hyposic-ischemic brain injury induces persistent production of striatal neurons from subventricular zone progenitors. *Developmental Neuroscience*. 29:331-340.

Zaidi, A.U., Bessert, D.A., Ong, J.E., Xu, H., Barks, J.D., Silverstein, F.S., Skoff, R.P. (2004). New oligodendrocytes are generated after neonatal hypoxic-ischemic brain injury in rodents. *Glia*. 46:380-390.

Zhu, X., Bergles, D.E., Nishiyama, A. (2008). Ng2 cells generate oligodendrocytes and gray matter astrocytes. *Development*. 135:145-157.

Zupanc, G. K. (1999). Neurogenesis, cell death and regeneration in the adult gymnotiform brain. *Journal of Experimental Biology*. 202, 1435-46.

Zupanc, G. K. (2008). Adult neurogenesis and neuronal regeneration in the brain of teleost fish. *Journal of Physiology Paris*. 102, 357-73.

15

Neural Stem Cells: Exogenous and Endogenous Promising Therapies for Stroke

M. Guerra-Crespo[1,2], A.K. De la Herrán-Arita[1], A. Boronat-García[1], G. Maya-Espinosa[1], J.R. García-Montes[1], J.H. Fallon[3] and R. Drucker-Colín[1,2]

[1]*Departamento de Neuropatología Molecular, Instituto de Fisiología Celular*
[2] *Grupo Células Troncales Adultas, Regeneración Neuronal y Enfermedad de Parkinson Universidad Nacional Autónoma de México*
[3]*Department of Psychiatry and Human Behavior, University of California, Irvine*
[1,2]*México*
[3]*USA*

1. Introduction

In the last three decades, neuroscience has been profoundly involved in stem cell research that focuses in attempting to develop mechanisms and strategies for secure therapy for different types of brain injury. The limited capacity for cellular regeneration in the highly specialized adult brain makes it particularly vulnerable to cellular damage produced by injuries, such as stroke, which results in a permanent loss of function. Stroke is a a vascular accident characterized by dramatic neuronal loss following a sudden cease of blood flow.

According to the World Health Organization, 15 million people worldwide suffer a stroke event each year. Of those 5 million pass away and another 5 million are permanently disabled (World Health Organization, 2007). Chronic stroke can have a devastating impact on the patient, family, and caregivers; but it also incurs in high economic costs to society. In the United States, stroke is the leading cause of adult disability with an estimated annual medical cost of approximately $40 billion.

High blood pressure contributes to the majority of stroke events, accounting for 12.7 million strokes over the world (World Health Organization, 2007). In developed countries, the incidence of stroke is declining, largely due to efforts in attempting to lower blood pressure and reduce smoking. However, the overall rate of stroke remains high due to the aging of the population.

There are two main types of stroke. One is ischemic stroke, which is caused by a blockage of an artery; the other is hemorrhagic stroke, which is the result of a tear in an artery's wall that produces a blood outflow into the brain. Ischemic stroke is the most common type, it accounts for about 85 percent of all stroke events (Foulkes *et al.*, 1988). In both cases, prompt treatment could mean the difference between life and death. Early treatment can also minimize damage to the brain and potential disability. The consequences of stroke on bodily functions and the severity of stroke depend on the affected area of the brain and the extent of the damage.

1.1 Current treatments for stroke

In general terms, a stroke event can be divided in acute, sub-acute and chronic phase, by taking in consideration the time course of the injury. The therapeutic strategies to each relative time point are different and centered to ameliorate different sorts of damage.

In the acute phase, the use of thrombolytic agents may dissolve the blood clot in order to slow down or prevent the cascading process that destroys nerve cells after a few hours of ischemic stroke. A currently available and successful intervention to reduce the size of the infarct is the employment of recombinant tissue plasminogen activator (t-PA). T-PA allows the dissolution of a blood clot occluding a cerebral vessel by converting plasminogen into plasmin, an important enzyme present in blood that degrades many blood plasma proteins, most notably, fibrin clots.

Administration of t-PA is approved only within 3 hours of the onset of ischemia, although optimal results are observed if given within 90 minutes (Hacke *et al.*, 2004). Unfortunately, due to this narrow time window as well as a number of contraindications, t-PA therapy is only available to about 5% of stroke patients evaluated in the emergency room. Of these, t-PA may be expected to yield an approximate of 30% increase in the number of patients avoiding long-term neurologic deficits (Ropper & Brown, 2005).

Surgical interventions have enhanced over time and proved to be effective in some cases; however, their success is still below a desired echelon. Although surgical decompression after a stroke event has proved to lessen mortality in severe cases, doubling the probability to survive in a favorable condition, the odds of surviving in a condition requiring assistance from others increases around 10 times (Vahedi *et al.*, 2007). Advances in endovascular techniques may improve recanalization sufficiently to improve patient or cell survival, but these have yet to be substantiated by randomized clinical trials (Burns *et al.*, 2008).

Given the narrow window of time in which thrombolytic drugs and surgical procedures are effective, current research is focused in developing neuroprotective agents that maintain the cellular viability of threatened neuronal tissue (ischemic penumbra) and reduce the secondary damage after ischemic stroke. Many drugs currently undergoing investigation target the excitatory amino acids, such as glycine and glutamate, released by dying neural cells, which are known to lead to downstream changes that destroy nerve cells several hours to several days after a stroke. Dejectedly, although a plethora of neuroprotective compounds have shown promise in animal models, currently their employment has not shown any effectiveness in clinical trials (Dirnagl, 2006).

This fact implicates that nowadays, not a single treatment has been successful in reversing the effects of the chronic stroke. Physical therapy is used to promote functional recovery in long-term stroke patients, but recovery is often incomplete. Therefore, reversal of symptoms after a chronic stroke is a daunting problem that requires the improvement of the patient's lost function achieved by the replacement of lost neurons and glia in the injured region, as well as the establishment of new functional connections. These requirements call for bold new treatments that induce new neural cells to differentiate and integrate into the circuitry that was damaged by the stroke, the cell replacement therapy.

Considering the large amount of data acquired over the past four decades, it has been confirmed that neural stem cells (NSCs) are present throughout life and that thousands of neurons are born on a daily basis in two specific zones of the brain, the subventricular zone (SVZ) and the hippocampus (For extensive review see Zhao *et al.*, 2008).

NSCs are endowed with a self-renewal capacity and are specified to give rise only to nervous tissue-specific cell types, including neurons, glia and oligodendroglia (Reynolds and Weiss, 1996). These features together with the recent finding of the NSCs endogenous response to certain types of insults, such as stroke, lead to the persistent pursue to replace the cellular loss that takes place in the central nervous system (CNS) after injury or neurodegenerative diseases. With the advent of neuroregeneration discipline, new insights have come to the management of stroke. In general, two broad approaches are currently in development for cell replacement therapy in stroke: the recruitment of endogenous neural stem cells and exogenous stem cells transplanted into the affected area.

For a successful therapy, both approaches require to follow a highly regulated process known as neurogenesis, defined as the birth or generation of new neurons from NSCs (Zhao *et al.*, 2008). Neurogenesis follows a course where important cellular steps such as proliferation, migration and differentiation (PMD reponse), as well as cell survival are taking place. However, without a doubt, in order to develop a successful cell therapy, more understanding about how the neurogenesis process is occurring is of foremost relevance.

To achieve this goal, some important questions emerge: what are the characteristics required from the microenvironment that allow the neurogenic process to persist throughout life in the adult brain? What are the underlying mechanisms of neurogenesis regulation? And what is the cellular and molecular process regulating neurogenesis under pathological conditions?

In this chapter we will describe the current knowledge about the cellular organization and the molecular regulation that takes place in the SVZ, mechanisms that could provide the basis for the development of cell therapy in stroke and other neurodegenerative diseases. Afterward, the chapter will depict the effects of several growth factors with towering therapeutic potential for their capacity to induce endogenous cell replacement. Finally, this chapter will depict the therapeutic potential of NSCs and other cell types that are suitable for transplantation and ergo, for regeneration therapy in human patients afflicted with cerebral ischemia.

2. Cellular and molecular regulation of adult NSCs in the SVZ

2.1 The SVZ and rostral migratory stream (RMS): A general view

The study of adult neurogenesis in mammalian CNS began in the 1960's with the pioneering observations made by Joseph Altman, who managed to observe cell proliferation in the adult brain with ^{3}H-thymidine, a recognized cell division marker (Altman 1963, 1965, 1969). In spite of the controversy, further studies corroborated the existence of brain areas with the potential to generate new neurons from NSCs and defined as neurogenic niches. Nowadays, it is well accepted that the main neurogenic areas in the adult mammalian brain are the SVZ located in the walls of the lateral ventricles and the subgranular layer of the dentate gyrus (DG) of the hippocampus (Figure 1).

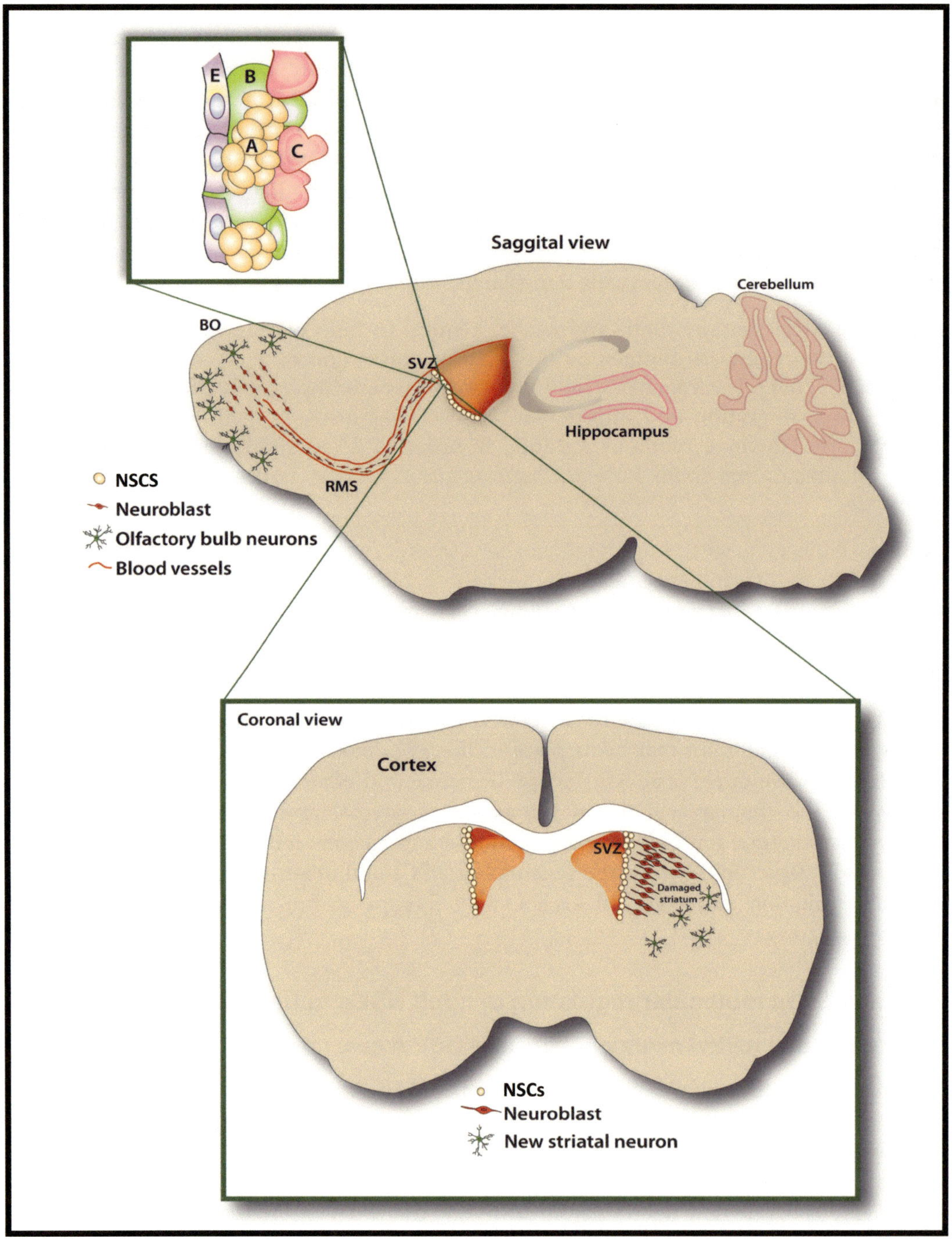

Fig. 1. Stroke and neurogenesis in the subventricular zone .

The SVZ contains many cell types, in addition to chemical and physical factors that create a special microenvironment, conducive to accurately regulate the self-renewal and multipotentiality properties of NSCs (Fuchs *et al.*, 2004). It also regulates the neurogenic processes of PMD as well as the less characterized cell integration.

The cellular composition and cytoarchitecture of the SVZ is remarkable peculiar and complex (Figure 1). There are at least five main different cell types integrating the SVZ: Astrocytes, also called B cells, divided in B1 (apical B) and B2 (tangential B); transit amplifying cells (C cells, the putative precursor); neuroblasts (A cells); tanycytes (D cells) and ependymal cells, divided in E1 and E2 (E cells) (Doetsch *et al.*, 1997; Mirzadeh *et al.*, 2008; Shen *et al.*, 2008). To give rise to new neuron, B cells generate C cells and their subsequent division generates A cells that migrate tangentially in clusters via the RMS pathway towards the OB (Figure 1) (Doetsch & Alvarez-Buylla, 1996; Lois *et al.*, 1996). In the OB, neuroblasts migrate radially to the granular and glomerular layers, where they differentiate as local interneurons and integrate into functional circuits (Belluzzi *et al.*, 2003; Carleton *et al.*, 2003; Kosaka *et al.*, 1995; Whitman *et al.*, 2007). Although uncertain, the functional relevance of the cell replacement that occurs in the OB throughout the lifespan of rodents is attributed to the olfactory adjustment to odor changes in the environment.

Without a doubt, the features shared by the many cellular types of the SVZ contributed to the debate about the true identity of NSCs (for review, Chojnacki *et al.*, 2009). Mitotic cells (astrocytes or B cells) have been generally considered to be the true NSCs, owing to their self-renew capacity and their ability to indefinitely produce neuronal and glial progeny (García-Verdugo *et al.*, 1998; Doetsch *et al.*, 1999). Nevertheless, *in vivo* observations suggest that ependymal cells function as NSCs (Johansson *et al.*, 1999). Nowadays, astrocytes are the current accepted NSCs in the adult SVZ. Nonetheless, it is well accepted that the combined stimuli of injury and a growth factor (e.g. transforming growth factor alpha,TGFα) induces a PMD response by NSCs in the ependymal layer and in the SVZ (Gleason *et al.*, 2008; Guerra *et al.*, 2009).

More recently, the presence of NSCs has been determined in the RMS (Gritti *et al.*, 2002), which was considered solely as a migratory pathway for neuroblasts that were migrating rostrally from the SVZ to their final destination, the OB (Lledo *et al.*, 2008). This cellular movement within the RMS is called "chain migration", a term established by Lois and coworkers in 1996; they showed that neuroblasts migrated in clusters without axonal guidance or radial glia regulation and instead, used a network of astrocytes that form "glial tubes" (Lois *et al.*, 1996). This unique type of migration is thought to enable cells to draw on neighboring cells as their scaffold for migration (Murase & Horwitz, 2002).

The close proximity to the striatum, shared by the SVZ and the RMS raised interest as a cellular replacement option for focal stroke occurring at the basal ganglia. Several molecules have been implicated in the highly regulated neurogenesis processes taking place in both areas. Some of them play multiple essential roles by regulating different levels of the PMD response. For the development of a thriving cell therapy, it is essential to control the processes of proliferation and migration, by means of increasing the number of proliferating cells and redirecting them to the injured area. Therefore, the mechanisms that will allow us to control neuronal fate are of great interest in the neurorepair field and represent a promising subject for the development of a factual clinical treatment.

In the following fraction, we describe the main molecules and cellular events that have been involved in the NSCs differentiation route. Table 1 summarizes the main findings.

2.2 Regulation of the NSCs development in the OB pathway

The tangential homotypic traverse of the cells that depart from the SVZ through the RMS is finely controlled by several molecules at different levels. Growth factors are the main mitogenic signals received by NSCs that trigger cell division. A subsequent generation of neuroblasts takes place after such signals; these neuroblasts are confined within the RMS via the combined effect of diffusible chemoattractants and chemorepellents that are flowing at a concentration gradient from the SVZ to the RMS. Noteworthy, the precise migration route is determined by cell-cell interaction mechanisms to ascertain their oriented organization into a continuous alignment. This migration configuration requires individual morphology arrangements in the wanderer neuroblasts, in order to make them suitable for migration.

2.2.1 Growth factors: Regulating multiple effects

Growth factors and neurotrophins have been implicated in the regulation of neurogenesis at early and postnatal development by controlling proliferation, migration and differentiation. Here, we describe some of these factors, including epidermal growth factor (EGF), fibroblast growth factor-2 (FGF-2), TGFα, brain derived neurotrophic factor (BDNF) and vascular endothelial growth factor (VEGF) for their critical role in these processes.

2.2.1.1 EGF, TGFα and FGF-2: Mitogenic signals

EGF and TGFα, two members of the EGF family, were the first growth factors to be associated to regulate SVZ proliferation through its binding to EGF receptor (EGFR). This tyrosine kynase receptor has been immunodetected in C cells (Doetsch *et al.*, 2002) and in astrocytes of the SVZ (Höglinger *et al.*, 2004). *In vitro,* studies have demonstrated that NSCs derived from the SVZ proliferate in the presence of EGF or FGF-2, as free-floating neuroespheres that have the potential to differentiate into neurons and glia under appropriate conditions (Doetsch *et al.*, 1999; Gritti *et al.*, 1999; Reynolds & Weiss, 1992). *In vivo,* intraventricular infusion of EGF and TFGα resulted in an increase of NSCs proliferation in the SVZ (Craig *et al.*, 1996; Khun *et al.*, 1997; Morshead *et al.*, 2003). Grippingly, only TGFα-knockout mice presented a significant reduction in proliferation and neuroblast migration in the SVZ and RMS (Tropepe *et al.*, 1997), pointing out its direct role in these two cellular processes.

Alternatively, FGF is a large family that is widely expressed in the central nervous system. Twenty-two members of the FGF family have been identified in humans (Itoh & Orniz, 2011) and have been related to several neuronal processes, for instance, development, adult neurogenesis and repair mechanisms (Reuss & von Bohlen, 2003). To accomplish its mentioned functions, FGF binds with different affinity to each one of the FGF receptor (FGFR) family, integrated by four different members (FGFR 1 to 4) (Ornitz *et al.*, 1996, Johnson & Williams, 1993).

From all the integrants of the FGF family, FGF-2 is the one commonly related to regulate adult neurogenesis. FGF-2 and its receptor, FGFR-1, have been described in the SVZ (Mudó *et al.*,

2007). *In vitro,* FGF-2 stimulates the proliferation of NSCs located in the SVZ and that these progenitors expressing nestin (neural progenitor marker) are able to differentiate into several brain cell types including neurons (Gritti *el al.*, 1996, 1999). *In vivo* studies demonstrated that a subcutaneous injection of FGF-2 increases mitotic activity in the adult SVZ (Wagner *et al.*, 1999) and restores the proliferative rates in aging rats (Fluxe *et al.*, 2008; Jin *et al.*, 2003). Nevertheless, the expression levels of FGF-2 and FGFR-1 in the SVZ do not change during aging, suggesting that although FGF-2/FGFR-1 interaction is involved in cell proliferation, it does not regulate the age-dependent declining of NSCs proliferation (Frinchi *et al.*, 2010).

2.2.1.2 BDNF: A multitask molecule

A neurotrophin recognized to undertake multiple roles in SVZ neurogenesis is BDNF. BDNF and its receptors of high and low affinity, TrkB and $p75^{NTR}$, respectively, are expressed all over the SVZ-RMS in a mutually exclusive pattern; with higher expression of BDNF and TrkB in the RMS when compared to the SVZ (Bath *et al.*, 2008; Chiaramello *et al.*, 2007; Galvão *et al.*, 2008; Gascon *et al.*, 2007; Giuliani *et al.*, 2004; Maisonpierre *et al.*, 1990; Snapyan *et al.*, 2009; Young *et al.*, 2007). In the adult RMS, BDNF mRNA expression was identified in endothelial cells of blood vessels, and absent in astrocytes as well as neuroblasts (Snapyan *et al.* 2009), whereas TrkB was found predominantly in astrocytes (Snapyan *et al.* 2009). Conversely, active TrkB (phosphorylated form) as well as the low affinity receptor $p75^{NTR}$ have been observed in migrating neuroblasts (Bath *et al.*, 2008, Galvão *et al.*, 2008; Snapyan *et al.*, 2009). $P75^{NTR}$ is also present in a small subset of GFAP-immunopositive cells and transit amplifying cells (C-cells) in the RMS (Bath *et al.*, 2008, Galvão *et al.*, 2008; Snapyan *et al.*, 2009). The BDNF-TrkB-p75 interaction is correlated to multiple cellular processes in the SVZ-RMS, in particular, those regarding survival, proliferation, migration and cell differentiation.

Luskin and Goldman's groups, who were working independently in a parallel fashion, were the firsts studying the effect of BDNF on adult neurogenesis. They both demonstrated that infusion or overexpression of BDNF into the lateral ventricles increased cell proliferation in the SVZ and migration through the default migratory pathway towards the OB (Benraiss *et al.*, 2001; Pencea *et al.*, 2001; Zigova *et al.*, 1998). Of equal relevance, was their discovery that a higher number of newborn neurons occur in the OB, highlighting its role in differentiation. Simultaneously, new heterotopic spiny neurons were found in the striatum; this finding raises the possibility to redirect the migration of neuroblasts after BDNF infusion, an astonishing breakthrough for endogenous replacement therapy (Benraiss *et al.*, 2001; Pencea *et al.*, 2001).

More recently, an additional function has been attributed to BDNF, the chemoattraction; defined as the movement of cells towards a chemical concentration gradient. Chiaramello *et al.*, tested the hypothesis that the lower expression pattern of BDNF in the RMS and a higher expression pattern in the OB is consistent with the chemoattraction property.

In explants cell cultures, the addition of this neurotrophin resulted essential for survival in addition to, a dose-dependent increase of migrating newly dividing cells. Interestingly, BDNF-induced motility on SVZ neuroblast explants was suppressed by blocking TrkB receptor autophosphorylation and by blocking BDNF action with neutralizing antibodies (Chiaramello *et al.*, 2007). Together, these results prove that BDNF has a chemoattractive function through an autocrine and/or paracrine signaling.

Despite the conclusive data concerning BDNF, the mechanisms responsible for maintaining the cells in proper formation while navigating towards the OB core still are unclear. New relevant information comes out from an interesting study from Snapyan and coworkers. They identified BDNF as a molecular signal released from endothelial cells of blood vessels (not from astrocytes or neuroblasts) in the SVZ-RMS, and demonstrated that it promotes neuronal migration via $p57^{NTR}$ activation on neuroblasts (Snapyan *et al.*, 2009). They proposed a vasculature-guided migration model in which migrating neuroblasts in the RMS are retained and migrate along this pathway, secondary to the presence of blood vessels that are oriented in a parallel fashion to the RMS (Snapyan *et al.*, 2009). This newly proposed model however, does not exclude the previously stated chemoatraction model, since the possibility that both mechanisms are operating is plausible (Snapyan *et al.*, 2009).

2.2.1.3 VEGF: Regulator of proliferation and a probable chemoattractive molecule in the SVZ

Another growth factor regulating cellular proliferation in the SVZ is VEGF, a glycoprotein known to be involved in angiogenesis and vasculogenesis. Intraventricular infusion of VEGF leads to an increase in cell proliferation of the SVZ (Jin *et al.*, 2002; Sun *et al.*, 2006). This finding has been confirmed by inducing the overexpression of VEGF in an ischemic model, where an increase in SVZ proliferation and migration of new cells into the ischemic injury was observed (Wang *et al.*, 2007a, 2007b).

The VEGF family ranges from VEGF-A to VEGF-D factors, which have an affinity to several tyrosine kinase receptors: VEGF receptor (VEGFR) 1 (Flt1), VEGFR2 (KDR/Flk1), VEGFR3 (Flt4) and neuropilin receptors (NP1/2) (Matsumoto & Claesson-Welsh, 2001). One of these receptors, VGFR2, has been demonstrated to mediate a chemotactic activity for VEGF in cell cultures (Zhang *et al.*, 2003). Its role as a chemoattractive guidance molecule for migrating neural progenitors arising from the SVZ only takes place when these cells are maintained under FGF-2 administration. These *in vitro* essays show that FGF-2 stimulates neural progenitors to express VGFR2, which confers them the capacity to respond to VEGF (Zhang *et al.*, 2003).

The chemoattractant function of VEGF in adult brain is still not clear. *In vivo*, intraventricular VEGF administration revealed that VEGFR2/Flk-1 receptors predominate in neurogenic niches such as the SVZ and co-localize with migrating neuroblasts as well as astroglial, endothelial and neuronal cells (Jin *et al.*, 2002; Schänzer *et al.*, 2004). VEGF expression is observed in astrocytes, being higher in astrocytes of the SVZ and RMS than in astrocytes from non-permissive regions (Balenci *et al.*, 2007). This expression represents an endogenous source of VEGF that can be related to neurogenesis and might have a direct effect on the migration of neural progenitors within neurogenic regions (Balenci *et al.*, 2007).

2.2.2 Chain migration: Molecules implicated

Neuroblast chain formation requires a myriad of specific and regulated interactions to constrain the migrating cells into a precisely organized shape of the RMS. Here, we are describing the main molecules orchestrating the cellular processes involved in chain formation.

2.2.2.1 Ephrin family: Driving proliferation and proper cell positioning in migrating chains

Ephrins are transmembrane-associated proteins that exert their actions through its binding to Ephrin (Eph) receptors; the largest family of tyrosine kynase receptors. Ligands and receptors are divided in two subclasses (A and B), based on their binding properties and structural homologies (Mosch *et al.*, 2010). In general, ephrin-A ligands (ephrin-A1 to A6) bind preferentially to EphA receptor (EphA1 to A9), whereas ephrin-B (ephrin-B1 to B3) ligands bind to EphB receptor (EphB1 to B6). The interaction triggers a bidirectional cascade signal, where ephrins mediate a "reverse" signal and receptors mediate a "forward" signal (Cowan & Henkemeyer, 2002).

Ephrins-Eph interaction during development has been implicated in multiple roles, including axonal growth and cell guidance (For review see Wilkinson, 2001). In the adult brain, this interaction has also been related to several roles during neurogenesis (Conover *et al.*, 2000; Holmberg *et al.*, 2005; Theus *et al.*, 2010).

The SVZ and RMS express Eph-B1-3 and EphA4 receptors (Conover *et al.*, 2000; Theus *et al.*, 2010). In the RMS, EphB2 receptor seems to be surrounding chains of migrating neuroblasts, whereas cells expressing EphB2 in the SVZ are yet still to be determined (astrocytes, neuroblasts or ependymal cells) (Conover *et al.*, 2000). On the other hand, ephrin-B ligands are expressed in astrocytes that envelop chains of migrating neuroblast along the RMS pathway. This ephrin/Eph complementary expression pattern seems to regulate the classic ephrin contact-mediated repulsive response to give a position to the cells in specific sites.

Interestingly, blocking Ephrin-B/EphB interactions originates an increase in astrocyte proliferation in the SVZ and promotes a disorganization of the chain network (Conover *et al.*, 2000). Therefore, ephrin-B/EphB interaction is a negative regulator of cell proliferation and controls spatial organization in the SVZ. Additionally, this result suggests that inhibition of cell proliferation is required to maintain the germinative niches homeostasis.

Recently, Theus and collaborators revealed the anti-proliferative effect of EphB3, another EphB receptor in the SVZ. EphB3 is expressed in neuronal stem progenitor cells and in neuroblasts. In the traumatic brain injury (TBI) model, where an increase in neurogenesis in the SVZ is induced, Theus's group observed a significant reduction of EphB3 expression, coincidently with enhanced NSCs precursors survival and proliferation post-injury. These findings were corroborated in both ephrin-B3 and EphB3 knockout mice. The two models showed a significant augmentation in SVZ proliferation. Interestingly, in ephrinB3$^{-/-}$ mice, cell division can be reverted by infusion of a soluble form of ephrinB3 (ephrinB3-Fc) in the lateral ventricle. Furthermore, its infusion also prevents TBI-induced neural stem progenitor cell proliferation (Theus *et al.*, 2010).

Studies made in ephrin subclass A receptors gave similar results to the ones observed on ephrin-B. The use of ephrin-A2 and ephA7 knockout mice showed that this ligand-receptor interaction is a key inhibitor of cell proliferation in adult brains (Holmberg *et al.*, 2005). Ephrin-A2$^{-/-}$ and EphA7$^{-/-}$ mice show an increase in SVZ proliferation concomitant to an increase in the number of new cells in the OB; suggesting that these cells migrate in a regular fashion to the OB. In this study, it was observed that Ephrin-A2 is expressed in neural progenitor cells and neuroblasts of the lateral ventricles, whereas EphA7 receptor was expressed in ependymal cells and astrocytes of the SVZ. Interestingly, once again it seems that they are localized in a mutually exclusive manner, proper to promote ligand-receptor interactions (Holmberg *et al.*, 2005).

Altogether, these findings indicate that ephrin signaling is involved in the regulation of at least two processes of adult neurogenesis, the proliferation and neuroblast guidance into migrating chains, although, the mechanisms controlling these events still have to be determined.

2.2.2.2 ErbB receptors and neuregulins: Chemoattractive implications in proliferation and migration

The ErbB family, also called EGFR family, is integrated by four related tyrosine kinase receptors (ErbB1-ErbB4); one of them involved in cell migration signaling. Anton and coworkers determined that ErbB4 is expressed in neuronal precursor cells (type A cells) residing in the SVZ and the RMS. Additionally, in a small subset of type B cells and type C cells (Anton *et al.*, 2004). Ghashghaei and collaborators detected the presence of ErbB4 in a subgroup of CD24+ ependymal cells (Ghashghaei *et al.*, 2006).

Experiments in mice lacking ErbB4 receptor, determined its role in the organization of the SVZ-RMS pathway (Anton *et al.*, 2004). ErbB4-null mice had a disorganized structure of the SVZ, whereas the neuronal precursors, typically organized in clusters of chains along the SVZ and RMS, were instead forming fragmented chains that migrated as individual cells with an altered orientation. Moreover, impaired placement of interneurons in the OB was observed. *In vitro*, explants from these mutant mice were also unable to form compact neuronal chains, suggesting that loss of ErbB4 disrupts the characteristic "glial" tubular organization in the RMS (Anton *et al.*, 2004).

In the SVZ, ErbB4 receptors can be activated by neuroregulins (NRGs), proteins belonging to the EGF family and directly involved in the migration process as chemoattractants (Anton *et al.*, 2004). NRGs are a family of four signaling proteins that mediate cell-cell interactions in different organs, including the brain. They have been related to the activation of intracellular signaling pathways that lead to specific cellular responses, including stimulation or inhibition of proliferation, apoptosis, migration, differentiation and adhesion (Yarden *et al.*, 2001).

Two NRG types, NRG1 and NRG2, are controlling adult neurogenesis, in particular, cell proliferation and migration. NRG1 type III isoform is highly expressed in the RMS and the OB in the early postnatal development period (Anton *et al.*, 2004). *In vitro*, NRG1 type III has been characterized as the preferred chemoattractive protein (compared to NRG1 type I), aiding the migration of neuronal precursor cells from the SVZ (Anton *et al.*, 2004).

NRG2 is expressed by immature neuroblasts and in a subset of astrocytes that are lining the ventricles of the SVZ (Ghashghaei *et al.*, 2006). Ghashghaei and collaborators demonstrated that NRG1 and NRG2 have different functions in the same niche. Intraventricular infusion of NRG1 induces the aggregation of proliferating precursors into clusters in the SVZ. This aggregation is fundamental for their proper migration, probably by a chemoattractive property. On the other hand, intraventricular infusion of NRG2 promotes astrocyte proliferation and a subsequent increase of neuroblast and GABAergic interneurons in the olfactory bulb (Ghashghaei *et al.*, 2006).

2.2.2.3 βI integrins: Receptors regulating chain formation and migration through laminin binding

Integrins are heterodimeric cell surface glycoproteins that regulate proliferation and cell adhesion. Their mechanism of action is mediated through the binding to ECM proteins (fibronectin, laminin), other proteins such as ADAMs (a desintegrin and metalloprotease) and Ig-superfamily cell surface counter-receptors (such as VCAM-1). Two different subunits have been identified, α (18) and β (8), which assemble at least into 24 distinct types of integrins (Hynes, 2002).

Integrins play an important role directing the migration of neuronal precursors demonstrated in both *in vivo* and *in vitro* experiments. At least 11 integrin subunits have been identified in the SVZ-RMS pathway (α1, α2, α3, α6, α7, αv, β1, β3, β5, β6 and β8) with modifications in the temporal expression pattern during development (Belvindrah *et al.*, 2007; Murase & Horwitz, 2002). The expression of α1, α6, and α7 integrin subunits in neuroblast of the RMS seems to be higher than the one observed in the SVZ (Belvindrah *et al.*, 2007) and grippingly, all of them are able to heterodimerize with a β1 subunit (Hynes, 2002). A model for NSCs-vascular niche adhesion in the SVZ proposes integrins as a receptor for laminin, a protein deposited between neuroblasts and mainly expressed by endothelial cells that surround the SVZ. Laminin is a chemoattractant molecule for SVZ-RMS neuronal precursors that has been proposed to promote cell-cell interactions for chain formation via binding to β1-integrin (Belvindrah *et al.*, 2007). As a counterpart, β5 integrin is dispensable for chain migration.

Another integrin, α6β1, is expressed in chain migrating cells along the rodent SVZ-OB pathway, mainly in cell-cell junctions (Belvindrah *et al.*, 2007; Emsley & Hagg, 2003; Jacques *et al.*, 1998). Laminin is the only known ligand for α6β1 integrin. It has been observed that laminin is recruited to the cell surface of migrating neuroblasts, where it induces chain formation in SVZ explants and the aggregation of neuronal precursors *in vivo*. In addition, it also induces chain migration *in vitro* and *in vivo* (Belvindrah *et al.*, 2007; Emsley & Hagg, 2003). Worthy of notice, laminin infusion, in regions where precursor cells are not normally seen, redirects neural precursors toward these locations, as it has been observed in the neostriatum (Emsley& Hagg, 2003). The evidence, strongly supports the idea that laminin-α6β1 interaction is involved in the directional guidance of neuronal precursors, and promotes the formation of cell chains in the adult RMS, as well as maintaining the integrity of glial tubes in the RMS (Belvindrah *et al.*, 2007; Emsley & Hagg, 2003).

2.2.2.4 Poly-sialated neural-cell adhesion (PSA-NCAM) molecule: Forming the chain migration process and controlling differentiation

NCAM is a cell surface glycoprotein, member of the immunoglobulin superfamily that mediates cell-cell homotypic interactions, while PSA is a polymer of neuroaminic acid residues added to the NCAM molecule that is highly expressed during development and persists in the adult nervous system. Interestingly, in the adult brain, its expression is particularly confined to neurogenic niches (Seki & Arai, 1993). Specifically, neuroblasts conforming the chain migration in the RMS express PSA-NCAM (Rousselot *et al.*, 1995).

In explants cultures, it has been observed, the absence of PSA produced by endoneuraminidase-N (endoN) treatment incites the dispersion of migrating chains, although neuroblasts are still able to migrate as single cells (Hu, 2000). Furthermore, *in vivo* essays demonstrate that these cells do not easily disperse, probably because the tubular glial structures that are ensheating the neuroblasts are confining the migrating cells (Hu, 2000). This clearly indicated that lack of PSA causes a decrease in cell-cell interactions in neural progenitor chains. Therefore, PSA is an important element for neuroblast chain formation (Hu, 2000).

Other studies have demonstrated that PSA is additionally controlling neuroblast differentiation (Battista *et al.*, 2010; Petridis *et al.*, 2004). Removal of PSA moieties secondary to endoN administration inhibits cell contact dependent differentiation *in vivo*. Furthermore, it promotes a minimal and ectopic dopaminergic differentiation of neural progenitors in the SVZ (Petridis *et al.*, 2004). PSA elimination can also result in a dispersal of progenitor cells from the SVZ to the surrounding regions such as cortex and striatum, where neuroblasts differentiate into a calretinin and GABAergic phenotypes (Battista *et al.*, 2010).

The effects of NCAM mutations are quite different compared to the ones observed after PSA depletion. NCAM mutations produce a dramatic size reduction of the OB (~30%), while disruption of PSA increases this effect (Cremer *et al.*, 1994). PSA-NCAM deficient mice have an altered RMS, where the normal borders are exceeded as a result of an altered accumulation of migrating cells (Chazal *et al.*, 2000).

Altogether, it seems that PSA-NCAM mutations lead neural precursors to fail in the tangential migration to the OB. On the other hand, the ability to migrate radially into the same structure remains unaltered under PSA removal. This was demonstrated after transplanting SVZ cells in the OB of normal and PSA depleted mice OBs (by endoN), in both cases, no differences in radial migration distance or cell morphology were detected, which indicate that PSA does not regulate radial migration (Hu, 1996).

2.2.2.5 Doublecortin (DCX): Cytoskeletal dynamics for cell translocation

DCX is a neuron-specific phosphoprotein associated with microtubules that is localized in cell bodies and leading processes. It is involved in the regulation of cytoskeletal dynamics. It is expressed in neuroblasts (type A cells) and immature neurons wherein induces polymerization, in addition to promote the stabilization of microtubules that contribute to cell locomotion (Francis *et al.*, 1999, Gleeson *et al.*, 1999).

DCX expression is downregulated in postmigratory neurons in the OB. Its co-distribution with PSA-NCAM characterizes early committed neurons within the RMS; therefore, DCX is selectively expressed in migratory cell populations within the SVZ, RMS and proximal OB (Moores *et al.*, 2004; Ocbina *et al.*, 2006).

This factor promotes and maintains a bipolar cell morphology, which allows nuclear translocation and therefore cellular migration. DCX deletion results in alterations of the RMS, instigating a thickening of the RMS by the double of its size; it also results in a multipolar neuroblast morphology that correlates to a pause in migration. Furthermore, the migratory cells suffer from an unpaired nuclear translocation towards the centrosome and undergo some defects in the length of leading processes, which suggests a failure in neurite stabilization (Koizumi *et al.*, 2006).

2.2.2.6 A Desintegrin And Metalloproteinase (ADAM) protein: Propelling migration

ADAM proteins are transmembrane proteins formed by metalloprotease and desintegrin domains. These molecules appear to be related to neuroblast migration, given the fact that other proteins involved in this process require cleavage-dependent activation promoted by ADAMs. The possible mechanism of regulation might be the high capacity of ADAMs to bind to integrins, which, as previously mentioned, are fundamental to this process (Yang *et al.*, 2006). ADAMs are widely expressed in the CNS and approximately 17 different ADAMs have been detected in the CNS, however, so far, only ADAM2 and ADAM21 are directly related to SVZ-RMS cell dynamics (Murase *et al.*, 2008; Yang *et al.*, 2005).

ADAM2 is a protein expressed in the RMS neuroblasts. Lack of ADAM2 results in defects in migration and morphological alterations of the RMS, which is caudally thicker and rostrally thinner, as seen in ADAM2-null mice. ADAM2 probably regulates migration by promoting polarized cell morphology that enables migration, since ADAM2-knockout mice present short leading cell processes and a slow cell migration rate compared with neuroblasts in wild type mice (Murase *et al.*, 2008).

ADAM21 is a protein expressed in ependymal and SVZ cells. The cell processes extending from the SVZ that express ADAM21 are surrounded by glial cells and project to blood vessels that course along the RMS. They are associated with integrin $\alpha6\beta1$ in neuroblasts and its location among neural progenitors and neuroblasts suggest that ADAM21 is involved in both cell proliferation and migration (Yang *et al.*, 2005).

2.2.2.7 Chondroitin Sulphate Proteoglycans (CSPGs): Confinement of neuroblast migration

Chondroitin sulfate is a sulfated glycosaminoglycan (GAG), composed of repeated disaccharide units of glucoronic acid (GlcA) and N-acetylgalactosamine (GalNAc). It is commonly attached to proteins as part of a proteoglycan and is a major component of the ECM that interacts with other proteins due to its negative charges (Viapiano & Matthews, 2006).

The adult brain is composed of a type of glia hallmarked by the expression of chondroitin sulfate proteoglycan NG2; which are called NG2 cells. These glial cells are widely distributed in the CNS, are present in multiple branched processes and have the capacity to differentiate into oligodendrocytes (Dawson *et al.*, 2003).

NG2 cells are aligned along the border of the SVZ and are more abundant as they are further distal to the RMS, being higher in the OB. This spatial pattern suggests a correlation between neuroblast maturity and the presence of NG2 cells. Furthermore, in the glomerular cell layer of the OB, these cells are in direct contact to cells expressing immature and mature neuronal markers (DCX, PSA-NCAM and the neuronal nuclear antigen, NeuN). The spatial organization of NG2 in SVZ-RMS raises the possibility that these cells confine the migration of neuroblasts within the RMS and prevents its dispersion (Komitova *et al.*, 2009) However, to make this assumption further research is required.

2.2.2.8 Slit and Robo proteins, chemorepulsive interactions for appropriate migration

Slit and Roundabout (Robo), ligand and receptor respectively, are evolutionarily conserved proteins in *Drosophila* and vertebrates. In mammals, Slit1-3 and Robo1-3 have been identified and implicated in axonal repulsion and cell guidance (Brose and Tessier-Lavigne 2000). Slit is a secreted protein that binds directly to Robo and functions as a chemorepellent for OB axons (Li *et al.*, 1999).

Slit and Robo function on the migratory pathway just started to be unraveled through *in vitro* experiments. Brain explants show that Slit 1 and Slit2 are expressed in the septum, where it repels progenitor cells rising from the SVZ and maintained along the RMS (Wu *et al.*, 1999). Moreover, Slit2 is expressed in the choroid plexus where it repels neural progenitors (Hu, 1999). Given the fact that Robo2 and Robo3 receptor expression has been determined in the SVZ and RMS (Marillat *et al.*, 2002), it is presumable that a Slit-Robo interaction is occurring in the migratory stream.

Slit1 deficient mice contributed to clarify the role of Slit1 in SVZ migration. Neuroblasts raised from the SVZ of knockout Slit1 mice migrate caudally to the corpus callosum, rather than to the RMS, which supports the idea that Slit1 plays an important role in directing migration. Moreover, Slit1 is also expressed in type A and type C cells within the SVZ and RMS, indicating that Slit presence is not limited to chemorepulsion activity, but possibly in parallel way, act as a individual cell migration inhibitor and might maintain chain migration integrity because Slit-mutant neurospheres migrate farther and in a disperse manner (Nguyen-Ba-Charvet *et al.*, 2004). Therefore, the evidence suggests that Slit1 and Slit2 are involved at least, in the beginning of the cell migration pathway from the SVZ towards the OB, orchestrating migration through a concentration gradient (Wu *et al.*, 1999).

2.2.2.9 Semaphorin-Neuropilin complex: Does it regulate SVZ-RMS migration?

Semaphorins are axonal guidance molecules with attractant or repellent activity that participate in early development, angiogenesis and cell migration (Tamagone & Comoglio, 2000). In vertebrates, semaphorins are divided in two groups; class 3 for secreted semaphorins and classes 4 to 7, which include transmembrane semaphorins (Raper, 2000). There is one receptor family manly involved in the regulation of semaphorin responses in the CNS: The neuropilins, NP1 and NP2 (De Wit & Verhaagen, 2003; Raper, 2000). Neuropilins are preserved throughout the entire adulthood (Giger *et al.*, 1998) and it is well known that these receptors maintain and stabilize neuronal connections and prevent axonal sprouting (Giger *et al.*, 2000; Wit & Verhaagen 2003).

Semaphorin 3A and its homodimer receptor NP1 are present along the entire RMS in the adult brain; they appear to be related to the regulation of neuroblast migration. NP1 is located in endothelial cells and binds to VEGF, an important angiogenic factor (Soker *et al.*, 1998). Therefore, it has been suggested that semaphorin 3A modulates angiogenesis. This might support the notion that the guidance of migrant neuroblasts chains could be regulated by semaphorin 3A through the indirect action of remodeling blood vessels (Melendez-Herrera *et al.*, 2008).

Complex/Molecule		Migration /Participate in proliferation(*)	Cell type expression: Receptors	Cell type expression: Ligands	Reference
Growth Factors	**EGF-EGFR**	Negative effect on neuroblast migration (*+)	EGFR:C cells, neuroblasts, and astrocytes	Neuronal precursor cells	Craig *et al.*, 1996; Doetsch *et al.*, 1999; Doetsch *et al.*, 2002; Gritti *et al.*, 1999; Kim *et al.* 2009; Morshead *et al.*, 2003; Reynolds & Weiss, 1992; Tropepe *et al.*, 1997
	EGF-TGFα	Neuroblast migration (*+)		TGFα:Astrocytes	
	BDNF-TrkB	Autophosphorylation induces motility on SVZ neuroblasts (*+)	TrkB:migrating neuroblasts TrkB-T1: astrocytes and ependymal cells	BDNF:Endothelial cells	Maisonpierre *et al.*, 1990; Giuliani *et al.*, 2004; Chiaramello *et al.*, 2007; Gascon *et al.*, 2007; Young *et al.*, 2007; Galvão *et al.*, 2008; Snapyan *et al.*, 2009
	BDNF-P75NTR	Promotes neuroblast migration	P75NTR: Neuroblasts, astrocytes and C-cells		
	FGF-2	Cell guidance with VEGF interaction (*+)	FGFR: Neural precursors (Nestin+)	FGF-2:Glial cells	Mudó *et al.*, 2007; Frinchi *et al.*, 2010
	VEGFR-VEGF	Neuroblasts guidance (*+)	VEGFR: Neuroblasts	VEGF:Astrocytes, endothelial cells	Jin *et al.*, 2002; Zhang *et al.*, 2003; Schänzer *et al.*, 2004; Schmidt *el al.*, 2009
Tyrosine kinase receptors	**Ephrins**	Chain organization (*-)	EphB3:C cells and neuroblasts EphA7:Ependymal cells and astrocytes	Ephrin B:Astrocytes Ephrin A2:Neuroblasts	Conover *et al.*, 2000; Holmberg *et al.*, 2005
	Erb	ErbB4:Chain organization NRG1:Neuroblast aggregation NRG2:SVZ-cell organization (*+)	ErbB4:Neuroblasts, C cells, ependymal cells	*NRG1*:Neuroblasts *NRG2*:Neuroblasts and astrocytes	Anton *et al.*, 2004; Ghashghaei *et al.*, 2006
Integrins	**Integrins**	Neuroblast aggregation and chain formation (*+)	α6β1-integrin: Neuroblasts and NSCs	Laminin:Recruited to the cell surface of neuroblasts Highly abundant around blood vessels in endothelial cells	Jacques *et al.*, 1998; Emsley & Hagg, 2003; Belvindrah *et al.*, 2007; Shen *et al.*, 2008

Complex/Molecule		Migration /Participate in proliferation(*)	Cell type expression		Reference
			Receptors	Ligands	
Immunoglobulin superfamily	**PSA-NCAM**	Chain formation. Cell adhesion for translocation	-	Neuroblasts and non-migrating glial progenitors	Seki and Arai 1993; Hu *et al.*, 1996; Chazal *et al.*, 2000; Petridis *et al.*, 2004
Microtubule-associated proteins	**DCX**	Stabilization of microtubules. Bipolar morphology for nuclear translocation	-	Neuroblasts	Francis *et al.*, 1999; Gleeson *et al.*, 1999; Moores *et al.*, 2004; Ocbina *et al.*, 2006
Extracelular matrix molecules	**ADAMs**	Chain formation. Maintains cell morphology	Integrins	ADAM2:Neuroblasts (RMS) ADAM21:Ependymal and SVZ cells	Komitova *et al.*, 2009; Murase *et al.*, 2008 Viapiano *et al.*, 2006; Yang *et al.*, 2006
	CSPGs	Neuroblast migration	Integrins	NG2 glial cells	Viapiano *et al.*, 2006; Komitova *et al.*, 2009
	Tenascin-R	Radial migration in OB	Not determined	Granular layer of the OB	Saghatelyan *et al.*, 2004
Slit-Robo	**Slit-Robo**	Chemorepulsion	Robo1:OB Robo2 and Robo3: SVZ-RMS neuroblasts	Slit1 and Slit2:Septum Slit1:Type A and C cells	Li *et al.*, 1999; Wu *et al.*, 1999; Marillat *et al.*, 2002; Nguyen-Ba-Charvet *et al.*, 2004
Semaphorin-NP	**Semaphorin-NP**	Suggested that modulates neuroblast migration	NP1:Endotelial cells (RMS)	Semaphorin 3A: Endothelial cells (RMS)	Tamagone & Comoglio 2000; Meléndez -Herrera *et al.*, 2008
Reelin/ApoER2-VLDLR	**Reelin/ ApoER2-VLDLR**	Radial migration in the OB by cell detachment	ApoER2-VLDLR: Neuroblasts	Reelin:Mitral cells of the OB	D´Arcangelo *et al.*, 1999; Hack *et al.*, 2002; Simó *et al.*, 2007
Prokineticin2	**PK2/Prokr2**	Cell detachment, radial migration in OB	Prokr2: OB Glomerular layer	PK2: OB Glomerular layer	Ng *et al.*, 2004; Prosser *et al.*, 2007

(*+) Increase cell proliferation, (*-) Decrease cell proliferation

Table 1. Regulation of NSCs proliferation and neuroblast migration in the SVZ-RMS pathway.

2.2.2.10 Radial migration in the OB: Reelin, Tenascin-R and Prokineticin2 as detachment signals

After completing chain migration through the RMS, neuroblasts finally arrive to the OB, where a shift in migration occurs, from a tangential to radial direction. To accomplish this cellular step, a neuroblast detachment signal is required. The most described ligand-receptor complex involved in this process is reelin/ApoER2-VLDLR. Reelin is an ECM secreted glycoprotein expressed in mitral cells of the OB. The surface receptors for reelin are the apolipoprotein E receptor 2 (ApoER2) and the very low-density lipoprotein receptor (VLDLR), both expressed in migrating neuroblasts. To allow the migratory switch, binding of reelin to its receptors induces phosphorylation of the intracellular adaptor protein disabled-1 (Dab1) and Src family kinases (SFK) (D´Arcangelo *et al.*, 1999, Hiesberger *et al.*, 1999).

In vitro, SVZ explants supplemented with reelin exhibit a loss of neuroblast chain formation, which gives raise to individual cell migration. Furthermore, *in vivo* observations of reeling-null mice reveal that neuroblasts fail to migrate radially in the OB and remain in clusters (Hack *et al.*, 2002). Additionally, the experimental overexpression of reelin by grafting reelin-expressing cells in the SVZ produces a dispersion of neuroblasts around the ventricular structures by chemokinetic activity and detachment (Courtes *et al.*, 2011).

Recently, another function of reelin has been hypothesized. It was demonstrated that reelin expression is reactivated after brain injury (focal demyelinization of corpus callosum) in mature neurons at the proximal damaged area. It is suggested that reelin enhances chemoattraction exerted by lesion-derived cytokines that contribute to neuroblast recruitment in the boundary of the damaged area. All together, these results support reelin's performance as a detachment and migration key factor (Courtes *et al.*, 2011).

There are other molecules involved in radial migration in the OB; one of these molecules is tenascin-R, an ECM glycoprotein expressed in the granular layer of the OB, where it initiates neuroblast chain detachment and radial migration. Tenascin-R-null mutant mice show a cell reduction in the granular layer, whereas neuroblasts remain in clusters in the OB (Saghatelyan *et al.*, 2004).

Finally, other molecule involved in OB migration is Prokineticin2 (PK2), which is expressed in the granular and periglomerular layers of the OB and acts as detachment signal and chemoattractant. PK2 receptor-null mice (Prokr2) exhibit a decrease in the volume of the OB and have an abnormal accumulation of neuroblasts around the olfactory ventricle. This suggests a deficiency in neuroblast migration and defects in chain migration detachment (Prosser *et al.*, 2007). Furthermore, *in vitro*, SVZ explants co-cultured with cells obtained from the glomerular layer of the OB, begin cell migration toward the glomerular layer of the explants; whereas cells from the glomerular layer of PK2-null mice do not exhibit chemotactic activity (Ng *et al.*, 2005).

3. Growth factors as an endogenous approach for neurorepair therapy

3.1 Endogenous response after stroke

We just described in the previous section the existence of a substantial number of molecules and its complex signaling regulating PMD response at the SVZ under basal conditions.

However, studies also reveal that neurogenesis could be triggered secondary to specific injury conditions, including stroke. Focal cerebral ischemia promotes neurogenesis in the DG of the hippocampus and in the SVZ (Arvidsson *et al.*, 2002; Jin *et al.*, 2001; Liu *et al.*, 1998; Parent *et al.*, 2002; Zhang *et al.*, 2001); both being a feature shared by rodent and the human brain (Jin *et al.*, 2006; Martí-Fabregas *et al.*, 2010). Moreover, these studies demonstrate that the default migratory pathway, followed by neuron precursors from the SVZ that supply the OB region, can be partially diverted to other destinations including the striatum and cortex after stroke injury. Recently, it was demonstrated that stroke enhances long-term neurogenesis, although, decreased in magnitude when compared to the acute phase (Thored *et al.*, 2006). These findings are of paramount relevance for their therapeutic potential in the field of neuronal damage reestablishment, by taking advantage of an endogenous cell source. However, Ardvisson *et al.*, in their seminal work, determined that within 6 weeks of transient focal stroke uniquely 0.2% of newborn neurons from the SVZ were integrated in the damaged striatum. Thus, neurogenesis occurring after cerebral ischemia represents an insufficient cell source for the purpose of neuronal replacement therapy (Ardvisson *et al.*, 2002).

3.2 Self-renewal induced by growth factors: The endogenous neuronal repair

The rationale for self-renewal induced by growth factors relies on the existence of neurogenic niches in the brain with the potential to modulate the proliferative response by either injury or growth factors. The purpose of this approach has been to determine whether growth factors are able to amplify the endogenous response of NSCs in a meaningful level that can account an increase of neuronal differentiation. The concept however, was not new in the field of stroke studies. One of us, Dr. James Fallon (Fallon *et al.*, 2000) employed the 6-hydroxydopamine lesion rodent model of Parkinson's disease, where he and collaborators demonstrated that proliferating cells significantly increased in the striatal SVZ ipsilateral to the injured side, however, this increase was exclusive in animals that also received an striatal infusion of TGF-α; neuroblasts then migrated in mass into the striatum to become neurons.

There is a long and increasing list of growth factors that has the potential to be implemented as a therapeutic tool in the recovery process after a stroke event, exploiting their neuroprotection and neurogenesis features. Here, we are focusing on studies that unveil that growth factors are modifying neurogenesis subsequent to the onset of focal cerebral ischemia, specifically in the SVZ-RMS. Table 2 represents a chronological compendium of such studies. The hematopoietic factors known to compel neurogenesis were omitted in our chapter, nonetheless, they are extensively reviewed by others (Greenberg & Jin, 2006).

The majority of the research described in this section of the chapter employs the most common method to study focal stroke, the middle cerebral artery occlusion (MCAO) model, which affects the striatum and/or cortex, resembling the injury that commonly occurs in stroke patients (Figure 1). To induce this kind of ischemia, an incision in the neck is made and after exposition of the common, external and internal carotid arteries, a monofilament suture is then carefully introduced via the external carotid artery through the lumen of the internal carotid artery until it reaches and occludes the middle cerebral artery (MCA).

3.2.1 EGF family of growth factors

It is important to highlight that in practically all the protocols regarding neurogenesis, the quintessential strategy to identify newborn cells is the administration of bromodeoxyuridine (BrdU), a marker of cell division, which intercalates into the DNA of cells that are undergoing proliferation. The neuronal (or glial) lineage of BrdU-labeled cells is determined by the co-staining with specific markers of immature or mature phenotypes, which may differs depending on the goal of the study.

3.2.1.1 EGF

Teramoto and collaborators reported in 2003 the first study about the effect of a growth factor on neurogenesis in the SVZ after an ischemic insult (Teramoto *et al.*, 2003). The antecedents implicating EGF as a factor that promotes NSCs proliferation and migration from the SVZ to the striatum lead them to test if EGF could induce a PMD response in the damaged brain after an ischemic event. They administered an intraventricular dose of EGF two days after stroke for a period of one week that allow a noteworthy neuronal replacement increase after week 13th post MCAO. Interestingly, Teramoto determined that the BrdU+/NeuN+ cells observed at the boundary zone of the stroke lesion were not DARPP-32 (dopamine and adenosine 3′,5′-monophosphate-regulated phosphoprotein of 32KD) positive, the prevailing neuronal population in the striatum. Rather than, they found 65% of the new cells maturated into aspiny parvalbumin-containing (PV) interneurons. This effect is intriguing, given the fact that *in vitro,* EGF stimulates the DARPP-32 phenotype (Reynolds & Weiss, 1992). Therefore, albeit EGF is probably directing the mechanisms related to neuronal commitment and fate specification, such claims need further study.

3.2.1.2 Heparin-binding epidermal growth factor-like growth factor (HB-EGF)

Another member of this family is the HB-EGF that also has a clear effect in increasing proliferation of NSCs in the SVZ. However, its effect on migration and differentiation is not fully disclosed. In one study, Jin and coworkers analyzed such effect 4 weeks after stroke and observed an increase in the number of BrdU cells (approximately 40%) and the number of BrdU cells co-localizing with the immature neuronal marker TUC-4 in the SVZ. Nonetheless, a concomitant decrease of approximately 60% of neuroblasts (DCX+) migrating towards the striatum was also observed. The authors suggested that this effect is probably due to its chemoattractant properties. In spite of the decrease in migration, a significant improvement in neurological outcome was accounted, probably due to a reduction in infarct size promoted by the growth factor (Jin *et al.*, 2004). Later on, another study conducted by Sugiura and coworkers determined that recombinant adenovirus-expressing HB-EGF promotes neurogenesis and angiogenesis in the SVZ. Nonetheless, opposite to the study developed by Jin *et al.*, no effect limiting migration of newborn cells toward the striatum was reported. Therefore, also in contrast, an increase of neuronal cells was observed in the ischemic striatum, going from 2 BrdU+/NeuN+ cells to 23 cells per mm^2 (Sugiura *et al.*, 2005) Additionally, a functional recovery was observed. The discrepancy in both results still remains to be discerned, however, it could be due to the different methods employed to deliver the growth factor, which impose differences in concentration and availability.

3.2.1.3 TGF-α

The first report of TGFα's effect on neurogenesis in rats that were subjected to transient MCAO, was made by our group (Guerra-Crespo *et al.*, 2009). The majority of stroke studies analyze the effect of growth factors in the acute phase of ischemia. However, we were interested in the delayed administration of the growth factor, since it resembles the typical situation in humans with preexisting stroke injuries, who might benefit from this type of therapy. Therefore, we infused TGFα directly in the striatum four weeks after injury, when the infarction area was no longer expanding and the cellular deterioration had stabilized, indicating that the acute phase was completed and the chronic phase had begun.

Eight weeks after MCAO, and four weeks after the onset of TGFα administration, we found a 4-fold increase of BrdU labeled cells arising from the ependymal layer and the SVZ. Many of the BrdU-labeled cells of the SVZ and others under migration were expressing the immature neuronal marker Meis2, a transcription factor that is strongly expressed in striatal precursors. Additionally, we found around the site of the infarction about a 7-fold increase of BrdU cells that were co-labeled with the neuronal-fate marker NeuN, whereas, several of the newborn cells co-labeled with DARPP-32, indicating that they differentiate into striatal neurons, which are typical for this brain region. These results indicate that TGFα treatment significantly increased the yield of neurons produced in the injury response. Although we did not examine the long-term survival of these neurons, an approximately 90% of behavioral recovery (corner and cylinder test) in the chronic animal suggests that many of them became functionally and integrated in the host's CNS (Guerra-Crespo *et al.*, 2009).

In the same year, another group (Leker *et al.*, 2009) reported an increase in neurogenesis as well as angiogenesis induced by TGFα. Leker *et al.*, administered an intraparenchymal dose of the growth factor 1 day after ischemia, for a consecutive period of 14 days and analyzed the long-term response. TGFα increased the number of BrdU cells and allow a 2-fold increase of neuroblasts in the ischemic hemisphere. Nonetheless, only a slight number of newborn neurons in proximity to the infarct border were observed, suggesting that under these specific experimental conditions, TGFα leads to a moderate but significant neuronal differentiation. Concomitant to neurogenesis, TGFα expanded (2.4-fold) the area covered by blood vessels in the ischemic border zone. The mechanism involved the recruitment of endothelial bone marrow-derived cells into newly formed cerebral blood vessels.

3.2.1.4 Intranasal infusion of TGFα

The employment of growth factors in patients is impeded by the facts that intracranial infusion is impractical and that many growth factors intravenously administrated are not able to cross the blood-brain barrier (BBB). Even though there are methods for bypassing the BBB, they typically consist of invasive neurosurgeries that would restrict clinical application to the most severe cases. Noninvasive techniques that are capable of delivering growth factors to the CNS represent a therapeutic alternative to surgery. A pioneer work was the intranasal administration of neurotrophic growth factor (NGF), which demonstrated that delivered growth factors could bypass the BBB. Since then, the prevalence of the intranasal administration technique for CNS treatment has grown considerably (Capsoni *et al.*, 2002; De Rosa *et al.*, 2005; Frey *et al.*, 1997; Liu *et al.*, 2004; Ma *et al.*, 2008). With this approach of administration, therapeutic molecules traverse the BBB through the olfactory pathway and the less-studied trigeminal neural pathway (Thorne *et al.*, 2004). The advantage of this method is that factors are delivered directly into the brain and thereby avoid adverse systemic effects. The simplicity of the intranasal administration of growth factors makes it an outstanding strategy. This non-surgical approach represents a potential therapeutic strategy for human patients.

Based in this knowledge, we delivered an intranasal dose of a pegylated form of TGFα (PEG-TGFα) to make it more stable for the nasal route. We found that intranasal delivery is a viable alternative because PEG-TGFα was able to induce the proliferation and migration of neural progenitors to the damaged striatum (in terms of BrdU incorporation and nestin expression), although of less magnitude compared to intracranial TGFα administration. This finding is associated with significant behavioral improvement in the MCAO model, measured by the corner and cylinder test. Therefore, intranasal delivery of PEG-TGFα holds great therapeutic potential.

3.2.2 FGF-2

From the FGF family, FGF-2 is the member which neurogenic properties have been demonstrated. The first study, performed in 2003, reported that, regardless of a 30% increase of BrdU+/DCX+ cells, no BrdU+/NeuN+ cells were observed (Wada *et al.*, 2003). In the same year, Matsuoka and collaborators also reported an increase (2.1 fold) of BrdU+ cells in the SVZ of a global ischemia rodent model, through FGF-2 gene delivery by an adenoviral vector. Although neuronal differentiation in the striatum was not analyzed, a small number of newborn cells (3%) were labeled positive for NeuN amongst the different layers of the cortex and by the proximity to the SVZ suggesting that such cells were derived from the SVZ (Matsuoka *et al.*, 2003). Matsuoka's research set a precedent on gene therapy field, since was the first time that a growth factor was administered through viral vector in a stroke model

Both former works were focused on the study of acute phase of stroke, where a short infusion period of the growth factor or a short time analysis after the growth factor-adenovirus transduction was made. More recently, Leker and coworkers employed an ischemia model with predominant cortical damage and an adenovirus delivery system that allowed overexpression of FGF-2 for a long time period. They were able to observe that FGF-2 increased proliferation and migration from the SVZ and that the immature neurons were localized in the border of damaged cortex when analyzed at 30 and 90 days after stroke. Additionally, the results shown that the group treated with FGF-2 presented a 2-fold increase of newborn cells expressing NeuN and even a higher increase (22 *versus* 1.6%) of cells labeled with the immature neuronal marker Hu. In either Matsuoka *et al.*, or Leker *et al.*, studies, no lineage analyses was performed to asseverate without a doubt, that such cells were originated in the SVZ, since they could be also produced in the cortex. Nevertheless, the evidence suggests that the majority of the cells arose from the SVZ (Leker *et al.*, 2007).

The previous studies disclose the enhancement achieved on endogenous neurogenesis in the SVZ when EGF or FGF-2 are administered independently. However, a more interesting fact would be to analyze the simultaneous effect of both growth factors in order to increase neurogenesis. Nakatomi *et al.*, in a breakthrough work addressed that question in a transient ischemia model with specific pyramidal hippocampal damage. The results demonstrated that endogenous proliferation and migration are enhanced by intraventricular co-administration of growth factors after stroke. However, that the growth factor induced neuronal regeneration of approximately 40% of the CA1 pyramidal layer, one month after stroke, was a remarkable finding (Nakatomi *et al.*, 2002). This study was centered in the analysis of the hippocampus; however, only a few years later, Baldauf and Reymann examined the combined effect of EGF/FGF-2 on the SVZ. They observed an increase (almost double) in the number of BrdU+/DCX+ cells in the ipsilateral striatum, in spite of a concomitant increase in the infarct volume. Unfortunately, neuronal fate and behavior were not analyzed; therefore, further analyses are required to conclusively discern the effect of the synchronic administration of EGF and FGF-2 for striatum neuronal replacement (Baldauf and Reymann 2005).

3.2.3 Neurotrophic factors: BDNF

The neuroprotective effects of the intraventricular infusion of BDNF after global ischemia were reported for the first time in 1994 (Beck *et al.*, 1994). However, the role of BDNF in neurogenesis was recently demonstrated. In one report (Gustafsson *et al.*, 2003), BDNF expression was successfully attained in the substantia nigra by means of an adenovirus; the objective was to transport BDNF anterogradely to the striatum and avoid damage in its cytoarchitecture. The transduction of the viral vector took place 4 to 5 weeks prior to the MCAO. With this strategy, they found a significant increase of neuroblast migration towards the striatum at 2 weeks after ischemia. Additionally, a higher number of BrdU cells co-laballed with early neuronal marker Hu and striatal neuronal marker Meis2 were observed. The significant relevance for replacement therapy cannot be established with these findings, because the transduction of the viral vector generated anomalous behavior patterns and therefore, the behavioral outcome cannot be correctly analyzed. Moreover, the high expression levels of BDNF aggravated the cellular death of cholinergic, PV and neuropeptide Y interneurons in the striatum, which could override the neuroprotective effects observed in former studies performed by the same research group (Andsberg *et al.*, 2002).

Some years later, in another attempt to determine the effect of BDNF on neurogenesis, Schavitz and coworkers induced a parietal cortical lesion, and after an intravenous injection of BDNF on the following 5 days, were able to observe precursor cells rising from the SVZ that were migrating toward the ipsilateral striatum. A substantial number of neuroblasts were recruited to the ischemic hemisphere 37 days after the last injection; still, neuronal replacement was not observed with this paradigm of cortical injury, since no BrdU+/NeuN+ cells were observed. Additionally, the authors were unable to detect cortical neurogenesis (Schavitz *et al.*, 2007).

In summary, a variety of growth factors have been tested in experimental stroke models and irrespective of their nature, practically all of them can induce proliferation on NSCs that reside in the SVZ. Most of them, enhance in a parallel fashion the recruitment of neuroblasts toward the peri-infarcted area, either striatum or cortex. However, extensive differences in neuronal differentiation have been observed in the damaged area (even when employing the same growth factor), ranging from a highly significant increase to a decrease in differentiation levels. The behavioral outcome, measured as an indirect index of functional integration, has been analyzed only in the minority of the studies, finding an important level of recovery in some of them.

The potential of growth factors for endogenous cell replacement is evident; however, clinical studies are not currently being undertaken, given the fact that growth factors encompass a strong and dangerous mitogenic effect.

Growth factor	Stroke Model	Delivery Method	Time period of infusion	Effect on Neurogenesis	Behavioral outcome	Reference
EGF Family						
EGF	Mouse Left MCAO	Minipump in lateral ventricle	21 days after ischemia for 7 days	- 65% of new PV interneurons 13 weeks after stroke - 100 fold of neuronal replacement	NA	Teramoto *et al.*, 2003
HB-EGF	Rat Right MCAO	Minipump in lateral ventricle	24 hours after ischemia for 3 days	- Increase of BrdU+ cells expressing TUC-4 -60% decrease in migrating neuroblasts 4 weeks after stroke	Improvement in neurological score	Jin *et al.*, 2004
HB-EGF	Rat Left MCAO	Adenoviral vector in lateral ventricle	3 days after ischemia	- 2-fold increase of vascular density (angiogenesis) - Increase of BrdU+/NeuN+ cells (2 vs 23 cells per mm^2) 28 days after ischemia	Improvement in rotarod test	Sugiura *et al.*, 2005
TGFα	Rat Left MCAO	Minipump in striatum	1 month after ischemia for 28 days (chronic treatment)	- 4-fold increase of BrdU+ cells - 7-fold increase of BrdU+/NeuN+ cells and several BrdU+/DARPP-32+ two months after stroke	90% of improvement in corner and cylinder test	Guerra-Crespo *et al.*, 2009
TGFα	Mouse Left MCAO	Minipump intra-parenchyma	1 day after ischemia for 14 days	- 2.4 fold increase in blood vessel cover area in the infarct border - Small number of newborn neurons in the infarct border 90 days post-infusion	NA	Leker *et al.*, 2009
TGFα	Rat Left MCAO	Intranasal delivery	4 intranasal doses per month (1 per week)	- Increased BrdU+ cells in SVZ and ischemic striatum - Increased neurogenesis (BrdU+/Nestin+ cells) in the peri-infarcted striatum	50% of improvement in corner and cylinder test	Guerra-Crespo *et al.*, 2010

Growth factor	Stroke Model	Delivery Method	Time period of infusion	Effect on Neurogenesis	Behavioral outcome	Reference
EGF Family						
FGF-2 (bFGF)	Rat Right MCAO (cortex)	Intracisternal injection	Injections 24 and 48 hrs after stroke	- 30% increase of BrdU+/DCX+ cells at day 7 and 2% at day 21 - No BrdU+/NeuN+ cells	NA	Wada *et al.*, 2003
FGF-2	Gerbil Global ischemia	Adenoviral vector in lateral ventricle	Transduction 3 hrs after stroke	- 2.1 fold increase of BrdU+ cells in the SVZ 7 days after ischemia - 3% of BrdU+/NeuN+ in cortex 30 days post-stroke	NA	Matsuoka *et al.*, 2003
FGF-2	Hypertensive Rat MCAO (cerebral cortex)	Adenoviral vector in the infarct border	Transduction starting the same day of ischemia (chronic treatment)	- Increase of BrdU+ cells expressing neural transcription factors MASH1 and Pax6 in the peri-infarcted area 30 days after stroke - Increase of BrdU+/Hu+ cells (22% vs 1.6%)	Improvement in motor disability score	Leker *et al.*, 2007
FGF-2+ EGF	Rat Left MCAO	Minipump in lateral ventricle	10 min after ischemia for 14 days	- Increased proliferation in the striatum - 2 fold increase of BrdU+/DCX+ cells in striatum 14 days after stroke	NA	Baldauf *et al.*, 2005
Neurotrophin						
BDNF	Rat MCAO	Adenoviral vector in SN transported anterogradely towards striatum	5 weeks previously to MCAO	- Increased neuroblasts (DCX+) - Increased neuronal death in the striatum - Increased BrdU+/Meis2+ cells 2 weeks post-ischemia	Vector-induced abnormal motor behavior	Gustafsson *et al.*, 2003
BDNF	Rat Photothrombotic ischemia (cortical stroke)	Intravenous injection	1 hour post-ischemia 2nd to 5th day	- Highly significant increase of neuroblasts in striatum 37 days after last injection - No BrdU+/NeuN+ cells	Improvement in neurological score and adhesive tape removal test	Schavitz *et al.*, 2007

Table 2. Growth factors induce SVZ neurogenesis in ischemic stroke.

4. NSCs transplant therapy: A new expectation

Neural transplantation is a promising strategy for treatment of several CNS pathologies that offers long-lasting improvement and the prospect of permanent cure. The most obvious possibility is to use neural transplantation as a technique for cell replacement therapy whereby the cells would occupy the place or the function of dead or degenerated cells. Potential advantages to this approach may include greater control over cell fate, the ability to deliver any desired number of cells, and reduced risks associated with mitogen infusion. A number of different cell types have been considered for cell transplantation with goals ranging from replacement of host circuitry to delivery of neuroprotective or immunomodulatory compounds.

The majority of studies to date have shown relatively limited cell replacement from endogenous NSCs. Further, the technology for mobilizing endogenous NSCs is relatively new. In contrast, work has been in progress for decades to replace lost neural cells by transplantation of either fetal brain tissue or more recently, NSCs.

Recent studies have highlighted the enormous potential of cell transplantation therapy for stroke. In this branch of the chapter, we will describe the experimental trials that utilized NSCs in the MCAO model, which placed the stepping stone for the first human trials of NSCs transplant therapy.

Fetal brain tissue transplants have been shown to produce some recovery in animal models of stroke (Mattsson *et al.*, 1999; Nishino *et al.*, 2000; Riolobos *et al.*, 2001), but ethical considerations and a short supply of human fetal tissue limited this approach. As a result, a variety of cell types have been tested in stroke models, they include human bone marrow cells, human umbilical cord blood cells (Chen *et al.*, 2001a; Chen *et al.*, 2001b; Savitz *et al.*, 2002; Zhao *et al.*, 2002), rat trophic factor-secreting kidney cells (Mattsson *et al.*, 1999; Nishino *et al.*, 2000; Riolobos *et al.*, 2001; Savitz *et al.*, 2002), and immortalized cell lines such as the human neuron-like NT2N (hNT) cells (Borlongan *et al.*, 1998; Saporta *et al.*, 1999) and MHP36, an embryonic murine immortalized neuroepithelial cell line (Modo *et al.*, 2002; Veizovic *et al.*, 2001). In spite of the vast types of transplanted cells employed, they yet need to demonstrate a significant behavioral recovery in animal models of stroke and a long-lasting survival of the grafted tissue (Table 3).

From the previously mentioned cell types employed for transplant, human cells that have been used in these studies fall into 3 categories: A) Neural stem/progenitor cells (NPCs) cultured from fetal tissue; B) immortalized neural cell lines, hematopoietic/endothelial progenitors and stromal cells isolated from bone marrow, umbilical cord blood, peripheral blood or C) adipose tissue. Even though transplanted human cells have shown promise, other sorts of cells have arisen to address the need for a quintessential cell source for transplant therapy.

Amongst them, NSCs have been proposed as a potential source of new cells to replace those lost due to central nervous system injury such as stroke, as well as a source of trophic molecules to minimize damage and promote recovery in clinical trials.

In the background of this imperative clinical need, hundreds of studies have recently published the therapeutic potential of either endogenous or transplanted NSCs in laboratory models of stroke. To their advantage, NSCs have the capacity to respond actively to their environment, migrate to areas of injury, and secrete neuroprotective compounds. Such properties may afford them therapeutic potential both in the acute phase and at later time points when the employment of conventional medical therapies would no longer be effective.

NSCs can be isolated from many regions of the CNS of embryonic as well as adult mammals. As mentioned in other section of the chapter, they can be propagated in culture in the presence of EGF and/or FGF-2 as proliferative clusters of cells termed neurospheres. Recent studies have demonstrated that rather than being homogeneous aggregates of stem cells, neurospheres actually represent a heterogeneous collection of cells including true stem cells, committed progenitors, and differentiated progeny. This is in contrast to embryonic stem cells (ESCs), which in the presence of appropriate signaling molecules can be maintained as a relatively homogeneous population of stem cells. NSCs also differ from ESCs in terms of the variety of neurons they can generate. Profiles of NSCs gene expression tend to point to NSCs expanded as neurospheres in EGF and FGF-2 as adopting a forebrain profile.

Consistent with this, attempts to differentiate NSCs into cells from other regions of the CNS, such as dopaminergic neurons, cerebellar Purkinje cells, or motoneurons have in most cases been unsuccessful. Nevertheless, NSCs can be successfully differentiated into representative cell types in parts of the brain most commonly affected by stroke, such as cortical projection neurons (Englund *et al.*, 2002), interneurons (Scheffler *et al.*, 2005) and hippocampal pyramidal neurons (Corti *et al.*, 2005). Retrograde labeling, synaptic integration, and action potential generation from NSCs-derived neurons has been demonstrated *in vivo* (Englund *et al.*, 2002).

Given the fact that NSCs have the capability to differentiate into neurons (Kelly *et al.*, 2004; Song *et al.*, 2002a; Song *et al.*, 2002b), astrocytes (Eriksson *et al.*, 2003; Herrera *et al.*, 1999; Winkler *et al.*, 1998), oligodendrocytes (Pluchino *et al.*, 2003; Yandava *et al.*, 1999), and perhaps endothelium (Wurmser *et al.*, 2004), advocates that conception that NSCs should be capable of replacing most of the cell types affected by an ischemic injury.

4.1 From theory to practice: NSCs transplant in rodent models of stroke

Actual results in preclinical studies, however, have been quite varied. NSCs, including human, can clearly survive after transplantation, have a tendency to migrate toward areas of infarct (Kelly *et al.*, 2004), and can generate functional neurons (Englund *et al.*, 2002) that may form connections with host cells (Park *et al.*, 2002). Although several studies have found NSCs to predominantly differentiate into glia after transplantation into normally non-neurogenic regions (Eriksson *et al.*, 2003; Herrera *et al.*, 1999; Winkler *et al.*, 1998), robust neural differentiation has been observed after transplantation of cells cultured on laminin (Wu *et al.*, 2002; Yan *et al.*, 2007).

Most studies have not observed substantial changes in infarct size after NSCs transplantation (Kelly *et al.*, 2004; Pollock *et al.*, 2006); however, neuroprotective (Lee *et al.*, 2007; Ourednik *et al.*, 2002) and immunomodulatory (Fujiwara *et al.*, 2004; Pluchino *et al.*, 2005) effects of NSCs in addition to their potential for at least some cell replacement (Sinden *et al.*, 1997), have collectively yielded beneficial effects in multiple animal models of neurodegeneration and brain injury, including stroke (Chu *et al.*, 2004; Pollock *et al.*, 2006; Sinden *et al.*, 1997).

NSCs transplantation enhances endogenous cell proliferation in the SVZ and promotes angiogenesis in the peri-infarct zone of adult rats, even if it is performed in the acute phase of ischemic injury. In addition, this transplanted NSCs managed to survive, migrate, differentiate, and also induce improvement in neurological functions (Zhang *et al.*, 2009a, 2009b; Zhang *et al.*, 2010). Grafted NSCs enhanced the number of BrdU-positive cells in ischemic ipsilateral SVZ at 7 days after transplantation, an effect that persisted to at least 14 days post-transplantation. These results revealed that NSCs transplantation increases cell proliferation in the SVZ and promotes angiogenesis in the peri-infarct zone after focal cerebral ischemia in adult rats. The reason that grafted NSCs increase endogenous NSPCs proliferation may be due to the production of certain growth factors or repression of inflammation and apoptosis.

In addition to promote proliferation, it has been shown that human fetal striatum derived NSCs, transplanted as neurospheres, survive in stroke-damaged rat striatum, migrate toward the site of the injury, and differentiate into mature neurons in the absence of tumor formation (Darsalia *et al.*, 2007).

Evidence of cell migration to the site of ischemic injury from administration by various routes has been seen, including intravenous, intraarterial, and intraparenchymal brain injection. Migration potential may differ according to cell type and route, but this has not been systematically explored for most cell lines. Animal studies with a number of NSCs lines have shown evidence of cell survival and in some studies improvement in behavioral outcomes after focal ischemic injury (Bacigaluppi *et al.*, 2008; Borlongan *et al.*, 1998; Chu *et al.*, 2004; Ishibashi *et al.*, 2004; Jiang *et al.*, 2006; Modo *et al.*, 2002; Pollock *et al.*, 2006; Wei *et al.*, 2005).

In a follow-up work, Darsalia *et al.* showed that transplantation shortly after stroke (48 hours) resulted in better cell survival than did transplantation 6 weeks after stroke, but the delayed transplantation did not influence the magnitude of migration, neuronal differentiation, and cell proliferation in the grafts. Additionally, transplanting greater numbers of grafted NSCs did not result in a greater number of surviving cells or increased neuronal differentiation. They observed a substantial number of activated microglia 48 hours after the insult in the injured striatum, but reached maximum levels 1 to 6 weeks after stroke (Darsalia *et al.*, 2011). Their findings show that the best survival of grafted human NSCs in stroke-damaged brain requires optimum numbers of cells to be transplanted in the early post stroke phase, before the inflammatory response is established.

In an attempt to improve transplant survival and behavioral outcome, Jin *et al.* found that intralesional transplantation of nestin/Sox2-immunopositive neuronal precursor cells (NPCs) derived from BG01 human embryonic stem cells 3 weeks after distal MCAO in rats reduced infarct volume and improved behavioral outcome 4–9 weeks post-transplant (Jin *et al.*, 2010a). In another study, they found that the beneficial effects of transplantation occurred in both young adult (3-month-old) and aged (24-month-old) rats (Jin *et al.*, 2010b).

NPCs express many factors known to influence neurite plasticity and thus have the potential to enhance structural plasticity after stroke. With that notion in mind, Andres *et al.* decided to analyze the effects of transplanted NPCs on structural plasticity and axonal transport in the ischemic rat brain. They found that NPCs transplant one week after the ischemic event enhanced dendritic plasticity in both the ipsi- and contralesional cortex. Moreover, stem cell-grafted rats demonstrated increased corticocortical, corticostriatal, corticothalamic and corticospinal axonal rewiring from the contralesional side; with the transcallosal and corticospinal axonal sprouting correlating with functional recovery (Andres *et al.*, 2011).

With the advent and enhancement of molecular biology technology, tumor-derived neuronal tissues, including immortal teratocarcinoma-derived cells such as NT2N and hNT cells have been tested in stroke models (Bacigaluppi *et al.*, 2008; Borlongan *et al.*, 1998). Human fetal cortex cells have been immortalized by genetic modification (e.g. insertion of transcription factor genes including v-myc (Cacci *et al.*, 2007) and c-myc (Pollock *et al.*, 2006).

Irrespective of cell type or route of administration, cell survival has been limited. Quantification of cell survival is poorly reported. Reports, from the few studies that have attempted to quantify cell survival, range from 1% (Hicks *et al.*, 2009) to 30% (Darsalia *et al.*, 2007). Site of transplantation may be relevant to cell survival, with pathological evidence of 33.4 ± 6.1% viable cells when human fetal neural stem cells (hNSCs) were transplanted in non-ischemic tissue medial to the ischemic lesion (Kelly *et al.*, 2004). In contrast 30–50% of hNT cells have been suggested to survive in and around the ischemic tissue (Bliss *et al.*, 2006). Inflammation following ischemic stroke may aid cell migration (Belmadani *et al.*, 2006), but a negative correlation between cell survival and inflammatory response has also been observed (Kelly *et al.*, 2004), and may be a factor to consider when timing cell administration.

Although transplanted NSCs can recover some of the function lost after stroke, recovery has shown to be incomplete and restoration of lost tissue is minimal in most of the cases. The challenge set was to provide transplanted cells with matrix support in order to optimize their ability to engraft the damaged tissue. Bible *et al.* demonstrated that plasma polymerised allylamine (ppAAm)-treated poly(D,L-lactic acid-co-glycolic acid) (PLGA) scaffold particles can act as a structural support for neural stem cells injected directly through a needle into the lesion cavity using magnetic resonance imaging-derived coordinates. Upon implantation, the neuro-scaffolds integrated efficiently within the rat host tissue forming a primitive neural tissue. These work demonstrated that neuro-scaffolds could be a more advanced method to enhance brain repair. This study provides a substantial step in the technology development required for the translation of this approach.

Other cell sources employed for this purpose are the induced pluripotent stem (iPS) cell and mesenchymal stem cells (MSCs). iPS cells can be produced with high reproduction ability and pluripotency to differentiate into various types of cells, making them a feasible resource for transplantation, with the additional benefit of obtaining these cells from the same patient. On the other hand, MSC are multipotent stem cells that can differentiate into a variety of cell types.

In one study, undifferentiated iPS cells were transplanted into the ipsilateral striatum of a MCAO model of stroke; the transplanted iPS cells expanded and formed larger tumors in the post-ischemic brain compared with the control condition. iPS cells formed a tridermal teratoma (Kawai *et al.*, 2010). Despite this finding, iPS cells are still a hopeful alternative to provide neural cells for ischemic brain injury; however, tumor formation still needs to be prevented and controlled.

In another study that employed human bone marrow-derived MSCs (hBMSCs) found a significant recovery of behavior in the hBMSCs-treated rats beginning at 14 days after MCAO compared with the control animals. High levels of BDNF, neurotrophin-3 (NT-3), and VEGF were detected in the hBMSCs-treated brain, as well as an increased proliferation of neuronal progenitor cells in the SVZ (Bao *et al.*, 2011). This indicates that it is unlikely that

MSCs replaced the damaged tissue and more likely that they secreted trophic factors that promoted functional recovery after stroke.

Altogether, these findings, therefore, placed the stepping stone for direct clinical implications. However, evidence indicates that NSCs transplantation may protect the CNS from inflammatory damage via a "bystander" mechanism rather than by direct cell replacement (Martino & Pluchino, 2006).

4.2 Human trials with stem cells

Results of NSCs transplantation in ischemic stroke patients have not been reported, although in 2010 ReNeuron, a company in the United Kingdom received an approval to start a clinical trial using expanded NSCs. Therefore, we will witness the first clinical results of NSCs transplants. Nonetheless, the potential of precursor cells as an exogenous source for transplant therapy has been already assessed in some clinical trials (summarized in table 3).

Neural cells derived from an immortalized human teratocarcinoma cell line (NT2N cells) underwent two small clinical studies after showing improved behavioral outcomes in rats (Borlongan *et al.*, 1998). In the first human safety study (Stilley *et al.*, 2004), twelve patients who had suffered ischemic stroke on average 27 months earlier, received doses of between 2 and 6 million cells by direct injection into the basal ganglia. No cell related adverse events were reported. Some motor improvement was reported in around half the subjects at 6 months based on one clinical stroke scale and an increase in 18F-flurodeoxyglucose uptake on brain positron emission tomography (PET) scans was also reported, although of unclear significance. Subsequently Kondziolka and colleagues (Kondziolka *et al.*, 2005) reported on an open-label randomized phase II efficacy trial of 5–10 million NT2N cells including 14 actively treated subjects with stroke between 1 and 6 years earlier. Patients were randomized to receive cells with rehabilitation or rehabilitation alone in a 7:2 ratio and two ascending dose arms. Ischemic and hemorrhagic stroke formed half of the subjects each. No cell related adverse events were noted, and the major adverse events reported (seizure and subdural hematoma) were considered to be procedure-related. Neurological motor scores that were stable 6 months prior to surgery were reported to improve in 6/14 subjects by 6 months while in 4/14 subjects the scores deteriorated. There was no indication of any dose-related effects.

Savitz and colleagues used the intra-parenchymal route for delivery of fetal porcine derived cells in five patients, who suffered stroke 3–10 months prior (Savitz *et al.*, 2005). The study was halted after the fourth and fifth patients had worsening motor deficits and seizures. Whether cell or procedure related complications were the cause, it still remains unclear.

Mesenchymal stem cells have been administered intravenous in a controlled trial (Bang *et al.*, 2005) and autologous cells have been transplanted intraparenchymally in an open study (Suárez-Monteagudo *et al.*, 2009). Neither trial reported any cell or procedure related adverse events up to one year follow-up. In the latter study, clinical scores showed minimal and insignificant changes. Bang and colleagues reported improvement in one functional score of activities of daily living (Barthel index scores measured at 3, 6 and 12 months after cell therapy) but not in other clinical measures of outcome (modified-Rankin scale and National Institutes of Health Stroke Scale scores) and the number of subjects was small (5 actively treated and 25 controls).

Cell Source	Stroke Area	Time period (after transplant)	Behavioral outcome	N° of surviving cells	Reference
NSCs transplant therapy in rodent models of stroke					
Human Neuron-Like NT2N	MCAO	24 weeks	3-fold improvement of Passive Avoidance Test	NA	Borlongan *et al.*, 1998
Rat Trophic Factor-Secreting Kidney Cells	MCAO	20 weeks	1 point augmentation in Prehensile Traction Test	NA	Mattsson *et al.*, 1999
Human NT Neurons	MCAO	12 weeks	50% of behavioral improvement in functional tests	24,217 ± 9,260	Saporta *et al.*, 1999
Human Umbilical Cord Blood Cells	MCAO	5 weeks	Two points reduction of Modified Neurological Severity Score	32 600 ± 1689	Chen *et al.*, 2001a
Rat Bone Marrow Cells	MCAO	2 weeks	50% of behavioral improvement in functional tests	~14% (Total of 4×10^5)	Chen *et al.*, 2001b
MHP36 Human Immortalized Cell Line	MCAO	8 weeks	2-Fold improvement in Water Maze Acquisition Test	~7500	Veizovic *et al.*, 2001
MHP36 Human Immortalized Cell Line	MCAO	4 weeks	30% of behavioral improvement in functional tests	~7500	Modo *et al.*, 2002
Human Bone Marrow Cells	MCAO	2-6 weeks	33.5 ± 8.7% of somatosensory asymmetries	NA	Zhao *et al.*, 2002
Human NSCs neurospheres	Distal MCAO	4 weeks	NA	100,147 ± 28,944	Kelly *et al.*, 2004
CTX0E03 Human Neural Stem Cells	MCAO	6-12 weeks	50% of behavioral improvement in functional tests	NA	Pollock *et al.*, 2006
iPS cells	MCAO	7 weeks	None	Tumor of $50mm^3$	Kawai *et al.*, 2010
Human Bone Marrow-Derived Mesenchymal	MCAO	4 weeks	Decrease of 4 points in the Modified Neurological Severity Scores	NA	Bao *et al.*, 2011
NSCs transplant therapy in stroke patients					
Cell Source	**Stroke Area**	**Time period (after transplant)**	**Behavioral outcome**	**N° of surviving cells**	**Reference**
Neural MSCs derived	MCA	48 weeks	Significant improvement for 12 weeks in Barthel Index (30 points augmentation)	NA	Bang *et al.*, 2005
Human Neuronal Cells	NA	96 weeks	Improvement of 6.9 points in the European Stroke Scale at 24 weeks	NA	Kondziolka *et al.*, 2005

Cell Source	Stroke Area	Time period (after transplant)	Behavioral outcome	N° of surviving cells	Reference
Neural Fetal Porcine Cells	Basal Ganglia	192 weeks	Improvement in speech, language and/or motor impairments in 40% of the patients	NA	Savitz *et al.*, 2005
Bone Marrow Stem Cells	NA	48 weeks	None	NA	Suárez-Monteagudo *et al.*, 2009

Table 3. Experimental and clinical NSCs transplant therapy in stroke.

4.3 The prospect of NSCs transplant therapy for stroke

NSCs transplantation therapy for stroke holds great promise. However, the mechanisms of recovery are not completely understood. It is very likely that more than one mechanism is involved in the processes of recovery, and it still remains to be answered.

To this end, some standardization of the basic research, especially for behavior, is needed so that direct comparisons can be made between studies. Furthermore, longer-term studies are required to determine whether the cell-enhanced recovery is sustained and also to determine the tumorigenic potential of the cells. Other challenges include ensuring appropriate characterization, manufacturing, and quality control of transplanted cells and rigorous testing of viral and adventitious agents. Clearly, more research is needed to understand the bidirectional interaction between the transplanted cells and the host to optimize the chances of success before proceeding to the clinic.

Although reconstructing normal brain circuitry following stroke via NSCs is not likely in the foreseeable future, and although great care must be taken to ensure safety before considering clinical trials, preliminary evidence supports the therapeutic potential of NSCs for treatment of ischemic brain injury in animal models.

Understanding the mechanism of action of human NPCs in the post-ischemic brain will be important for the successful translation of cell transplantation strategies to the clinic. For example, if modulation of host brain plasticity is a major human NPCs mechanism of action, this could dictate the best time to transplant cells after stroke; "network relearning" occurs within weeks of stroke and continues for several months, making it a good therapeutic target with a large time window of intervention. Furthermore, knowing what changes the human NPCs elicit in the brain offers useful surrogate indicators of transplanted cell activity.

5. Conclusions

Under normal and pathological conditions, the adult brain is able to preserve regions with regenerative potential. Current research of neurogenic niches is revealing their complex homeostatic process, but at the same time, is bestowing with expectation that unraveling the characteristics of the unique molecular environment of the SVZ and the understanding of the underlying mechanisms that regulate the creation of new neurons in the adult brain will allow us to manipulate NSCs to yield a significant number of neurons capable of integration into human functional brain circuitries that were damaged and improve the motor deficits secondary to stroke or neurodegenerative diseases.

Long-term studies reviewed here support the persistence of an attenuate plasticity process residing within the neurogenic SVZ and RMS throughout life, which can be modulated in major extent by the action of some growth factors. The employment of growth factors could circumvent the technical and ethical constraints by using stem cells for transplant therapy. In addition, autologous transplantation of NSCs expanded *in vitro* could also avoid these concerns, and for this reason, it can be considered as a promising alternative. The first transplant of NSCs in stroke patients is currently in progress.

Altogheter, the large body of evidence supports the manipulation of endogenous NSCs and employment of grafted stem cells as future treatments for acute and chronic stroke. In spite of current efforts, the effectiveness and safety of both approaches are still being developed. It is clear that further investigation is necessary before such methods can be applied for human treatment and in our opinion, successful cell therapy for stroke patients, is still in a distant future.

6. Acknowledgments

We thank to Diana Millán-Aldaco and Marcela Palomero-Rivero for their relevant critical input on the manuscript. This work was supported by IMPULSA of the Universidad Nacional Autónoma de México (UNAM); by the Iniciativa de Apoyo complementario a la realización de las obras determinadas (IACOD), UNAM Grant No. I1201911 and by Programa de Apoyo a Proyectos de Investigación e Innovación Tecnológica (PAPIIT), UNAM Grant No. IN225209-3.

7. References

Altman, J. (1963) Autoradiographic investigation of cell proliferation in the brains of rats and cats. *The Anatomical record,* 145: 573–91.

Altman, J., Das, G.D. (1965) Autoradiographic and histological evidence of postnatal hippocampal neurogenesis in rats. *J Comp Neurol.,* 124:319–335.

Altman, J. (1969) Autoradiographic and histological studies of postnatal neurogenesis IV. Cell proliferation and migration in the anterior forebrain, with special reference to persisting neurogenesis in the olfactory bulb. *J Comp Neurol.,* I137:433–458.

American Heart Association. (2007) International Cardiovascular Disease Statistics; www.americanheart.org.

Andres, R.H., Horie, N., Slikker, W., Keren-Gill, H., Zhan, K., Sun, G., Manley, N.C., Pereira, M.P., Sheikh, L.A., McMillan, E.L., Schaar, B.T., Svendsen, C.N., Bliss, T.M., Steinberg, G.K. (2011) Human neural stem cells enhance structural plasticity and axonal transport in the ischaemic brain. *Brain,* 134:1777-1789.

Andsberg, G., Kokaia, Z., Klein, R.L., Muzyczka, N., Lindvall, O., Mandel, R.J. (2002) Neuropathological and behavioral consequences of adeno-associated viral vector-mediated continuous intrastriatal neurotrophin delivery in a focal ischemia model in rats. *Neurobiol Dis.,* 9:187-204.

Anton, E.S., Ghashghaei, H.T., Weber, J.L., McCann, C., Fischer, T.M., Cheung, I.D., Gassmann, M., Messing, A., Klein, R., Loyd, K.C.K., Lai, C. (2004) Receptor tyrosine kinase ErbB4 modulates neuroblast migration and placement in the adult forebrain. *Nat Neurosci.,* 7:1319-1328.

Arvidsson, A., Collin, T., Kirik, D., Kokaia, Z., Lindvall, O. (2002) Neuronal replacement from endogenous precursors in the adult brain after stroke. *Nat Med.*, 8:963-70.

Bacigaluppi, M., Pluchino, S., Martino, G., Kilic, E., Hermann, D.M. (2008) Neural stem/precursor cells for the treatment of ischemic stroke. *J Neurol Sci.*, 265:73-77.

Baldauf, K., Reymann, K.G. (2005) Influence of EGF/bFGF treatment on proliferation, early neurogenesis and infarct volume after transient focal ischemia. *Brain Res.*, 1056:158-167.

Balenci, L., Saoudi, Y., Grunwald, D., Deloulme, J.C., Bouron, A., Bernards, A., Baudier, J. (2007) IAGAP1 regulates adult neural progenitors in vivo and vascular endothelial growth factor-triggered neural progenitor migration in vitro. J Neurosci., 27:4716–4724.

Bang, O.Y., Lee, J.S., Lee, P.H, Lee, G. (2005) Autologous mesenchymal stem cell transplantation in stroke patients. Ann Neurol., 57:874-882.

Bao, X., Wei, J., Feng, M., Lu, S., Li, G., Dou, W., Ma, W., Ma, S., An, Y., Qin, C., Zhao, R.C., Wang, R. (2011) Transplantation of human bone marrow-derived mesenchymal stem cells promotes behavioral recovery and endogenous neurogenesis after cerebral ischemia in rats. Brain Res., 1367:103-113.

Bath, K.G., Mandairon, N., Jing, D., Rajagopal, R., Kapoor, R., Chen, Z.Y., Khan, T., Proenca, C.C., Kraemer, R., Cleland, T.A., Hempstead, B.L., Chao, M.V., Lee, F.S. (2008) Variant brain-derived neurotrophic factor (Val66Met) alters adult olfactory bulb neurogenesis and spontaneous olfactory discrimination. J Neurosci., 28:2383–2393.

Battista, D., Rutishauser, U. (2010) removal of polysialic acid triggers dispersion of subventricularly derived neuroblasts into surrounding CNS tissues. J Neurosci., 11:3995- 4003.

Belluardo, N., Mudó, G., Bonomo, A., Di Liberto, V., Frinchi, M., Fluxe, K. (2008). Nicotine induced fibroblast growth factor-2 restores the age-related decline of precursor cell proliferation in the subventricular zone of rat brain. Brain Res., 1193:12-24.

Belluzzi, O., Benedusi, M., Ackman, J., LoTurco, J.J. (2003) Electrophysiological differentiation of new neurons in the olfactory bulb. *J Neurosci.*, 23, 10411–10418.

Belmadani, A., Tran, P.B., Ren, D., Miller, R.J. (2006) Chemokines regulate the migration of neural progenitors to sites of neuroinflammation. *J Neurosci.*, 26:3182-3191.

Belvindrah, R., Hankel, S., Walker, J., Patton, B.L., Müller, U. (2007) Integrins Control the Formation of Cell Chains in the Adult Rostral Migratory Stream. *J Neurosci.*, 27:2704–2717.

Benraiss, A., Chmielnicki, E., Lerner, K., Roh, D., Goldman, S.A. (2001) Adenoviral brain-derived neurotrophic factor induces both neostriatal and olfactory neuronal recruitment from endogenous progenitor cells in the adult forebrain. *J Neurosci.*, 21, 6718–6731.

Bible, E., Chau, D.Y., Alexander, M.R., Price, J., Shakesheff, K.M., Modo, M. (2009) The support of neural stem cells transplanted into stroke-induced brain cavities by PLGA particles. *Biomaterials.*, 30:2985-9294.

Bliss, T.M., Kelly, S., Shah, A.K., Foo, W.C., Kohli, P., Stokes, C., Sun, G.H., Ma, M., Masel, J., Kleppner, S.R., Schallert, T., Palmer, T., Steinberg, G.K. (2006) Transplantation of hNT neurons into the ischemic cortex: cell survival and effect on sensorimotor behavior. *J Neurosci Res.*, 83:1004-1014.

Borlongan, C.V., Tajima, Y., Trojanowski, J.Q., Lee, V.M., Sanberg, P.R. (1998) Transplantation of cryopreserved human embryonal carcinoma-derived neurons (NT2N cells) promotes functional recovery in ischemic rats *Exp. Neurol.* 149:310–321.

Brose, K., Tessier-Lavigne, M. (2000) Slit proteins: key regulators of axon guidance, axonal branching, and cell migration. *Curr Opin Neurobiol.*, 10:95–102.

Burns, T.C., Rodriguez, G.J., Patel, S., Hussein, H.M., Georgiadis, A.L., Lakshminarayan, K., Qureshi, A.I. (2008) Endovascular interventions following intravenous thrombolysis may improve survival and recovery in patients with acute ischemic stroke: a case control study. *Am J Neuroradiol.*, 29:1918-1924.

Cacci, E., Villa, A., Parmar, M., Cavallaro, M., Mandahl, N., Lindvall, O., Martinez-Serrano, A., Kokaia, Z. (2007) Generation of human cortical neurons from a new immortal fetal neural stem cell line. *Exp Cell Res.*, 313:588-601.

Capsoni, S., Giannotta, S., Cattaneo, A. (2002) Nerve growth factor and galantamine ameliorate early signs of neurodegeneration in anti-nerve growth factor mice. *Proc Natl Acad Sci USA*, 99:12432-7.

Carleton, A., Petreanu, L.T., Lansford, R., Alvarez-Buylla, A., Lledo, P.M. (2003) Becoming a new neuron in the adult olfactory bulb. *Nat Neurosci.*, 6:507–518.

Chazal, G., Durbec, P., Jankovski, A., Rougon, G., Cremer, H. (2000) Consequences of neural cell adhesion molecule deficiency on cell migration in the rostral migratory stream of the mouse. *J Neurosci.*, 20:1446–1457.

Chen, J., Li, Y., Wang, L., Lu, M., Zhang, X., Chopp, M. (2001b) Therapeutic benefit of intracerebral transplantation of bone marrow stromal cells after cerebral ischemia in rats. *J. Neurol. Sci.*, 189:49–57.

Chen, J., Sanberg, P.R., Li, Y., Wang, L., Lu, M., Willing, A.E., Sanchez-Ramos, J., Chopp, M. (2001a) Intravenous administration of human umbilical cord blood reduces behavioral deficits after stroke in rats. *Stroke*, 32:2682–2688.

Chiaramello, S., Dalmasso, G., Bezin, L., Marcel, D., Jourdan, F., Peretto, P., Fasolo, A., De Marchis, S. (2007) BDNF/TrkB interaction regulates migration of SVZ precursor cells via PI3-K and MAP-K signalling pathways. *Eur J Neurosci.*, 26:1780–1790.

Chojnacki, A.K., Mak, G.K., Weiss, S. (2009) Identity crisis for adult periventricular neural stem cells: subventricular zone astrocytes, ependymal cells or both?. *Nat Rev Neurosci.*, 10:153-63.Chu, K., Kim, M., Park, K.I., Jeong, S.W., Park, H.K., Jung, K.H., Lee, S.T., Kang, L., Lee, K., Park, D.K., Kim, S.U., Roh, J.K. (2004) Human neural stem cells improve sensorimotor deficits in the adult rat brain with experimental focal ischemia. *Brain Res.*, 1016:145–153.

Conover, J.C., Doetsch, F., Garcia-Verdugo, J.M., Gale, N.W., Yancopoulos, G.D., Alvarez-Buylla, A. (2000) Disruption of Eph/ephrin signaling affects migration and proliferation in the adult subventricular zone. *Nat Neurosci.*, 3:1091-1097.

Corti, S., Locatelli, F., Papadimitriou, D., Donadoni, C., Del Bo, R., Fortunato, F., Strazzer, S., Salani, S., Bresolin, N., Comi, G.P. (2005) Multipotentiality, homing properties, and pyramidal neurogenesis of CNS-derived LeX(ssea-1)_/CXCR4_ stem cells. *FASEB J.*, 19:1860 -1862.

Courtès, S., Vernerey, J., Pujadas, L., Magalon, K., Cremer, H., Soriano, E., Durbec, P., Cayre, M. (2011) reelin controls progenitor cell migration in the healthy and pathological adult mouse brain. *PLoS ONE*, 6:e20430.

Cowan, C.A., Henkemeyer, M. (2002). Ephrins in reverse, park and drive. *Trends Cell Biol.*, 12: 339–346.

Craig, C.G., Tropepe, V., Morshead, C.M., Reynolds, B.A., Weiss, S., van der Kooy, D. (1996) In vivo growth factor expansion of endogenous subependymal neural precursor cell populations in the adult mouse brain. *J Neurosci.*, 16:2694-2658.

Cremer, H., Lange, R., Christoph, A., Plomann, M., Vopper, G., Roes, J., Brown, R., Baldwin, S., Kraemer, P., Scheff, S., Barthels, D., Rajewsky, K., Wille, W. (1994) Inactivation of the N-CAM gene in mice results in size reduction of the olfactory bulb and deficits in spatial learning. *Nature*, 367:455–459.

D'Arcangelo, G., Miao, G.G., Chen, S.C., Soares, H.D., Morgan, J.I., Curran, T. (1995) A protein related to extracelular matrix proteins deleted in the mouse mutant reeler. *Nature*, 374:719–723.

Darsalia, V., Kallur, T., Kokaia, Z. (2007) Survival, migration and neuronal differentiation of human fetal striatal and cortical neural stem cells grafted in stroke-damaged rat striatum. *Eur J Neurosci.*, 26:605–14.

Darsalia, V., Allison, S.J., Cusulin, C., Monni, E., Kuzdas, D., Kallur, T., Lindvall, O., Kokaia, Z. (2011) Cell number and timing of transplantation determine survival of human neural stem cell grafts in stroke-damaged rat brain. *J Cereb Blood Flow Metab.*, 31:235-42.

Dawson, M.R., Polito, A., Levine, J.M., Reynolds, R. (2003) NG2-expressing glial progenitor cells: an abundant and widespread population of cycling cells in the adult rat CNS. *Mol Cell Neurosci.*, 24:476–488.

De Rosa, R., Garcia, A.A., Braschi, C., Capsoni, S., Maffei, L., Berardi, N., Cattaneo, A. (2005) Intranasal administration of nerve growth factor (NGF) rescues recognition memory deficits in AD11 anti-NGF transgenic mice. *Proc Natl Acad Sci USA*, 102:3811-6.

De Wit, J., Verhaagen, J. (2003) Role of semaphorins in the adult nervous system. *Prog Neurobiol.*, 71:249-67.

Dirnagl U. (2006) Bench to bedside: the quest for quality in experimental stroke research. *J Cereb Blood Flow Metab.*, 26:1465–1478.

Doetsch, F., Alvarez-Buylla, A. (1996) Network of tangentially phatways for neuronal migration in adult mammalian brain. *Proc Natl Acad Sci USA*., 93:14895–14900.

Doetsch, F., Garcia-Verdugo, J. M., Alvarez-Buylla, A. (1997) Cellular composition and three-dimensional organization of the subventricular germinal zone in the adult mammalian brain. *J Neurosci.*, 17:5046–5061.

Doetsch, F., Caille, I., Lim, D. A., Garcia-Verdugo, J. M., Alvarez-Buylla, A. (1999) Subventricular zone astrocytes are neural stem cells in the adult mammalian brain. *Cell*, 97:703–716.

Doetsch, F., Petreanu, L., Caille, I., Garcia-Verdugo, J.M., Alvarez-Buylla, A. (2002) EGF converts transit-amplifying neurogenic precursors in the adult brain into multipotent stem cells. *Neuron.*, 36:1021-1034.

Drescher, U. (1997) The Eph family in the patterning of neural development. *Curr Biol.*, 7:799–807.

Emsley, J.G., Hagg, T. (2003) alpha6beta1 integrin directs migration of neuronal precursors in adult mouse forebrain. *Exp. Neurol.*, 183:273–285.

Englund, U., Bjorklund, A., Wictorin, K., Lindvall, O., Kokaia, M. (2002) Grafted neural stem cells develop into functional pyramidal neurons and integrate into host cortical circuitry. *Proc Natl Acad Sci USA*, 99:17089-17094.

Eriksson, P.S., Perfilieva, E., Bjork-Eriksson, T., Alborn, A.M., Nordborg, C., Peterson, D.A., Gage, F.H. (1998) Neurogenesis in the adult human hippocampus. *Nat Med.*, 4:1313-1317.

Fallon, J., Reid, S., Kinyamu, R., Opole, I.., Opole, R., Baratta, J., Korc, M., Endo, T.L., Duong, A., Nguyen, G., Karkehabadhi, M., Twardzik, D., Patel, S., Loughlin, S. (2000) *In vivo* induction of massive proliferation, directed migration, and differentiation of neural cells in the adult mammalian brain. *Proc Natl Acad Sci USA*, 97:14686-91.

Foulkes, M.A., Wolf, P.A., Price, T.R., Mohr, J.P., Hier, D.B. (1988) The Stroke Data Bank: design, methods, and baseline characteristics. *Stroke*, 19:547-54.

Francis, F., Koulakoff, A., Boucher, D., Chafey, P., Schaar, B., Vinet, M.C., Friocourt, G., McDonnell, N., Reiner, O., Kahn, A., McConnell, S.K., Berwald-Netter, Y., Denoulet, P., Chelly, J. (1999) Doublecortin is a developmentally regulated, microtubule-associated protein expressed in migrating and differentiating neurons. *Neuron*, 23:247-56.

Frey, W.H. 2nd, Liu, J., Chen, X., Thorne, R.G., Fawcett, J.R., Ala, T.A., Rahman, Y.E. (1997) Delivery of 125I-NGF to the brain via the olfactory route. *Drug Delivery*, 4:87-92.

Frinchi, M., Di Liberto, V., Olivieri, M., Fuxe, K., Belluardo, N., Mudò, G. (2010) FGF-2/FGFR1 neurotrophic system expressionlevel and its basal activation do not account for the age-dependent decline of precursor cell proliferation in the subventricular zone of rat brain. *Brain Res.*, 1358:39-45.

Fuchs, E., Tumbar, T., Guasch, G. (2004) Socializing with the neighbors: stem cells and their niche. *Cell*, 116:769-778.

Fujiwara, Y., Tanaka, N., Ishida, O., Fujimoto, Y., Murakami, T., Kajihara, H., Yasunaga, Y., Ochi, M. (2004) Intravenously injected neural progenitor cells of transgenic rats can migrate to the injured spinal cord and differentiate into neurons, astrocytes and oligodendrocytes. *Neurosci Lett.*, 366:287-291.

Galvão, R.P., Garcia-Verdugo, J.M., Alvarez-Buylla, A. (2008) Brain-derived neurotrophic factor signaling does not stimulate subventricular zone neurogenesis in adult mice and rats. *J Neurosci.*, 28: 13368-13383.

García-Verdugo, J.M., Doetsch, F., Wichterle, H., Lim, D.A., Alvarez-Buylla, A. (1998) Architecture and cell types of the adult subventricular zone: in search of the stem cells. *J Neurobiol.*, 36:234-248.

Gascon, E., Vutskits, L., Jenny, B., Durbec, P., Kiss, J.Z. (2007) PSA-NCAM in postnatally generated immature neurons of the olfactory bulb: a crucial role in regulating p75 expression and cell survival. *Development*, 134:1181-1190.

Ghashghaei, H.T., Weber, J., Pevny, L., Schmid, R., Schwab, M.H., Lloyd, K.C.K., Eisenstat, D.D., Lai, C., Anton, E.S. (2006) The role of neuregulin-ErbB4 interactions on the proliferation and organization of cells in the subventricular zone. *Proc Natl Acad Sci USA*, 103:1930-1935.

Giger, R.J., Pasterkamp, R.J., Heijnen, S., Holtmaat, A.J., Verhaagen, J. (1998) Anatomical distribution of the chemorepellent semaphorin III/collapsin-1 in the adult rat and human brain: predominant expression in structures of the olfactory-hippocampal pathway and the motor system. *J Neurosci Res.*, 52:27-42.

Giger, R.J., Cloutier, J.F., Sahay, A., Prinjha, R.K., Levengood, D.V., Moore, S.E., Pickering, S., Simmons, D., Rastan, S., Walsh, F.S., Kolodkin, A.L., Ginty, D.D., Geppert, M. (2000) Neuropilin-2 is required in vivo for selective axon guidance responses to secreted semaphorins. *Neuron,* 25:29–41.

Giuliani, A., D'Intino, G., Paradisi, M., Giardino, L., Calza, L. (2004) p75(NTR)-immunoreactivity in the subventricular zone of adult male rats: expression by cycling cells. *J Mol Histol.,* 35:749–758

Gleason, D., Fallon, J.H., Guerra, M., Liu, J.C., Bryant, P.J. (2008) Ependymal stem cells divide asymmetrically and transfer progeny into the subventricular zone when activated by injury. *Neuroscience,* 156:81-8.

Gleeson, J.G., Lin, P.T., Flanagan, L.A., Walsh, C.A. (1999) Doublecortin is a microtubule-associated protein and is expressed widely by migrating neurons. *Neuron,* 23:257-71.

Greenberg, D.A., Jin, K. (2006) Growth factors and stroke. *NeuroRx,* 3:458-65.

Gritti, A., Parati, E.A., Cova, L., Frolichsthal, P., Galli, R., Wanke, E., Faravelli, L., Morassutti, D.J., Roisen, F., Nickel, D.D., Vescovi, A.L. (1996) Multipotential stem cells from the adult mouse brain proliferate and self-renew in response to basic fibroblast growth factor. *J Neurosci.,* 16:1091–1100.

Gritti, A., Frölichsthal-Schoeller, P., Galli, R., Parati, E.A., Cova, L., Pagano, S.F., Bjornson, C.R., Vescovi, A.L. (1999) Epidermal and fibroblast growth factors behave as mitogenic regulators for a single multipotent stem cell-like population from the subventricular region of the adult mouseforebrain. *J Neurosci.,* 19:3287-3297.

Gritti, A., Bonfanti, L., Doetsch, F., Caille, I., Alvarez-Buylla, A., Lim D., Galli R., Garcia-Verdugo J.M., Herrera, D.G., Vescovi, A.L. (2002) multipotent neural stem cells reside into the rostral extension and olfactory bulb of adult rodents. *J Neurosci.,* 22:437-445.

Guerra-Crespo, M., Gleason, D., Sistos, A., Toosky, T., Solaroglu, I., Zhang, J.H., Bryant, P.J., Fallon, J.H. (2009) Transforming growth factor-alpha induces neurogenesis and behavioral improvement in a chronic stroke model. *Neuroscience,* 160:470-83.

Guerra-Crespo, M., Sistos, A., Gleason, D., Fallon, J.H. (2010) Intranasal administration of PEGylated transforming growth factor-alpha improves behavioral deficits in a chronic stroke model. *J Stroke Cerebrovasc Dis.,* 19:3-9.

Gustafsson, E., Andsberg, G., Darsalia, V., Mohapel, P., Mandel, R.J., Kirik, D., Lindvall, O., Kokaia, Z. (2003) Anterograde delivery of brain-derived neurotrophic factor to striatum via nigral transduction of recombinant adeno-associated virus increases neuronal death but promotes neurogenic response following stroke. *Eur J Neurosci,* 17:2667-2678.

Hack, I., Bancila, M., Loulier, K., Carroll, P., Cremer, H. (2002) Reelin is a detachment signal in tangential chain-migration during postnatal neurogenesis. *Nat Neurosci.,* 5: 939–945.

Hacke, W., Donnan, G., Fieschi, C., Kaste, M., von Kummer, R., Broderick, J.P., Brott T, Frankel M, Grotta JC, Haley EC Jr, Kwiatkowski T, Levine SR, Lewandowski C, Lu M, Lyden P, Marler JR, Patel S, Tilley BC, Albers G, Bluhmski E, Wilhelm M, Hamilton S, ATLANTIS Trials Investigators, ECASS Trials Investigators, NINDS rt-PA Study Group Investigators. (2004) Association of outcome with early stroke treatment: pooled analysis of ATLANTIS, ECASS, and NINDS rt-PA stroke trials. *Lancet,* 363:768–774.

Herrera, D.G., Garcia-Verdugo, J.M., Alvarez-Buylla, A. (1999) Adult-derived neural precursors transplanted into multiple regions in the adult brain. *Ann Neurol.,* 46:867–877.

Hicks, A.U., Lappalainen, R.S., Narkilahti, S., Suuronen, R., Corbett, D., Sivenius, J., Hovatta, O., Jolkkonen, J. (2009) Transplantation of human embryonic stem cell-derived neural precursor cells and enriched environment after cortical stroke in rats: cell survival and functional recovery. *Eur J Neurosci.,* 29:562-574.

Hiesberger, T., Trommsdorff, M., Howell, B.W., Goffinet, A., Mumby, M.C., Cooper, J.A., Herz, J. (1999) Direct binding of Reelin to VLDL receptor ApoE receptor 2 induces tyrosine phosphorylation of disabled-1 and modulates tau phosphorylation. *Neuron,* 24:481--489.

Höglinger, G.U., Rizk, P., Muriel, M.P., Duyckaerts, C., Oertel, W.H., Caille, I., Hirsch, E.C. (2004) Dopamine depletion impairs precursor cell proliferation in Parkinson disease. *Nat Neurosci.,* 7:726-7235.

Holmberg, J., Armulik, A., Senti, K.A., Edoff, K., Spalding, K., Momma, S., Cassidy, R., Flanagan, J.G., Frisén, J. (2005) Ephrin-A2 reverse signaling negatively regulates neural progenitor proliferation and neurogenesis *Genes Dev.,* 19:462-471.

Hu, H., Tomasiewicz, H., Magnuson, T., Rutishauser, U. (1996) The role of polysialic acid in migration of olfactory bulb interneuron precursors in the subventricular zone. *Neuron,* 16:735–743.

Hu, H. (1999) Chemorepulsion of neuronal migration by Slit2 in the developing mammalian forebrain. *Neuron,* 23:703–711.

Hu, H. (2000) Polysialic acid regulates chain formation by migrating olfactory interneuron precursors. *J Neurosci Res.,* 61:480–492.

Hynes, R.O. (2002) Integrins: Bidirectional, Allosteric Signaling Machines. *Cell,* 110:673–687.

Ishibashi, S., Sakaguchi, M., Kuroiwa, T., Yamasaki, M., Kanemura, Y., Shizuko, I., Shimazaki, T., Onodera, M., Okano, H., Mizusawa, H. (2004) Human neural stem/progenitor cells, expanded in long-term neurosphere culture, promote functional recovery after focal ischemia in Mongolian gerbils. *J Neurosci Res.,* 78:215-223.

Itoh, N., Ornitz, D.M. (2011) Fibroblast growth factors: from molecular evolution to roles in development, metabolism and disease. *J Biochem.,* 149:121–130.

Jacques, T.S., Relvas, J.B., Nishimura, S., Pytela, R., Edwards, G.M., Streuli, C.H., ffrench-Constant, C. (1998) Neural precursor cell chain migration and division are regulated through different beta1 integrins. *Development,* 125:3167-3177.

Jiang, Q., Zhang, Z.G., Ding, G.L., Silver, B., Zhang, L., Meng, H., Lu, M., Pourabdillah-Nejed-D, S., Wang, L., Savant-Bhonsale, S., Li, L., Bagher-Ebadian, H., Hu, J., Arbab, A.S., Vanguri, P., Ewing, J.R., Ledbetter, K.A., Chopp, M. (2006) MRI detects white matter reorganization after neural progenitor cell treatment of stroke. Neuroimage, 32:1080-1089.

Jin, K., Minami, M., Lan, J.Q., Mao, X.O., Batteur, S., Simon, R.P., Greenberg, D.A. (2001) Neurogenesis in dentate subgranular zone and rostral subventricular zone after focal cerebral ischemia in the rat. *Proc Natl Acad Sci USA,* 98:4710-5.

Jin, K., Zhu, Y., Sun, Y., Mao, X.O., Xie, L., Greenberg, D.A. (2002) Vascular endothelial growth factor (VEGF) stimulates neurogenesis in vitro and in vivo. *Proc Natl Acad Sci USA*, 99:11946–11950.

Jin, K., Sun, Y., Xie, L., Batteur, S., Mao, X.O., Smelick, C., Logvinova, A., Greenberg, D.A. (2003) Neurogenesis and aging: FGF-2 and HB-EGF restore neurogenesis in hippocampus and subventricular zone of aged mice. *Aging Cell*, 2:175-183.

Jin, K., Sun, Y., Xie, L., Childs, J., Mao, X.O., Greenberg, D.A. (2004) Post-ischemic administration of heparin-binding epidermal growth factor-like growth factor (HB-EGF) reduces infarct size and modifies neurogenesis after focal cerebral ischemia in the rat. *J Cereb Blood Flow Metab.*, 24:399-408.

Jin, K., Wang, X., Xie, L., Mao, X.O., Zhu, W., Wang, Y., Shen, J., Mao, Y., Banwait, S., Greenberg, D.A. (2006) Evidence for stroke-induced neurogenesis in the human brain. *Proc Natl Acad Sci USA*, 103:13198-202.

Jin, K., Mao, X., Xie, L., Galvan, V., Lai, B., Wang, Y., Gorostiza, O., Wang, X., Greenberg, D.A. (2010a) Transplantation of human neural precursor cells in Matrigel scaffolding improves outcome from focal cerebral ischemia after delayed postischemic treatment in rats. *J. Cereb. Blood Flow Metab.*, 30:534–544.

Jin, K., Mao, X., Xie, L., Greenberg, R.B., Peng, B., Moore, A., Greenberg, M.B., Greenberg, D.A. (2010b) Delayed transplantation of human neural precursor cells improves outcome from focal cerebral ischemia in aged rats. *Aging Cell*, 9:1076–108.

Johansson, C.B., Momma, S., Clarke, D.L., Risling, M., Lendahl, U., Frisén, J. (1999) Identification of a neural stem cell in the adult mammalian central nervous system. *Cell*, 96: 25–34.

Johnson, D.E., Williams, L.T. (1993) Structural and functional diversity in the FGF receptor multigene family. *Adv. Cancer Res.*, 60:1–41.

Kawai, H., Yamashita, T., Ohta, Y., Deguchi, K., Nagotani, S., Zhang, X., Ikeda, Y., Matsuura, T., Abe, K. (2010) Tridermal tumorigenesis of induced pluripotent stem cells transplanted in ischemic brain. *J Cereb Blood Flow Metab.*, 30:1487-1493.

Kelly, S., Bliss, T.M., Shah, A.K., Sun, G.H., Ma, M., Foo, W.C., Masel, J., Yenari, M.A., Weissman, I.L., Uchida, N., Palmer, T., Steinberg, G.K. (2004) Transplanted human fetal neural stem cells survive, migrate, and differentiate in ischemic rat cerebral cortex. *Proc Natl Acad Sci., U S A*, 101:11839–11844.

Koizumi, H., Higginbotham, H., Poon, T., Tanaka, T., Brinkman, B.C., Gleeson, J.G. (2006) Doublecortin maintains bipolar shape and nuclear translocation during migration in the adult forebrain. *Nat Neurosci.*, 9:779–786.

Komitova, M., Zhu, X., Serwanski, D.R., Nishiyama, A. (2009) NG2 cells are distinct from neurogenic cells in the postnatal mouse subventricular zone. *J Comp Neurol.*, 512:702-16.

Kondziolka, D., Steinberg, G.K., Wechsler, L., Meltzer, C.C., Elder, E., Gebel, J., Decesare, S., Jovin, T., Zafonte, R., Lebowitz, J., Flickinger, J.C., Tong, D., Marks, M.P., Jamieson, C., Luu, D., Bell-Stephens, T., Teraoka, J. (2005) Neurotransplantation for patients with subcortical motor stroke: a phase 2 randomized trial. *J Neurosurg.*, 103:38-45.

Kuhn, H.G., Winkler, J., Kempermann, G., Thal, L.J., Gage, F.H. (1997) Epidermal growth factor and fibroblast growth factor-2 have differenteffects on neural progenitors in the adult rat brain. *J Neurosci.*, 17:5820-5828.

Lee, J.P., Jeyakumar, M., Gonzalez, R., Takahashi, H., Lee, P.J., Baek, R.C., Clark, D., Rose, H., Fu, G., Clarke, J., McKercher, S., Meerloo, J., Muller, F.J., Park, K.I., Butters, T.D., Dwek, R.A., Schwartz, P., Tong, G., Wenger, D., Lipton, S.A., Seyfriend, T.N., Platt, F.M., Snyder, E.Y. (2007) Stem cells act through multiple mechanisms to benefit mice with neurodegenerative metabolic disease. *Nat Med.,* 13:439–447.

Leker, R.R., Soldner, F., Velasco, I., Gavin, D.K., Androutsellis-Theotokis, A., McKay, R.D. (2007) Long-lasting regeneration after ischemia in the cerebral cortex. *Stroke,* 38:153-161.

Leker, R.R., Toth, Z.E., Shahar, T., Cassiani-Ingoni, R., Szalayova, I., Key, S., Bratincsák, A., Mezey, E. (2009) Transforming growth factor alpha induces angiogenesis and neurogenesis following stroke. *Neuroscience,* 163:233-43.

Li, H.S., Chen, J.H., Wu, W., Fagaly, T., Zhou, L., Yuan, W., Dupuis, S., Jiang, Z.H., Nash, W., Gick, C. (1999) Vertebrate slit, a secreted ligand for the transmembrane protein roundabout, is a repellent for olfactory bulb axons. *Cell,* 96:807-818.

Liu, J., Solway, K., Messing, R.O., Sharp, F.R. (1998) Increased neurogenesis in the dentate gyrus after transient global ischemia in gerbils. *J Neurosci.,* 18:7768-78.

Liu, X.F., Fawcett, J.R., Hanson, L.R., Frey, W.H. 2nd. (2004) The window of opportunity for treatment of focal cerebral ischemic damage with noninvasive intranasal insulin-like growth factor-I in rats. *J Stroke Cerebrovasc Dis.,* 13:16-23.

Lledo, P.M., Merkle, F.T., Alvarez-Buylla, A. (2008) Origin and function of olfactory bulb interneuron diversity. *Trends Neurosci.,* 31:392-400.

Lois, C., Alvarez-Buylla, A. (1994), Long distance neuronal migration in the adult mammalian brain. *Science,* 264:1145–1148.

Lois, C., Garcia-Verdugo, J.M., Alvarez-Buylla, A. (1996) Chain migration of neuronal precursors. *Science,* 271:978–981.

Ma, M., Ma, Y., Yi, X., Guo, R., Zhu, W., Fan, X., Xu, G., Frey, W.H. 2nd, Liu, X. (2008) Intranasal delivery of transforming growth factor-beta1 in mice after stroke reduces infarct volume and increases neurogenesis in the subventricular zone. *BMC Neurosci.,* 9:117.

Maisonpierre, P.C., Belluscio, L., Friedman, B., Alderson, R.F., Wiegand, S.J., Furth, M.E., Lindsay, R.M., Yancopoulos, G.D. (1990) NT-3, BDNF, and NGF in the developing rat nervous system: parallel as well as reciprocal patterns of expression. *Neuron,* 5:501-509.

Marillat, V., Cases, O., Nguyen-Ba-Charvet, K.T., Tessier-Lavigne, M., Sotelo, C., Chedotal, A. (2002) Spatiotemporal expression patterns of slit and robo genes in the rat brain. *J Comp Neurol.,* 442:130–155.

Martí-Fàbregas, J., Romaguera-Ros, M., Gómez-Pinedo, U., Martínez-Ramírez, S., Jiménez-Xarrié, E., Marín, R., Martí-Vilalta, J.L., García-Verdugo, J.M. (2010) Proliferation in the human ipsilateral subventricular zone after ischemic stroke. *Neurology,* 74:357-65.

Martino,. G, Pluchino, S. (2006) The therapeutic potential of neural stem cells. *Nat Rev Neurosci.,* 7:395–406.

Matsumoto, T., Claesson-Welsh, L. (2001) VEGF receptor signal transduction. *Sci. STKE,* 2001:Re21.

Matsuoka, N., Nozaki, K., Takagi, Y., Nishimura, M., Hayashi, J., Miyatake, S., Hashimoto, N. (2003) Adenovirus-mediated gene transfer of fibroblast growth factor-2 increases BrdU-positive cells after forebrain ischemia in gerbils. *Stroke*, 34:1519-25.

Mattsson, B., Sorensen, J.C., Zimmer, J., Johansson, B.B. (1997) Neural grafting to experimental neocortical infarcts improves behavioral outcome and reduces thalamic atrophy in rats housed in enriched but not in standard environments. *Stroke*, 28:1225–1231.

Meléndez-Herrera, E., Colín-Castelán, D., Varela-Echavarría, A., Gutiérrez-Ospina, G. (2008) Semaphorin-3A and its receptor neuropilin-1 are predominantly expressed in endothelial cells along the rostral migratory stream of young and adult mice. *Cell Tissue Res.*, 333:175-84.

Mirzadeh, Z., Merkle, F. T., Soriano-Navarro, M., Garcia-Verdugo, J. M., Alvarez-Buylla, A. (2008) Neural stem cells confer unique pinwheel architecture to the ventricular surface in neurogenic regions of the adult brain. *Cell Stem Cell*, 3:265–278.

Modo, M., Stroemer, R.P., Tang, E., Patel, S., Hodges, H. (2002) Effects of implantation site of stem cell grafts on behavioral recovery from stroke damage. *Stroke*, 33:2270–2278.

Moores, C.A., Perderiset, M., Francis, F., Chelly, J., Houdusse, A., Milligan, R.A. (2004) Mechanism of microtubule stabilization by doublecortin. *Mol Cell*, 14:833-9.

Morshead, C.M., Garcia, A.D., Sofroniew, M.V., van Der Kooy, D. (2003) The ablation of glial fibrillary acidic proteinpositive cells from the adult central nervous system results in the loss of forebrain neural stem cells but not retinal stem cells. *Eur J Neurosci.*, 18:76–84.

Mosch, B., Reissenweber, B., Neuber, C., Pietzsch, J. (2010) Eph receptors and ephrin ligands: Important players in angiogenesis and tumor angiogenesis. *J Onc.*, 2010:1-12.

Mudó, G., Belluardo, N., Fuxe, K. (2007) Acute intermittent nicotine treatment induces fibroblast growth factor-2 in the subventricular zone of the adult rat brain and enhances neuronal precursor cell proliferation. *Neuroscience*, 145:470-83.

Murase, S., Horwitz, A. F. (2002) deleted in colorectal carcinoma and differentially expressed integrins mediate the directional migration of neural precursors in the rostral migratory stream. *J Neurosci.*, 22:3568–3579.

Murase, S., Cho, C., White, J.M., Horwitz, A.F. (2008) ADAM2 promotes migration of neuroblasts in the rostral migratory stream to the olfactory bulb. *Eur J Neurosci.*, 27:1585-95.

Nakatomi, H., Kuriu, T., Okabe, S., Yamamoto, S., Hatano, O., Kawahara, N., Tamura, A., Kirino, T., Nakafuku, M. (2002) Regeneration of hippocampal pyramidal neurons after ischemic brain injury by recruitment of endogenous neural progenitors. *Cell*, 110:429-41.

Ng, K.L., Li, J.D., Cheng, M.Y., Leslie, F.M., Lee, A.G., Zhou, Q.Y. (2005) Dependence of olfactory bulb neurogenesis on prokineticin 2 signaling. *Science*, 308:1923–1927.

Nguyen-Ba-Charvet, K.T., Picard-Riera, N., Tessier-Lavigne, M., Baron-Van, Evercooren, A., Sotelo, C., Chedotal, A. (2004) Multiple roles for slits in the control of cell migration in the rostral migratory stream. *J Neurosci.*, 24:1497–1506.

Nishino, H., Borlongan, C.V. (2000) Restoration of function by neural transplantation in the ischemic brain. *Prog. Brain Res.*, 127:461–476.

Ocbina, P.J., Dizon, M.L., Shin, L., Szele, F.G. (2006) Doublecortin is necessary for the migration of adult subventricular zone cells from neurospheres. *Mol Cell Neurosci.*, 33:126-35.

Ornitz, D.M., Xu, J., Colvin, J.S., McEwen, D.G., MacArthur, C.A., Coulier, F., Gao, G., Goldfarb, M. (1996) Receptor specificity of the fibroblast growth factor family. *J Biol Chem.*, 271:15292–15297.

Ourednik, J., Ourednik, V., Lynch, W.P., Schachner, M., Snyder, E.Y. (2002) Neural stem cells display an inherent mechanism for rescuing dysfunctional neurons. *Nat Biotechnol.*, 20:1103–1110.

Parent, J.M., Vexler, Z.S., Gong, C., Derugin, N., Ferriero, D.M. (2002) Rat forebrain neurogenesis and striatal neuron replacement after focal stroke. *Ann Neurol.*, 52:802-13.

Park, K.I., Lachyankar, M., Nissim, S., Snyder, E.Y. (2002) Neural stem cells for CNS repair: state of the art and future directions. *Adv Exp Med Biol.*, 506:1291-1296.

Pencea, V., Bingaman, K.D., Wiegand, S.J., Luskin, M.B. (2001) Infusion of brain-derived neurotrophic factor into the lateral ventricle of the adult rat leads to new neurons in the parenchyma of the striatum, septum, thalamus, and hypothalamus. *J Neurosci.*, 21:6706–6717.

Petridis, A.K., El-Maarouf, A., Rutishauser, U. (2004) Polysialic acid regulates cell contact-dependent neuronal differentiation of progenitor cells from the subventricular zone. *Dev Dyn.*, 230:675–684.

Pluchino, S., Quattrini, A., Brambilla, E., Gritti, A., Salani, G., Dina, G., Galli, R., Del Carro, U., Amadio, S., Bergami, A., Furlan, R., Comi, G., Vescovi, A.L., Martino, G. (2003) Injection of adult neurospheres induces recovery in a chronic model of multiple sclerosis. *Nature*, 422:688–694.

Pollock, K., Stroemer, P., Patel, S., Stevanato, L., Hope, A., Miljan, E., Dong, Z., Hodges, H., Price, J., Sinden, J.D. (2006) A conditional immortal clonal stem line from human cortical neuroepithelium for the treatment of ischemic stroke. *Exp Neurol.*, 199:143–155.

Prosser, H.M., Bradley, A., Caldwell, M.A. (2007) Olfactory bulb hypoplasia in Prokr2 null mice stems from defective neuronal progenitor migration and differentiation. *Eur J Neurosci.*, 12: 3339–3344.

Raper, J.A. (2000) Semaphorins and their receptors in vertebrates and invertebrates. *Curr Op Neurobiol.*, 10:88–94.

Reuss, B., Von, Bohlen., Halbach, O. (2003) Fibroblast growth factors and their receptors in the central nervous system. *Cell Tissue Res.*, 313:139–157

Reynolds, B.A., Weiss, S. (1992) Generation of neurons and astrocytes from isolated cells of the adult mammalian central nervous system. *Science*, 255:1707-1710.

Reynolds, B.A., Weiss, S. (1996) Clonal and population analyses demonstrate that an EGF-responsive mammalian embryonic CNS precursor is a stem cell. *Dev Biol.*, 175:1–13.

Riolobos, A. S., Heredia, M., de la Fuente, J.A., Criado, J.M., Yajeya, J., Campos, J., Santacana, M. (2001) Functional recovery of skilled forelimb use in rats obliged to use the impaired limb after grafting of the frontal cortex lesion with homotopic fetal cortex. *Neurobiol Learn Mem.*, 75:274–292.

Ropper, A.H., Brown, R.H. (2009) Adams and Victor's principles of neurology, 9th Ed. New York: McGraw-Hill.

Rousselot, P., Lois, C., Alvarez-Buylla, A. (1995) Embryonic (PSA) N-CAM reveals chains of migrating neuroblasts between the lateral ventricle and the olfactory bulb of adult mice. *J Comp Neurol.*, 351:51-61.

Saghatelyan, A., de Chevigny, A., Schachner, M., Lledo, P.M. (2004). Tenascin-R mediates activity-dependent recruitment of neuroblasts in the adult mouse forebrain. *Nat Neurosci.*, 7:347–356.

Saporta, S., Borlongan, C.V., Sanberg, P.R. (1999) Neural transplantation of human neuroteratocarcinoma (hNT) neurons into ischemic rats. A quantitative dose-response analysis of cell survival and behavioral recovery. *Neuroscience*, 91:519–525.

Savitz, S.I., Rosenbaum, D.M., Dinsmore, J.H., Weschler, L.R., Caplan, L.R. (2002) Cell transplantation for stroke *Ann. Neurol.*, 53:266–275.

Savitz, S.I., Dinsmore, J., Wu, J., Henderson, G.V., Stieg, P., Caplan, L.R. (2005) Neurotransplantation of fetal porcine cells in patients with basal ganglia infarcts: a preliminary safety and feasibility study. *Cerebrovasc Dis.*, 20:101-107.

Schabitz, W.R., Steigleder, T., Cooper-Kuhn, C.M., Schwab, S., Sommer, C., Schneider, A., Kuhn, H.G. (2007) Intravenous brain-derived neurotrophic factor enhances poststroke sensorimotor recovery and stimulates neurogenesis. *Stroke*, 38:2165–2172.

Schänzer, A., Wachs, F.P., Wilhelm, D., Acker, T., Cooper-Kuhn, C., Beck, H., Winkler, J., Aigner, L., Plate, KH., Kuhn, H.G. (2004) Direct stimulation of adult neural stem cells in vitro and neurogenesis in vivo by vascular endothelial growth factor. *Brain Pathol.*, 14:237-48.

Scheffler, B., Walton, N.M., Lin, D.D., Goetz, A.K., Enikolopov, G., Roper, S.N., Steindler, D.A. (2005) Phenotypic and functional characterization of adult brain neuropoiesis. *Proc Natl Acad Sci., U S A*, 102:9355–9358.

Seki, T., Arai, Y. (1993) Highly polysialylated neural cell adhesion molecule (NCAM-H) is expressed by newly generated granule cells in the dentate gyrus of the adult rat. *J Neurosci.*, 13:2351–2358.

Shen, Q., Wang, Y., Kokovay, E., Lin, G., Chuang, S.M., Goderie, S.K., Roysam, B., Temple, S. (2008) Adult SVZ stem cells lie in a vascular niche: a quantitative analysis of niche cell-cell interactions. *Cell Stem Cell*, 3:289–300.

Sinden, J.D., Rashid-Doubell, F., Kershaw, T.R., Nelson, A., Chadwick, A., Jat, P.S., Noble, M.D., Hodges, H., Gray, J.A. (1997) Recovery of spatial learning by grafts of a conditionally immortalized hippocampal neuroepithelial cell line into the ischaemia-lesioned hippocampus. *Neuroscience*, 81:599–608.

Snapyan, M., Lemasson, M., Brill, M.S., Blais, M., Massouh, M., Ninkovic, J., Gravel, C., Berthod, F., Götz, M., Barker, P.A., Parent, A.,Saghatelyan, A. (2009) Vasculature guides migrating neuronal precursors in the adult mammalian forebrain via brain-derived neurotrophic factor signaling. *J Neurosci.*, 29:4172-4188.

Soker, S., Takashima, S., Miao, H.Q., Neufeld, G., Klagsbrun, M. (1998) Neuropilin-1 is expressed by endothelial and tumor cells as an isoformspecific receptor for vascular endothelial growth factor. *Cell*, 92:735-745.

Song, H.J., Stevens, C.F., Gage, F.H. (2002a) Astroglia induce neurogenesis from adult neural stem cells. *Nature*, 417:39–44.

Song, H.J., Stevens, C.F., Gage, F.H. (2002b) Neural stem cells from adult hippocampus develop essential properties of functional CNS neurons. *Nat Neurosci.*, 5:438–445.

Stilley, C.S., Ryan, C.M., Kondziolka, D., Bender, A., DeCesare, S., Wechsler, L. (2004) Changes in cognitive function after neuronal cell transplantation for basal ganglia stroke. *Neurology,* 63:1320-1322.

Suárez-Monteagudo, C., Hernández-Ramírez, P., Alvarez-González, L., García-Maeso, I., de la Cuétara-Bernal, K., Castillo-Díaz, L., Bringas-Vega, M.L., Martínez-Aching, G., Morales-Chacón, L.M., Báez-Martín, M.M., Sánchez-Catasús, C., Carballo-Barreda, M., Rodríguez-Rojas, R., Gómez-Fernández, L., Alberti-Amador, E., Macías-Abraham, C., Balea, E.D., Rosales, L.C., Del Valle Pérez, L., Ferrer, B.B., González, R.M., Bergado, J.A. (2009) Autologous bone marrow stem cell neurotransplantation in stroke patients. An open study. *Restor Neurol Neurosci.,* 27:151-161.

Sugiura, S., Kitagawa, K., Tanaka, S., Todo, K., Omura-Matsuoka, E., Sasaki, T., Mabuchi, T., Matsushita, K., Yagita, Y., Hori, M. (2005) Adenovirus-mediated gene transfer of heparin-binding epidermal growth factor-like growth factor enhances neurogenesis and angiogenesis after focal cerebral ischemia in rats. *Stroke,* 36:859-864.

Sun, Y., Jin, K., Childs, J.T., Xie, L., Mao, X.O., Greenberg, D.A. (2006) Vascular endothelial growth factor-B (VEGFB) stimulates neurogenesis: Evidence from knockout mice and growth factor administration. *Dev Biol.,* 289:329–33.

Tamagnone, L., Comoglio, P.M. (2000) Signaling by semaphorin receptors: cell guidance and beyond. *Trends Cell Biol.,* 10, 377–383.

Teramoto, T., Qiu, J., Plumier, J.C., Moskowitz, M.A. (2003) EGF amplifies the replacement of parvalbumin-expressing striatal interneurons after ischemia. *J Clin Invest.,* 111:1125-1132.

Theus, M.G., Ricard, J., Bethea, J.R., Liebl, D.J. (2010) EphB3 limits the expansion of neural progenitor cells in the subventricular zone by regulatinf p53 during homeostasis and following traumatic brain injury. *Stem Cells,* 28:1231-1242.

Thored, P., Arvidsson, A., Cacci, E., Ahlenius, H., Kallur, T., Darsalia, V., Ekdahl, C.T., Kokaia, Z., Lindvall, O. (2006) Persistent production of neurons from adult brain stem cells during recovery after stroke. *Stem Cells,* 24:739-47.

Thorne, R.G., Pronk, G.J., Padmanabhan, V., Frey, W.H. 2nd. (2004) Delivery of insulin-like growth factor-I to the rat brain and spinal cord along olfactory and trigeminal pathways following intranasal administration. *Neuroscience,* 127:481-96.

Tropepe, V., Craig, C.G., Morshead, C.M., van der Kooy, D. (1997) Transforming growth factor-alpha null and senescent mice show decreased neural progenitor cell proliferation in the forebrain subependyma. *J Neurosci.,* 17:7850–7859.

Vahedi, K., Hofmeijer, J., Juettler, E., Vicaut, E., George, B., Algra, A., Amelink, G.J., Schmiedeck, P., Schwab, S., Rothwell, P.M., Bousser, M.G., van der Worp, H.B., Hacke, W., DECIMAL, DESTINY, and HAMLET investigators. (2007) Early decompressive surgery in malignant infarction of the middle cerebral artery: a pooled analysis of three randomised controlled trials. *Lancet Neuro.,* 6:215–222.

Veizovic, T., Beech, J.S., Stroemer, R.P., Watson, W.P., Hodges, H. (2001) Resolution of stroke deficits following contralateral grafts of conditionally immortal neuroepithelial stem cells. *Stroke,* 32:1012–1019.

Viapiano, M.S., Matthews, R.T. (2006) From barriers to bridges: chondroitin sulfate proteoglycans in neuropathology. *Trends Mol Med.,* 12:488-496.

Wada, K., Sugimori, H., Bhide, P.G., Moskowitz, M.A., Finklestein, S.P. (2003) Effect of basic fibroblast growth factor treatment on brain progenitor cells after permanent focal ischemia in rats. *Stroke,* 34:2722-2728.

Wagner, J.P., Black, I.B., DiCicco-Bloom, E. (1999) Stimulation of neonatal and adult brain neurogenesis by subcutaneous injection of basic fibroblast growth factor. *J Neurosci.,* 19:6006–6016.

Wang, Y.Q., Jin, K., Mao, X.O., Xie, L., Banwait, S., Marti, H.H., Greenberg, D.A. (2007a) VEGF-overexpressing transgenic mice show enhanced post-ischemic neurogenesis and neuromigration. *J Neurosci Res.,* 85:740–747.

Wang, Y.Q., Guo, X., Qiu, M.H., Feng, X.Y., Sun, F.Y. (2007b) VEGF overexpression enhances striatal neurogenesis in brain of adult rat after a transient middle cerebral artery occlusion. *J Neurosci Res.,* 85:73–82.

Wei, L., Cui, L., Snider, B.J., Rivkin, M., Yu, S.S., Lee, C.S., Adams, L.D., Gottlieb, D.I., Johnson, E.M. Jr Yu, S.P., Choi, D.W. (2005) Transplantation of embryonic stem cells overexpressing Bcl-2 promotes functional recovery after transient cerebral ischemia. *Neurobiol Dis.,* 19:183-193.

Weiss, S., Dunne, C., Hewson, J., Wohl, C., Wheatley, M., Peterson, A.C., Reynolds, B.A. (1996) Multipotent CNS stem cells are present in the adult mammalian spinal cord and ventricular neuroaxis. *J Neurosci.,* 16:7599–7609.

Whitman, M.C., Greer, C.A. (2007) Adult-generated neurons exhibit diverse developmental fates. *Dev. Neurobiol.,* 67:1079–1093.

Wilkinson, D.G. (2001) Multiple roles of EPH receptors and ephrins in neural development. Multiple roles of EPH receptors and ephrins in neural development. *Nat Rev Neurosci.,* 2:155-64.

Winkler, C., Fricker, R.A., Gates, M.A., Olsson, M., Hammang, J.P., Carpenter, M.K., Bjorklund, A. (1998) Incorporation and glial differentiation of mouse EGF-responsive neural progenitor cells after transplantation into the embryonic rat brain. *Mol Cell Neurosci.,* 11:99 -116.

World Health Organization (2007) World Health Report ; www.who.int.

Wu, P., Tarasenko, Y.I., Gu, Y., Huang, L.Y., Coggeshall, R.E., Yu, Y. (2002) Region-specific generation of cholinergic neurons from fetal human neural stem cells grafted in adult rat. *Nat Neurosci.,* 5:1271–1278.

Wu, W., Wong, K., Chen, J., Jiang, Z., Dupuis, S., Wu, J.Y., Rao, Y. (1999) Directional guidance of neuronal migration in the olfactory system by the protein Slit. *Nature,* 400:331–336.

Wurmser, A.E., Nakashima, K., Summers, R.G., Toni, N., D'Amour, K.A., Lie, D.C., Gage, F.H. (2004) Cell fusion-independent differentiation of neural stem cells to the endothelial lineage. *Nature,* 430:350 -356.

Yan, J., Xu, L., Welsh, A.M., Hatfield, G., Hazel, T., Johe, K., Koliatsos, V.E. (2007) Extensive neuronal differentiation of human neural stem cell grafts in adult rat spinal cord. *PLoS Med.,* 4:e39.

Yandava, B.D., Billinghurst, L.L., Snyder, E.Y. (1999) "Global" cell replacement is feasible via neural stem cell transplantation: evidence from the dysmyelinated shiverer mouse brain. *Proc Natl Acad Sci U S A,* 96:7029–7034.

Yang, P., Baker, K.A., Hagg, T. (2005) A disintegrin and metalloprotease 21 (ADAM21) is associated with neurogenesis and axonal growth in developing and adult rodent CNS. *J Comp Neurol.*, 490:163–179.

Yang, P., K., Baker, A., Hagg, T. (2006) The ADAMs family: Coordinators of nervous system development, plasticity and repair. *Prog Neurobiol.*, 79:73–94.

Yarden, Y., Sliwkowski, M.X. (2001) Untangling the ErbB signaling network, *Nat Rev Mol Cell Biol.*, 2:127–137.

Young, K.M., Merson, T.D., Sotthibundhu, A., Coulson, E.J., Bartlett, P.F. (2007) p75 neurotrophin receptor expression defines a population of BDNF- responsive neurogenic precursor cells. *J Neurosci.*, 27:5146–5155.

Zhang, H., Vutskits, L., Pepper, M.S., Kiss, J.Z. (2003) VEGF is a chemoattractant for FGF-2-stimulated neural progenitors *J Cell Biol.*, 163:1375–1384.

Zhang, P., Li, J., Liu, Y., Chen, X., Kang, Q. (2009a) Transplanted human embryonic neural stem cells survive, migrate, differentiate and increase endogenous nestin expression in adult rat cortical peri-infarction zone. *Neuropathology,* 29:410–421.

Zhang, P., Li, J., Liu, Y., Chen, X., Kang, Q., Zhao, J., Li, W. (2009b) Human neural stem cell transplantation attenuates apoptosis and improves neurological functions after cerebral ischemia in rats. *Acta Anaesthesiol Scand,* 53:1184-1191.

Zhang, P., Li, J., Liu, Y., Chen, X., Lu, H., Kang, Q., Li, W., Gao, M. (2010) Human embryonic neural stem cell transplantation increases subventricular zone cell proliferation and promotes peri-infarct angiogenesis after focal cerebral ischemia. *Neuropathology,* 39:1440-1789.

Zhang, R.L., Zhang, Z.G., Zhang, L., Chopp, M. (2001) Proliferation and differentiation of progenitor cells in the cortex and the subventricular zone in the adult rat after focal cerebral ischemia. *Neuroscience,* 105:33-41.

Zhao, C., Deng, W., Gage, F.H. (2008) Mechanisms and functional implications of adult neurogenesis. *Cell,* 132:645-60.

Zhao, L.R., Duan, W.M., Reyes, M., Keene, C.D., Verfaillie, C.M., Low, W.C. (2002) Human bone marrow stem cells exhibit neural phenotypes and ameliorate neurological deficits after grafting into the ischemic brain of rats *Exp. Neurol.,* 174:11–20.

Zigova, T., Pencea, V., Wiegand, S.J., Luskin, M.B. (1998) Intraventricular administration of BDNF increases the number of newly generated neurons in the adult olfactory bulb. *Mol Cell Neurosci.*, 11:234–245.

16

Ischemia-Induced Neural Stem/Progenitor Cells Within the Post-Stroke Cortex in Adult Brains

Takayuki Nakagomi and Tomohiro Matsuyama
Institute for Advanced Medical Sciences,
Hyogo College of Medicine
Hyogo,
Japan

1. Introduction

Stroke is one of the major causes of death and disability in developed countries. The central nervous system (CNS) is known for its limited reparative capacity, but several studies demonstrated that the CNS has some reparative potential and cerebral ischemia is followed by activation of endogenous neurogenesis (Nakatomi *et al.*, 2002; Taguchi *et al.*, 2004). It is well-known that new neurons are continuously generated in specific brain regions such as the subventricular zones (SVZ) (Alvarez-Buylla *et al.*, 2002) and the subgranular zone within the dentate gyrus of the hippocampus (SGZ) (Kuhn *et al.*, 1996). Although adult cerebral cortical neurogenesis remains controversial, accumulating evidence has shown that under pathological conditions, new neurons are generated in the adult mammalian cerebral cortex (Magavi *et al.*, 2000; Jiang *et al.*, 2001; Jin *et al.*, 2006; Yang *et al.*, 2007). This suggests that neural stem/progenitor cells (NSPCs) can be activated in the cortex by brain injury such as ischemic stroke. In support of this notion, we demonstrated that NSPCs develop in the post-stroke area of the cortex in the adult murine (Nakagomi *et al.*, 2009a; Nakagomi *et al.*, 2009b; Nakano-Doi *et al.*, 2010; Saino *et al.*, 2010) and human brain (Nakayama *et al.*, 2010), and we referred to these as ischemia/injury-induced NSPCs (iNSPCs). These cells express markers of NSPCs, such as nestin and Sox2. They also form neurospheres that have the capacity for self-renewal, and differentiate into electrophysiologically functional neurons, astrocytes, and myelin-producing oligodendrocytes (Nakagomi *et al.*, 2009a; Nakagomi *et al.*, 2009b; Nakano-Doi *et al.*, 2010; Clausen *et al.*, 2011). In addition, we demonstrated that iNSPCs originate, at least in part, from within the cerebral cortex, but not from SVZ cells (Nakagomi *et al.*, 2009b). However, the detailed origin and identity of the iNSPCs remains unclear. In this chapter, we introduce the characterization and possible origin of iNSPCs based on our reports and recent viewpoint, and compare them to other previously reported types of CNS stem/progenitor cells, including SVZ astrocytes (Doetsch *et al.*, 1999), ependymal cells (Moreno-Manzano *et al.*, 2009), reactive astrocytes (Shimada *et al.*, 2010), resident glia (Zawadzka *et al.*, 2010), and oligodendrocyte precursor cells (OPCs) (Kondo *et al.*, 2000). We also refer to the possible cortical neurogenesis by iNSPCs and to the therapeutic potential of iNSPC transplantation in stroke patients.

2. NSPCs in the adult cortex

In the CNS of adult mammals, it is well-known that NSPCs are present in the SVZ and SVG, and that ongoing neurogenesis is retained in these two zones. However, accumulating evidence suggests that NSPCs reside in many parts of the adult brain including the cortex (Arsenijevic *et al.*, 2001; Joh *et al.*, 2005; Kallur *et al.*, 2006; Jiao *et al.*, 2008; Willaime-Morawek *et al.*, 2008), striatum (Kallur *et al.*, 2006; Willaime-Morawek *et al.*, 2008), subcortical white matter (Nunes *et al.*, 2003), and spinal cord (Weiss *et al.*, 1996; Parr *et al.*, 2008). These observations suggest that NSPCs are widely distributed throughout the adult CNS. In this chapter, we introduce iNSPCs, which are induced within the post-stroke cortex after brain injury/ischemia in adult brains.

2.1 Cortical development in the embryonic stage: comparison to iNSPCs in the cortex

In the embryonic stage, neurogenesis was observed throughout the CNS including the cortex. Mignone and colleagues traced nestin-expressing NSPCs, and showed that green fluorescent protein (GFP) expression in developing transgenic nestin-GFP mice was evident on as early as day 7 of embryonic development (e7). At e8, a GFP signal was observed predominantly in the neural plate, and by e10 intense GFP fluorescence was observed throughout the neuroepithelium. At e10 to e12, GFP signals marked the entire thickness of the cerebral wall, but GFP expression became weaker near the pial surface and stronger in the ventricular zones starting from e12. Finally, in the adult brain, GFP was selectively expressed in the SVZ and SGZ in areas related to continuous neurogenesis (Mignone *et al.*, 2004). Thus, in the postnatal CNS, constitutive neurogenesis is known to be retained in only two regions the SVZ (Alvarez-Buylla *et al.*, 2002) and SGZ (Kuhn *et al.*, 1996). However, under pathological conditions, neurogenesis may occur again in the adult cerebral cortex (Magavi *et al.*, 2000; Jiang *et al.*, 2001; Jin *et al.*, 2006; Yang *et al.*, 2007). Supporting their observations, nestin-positive NSPCs were observed after brain injury/ischemia in nonconventional neurogenic zones, such as the cortex (Nakagomi *et al.*, 2009b; Nakayama *et al.*, 2010). Because they were rarely observed in the absence of brain injury (Nakagomi *et al.*, 2009b), cortical neurogenesis may reoccur only in the case of brain injury. These findings suggest that in adult mammalian brains, NSPC activation and neuronal homeostasis are maintained under physiological conditions, at least in part, in specific brain regions, such as the SVZ and SGZ. However, after brain injury, it appears that regional NSPCs are mobilized to accelerate tissue repair by a mechanism similar to embryonic neurogenesis. Taken together, these observations suggest that ischemia/hypoxia is essential for the induction of NSPCs in the adult cortex, although we remain unaware of the required signaling and/or factors.

2.1.1 Characteristics of iNSPCs from the post-stroke cortex

To confirm the possible adult neurogenesis induced by brain injury, we have sought to isolate NSPCs from the injured area of the post-stroke cortex. Previously, we established a highly reproducible murine model of cortical infarction using CB-17/Icr-+/+Jcl and CB-17/Icr-Scid/scid Jcl mice. The infarct area in mice of this background has been limited to the ipsilateral cerebral cortex of the territory occupied by the middle cerebral artery (MCA) (Taguchi *et al.*, 2004; Taguchi *et al.*, 2007; Nakagomi *et al.*, 2009a; Nakagomi *et al.*, 2009b;

Nakano-Doi *et al.*, 2010; Saino *et al.*, 2010; Taguchi *et al.*, 2010). Following MCA occlusion, abundant nestin-positive cells emerged within the post-stroke cortex, although they were rarely observed in the non-ischemic cortex (Nakagomi *et al.*, 2009b; Nakano-Doi *et al.*, 2010; Saino *et al.*, 2010). To examine whether these cells showed stem cell-characteristics, we cultured cells isolated from the post-stroke cortex under conditions that promoted the formation of neurospheres (Reynolds *et al.*, 1992). In brief, tissue from the ischemic core of the post-infarct cerebral cortex was obtained on day 7 after MCA occlusion. Cells were dissociated by passage through 23 and 27 gauge needles, and cell suspensions were incubated in tissue culture flasks with DMEM containing epidermal growth factor (EGF), basic fibroblast growth factor (bFGF) and N2 supplement. This procedure allowed us to obtain nestin-positive neurosphere-like cell clusters (iNSPCs) (Nakagomi *et al.*, 2009a; Nakagomi *et al.*, 2009b; Nakano-Doi *et al.*, 2010) (Fig. 1).

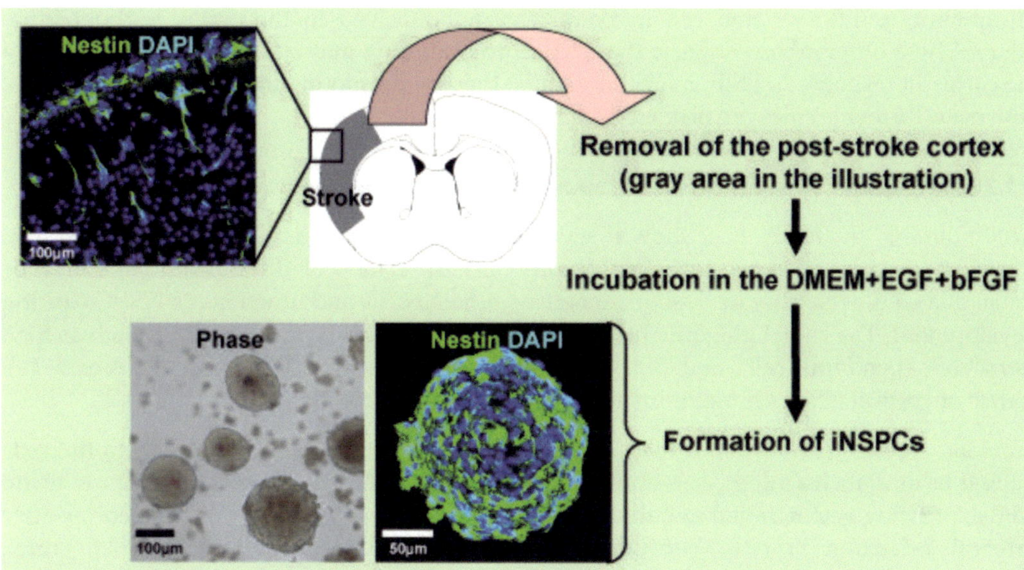

Fig. 1. Isolation of nestin-positive iNSPCs developing within the post-stroke cortex

However, iNSPCs were rarely obtained in the absence of brain injury. Notably, because we could not obtain iNSPCs from the peri-stroke cortex, it is possible that these cells are generated within the degenerating cortical tissue after ischemic stroke.

Uptake of 5-bromo-2'-deoxyuridine (BrdU) by iNSPCs was confirmed *in vivo* (Nakano-Doi *et al.*, 2010; Saino *et al.*, 2010) and *in vitro* (Nakagomi *et al.*, 2009a; Nakagomi *et al.*, 2009b), showing that they have the proliferative activity. They possessed self-renewal capacity, which was confirmed by a clonal assay. However, in contrast to the embryonic stem cells, the cluster formation in the same medium at a clonal density was limited to between three and five cell passages (Nakagomi *et al.*, 2009b; Nakano-Doi *et al.*, 2010), consistent with other adult candidates of stem cells, such as neurospheres derived from hippocampus (Bull *et al.*, 2005) and subcortical white matter (Nunes *et al.*, 2003). These observations suggest that cortex-derived iNSPCs are more likely to be neural progenitors than neural stem cells (NSCs). However, they certainly differentiated into electrophysiologically functional neurons, astrocytes, and myelin-producing oligodendrocytes (Nakagomi *et al.*, 2009a; Nakagomi *et al.*, 2009b; Nakano-Doi *et al.*, 2010), indicating that iNSPCs have a stemness-capacity similar to other adult NSPCs. Interestingly, they predominantly differentiated into neurons (approximately 35%) and oligodendrocytes (approximately 30%) rather than astrocytes (approximately 5%) (Nakagomi *et al.*, 2009a; Nakano-Doi *et al.*, 2010; Nakagomi *et al.*, 2011) with characteristics discriminating from other adult NSPCs such as SVZ astrocytes, most of which are known to differentiate into astrocytes. These findings suggest that iNSPCs have a strong potential of contributing to cortical neurogenesis compared to NSPCs derived from other origins, especially under the conditions of brain injury.

Consistent with the SVZ-derived NSPCs (Kim *et al.*, 2009b), cortical iNSPCs expressed several pluripotent/undifferentiated cell markers, including Sox2, Klf4, c-myc and Nanog (Nakagomi *et al.*, 2009b; Nakagomi *et al.*, 2011). However, expression of various pluripotent/undifferentiated cell markers was not observed in the cortex without brain injury. These observations suggest that cell reprogramming may occur in unknown cells of the cortex in response to brain injury/ischemia, thereby promoting the induction of iNSPCs. However, further studies are needed to clarify this hypothesis.

2.1.2 Comparison to other types of reported CNS stem/progenitor cells

Accumulating evidence has shown several candidates for adult NSPCs, which can contribute to adult neurogenesis in the cerebral cortex. One of these candidates may be radial glia cells, which are derived from neuroepithelial cells and functions as NSPCs during development. The radial glia are able to develop into several types of NSPCs, such as SVZ astrocytes, ependymal cells, and OPCs in adult (Kriegstein *et al.*, 2009). However, precise cell source of cortical NSPCs remains unclear, especially in the injured brain.

Previous studies demonstrated that SVZ astrocytes have the capacity to migrate towards injured lesions, including the cerebral cortex (Goings *et al.*, 2004). However, our study using GFP-expressing vector, failed to demonstrate cell migration from the SVZ to the cortex after cerebral infarction *in vivo*, but demonstrated that iNSPCs in the post-stroke cortex originated, at least in part, from the cerebral cortex (Nakagomi *et al.*, 2009b). Consistently, subsequent studies showed that NSPCs developing within and around the post-stroke cortex are derived from locally activated stem/progenitor cells, but not from SVZ cells (Ohira *et al.*, 2010; Shimada *et al.*, 2010). To answer which cells can be activated by cerebral injury, some studies proposed the reactive astrocytes as a source of injury-induced NSPCs (Oki *et al.*, 2010; Shimada *et al.*, 2010), because NSPCs express the astrocyte marker, GFAP. However, we could not detect GFAP- (Nakagomi *et al.*, 2011) and S100β-positive astrocytes within the post-stroke cortex (Nakagomi *et al.*, 2009b). Eventually, the isolated iNSPCs from the infarct cortex rarely expressed GFAP and developed few astrocytic traits even after differentiation (Nakagomi *et al.*, 2011). In addition, although we found some nestin and GFAP double-positive reactive astrocyte-like cells in the peri-infarct area, we could not obtain neurospheres from these areas. These findings strongly suggest that the source of iNSPCs within the infarct cortex is distinct from reactive astrocytes.

Currently, it is still highly controversial whether periventricular NSPCs can be derived from SVZ astrocytes, ependymal cells, or both (Chojnacki *et al.*, 2009). Ependymal cells were originally considered to be the resident stem cell population in the wall of the lateral ventricle, in which they locate nearby perivascular cells (Pfenninger *et al.*, 2007; Coskun *et al.*, 2008). Although it is controversial whether ependymal cells have NSPC activity or not, recent studies confirmed that ependymal cells do not play a role in adult neurogenesis under normal conditions, but do possess NSPC activity and can differentiate into neurons, astrocytes, and oligodendrocytes in response to the CNS injuries including ischemic stroke (Carlen *et al.*, 2009; Moreno-Manzano *et al.*, 2009). Furthermore, ependymal cells express PDGFRα (Danilov *et al.*, 2009) and NG2 (Moreno-Manzano *et al.*, 2009), and have the structure of lipid droplets, microvilli, and cilia (Coskun *et al.*, 2008; Danilov *et al.*, 2009). Consistent with the traits of ependymal cells, iNSPCs express PDGFRα and NG2, but do not possess microvilli-like structures (Nakagomi *et al.*, 2011). These findings indicate that iNSPCs do not have completely identical characteristics to those of ependymal cells.

Adult OPCs comprise approximately 5%–8% of the glial cell population in the CNS. Their function in the CNS remains unknown, although accumulating evidence has shown that they have NSPC activity (Kondo *et al.*, 2000; Gaughwin *et al.*, 2006), in addition to myelin-producing abilities (Sundberg *et al.*, 2010). OPCs are known to express NG2 (Ulrich *et al.*, 2008) and PDGFRα (Hall *et al.*, 1996), and OPCs expressing A2B5 have NSPC activity (Kondo *et al.*, 2000; Gaughwin *et al.*, 2006). To investigate whether iNSPCs are derived from OPCs, we analyzed OPC markers expressed by iNSPCs *in vivo* and *in vitro*. Although iNSPCs express some OPC markers such as NG2 and PDGFRα, they do not possess A2B5 or even Olig2 (another OPC marker) (Billon *et al.*, 2002). These observations indicate that iNSPCs are different from reported multipotent OPCs (Kondo *et al.*, 2000; Gaughwin *et al.*, 2006). However, when iNSPCs were incubated in OPC-promoting medium (Chen *et al.*, 2007), they began to express Olig2. In addition, almost all cells developed from iNSPCs in this medium differentiated into O4- and/or myelin-associated glycoprotein (MAG)-positive oligodendrocytes (Nakagomi *et al.*, 2011). These findings suggest that iNSPCs express some OPC markers during their development/differentiation.

It is well-known that NG2 is not only the marker of OPC, but is also the marker of resident glial cells/glial progenitors (Stallcup *et al.*, 1987). More recently, NG2-positive resident glia was reported to develop NSPC activity after brain injury (Yokoyama *et al.*, 2006; Zawadzka *et al.*, 2010). We demonstrated that cortical iNSPCs express NG2 and PDGFRα in a similar manner to resident glial/progenitor cells. However, neuronal differentiation from NG2- and/or PDGFRα-positive glial cells is rarely observed (Zawadzka *et al.*, 2010; Richardson *et al.*, 2011), suggesting that iNSPCs may be different from these glial cells or belonging to unknown cell type, which expresses some glial markers.

2.1.3 What is the origin of iNSPCs in the cortex?

So far, it seems possible that iNSPCs are different from previously proposed CNS stem/progenitor cells such as SVZ astrocytes, reactive astrocytes, ependymal cells, or OPCs. The essential difference of these cells may be their induction pattern and localization, because iNSPCs were found only after ischemic insult, and in close association with the blood vessels in the cortex. This unique localization allowed us to examine the characteristics of cells nearby blood vessels as a candidate of iNSPCs.

Our studies showed that the nestin-positive iNSPCs developed in the perivascular regions of the post-stroke cortex (Nakano-Doi *et al.*, 2010; Nakayama *et al.*, 2010), where nestin-positive cells express NG2 and PDGFRβ (both of which are the pericyte marker), suggesting that the iNSPCs are derived from pericytes. Pericytes with multipotent progenitor activity have been indentified in various organs (Crisan *et al.*, 2009) as well as in the CNS (Dore-Duffy *et al.*, 2006). In addition, Dore-Duffy and colleagues (Dore-Duffy *et al.*, 2006) showed that pericyte-derived NSPCs can be isolated from the CNS of non-injured animals. However, we hardly obtained iNSPCs from the nonischemic cortex (Nakagomi *et al.*, 2009b), suggesting that pericytes in the cortical tissues increase their stemness activity during the progression of cerebral injury.

Increasing evidence has shown that ischemic insult promotes stem cell activity, and NSPCs (Sirko *et al.*, 2009; Xue *et al.*, 2009) and neuronal progenitors (Ohira *et al.*, 2010) are also induced in response to cortical ischemic injury. These cortical NSPCs are frequently observed at the subpial/cortical layer 1 regions, suggesting that NSPCs can be activated preferentially in the cortical surface. Independent of these studies, we found nestin/Sox2-positive iNSPCs proliferating in the pia mater, which covers the surface of the post-ischemic cortex (Nakagomi *et al.*, 2011). Pia mater is widely distributed throughout the CNS, and is closely associated with the blood vessels. It has been reported that leptomeninges (including pia mater and arachnoid membrane) regulate NSPCs (Sockanathan *et al.*, 2009) and cortical neuron generation (Siegenthaler *et al.*, 2009) in embryonic cortical formation, and function as a niche for stem/progenitor cells with neuronal differentiation potential (Bifari *et al.*, 2009). These findings suggest that pia mater contains NSPCs at embryonic stage. The pial iNSPCs, which we found in the adult brain, partially spread into the cortical parenchyma as perivascular cells/pericytes with expression of pericyte markers such as NG2 and PDGFRβ. In addition, cells isolated from the infarcted area including pia mater and cortex and sorted by magnetic cell sorting (MACS) with a pericyte marker (PDGFRβ) had NSPC activity and differentiated into neurons (Nakagomi *et al.*, 2011). These findings indicate that the microvascular pericytes that distribute from the pia mater to the cortex are a potential source of the iNSPCs (Fig. 2).

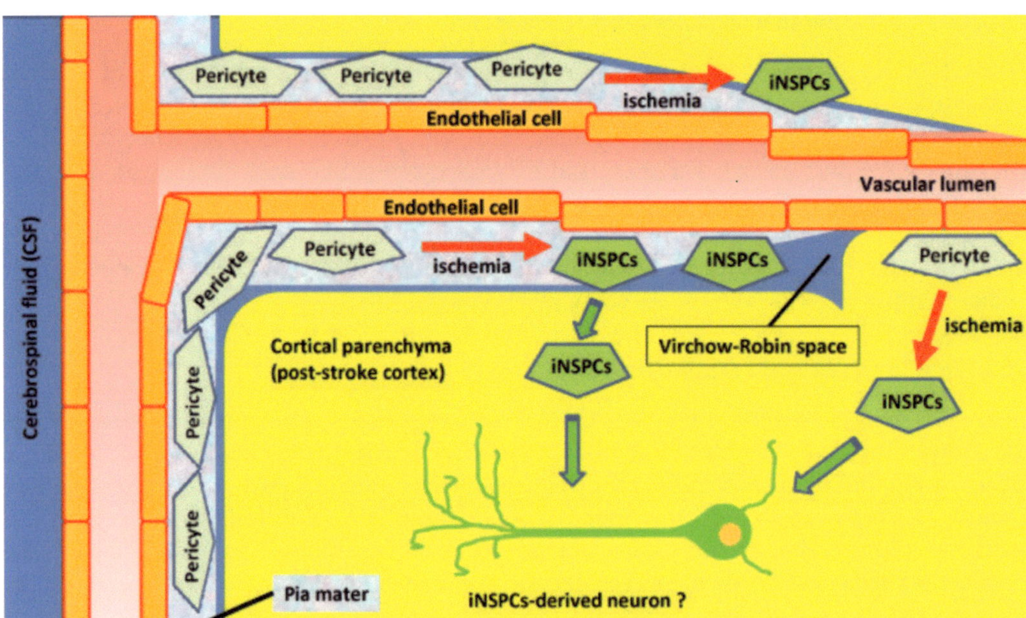

Fig. 2. Schematic representation for the fate of iNSPCs following cortical infarction

Thus, our recent study suggests that pia mater may have the potential to generate NSPCs even in the adult brain. Until now, it has been demonstrated that pia mater, as well as some CNS pericytes, originate from the neural crest (Morse *et al.*, 1984; Etchevers *et al.*, 2001). We recently demonstrated that pial iNSPCs express various neural crest markers, such as Sox9, Sox10, Snail, Slug, and Twist (Aihara *et al.*, 2010; Nakagomi *et al.*, 2011), suggesting that pial iNSPCs are neural crest derivatives. This may provide a novel concept that neural crest-derived cells play a crucial role in the CNS repair following cortical infarction by a similar mechanism to the CNS formation in development. Considering that the neural crest has stem cell potential (neural crest-derived stem cells) (Teng *et al.*, 2006) with differentiation into a variety of cell types including neurons and glia (Nagoshi *et al.*, 2009), this hypothesis would not be surprising. In addition, this might explain why Schwann cells, which have neural crest origin, are induced in the injured CNS (Zawadzka *et al.*, 2010). However, the precise source, lineage, and traits of iNSPCs warrants further investigation. This may be clarified through experiment of lineage labeling by genetic means.

2.2 Potential contribution of endogenous iNSPCs to cortical neurogenesis

Although iNSPCs are generated within the post-stroke area following cortical infarction, almost all of them can undergo apoptotic cell death (Saino *et al.*, 2010). Subsequently, appropriate support for survival of iNSPCs is essential in maintaining post-stroke neurogenesis. Because iNSPCs developed in close association with the blood vessels from the pia mater to the cortex, they must be influenced by the vascular microenvironment, consisting of endothelial cells (ECs) (Palmer *et al.*, 2000; Louissaint *et al.*, 2002) and inflammatory cells infiltrated after cerebral injury (Saino *et al.*, 2010).

ECs are a component of the blood brain barrier (BBB) and also function as a vascular niche (Shen *et al.*, 2008). It has been reported that although inflammation exacerbates post-stroke neuronal damage, inflammation is a strong stimulus for activation of neurogenesis. Such inflammatory reactions may happen in perivascular (Virchow-Robin) spaces (Hutchings *et al.*, 1986), in which inflammatory cells such as macrophage and lymphocytes infiltrate and may affect angiogenesis and neurogenesis after brain injury. These factors should be considered when observing cortical neurogenesis through iNSPCs after ischemic stroke.

NSPCs reside in a vascular niche and the vasculature is regarded as a key element, especially in the adult SVZ (Tavazoie *et al.*, 2008). ECs are believed to make valuable contribution to this vascular microenvironment (Palmer *et al.*, 2000; Louissaint *et al.*, 2002). In support of this viewpoint, co-culture experiments showed that ECs increase proliferation of NSPCs derived from the adult SVZ (Shen *et al.*, 2004; Teng *et al.*, 2008). Furthermore, we demonstrated both *in vitro* and *in vivo*, that the presence of ECs enhances survival, proliferation, migration, and differentiation of iNSPCs (Nakagomi *et al.*, 2009a), indicating that augmentation of ECs (e.g., proliferation of ECs [angiogenesis]) can promote neurogenesis by enhancing the proliferation of endogenous iNSPCs.

Thus, therapeutic angiogenesis may enhance endogenous neurogenesis even after cerebral injury (Hamano *et al.*, 2000; Chen *et al.*, 2003). It has been reported that bone marrow cells (BMCs) such as bone marrow mononuclear cells (BMMCs) (Li *et al.*, 2006; Kim *et al.*, 2009a; Ribeiro-Resende *et al.*, 2009) and mesenchymal stem cells (MSCs) (Labouyrie *et al.*, 1999; Mahmood *et al.*, 2004; Kurozumi *et al.*, 2005) induce angiogenic effects by secreting multiple growth factors including vascular endothelial growth factor (VEGF), glia-derived neurotrophic factor (GDNF), brain-derived neurotrophic factor (BDNF), nerve growth factor (NGF), and hepatocyte growth factor (HGF). We showed that BMMCs can contribute to the proliferation of endogenous iNSPCs through vascular niche regulation, which includes EC proliferation following cortical infarction (Nakano-Doi *et al.*, 2010).

In addition to ECs, astrocytes are also reported to be important niche cells for NSPCs in the SVZ (Song *et al.*, 2002), SGZ (Lim *et al.*, 1999) and cortex (Jiao *et al.*, 2008). Our study already showed that astrocytes, as well as ECs, promote the proliferation of iNSPCs (Nakagomi *et al.*, 2009a), suggesting that astrocytes function as a niche for cortex-derived iNSPCs. Although astrocytes were not observed within the post-stroke cortex after permanent ischemia (Nakagomi *et al.*, 2011), astrocytes are resistant to hypoxia/ischemia and they can still survive after transient ischemia (Li *et al.*, 1995). These findings might explain the reason why new-born neurons are frequently found in the post-stroke cortex after mild transient ischemia (Ohira *et al.*, 2010), but are not seen after severe permanent ischemia (Nakagomi *et al.*, 2009b).

Regulation of the immune system has also been proposed as one of the key factors in enhancing neurogenesis and functional recovery after stroke. Our studies showed that T lymphocytes, mainly CD4- but not CD8-positive T cells, induce apoptosis in iNSPCs (Saino *et al.*, 2010; Takata *et al.*, 2011). The details of the mechanism are still under investigation, but these findings suggest that the immune response and/or enhanced inflammation triggered by CD4-positive T cells, are major deteriorating modulators of post-stroke neurogenesis. These findings, at least in part, are consistent with previous results demonstrating that transplantation of mesenchymal cells accelerates endogenous neurogenesis after stroke (Li *et al.*, 2008; Yoo *et al.*, 2008), because such treatment is known to suppress the immune response in graft-versus-host disease.

2.3 Exogenous iNSPC transplantation after cerebral infarction

Compared to the strategy focusing on enhanced endogenous neurogenesis, exogenous NSPC transplantation may have some advantages in treating stroke patients; this therapy allows a longer therapeutic time window to administer larger numbers of stem cells, and to repeat the treatment. The therapeutic time window to enhance the endogenous neurogenesis may be limited, because we observed that neurogenesis peaks for several days and ends within a few weeks after stroke onset in patients (Nakayama *et al.*, 2010).

Until now, various cell sources for exogenous NSPC transplantation have been proposed; e.g., fetal brain (Ishibashi *et al.*, 2004; Kelly *et al.*, 2004; Cayre *et al.*, 2006; Darsalia *et al.*, 2007), adult brain tissue obtained from the SVZ (Cayre *et al.*, 2006; Hicks *et al.*, 2007; Kameda *et al.*, 2007), gene transfected bone marrow cells (Dezawa *et al.*, 2004), immortalized tumor cell lines (Staines *et al.*, 1994), embryonic stem (ES) cells/induced pluripotent stem cell (iPS) cells (Bjorklund *et al.*, 2002; Wei *et al.*, 2005; Buhnemann *et al.*, 2006) and *ex vivo* expanded cortex-derived iNSPCs (Nakagomi *et al.*, 2009a; Nakagomi *et al.*, 2009b; Nakano-Doi *et al.*, 2010). Transplantation of exogenous NSPCs can be performed even at the chronic stage of post-stroke. In experimental models of stroke using fetal NSPCs, transplanted cells were reported to survive within the host brain, migrate into the injured area, and maintain their multipotency (Ishibashi *et al.*, 2004; Kelly *et al.*, 2004; Darsalia *et al.*, 2007). However, there are some issues to be solved for clinical application of exogenous NSPC transplantation in stroke patients; e.g., survival, safety and suitability of transplanted cells, and their capacity to repair injured adult brain. Indeed, the other lines of experiment using NSPCs derived from adult mammalian brains showed that only a small population of grafted cells can survive in the injured brain (Toda *et al.*, 2001; Hicks *et al.*, 2007; Kameda *et al.*, 2007; Takahashi *et al.*, 2008). Consistent with these reports, we showed that the majority of transplanted iNSPCs, which are derived from the adult cortex, cannot survive in the injured cortex (Nakagomi *et al.*, 2009a).

A higher survival rate of transplanted NSPCs carrying the property of neoplasm (such as ES/iPS-derived NSPCs) can be expected, because survival is often attributed to a lack of apoptotic signaling. However, this property may be directly linked to a high risk of tumorigenesis. Whether the transplanted fetal NSPCs will be able to contribute to reconstitution of the adult brain is also an issue to be addressed, because they are the cells destined to form the infant brain. Therefore, we must achieve significant recovery of impaired neurological functions of the adult brain to determine the suitability of transplanted cells.

Our study showed that iNSPCs from the injured cortex differentiate into functional neurons with less tumorigenesis, suggesting that these cells are one of the most suitable NSPCs for transplantation. Therefore, we may choose alternative ways to continue the survival of transplanted iNSPCs. Recently, we reported that co-transplantation of iNSPCs with ECs as a vascular niche, enhances functional recovery after cortical infarction with longer survival of transplanted cells (Nakagomi *et al.*, 2009a). This suggests that the microenvironment around the transplants has to be considered for cell therapy. From another point of view, as differentiated cells are more resistant to apoptotic cell death, enhancing differentiation of NSPCs into mature neurons may be a choice in maintaining the transplant. Recent studies showed that transplantation of NSPCs with valproic acid, which inhibits proliferation but enhances differentiation of transplanted stem cells to functional neurons, significantly improves motor function in a spinal cord injury model (Abematsu *et al.*, 2010). These results may indicate a future direction for the clinical application of exogenous NSPC transplantation for patients after cerebral infarction.

Another problem regarding cell transplantation is the difficulty in regulating the differentiation of transplanted NSPCs *in vivo*. It is well-known that a variety of chemical mediators/cytokines are produced/activated at the site of brain injury, and among these, IL-6, CNTF, and BMPs promote differentiation of NSPCs into the astrocytic phenotype (Nakashima *et al.*, 1999; Okada *et al.*, 2004). Our previous studies showed that transplanted iNSPCs largely differentiated into glial cells *in vivo*, although they predominantly differentiated into neuronal cells *in vitro* (Nakagomi *et al.*, 2009a). These results suggest that the neurogenesis-oriented regulation of transplanted iNSPCs might accomplish a real functional restoration of stroke patients in the future.

3. Conclusion

In conclusion, we demonstrated that iNSPCs, which are the potential cell sources for neocortical neurogenesis, develop in the murine post-stroke cortex (Nakagomi *et al.*, 2009a; Nakagomi *et al.*, 2009b; Nakano-Doi *et al.*, 2010; Saino *et al.*, 2010). Furthermore, we demonstrat that iNSPCs develop within the post-stroke pia mater, suggesting that pia mater is an important target for cortical neurogenesis. In the field of cardiology, accumulating evidence has shown that cardiac stem/progenitor cells reside in epicardium, termed as "epicardial progenitor cells" (Zhou *et al.*, 2008; Smart *et al.*, 2011). These findings may raise a possibility that stem/progenitor cells are present in the surface of multiple organs as well as those observed in the brain and heart. In addition, deposition of several materials including cell sheets onto infarcted heart could improve cardiac repair and functions after myocardial infarction (Zakharova *et al.*; Miyahara *et al.*, 2006; Derval *et al.*, 2008). Thus, patches of cell sheets carrying bioactive substances on post-stroke pia mater may promote cortical repair/neurogenesis without parenchymal damage, due to the intracerebral approach of cell transplantation.

In the past, cerebral infarction was believed to be a region occupied only by necrotic tissue and inflammatory cells. However, we detected viable cells with the capacity for proliferation, differentiation, and multipotency within the post-stroke cortex and pia mater in an experimental murine model of ischemic stroke. Because similar iNSPCs were detected in the post-stroke human cortex (Nakayama *et al.*, 2010), further investigation will establish novel therapeutic neurogenesis for stroke patients by iNSPCs.

4. Acknowledgement

Our series of studies were partially supported by a Grant-in-Aid for Scientific Research from the Ministry of Education, Culture, Sports, Science and Technology [(18800071 and 21700363; to T. Nakagomi) and (21500359; to T. Matsuyama)], Hyogo Science and Technology Association (2009), Takeda Science Foundation (2009) and Grant-in-Aid for Researchers, Hyogo College of Medicine (2006 and 2011). We would like to thank Dr. A. Nakano-Doi, Y. Okinaka, Y. Tanaka, Y. Tatsumi, Dr. O. Saino, Dr. Y. Momota, Dr. M. Takata, Dr. N. Nakagomi, Y. Kasahara, Dr. A. Taguchi and Prof. Z. Molnár for helpful assistances.

5. References

Abematsu, M., Tsujimura, K., Yamano, M., et al (2010) Neurons derived from transplanted neural stem cells restore disrupted neuronal circuitry in a mouse model of spinal cord injury. *J Clin Invest*, 120, 3255-3266.

Aihara, Y., Hayashi, Y., Hirata, M., et al (2010) Induction of neural crest cells from mouse embryonic stem cells in a serum-free monolayer culture. *Int J Dev Biol*, 54, 1287-1294.

Alvarez-Buylla, A. & Garcia-Verdugo, J.M. (2002) Neurogenesis in adult subventricular zone. *J Neurosci*, 22, 629-634.

Arsenijevic, Y., Villemure, J.G., Brunet, J.F., et al (2001) Isolation of multipotent neural precursors residing in the cortex of the adult human brain. *Exp Neurol*, 170, 48-62.

Bifari, F., Decimo, I., Chiamulera, C., et al (2009) Novel stem/progenitor cells with neuronal differentiation potential reside in the leptomeningeal niche. *J Cell Mol Med*, 13, 3195-3208.

Billon, N., Jolicoeur, C., Ying, Q.L., et al (2002) Normal timing of oligodendrocyte development from genetically engineered, lineage-selectable mouse ES cells. *J Cell Sci*, 115, 3657-3665.

Bjorklund, L.M., Sanchez-Pernaute, R., Chung, S., et al (2002) Embryonic stem cells develop into functional dopaminergic neurons after transplantation in a Parkinson rat model. *Proc Natl Acad Sci U S A*, 99, 2344-2349.

Buhnemann, C., Scholz, A., Bernreuther, C., et al (2006) Neuronal differentiation of transplanted embryonic stem cell-derived precursors in stroke lesions of adult rats. *Brain*, 129, 3238-3248.

Bull, N.D. & Bartlett, P.F. (2005) The adult mouse hippocampal progenitor is neurogenic but not a stem cell. *J Neurosci*, 25, 10815-10821.

Carlen, M., Meletis, K., Goritz, C., et al (2009) Forebrain ependymal cells are Notch-dependent and generate neuroblasts and astrocytes after stroke. *Nat Neurosci*, 12, 259-267.

Cayre, M., Bancila, M., Virard, I., et al (2006) Migrating and myelinating potential of subventricular zone neural progenitor cells in white matter tracts of the adult rodent brain. *Mol Cell Neurosci*, 31, 748-758.

Chen, J., Li, Y., Katakowski, M., et al (2003) Intravenous bone marrow stromal cell therapy reduces apoptosis and promotes endogenous cell proliferation after stroke in female rat. *J Neurosci Res*, 73, 778-786.

Chen, Y., Balasubramaniyan, V., Peng, J., et al (2007) Isolation and culture of rat and mouse oligodendrocyte precursor cells. *Nat Protoc*, 2, 1044-1051.

Chojnacki, A.K., Mak, G.K. & Weiss, S. (2009) Identity crisis for adult periventricular neural stem cells: subventricular zone astrocytes, ependymal cells or both? *Nat Rev Neurosci*, 10, 153-163.

Clausen, M., Nakagomi, T., Nakano-Doi, A., et al. (2011) Ischemia-induced neural stem/progenitor cells express pyramidal cells markers. Neuroreport, 22, 789-794.

Coskun, V., Wu, H., Blanchi, B., et al (2008) CD133+ neural stem cells in the ependyma of mammalian postnatal forebrain. *Proc Natl Acad Sci U S A*, 105, 1026-1031.

Crisan, M., Chen, C.W., Corselli, M., et al (2009) Perivascular multipotent progenitor cells in human organs. *Ann N Y Acad Sci*, 1176, 118-123.

Danilov, A.I., Gomes-Leal, W., Ahlenius, H., et al (2009) Ultrastructural and antigenic properties of neural stem cells and their progeny in adult rat subventricular zone. *Glia*, 57, 136-152.

Darsalia, V., Kallur, T. & Kokaia, Z. (2007) Survival, migration and neuronal differentiation of human fetal striatal and cortical neural stem cells grafted in stroke-damaged rat striatum. *Eur J Neurosci*, 26, 605-614.

Derval, N., Barandon, L., Dufourcq, P., et al (2008) Epicardial deposition of endothelial progenitor and mesenchymal stem cells in a coated muscle patch after myocardial infarction in a murine model. *Eur J Cardiothorac Surg*, 34, 248-254.

Dezawa, M., Kanno, H., Hoshino, M., et al (2004) Specific induction of neuronal cells from bone marrow stromal cells and application for autologous transplantation. *J Clin Invest*, 113, 1701-1710.

Doetsch, F., Caille, I., Lim, D.A., et al (1999) Subventricular zone astrocytes are neural stem cells in the adult mammalian brain. *Cell*, 97, 703-716.

Dore-Duffy, P., Katychev, A., Wang, X., et al (2006) CNS microvascular pericytes exhibit multipotential stem cell activity. *J Cereb Blood Flow Metab*, 26, 613-624.

Etchevers, H.C., Vincent, C., Le Douarin, N.M., et al (2001) The cephalic neural crest provides pericytes and smooth muscle cells to all blood vessels of the face and forebrain. *Development*, 128, 1059-1068.

Gaughwin, P.M., Caldwell, M.A., Anderson, J.M., et al (2006) Astrocytes promote neurogenesis from oligodendrocyte precursor cells. *Eur J Neurosci*, 23, 945-956.

Goings, G.E., Sahni, V. & Szele, F.G. (2004) Migration patterns of subventricular zone cells in adult mice change after cerebral cortex injury. *Brain Res*, 996, 213-226.

Hall, A., Giese, N.A. & Richardson, W.D. (1996) Spinal cord oligodendrocytes develop from ventrally derived progenitor cells that express PDGF alpha-receptors. *Development*, 122, 4085-4094.

Hamano, K., Li, T.S., Kobayashi, T., et al (2000) Angiogenesis induced by the implantation of self-bone marrow cells: a new material for therapeutic angiogenesis. *Cell Transplant*, 9, 439-443.

Hicks, A.U., Hewlett, K., Windle, V., et al (2007) Enriched environment enhances transplanted subventricular zone stem cell migration and functional recovery after stroke. *Neuroscience*, 146, 31-40.

Hutchings, M. & Weller, R.O. (1986) Anatomical relationships of the pia mater to cerebral blood vessels in man. *J Neurosurg*, 65, 316-325.

Ishibashi, S., Sakaguchi, M., Kuroiwa, T., et al (2004) Human neural stem/progenitor cells, expanded in long-term neurosphere culture, promote functional recovery after focal ischemia in Mongolian gerbils. *J Neurosci Res*, 78, 215-223.

Jiang, W., Gu, W., Brannstrom, T., et al (2001) Cortical neurogenesis in adult rats after transient middle cerebral artery occlusion. *Stroke*, 32, 1201-1207.

Jiao, J. & Chen, D.F. (2008) Induction of neurogenesis in nonconventional neurogenic regions of the adult central nervous system by niche astrocyte-produced signals. *Stem Cells*, 26, 1221-1230.

Jin, K., Wang, X., Xie, L., et al (2006) Evidence for stroke-induced neurogenesis in the human brain. *Proc Natl Acad Sci U S A*, 103, 13198-13202.

Joh, T., Sasaki, M., Kataoka, H., et al (2005) Helicobacter pylori eradication decreases the expression of glycosylphosphatidylinositol-anchored complement regulators, decay-accelerating factor and homologous restriction factor 20, in human gastric epithelium. *J Gastroenterol Hepatol*, 20, 1344-1351.

Kallur, T., Darsalia, V., Lindvall, O., et al (2006) Human fetal cortical and striatal neural stem cells generate region-specific neurons in vitro and differentiate extensively to neurons after intrastriatal transplantation in neonatal rats. *Journal of neuroscience research*, 84, 1630-1644.

Kameda, M., Shingo, T., Takahashi, K., et al (2007) Adult neural stem and progenitor cells modified to secrete GDNF can protect, migrate and integrate after intracerebral transplantation in rats with transient forebrain ischemia. *Eur J Neurosci*, 26, 1462-1478.

Kelly, S., Bliss, T.M., Shah, A.K., et al (2004) Transplanted human fetal neural stem cells survive, migrate, and differentiate in ischemic rat cerebral cortex. *Proc Natl Acad Sci U S A*, 101, 11839-11844.

Kim, H., Park, J.S., Choi, Y.J., et al (2009a) Bone Marrow Mononuclear Cells Have Neurovascular Tropism and Improve Diabetic Neuropathy. *Stem Cells*, 27, 1686-1696.

Kim, J.B., Sebastiano, V., Wu, G., et al (2009b) Oct4-induced pluripotency in adult neural stem cells. *Cell*, 136, 411-419.

Kondo, T. & Raff, M. (2000) Oligodendrocyte precursor cells reprogrammed to become multipotential CNS stem cells. *Science*, 289, 1754-1757.

Kriegstein, A. & Alvarez-Buylla, A. (2009) The glial nature of embryonic and adult neural stem cells. *Annu Rev Neurosci*, 32, 149-184.

Kuhn, H.G., Dickinson-Anson, H. & Gage, F.H. (1996) Neurogenesis in the dentate gyrus of the adult rat: age-related decrease of neuronal progenitor proliferation. *J Neurosci*, 16, 2027-2033.

Kurozumi, K., Nakamura, K., Tamiya, T., et al (2005) Mesenchymal stem cells that produce neurotrophic factors reduce ischemic damage in the rat middle cerebral artery occlusion model. *Mol Ther*, 11, 96-104.

Labouyrie, E., Dubus, P., Groppi, A., et al (1999) Expression of neurotrophins and their receptors in human bone marrow. *Am J Pathol*, 154, 405-415.

Li, T.S., Furutani, A., Takahashi, M., et al (2006) Impaired potency of bone marrow mononuclear cells for inducing therapeutic angiogenesis in obese diabetic rats. *Am J Physiol Heart Circ Physiol*, 290, H1362-1369.

Li, W.Y., Choi, Y.J., Lee, P.H., et al (2008) Mesenchymal stem cells for ischemic stroke: changes in effects after ex vivo culturing. *Cell Transplant*, 17, 1045-1059.

Li, Y., Chopp, M., Zhang, Z.G., et al (1995) Expression of glial fibrillary acidic protein in areas of focal cerebral ischemia accompanies neuronal expression of 72-kDa heat shock protein. *J Neurol Sci*, 128, 134-142.

Lim, D.A. & Alvarez-Buylla, A. (1999) Interaction between astrocytes and adult subventricular zone precursors stimulates neurogenesis. *Proc Natl Acad Sci U S A*, 96, 7526-7531.

Louissaint, A., Jr., Rao, S., Leventhal, C., et al (2002) Coordinated interaction of neurogenesis and angiogenesis in the adult songbird brain. *Neuron*, 34, 945-960.

Magavi, S.S., Leavitt, B.R. & Macklis, J.D. (2000) Induction of neurogenesis in the neocortex of adult mice. *Nature*, 405, 951-955.

Mahmood, A., Lu, D. & Chopp, M. (2004) Intravenous administration of marrow stromal cells (MSCs) increases the expression of growth factors in rat brain after traumatic brain injury. *J Neurotrauma*, 21, 33-39.

Mignone, J.L., Kukekov, V., Chiang, A.S., et al (2004) Neural stem and progenitor cells in nestin-GFP transgenic mice. *J Comp Neurol*, 469, 311-324.

Miyahara, Y., Nagaya, N., Kataoka, M., et al (2006) Monolayered mesenchymal stem cells repair scarred myocardium after myocardial infarction. *Nat Med*, 12, 459-465.

Moreno-Manzano, V., Rodriguez-Jimenez, F.J., Garcia-Rosello, M., et al (2009) Activated spinal cord ependymal stem cells rescue neurological function. *Stem Cells*, 27, 733-743.

Morse, D.E. & Cova, J.L. (1984) Pigmented cells in the leptomeninges of the cat. *Anat Rec*, 210, 125-132.

Nagoshi, N., Shibata, S., Nakamura, M., et al (2009) Neural crest-derived stem cells display a wide variety of characteristics. *J Cell Biochem*, 107, 1046-1052.

Nakagomi, N., Nakagomi, T., Kubo, S., et al (2009a) Endothelial cells support survival, proliferation, and neuronal differentiation of transplanted adult ischemia-induced neural stem/progenitor cells after cerebral infarction. *Stem Cells*, 27, 2185-2195.

Nakagomi, T., Molnár, Z., Nakano-Doi, A., et al (2011) Ischemia-induced neural stem/progenitor cells in the pia mater following cortical infarction. *Stem Cells and Development*, 20, 2037-2051.

Nakagomi, T., Taguchi, A., Fujimori, Y., et al (2009b) Isolation and characterization of neural stem/progenitor cells from post-stroke cerebral cortex in mice. *Eur J Neurosci*, 29, 1842-1852.

Nakano-Doi, A., Nakagomi, T., Fujikawa, M., et al (2010) Bone Marrow Mononuclear Cells Promote Proliferation of Endogenous Neural Stem Cells Through Vascular Niches After Cerebral Infarction. *Stem Cells*, 28, 1292-1302.

Nakashima, K., Yanagisawa, M., Arakawa, H., et al (1999) Synergistic signaling in fetal brain by STAT3-Smad1 complex bridged by p300. *Science*, 284, 479-482.

Nakayama, D., Matsuyama, T., Ishibashi-Ueda, H., et al (2010) Injury-induced neural stem/progenitor cells in post-stroke human cerebral cortex. *Eur J Neurosci*, 31, 90-98.

Nunes, M.C., Roy, N.S., Keyoung, H.M., et al (2003) Identification and isolation of multipotential neural progenitor cells from the subcortical white matter of the adult human brain. *Nat Med*, 9, 439-447.

Ohira, K., Furuta, T., Hioki, H., et al (2010) Ischemia-induced neurogenesis of neocortical layer 1 progenitor cells. *Nat Neurosci*, 13, 173-179.

Okada, S., Nakamura, M., Mikami, Y., et al (2004) Blockade of interleukin-6 receptor suppresses reactive astrogliosis and ameliorates functional recovery in experimental spinal cord injury. *J Neurosci Res*, 76, 265-276.

Oki, K., Kaneko, N., Kanki, H., et al (2010) Musashi1 as a marker of reactive astrocytes after transient focal brain ischemia. *Neurosci Res*, 66, 390-395.

Palmer, T.D., Willhoite, A.R. & Gage, F.H. (2000) Vascular niche for adult hippocampal neurogenesis. *J Comp Neurol*, 425, 479-494.

Parr, A.M., Kulbatski, I., Zahir, T., et al (2008) Transplanted adult spinal cord-derived neural stem/progenitor cells promote early functional recovery after rat spinal cord injury. *Neuroscience*, 155, 760-770.

Pfenninger, C.V., Roschupkina, T., Hertwig, F., et al (2007) CD133 is not present on neurogenic astrocytes in the adult subventricular zone, but on embryonic neural stem cells, ependymal cells, and glioblastoma cells. *Cancer Res*, 67, 5727-5736.

Reynolds, B.A. & Weiss, S. (1992) Generation of neurons and astrocytes from isolated cells of the adult mammalian central nervous system. *Science*, 255, 1707-1710.

Ribeiro-Resende, V.T., Pimentel-Coelho, P.M., Mesentier-Louro, L.A., et al (2009) Trophic activity derived from bone marrow mononuclear cells increases peripheral nerve regeneration by acting on both neuronal and glial cell populations. *Neuroscience*, 159, 540-549.

Richardson, W.D., Young, K.M., Tripathi, R.B., et al (2011) NG2-glia as Multipotent Neural Stem Cells: Fact or Fantasy? *Neuron*, 70, 661-673.

Saino, O., Taguchi, A., Nakagomi, T., et al (2010) Immunodeficiency reduces neural stem/progenitor cell apoptosis and enhances neurogenesis in the cerebral cortex after stroke. *J Neurosci Res*, 88, 2385-2397.

Shen, Q., Goderie, S.K., Jin, L., et al (2004) Endothelial cells stimulate self-renewal and expand neurogenesis of neural stem cells. *Science*, 304, 1338-1340.

Shen, Q., Wang, Y., Kokovay, E., et al (2008) Adult SVZ stem cells lie in a vascular niche: a quantitative analysis of niche cell-cell interactions. *Cell Stem Cell*, 3, 289-300.

Shimada, I.S., Peterson, B.M. & Spees, J.L. (2010) Isolation of locally derived stem/progenitor cells from the peri-infarct area that do not migrate from the lateral ventricle after cortical stroke. *Stroke*, 41, e552-560.

Siegenthaler, J.A., Ashique, A.M., Zarbalis, K., et al (2009) Retinoic acid from the meninges regulates cortical neuron generation. *Cell*, 139, 597-609.

Sirko, S., Neitz, A., Mittmann, T., et al (2009) Focal laser-lesions activate an endogenous population of neural stem/progenitor cells in the adult visual cortex. *Brain*, 132, 2252-2264.

Smart, N., Bollini, S., Dube, K.N., et al (2011) De novo cardiomyocytes from within the activated adult heart after injury. *Nature*, 474, 640-644.

Sockanathan, S. & Gaiano, N. (2009) Preview. Meninges Play a RAdical Role in Embryonic Neural Stem Cell Regulation. *Cell Stem Cell*, 5, 455-456.

Song, H., Stevens, C.F. & Gage, F.H. (2002) Astroglia induce neurogenesis from adult neural stem cells. *Nature*, 417, 39-44.

Staines, W.A., Morassutti, D.J., Reuhl, K.R., et al (1994) Neurons derived from P19 embryonal carcinoma cells have varied morphologies and neurotransmitters. *Neuroscience*, 58, 735-751.

Stallcup, W.B. & Beasley, L. (1987) Bipotential glial precursor cells of the optic nerve express the NG2 proteoglycan. *J Neurosci*, 7, 2737-2744.

Sundberg, M., Skottman, H., Suuronen, R., et al (2010) Production and isolation of NG2+ oligodendrocyte precursors from human embryonic stem cells in defined serum-free medium. *Stem Cell Res*, 5, 91-103.

Taguchi, A., Kasahara, Y., Nakagomi, T., et al (2010) A Reproducible and Simple Model of Permanent Cerebral Ischemia in CB-17 and SCID Mice. *J Exp Stroke Transl Med*, 3, 28-33.

Taguchi, A., Soma, T., Tanaka, H., et al (2004) Administration of CD34+ cells after stroke enhances neurogenesis via angiogenesis in a mouse model. *J Clin Invest*, 114, 330-338.

Taguchi, A., Wen, Z., Myojin, K., et al (2007) Granulocyte colony-stimulating factor has a negative effect on stroke outcome in a murine model. *Eur J Neurosci*, 26, 126-133.

Takahashi, K., Yasuhara, T., Shingo, T., et al (2008) Embryonic neural stem cells transplanted in middle cerebral artery occlusion model of rats demonstrated potent therapeutic effects, compared to adult neural stem cells. *Brain Res*, 1234, 172-182.

Takata, M., Nakagomi, T., Kashiwamura, S., et al. (2011) Glucocorticoid-induced TNF receptor-triggered T cells are key modulators for survival/death of neural stem/progenitor cells induced by ischemic stroke. Cell death and Differentiation, in press.

Tavazoie, M., Van der Veken, L., Silva-Vargas, V., et al (2008) A specialized vascular niche for adult neural stem cells. *Cell stem cell*, 3, 279-288.

Teng, H., Zhang, Z.G., Wang, L., et al (2008) Coupling of angiogenesis and neurogenesis in cultured endothelial cells and neural progenitor cells after stroke. *J Cereb Blood Flow Metab*, 28, 764-771.

Teng, L. & Labosky, P.A. (2006) Neural crest stem cells. *Adv Exp Med Biol*, 589, 206-212.

Toda, H., Takahashi, J., Iwakami, N., et al (2001) Grafting neural stem cells improved the impaired spatial recognition in ischemic rats. *Neurosci Lett*, 316, 9-12.

Ulrich, R., Seeliger, F., Kreutzer, M., et al (2008) Limited remyelination in Theiler's murine encephalomyelitis due to insufficient oligodendroglial differentiation of nerve/glial antigen 2 (NG2)-positive putative oligodendroglial progenitor cells. *Neuropathol Appl Neurobiol*, 34, 603-620.

Wei, L., Cui, L., Snider, B.J., et al (2005) Transplantation of embryonic stem cells overexpressing Bcl-2 promotes functional recovery after transient cerebral ischemia. *Neurobiol Dis*, 19, 183-193.

Weiss, S., Dunne, C., Hewson, J., et al (1996) Multipotent CNS stem cells are present in the adult mammalian spinal cord and ventricular neuroaxis. *J Neurosci*, 16, 7599-7609.

Willaime-Morawek, S. & van der Kooy, D. (2008) Cortex- and striatum- derived neural stem cells produce distinct progeny in the olfactory bulb and striatum. *The European journal of neuroscience*, 27, 2354-2362.

Xue, J.H., Yanamoto, H., Nakajo, Y., et al (2009) Induced spreading depression evokes cell division of astrocytes in the subpial zone, generating neural precursor-like cells and new immature neurons in the adult cerebral cortex. *Stroke*, 40, e606-613.

Yang, Z., Covey, M.V., Bitel, C.L., et al (2007) Sustained neocortical neurogenesis after neonatal hypoxic/ischemic injury. *Ann Neurol*, 61, 199-208.

Yokoyama, A., Sakamoto, A., Kameda, K., et al (2006) NG2 proteoglycan-expressing microglia as multipotent neural progenitors in normal and pathologic brains. *Glia*, 53, 754-768.

Yoo, S.W., Kim, S.S., Lee, S.Y., et al (2008) Mesenchymal stem cells promote proliferation of endogenous neural stem cells and survival of newborn cells in a rat stroke model. *Exp Mol Med*, 40, 387-397.

Zakharova, L., Mastroeni, D., Mutlu, N., et al Transplantation of cardiac progenitor cell sheet onto infarcted heart promotes cardiogenesis and improves function. *Cardiovasc Res.*

Zawadzka, M., Rivers, L.E., Fancy, S.P., et al (2010) CNS-resident glial progenitor/stem cells produce Schwann cells as well as oligodendrocytes during repair of CNS demyelination. *Cell Stem Cell*, 6, 578-590.

Zhou, B., Ma, Q., Rajagopal, S., et al (2008) Epicardial progenitors contribute to the cardiomyocyte lineage in the developing heart. *Nature*, 454, 109-113.

17

Mesenchymal Stromal Cells and Neural Stem Cells Potential for Neural Repair in Spinal Cord Injury and Human Neurodegenerative Disorders

Dasa Cizkova*

Institute of Neurobiology, Center of Excellence for Brain Research, Slovak Academy of Sciences, Kosice, Slovakia

1. Introduction

Spinal cord injury represents a serious neurodegenerative condition mostly characterized by inflammation, demyelination, loss of neurons and glial cells. Patients who suffer from spinal cord trauma show limited functional recovery, which frequently leads to deficit of multiple sensory, motor and autonomic systems resulting to clinical signs of partial or complete paralysis with prominent spasticity and rigidity (Cizkova et al. 2007). Because of the limited regenerative capacity of the adult CNS due to the inhibitory molecules, decrease of trophic factor support and scar tissue formation, the current functional treatments for SCI are not successful (Rowland et al. 2008). However, emerging research evidences on regenerative medicine involving adult and neural stem cells has put much attention on the development of cell based therapies which could promote regeneration of lesioned CNS (Barnabe-Heider & Frisen, 2008; Goldman, 2005). One of the most important factors for the stem cells candidates that are being used in transplantation strategies, is their compatibility with the host tissue. Therefore, preferential criteria for stem cells transplantation in clinical trials are their ability to be used as autologous transplant to avoid moral and ethical dilemma as well as immunosuppressive therapy. Mesenchymal stem cells (MSCs) fulfill all these criteria and can be easily isolated from patient's bone marrow or adipose tissue. However, in many cases their beneficial effect in regard to the treatment of neurodegenerative disorders is most likely due to paracrine (Zacharek et al. 2007) or immunomodulatory effects (Djouad et al. 2003), rather than by direct cell replacement (Jorgensen, 2009). Therefore, other sources of autologous stem cells, such as „Schwann cells" derived from peripheral nerve, „Olfactory ensheating cells" (OECs) (Papastefanaki et al. 2007; Raisman et al. 2011) from olfactory bulb, or even allogenic

*Norbert Zilka[3], Zuzana Kazmerova[3], Lucia Slovinska[1], Ivo Vanicky[1], Ivana Novotna-Grulova[1], Viera Cigankova[2], Milan Cizek[2] and Michal Novak[3]

[1]*Institute of Neurobiology, Center of Excellence for Brain Research, Slovak Academy of Sciences, Kosice, Slovakia*

[2]*University of Veterinary Medicine and Pharmacy , Kosice, Slovakia*

[3]*Institute of Neuroimunology, Center of Excellence for Brain Research, Slovak Academy of Sciences, Bratislava, Slovakia*

embryonic or neural stem cells have been involved in different studies to replace lost/impaired neural population. Particularly, rapidly improving neural stem cells (NSCs) research has been providing encouraging evidence that stem cells derived from nerve tissue can repair CNS structure and perhaps even function which is impaired by various neurodegenerative disorders. NSCs that can self-renew, are multipotent cells committed to generate a neural phenotype, thus making them easier to differentiate into the desired sources of neuronal or pro-oligodendroglial cells that may be applied for further transplantation strategies. The accuracy of both autologous vs allogenic cell based approaches was confirmed in recent studies where application of adult and neural stem cells into injured spinal cord or to a wide variety of neurodegenerative diseases led to improvement of functional outcome in animal models through replacement of damaged or dead motor neurons and thereby remyelination of spared axons and modulation of inflammation (Louro & Pearse, 2008; Kim & de Vellis, 2009; Nandoe Tewarie et al. 2009). As with any cell therapy in CNS, it is important to realize that more complex issues need to be considered, such as: the selection of cell source, effective delivery strategies, optimal dosing of stem cells, proper timing and safety guarantees of stem cells based treatment.

Here we have tried to outline the most important basic issues of MSC, NSC research in regard to their therapeutic potential to repair or enhance plasticity in neurodegenerative disorders, with main focus on SCI. The following sections summarize the MSCs and NSCs fundamental biological properties, their potential sources and perspective advantages in cell-based therapies.

2. Mesenchymal stem cells

Mesenchymal stem cells, also called bone marrow stromal cells represent a heterogeneous population of the cells derived from the non-blood forming fraction of bone marrow . They are able to differentiate into bone, tendon, cartilage and fat (Pittenger et al. 1999) or under specific condition into neuronal, muscle, liver cells (Keilhoff et al. 2006; Yu et al. 2007; Greco & Rameshwar, 2008) as well as epithelial cells of lung, skin, kidney and gastrointestinal tract (Herzog et al. 2003). The first evidence for the existence of non-hematopoietic stem cells derived from bone marrow has been available from Friedenstein's work in 1970s (Friedenstein et al. 1976). Friedenstein isolated cells from bone marrow and plated them on plastic culture dish. After 4 hours, he removed the medium with non-adherent cells (mostly containing hematopoietic stem cells) and observed that a small number of cells with spindle-shape morphology remained adhered to the Petri dish and form foci of two or four cells. After the 2-4 days, the adherent cells started to multiply and attained spindle-shaped morphology (Friedenstein et al. 1976). From a physiological point of view, MSCs represent a major population of bone marrow stromal cells, that by the continuous release of EPO (erythropoietin-EPO) and granulocyte-colony formation stimulating factor (granulocyte colony stimulating factor G-CSF), promote survival, division and differentiation of hematopoietic precursor/stem cells (Cui et al. 2009). Since then non-hematopoietic stem cells have been identified in many other organs and tissues including skin, skeletal muscle, teeth, adipose tissue, testis, gut, liver and ovarian epithelium (Kerkis et al. 2006; Guan et al. 2006; Zuk et al. 2002).

2.1 Isolation of MSCs from bone marrow and adipose tissue

MSCs can be isolated by aspiration of bone marrow from the diaphysis of the tibia or femur in rats, mice, which represent only 0,001-0,01% of the total population of nucleated cells (Pittenger et al. 1999). In humans, bone marrow derived MSCs (BM-MSCs) are mainly obtained from superior iliac crest of pelvis (Digirolamo et al. 1999). *In-vitro* cultivation of MSCs is very simple because of their plastic adherence, their extensive proliferative capacity and ability to create single-cell-derived colonies (Colter et al. 2000). There is a possibility for MSCs exploitation in autologous transplantations to prevent immunological response or rejection of implanted cells. Compared to embryonic stem cells, MSCs have reportedly low tumorigenic potential and they are capable to migrate toward tumors (Loebinger et al. 2009) and into the sites of neural lesions (Chen et al. 2008). Another source of mesenchymal stem cells represents the adipose tissue. Adipose tissue-derived mesenchymal stem cells (AT-MSCs) are also multipotent, plastic adherent, have similar CD markers as BM-MSCs and under specific condition they are able to differentiate into cells of the mesodermal, osteogenic, chondrogenic, adipogenic and myogenic lineages and even into cells with neuron-like morphology (Zuk et al. 2002). Moreover, isolation of AT-MSCs is easier (by liposuction); less painful and number of obtained cells is much higher in comparison to BM-MSCs (Lin et al. 2008). In spite of this, MSCs obtained from bone marrow represent the main source of stem cells in preclinical and clinical studies until now.

2.1.1 Morphology and phenotype of MSCs

According to the morphology, MSCs are classified into two groups: spindle-shaped type, also called very small rapidly self-renewal round cells (RSCs) (Colter et al. 2001) and flattened type (Mets & Verdonk, 1981) known as a mature MSCs (mMSCs). RSCs are characterised by rapid rate of replication after low density plating, potential for multilineage differentiation and by the presence of specific cell surface epitopes which are not found at mMSCs stage, such as: vascular endothelial growth factor receptor-2 (FLK-1), TRK (a nerve growth factor receptor), transferrin receptor and annexin II (lipocortine 2) (Colter et al. 2001). Unlike, mMSCs are characterised by large-scale and flatted morphology, lower property of replication and higher ratio of cytoplasm-to-nucleus when compared to RSCs. Moreover, MSCs express several positive cell surface molecules that allow us to distinguish them from the hematopoietic stem cells such as: β-integrins (CD29), CD44, α-integrins (CD49a, CD49b), CD61, P-selectin (CD62), CD90 (thy-1), CD105, CD106 (VCAM-1) and CD166 (Majumdar et al. 2003; Docheva et al. 2007), collagen type I and IV, laminin, fibronectin; chemokine receptors: CXCR5,6-R, CCR1,7,9-R; CX3CL1-R; growth factor receptors: TGFβ-R, PDGF-R, NGF-R, FGF-R; and cytokine receptors: IL1,3,4,6,7,15-R, TNFα-R (Dominici et al. 2006; Stagg, 2007). The immune phenotype of cultured MSCs is described as MHC class I+, MHC class II-, CD40-, CD80- and CD86-. This phenotype is regarded as non-immunogenic and suggests that MSCs might be effective in inducing tolerance (Javazon et al. 2001). It has been documented that during aging, MSCs undergo several changes and thereby lose their differentiation capacity and decrease production of specific proteins and factors responsible for cell differentiation such as bone morphogenic protein (BMP-7), alkaline phosphatase, G-CSF (granulocyte colony-stimulating factor), LIF (leukemia inhibitory factor) and stem cell factor (SCF). Moreover, differentiation potential of MSCs is down regulated from the 6th passage on and the mean length of telomeres is shortened after 9th passage revealing morphological abnormalities typical of the Hayflick model of cellular aging (Bonab et al. 2006). According to these evidences it is very important to realize the fact that mesenchymal stem cells which are applied in regenerative medicine should be used in early passages where currently their rapid proliferation and increased differentiation capacity are utilized.

2.1.2 Trophic properties of MSCs

Several reports suggest that application of MSCs in neurodegenerative disorders led to neuroprotective effect and to the replacement of diseased and damaged cells and tissues in the most affected area. Profuse scientific investigations revealed that the main effect of the neuroprotection and neuroregeneration is mediated by specific neurotrophic molecules and cytokines that are directly produced by MSCs. It has been also shown that these factors can support neuronal cell survival and regenerate nerve fibers at the lesion sites (Mahmood et al. 2004). *In vitro* studies have confirmed the presence of various neurotrophic factors produced by MSCs, including nerve growth factor (NGF), brain-derived neurotrophic factor (BDNF), glial cell line-derived neurotrophic factor (GDNF), ciliary neurotrophic factor (CNTF) and neurotrophin-3 (NT-3) (Chen et al. 2005; Kurozumi et al. 2005). Measurement of 56 separate subclones derived from human MSCs showed that differences in neurotrophin's production between single cell clones can vary in a huge range (from 167 to 2000-fold) and expression of these neuro-regulatory molecules was able to promote survival and neurite outgrowth in the SH-SY5Y neuroblastoma cell line. Consecutive selection of the most producing single cell derived clones can lead to better exploitation of MSCs in regenerative and cell replacement medicine (Crigler et al. 2006). Moreover, MSCs also constitutively express several interleukins including IL-6, IL-7, IL-8, IL-11, IL-12, IL-14, IL-15, macrophage or granulocyte-macrophage colony stimulating factor (M-CSF, GM-CSF), stromal cell-derived factor 1α (SDF-1α) (Crigler et al. 2006), Flt-3 ligand and stem cell factor (Majumdar et al. 1998) that can play an important role in immunomodulatory processes.

2.1.3 Immunomodulatory effect of MSCs

Recent studies demonstrate that MSCs command with the ability to modulate an immune response depending on the stimulus to which they are exposed. Their dual ability, to suppress and/or activate immune responses, can lead to modulation of the reaction of broad range of immune cells, including T cells, B cells, NK cells and antigen-presenting cells (Stagg, 2007). It is assumed that the main effect of immunosuppresion is evoked by soluble factors that are produced by MSCs or immune cells, such as: hepatocyte growth factor, indoleamine 2, 3-dioxygenase (IDO), prostaglandin E2, TGF-β1, nitric oxide and IL10. It has been also observed that MSCs use different mechanisms that are responsible for inhibition of function and proliferation of immune cells (Nauta & Fibbe, 2007). INFγ play a crucial role in regulation of MSC-mediated immunosuppresion. INFγ induce MSCs to release prostaglandins and IDO, which causes depletion of tryptophan, an essential factor for lymphocyte proliferation (Aggarwal & Pittenger, 2005). The similar suppressive effect on T-cell proliferation was also suggested in the presence of TGF-β and hepatocyte growth factor, which are constitutively produced by MSCs (Di Nicola et al. 2002). Cocultivation of MSCs with lymphocytes revealed that MSCs don't constitutively secrete suppressive factors but provide a dynamic cross-talk between MSCs and lymphocytes (Augello et al. 2005). MSCs can interfere with dendritic cells (DCs) differentiation, maturation and function. It has been observed that MSCs had an inhibitory effect on differentiation of monocytes and CD34+ progenitors into CD1a+-DCs by skewing of their differentiation property toward macrophages (Nauta & Fibbe, 2007). At the same time, immature DCs were unable to induce T cells activation in the presence of MSCs. Cocultivation of MSCs with NK cells showed that allogeneic MSCs could inhibit IL-2 and IL-15-induced proliferation of resting NK cells and either MSCs are able to suppress the proliferation and cytokine production of IL-15 stimulated NK cells via soluble factors. Suggesting that there is also the existence of different mechanisms for MSC-mediated NK cell suppression demonstrated experiments where after inhibition of both soluble factors - PGE2 and TGF-β produced by MSCs complete restoration of proliferation capacity of NK cells was observed (Sotiropoulou et al. 2006).

3. Application of MSCs in neurodegenerative diseases

Transplantation of autologous or allogenic mesenchymal stem cells has been considered as a potential therapeutic approach to a wide variety of neurodegenerative diseases such as Alzheimer's disease (AD), Parkinson disease (PD), sclerosis multiplex (SM), amyotrophic lateral sclerosis (ALS), spinal cord injury (SCI) or stroke.

3.1 Utilization of MSCs in therapy for SCI

Traumatic injury to the spinal cord initiates a cascade of reactive changes, which results in permanent damage and loss of neurological function below the lesion site (Rowland et al., 2008). The inflammatory events, together with ischemia, Ca^{2+} influx into cells, edema, and progressive hemorrhagic necrosis significantly contribute to secondary injury, which causes progressive cavitation and loss of spinal tissue (Kwon et al. 2010). The expression of adverse neurite growth-inhibitory molecules in the extracellular matrix (Fawcett, 2006; Schwab, 2004) together with lack of trophic factor support and the discontinuity of axonal projections caused by progressive tissue cyst formation pose multifactor obstacles contributing to the loss of spinal cord regeneration and inability to find an effective therapy (Nagahara & Tuszynski, 2011). However, by addressing aspects, such as neutralization of growth inhibitors Nogo-A, CSPGs, delivery of various trophic factors or utilizing stem cells/progenitors, a considerable progress has been made in enhancing the growth of injured adult axons (Bradbury et al. 2002). The widespread use of stem cell therapy has shown that transplantation of MSCs can improve recovery after stroke (Chopp & Li, 2002), promote remyelinization (Akiyama et al. 2002), as well as contribute to partial recovery of locomotor function in animal models of spinal cord injury (SCI) (Cizkova et al. 2006) (Sykova & Jendelova, 2005; Arboleda et al. 2011; Forostyak et al. 2011). Thus, achieved progress in animal SCI models utilizing MSCs made it possible for translating preclinical findings to human clinical trials. For example, transplantation of unmanipulated autologous bone marrow in patients with subacute and chronic SCI resulted into improvement of motor/or sensory functions within 3 months. Although, implantation of autologous bone marrow cells appears to be safe, it is necessary to follow up patients outcome data, for more than 2 years (Sykova et al. 2006; Pal et al. 2009; Moviglia et al. 2009). While there is evidence that MSCs can give rise to cells with neural characteristics in vitro (Kim et al. 2002) and in vivo (Jendelova et al. 2004), it is more likely that production of neurotrophic or vascular factors (Zhong et al. 2003; Hamano et al. 2000) together with immunomodulatory effects (Aggarwal & Pittenger, 2005) have a dominant influence on recovery of function following spinal cord trauma. Particularly, suggested hypoimmunologic nature of MSCs, imply for unique MSCs immunomodulatory approaches, that could be used for immunosuppression to induce allotransplantation tolerance or even to attenuate autoimmune, inflammatory responses (Le Blanc & Ringden, 2005). Although some experimental studies in animals or pre-clinical human studies demonstrate the effectiveness and safety of MSCs therapy, there are still many questions to be answered regarding the mechanisms of engraftment, homing, inter-cellular interactions, immunological profiles, in vivo differentiation as well as long-term safety.

3.1.1 MSCs therapy for Parkinson disease

Parkinson disease (PD) is the second most common neurodegenerative disorder in the world characterized by progressive loss of nigrostriatal dopaminergic neurons leading to deficiency of dopamine in striatum which is responsible for control of movement. The characteristic symptoms in patients suffering from PD are rigidity, akinesia, tremor and balance problems (Pechadre et al. 1976). Number of studies investigated whether transplantation of human mesenchymal stem cells (hMSCs) can lead to protective effect on progressive dopaminergic neuronal loss in vitro or in vivo conditions. Intravenous injection of hMSCs into the PD transgenic rat models showed strong protective effect on progressive loss of dopaminergic neurons in substantia nigra. Human MSCs reduced the caspase-3 activity and increased survival of TH-immunoreactive cells in substantia nigra in comparison with the control group. Moreover, a significant improvement in behavioral motor tests in hMSCs treated group has also been observed (Park et al. 2008). *In vitro* study demonstrated that SDF-α-1, chemokine constitutively produced by MSCs increased dopamine release and led to suppression of cell death induced by 6-OHDA administration compared to untreated group (Wang et al. 2010). Neuroprotective effect of hMSCs on dopaminergic neurons mediated by anti-inflammatory properties of MSCs and their modulation of microglial activation were uncovered (Kim et al. 2009). Transplantation of GDNF-transduced MSCs into the PD animal model supported the evidence, that they are capable to induce a local trophic effect in the denervated striatum and sprouting from remaining dopaminergic terminals toward neurotrophic milieu. Exploitation of new optogenetic technique demonstrated for the first time that intrastriatally grafted stem cell-derived dopamine neurons become functionally integrated in the dopamine-denervated striatum (Tonnesen et al. 2011). Noninvasive intranasal delivery of MSCs to the unilaterally 6-hydroxydopamine - lesioned rat brains showed decreasing concentrations of inflammatory cytokines, increasing of tyrosine hydroxylase level in the lesioned ipsilateral striatum and substancia nigra, and prevented any decrease of dopamine in the lesioned hemisphere. Simultaneously, significant improvement of motor function of forepaw in PD rat model was observed (Danielyan et al. 2011).

3.1.2 MSCs therapy for Alzheimer disease

Alzheimer disease (AD), the most common form of dementia, is characterized as a progressive neurodegenerative disorder (Berchtold & Cotman, 1998). Degeneration and dysfunction of the neurons and decline of synaptic function and plasticity mostly in brain regions responsible for memory and learning, as hippocampus, entorhinal cortex, basal forebrain and neocortical association cortices, are the most incident symptoms that generally characterize AD (DeKosky et al. 1996). There is no cure or early preclinical diagnostic assay available for Alzheimer's disease. Currently, most prevalent is symptomatic therapy, which is not able to stop the progression of the disease. Therefore, Alzheimer's disease is still being recognized as an unmet medical need. In 1906, Dr. Alois Alzheimer, identified two specific features that are mostly figured in AD human brain, neurofibrillary tangles and amyloid plaques. Deep investigation in the study of the main structural components responsible for the creation of two pathological hallmarks in AD brain, uncovered inherence of tau protein in NFT and amyloid beta peptide in amyloid plaques. Several years later, it was demonstrated that strong neuroinflammation occurs in AD brain (Novak et al. 1993) (Dickson et al. 1988; Zilka et al. 2006; Zilkova et al. 2006).

Application of stem cells in AD preclinical studies brought in last years several positive results. Taking advantage of stem cells immunomodulatory and trophic properties and their transplantation into AD transgenic animal models showed that they are the most appropriate tool for the achievement of functional restoration of damaged cells and in the same manner for the replacement by healthy one (Blurton-Jones et al. 2009; Hampton et al. 19 2010; Lee et al. 2010). Recent developments in stem cell technology raise the prospect of cell therapy for human neurodegenerative tauopathies. Transplantation of the neural stem cells or administration of mesenchymal stem cells isolated either from human umbilical cord or from the bone marrow has produced beneficial effects in several independent animal models of AD (Blurton-Jones et al. 2009). Above mentioned reports have shown that the neuroprotective effect of stem cells may be mediated 1) by their ability to produce various trophic factors that contribute to functional recovery or 2) by activation of neuroinflammatory pathways. *In vitro* studies show that MSCs can prevent tau mediated cell death in the Alzheimer's cell model. It has been confirmed that MSCs have significant impact on tau cell death cascade and can ameliorate toxic effect of misfolded truncated tau that is considered to be driving force behind neurofibrillary degeneration. Therefore it may be suggested that the cell neuroprotective therapy rather than cell replacement therapy represents prospective strategy for treatment of Alzheimer's disease and related tauopathies (Zilka et al. 2011).

4. Neural stem cells

The human brain contains roughly 100 billion neurons, of which several thousands die every day, representing the loss of millions of nerve cells across the life span. For this reason, it has been believed for a long time, that adult mammalian central nervous system (CNS) is rather rigid structure, unable to repair itself following diseases or injury. However, in some brain regions dead neurons could be replaced and potentially could contribute to the regeneration of damaged nerve tissue (Graziadei & Graziadei, 1979). Therefore, a number of controversial issues concerning possible CNS plasticity was raised and broadly discussed. Finally, in the 1960s and 1970s, most of the uncertainties were addressed and neuroscience's central tenets the 'no new neurons' doctrine, was reconsidered following the key-revolutionary discovery of Joseph Altman (Altman, 1962; Altman & Das, 1965), documenting thymidine-H_3-labelled neurons and neuroblasts in the adult rat brain. From now on a huge effort has gone into unraveling and understanding the fundamental mechanisms of adult CNS regeneration in mammals.

4.1 Neural stem cells definition and origin

It took almost twenty years of dedicated research involving a large number of scientific experiments which clearly confirmed ongoing neurogenesis not only in songbirds (Nottebohm, 1981), but also in rodents, non-human primates and humans, in whom new imaging techniques, such as bromodeoxyuridin (BrdU) labeling, etc, enabling identification of proliferating cells were applied (Eriksson et al. 1998). All these studies jointly confirmed that new functional neurons are generated in the adult mammalian, including human CNS in two discrete areas: i) in the hippocampus, the subgranular zone (SGZ) of the dentate gyrus, which is an important center of our memory (Gage, 2000; Alvarez-Buylla et al. 2002) and, ii) in subventricular zone (SVZ), representing a thin layer of cells lining along the lateral cerebral ventricles, where a nerve cells essential for olfaction are generated (Gage, 2000; Lledo et al. 2006). In both areas, neurogenesis progresses as a complex multi-stage process, which starts with the proliferation, followed by migration and terminal differentiation (Abrous et al. 2005). The current knowledge of self-renewing and multipotent neural stem cells is largely defined by in vitro, as well as *in vivo* evidences documenting their ability to generate the main progeny of the nervous system: neurons, astrocytes and oligodendrocytes (Gage, 2000). NSCs reside in specific anatomical microenvironments that are called neurogenic niches; small islands where neurons and glial cells are continuously generated (Doetsch et al. 1999). However, neurogenic regions (SVZ, SGZ) must meet following criteria: 1) contain neural precursors (NPCs) that are generated in, 2) neurogenic niches, providing cell-cell contacts and diffusible factors for terminal neural differentiation, and 3) provide neurogenic potential (thus, ability of NPCs that are implanted in a neurogenic areas to generate neurons, while when implanted into other brain location they give rise to glia). Another interesting pool of neural precursor cells is represented by astrocytes found within the germinal layers of the adult brain. It has been broadly documented that these astrocytes retain the stem cell properties throughout the life span, and are involved in both neuro- and glio-genesis (Alvarez-Buylla et al. 2001; Gotz & Huttner, 2005; Mori et al. 2005).

4.1.1 Neurogenesis mediated by pathological conditions; Properties of non-neurogenic areas

Normal adult neurogenesis produces a limited number of newly generated functional cells that primarily serves to maintain physiological tissue homeostasis in specific CNS systems. Initially, the neurogenic processes have been expected to be rather stable, moreover insensitive to external stimuli. However, this view has been changed, due to the growing evidence documenting that SVZ and SGZ are responding to a various local or global signals generated from nerve tissue damage. For example, neurogenesis in both neurogenic zones is increased in animal experimental models of ischemia/stroke (Zhang et al. 2008) as well as in humans suffering from stroke (Curtis et al. 2007), epileptic seizures (Grote & Hannan, 2007) and multiple sclerosis (Nait-Oumesmar et al. 2007). Furthermore, neurogenesis is increased in human cases and animal models of Huntington's disease while it is reduced in Alzheimer's and Parkinson's disease as well as in depression and stress (Elder et al. 2006; Grote & Hannan, 2007). Stem cells with the potential to generate new neurons that could replace dying neurons in neurodegenerative diseases or CNS injuries reside also in other areas of the adult CNS, indicating to the possibility that endogenous sources of NSCs can be mobilized also from non-neurogenic regions (Minger, 2007). These NSCs have been demonstrated in brain areas such as septum, striatum or even in the spinal cord, but so far it was not clearly established whether these stem cells are capable of differentiation to the final functional neurons (Liu & Martin, 2003; Wiltrout et al. 2007). Furthermore, it has been suggested that ependymal cells (ECs) adjacent to the SVZ of the lateral ventricles, may mimic the characteristics of NSCs (Johansson et al. 1999; Doetsch et al. 1999). A study by Coskun et al. (Coskun et al. 2008) documented that this may be the case, because the subpopulation of ependymal cells, CD133+/CD24-, exhibited features of quiescent NSCs *in vitro*, i.e., self-renewal and multipotency as well as participation in neurogenesis *in vivo* after injury. In this relation, the occurrence of ependymal cell layer covering CNS ventricular system including the areas around the third, fourth ventricles, and the central canal (CC) of the spinal cord supports suggestion, that also these regions may retain similar quiescent NSCs as those which were identified in the lateral ventricles (Weiss et al. 1996).

4.1.2 Neurogenic potential in the spinal cord and stimulatory factors

There is increasing evidence that the CC ependymal cell region, which is regarded as presumptive neurogenic area of adult spinal cord, contains a limited number of neural stem cells. Once implanted in the animals, they differentiate into oligodendrocytes and astrocytes (Mothe & Tator, 2005) while, under *in vitro* conditions, they give rise to both neurons and glia (Yamamoto et al. 2001). On the other hand, neuronal or glial fate of grafted ECs is highly depended on the host neurogenic/non-neurogenic microenvironment (Shihabuddin et al. 2000). These contradictory findings are often explained in regard to beneficial (*in vitro*) or inhibitory (*in vivo*) conditions directly influencing neuronal or glial fate (Weiss et al. 1996). Furthermore, after pathological condition such as spinal cord injury, most of the newly dividing intrinsic ependymal stem cells migrate toward damaged tissue, where they develop into macroglial cells, while only few cells retain primitive nestin-like phenotype (Johansson et al. 1999; Cizkova et al. 2009a). Likewise, a significant number of neural progenitors could be activated also in other regions of the parenchyma (Horner et al. 2000) (Kehl et al. 1997). However, it remains unclear whether these progenitors develop into functional neurons.

A stimulatory effect on spinal progenitors may be obtained also after physiological stimulation, when experimental animals are exposed to an enrichment environment or physical activity. Previous experiments have shown that mice providing systematic exercise in a running wheel had twice more new hippocampal neurons than controls (Gomez-Pinilla et al. 2001). Beside this, it has been confirmed that voluntary exercise can increase levels of brain-derived neurotrophic factor (BDNF) and other growth factors, which stimulate neurogenesis, improve learning, mental performance (Gomez-Pinilla et al. 2001) and may mobilize gene expression profiles that could be beneficial for CNS plasticity processes (Neeper et al. 1995). These data were further confirmed in latter studies showing that enhanced physical activity in adult rats induces an endogenous ependymal cell response leading to increased proliferation, although in more attenuated manner if compared with

SCI (Cizkova et al. 2009b) (Fig.1). Indeed, there is one group of studies that favor the fact that ECs might contribute to *de novo* neuronal differentiation following CNS injury (Ke et al. 2006; Danilov et al. 2006), while others refuse this suggestion (Zai & Wrathall, 2005). Based on these findings, it is un-doubtful that the adult spinal cord retain a certain reservoir of neural precursors, which can under various specific conditions stimulate and promote the recovery of injured spinal cord.

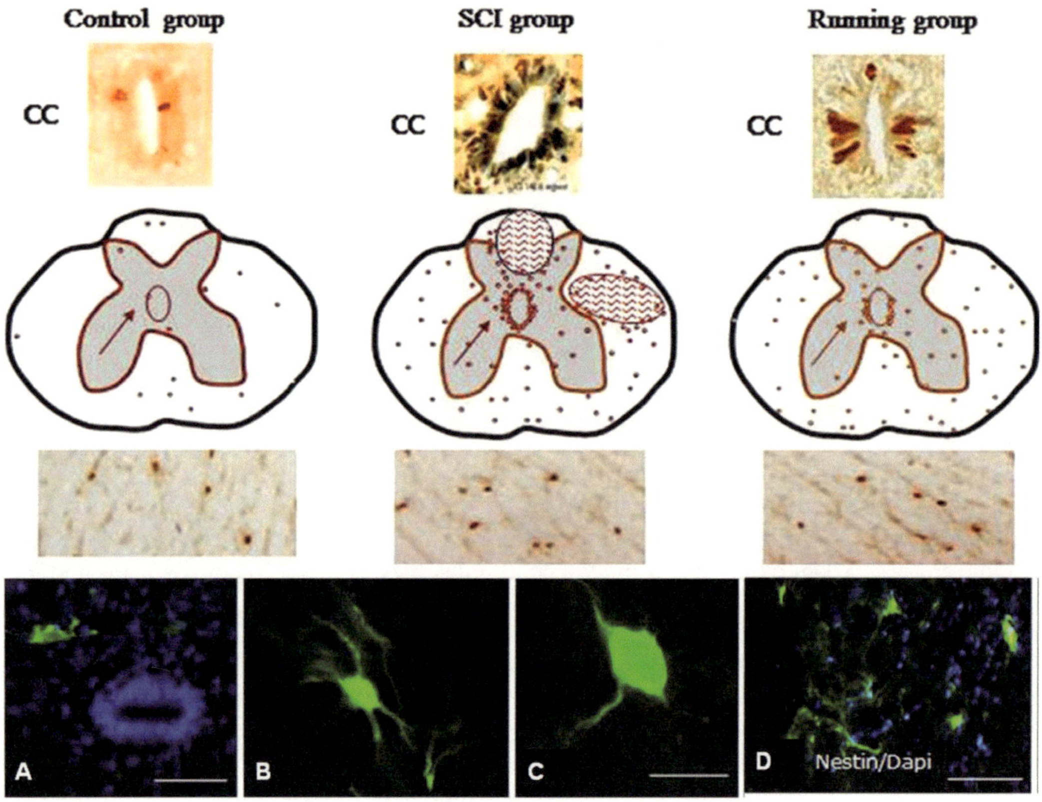

Fig. 1. Schematic illustration of BrdU IR in the thoracic spinal cord section (Th8) of the control, SCI or Running group. Note, the highest BrdU expression in the CC canal, and around the lesion site of SCI group, different distribution patterns of BrdUpositive nuclei in the ependyma between SCI and Running group, and increased BrdU response in the parenchyma of the SC in both groups. Below each schematic drawing, a panel revealing BrdU-IR in the corresponding ventral white matter is performed. (A-D) Fluorescence microscopy images of occasionally occurring nestin-positive cell bodies (green) with processes, found in the close vicinity to the CC gray matter, dorsal horn or adjacent to lesion site.

4.1.3 Molecular mechanisms of neurogenesis

Neurogenesis is understood as a complex process that is regulated by a wide variety of important signaling molecules such as: growth factors, cytokines, and neurotransmitters.

Their primary function is to mediate a balance between proliferation, migration and survival of NSCs within the neurogenic niche. The most important growth factors affecting cell division are: FGF (fibroblast growth factor), VEGF (vascular endothelial growth factor), EGF (epidermal growth factor / epidermal growth factor), PDGF (platelet-derived growth factor) and BDNF (brain derived neurotrophic factor). Therefore, endogenous neurogenesis can be stimulated by intraventricular infusion of mitogenic factors such as EGF, bFGF, TGFβ (transforming growth factor β) that stimulate the proliferation activity in the SVZ and thus restore the nervous tissue (Kuhn et al. 1997). Nitric oxide (NO), erythropoietin, bone morphogenetic protein (BMP Bone Morphogenetic Protein) and Wnt proteins (Wiltrout et al. 2007) also play an important role in regulating neurogenesis. BMP and its receptor that are expressed by the SVZ cells promoting differentiation of the NSCs toward glial phenotype are blocked by Noggin, which is produced by ECs and in contrast drives differentiation into neurons (Lim et al. 2000). The most important regulatory neurotransmitters include GABA (γ-aminobutyric acid) and glutamate, which maintain homeostasis of newly formed neurons (Platel et al. 2007). GABA decreases the proliferation of neuroblasts and NSCs, whereas glutamate stimulates their division. It is noteworthy that in all types of damaged nerve tissue which is associated with glutamate excitotoxicity an increased neurogenesis, is documented. GABA is synthesized and released by neuroblasts and activates $GABA_A$ receptor, causing loss of proliferation of neuroblasts and astrocytes. We can conclude that GABA acts as a negative modulator inhibiting cell division, which means that with increased number of neuroblasts there is a higher amount of released of GABA and more $GABA_A$ receptors are activated (Bordet et al. 2007).

5. Transplantation strategies utilizing NSCs

Neural progenitors isolated from vertebrate central nervous system (CNS) represent valuable source of cells that hold particular promise for treating a variety of human neurological diseases such as spinal cord injury (Goldman, 2005). Due to the pathological events and limited ability of the spinal cord to repair itself, therapeutic approaches are focused either on: i) stimulation of endogenous neuronal plasticity and mobilization of oligodendoglial progenitors (Azari et al. 2005; Fawcett, 2006; Yang et al. 2006) or ii) development of an effective cell selection techniques to gain desired NSCs progeny used for cell-replacement therapy (Faulkner & Keirstead, 2005; Hofstetter et al. 2005; Keirstead et al. 2005). However, an important issue due to the pathological nature of spinal cord damage it is important to select the most convenient strategy involving desired cellular pools for transplantation. For example, spinal ischemia-induced spastic paraplegia which is associated with a selective loss of small inhibitory interneurons, would necessarily involve implantation of neuronal progenitors. On the other hand, diseases or spinal cord trauma, with different pathological outcome, resulting in demyelination of axons followed by destruction of long descending tracts would rather require transplantation of myelin-producing cells such as oligodendroglial cells, Schwann cells or Olfactory ensheating cells (Keirstead et al. 2005; Keilhoff et al. 2006; Pearse et al. 2007; Raisman, 2007). Since a well-documented repertoire of specific surface markers for cells of NSCs at different developmental stages have been identified, it may be possible to identify factors which affect their commitment to oligodendroglial cells or neurons and combine this with optimal sorting methods (Deng & Poretz, 2003; Pruszak et al. 2007; Uchida et al. 2000). In particular, magnetic cell separation using specific monoclonal antibodies (e.g. A2B5, PSA-NCAM) conjugated to nanoparticles allowing positive retention or negative dilution of selected cells provide a feasible approach for experimental cell enrichment of desired oligodendroglial progeny, which may be used in future trials for cell-based therapies to treat spinal cord injury (Cizkova et al. 2009a). These studies have shown that MACs technology enable us to gain about a 5 to 9 fold increase of immature, mature oligodendrocytes content (NG2+, RIP+, MBP+) when compared to amount of oligodendroglial cells acquired from unseparated population (Fig.2). A great deal of attention has been given to NSCs isolated from various regions of CNS, including embryonic and adult spinal cord, that could differentiate into desired oligodendrocytes and myelinate host axons in various pre-clinical animal models of SCI (Tarasenko et al. 2007; Kakinohana et al. 2004). For example, NSCs derived from human fetal brain improved recovery after contusion SCI either in severe combined immunodeficiency (SCID) or myelin-deficient shivered mice (Cummings et al. 2005). Highly purified oligodendrocyte progenitors could be generated also from human embryonic stem cells (hESCs) (Nistor et al. 2005; Cloutier et al. 2006). Based on their remyelination properties described in preclinical animal SCI models, the Geron Corporation has initiated a first clinical trial (Phase I) by transplanting hESC-derived oligodendrocyte

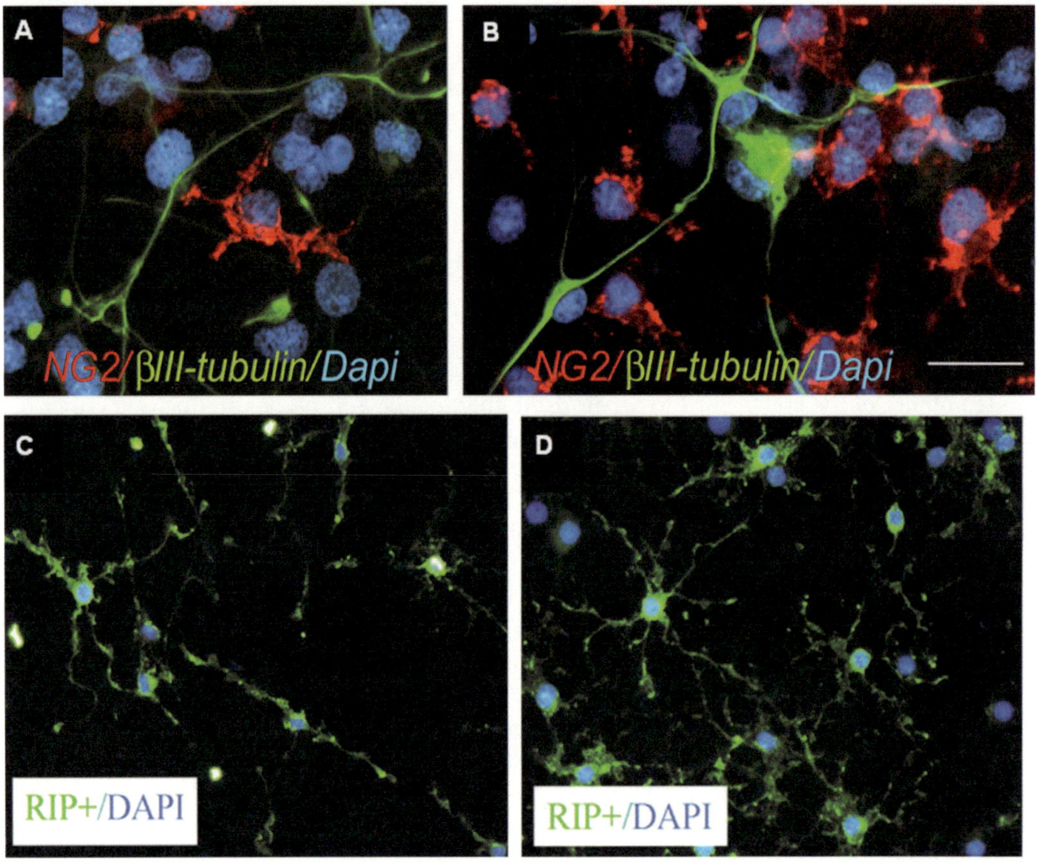

Fig. 2. Immature neurons expressing βIII-tubulin (green) occurred in both, unseparated (A) and separated NSC population (B), but higher number of immature NG2+ oligodendrocytes (red, A, B) and mature RIP+ oligodendrocytes (green C, D) was found after MACs (B, D) (compare A with B and C with D).

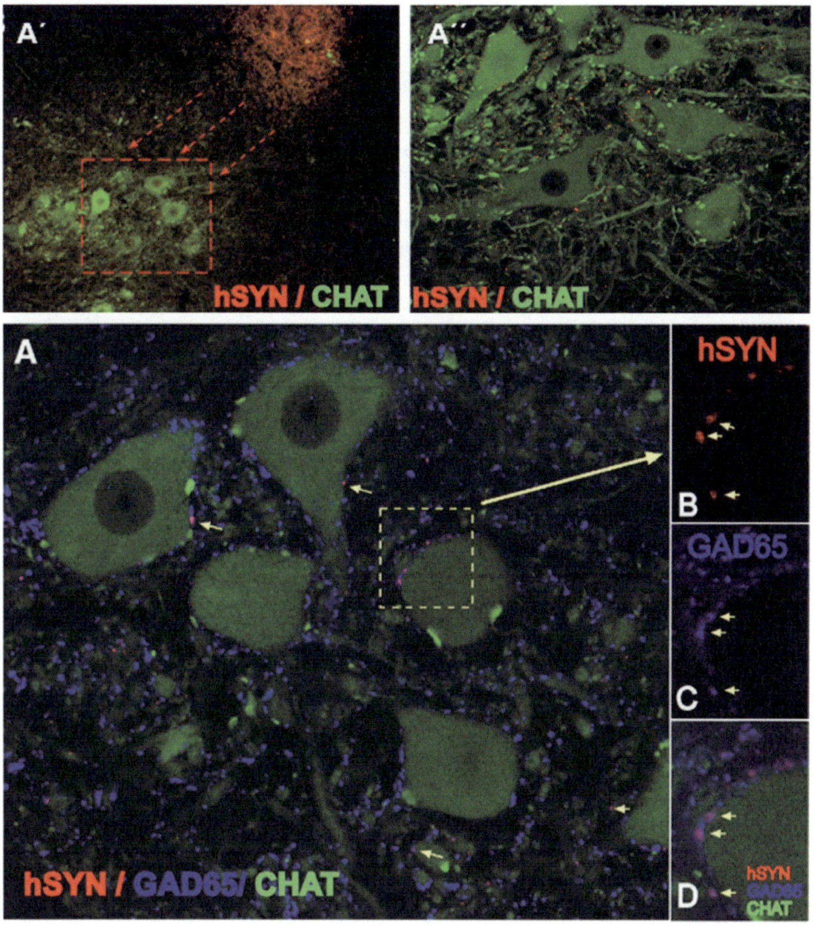

Fig. 3. Fluorescent microscopy images (A´, A´´) and single confocal optical images (A-D) of transverse spinal cord sections taken at 3 months after grafting and stained with human specific hSYN antibody (B, red), CHAT antibody (A, green) and GAD65 antibody (C, blue) antibodies. The majority of hSYN terminals showed co-localization with GAD65 (B–D, yellow arrows).

Progenitor cells in patients with spinal cord injuries. Their preliminary data showed a very good safety profile, with no serious adverse events, no evidence of cavitation at the injury site and no immune responses to the transplanted cells even after complete withdrawal of immunesuppression. One of the most important properties of NSCs is their ability to generate functional neurons, which could potentially rebuild altered local neuronal network following spinal injury. Thus, implanting NSCs-derived neuronal pools in animals subjected to spinal ischemia-induced paraplegia, where selective loss of small local inhibitory interneurons, with persisting α-motoneurons occurs, could meet the needs and expectations to reconstruct impaired local inhibitory neuronal circuits. Although, the precise mechanism leading to spastic paraplegia and rigidity is not certain, the neuropathological features of a selective degeneration of GABA, GAD immunopositive inhibitory neurons are well defined. In addition, the loss of these specific inhibitory pools localized in the intermediate zone of the spinal grey matter, ultimately leads to an increase in the monosynaptic reflex and near-complete loss in spinal polysynaptic activity. A challenging study done in collaboration with anesthesiology research laboratory at University of California San Diego, has shown that NSCs derived from human fetal spinal cord grafted into a rat model of ischemic spastic paraplegia resulted into a progressive recovery of motor function with correlative improvement in motor evoked potentials (Cizkova et al. 2007). Of note, transplanted NSCs became integrated into host neuronal circuits and displayed an extensive axo-dendritic outgrowth and active rostrocaudal/dorsoventral migration for about 8-12 weeks. Furthermore, intense hSYN immunoreactivity was identified within the grafts and in the vicinity of persisting α-motoneurons. These hSYN immunoreactive synaptic terminals expressed GAD65 immunoreactivity in 40-45% of human grafted cells, referring to their inhibitory fate (Fig. 3). All together, these data conclude that functional recovery was associated with long term survival of grafted neurons with GABAergic phenotype that most probably contributed to suppression of spasticity (Cizkova et al. 2007). Similarly, human hNT neurons (teratocarcinoma cell line) or rat spinal neuronal precursors (SNPs), grafted into ischemic spinal segments depleted of inhibitory neurons, restore local inhibitory tone and ameliorate spasticity (Marsala et al. 2004). In addition, when human derived NSCs were treated with a cocktail of growth factors and later transplanted into the injured spinal cord, they differentiated preferentially into cholinergic neurons (Wu et al. 2009). Although, it seems that NSCs are a powerful source of neural progenitors that are constitutively secreting a variety of growth stimulating factors (NGF, BDNF, GDNF), they are often genetically modified to further enhance their potential and secrete additional factors such as neurotrophin 3 (NT-3), or are combined with antibodies that neutralize ciliary neurothrophic factor (CNTF), in an attempt to attenuate astrocytic differentiation (Ishii et al. 2006).

6. Acknowledgments

We acknowledge the financial support given by MVTS-COST BM-1002- NANONET, the Grant Agency of the Slovak Academy of Sciences VEGA 2/0114/11 and the Center of Excellence for Brain Research Slovak Academy of Sciences and the project No 26110230036 University of Veterinary Medicine and Pharmacy (Nove studijne programy a vzdelavanie na Univerzite veterinarneho lekarstva v Kosiciach, Operacny program Vzdelavanie).

7. References

Abrous, D. N.; Koehl, M.& Le Moal, M. (2005). Adult neurogenesis: from precursors to network and physiology. *Physiol Rev,* Vol. 85, No. 2, (April 2005), pp 523-569, ISSN 0031-9333

Aggarwal, S .& Pittenger, M. F. (2005). Human mesenchymal stem cells modulate allogeneic immune cell responses. *Blood,* Vol. 105, No. 4, (February 2005), pp 1815-1822, ISSN 0006-4971

Akiyama, Y.; Radtke, C.; Honmou, O.& Kocsis, J. D. (2002). Remyelination of the spinal cord following intravenous delivery of bone marrow cells. *Glia,* Vol. 39, No. 3, (September 2002), pp 229-236, ISSN 0894-1491

Altman, J. (1962). Are new neurons formed in the brains of adult mammals? *Science,* Vol. 135, (March 1962), pp 1127-1128, ISSN 0036-8075

Altman, J.&Das, G. D. (1965). Autoradiographic and histological evidence of postnatal hippocampal neurogenesis in rats. *J Comp Neurol,* Vol. 124, No. 3, pp 319-335, ISSN 0021-9967

Alvarez-Buylla, A.; Garcia-Verdugo, J. M.& Tramontin, A. D. (2001). A unified hypothesis on the lineage of neural stem cells. *Nat Rev Neurosci,* Vol. 2, No. 4, (April 2001), pp 287-293, ISSN 1471-0048

Alvarez-Buylla, A.; Seri, B.& Doetsch, F. (2002). Identification of neural stem cells in the adult vertebrate brain. *Brain Res Bull,* Vol. 57, No. 6, (April 2002), pp 751-758, ISSN 0361-9230

Arboleda, D.; Forostyak, S.; Jendelova, P.; Marekova, D.; Amemori, T.; Pivonkova, H.; Masinova, K.& Sykova, E. (2011). Transplantation of Predifferentiated Adipose-Derived Stromal Cells for the Treatment of Spinal Cord Injury. *Cell Mol Neurobiol,* (June 2011), ISSN 0272-4340

Augello, A.; Tasso, R.; Negrini, S. M.; Amateis, A.; Indiveri, F.; Cancedda, R.& Pennesi, G. (2005). Bone marrow mesenchymal progenitor cells inhibit lymphocyte proliferation by activation of the programmed death 1 pathway. *Eur J Immunol,* Vol. 35, No. 5, (May 2005), (May 2005), pp 1482-1490, ISSN 0014-2980

Azari, M. F.; Profyris, C.; Zang, D. W.; Petratos, S.& Cheema, S. S. (2005). Induction of endogenous neural precursors in mouse models of spinal cord injury and disease. *Eur J Neurol,* Vol. 12, No. 8, (August 2005), pp 638-648, ISSN 1351-5101

Barnabe-Heider, F. & Frisen, J. (2008). Stem cells for spinal cord repair. *Cell Stem Cell,* Vol. 3, No. 1, (Jul 2008), pp 16-24, ISSN 1875-9777

Berchtold, N. C. & Cotman, C. W. (1998). Evolution in the conceptualization of dementia and Alzheimer's disease: Greco-Roman period to the 1960s. *Neurobiol Aging,* Vol. 19, No. 3, (May-June 1998), pp 173-189, ISSN 0197-4580

Bjorklund, A. & Svendsen, C. (1999). Stem cells. Breaking the brain-blood barrier. *Nature,* Vol. 397, No. 6720, pp 569-570, ISSN 0028-0836

Blurton-Jones, M.; Kitazawa, M.; Martinez-Coria, H.; Castello, N. A.; Muller, F. J.; Loring, J. F.; Yamasaki, T. R.; Poon, W. W.; Green, K. N.& LaFerla, F. M. (2009). Neural stem cells improve cognition via BDNF in a transgenic model of Alzheimer disease. *Proc Natl Acad Sci U S A,* Vol. 106, No. 32, (August 2009), pp 13594-13599, ISSN 1091-6490

Bonab, M. M.; Alimoghaddam, K.; Talebian, F.; Ghaffari, S. H.; Ghavamzadeh, A.& Nikbin, B. (2006). Aging of mesenchymal stem cell in vitro. *BMC Cell Biol,* Vol. 7, (March 2006), pp 14, ISSN 1471-2121

Bordet, R.; Lestage, P.& Onteniente, B. (2007). [The concept of neuroprotective agents as a treatment modulator in the development of brain diseases]. *Therapie,* Vol. 62, No. 6, (November/December 2007), pp 463-472, ISSN 0040-5957

Bradbury, E. J.; Moon, L. D.; Popat, R. J.; King, V. R.; Bennett, G. S.; Patel, P. N.; Fawcett, J. W.& McMahon, S. B. (2002). Chondroitinase ABC promotes functional recovery after spinal cord injury. *Nature,* Vol. 416, No. 6881, (April 2002), pp 636-640, ISSN 0028-0836

Cizkova, D.; Cizek, M.; Nagyova, M.; Slovinska, L.; Novotna, I.; Jergova, S.; Radonak, J.; Hlucilova, J.& Vanicky, I. (2009a). Enrichment of rat oligodendrocyte progenitor cells by magnetic cell sorting. *J Neurosci Methods,* Vol. 184, No. 1, (October 2009), pp 88-94, ISSN 1872-678X

Cizkova, D.; Kakinohana, O.; Kucharova, K.; Marsala, S.; Johe, K.; Hazel, T.; Hefferan, M. P.& Marsala, M. (2007). Functional recovery in rats with ischemic paraplegia after spinal grafting of human spinal stem cells. *Neuroscience,* Vol. 147, No. 2, (June 2007), pp 546-560, ISSN 0306-4522

Cizkova, D.; Nagyova, M.; Slovinska, L.; Novotna, I.; Radonak, J.; Cizek, M.; Mechirova, E.; Tomori, Z.; Hlucilova, J.; Motlik, J.; Sulla, I., Jr. & Vanicky, I. (2009b). Response of ependymal progenitors to spinal cord injury or enhanced physical activity in adult rat. *Cell Mol Neurobiol,* Vol. 29, No. 6-7, (September 2009), pp 999-1013, ISSN 1573-6830

Cizkova, D.; Rosocha, J.; Vanicky, I.; Jergova, S.& Cizek, M. (2006). Transplants of human mesenchymal stem cells improve functional recovery after spinal cord injury in the rat. *Cell Mol Neurobiol,* Vol. 26, No. 7-8, (November 2006), pp 1165-1178, ISSN 0272-4340

Cloutier, F.; Siegenthaler, M. M.; Nistor, G.& Keirstead, H. S. (2006). Transplantation of human embryonic stem cell-derived oligodendrocyte progenitors into rat spinal cord injuries does not cause harm. *Regen Med,* Vol. 1, No. 4, (July 2006), pp 469-479, ISSN 1746-0751

Colter, D. C.; Class, R.; DiGirolamo, C. M.& Prockop, D. J. (2000). Rapid expansion of recycling stem cells in cultures of plastic-adherent cells from human bone marrow. *Proc Natl Acad Sci U S A,* Vol. 97, No. 7, (March 2000), pp 3213-3218, ISSN 0027-8424

Colter, D. C.; Sekiya, I.& Prockop, D. J. (2001). Identification of a subpopulation of rapidly self-renewing and multipotential adult stem cells in colonies of human marrow stromal cells. *Proc Natl Acad Sci U S A,* Vol. 98, No. 14, (July 2001), pp 7841-7845, ISSN 0027-8424

Coskun, V.; Wu, H.; Blanchi, B.; Tsao, S.; Kim, K.; Zhao, J.; Biancotti, J. C.; Hutnick, L.; Krueger, R. C., Jr.; Fan, G.; de Vellis, J.& Sun, Y. E. (2008). CD133+ neural stem cells in the ependyma of mammalian postnatal forebrain. *Proc Natl Acad Sci U S A,* Vol. 105, No. 3, (January 2008), pp 1026-1031, ISSN 1091-6490

Crigler, L.; Robey, R. C.; Asawachaicharn, A.; Gaupp, D.& Phinney, D. G. (2006). Human mesenchymal stem cell subpopulations express a variety of neuro-regulatory molecules and promote neuronal cell survival and neuritogenesis. *Exp Neurol,* Vol. 198, No. 1, (March 2006), (March 2006), pp 54-64, ISSN 0014-4886

Cui, X.; Chopp, M.; Zacharek, A.; Roberts, C.; Lu, M.; Savant-Bhonsale, S. & Chen, J. (2009). Chemokine, vascular and therapeutic effects of combination Simvastatin and BMSC treatment of stroke. *Neurobiol Dis,* Vol. 36, No. 1, (October 2009), pp 35-41, ISSN 09699961

Cummings, B. J.; Uchida, N.; Tamaki, S. J.; Salazar, D. L.; Hooshmand, M.; Summers, R.; Gage, F. H.& Anderson, A. J. (2005). Human neural stem cells differentiate and promote locomotor recovery in spinal cord-injured mice. *Proc Natl Acad Sci U S A,* Vol. 102, No. 39, (September 2005), pp 14069-14074, ISSN 0027-8424

Curtis, M. A.; Kam, M.; Nannmark, U.; Anderson, M. F.; Axell, M. Z.; Wikkelso, C.; Holtas, S.; van Roon-Mom, W. M.; Bjork-Eriksson, T.; Nordborg, C.; Frisen, J.; Dragunow, M.; Faull, R. L.& Eriksson, P. S. (2007). Human neuroblasts migrate to the olfactory bulb via a lateral ventricular extension. *Science,* Vol. 315, No. 5816, (March 2007), pp 1243-1249, ISSN 0036-8075

Danielyan, L.; Schafer, R.; von Ameln-Mayerhofer, A.; Bernhard, F.; Verleysdonk, S.; Buadze, M.; Lourhmati, A.; Klopfer, T.; Schaumann, F.; Schmid, B.; Koehle, C.; Proksch, B.; Weissert, R.; Reichardt, H. M.; van den Brandt, J.; Buniatian, G. H.; Schwab, M.; Gleiter, C. H.& Frey, W. H., 2nd. (2011). Therapeutic efficacy of intranasally delivered mesenchymal stem cells in a rat model of Parkinson disease. *Rejuvenation Res,* Vol. 14, No. 1, (February 2011), pp 3-16, ISSN 1557-8577

Danilov, A. I.; Covacu, R.; Moe, M. C.; Langmoen, I. A.; Johansson, C. B.; Olsson, T.& Brundin, L. (2006). Neurogenesis in the adult spinal cord in an experimental model of multiple sclerosis. Eur J Neurosci, Vol. 23, No. 2, (January 2006), pp 394-400, ISSN 0953-816X

DeKosky, S. T.; Scheff, S. W.& Styren, S. D. (1996). Structural correlates of cognition in dementia: quantification and assessment of synapse change. *Neurodegeneration,* Vol. 5, No. 4, (December 1996), pp 417-421, ISSN 1055-8330

Deng, W.&Poretz, R. D. (2003). Oligodendroglia in developmental neurotoxicity. *Neurotoxicology,* Vol. 24, No. 2, (March 2003), pp 161-178, ISSN 0161-813X

Di Nicola, M.; Carlo-Stella, C.; Magni, M.; Milanesi, M.; Longoni, P. D.; Matteucci, P.; Grisanti, S.& Gianni, A. M. (2002). Human bone marrow stromal cells suppress T-lymphocyte proliferation induced by cellular or nonspecific mitogenic stimuli. *Blood,* Vol. 99, No. 10, (May 2002), pp 3838-3843, ISSN 0006-4971

Dickson, D. W.; Farlo, J.; Davies, P.; Crystal, H.; Fuld, P.& Yen, S. H. (1988). Alzheimer's disease. A double-labeling immunohistochemical study of senile plaques. *Am J Pathol,* Vol. 132, No. 1, (July 1988), pp 86-101, ISSN 0002-9440

Digirolamo, C. M.; Stokes, D.; Colter, D.; Phinney, D. G.; Class, R.& Prockop, D. J. (1999). Propagation and senescence of human marrow stromal cells in culture: a simple colony-forming assay identifies samples with the greatest potential to propagate and differentiate. *Br J Haematol,* Vol. 107, No. 2, (November 1999), pp 275-281, ISSN 0007-1048

Djouad, F.; Plence, P.; Bony, C.; Tropel, P.; Apparailly, F.; Sany, J.; Noel, D.& Jorgensen, C., 2003. Immunosuppressive effect of mesenchymal stem cells favors tumor growth in allogeneic animals. *In: Blood,* vol. 102. no. 10, pp. 3837-3844, ISSN 0006-4971

Doetsch, F.; Caille, I.; Lim, D. A.; Garcia-Verdugo, J. M.& Alvarez-Buylla, A. (1999). Subventricular zone astrocytes are neural stem cells in the adult mammalian brain. *Cell,* Vol. 97, No. 6, pp 703-716, ISSN 0092-8674

Docheva, D.; Popov, C.; Mutschler, W.& Schieker, M. (2007). Human mesenchymal stem cells in contact with their environment: surface characteristics and the integrin system. *J Cell Mol Med,* Vol. 11, No. 1, (January/February 2007), pp 21-38, ISSN 1582-1838

Dominici, M.; Le Blanc, K.; Mueller, I.; Slaper-Cortenbach, I.; Marini, F.; Krause, D.; Deans, R.; Keating, A.; Prockop, D.& Horwitz, E. (2006). Minimal criteria for defining multipotent mesenchymal stromal cells. The International Society for Cellular Therapy position statement. *Cytotherapy,* Vol. 8, No. 4, pp 315-317, ISSN 1465-3249

Elder, G. A.; De Gasperi, R.& Gama Sosa, M. A. (2006). Research update: neurogenesis in adult brain and neuropsychiatric disorders. *Mt Sinai J Med,* Vol. 73, No. 7, (November 2006), pp 931-940, ISSN 0027-2507

Eriksson, P. S.; Perfilieva, E.; Bjork-Eriksson, T.; Alborn, A. M.; Nordborg, C.; Peterson, D. A.& Gage, F. H. (1998). Neurogenesis in the adult human hippocampus. *Nat Med,* Vol. 4, No. 11, (November 1998), pp 1313-1317, ISSN 1078-8956

Faulkner, J.&Keirstead, H. S. (2005). Human embryonic stem cell-derived oligodendrocyte progenitors for the treatment of spinal cord injury. *Transpl Immunol,* Vol. 15, No. 2, (December 2005), pp 131-142, ISSN 0966-3274

Fawcett, J. W. (2006). Overcoming inhibition in the damaged spinal cord. *J Neurotrauma,* Vol. 23, No. 3-4, (March/April 2006), pp 371-383, 0897-7151 ISSN 0897-7151

Forostyak, S.; Jendelova, P.; Kapcalova, M.; Arboleda, D.& Sykova, E. (2011). Mesenchymal stromal cells prolong the lifespan in a rat model of amyotrophic lateral sclerosis. *Cytotherapy,* (July 2011), ISSN 1477-2566

Friedenstein, A. J.; Gorskaja, J. F.& Kulagina, N. N. (1976). Fibroblast precursors in normal and irradiated mouse hematopoietic organs. *Exp Hematol,* Vol. 4, No. 5, (September 1976), pp 267-274, ISSN 0301-472X

Gage, F. H. (2000). Mammalian neural stem cells. *Science,* Vol. 287, No. 5457, (February 2000), pp 1433-1438, ISSN 0036-8075

Goldman, S. (2005). Stem and progenitor cell-based therapy of the human central nervous system. *Nat Biotechnol,* Vol. 23, No. 7, (July,2005), pp 862-871, ISSN 1087-0156

Gomez-Pinilla, F.; Ying, Z.; Opazo, P.; Roy, R. R.& Edgerton, V. R. (2001). Differential regulation by exercise of BDNF and NT-3 in rat spinal cord and skeletal muscle. *Eur J Neurosci,* Vol. 13, No. 6, (March 2001), pp 1078-1084, ISSN 0953-816X

Gotz, M.&Huttner, W. B. (2005). The cell biology of neurogenesis. *Nat Rev Mol Cell Biol,* Vol. 6, No. 10, (Pctober 2005), pp 777-788, ISSN 1471-0072

Graziadei, P. P.&Graziadei, G. A. (1979). Neurogenesis and neuron regeneration in the olfactory system of mammals. I. Morphological aspects of differentiation and structural organization of the olfactory sensory neurons. *J Neurocytol,* Vol. 8, No. 1, (February 1979), pp 1-18, ISSN 0300-4864

Greco, S. J.&Rameshwar, P. (2008). Microenvironmental considerations in the application of human mesenchymal stem cells in regenerative therapies. *Biologics,* Vol. 2, No. 4, pp 699-705, ISSN 1177-5475

Grote, H. E.&Hannan, A. J. (2007). Regulators of adult neurogenesis in the healthy and diseased brain. *Clin Exp Pharmacol Physiol,* Vol. 34, No. 5-6, (May/June 2007), pp 533-545, ISSN 0305-1870

Guan, K.; Nayernia, K.; Maier, L. S.; Wagner, S.; Dressel, R.; Lee, J. H.; Nolte, J.; Wolf, F.; Li, M.; Engel, W.& Hasenfuss, G. (2006). Pluripotency of spermatogonial stem cells from adult mouse testis. *Nature,* Vol. 440, No. 7088, (April 2006), pp 1199-1203, ISSN 1476-4687

Hamano, K.; Li, T. S.; Kobayashi, T.; Kobayashi, S.; Matsuzaki, M.& Esato, K. (2000). Angiogenesis induced by the implantation of self-bone marrow cells: a new material for therapeutic angiogenesis. *Cell Transplant,* Vol. 9, No. 3, (May/June 2000), pp 439-443, ISSN 0963-6897

Hampton, D. W.; Webber, D. J.; Bilican, B.; Goedert, M.; Spillantini, M. G.& Chandran, S. (2010). Cell-mediated neuroprotection in a mouse model of human tauopathy. J *Neurosci,* Vol. 30, No. 30, (July 2010), pp 9973-9983, ISSN 1529-2401

Herzog, E. L.; Chai, L.& Krause, D. S. (2003). Plasticity of marrow-derived stem cells. *Blood,* Vol. 102, No. 10, (November 2003), pp 3483-3493, ISSN 0006-4971

Hofstetter, C. P.; Holmstrom, N. A.; Lilja, J. A.; Schweinhardt, P.; Hao, J.; Spenger, C.; Wiesenfeld-Hallin, Z.; Kurpad, S. N.; Frisen, J.& Olson, L. (2005). Allodynia limits the usefulness of intraspinal neural stem cell grafts; directed differentiation improves outcome. *Nat Neurosci,* Vol. 8, No. 3, (March 2005), pp 346-353, ISSN 1097-6256

Horner, P. J.; Power, A. E.; Kempermann, G.; Kuhn, H. G.; Palmer, T. D.; Winkler, J.; Thal, L. J.& Gage, F. H. (2000). Proliferation and differentiation of progenitor cells throughout the intact adult rat spinal cord. *J Neurosci,* Vol. 20, No. 6, (March 2000), pp 2218-2228, ISSN 1529-2401

Chen, J. R.; Cheng, G. Y.; Sheu, C. C.; Tseng, G. F.; Wang, T. J.& Huang, Y. S. (2008). Transplanted bone marrow stromal cells migrate, differentiate and improve motor function in rats with experimentally induced cerebral stroke. *J Anat,* Vol. 213, No. 3, (September 2008), pp 249-258, ISSN 1469-7580

Chen, Q.; Long, Y.; Yuan, X.; Zou, L.; Sun, J.; Chen, S.; Perez-Polo, J. R.& Yang, K. (2005). Protective effects of bone marrow stromal cell transplantation in injured rodent brain: synthesis of neurotrophic factors. *J Neurosci Res,* Vol. 80, No. 5, (June 2005), pp 611-619, ISSN 0360-4012

Chopp, M.&Li, Y. (2002). Treatment of neural injury with marrow stromal cells. *Lancet Neurol,* Vol. 1, No. 2, (June 2002), pp 92-100, ISSN 1474-4422

Ishii, K.; Nakamura, M.; Dai, H.; Finn, T. P.; Okano, H.; Toyama, Y.& Bregman, B. S. (2006). Neutralization of ciliary neurotrophic factor reduces astrocyte production from transplanted neural stem cells and promotes regeneration of corticospinal tract fibers in spinal cord injury. *J Neurosci Res,* Vol. 84, No. 8, (December 2006), pp 1669-1681, ISSN 0360-4012

Javazon, E. H.; Colter, D. C.; Schwarz, E. J.& Prockop, D. J. (2001). Rat marrow stromal cells are more sensitive to plating density and expand more rapidly from single-cell-derived colonies than human marrow stromal cells. *Stem Cells,* Vol. 19, No. 3, (May 2001), pp 219-225, ISSN 1066-5099

Jendelova, P.; Herynek, V.; Urdzikova, L.; Glogarova, K.; Kroupova, J.; Andersson, B.; Bryja, V.; Burian, M.; Hajek, M.& Sykova, E. (2004). Magnetic resonance tracking of transplanted bone marrow and embryonic stem cells labeled by iron oxide nanoparticles in rat brain and spinal cord. *J Neurosci Res,* Vol. 76, No. 2, (April 2004), pp 232-243, ISSN 0360-4012

Johansson, C. B.; Momma, S.; Clarke, D. L.; Risling, M.; Lendahl, U.& Frisen, J. (1999). Identification of a neural stem cell in the adult mammalian central nervous system. *Cell,* Vol. 96, No. 1, (January 1999), pp 25-34, ISSN 0092-8674

Jorgensen, C. (2009). Link between cancer stem cells and adult mesenchymal stromal cells: implications for cancer therapy. *Regen Med,* Vol. 4, No. 2, (March 2009), pp 149-152, ISSN 1746-076X

Kakinohana, O.; Cizkova, D.; Tomori, Z.; Hedlund, E.; Marsala, S.; Isacson, O.& Marsala, M. (2004). Region-specific cell grafting into cervical and lumbar spinal cord in rat: a qualitative and quantitative stereological study. *Exp Neurol,* Vol. 190, No. 1, (November 2004), pp 122-132, ISSN 0014-4886

Ke, Y.; Chi, L.; Xu, R.; Luo, C.; Gozal, D.& Liu, R. (2006). Early response of endogenous adult neural progenitor cells to acute spinal cord injury in mice. *Stem Cells,* Vol. 24, No. 4, (April 2006), pp 1011-1019, ISSN 1066-5099

Kehl, L. J.; Fairbanks, C. A.; Laughlin, T. M.& Wilcox, G. L. (1997). Neurogenesis in postnatal rat spinal cord: a study in primary culture. *Science,* Vol. 276, No. 5312, (April 1997), pp 586-589, ISSN 0036-8075

Keilhoff, G.; Goihl, A.; Langnase, K.; Fansa, H.& Wolf, G. (2006). Transdifferentiation of mesenchymal stem cells into Schwann cell-like myelinating cells. *Eur J Cell Biol,* Vol. 85, No. 1, (January 2006), pp 11-24, ISSN 0171-9335

Keirstead, H. S.; Nistor, G.; Bernal, G.; Totoiu, M.; Cloutier, F.; Sharp, K.& Steward, O. (2005). Human embryonic stem cell-derived oligodendrocyte progenitor cell transplants remyelinate and restore locomotion after spinal cord injury. *J Neurosci,* Vol. 25, No. 19, (May 2005), pp 4694-4705, ISSN 1529-2401

Kerkis, I.; Kerkis, A.; Dozortsev, D.; Stukart-Parsons, G. C.; Gomes Massironi, S. M.; Pereira, L. V.; Caplan, A. I.& Cerruti, H. F. (2006). Isolation and characterization of a population of immature dental pulp stem cells expressing OCT-4 and other embryonic stem cell markers. *Cells Tissues Organs,* Vol. 184, No. 3-4, pp 105-116, ISSN 1422-6421

Kim, B. J.; Seo, J. H.; Bubien, J. K.& Oh, Y. S. (2002). Differentiation of adult bone marrow stem cells into neuroprogenitor cells in vitro. *Neuroreport,* Vol. 13, No. 9, pp 1185-1188, ISSN 0959-4965

Kim, S. U.&de Vellis, J. (2009). Stem cell-based cell therapy in neurological diseases: a review. *J Neurosci Res,* Vol. 87, No. 10, (August 2009), pp 2183-2200, ISSN 1097-4547

Kim, Y. J.; Park, H. J.; Lee, G.; Bang, O. Y.; Ahn, Y. H.; Joe, E.; Kim, H. O.& Lee, P. H. (2009). Neuroprotective effects of human mesenchymal stem cells on dopaminergic neurons through anti-inflammatory action. *Glia,* Vol. 57, No. 1, (January 2009), pp 13-23, ISSN 1098-1136

Krampera, M.; Cosmi, L.; Angeli, R.; Pasini, A.; Liotta, F.; Andreini, A.; Santarlasci, V.; Mazzinghi, B.; Pizzolo, G.; Vinante, F.; Romagnani, P.; Maggi, E.; Romagnani, S.& Annunziato, F. (2006). Role for interferon-gamma in the immunomodulatory activity of human bone marrow mesenchymal stem cells. *Stem Cells,* Vol. 24, No. 2, (February 2006), pp 386-398, ISSN 1066-5099

Kuhn, H. G.; Winkler, J.; Kempermann, G.; Thal, L. J.& Gage, F. H. (1997). Epidermal growth factor and fibroblast growth factor-2 have different effects on neural progenitors in the adult rat brain. J Neurosci, Vol. 17, No. 15, (August 1997), pp 5820-5829, ISSN 0270-6474

Kurozumi, K.; Nakamura, K.; Tamiya, T.; Kawano, Y.; Ishii, K.; Kobune, M.; Hirai, S.; Uchida, H.; Sasaki, K.; Ito, Y.; Kato, K.; Honmou, O.; Houkin, K.; Date, I.& Hamada, H. (2005). Mesenchymal stem cells that produce neurotrophic factors reduce ischemic damage in the rat middle cerebral artery occlusion model. *Mol Ther,* Vol. 11, No. 1, (January 2005), pp 96-104, ISSN 1525-0016

Kwon, B. K.; Okon, E. B.; Plunet, W.; Baptiste, D.; Fouad, K.; Hillyer, J.; Weaver, L. C.; Fehlings, M. G.& Tetzlaff, W. (2010). A systematic review of directly applied biologic therapies for acute spinal cord injury. *J Neurotrauma,* Vol. 28, No. 8, (August 2010), pp 1589-1610, ISSN 1557-9042

Le Blanc, K.&Ringden, O. (2005). Immunobiology of human mesenchymal stem cells and future use in hematopoietic stem cell transplantation. *Biol Blood Marrow Transplant,* Vol. 11, No. 5, (May 2005), pp 321-334, ISSN 1083-8791

Lee, J. K.; Jin, H. K.; Endo, S.; Schuchman, E. H.; Carter, J. E.& Bae, J. S. (2010). Intracerebral transplantation of bone marrow-derived mesenchymal stem cells reduces amyloid-beta deposition and rescues memory deficits in Alzheimer's disease mice by modulation of immune responses. *Stem Cells,* Vol. 28, No. 2, (February 2010), pp 329-343, ISSN 1549-4918

Lim, D. A.; Tramontin, A. D.; Trevejo, J. M.; Herrera, D. G.; Garcia-Verdugo, J. M.& Alvarez-Buylla, A. (2000). Noggin antagonizes BMP signaling to create a niche for adult neurogenesis. *Neuron,* Vol. 28, No. 3, (December 2000), pp 713-726, ISSN 0896-6273

Lin, G.; Garcia, M.; Ning, H.; Banie, L.; Guo, Y. L.; Lue, T. F.& Lin, C. S. (2008). Defining stem and progenitor cells within adipose tissue. *Stem Cells Dev,* Vol. 17, No. 6, (December 2008), pp 1053-1063, ISSN 1557-8534

Liu, Z.&Martin, L. J. (2003). Olfactory bulb core is a rich source of neural progenitor and stem cells in adult rodent and human. *J Comp Neurol,* Vol. 459, No. 4, (May 2003), pp 368-391, ISSN 0021-9967

Lledo, P. M.; Alonso, M.& Grubb, M. S. (2006). Adult neurogenesis and functional plasticity in neuronal circuits. Nat Rev Neurosci, Vol. 7, No. 3, (March 2006), pp 179-193, ISSN 1471-003X

Loebinger, M. R.; Eddaoudi, A.; Davies, D.& Janes, S. M. (2009). Mesenchymal stem cell delivery of TRAIL can eliminate metastatic cancer. *Cancer Res,* Vol. 69, No. 10, (May 2009), pp 4134-4142, ISSN 1538-7445

Louro, J.&Pearse, D. D. (2008). Stem and progenitor cell therapies: recent progress for spinal cord injury repair. *Neurol Res,* Vol. 30, No. 1, (February 2008), pp 5-16, ISSN 0161-6412

Mahmood, A.; Lu, D.& Chopp, M. (2004). Marrow stromal cell transplantation after traumatic brain injury promotes cellular proliferation within the brain. *Neurosurgery,* Vol. 55, No. 5, (November 2004), pp 1185-1193, ISSN 1524-4040

Majumdar, M. K.; Keane-Moore, M.; Buyaner, D.; Hardy, W. B.; Moorman, M. A.; McIntosh, K. R.& Mosca, J. D. (2003). Characterization and functionality of cell surface molecules on human mesenchymal stem cells. *J Biomed Sci,* Vol. 10, No. 2, (March/April 2003), pp 228-241, ISSN 1021-7770

Majumdar, M. K.; Thiede, M. A.; Mosca, J. D.; Moorman, M.& Gerson, S. L. (1998). Phenotypic and functional comparison of cultures of marrow-derived mesenchymal stem cells (MSCs) and stromal cells. *J Cell Physiol,* Vol. 176, No. 1, (July 1998), pp 57-66, ISSN 0021-9541

Marsala, M.; Kakinohana, O.; Yaksh, T. L.; Tomori, Z.; Marsala, S.& Cizkova, D. (2004). Spinal implantation of hNT neurons and neuronal precursors: graft survival and functional effects in rats with ischemic spastic paraplegia. *Eur J Neurosci,* Vol. 20, No. 9, (November 2004), pp 2401-2414, ISSN 0953-816x

Mets, T.&Verdonk, G. (1981). In vitro aging of human bone marrow derived stromal cells. *Mech Ageing Dev,* Vol. 16, No. 1, (May 1981), pp 81-89, ISSN 0047-6374

Minger, S. (2007). Interspecies SCNT-derived human embryos--a new way forward for regenerative medicine. *Regen Med,* Vol. 2, No. 2, (March 2007), pp 103-106, ISSN 1746-076X

Mori, H.; Kanemura, Y.; Onaya, J.; Hara, M.; Miyake, J.; Yamasaki, M.& Kariya, Y. (2005). Effects of heparin and its 6-O-and 2-O-desulfated derivatives with low anticoagulant activity on proliferation of human neural stem/progenitor cells. *J Biosci Bioeng,* Vol. 100, No. 1, (July 2005), pp 54-61, ISSN 1389-1723

Mothe, A. J.&Tator, C. H. (2005). Proliferation, migration, and differentiation of endogenous ependymal region stem/progenitor cells following minimal spinal cord injury in the adult rat. *Neuroscience,* Vol. 131, No. 1, pp 177-187, ISSN 0306-4522

Moviglia, G. A.; Varela, G.; Brizuela, J. A.; Moviglia Brandolino, M. T.; Farina, P.; Etchegaray, G.; Piccone, S.; Hirsch, J.; Martinez, G.; Marino, S.; Deffain, S.; Coria, N.; Gonzales, A.; Sztanko, M.; Salas-Zamora, P.; Previgliano, I.; Aingel, V.; Farias, J.; Gaeta, C. A.; Saslavsky, J.& Blasseti, N. (2009). Case report on the clinical results of a combined cellular therapy for chronic spinal cord injured patients. *Spinal Cord,* Vol. 47, No. 6, (June 2009), pp 499-503, ISSN 1476-5624

Nagahara, A. H.&Tuszynski, M. H. (2011). Potential therapeutic uses of BDNF in neurological and psychiatric disorders. *Nat Rev Drug Discov,* Vol. 10, No. 3, (March 2011), pp 209-219, ISSN 1474-1784

Nait-Oumesmar, B.; Picard-Riera, N.; Kerninon, C.; Decker, L.; Seilhean, D.; Hoglinger, G. U.; Hirsch, E. C.; Reynolds, R.& Baron-Van Evercooren, A. (2007). Activation of the subventricular zone in multiple sclerosis: evidence for early glial progenitors. Proc *Natl Acad Sci U S A,* Vol. 104, No. 11, (March 2007), pp 4694-4699, ISSN 0027-8424

Nandoe Tewarie, R. S.; Hurtado, A.; Bartels, R. H.; Grotenhuis, A.& Oudega, M. (2009). Stem cell-based therapies for spinal cord injury. *J Spinal Cord Med,* Vol. 32, No. 2, pp 105-114, ISSN 1079-0268

Nauta, A. J.&Fibbe, W. E. (2007). Immunomodulatory properties of mesenchymal stromal cells. *Blood,* Vol. 110, No. 10, (November 2007), pp 3499-3506, ISSN 0006-4971

Neeper, S. A.; Gomez-Pinilla, F.; Choi, J.& Cotman, C. (1995). Exercise and brain neurotrophins. *Nature,* Vol. 373, No. 6510, (January 1995), pp 109, ISSN 0028-0836

Nistor, G. I.; Totoiu, M. O.; Haque, N.; Carpenter, M. K.& Keirstead, H. S. (2005). Human embryonic stem cells differentiate into oligodendrocytes in high purity and myelinate after spinal cord transplantation. *Glia,* Vol. 49, No. 3, (February 2005), pp 385-396, ISSN 0894-1491

Nottebohm, F. (1981). A brain for all seasons: cyclical anatomical changes in song control nuclei of the canary brain. Science, Vol. 214, No. 4527, (December 1981), pp 1368-1370, ISSN 0036-8075

Novak, M.; Kabat, J.& Wischik, C. M. (1993). Molecular characterization of the minimal protease resistant tau unit of the Alzheimer's disease paired helical filament. *Embo J,* Vol. 12, No. 1, (January 1993), pp 365-370, ISSN 0261-4189

Pal, R.; Venkataramana, N. K.; Bansal, A.; Balaraju, S.; Jan, M.; Chandra, R.; Dixit, A.; Rauthan, A.; Murgod, U.& Totey, S. (2009). Ex vivo-expanded autologous bone marrow-derived mesenchymal stromal cells in human spinal cord injury/paraplegia: a pilot clinical study. *Cytotherapy,* Vol. 11, No. 7, pp 897-911, ISSN 1477-2566

Papastefanaki, F.; Chen, J.; Lavdas, A. A.; Thomaidou, D.; Schachner, M.& Matsas, R. (2007). Grafts of Schwann cells engineered to express PSA-NCAM promote functional recovery after spinal cord injury. *Brain,* Vol. 130, No. Pt 8, (August 2007), pp 2159-2174, ISSN 1460-2156

Park, H. J.; Lee, P. H.; Bang, O. Y.; Lee, G.& Ahn, Y. H. (2008). Mesenchymal stem cells therapy exerts neuroprotection in a progressive animal model of Parkinson's disease. *J Neurochem,* Vol. 107, No. 1, (October 2008), pp 141-151, ISSN 1471-4159

Pearse, D. D.; Sanchez, A. R.; Pereira, F. C.; Andrade, C. M.; Puzis, R.; Pressman, Y.; Golden, K.; Kitay, B. M.; Blits, B.; Wood, P. M.& Bunge, M. B. (2007). Transplantation of Schwann cells and/or olfactory ensheathing glia into the contused spinal cord: Survival, migration, axon association, and functional recovery. *Glia,* Vol. 55, No. 9, (July 2007), pp 976-1000, ISSN 0894-1491

Pechadre, J. C.; Larochelle, L.& Poirier, L. J. (1976). Parkinsonian akinesia, rigidity and tremor in the monkey. Histopathological and neuropharmacological study. *J Neurol Sci,* Vol. 28, No. 2, (June 1976), pp 147-157, ISSN 0022-510X

Pittenger, M. F.; Mackay, A. M.; Beck, S. C.; Jaiswal, R. K.; Douglas, R.; Mosca, J. D.; Moorman, M. A.; Simonetti, D. W.; Craig, S.& Marshak, D. R. (1999). Multilineage potential of adult human mesenchymal stem cells. *Science,* Vol. 284, No. 5411, (April 1999), pp 143-147, ISSN 0036-8075

Platel, J. C.; Lacar, B.& Bordey, A. (2007). GABA and glutamate signaling: homeostatic control of adult forebrain neurogenesis. *J Mol Histol,* Vol. 38, No. 4, (August 2007), pp 303-311, ISSN 1567-2379

Prockop, D. J. (1997). Marrow stromal cells as stem cells for nonhematopoietic tissues. *Science,* Vol. 276, No. 5309, (April 1997), pp 71-74, ISSN 0036-8075

Pruszak, J.; Sonntag, K. C.; Aung, M. H.; Sanchez-Pernaute, R.& Isacson, O. (2007). Markers and methods for cell sorting of human embryonic stem cell-derived neural cell populations. *Stem Cells,* Vol. 25, No. 9, (September 2007), pp 2257-2268, ISSN 1549-4918

Raisman, G. (2007). Repair of spinal cord injury by transplantation of olfactory ensheathing cells. *C R Biol,* Vol. 330, No. 6-7, (June/July 2007), pp 557-560, ISSN 1631-0691

Raisman, G.; Carlstedt, T.; Choi, D.& Li, Y. (2011). Clinical prospects for transplantation of OECs in the repair of brachial and lumbosacral plexus injuries: Opening a door. *Exp Neurol,* (May2011), ISSN 1090-2430

Rowland, J. W.; Hawryluk, G. W.; Kwon, B.& Fehlings, M. G. (2008). Current status of acute spinal cord injury pathophysiology and emerging therapies: promise on the horizon. *Neurosurg Focus,* Vol. 25, No. 5, pp E2, ISSN 1092-0684

Shihabuddin, L. S.; Horner, P. J.; Ray, J.& Gage, F. H. (2000). Adult spinal cord stem cells generate neurons after transplantation in the adult dentate gyrus. *J Neurosci,* Vol. 20, No. 23, (December 200), pp 8727-8735, ISSN 1529-2401

Schwab, M. E. (2004). Nogo and axon regeneration. *Curr Opin Neurobiol,* Vol. 14, No. 1, (February 2004), pp 118-124, ISSN 0959-4388

Sotiropoulou, P. A.; Perez, S. A.; Gritzapis, A. D.; Baxevanis, C. N.& Papamichail, M. (2006). Interactions between human mesenchymal stem cells and natural killer cells. *Stem Cells,* Vol. 24, No. 1, (January 2006), pp 74-85, ISSN 1066-5099

Stagg, J. (2007). Immune regulation by mesenchymal stem cells: two sides to the coin. *Tissue Antigens,* Vol. 69, No. 1, (January 2007), pp 1-9, ISSN 0001-2815

Sykova, E.; Homola, A.; Mazanec, R.; Lachmann, H.; Konradova, S. L.; Kobylka, P.; Padr, R.; Neuwirth, J.; Komrska, V.; Vavra, V.; Stulik, J.& Bojar, M. (2006). Autologous bone marrow transplantation in patients with subacute and chronic spinal cord injury. *Cell Transplant,* Vol. 15, No. 8-9, (January 2006), pp 675-687, ISSN 0963-6897

Sykova, E. &J endelova, P. (2005). Magnetic resonance tracking of implanted adult and embryonic stem cells in injured brain and spinal cord. *Ann N Y Acad Sci,* Vol. 1049, (May 2005), pp 146-160, ISSN 0077-8923

Tarasenko, Y. I.; Gao, J.; Nie, L.; Johnson, K. M.; Grady, J. J.; Hulsebosch, C. E.; McAdoo, D. J.& Wu, P. (2007). Human fetal neural stem cells grafted into contusion-injured rat spinal cords improve behavior. *J Neurosci Res,* Vol. 85, No. 1, pp 47-57, ISSN 0360-4012

Tonnesen, J.; Parish, C. L.; Sorensen, A. T.; Andersson, A.; Lundberg, C.; Deisseroth, K.; Arenas, E.; Lindvall, O.& Kokaia, M. (2011). Functional integration of grafted neural stem cell-derived dopaminergic neurons monitored by optogenetics in an in vitro Parkinson model. *PLoS One,* Vol. 6, No. 3, (March 2011), pp e17560, ISSN 1932-6203

Uchida, N.; Buck, D. W.; He, D.; Reitsma, M. J.; Masek, M.; Phan, T. V.; Tsukamoto, A. S.; Gage, F. H.& Weissman, I. L. (2000). Direct isolation of human central nervous system stem cells. *Proc Natl Acad Sci U S A,* Vol. 97, No. 26, (December 2000), pp 14720-14725, ISSN 0027-8424

Wang, F.; Yasuhara, T.; Shingo, T.; Kameda, M.; Tajiri, N.; Yuan, W. J.; Kondo, A.; Kadota, T.; Baba, T.; Tayra, J. T.; Kikuchi, Y.; Miyoshi, Y.& Date, I. (2010). Intravenous administration of mesenchymal stem cells exerts therapeutic effects on parkinsonian model of rats: focusing on neuroprotective effects of stromal cell-derived factor-1alpha. *BMC Neurosci,* Vol. 11, pp 52, ISSN 1471-2202

Weiss, S.; Dunne, C.; Hewson, J.; Wohl, C.; Wheatley, M.; Peterson, A. C.& Reynolds, B. A. (1996). Multipotent CNS stem cells are present in the adult mammalian spinal cord and ventricular neuroaxis. *J Neurosci,* Vol. 16, No. 23, (December 1996), pp 7599-7609, ISSN

Wiltrout, C.; Lang, B.; Yan, Y.; Dempsey, R. J.& Vemuganti, R. (2007). Repairing brain after stroke: a review on post-ischemic neurogenesis. *Neurochem Int,* Vol. 50, No. 7-8, (June 2007), pp 1028-1041, ISSN 0270-6474

Wu, Y.; Shatapathy, C. C.& Minger, S. L. (2009). Isolation, in vitro cultivation and characterisation of foetal liver cells. *Methods Mol Biol,* Vol. 481, pp 181-192, ISSN 1064-3745

Yamamoto, S.; Yamamoto, N.; Kitamura, T.; Nakamura, K.& Nakafuku, M. (2001). Proliferation of parenchymal neural progenitors in response to injury in the adult rat spinal cord. *Exp Neurol,* Vol. 172, No. 1, (November 2001), pp 115-127, ISSN 0014-4886

Yang, H.; Lu, P.; McKay, H. M.; Bernot, T.; Keirstead, H.; Steward, O.; Gage, F. H.; Edgerton, V. R.& Tuszynski, M. H. (2006). Endogenous neurogenesis replaces oligodendrocytes and astrocytes after primate spinal cord injury. *J Neurosci,* Vol. 26, No. 8, (February 2006), pp 2157-2166, ISSN 1529-2401

Yu, Y.; Yao, A. H.; Chen, N.; Pu, L. Y.; Fan, Y.; Lv, L.; Sun, B. C.; Li, G. Q.& Wang, X. H. (2007). Mesenchymal stem cells over-expressing hepatocyte growth factor improve small-for-size liver grafts regeneration. *Mol Ther,* Vol. 15, No. 7, (July 2007), pp 1382-1389, ISSN 1525-0024

Zacharek, A.; Chen, J.; Cui, X.; Li, A.; Li, Y.; Roberts, C.; Feng, Y.; Gao, Q.& Chopp, M. (2007). Angiopoietin1/Tie2 and VEGF/Flk1 induced by MSC treatment amplifies angiogenesis and vascular stabilization after stroke. *J Cereb Blood Flow Metab,* Vol. 27, No. 10, (October 2007), pp 1684-1691, ISSN 0271-678X

Zai, L. J.&Wrathall, J. R. (2005). Cell proliferation and replacement following contusive spinal cord injury. *Glia,* Vol. 50, No. 3, (May 2005), pp 247-257, ISSN 0894-1491

Zhang, R. L.; Zhang, Z. G.& Chopp, M. (2008). Ischemic stroke and neurogenesis in the subventricular zone. *Neuropharmacology,* Vol. 55, No. 3, (September 2008), pp 345-352, ISSN 0028-3908

Zhong, C.; Qin, Z.; Zhong, C. J.; Wang, Y.& Shen, X. Y. (2003). Neuroprotective effects of bone marrow stromal cells on rat organotypic hippocampal slice culture model of cerebral ischemia. *Neurosci Lett,* Vol. 342, No. 1-2, (May 2003), pp 93-96, ISSN 0304-3940

Zilka, N.; Ferencik, M.& Hulin, I. (2006). Neuroinflammation in Alzheimer's disease: protector or promoter? *Bratisl Lek Listy,* Vol. 107, No. 9-10, pp 374-383, ISSN 0006-9248

Zilka, N.; Zilkova, M.; Kazmerova, Z.; Sarissky, M.; Cigankova, V.& Novak, M. (2011). Mesenchymal stem cells rescue the Alzheimer's disease cell model from cell death induced by misfolded truncated tau. Neuroscience, Vol. 193, No (July 14), pp, 330–337, ISSN 1873-7544

Zilkova, M.; Koson, P.& Zilka, N. (2006). The hunt for dying neurons: insight into the neuronal loss in Alzheimer's disease. *Bratisl Lek Listy,* Vol. 107, No. 9-10, (2006), pp 366-373, ISSN 0006-9248

Zuk, P. A.; Zhu, M.; Ashjian, P.; De Ugarte, D. A.; Huang, J. I.; Mizuno, H.; Alfonso, Z. C.; Fraser, J. K.; Benhaim, P.& Hedrick, M. H. (2002). Human adipose tissue is a source of multipotent stem cells. *Mol Biol Cell,* Vol. 13, No. 12, (November 2002), pp 4279-4295, ISSN 1059-1524

18

Assessing the Influence of Neuroinflammation on Neurogenesis: *In Vitro* Models Using Neural Stem Cells and Microglia as Valuable Research Tools

Bruno P. Carreira[1], Maria Inês Morte[1],
Caetana M. Carvalho[1] and Inês M. Araújo[1,2]
[1]*Center for Neuroscience and Cell Biology, Neuroendocrinology and Neurogenesis Group, University of Coimbra, Coimbra,*
[2]*Regenerative Medicine Program, Department of Biomedical Sciences and Medicine University of Algarve, Faro, Portugal*

1. Introduction

1.1 Neural stem cells

Neural stem cells are localized in two limited regions of the adult mammalian brain: the subgranular zone of the dentate gyrus (DG) of the hippocampus, a cell layer located between the granule cell layer and the hilus (Eriksson *et al.*, 1998; Limke and Rao, 2002), and the subventricular zone (SVZ), located next to the ependyma of the lateral walls of the lateral ventricles (Doetsch and Scharff, 2001; Curtis *et al.*, 2007). These regions are thought to provide a specific microenvironment, the stem cell niche, characterized by the presence of several agents involved in the maintenance of self-renewal and/or multipotency of neural stem cells (Alvarez-Buylla and Lim, 2004).

Although neurogenesis has been intensively studied over the past decades, only recently it has been established that newly formed neurons in the adult mammalian brain are functional and integrate into the existing neuronal network (Carlen *et al.*, 2002). The several stages of adult neurogenesis include proliferation of adult neural stem cells, fate determination, migration, integration and maturation of the newborn neurons. Using specific cell markers it is possible to independently investigate the different phases of development. Hippocampal neurogenesis plays an important role in normal hippocampal function, learning and memory (Gould *et al.*, 1999a; Shors *et al.*, 2001; Drapeau *et al.*, 2007). Newborn cells emerging from the SVZ migrate through the rostral migratory stream and integrate into the neuronal network of the olfactory bulb, establish functional synaptic connections and develop electrophysiological properties of mature neurons (Carlen *et al.*, 2002; Petreanu and Alvarez-Buylla, 2002; Belluzzi *et al.*, 2003). Furthermore, neurogenesis in the olfactory bulb is involved in important functions such as odor memory and discrimination (Gheusi *et al.*, 2000; Rochefort *et al.*, 2002; Shingo *et al.*, 2003). Under physiological conditions, neural stem cells are tightly controlled contributing for the maintenance of brain homeostasis (Morshead *et al.*, 1994; Morshead *et al.*, 1998), however they seem to be also involved in neuronal replacement in response to pathophysiological conditions, particularly in conditions associated with neuroinflammation. Although little is known about the molecular mechanisms involved in the regulation of neural stem cells, several factors, both intrinsic and extrinsic, have been described to modulate the neurogenic process, such as hormones, trophic factors, neurotransmitters, neuromodulators and glial cells (for review see Ming and Song, 2005).

The existence of neurogenesis in areas beyond the SVZ and the DG of the adult mammalian brain have also been reported, namely in the neocortex (Gould *et al.*, 1999b; Dayer *et al.*, 2005), striatum, amygdala (Bernier *et al.*, 2002), hypothalamus (Gould *et al.*, 2001; Xu *et al.*, 2005), mesencephalon (Zhao *et al.*, 2003) and spinal cord (Yamamoto *et al.*, 2001). However, these findings need further experimental support, thus more studies need to be conducted.

1.2 Neuroinflammation

The central nervous system (CNS) was considered an immunologically privileged site, not susceptible to immune activation, due to its protection by the blood-brain barrier, which selectively allows certain inflammatory agents to enter and/or exit (Lucas *et al.*, 2006). Nowadays it is well established that immune surveillance takes place in the CNS due to the selective permeability of the blood-brain barrier to immune cells such as T cells, macrophages and dendritic cells (Hickey, 1999). Following injury or exposure to pathogens, an inflammatory response is driven by the activation of two types of immune cells: CNS resident cells, such as microglial cells and astrocytes, and CNS infiltrating cells, such as lymphocytes, monocytes and macrophages from the hematopoietic system (Stoll and Jander, 1999; Streit *et al.*, 1999). The activation of immune cells leads to the production and release of a plethora of regulatory substances, like cytokines, chemokines, neurotransmitters, reactive oxygen species and reactive nitrogen species (reviewed by Whitney *et al.*, 2009). These inflammatory mediators are essential for the recruitment of immune cells, particularly microglial cells, but also for changing the permeability of the blood-brain barrier and recruitment of monocytes and lymphocytes from the hematopoietic system to the compromised area (Hickey, 1999; Lossinsky and Shivers, 2004; Taupin, 2008), which creates a positive feedback loop to the inflammatory response.

Microglia, frequently referred to as the resident macrophages of the brain parenchyma, play a central role in the inflammatory response. Unlike astrocytes, oligodendrocytes and ependymal cells, microglial cells derive from the mesodermal germ layer. During adult life, the microglial cell pool is renewed by division of CNS resident cells. Moreover, microglia are distributed throughout the CNS with distinct densities (Lawson *et al.*, 1990). In the healthy brain, microglia are present in a resting state assuming a typical and dynamic morphology, whose function has been clarified by different studies (Davalos *et al.*, 2005; Nimmerjahn *et al.*, 2005; Davalos *et al.*, 2008). This resting state consists of a constant surveillance activity of the brain parenchyma, which enables microglial cells to screen different brain regions without disturbing the neuronal network (Hanisch and Kettenmann, 2007). Therefore, microglial cells can rapidly react to subtle homeostatic variations by changing morphology and acquiring an array of functions that allow the targeted migration into a site of injury and release of inflammatory mediators (Gehrmann, 1996; Kreutzberg, 1996; Haynes *et al.*, 2006). Reactive microglia have the ability to rapidly upregulate a large number of receptor types, like cytokine receptors, toll-like receptors or cell adhesion molecules, but also to release a plethora of inflammatory agents (for review see Block and Hong, 2005). In fact, chemokines released by reactive microglial cells attract more microglia that, following activation, contribute to further propagate the neuroinflammatory event (Whitney *et al.*, 2009).

Astrocytes constitute the majority of glial cells in the CNS, and play an important structural function, providing support for neurons, playing also regulatory functions, including maintenance of extracellular ion balance, signaling to neurons, repair and scarring process of the CNS (Svendsen, 2002). During inflammation, astrocytes also become activated and release inflammatory factors, growth factors and excitatory amino acids, such as glutamate, which are involved in the regulation of the inflammatory response (Song *et al.*, 2002).

1.3 Neuroinflammation and neurogenesis

Neuroinflammation is a complex event with different outcomes in the neurogenic process, which can therefore enhance or suppress neurogenesis. The secreted products during inflammation have been shown to act as pro- or anti-neurogenic agents, contributing to beneficial or detrimental outcomes of neuroinflammation on the different steps of neurogenesis. Moreover, these effects seem to be particularly dependent on how and for how long microglial cells are activated. Inflammation and microglia activation were initially thought to inhibit adult neurogenesis (Ekdahl *et al.*, 2003; Monje *et al.*, 2003), while recent evidence indicates that microglia under certain circumstances can support neurogenic events (reviewed by Hanisch and Kettenmann, 2007). It has been suggested that mediators released by reactive microglia, such as cytokines and nitric oxide (NO), can inhibit adult neurogenesis in inflammatory conditions (Vallieres *et al.*, 2002; Monje *et al.*, 2003; Liu *et al.*, 2006). On the other hand, neurogenesis seems to be induced by microglial cells activated by IL-4 or low level of IFN-gamma, which has been associated with increased neuroprotection (Wong *et al.*, 2004; Song *et al.*, 2005; Baron *et al.*, 2008). Moreover, some inflammatory mediators like NO seem to have opposite roles in regulating neurogenesis in inflammatory conditions (Carreira *et al.*, 2010). Apparently, microglial cells and the factors they release play a dual role in neurogenesis acting as antiproliferative or proliferative agents. Indeed, self-renewal, proliferation, migration, differentiation, integration and, more importantly, survival of newborn neurons is modulated by the local microenvironment characterizing the neuroinflammatory response. Neural stem cells become "activated" following brain injury and migrate into the lesioned areas, which suggests that mediators present in the inflammatory microenvironment can guide the migration of newborn cells (Arvidsson *et al.*, 2002; Nakatomi *et al.*, 2002).

The role of neuroinflammation in regulating neurogenesis and neuroprotection is not clear yet, and is the subject of numerous studies (for comprehensive review see Whitney *et al.*, 2009; and Gonzalez-Perez *et al.*, 2010). There is, however, evidence for some of the most important mediators of the inflammatory response in their role in the regulation of neurogenesis and neuroprotection (Table 1).

Inflammatory factor	Neurogenesis	Neuroprotection	References
IFN-gamma	Pro-neurogenic	Decreased	(Ben-Hur *et al.*, 2003; Wong *et al.*, 2004; Butovsky *et al.*, 2006; Johansson *et al.*, 2008)
Interleukin-6 Interleukin-18	Anti-neurogenic	Decreased	(Ekdahl *et al.*, 2003; Liu *et al.*, 2005; Nakanishi *et al.*, 2007; Koo and Duman, 2008; Bauer, 2009; Islam *et al.*, 2009)
Nitric oxide	Anti-neurogenic (nNOS) Pro-astrogliogenic (iNOS)	Decreased	(Contestabile *et al.*, 2003; Moreno-Lopez *et al.*, 2004; Matarredona *et al.*, 2005; Ciani *et al.*, 2006; Covacu *et al.*, 2006; Fritzen *et al.*, 2007; Luo *et al.*, 2007; Carreira *et al.*, 2010)
TNF-alpha	Anti-neurogenic (TNF-R1) Pro-neurogenic (TNF-R2)	Decreased or Increased	(Ben-Hur *et al.*, 2003; Wong *et al.*, 2004; Cacci *et al.*, 2005; Heldmann *et al.*, 2005; Liu *et al.*, 2005; Iosif *et al.*, 2006; Bernardino *et al.*, 2008)

Table 1. Effect of some inflammatory factors on neurogenesis and their neuroprotective role.

We are only beginning to understand how inflammatory factors and microglial cells influence neurogenesis in an inflammatory scenario, and the mechanisms, function and modulation of neurogenesis during inflammation require further investigation. This field of work is of particular interest for a better understanding of the mechanisms underlying the effects of neuroinflammation on neurogenesis, and further studies need to be conducted to increase the potential therapeutic value of regulating neuroinflammation in cellular regeneration in the diseased brain.

1.4 Brain repair and stem cell based therapies

Repair of damaged tissues is essential for the survival of living organisms. Each tissue or organ has an intrinsic, albeit limited ability for the replacement of dead cells, and correct integration of the newborn cells that, ideally, should restore the original structure. Cell replacement and correct integration of the newborn cells in the CNS is not so efficient as in other tissues such as skin or bone, which present a higher cell turnover. The CNS, on the other hand, has weak capabilities for both endogenous cell replacement and pattern repair. Some approaches have been used to attempt to develop therapeutic strategies for brain repair, namely transplantation of neural stem cells, stimulation of endogenous neurogenesis, neuroprotective strategies and anti-inflammatory approaches.

Transplantation of neural stem cells is one of the promising methods in study to be used in the reconstruction of neuronal circuits. However, the cells to be transplanted should be phenotypically plastic and able to proliferate *ex vivo* in response to external stimulus (Wang *et al.*, 1998; Sheen *et al.*, 1999). Intracerebral transplantation of SVZ-derived neural stem cells has been successfully used in experimental models of Parkinson's disease (Zigova *et al.*, 1998; Richardson *et al.*, 2005), Huntington's disease (Vazey *et al.*, 2006), and in Multiple Sclerosis (Cayre *et al.*, 2006). Cell replacement could also be achieved by inducing endogenous neural stem cells to differentiate into neurons in the adult CNS, which consists in a less invasive strategy when compared to cell transplantation.

Indeed, *in situ* stimulation of endogenous adult neural stem cells and modulation of injury-induced neurogenesis is a therapeutic strategy, developed to upregulate endogenous neurogenesis, for instance through the control of the inflammatory response in a safe and efficient way. This approach seems to be a more advantageous strategy for multifocal diseases such as Alzheimer's disease, when compared to grafting strategies. Therefore, increased neurogenesis has been achieved by different strategies, such as administration of mitotic agents or trophic factors (Craig *et al.*, 1996; Kuhn *et al.*, 1997; Zigova *et al.*, 1998), treatment with neuroleptics like olanzepine (Green *et al.*, 2006), administration of NO donors or 5-phosphodiesterase inhibitors (Zhang *et al.*, 2003; Imitola *et al.*, 2004; Sun *et al.*, 2004; Sun *et al.*, 2006).

Other strategies designed to improve brain repair are being investigated, such as neuroprotective approaches consisting in the administration of radical scavengers, apoptosis inhibitors, neurotrophic agents, metal ions chelators and gene therapy, which seem to be useful to limit injury-induced lesion, but also for the enhancement of the survival of newborn cells (Polazzi and Monti, 2010). The use of anti-inflammatory drugs as a strategy to promote neurogenesis has also been explored and, although the chronic use of nonsteroidal anti-inflammatory drugs is detrimental for the gastrointestinal tract, it has also been associated with a decreased risk for neurodegenerative diseases (McGeer and McGeer, 1995; Lim *et al.*, 2000; Chen *et al.*, 2003). In fact, control of the inflammatory response seems to be an important strategy to increase proliferation of neural stem cells and/or differentiation of newborn neurons.

Strategies to promote regeneration of lesioned areas or cell replacement therapies will have to take into account the effects of inflammation on the formation and survival of newly generated neurons, either from the brain's own pool of neural stem cells, or from transplanted neural stem cells. Thus, the understanding of the mechanisms underlying the effect of neuroinflammation in proliferation, fate determination, migration and differentiation of neural stem cells is the first step in the development of specific strategies that could target the deleterious effect of inflammation in neurogenesis. Since the neuroinflammatory event is mostly characterized by the activation of resident microglial cells, the use of *in vitro* models that allow the study of the effects of microglia activation in the modulation of neural stem cells proliferation, fate determination, migration and differentiation into neurons is of high importance for the development of therapeutic strategies.

2. *In vitro* models to assess the crosstalk of neurogenesis and neuroinflammation

In vitro culture systems are critical tools for the study of various aspects related to the mechanisms that regulate biological functions. The removal of cells from their native microenvironment allows the study in a more focused way without the restrictions or control of other cell types. When using *in vitro* systems it is essential to recognize that some of the isolated cells must be studied within a short period of time following isolation, or instead, the experimental model must reproduce the microenvironment of the CNS from where cells were isolated. These limitations can, however, be useful to investigate the factors that regulate the phenotype of isolated cells. Different *in vitro* models using neural stem cells and microglial cells may be used, to better understand how inflammation affects the formation of new neurons from neural stem cells.

2.1 Neural stem cell cultures

Reynolds and collaborators performed the first adult neural stem cell culture in the 90's (Reynolds *et al.*, 1992; Reynolds and Weiss, 1992), as free floating cell clusters, commonly referred to as neurospheres. These adult neural stem cells found *in vivo* were dissociated *in vitro* and kept their main properties: self-renewal capacity and multipotency, when in presence of mitogens such as basic fibroblast growth factor (bFGF) and epidermal growth factor (EGF). This cell culture system is extensively used by researchers in neural stem cell biology, and models based on adherent adult neural stem cells cultured in a monolayer on matrix are also widely used (Pollard *et al.*, 2006).

The neural stem cell cultures can be obtained from different regions from the neuroaxis of the adult mammalian CNS, from the olfactory bulb to the spinal cord, and kept in uncoated dishes under serum-free conditions plus mitogens and other essential supplements (Golmohammadi *et al.*, 2008). These adult neural stem cells can be identified based on the expression of specific protein markers such as the transcription factor Sox2, nestin, musashi-1 and the EGF receptor, among others (Kaneko *et al.*, 2000; Ming and Song, 2005). After removal of mitogens these cells can give rise to three different cell types, namely neurons, astrocytes and oligodendrocytes (Levison and Goldman, 1997; Luskin *et al.*, 1997; Palmer *et al.*, 2001; Sanai *et al.*, 2004). Thus, in cultures we can find cells expressing the referred markers but also cells expressing other specific markers, such as glial fibrillary acidic protein (GFAP), polysialylated-neural cell adhesion molecule (PSA-NCAM) and beta-IIII tubulin (Suslov *et al.*, 2002; Ming and Song, 2011).

It is believed that the neurosphere culture may closer resemble the *in vivo* architecture than adherent cultures since it is believed that the stem cell niche is created by clustered cells. On the other hand, the sphere size can be a limitation of this culture in comparison to adherent neural stem cell cultures since the cells that are in the sphere core can have lower access to the nutrients and oxygen, thus undergoing cell death (Ostenfeld *et al.*, 2002; Bez *et al.*, 2003).

Adult neural stem cell culturing systems have been a relevant tool in the study of biological processes within the mammalian nervous system such as neurogenesis and their distinct phases. Cultures are good platforms for expansion of adult neural stem cells, being easily manipulated without loss of function. Additionally, they can be used as experimental models for the study of differentiation and intrinsic specification, and also for screening of drugs with the potential to enhance neurogenesis. However, further investigation should be performed for characterization of stem cells in these models, since a specific marker for neural stem cells is still lacking.

On the other hand, adult stem cell cultures have some limitations, as described next. Cells are sensitive to the culturing protocols, namely the overall number of passages, mitogen concentration and also to the methodology adopted to dissociate spheres - mechanically or by enzymatic digestion (Caldwell, 2001; Caldwell *et al.*, 2001; Morshead *et al.*, 2002; Irvin *et al.*, 2003). The overall size of spheres has been linked to the heterogeneity of sphere composition, since it increases with sphere size, the artificiality of the cell cultures, since cells propagate without instructions of their niche, and the fact that all dividing cells propagate resulting in a mixture of different cell types, are all limitations of the neurosphere culture (Reynolds and Weiss, 1996; Suslov *et al.*, 2002; Parmar *et al.*, 2003). Moreover, the non-limited expansion of cultures could be a disadvantage once the proliferative capacity could be lost by fast dividing cells over multipotent cells or by loss of stem cell capacity over the number of passages. This situation may occur at the expense of differentiation. Moreover, long-term culturing emphasizes the tendency for neural stem cells to adopt an astrocytic phenotype, with reduced capacity to generate oligodendrocytes and neurons (Chang *et al.*, 2004; Vukicevic *et al.*, 2010). Despite these limitations, free floating neural stem cell culturing systems have several advantages and are by far the most used tool concerning the study of neural stem cell biology. The use of neural stem cell cultures allows the easy access to different stages of adult neurogenesis, including proliferation of neural stem cells or progenitors, differentiation and fate determination of progenitor cells, migration of newborn cells and cell survival. By choosing the right tools and correct techniques, these different stages can be independently studied *in vitro*.

Adult neurogenesis was initially reported *in vivo* using autoradiography to track tritiated ([^{3}H])-thymidine. [^{3}H]-thymidine is incorporated in the DNA of dividing cells, thus proving evidence for the existence of newborn cells in the hippocampus (Altman and Das, 1965) and later, in the olfactory bulb (Altman, 1969). Proliferation of neural stem cells, the first stage of neurogenesis, can be also detected *in vitro*. Different methods have been developed since, such as the evaluation of 5-bromo-2'deoxyuridine (BrdU) incorporation, a thymidine analogue that can be incorporated by S-phase cells during DNA synthesis, to detect cell proliferation instead of [^{3}H]-thymidine (Gratzner, 1982; Nowakowski *et al.*, 1989). BrdU has been the golden standard in the detection of cell proliferation for the last 20 years both *in vivo* and *in vitro*. Detection of BrdU can be easily performed with antibodies, either by immunocytochemistry, microplate assay or by flow cytometry. However, BrdU detection requires aggressive treatment for DNA denaturation, in order to allow exposure of the incorporated BrdU to antibodies. Such harsh treatment can be a major drawback in the technique, as head or acid treatment can destroy several epitopes, thus precluding multiplex labeling with other antibodies, and DNA denaturation causes the loss of binding sites for cell cycle dyes.

The use of 5-ethynyl-2'-deoxyuridine (EdU) has recently been proposed as an alternative to BrdU, since EdU detection does not require DNA denaturation, thus improving DNA structural preservation (Salic and Mitchison, 2008). EdU is also a thymidine analog that is incorporated into DNA by dividing cells during active DNA synthesis, and can be used *in vitro* as well as *in vivo* (Rostovtsev *et al.*, 2002). EdU detection is based on click chemistry, via the copper-mediated covalent coupling of the ethynyl group of EdU to a fluorescent dye-conjugated azide (Rostovtsev *et al.*, 2002). Detection can be performed by microscopy, high-throughput analysis equipment or flow cytometry. Particularly, flow cytometry is extremely useful for fast cell cycle analysis together with detection of EdU incorporation, while at same time it is possible to co-label the proliferative cells with other cell-type specific markers. The use of cell cycle markers (described next) complement detection of proliferation by ^{3}H-thymidine, BrdU or EdU, allowing for a more accurate timing of the birth of newborn cells

(Eisch and Mandyam, 2007). Other thymidine analogues that can be detected with antibodies are also available, such as iododeoxyuridine (IdU) and chlorodeoxyuridine (CldU).

Proteins related to the cell cycle have different expression patterns in the neurogenic regions accordingly to the phases of the cell cycle: retinoblastoma protein (Rb), a nuclear protein involved in the control of cell cycle progression, has a functional domain that binds to transcription factors and is expressed mostly in late G1 phase (Yoshikawa, 2000). Proliferating cell nuclear antigen (PCNA), a catalytic nuclear protein associated with DNA polymerase δ, is detected throughout all four phases of the cell cycle, however it is most abundant at late G1 and early S and scarce during G2 and M (Kawabe *et al.*, 2002). Ki-67, a nonhistone nuclear protein, is present during G1, S, G2 and M phase (Gerdes *et al.*, 1984). Cyclin-dependent kinase 1 (CDK1) or Cdc2 (the p34cdc2) is one of the mitosis-promoting factors and has an important role in the initiation of mitosis (Draetta *et al.*, 1988; Okano *et al.*, 1993).

Multi-labeling cells with specific cell markers and proliferation makers could easily identify newly generated neurons and glial cells, such as astrocytes and oligodendrocytes, which allows the distinction between these cell types. Proteins such as RNA-binding protein Hu and musashi-1 are exclusively expressed in mitotic active neural precursor cells, and they are absent in fully differentiated neuronal cells (Sakakibara *et al.*, 1996; Akamatsu *et al.*, 1999). The expression pattern of these markers can be detected by immunolabeling or quantitative real-time PCR (qRT-PCR). Mature neurons can be identified by assessing the presence of markers such as beta-III-tubulin, which contributes to microtubule stability in neuronal cell bodies and axons (Lee *et al.*, 1990; Memberg and Hall, 1995), or by evaluating the presence of neuronal nuclear antigen (NeuN) (Mullen *et al.*, 1992). Also the transcription factor NeuroD can be used since it is expressed throughout maturation until new neurons develop dendrites (Seki, 2002). Other markers that are commonly used can also be found in non-neuronal cells, namely PSA-NCAM (Seki and Arai, 1993; Kiss and Rougon, 1997); nestin, which is expressed in newly generated cells that still have the capacity to divide and differentiate into neurons or astrocytes (Reynolds and Weiss, 1992; Daniel *et al.*, 2008); Sox2, a transcription factor essential to maintain self-renewal of stem cells (Pevny and Placzek, 2005); and doublecortin (DCX) which has a transient expression in proliferating progenitor cells and newly generated neuroblasts or glial cells (Brown *et al.*, 2003; Kempermann *et al.*, 2003; Rao and Shetty, 2004). Oligodendrocytes are easily identified by imunolabeling against 2′, 3′-cyclic nucleotide 3′-phosphodiesterase (CNPase), APC or O4 (Vernadakis *et al.*, 1984; Wu *et al.*, 2008; Girolamo *et al.*, 2010), while astrocytes can be identified by immunolabeling against GFAP, a specific protein for astrocytes (Bock *et al.*, 1977).

Concerning the migration of newly formed cells, it has been extensively studied *in vivo* (Kempermann *et al.*, 2003; Rao and Shetty, 2004), but also *in vitro*, by measuring DCX immunoreactivity (Francis *et al.*, 1999; Cohen *et al.*, 2008). DCX is a microtubule-associated protein having an important role in neuronal migration, by stabilizing microtubules and causing bundling (Sapir *et al.*, 2000). While immunolabeling is currently used, other assays have been developed in order to evaluate migration and simultaneously the mechanisms controlling cell migration, cell protrusion and cell polarization, such as the scratch-wound migration assay (Etienne-Manneville, 2006). Additionally, Durbec and collaborators compared three different assays to evaluate migration of neural stem cells *in vitro*: matrigel, a three-dimensional substrate mimicking the *in vivo* extracellular matrix, detection of soluble factors influencing radial migration and the chemotaxis chamber assay, where the researcher can evaluate whether the cells prefer or not a chemical factor (Durbec *et al.*, 2008).

When mature, not all neurons in culture are functional or survive. It is important to check their viability, namely identify functional synapses by morphological, electrophysiological and immunological characterization (Hartley *et al.*, 1999). Several methods have been used, including immunocytochemical assays, Western blotting and qRT-PCR which allow identification and quantification of proteins, neurotransmitters, neurotrophic factors, among others, involved in neuronal or glial neurotransmitter systems (Hartley *et al.*, 1999; Elmariah *et al.*, 2005; Goodfellow *et al.*, 2011). Using patch-clamp techniques *in vitro* the electrophysiological characterization of neural stem cell cultures can be performed by evaluating the formation of action potentials and activity patterns (Li *et al.*, 2008; Cheyne *et al.*, 2011). Also single-cell calcium currents may be evaluated to discriminate neuronal profile and viability in response to different stimuli, as reported by Bernardino and collaborators (Bernardino *et al.*, 2008).

2.2 Microglial cell cultures

Microglial cells may be obtained for culturing by several methods. One of the most used models for the study of microglial cell function consists in the isolation and expansion of microglia from the neonatal brain. However, there are several limitations and criticisms to this approach since it consists in the isolation of microglial cells from the neonatal brain, not the adult brain. One of the main problems associated with the use of microglial cells *in vitro* is related to the characterization of microglia phenotype. Since there are no truly, unique and specific microglial cell markers, microglia phenotype is defined through a combined analysis of morphology and presence or absence of certain antigens. Several works lack a proper evaluation of microglia phenotype that would allow to distinguish microglia from macrophages. In most studies, the presence of microglial cell markers is excluded from cells that are positive for astrocytic or neuronal markers, but do not distinguish between microglia or macrophages. One of the most used immunocytochemical marker of microglial cells that is the ionized calcium binding adapter molecule 1 (Iba1) (Ito *et al.*, 1998). Other markers that have been identified include the beta-integrin marker CD11b (Ling and Wong, 1993; Gonzalez-Scarano and Baltuch, 1999), the glucose transporter 5 (GLUT5) (Sasaki *et al.*, 2004), CD163 (Roberts *et al.*, 2004; Borda *et al.*, 2008), CCR2 (Albright *et al.*, 1999; Zhang *et al.*, 2007), CD34 (Asheuer *et al.*, 2004; Ladeby *et al.*, 2005) and C-type lectin CD209b (Park *et al.*, 2009). Toll-like receptor 2 (TLR2) and Toll-like receptor 4 (TLR4) have been also used as markers of microglial cells as they appear to be involved in determining the phenotype and function of microglia (Li *et al.*, 2009). A combination of several of these markers would allow for a better characterization of microglia phenotype, rather than the use of a single marker, which is the current standard. The use of multiplex detection systems would be the best approach for a full molecular characterization of microglia (Albright and Gonzalez-Scarano, 2004; Duke *et al.*, 2004; Gebicke-Haerter, 2005; Glanzer *et al.*, 2007; Moran *et al.*, 2007).

The most popular protocol to isolate microglial cells is the shaking method described by Guilian and Baker (Giulian and Baker, 1986) and Frei and colleagues (Frei *et al.*, 1986). In this method, microglial cells are separated from confluent primary mixed glial cultures, isolated from the rodent neonatal cortex, by agitation in an orbital shaker. Although this method allows the preparation of highly pure microglial cultures, the yield of this protocol is low. Saura and colleagues described a method to isolate microglial cells from primary mixed glial cultures of rodent brain by a mild trypsinization protocol, which allows the preparation of high purity microglial cultures, with a higher yield when compared to the shaking method (Saura *et al.*, 2003). Similarly to the shaking method, several works describe the isolation of microglia from adult rodents, and the large majority of these studies take advantage from the astrocyte-microglia interaction for the success of cell cultures (Rosenstiel *et al.*, 2001; Ponomarev *et al.*, 2005). These studies showed that microglial cells, when grown on a monolayer of astrocytes, develop a highly branched morphology which seems to be associated with the downregulation of the nuclear factor kappa B (NF-kappaB) (Rosenstiel *et al.*, 2001). It has been shown that microglial cells isolated from the neonatal or adult brain are sensitive to the treatment with granulocyte macrophage colony-stimulating factor (GM-CSF), which induced a differentiation into a phenotype more similar to those of dendritic cells (Suzumura *et al.*, 1990; Aloisi, 2001). On the other hand, the isolation of adult microglial cells and subsequent culture with low concentrations of macrophage colony-stimulating factor (M-CSF) leads to increased proliferation and survival of cells that persists for several weeks (Suzumura *et al.*, 1990; Ponomarev *et al.*, 2005). M-CSF seems to be a key factor for the maintenance and survival of microglial cells *in vitro*, and has been used in several works (Wegiel *et al.*, 1998; Ponomarev *et al.*, 2005; Carreira *et al.*, 2010). Other methods are also described for the isolation of microglial cells, which include isolation from CNS tissue by Percoll gradient (Dick *et al.*, 1995; Ford *et al.*, 1995), isolation from primary cultures by nutritional deprivation (Hao *et al.*, 1991) or by collecting floating cells in mixed glial cultures (Ganter *et al.*, 1992), but the yield is generally very low.

The use of *in vitro* models allows for the understanding of many aspects of the dynamics associated with the biological functions of microglial cells in a quick and simple manner. However, one cannot overlook that the relevance of the observations obtained can only be extrapolated following *in vivo* studies. Several groups work with microglial cell lines, such as BV-2, HAPI or N9, however the use of microglial cell lines should be carefully considered since immortalization could significantly affect cell biology when compared to the use of primary microglial cultures (Corradin *et al.*, 1993; Lockhart *et al.*, 1998; Horvath *et al.*, 2008).

Concerning primary cultures of microglial cells it is always important to assess the purity of the cultures, this parameter being intrinsically linked to the method of isolation adopted. The isolation method described by Saura and collaborators is, therefore, one of the methods that seems to offer the best value yield/purity (Saura *et al.*, 2003). We favor the isolation of microglial cells by shaking from mixed glial cultures treated with low levels of M-CSF as an alternative to the method of Saura (Saura *et al.*, 2003), with a high purity of the microglia obtained (>90%) and, unlike previous methods, with a high yield (Carreira *et al.*, 2010).

When microglial cells become activated in response to immunologic stimuli or brain injury, activation is characterized by changes in microglia morphology (Streit *et al.*, 1988; Kreutzberg, 1996; Streit *et al.*, 1999; Liu and Hong, 2003), from resting ramified into activated amoeboid microglia (Kreutzberg, 1996). There is also a complex cellular response after activation of microglial cells, which is characterized by upregulation of surface molecules, such as complement receptors and major histocompatibility complex molecules (Oehmichen and Gencic, 1975; Graeber *et al.*, 1988). In addition, activated microglia release a large variety of soluble factors, with a pro- or anti- inflammatory nature and potentially cytotoxic (for review see Block and Hong, 2005). It is therefore important, when establishing primary cultures of microglia, to assess whether microglial cells *in vitro* are also responsive to inflammatory stimuli similarly to what occurs *in vivo*. Microglial cells can be challenged with different stimuli *in vitro*, and by far the most widely used stimulus in primary cultures of microglia isolated from rodents is the bacterial endotoxin lipopolysaccharide (LPS) (Qin *et al.*, 2005a; Qin *et al.*, 2005b; Pei *et al.*, 2007). LPS mimics the infection by Gram-negative bacteria, which induces an increase in the synthesis of inflammatory mediators, namely cytokines, such as IL-1, IL-6 and tumor necrosis factor-alpha (TNF-alpha), chemokines, such as stromal derived factor-1 alpha (SDF-1alpha), free radicals and nitric oxide (Block and Hong, 2005). Other stimuli may consist in the use of ATP, interleukins, IFN-gamma or LPS plus IFN-gamma (Wollmer *et al.*, 2001; Saura *et al.*, 2003).

To characterize the activation of microglial cells after an inflammatory stimulus, we suggest to define at least three parameters to evaluate the activation of microglial cells following exposure to an inflammatory stimulus, including: change to an amoeboid morphology (Suzumura *et al.*, 1991; Wollmer *et al.*, 2001), the expression of NF-kappaB (Heyen *et al.*, 2000; Wollmer *et al.*, 2001), expression of the inducible nitric oxide synthase (iNOS) and subsequent evaluation of the production of NO (Boje and Arora, 1992; Chao *et al.*, 1992b), or the release of TNF-alpha (Sawada *et al.*, 1989; Chao *et al.*, 1992a). The various mechanisms by which microglial cells are activated and the identity of the inflammatory factors released by microglia have been studied and characterized, but there still is a great controversy whether these factors are neuroprotective or neurotoxic when released. The hypothesis that seems to be more acceptable is that, depending on the aggressiveness of the inflammatory response, the activation of microglial cells may shift from a beneficial to a harmful outcome for neurogenesis.

2.3 Combination of neural stem cells and microglial cell cultures

The study of the link between brain inflammation and neurogenesis, in particular the role of microglia in the modulation of the various steps of the neurogenic process, is of particular relevance. In order to operate at a therapeutic level there is an urgent need to understand the crosstalk between microglia and neural stem cells and the implications of the inflammatory response for the neurogenic outcome. Several studies *in vivo* have been developed in recent years, but the potential of *in vitro* studies becomes indisputable when the aim is to study the effect of a particular inflammatory factor or a very specific parameter related to the inflammatory response and its effect on neurogenesis. Whether the function of microglial cells is pro- or anti-neurogenic and whether it is possible to control microglial activation in order to reach a beneficial effect are important questions that need to be answered. Thus, the development of basic models for the *in vitro* study of these issues is an asset to the studies in this area. The use of combined primary neuronal and microglial cell cultures has been a very useful tool in studying the effect of the inflammatory response on neurons from different brain regions. In fact, there are numerous published studies where different approaches have been adopted for the study of the crosstalk between microglial cells and neurons *in vitro* (Boje and Arora, 1992; Lambertsen *et al.*, 2009). Here we describe the use of three different *in vitro* models, which address different aspects of the effects of inflammatory factors released by microglial cells in the neurogenic process.

2.3.1 Co-cultures of neural stem cells with microglia

The inflammatory response has been identified as responsible for the down-regulation of neurogenesis. This hypothesis has been supported by several studies *in vivo* (Ekdahl *et al.*, 2003; Monje *et al.*, 2003), but also by *in vitro* studies where the survival of new neurons is compromised when these are co-cultured with microglial cells activated by LPS (Monje *et al.*, 2003; Cacci *et al.*, 2005; Liu *et al.*, 2005; Cacci *et al.*, 2008). Co-cultures of neural stem cells with microglia, without physical contact between the two cell types, is an experimental model that allows the researcher to assess the role of soluble neuroinflammatory factors using co-cultures of microglial cells seeded in membrane inserts placed on top of multiwell plates containing neural stem cells. The use of techniques of immunodepletion, but also the use of genetically modified animals, allowed to correlate this anti-neurogenic inflammatory response to different interleukins produced during the activation of microglial cells, including IL-6 and IL-1beta (Vallieres *et al.*, 2002; Monje *et al.*, 2003; Nakanishi *et al.*, 2007; Goshen *et al.*, 2008; Koo and Duman, 2008; Spulber *et al.*, 2008). Other factors involved in the inflammatory response appear to contribute to the inhibition of neurogenesis. For example, the increased production of TNF-alpha by microglial cells appears to reduce the survival and differentiation of neural stem cells (Vezzani *et al.*, 2002; Monje *et al.*, 2003; Liu *et al.*, 2005; Iosif *et al.*, 2006).

Although some studies have described IFN-gamma as having a deleterious effect on neurogenesis, it has been demonstrated that microglia stimulated with low levels of IFN-gamma can support the neurogenic process, promoting neuronal differentiation *in vitro* (Butovsky *et al.*, 2006). In other studies it was observed that IFN-gamma is involved in the modulation of proliferation and differentiation of neural stem cells into neurons (Wong *et al.*, 2004; Song *et al.*, 2005; Baron *et al.*, 2008). Recent *in vitro* studies based on the establishment of co-cultures of microglia and neural stem cells, without physical contact between cells, reported that microglia might have a more complex role in neurogenesis contrarily to initial thoughts. Microglia seems to play a dual role in adult neurogenesis, being detrimental or beneficial and support the different steps in neurogenesis, such as stem cell proliferation, differentiation, migration and survival (reviewed in Ekdahl *et al.*, 2009). This dual effect becomes associated to different soluble factors produced by activated microglial cells, such as TNF-alpha or nitric oxide.

The establishment of experimental models such as co-cultures of microglia and neural stem cells allows to mimic the chemical microenvironment that surrounds the SVZ and/or the DG during inflammatory conditions when microglial cells are recruited and activated. On the other hand, the fact that both cell types share the same culture environment is important to determine the effect of factors produced by microglial cells on neural stem cells. The fact that this is a system without physical contact between the two cell types also allows determining more quickly, and using more economic approaches, the modulation of the multistep neurogenic process mediated by the inflammatory response. Thus, experimental approaches to determine cell proliferation and cell cycle, such as flow cytometry, cell migration, could be performed without the need for prior characterization to distinguish neural stem cells from microglial cells as in mixed cultures. Moreover, signaling pathways present in both cell types can be studied this way, as is the case of TLR4 that directly modulates self-renewal and the decision-cell-fate in neural stem cells (Rolls *et al.*, 2007) and in microglial cells is involved in its activation, particularly in the regulation of gene expression of iNOS (Graeber and Streit, 2010).

However, there are also some disadvantages associated with the use of this experimental methodology. Firstly, the fact that it does not allow an easy processing of microglia cells, which are placed in membrane inserts, after experimental treatment. In fact, simple experimental procedures such as protein, RNA or DNA extraction from microglial cells becomes difficult to perform. On the other hand, it is not possible to perform immunostaining techniques for subsequent microscopic analysis of microglial cells plated in inserts. In addition, this model does not answer a question that seems to be increasingly important which is the influence of cell-to-cell contact in the modulation of neurogenesis by the inflammatory response (Song *et al.*, 2002; Aarum *et al.*, 2003; Alvarez-Buylla and Lim, 2004). Despite these disadvantages, the use of co-cultures of neural stem cells with microglia, without physical contact between the two cell types, is a good approach for some studies.

2.3.2 Neural stem cell cultures exposed to microglia-conditioned medium

The production of cytokines and other molecules by activated microglial cells with implications in cellular processes has been demonstrated in many studies based on *in vitro* models (Banati *et al.*, 1993; Minghetti and Levi, 1998; Gebicke-Haerter *et al.*, 2001; Hanisch, 2002; Hausler *et al.*, 2002). However, there is still much to be learned about how cellular pathways in neural stem cells are regulated by these soluble factors from microglial origin. It is therefore important to assess how these diffusible factors influence phenomena as diverse as proliferation, differentiation, migration or cell survival.

Culturing neural stem cells with microglia conditioned medium, obtained from a separate microglia culture, allows the isolation of the unidirectional communication between activated microglia and neural stem cells, with further investigation of soluble inflammatory factors. According to studies using this experimental model, the conditioned medium of microglial cells acutely challenged with LPS reduced the survival of neural stem cells, preventing their differentiation into neurons (Monje *et al.*, 2003; Cacci *et al.*, 2008). One of the inflammatory agents reported to be responsible for this antineurogenic effect is the cytokine IL-6, as evidenced by the works of Monje and collaborators or Nakanishi and colleagues that by using a specific antibody against IL-6 rescued neurogenesis (Monje *et al.*, 2003; Nakanishi *et al.*, 2007). On the other hand, several *in vitro* studies described a pro-neurogenic effect of microglial cells and their conditioned medium, in which neural stem cells grow (Aarum *et al.*, 2003; Morgan *et al.*, 2004; Walton *et al.*, 2006; Nakanishi *et al.*, 2007).

Despite the advantages of this experimental model, namely the fact that it allows a study of the unidirectional effect of microglia on neural stem cells, there are also some disadvantages. This model does not allow inferring any conclusion about the influence of cell-to-cell contact between microglia and neural stem cells, an event that has been described to occur between glial cells and neural stem cells (Song *et al.*, 2002; Aarum *et al.*, 2003; Alvarez-Buylla and Lim, 2004). On the other hand, this model completely neglects the fact that some of the factors released by microglial cells have physical characteristics that do not allow their study in a conditioned medium transferred from a cell culture to another. Particularly nitric oxide, a gaseous molecule with a short half-life, cannot be studied because it is highly reactive in aqueous solution at 37 °C and physiological pH (pH = 7.4). Thus, although stable end products of NO can be detected in conditioned medium from activated microglial cell cultures, the effect of NO in the neural stem cells cannot be analyzed. These are negative aspects that must be taken into account when a researcher decides to select this experimental model. Despite these aspects, the use of conditioned medium of microglia in cultures of neural stem cells is a good model to further study the influence of inflammation on neurogenesis. This model is useful to complement other *in vitro* approaches, including co-cultures of microglia and neural stem cells, with or without physical contact.

2.3.3 Mixed cultures of neural stem cells with microglia

The progression of the neurogenic process until the differentiation of neural stem cells into neurons appears to be regulated by the inflammatory microenvironment but also by cell-to-cell interactions involved (Arvidsson *et al.*, 2002; Nakatomi *et al.*, 2002; Ben-Hur *et al.*, 2003; Thored *et al.*, 2006; Thored *et al.*, 2009). Therefore, the optimization of an *in vitro* system that allows the study of physical interactions between microglia and neural stem cells is of great interest to understand how both cell types crosstalk in inflammatory conditions.

Mixed cultures are co-cultures of neural stem cells with microglia with physical contact between the two cell types. In this culture model, the role of physical contacts between microglia and neural stem cells can be studied. The mixed culture system is, probably, the *in vitro* approach that more closely mimics what happens *in vivo*, where microglial cells physically contact with the neural stem cells from neurogenic areas. Adopting this experimental model, the researcher can study the influence of the inflammatory response on the several steps of the neurogenic process, but also cell-cell interactions, which is an advantage compared to the *in vitro* models already described. An example of a mixed culture of neural stem cells cultured together with forebrain microglia is shown in Fig. 1. Enhanced green fluorescent protein (EGFP)-positive SVZ cells were isolated from the SVZ of postnatal day 1-3 actin-EGFP C57Bl6 mice, thus being readily distinguishable from microglia isolated from wild-type mice (Fig. 1A).

The mixed culture model allows simultaneous evaluation of microglia and neural stem cells. Thus, following stimulation of microglial cells, the researcher can evaluate the activation of these cells as well as several biological processes of neural stem cells, such as proliferation, differentiation and/or survival. Moreover, multi-labeling experiments of proliferation markers, such as BrdU or EdU (Fig. 1B), with microglia-specific (Iba-1 or CD11b), neuron-specific (NeuN or Tuj-1) or glia-specific (GFAP) proteins by confocal microscopy or flow cytometry are a good way to determine the phenotype of proliferating cells (Nixon and Crews, 2004). In addition, it is also possible to evaluate the effect of diffusible factors that are produced following activation of microglial cells. Separation of the two cell populations for posterior analysis (e.g. of protein or nucleic acids) is possible using a cell sorter. The researcher can confirm whether the effects observed in mixed cultures are caused by physical interactions or by diffusible factors released by microglial cells by combining such experiments with a comparative study using co-cultured cells without physical contact.

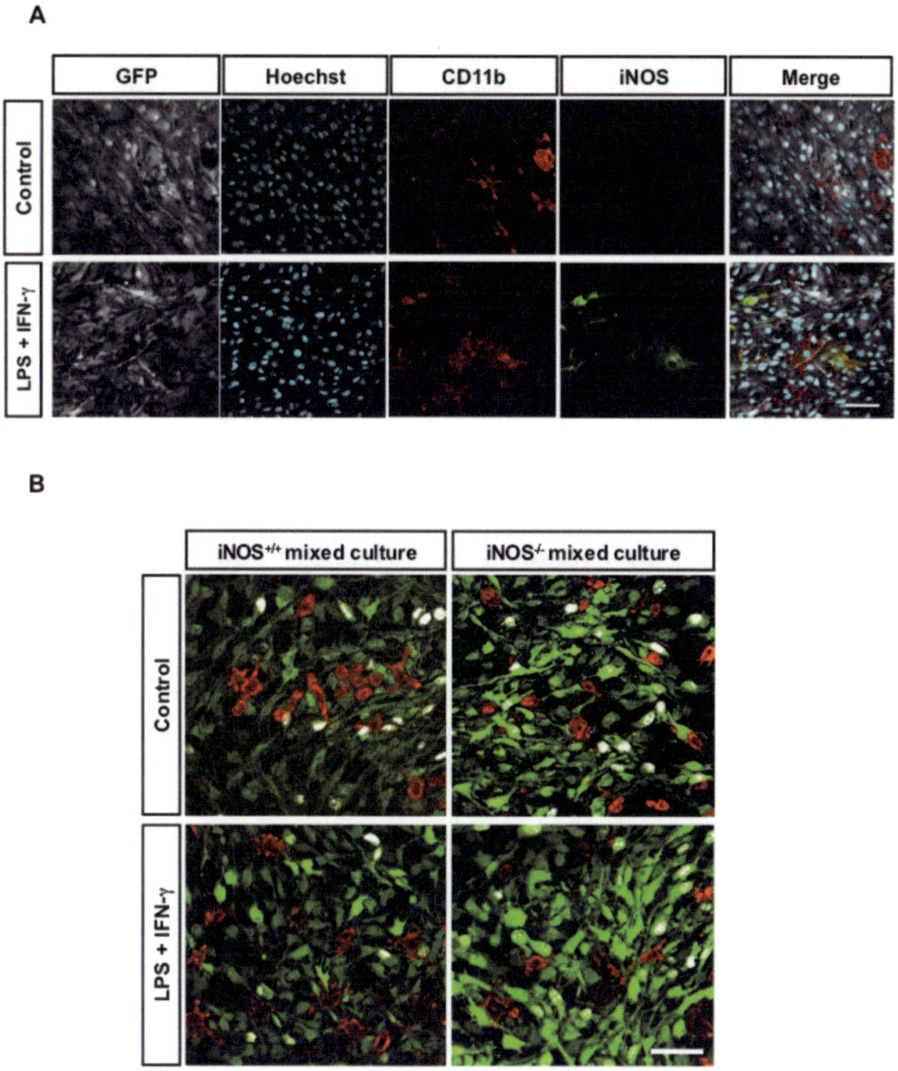

Fig. 1. Mixed cultures of primary microglial cells and subventricular zone (SVZ)-derived neural stem cells. SVZ cells (isolated from transgenic mice expressing green fluorescence protein (GFP) under the actin promoter (shown in white) are readily distinguishable from CD11b-positive microglia (red) (A). Microglia (red) cultured with GFP-positive SVZ cells (white) show immunoreactivity for inducible nitric oxide synthase (iNOS, green), following treatment with lipopolysaccharide (LPS; 100 ng/ml) plus interferon-gamma (IFN-gamma; 0.5 ng/ml), for 24 h. Nuclei are labeled with Hoechst 33342 (blue). Scale bar: 20 μm. B) Stimulation with LPS plus IFN-gamma decreases the proliferation of GFP-positive SVZ-derived neural stem cells (green), in mixed cultures of SVZ and microglia obtained from wild type mice (iNOS$^{+/+}$), which are CD11b-positive (red). Cell proliferation was assessed by 5-ethynyl-2′-deoxyuridine (EdU) incorporation (white). The antiproliferative effect of LPS plus IFN-gamma on EdU incorporation is abolished in mixed cultures in which the microglia was obtained from iNOS-knockout mice (iNOS$^{-/-}$). Scale bar: 20 μm.

3. Summary and future directions

Microglial cells may cause different effects on the neurogenic process, promoting or inhibiting it. Experimental evidence has been presented indicating that microglia, depending on their activation status and phenotype, could favor or hinder adult neurogenesis, in physiological or pathophysiological conditions. In fact, microglia can have a dual role in different steps of the neurogenic process, namely in the formation, maturation and integration of newly formed neurons. Therefore the need to explore in more detail how microglia regulate adult neurogenesis in physiological and pathophysiological conditions is of particular importance (Graeber and Streit, 2010).

Genetic mouse models in which the researcher can selectively ablate genes have already been described as useful strategies to study the involvement of particular effectors of the neuroinflammatory response on neural stem cells. Experimental models may have as an objective the determination of how modulation of microglial cell activation can be used as a therapeutic target to regulate neurogenesis in the adult brain (Ekdahl *et al.*, 2009; Whitney *et al.*, 2009; Polazzi and Monti, 2010). These models are suitable to evaluate the neurogenic potential of anti-inflammatory drugs or identify pro-neurogenic targets. Thus, these experimental approaches will allow the design of therapeutic strategies to enhance the formation, proper migration, differentiation, integration and survival of new neuronal cells in the injured nervous system. Moreover, all culture models are suitable for pharmacological or genetic manipulation, including obtaining the cells used in the cultures from wild-type or genetically modified animals, and can be adapted for high-throughput analysis and drug screening. The use of anti-inflammatory drugs with a selective mechanism of action at the level of microglial cells, or the use of anti-inflammatory drugs which may release molecules that may enhance the neurogenesis are strategies under investigation (Keeble and Moore, 2002; Napoli and Ignarro, 2003; Ajmone-Cat *et al.*, 2008; Koc and Kucukguzel, 2009). In order to develop more specific therapeutic interventions in the future, it is necessary to identify the mechanisms and factors that regulate the switch between the enhancing or detrimental effect of the inflammatory response on neurogenic events. The *in vitro* strategies discussed here are important as a first step in identifying and characterizing these events (Table 2).

Experimental model	**Parameters evaluated**			
	Diffusible/soluble factors	**Cell-to-cell interaction**	**Cellular characterization**	**Protein, RNA and DNA content**
Co-culture	Very Good	-	Very Good	Very Good
Conditioned medium	Good	-	Very Good	Very Good
Mixed culture	Very Good	Very Good	Good (requires multiplex analysis)	Good (requires cell sorting)

Table 2. Evaluation of experimental *in vitro* models using neural stem cells and microglial cells as research tools to evaluate the effect of neuroinflammation in the neurogenesis.

4. References

Aarum J, Sandberg K, Haeberlein SL and Persson MA (2003). "Migration and differentiation of neural precursor cells can be directed by microglia." *Proc Natl Acad Sci U S A* 100(26): 15983-8.

Ajmone-Cat MA, Cacci E and Minghetti L (2008). "Non steroidal anti-inflammatory drugs and neurogenesis in the adult mammalian brain." *Curr Pharm Des* 14(14): 1435-42.

Akamatsu W, Okano HJ, Osumi N, Inoue T, Nakamura S, Sakakibara S, Miura M, Matsuo N, Darnell RB and Okano H (1999). "Mammalian ELAV-like neuronal RNA-binding proteins HuB and HuC promote neuronal development in both the central and the peripheral nervous systems." *Proc Natl Acad Sci U S A* 96(17): 9885-90.

Albright AV and Gonzalez-Scarano F (2004). "Microarray analysis of activated mixed glial (microglia) and monocyte-derived macrophage gene expression." *J Neuroimmunol* 157(1-2): 27-38.

Albright AV, Shieh JT, Itoh T, Lee B, Pleasure D, O'Connor MJ, Doms RW and Gonzalez-Scarano F (1999). "Microglia express CCR5, CXCR4, and CCR3, but of these, CCR5 is the principal coreceptor for human immunodeficiency virus type 1 dementia isolates." *J Virol* 73(1): 205-13.

Aloisi F (2001). "Immune function of microglia." *Glia* 36(2): 165-79.

Altman J (1969). "Autoradiographic and histological studies of postnatal neurogenesis. IV. Cell proliferation and migration in the anterior forebrain, with special reference to persisting neurogenesis in the olfactory bulb." *J Comp Neurol* 137(4): 433-57.

Altman J and Das GD (1965). "Autoradiographic and histological evidence of postnatal hippocampal neurogenesis in rats." *J Comp Neurol* 124(3): 319-35.

Alvarez-Buylla A and Lim DA (2004). "For the long run: maintaining germinal niches in the adult brain." *Neuron* 41(5): 683-6.

Arvidsson A, Collin T, Kirik D, Kokaia Z and Lindvall O (2002). "Neuronal replacement from endogenous precursors in the adult brain after stroke." *Nat Med* 8(9): 963-70.

Asheuer M, Pflumio F, Benhamida S, Dubart-Kupperschmitt A, Fouquet F, Imai Y, Aubourg P and Cartier N (2004). "Human CD34+ cells differentiate into microglia and express recombinant therapeutic protein." *Proc Natl Acad Sci U S A* 101(10): 3557-62.

Banati RB, Gehrmann J, Schubert P and Kreutzberg GW (1993). "Cytotoxicity of microglia." *Glia* 7(1): 111-8.

Baron R, Nemirovsky A, Harpaz I, Cohen H, Owens T and Monsonego A (2008). "IFN-gamma enhances neurogenesis in wild-type mice and in a mouse model of Alzheimer's disease." *Faseb J* 22(8): 2843-52.

Bauer S (2009). "Cytokine control of adult neural stem cells." *Ann N Y Acad Sci* 1153: 48-56.

Belluzzi O, Benedusi M, Ackman J and LoTurco JJ (2003). "Electrophysiological differentiation of new neurons in the olfactory bulb." *J Neurosci* 23(32): 10411-8.

Ben-Hur T, Ben-Menachem O, Furer V, Einstein O, Mizrachi-Kol R and Grigoriadis N (2003). "Effects of proinflammatory cytokines on the growth, fate, and motility of multipotential neural precursor cells." *Mol Cell Neurosci* 24(3): 623-31.

Bernardino L, Agasse F, Silva B, Ferreira R, Grade S and Malva JO (2008). "Tumor necrosis factor-alpha modulates survival, proliferation, and neuronal differentiation in neonatal subventricular zone cell cultures." *Stem Cells* 26(9): 2361-71.

Bernier PJ, Bedard A, Vinet J, Levesque M and Parent A (2002). "Newly generated neurons in the amygdala and adjoining cortex of adult primates." *Proc Natl Acad Sci U S A* 99(17): 11464-9.

Bez A, Corsini E, Curti D, Biggiogera M, Colombo A, Nicosia RF, Pagano SF and Parati EA (2003). "Neurosphere and neurosphere-forming cells: morphological and ultrastructural characterization." *Brain Res* 993(1-2): 18-29.

Block ML and Hong JS (2005). "Microglia and inflammation-mediated neurodegeneration: multiple triggers with a common mechanism." *Prog Neurobiol* 76(2): 77-98.

Bock E, Moller M, Nissen C and Sensenbrenner M (1977). "Glial fibrillary acidic protein in primary astroglial cell cultures derived from newborn rat brain." *FEBS Lett* 83(2): 207-11.

Boje KM and Arora PK (1992). "Microglial-produced nitric oxide and reactive nitrogen oxides mediate neuronal cell death." *Brain Res* 587(2): 250-6.

Borda JT, Alvarez X, Mohan M, Hasegawa A, Bernardino A, Jean S, Aye P and Lackner AA (2008). "CD163, a marker of perivascular macrophages, is up-regulated by microglia in simian immunodeficiency virus encephalitis after haptoglobin-hemoglobin complex stimulation and is suggestive of breakdown of the blood-brain barrier." *Am J Pathol* 172(3): 725-37.

Brown JP, Couillard-Despres S, Cooper-Kuhn CM, Winkler J, Aigner L and Kuhn HG (2003). "Transient expression of doublecortin during adult neurogenesis." *J Comp Neurol* 467(1): 1-10.

Butovsky O, Ziv Y, Schwartz A, Landa G, Talpalar AE, Pluchino S, Martino G and Schwartz M (2006). "Microglia activated by IL-4 or IFN-gamma differentially induce neurogenesis and oligodendrogenesis from adult stem/progenitor cells." *Mol Cell Neurosci* 31(1): 149-60.

Cacci E, Ajmone-Cat MA, Anelli T, Biagioni S and Minghetti L (2008). "In vitro neuronal and glial differentiation from embryonic or adult neural precursor cells are differently affected by chronic or acute activation of microglia." *Glia* 56(4): 412-25.

Cacci E, Claasen JH and Kokaia Z (2005). "Microglia-derived tumor necrosis factor-alpha exaggerates death of newborn hippocampal progenitor cells in vitro." *J Neurosci Res* 80(6): 789-97.

Caldwell MA (2001). "Recent advances in neuralstem cell technologies." *Trends Neurosci* 24(2): 72-4.

Caldwell MA, He X, Wilkie N, Pollack S, Marshall G, Wafford KA and Svendsen CN (2001). "Growth factors regulate the survival and fate of cells derived from human neurospheres." *Nat Biotechnol* 19(5): 475-9.

Carlen M, Cassidy RM, Brismar H, Smith GA, Enquist LW and Frisen J (2002). "Functional integration of adult-born neurons." *Curr Biol* 12(7): 606-8.

Carreira BP, Morte MI, Inacio A, Costa G, Rosmaninho-Salgado J, Agasse F, Carmo A, Couceiro P, Brundin P, Ambrosio AF, Carvalho CM and Araujo IM (2010). "Nitric oxide stimulates the proliferation of neural stem cells bypassing the epidermal growth factor receptor." *Stem Cells* 28(7): 1219-30.

Cayre M, Bancila M, Virard I, Borges A and Durbec P (2006). "Migrating and myelinating potential of subventricular zone neural progenitor cells in white matter tracts of the adult rodent brain." *Mol Cell Neurosci* 31(4): 748-58.

Chang MY, Park CH, Lee SY and Lee SH (2004). "Properties of cortical precursor cells cultured long term are similar to those of precursors at later developmental stages." *Brain Res Dev Brain Res* 153(1): 89-96.

Chao CC, Hu S, Close K, Choi CS, Molitor TW, Novick WJ and Peterson PK (1992a). "Cytokine release from microglia: differential inhibition by pentoxifylline and dexamethasone." *J Infect Dis* 166(4): 847-53.

Chao CC, Hu S, Molitor TW, Shaskan EG and Peterson PK (1992b). "Activated microglia mediate neuronal cell injury via a nitric oxide mechanism." *J Immunol* 149(8): 2736-41.

Chen H, Zhang SM, Hernan MA, Schwarzschild MA, Willett WC, Colditz GA, Speizer FE and Ascherio A (2003). "Nonsteroidal anti-inflammatory drugs and the risk of Parkinson disease." *Arch Neurol* 60(8): 1059-64.

Cheyne JE, Grant L, Butler-Munro C, Foote JW, Connor B and Montgomery JM (2011). "Synaptic integration of newly generated neurons in rat dissociated hippocampal cultures." *Mol Cell Neurosci* 47(3): 203-14.

Ciani E, Calvanese V, Crochemore C, Bartesaghi R and Contestabile A (2006). "Proliferation of cerebellar precursor cells is negatively regulated by nitric oxide in newborn rat." *J Cell Sci* 119(Pt 15): 3161-70.

Cohen D, Segal M and Reiner O (2008). "Doublecortin supports the development of dendritic arbors in primary hippocampal neurons." *Dev Neurosci* 30(1-3): 187-99.

Contestabile A, Monti B and Ciani E (2003). "Brain nitric oxide and its dual role in neurodegeneration/neuroprotection: understanding molecular mechanisms to devise drug approaches." *Curr Med Chem* 10(20): 2147-74.

Corradin SB, Mauel J, Donini SD, Quattrocchi E and Ricciardi-Castagnoli P (1993). "Inducible nitric oxide synthase activity of cloned murine microglial cells." *Glia* 7(3): 255-62.

Covacu R, Danilov AI, Rasmussen BS, Hallen K, Moe MC, Lobell A, Johansson CB, Svensson MA, Olsson T and Brundin L (2006). "Nitric oxide exposure diverts neural stem cell fate from neurogenesis towards astrogliogenesis." *Stem Cells* 24(12): 2792-800.

Craig CG, Tropepe V, Morshead CM, Reynolds BA, Weiss S and van der Kooy D (1996). "In vivo growth factor expansion of endogenous subependymal neural precursor cell populations in the adult mouse brain." *J Neurosci* 16(8): 2649-58.

Curtis MA, Eriksson PS and Faull RL (2007). "Progenitor cells and adult neurogenesis in neurodegenerative diseases and injuries of the basal ganglia." *Clin Exp Pharmacol Physiol* 34(5-6): 528-32.

Daniel C, Albrecht H, Ludke A and Hugo C (2008). "Nestin expression in repopulating mesangial cells promotes their proliferation." *Lab Invest* 88(4): 387-97.

Davalos D, Grutzendler J, Yang G, Kim JV, Zuo Y, Jung S, Littman DR, Dustin ML and Gan WB (2005). "ATP mediates rapid microglial response to local brain injury in vivo." *Nat Neurosci* 8(6): 752-8.

Davalos D, Lee JK, Smith WB, Brinkman B, Ellisman MH, Zheng B and Akassoglou K (2008). "Stable in vivo imaging of densely populated glia, axons and blood vessels in the mouse spinal cord using two-photon microscopy." *J Neurosci Methods* 169(1): 1-7.

Dayer AG, Cleaver KM, Abouantoun T and Cameron HA (2005). "New GABAergic interneurons in the adult neocortex and striatum are generated from different precursors." *J Cell Biol* 168(3): 415-27.

Dick AD, Ford AL, Forrester JV and Sedgwick JD (1995). "Flow cytometric identification of a minority population of MHC class II positive cells in the normal rat retina distinct from CD45lowCD11b/c+CD4low parenchymal microglia." *Br J Ophthalmol* 79(9): 834-40.

Doetsch F and Scharff C (2001). "Challenges for brain repair: insights from adult neurogenesis in birds and mammals." *Brain Behav Evol* 58(5): 306-22.

Draetta G, Brizuela L, Moran B and Beach D (1988). "Regulation of the vertebrate cell cycle by the cdc2 protein kinase." *Cold Spring Harb Symp Quant Biol* 53 Pt 1: 195-201.

Drapeau E, Montaron MF, Aguerre S and Abrous DN (2007). "Learning-induced survival of new neurons depends on the cognitive status of aged rats." *J Neurosci* 27(22): 6037-44.

Duke DC, Moran LB, Turkheimer FE, Banati R and Graeber MB (2004). "Microglia in culture: what genes do they express?" *Dev Neurosci* 26(1): 30-7.

Durbec P, Franceschini I, Lazarini F and Dubois-Dalcq M (2008). "In vitro migration assays of neural stem cells." *Methods Mol Biol* 438: 213-25.

Eisch AJ and Mandyam CD (2007). "Adult neurogenesis: can analysis of cell cycle proteins move us "Beyond BrdU"?" *Curr Pharm Biotechnol* 8(3): 147-65.

Ekdahl CT, Claasen JH, Bonde S, Kokaia Z and Lindvall O (2003). "Inflammation is detrimental for neurogenesis in adult brain." *Proc Natl Acad Sci U S A* 100(23): 13632-7.

Ekdahl CT, Kokaia Z and Lindvall O (2009). "Brain inflammation and adult neurogenesis: the dual role of microglia." *Neuroscience* 158(3): 1021-9.

Elmariah SB, Oh EJ, Hughes EG and Balice-Gordon RJ (2005). "Astrocytes regulate inhibitory synapse formation via Trk-mediated modulation of postsynaptic GABAA receptors." *J Neurosci* 25(14): 3638-50.

Eriksson PS, Perfilieva E, Bjork-Eriksson T, Alborn AM, Nordborg C, Peterson DA and Gage FH (1998). "Neurogenesis in the adult human hippocampus." *Nat Med* 4(11): 1313-7.

Etienne-Manneville S (2006). "In vitro assay of primary astrocyte migration as a tool to study Rho GTPase function in cell polarization." *Methods Enzymol* 406: 565-78.

Ford AL, Goodsall AL, Hickey WF and Sedgwick JD (1995). "Normal adult ramified microglia separated from other central nervous system macrophages by flow cytometric sorting. Phenotypic differences defined and direct ex vivo antigen presentation to myelin basic protein-reactive CD4+ T cells compared." *J Immunol* 154(9): 4309-21.

Francis F, Koulakoff A, Boucher D, Chafey P, Schaar B, Vinet MC, Friocourt G, McDonnell N, Reiner O, Kahn A, McConnell SK, Berwald-Netter Y, Denoulet P and Chelly J (1999). "Doublecortin is a developmentally regulated, microtubule-associated protein expressed in migrating and differentiating neurons." *Neuron* 23(2): 247-56.

Frei K, Bodmer S, Schwerdel C and Fontana A (1986). "Astrocyte-derived interleukin 3 as a growth factor for microglia cells and peritoneal macrophages." *J Immunol* 137(11): 3521-7.

Fritzen S, Schmitt A, Koth K, Sommer C, Lesch KP and Reif A (2007). "Neuronal nitric oxide synthase (NOS-I) knockout increases the survival rate of neural cells in the hippocampus independently of BDNF." *Mol Cell Neurosci* 35(2): 261-71.

Ganter S, Northoff H, Mannel D and Gebicke-Harter PJ (1992). "Growth control of cultured microglia." *J Neurosci Res* 33(2): 218-30.

Gebicke-Haerter PJ (2005). "Microarrays and expression profiling in microglia research and in inflammatory brain disorders." *J Neurosci Res* 81(3): 327-41.

Gebicke-Haerter PJ, Spleiss O, Ren LQ, Li H, Dichmann S, Norgauer J and Boddeke HW (2001). "Microglial chemokines and chemokine receptors." *Prog Brain Res* 132: 525-32.

Gehrmann J (1996). "Microglia: a sensor to threats in the nervous system?" *Res Virol* 147(2-3): 79-88.

Gerdes J, Lemke H, Baisch H, Wacker HH, Schwab U and Stein H (1984). "Cell cycle analysis of a cell proliferation-associated human nuclear antigen defined by the monoclonal antibody Ki-67." *J Immunol* 133(4): 1710-5.

Gheusi G, Cremer H, McLean H, Chazal G, Vincent JD and Lledo PM (2000). "Importance of newly generated neurons in the adult olfactory bulb for odor discrimination." *Proc Natl Acad Sci U S A* 97(4): 1823-8.

Girolamo F, Strippoli M, Errede M, Benagiano V, Roncali L, Ambrosi G and Virgintino D (2010). "Characterization of oligodendrocyte lineage precursor cells in the mouse cerebral cortex: a confocal microscopy approach to demyelinating diseases." *Ital J Anat Embryol* 115(1-2): 95-102.

Giulian D and Baker TJ (1986). "Characterization of ameboid microglia isolated from developing mammalian brain." *J Neurosci* 6(8): 2163-78.

Glanzer JG, Enose Y, Wang T, Kadiu I, Gong N, Rozek W, Liu J, Schlautman JD, Ciborowski PS, Thomas MP and Gendelman HE (2007). "Genomic and proteomic microglial profiling: pathways for neuroprotective inflammatory responses following nerve fragment clearance and activation." *J Neurochem* 102(3): 627-45.

Golmohammadi MG, Blackmore DG, Large B, Azari H, Esfandiary E, Paxinos G, Franklin KB, Reynolds BA and Rietze RL (2008). "Comparative analysis of the frequency and distribution of stem and progenitor cells in the adult mouse brain." *Stem Cells* 26(4): 979-87.

Gonzalez-Perez O, Jauregui-Huerta F and Galvez-Contreras AY (2010). "Immune system modulates the function of adult neural stem cells." *Curr Immunol Rev* 6(3): 167-173.

Gonzalez-Scarano F and Baltuch G (1999). "Microglia as mediators of inflammatory and degenerative diseases." *Annu Rev Neurosci* 22: 219-40.

Goodfellow CE, Graham SE, Dragunow M and Glass M (2011). "Characterization of NTera2/D1 cells as a model system for the investigation of cannabinoid function in human neurons and astrocytes." *J Neurosci Res.*

Goshen I, Kreisel T, Ben-Menachem-Zidon O, Licht T, Weidenfeld J, Ben-Hur T and Yirmiya R (2008). "Brain interleukin-1 mediates chronic stress-induced depression in mice via adrenocortical activation and hippocampal neurogenesis suppression." *Mol Psychiatry* 13(7): 717-28.

Gould E, Beylin A, Tanapat P, Reeves A and Shors TJ (1999a). "Learning enhances adult neurogenesis in the hippocampal formation." *Nat Neurosci* 2(3): 260-5.

Gould E, Reeves AJ, Graziano MS and Gross CG (1999b). "Neurogenesis in the neocortex of adult primates." *Science* 286(5439): 548-52.

Gould E, Vail N, Wagers M and Gross CG (2001). "Adult-generated hippocampal and neocortical neurons in macaques have a transient existence." *Proc Natl Acad Sci U S A* 98(19): 10910-7.

Graeber MB and Streit WJ (2010). "Microglia: biology and pathology." *Acta Neuropathol* 119(1): 89-105.

Graeber MB, Streit WJ and Kreutzberg GW (1988). "The microglial cytoskeleton: vimentin is localized within activated cells in situ." *J Neurocytol* 17(4): 573-80.

Gratzner HG (1982). "Monoclonal antibody to 5-bromo- and 5-iododeoxyuridine: A new reagent for detection of DNA replication." *Science* 218(4571): 474-5.

Green W, Patil P, Marsden CA, Bennett GW and Wigmore PM (2006). "Treatment with olanzapine increases cell proliferation in the subventricular zone and prefrontal cortex." *Brain Res* 1070(1): 242-5.

Hanisch UK (2002). "Microglia as a source and target of cytokines." *Glia* 40(2): 140-55.

Hanisch UK and Kettenmann H (2007). "Microglia: active sensor and versatile effector cells in the normal and pathologic brain." *Nat Neurosci* 10(11): 1387-94.

Hao C, Richardson A and Fedoroff S (1991). "Macrophage-like cells originate from neuroepithelium in culture: characterization and properties of the macrophage-like cells." *Int J Dev Neurosci* 9(1): 1-14.

Hartley RS, Margulis M, Fishman PS, Lee VM and Tang CM (1999). "Functional synapses are formed between human NTera2 (NT2N, hNT) neurons grown on astrocytes." *J Comp Neurol* 407(1): 1-10.

Hausler KG, Prinz M, Nolte C, Weber JR, Schumann RR, Kettenmann H and Hanisch UK (2002). "Interferon-gamma differentially modulates the release of cytokines and chemokines in lipopolysaccharide- and pneumococcal cell wall-stimulated mouse microglia and macrophages." *Eur J Neurosci* 16(11): 2113-22.

Haynes SE, Hollopeter G, Yang G, Kurpius D, Dailey ME, Gan WB and Julius D (2006). "The P2Y12 receptor regulates microglial activation by extracellular nucleotides." *Nat Neurosci* 9(12): 1512-9.

Heldmann U, Thored P, Claasen JH, Arvidsson A, Kokaia Z and Lindvall O (2005). "TNF-alpha antibody infusion impairs survival of stroke-generated neuroblasts in adult rat brain." *Exp Neurol* 196(1): 204-8.

Heyen JR, Ye S, Finck BN and Johnson RW (2000). "Interleukin (IL)-10 inhibits IL-6 production in microglia by preventing activation of NF-kappaB." *Brain Res Mol Brain Res* 77(1): 138-47.

Hickey WF (1999). "Leukocyte traffic in the central nervous system: the participants and their roles." *Semin Immunol* 11(2): 125-37.

Horvath RJ, Nutile-McMenemy N, Alkaitis MS and Deleo JA (2008). "Differential migration, LPS-induced cytokine, chemokine, and NO expression in immortalized BV-2 and HAPI cell lines and primary microglial cultures." *J Neurochem* 107(2): 557-69.

Imitola J, Raddassi K, Park KI, Mueller FJ, Nieto M, Teng YD, Frenkel D, Li J, Sidman RL, Walsh CA, Snyder EY and Khoury SJ (2004). "Directed migration of neural stem cells to sites of CNS injury by the stromal cell-derived factor 1alpha/CXC chemokine receptor 4 pathway." *Proc Natl Acad Sci U S A* 101(52): 18117-22.

Iosif RE, Ekdahl CT, Ahlenius H, Pronk CJ, Bonde S, Kokaia Z, Jacobsen SE and Lindvall O (2006). "Tumor necrosis factor receptor 1 is a negative regulator of progenitor proliferation in adult hippocampal neurogenesis." *J Neurosci* 26(38): 9703-12.

Irvin DK, Dhaka A, Hicks C, Weinmaster G and Kornblum HI (2003). "Extrinsic and intrinsic factors governing cell fate in cortical progenitor cultures." *Dev Neurosci* 25(2-4): 162-72.

Islam O, Gong X, Rose-John S and Heese K (2009). "Interleukin-6 and neural stem cells: more than gliogenesis." *Mol Biol Cell* 20(1): 188-99.

Ito D, Imai Y, Ohsawa K, Nakajima K, Fukuuchi Y and Kohsaka S (1998). "Microglia-specific localisation of a novel calcium binding protein, Iba1." *Brain Res Mol Brain Res* 57(1): 1-9.

Johansson S, Price J and Modo M (2008). "Effect of inflammatory cytokines on major histocompatibility complex expression and differentiation of human neural stem/progenitor cells." *Stem Cells* 26(9): 2444-54.

Kaneko Y, Sakakibara S, Imai T, Suzuki A, Nakamura Y, Sawamoto K, Ogawa Y, Toyama Y, Miyata T and Okano H (2000). "Musashi1: an evolutionally conserved marker for CNS progenitor cells including neural stem cells." *Dev Neurosci* 22(1-2): 139-53.

Kawabe T, Suganuma M, Ando T, Kimura M, Hori H and Okamoto T (2002). "Cdc25C interacts with PCNA at G2/M transition." *Oncogene* 21(11): 1717-26.

Keeble JE and Moore PK (2002). "Pharmacology and potential therapeutic applications of nitric oxide-releasing non-steroidal anti-inflammatory and related nitric oxide-donating drugs." *Br J Pharmacol* 137(3): 295-310.

Kempermann G, Gast D, Kronenberg G, Yamaguchi M and Gage FH (2003). "Early determination and long-term persistence of adult-generated new neurons in the hippocampus of mice." *Development* 130(2): 391-9.

Kiss JZ and Rougon G (1997). "Cell biology of polysialic acid." *Curr Opin Neurobiol* 7(5): 640-6.

Koc and Kucukguzel SG (2009). "Medicinal chemistry and anti-inflammatory activity of nitric oxide-releasing NSAI drugs." *Mini Rev Med Chem* 9(5): 611-9.

Koo JW and Duman RS (2008). "IL-1beta is an essential mediator of the antineurogenic and anhedonic effects of stress." *Proc Natl Acad Sci U S A* 105(2): 751-6.

Kreutzberg GW (1996). "Microglia: a sensor for pathological events in the CNS." *Trends Neurosci* 19(8): 312-8.

Kuhn HG, Winkler J, Kempermann G, Thal LJ and Gage FH (1997). "Epidermal growth factor and fibroblast growth factor-2 have different effects on neural progenitors in the adult rat brain." *J Neurosci* 17(15): 5820-9.

Ladeby R, Wirenfeldt M, Dalmau I, Gregersen R, Garcia-Ovejero D, Babcock A, Owens T and Finsen B (2005). "Proliferating resident microglia express the stem cell antigen CD34 in response to acute neural injury." *Glia* 50(2): 121-31.

Lambertsen KL, Clausen BH, Babcock AA, Gregersen R, Fenger C, Nielsen HH, Haugaard LS, Wirenfeldt M, Nielsen M, Dagnaes-Hansen F, Bluethmann H, Faergeman NJ, Meldgaard M, Deierborg T and Finsen B (2009). "Microglia protect neurons against ischemia by synthesis of tumor necrosis factor." *J Neurosci* 29(5): 1319-30.

Lawson LJ, Perry VH, Dri P and Gordon S (1990). "Heterogeneity in the distribution and morphology of microglia in the normal adult mouse brain." *Neuroscience* 39(1): 151-70.

Lee MK, Rebhun LI and Frankfurter A (1990). "Posttranslational modification of class III beta-tubulin." *Proc Natl Acad Sci U S A* 87(18): 7195-9.

Levison SW and Goldman JE (1997). "Multipotential and lineage restricted precursors coexist in the mammalian perinatal subventricular zone." *J Neurosci Res* 48(2): 83-94.

Li T, Jiang L, Chen H and Zhang X (2008). "Characterization of excitability and voltage-gated ion channels of neural progenitor cells in rat hippocampus." *J Mol Neurosci* 35(3): 289-95.

Li W, Gao G, Guo Q, Jia D, Wang J, Wang X, He S and Liang Q (2009). "Function and phenotype of microglia are determined by toll-like receptor 2/toll-like receptor 4 activation sequence." *DNA Cell Biol* 28(10): 493-9.

Lim GP, Yang F, Chu T, Chen P, Beech W, Teter B, Tran T, Ubeda O, Ashe KH, Frautschy SA and Cole GM (2000). "Ibuprofen suppresses plaque pathology and inflammation in a mouse model for Alzheimer's disease." *J Neurosci* 20(15): 5709-14.

Limke TL and Rao MS (2002). "Neural stem cells in aging and disease." *J Cell Mol Med* 6(4): 475-96.

Ling EA and Wong WC (1993). "The origin and nature of ramified and amoeboid microglia: a historical review and current concepts." *Glia* 7(1): 9-18.

Liu B and Hong JS (2003). "Role of microglia in inflammation-mediated neurodegenerative diseases: mechanisms and strategies for therapeutic intervention." *J Pharmacol Exp Ther* 304(1): 1-7.

Liu BF, Gao EJ, Zeng XZ, Ji M, Cai Q, Lu Q, Yang H and Xu QY (2006). "Proliferation of neural precursors in the subventricular zone after chemical lesions of the nigrostriatal pathway in rat brain." *Brain Res* 1106(1): 30-9.

Liu YP, Lin HI and Tzeng SF (2005). "Tumor necrosis factor-alpha and interleukin-18 modulate neuronal cell fate in embryonic neural progenitor culture." *Brain Res* 1054(2): 152-8.

Lockhart BP, Cressey KC and Lepagnol JM (1998). "Suppression of nitric oxide formation by tyrosine kinase inhibitors in murine N9 microglia." *Br J Pharmacol* 123(5): 879-89.

Lossinsky AS and Shivers RR (2004). "Structural pathways for macromolecular and cellular transport across the blood-brain barrier during inflammatory conditions. Review." *Histol Histopathol* 19(2): 535-64.

Lucas SM, Rothwell NJ and Gibson RM (2006). "The role of inflammation in CNS injury and disease." *Br J Pharmacol* 147 Suppl 1: S232-40.

Luo CX, Zhu XJ, Zhou QG, Wang B, Wang W, Cai HH, Sun YJ, Hu M, Jiang J, Hua Y, Han X and Zhu DY (2007). "Reduced neuronal nitric oxide synthase is involved in ischemia-induced hippocampal neurogenesis by up-regulating inducible nitric oxide synthase expression." *J Neurochem* 103(5): 1872-82.

Luskin MB, Zigova T, Soteres BJ and Stewart RR (1997). "Neuronal progenitor cells derived from the anterior subventricular zone of the neonatal rat forebrain continue to proliferate in vitro and express a neuronal phenotype." *Mol Cell Neurosci* 8(5): 351-66.

Matarredona ER, Murillo-Carretero M, Moreno-Lopez B and Estrada C (2005). "Role of nitric oxide in subventricular zone neurogenesis." *Brain Res Brain Res Rev* 49(2): 355-66.

McGeer PL and McGeer EG (1995). "The inflammatory response system of brain: implications for therapy of Alzheimer and other neurodegenerative diseases." *Brain Res Brain Res Rev* 21(2): 195-218.

Memberg SP and Hall AK (1995). "Dividing neuron precursors express neuron-specific tubulin." *J Neurobiol* 27(1): 26-43.

Ming GL and Song H (2005). "Adult neurogenesis in the mammalian central nervous system." *Annu Rev Neurosci* 28: 223-50.

Ming GL and Song H (2011). "Adult neurogenesis in the mammalian brain: significant answers and significant questions." *Neuron* 70(4): 687-702.

Minghetti L and Levi G (1998). "Microglia as effector cells in brain damage and repair: focus on prostanoids and nitric oxide." *Prog Neurobiol* 54(1): 99-125.

Monje ML, Toda H and Palmer TD (2003). "Inflammatory blockade restores adult hippocampal neurogenesis." *Science* 302(5651): 1760-5.

Moran LB, Duke DC and Graeber MB (2007). "The microglial gene regulatory network activated by interferon-gamma." *J Neuroimmunol* 183(1-2): 1-6.

Moreno-Lopez B, Romero-Grimaldi C, Noval JA, Murillo-Carretero M, Matarredona ER and Estrada C (2004). "Nitric oxide is a physiological inhibitor of neurogenesis in the adult mouse subventricular zone and olfactory bulb." *J Neurosci* 24(1): 85-95.

Morgan SC, Taylor DL and Pocock JM (2004). "Microglia release activators of neuronal proliferation mediated by activation of mitogen-activated protein kinase, phosphatidylinositol-3-kinase/Akt and delta-Notch signalling cascades." *J Neurochem* 90(1): 89-101.

Morshead CM, Benveniste P, Iscove NN and van der Kooy D (2002). "Hematopoietic competence is a rare property of neural stem cells that may depend on genetic and epigenetic alterations." *Nat Med* 8(3): 268-73.

Morshead CM, Craig CG and van der Kooy D (1998). "In vivo clonal analyses reveal the properties of endogenous neural stem cell proliferation in the adult mammalian forebrain." *Development* 125(12): 2251-61.

Morshead CM, Reynolds BA, Craig CG, McBurney MW, Staines WA, Morassutti D, Weiss S and van der Kooy D (1994). "Neural stem cells in the adult mammalian forebrain: a relatively quiescent subpopulation of subependymal cells." *Neuron* 13(5): 1071-82.

Mullen RJ, Buck CR and Smith AM (1992). "NeuN, a neuronal specific nuclear protein in vertebrates." *Development* 116(1): 201-11.

Nakanishi M, Niidome T, Matsuda S, Akaike A, Kihara T and Sugimoto H (2007). "Microglia-derived interleukin-6 and leukaemia inhibitory factor promote astrocytic differentiation of neural stem/progenitor cells." *Eur J Neurosci* 25(3): 649-58.

Nakatomi H, Kuriu T, Okabe S, Yamamoto S, Hatano O, Kawahara N, Tamura A, Kirino T and Nakafuku M (2002). "Regeneration of hippocampal pyramidal neurons after ischemic brain injury by recruitment of endogenous neural progenitors." *Cell* 110(4): 429-41.

Napoli C and Ignarro LJ (2003). "Nitric oxide-releasing drugs." *Annu Rev Pharmacol Toxicol* 43: 97-123.

Nimmerjahn A, Kirchhoff F and Helmchen F (2005). "Resting microglial cells are highly dynamic surveillants of brain parenchyma in vivo." *Science* 308(5726): 1314-8.

Nixon K and Crews FT (2004). "Temporally specific burst in cell proliferation increases hippocampal neurogenesis in protracted abstinence from alcohol." *J Neurosci* 24(43): 9714-22.

Nowakowski RS, Lewin SB and Miller MW (1989). "Bromodeoxyuridine immunohistochemical determination of the lengths of the cell cycle and the DNA-synthetic phase for an anatomically defined population." *J Neurocytol* 18(3): 311-8.

Oehmichen W and Gencic M (1975). "Experimental studies on kinetics and functions of monuclear phagozytes of the central nervous system." *Acta Neuropathol Suppl* Suppl 6: 285-90.

Okano HJ, Pfaff DW and Gibbs RB (1993). "RB and Cdc2 expression in brain: correlations with 3H-thymidine incorporation and neurogenesis." *J Neurosci* 13(7): 2930-8.

Ostenfeld T, Joly E, Tai YT, Peters A, Caldwell M, Jauniaux E and Svendsen CN (2002). "Regional specification of rodent and human neurospheres." *Brain Res Dev Brain Res* 134(1-2): 43-55.

Palmer TD, Schwartz PH, Taupin P, Kaspar B, Stein SA and Gage FH (2001). "Cell culture. Progenitor cells from human brain after death." *Nature* 411(6833): 42-3.

Park JY, Choi HJ, Prabagar MG, Choi WS, Kim SJ, Cheong C, Park CG, Chin CY and Kang YS (2009). "The C-type lectin CD209b is expressed on microglia and it mediates the uptake of capsular polysaccharides of Streptococcus pneumoniae." *Neurosci Lett* 450(3): 246-51.

Parmar M, Sjoberg A, Bjorklund A and Kokaia Z (2003). "Phenotypic and molecular identity of cells in the adult subventricular zone. in vivo and after expansion in vitro." *Mol Cell Neurosci* 24(3): 741-52.

Pei Z, Pang H, Qian L, Yang S, Wang T, Zhang W, Wu X, Dallas S, Wilson B, Reece JM, Miller DS, Hong JS and Block ML (2007). "MAC1 mediates LPS-induced production of superoxide by microglia: the role of pattern recognition receptors in dopaminergic neurotoxicity." *Glia* 55(13): 1362-73.

Petreanu L and Alvarez-Buylla A (2002). "Maturation and death of adult-born olfactory bulb granule neurons: role of olfaction." *J Neurosci* 22(14): 6106-13.

Pevny L and Placzek M (2005). "SOX genes and neural progenitor identity." *Curr Opin Neurobiol* 15(1): 7-13.

Polazzi E and Monti B (2010). "Microglia and neuroprotection: from in vitro studies to therapeutic applications." *Prog Neurobiol* 92(3): 293-315.

Pollard SM, Conti L, Sun Y, Goffredo D and Smith A (2006). "Adherent neural stem (NS) cells from fetal and adult forebrain." *Cereb Cortex* 16 Suppl 1: i112-20.

Ponomarev ED, Novikova M, Maresz K, Shriver LP and Dittel BN (2005). "Development of a culture system that supports adult microglial cell proliferation and maintenance in the resting state." *J Immunol Methods* 300(1-2): 32-46.

Qin H, Wilson CA, Lee SJ, Zhao X and Benveniste EN (2005a). "LPS induces CD40 gene expression through the activation of NF-kappaB and STAT-1alpha in macrophages and microglia." *Blood* 106(9): 3114-22.

Qin L, Li G, Qian X, Liu Y, Wu X, Liu B, Hong JS and Block ML (2005b). "Interactive role of the toll-like receptor 4 and reactive oxygen species in LPS-induced microglia activation." *Glia* 52(1): 78-84.

Rao MS and Shetty AK (2004). "Efficacy of doublecortin as a marker to analyse the absolute number and dendritic growth of newly generated neurons in the adult dentate gyrus." *Eur J Neurosci* 19(2): 234-46.

Reynolds BA, Tetzlaff W and Weiss S (1992). "A multipotent EGF-responsive striatal embryonic progenitor cell produces neurons and astrocytes." *J Neurosci* 12(11): 4565-74.

Reynolds BA and Weiss S (1992). "Generation of neurons and astrocytes from isolated cells of the adult mammalian central nervous system." *Science* 255(5052): 1707-10.

Reynolds BA and Weiss S (1996). "Clonal and population analyses demonstrate that an EGF-responsive mammalian embryonic CNS precursor is a stem cell." *Dev Biol* 175(1): 1-13.

Richardson RM, Broaddus WC, Holloway KL and Fillmore HL (2005). "Grafts of adult subependymal zone neuronal progenitor cells rescue hemiparkinsonian behavioral decline." *Brain Res* 1032(1-2): 11-22.

Roberts ES, Masliah E and Fox HS (2004). "CD163 identifies a unique population of ramified microglia in HIV encephalitis (HIVE)." *J Neuropathol Exp Neurol* 63(12): 1255-64.

Rochefort C, Gheusi G, Vincent JD and Lledo PM (2002). "Enriched odor exposure increases the number of newborn neurons in the adult olfactory bulb and improves odor memory." *J Neurosci* 22(7): 2679-89.

Rolls A, Shechter R, London A, Ziv Y, Ronen A, Levy R and Schwartz M (2007). "Toll-like receptors modulate adult hippocampal neurogenesis." *Nat Cell Biol* 9(9): 1081-8.

Rosenstiel P, Lucius R, Deuschl G, Sievers J and Wilms H (2001). "From theory to therapy: implications from an in vitro model of ramified microglia." *Microsc Res Tech* 54(1): 18-25.

Rostovtsev VV, Green LG, Fokin VV and Sharpless KB (2002). "A stepwise huisgen cycloaddition process: copper(I)-catalyzed regioselective "ligation" of azides and terminal alkynes." *Angew Chem Int Ed Engl* 41(14): 2596-9.

Sakakibara S, Imai T, Hamaguchi K, Okabe M, Aruga J, Nakajima K, Yasutomi D, Nagata T, Kurihara Y, Uesugi S, Miyata T, Ogawa M, Mikoshiba K and Okano H (1996). "Mouse-Musashi-1, a neural RNA-binding protein highly enriched in the mammalian CNS stem cell." *Dev Biol* 176(2): 230-42.

Salic A and Mitchison TJ (2008). "A chemical method for fast and sensitive detection of DNA synthesis in vivo." *Proc Natl Acad Sci U S A* 105(7): 2415-20.

Sanai N, Tramontin AD, Quinones-Hinojosa A, Barbaro NM, Gupta N, Kunwar S, Lawton MT, McDermott MW, Parsa AT, Manuel-Garcia Verdugo J, Berger MS and Alvarez-Buylla A (2004). "Unique astrocyte ribbon in adult human brain contains neural stem cells but lacks chain migration." *Nature* 427(6976): 740-4.

Sapir T, Horesh D, Caspi M, Atlas R, Burgess HA, Wolf SG, Francis F, Chelly J, Elbaum M, Pietrokovski S and Reiner O (2000). "Doublecortin mutations cluster in evolutionarily conserved functional domains." *Hum Mol Genet* 9(5): 703-12.

Sasaki A, Yamaguchi H, Horikoshi Y, Tanaka G and Nakazato Y (2004). "Expression of glucose transporter 5 by microglia in human gliomas." *Neuropathol Appl Neurobiol* 30(5): 447-55.

Saura J, Tusell JM and Serratosa J (2003). "High-yield isolation of murine microglia by mild trypsinization." *Glia* 44(3): 183-9.

Sawada M, Kondo N, Suzumura A and Marunouchi T (1989). "Production of tumor necrosis factor-alpha by microglia and astrocytes in culture." *Brain Res* 491(2): 394-7.

Seki T (2002). "Hippocampal adult neurogenesis occurs in a microenvironment provided by PSA-NCAM-expressing immature neurons." *J Neurosci Res* 69(6): 772-83.

Seki T and Arai Y (1993). "Highly polysialylated neural cell adhesion molecule (NCAM-H) is expressed by newly generated granule cells in the dentate gyrus of the adult rat." *J Neurosci* 13(6): 2351-8.

Sheen VL, Arnold MW, Wang Y and Macklis JD (1999). "Neural precursor differentiation following transplantation into neocortex is dependent on intrinsic developmental state and receptor competence." *Exp Neurol* 158(1): 47-62.

Shingo T, Gregg C, Enwere E, Fujikawa H, Hassam R, Geary C, Cross JC and Weiss S (2003). "Pregnancy-stimulated neurogenesis in the adult female forebrain mediated by prolactin." *Science* 299(5603): 117-20.

Shors TJ, Miesegaes G, Beylin A, Zhao M, Rydel T and Gould E (2001). "Neurogenesis in the adult is involved in the formation of trace memories." *Nature* 410(6826): 372-6.

Song H, Stevens CF and Gage FH (2002). "Astroglia induce neurogenesis from adult neural stem cells." *Nature* 417(6884): 39-44.

Song JH, Wang CX, Song DK, Wang P, Shuaib A and Hao C (2005). "Interferon gamma induces neurite outgrowth by up-regulation of p35 neuron-specific cyclin-dependent kinase 5 activator via activation of ERK1/2 pathway." *J Biol Chem* 280(13): 12896-901.

Spulber S, Oprica M, Bartfai T, Winblad B and Schultzberg M (2008). "Blunted neurogenesis and gliosis due to transgenic overexpression of human soluble IL-1ra in the mouse." *Eur J Neurosci* 27(3): 549-58.

Stoll G and Jander S (1999). "The role of microglia and macrophages in the pathophysiology of the CNS." *Prog Neurobiol* 58(3): 233-47.

Streit WJ, Graeber MB and Kreutzberg GW (1988). "Functional plasticity of microglia: a review." *Glia* 1(5): 301-7.

Streit WJ, Walter SA and Pennell NA (1999). "Reactive microgliosis." *Prog Neurobiol* 57(6): 563-81.

Sun L, Lee J and Fine HA (2004). "Neuronally expressed stem cell factor induces neural stem cell migration to areas of brain injury." *J Clin Invest* 113(9): 1364-74.

Sun Y, Jin K, Childs JT, Xie L, Mao XO and Greenberg DA (2006). "Vascular endothelial growth factor-B (VEGFB) stimulates neurogenesis: evidence from knockout mice and growth factor administration." *Dev Biol* 289(2): 329-35.

Suslov ON, Kukekov VG, Ignatova TN and Steindler DA (2002). "Neural stem cell heterogeneity demonstrated by molecular phenotyping of clonal neurospheres." *Proc Natl Acad Sci U S A* 99(22): 14506-11.

Suzumura A, Marunouchi T and Yamamoto H (1991). "Morphological transformation of microglia in vitro." *Brain Res* 545(1-2): 301-6.

Suzumura A, Sawada M, Yamamoto H and Marunouchi T (1990). "Effects of colony stimulating factors on isolated microglia in vitro." *J Neuroimmunol* 30(2-3): 111-20.

Svendsen CN (2002). "The amazing astrocyte." *Nature* 417(6884): 29-32.

Taupin P (2008). "Adult neurogenesis, neuroinflammation and therapeutic potential of adult neural stem cells." *Int J Med Sci* 5(3): 127-32.

Thored P, Arvidsson A, Cacci E, Ahlenius H, Kallur T, Darsalia V, Ekdahl CT, Kokaia Z and Lindvall O (2006). "Persistent production of neurons from adult brain stem cells during recovery after stroke." *Stem Cells* 24(3): 739-47.

Thored P, Heldmann U, Gomes-Leal W, Gisler R, Darsalia V, Taneera J, Nygren JM, Jacobsen SE, Ekdahl CT, Kokaia Z and Lindvall O (2009). "Long-term accumulation of microglia with proneurogenic phenotype concomitant with persistent neurogenesis in adult subventricular zone after stroke." *Glia* 57(8): 835-49.

Vallieres L, Campbell IL, Gage FH and Sawchenko PE (2002). "Reduced hippocampal neurogenesis in adult transgenic mice with chronic astrocytic production of interleukin-6." *J Neurosci* 22(2): 486-92.

Vazey EM, Chen K, Hughes SM and Connor B (2006). "Transplanted adult neural progenitor cells survive, differentiate and reduce motor function impairment in a rodent model of Huntington's disease." *Exp Neurol* 199(2): 384-96.

Vernadakis A, Mangoura D, Sakellaridis N and Linderholm S (1984). "Glial cells dissociated from newborn and aged mouse brain." *J Neurosci Res* 11(3): 253-62.

Vezzani A, Moneta D, Richichi C, Aliprandi M, Burrows SJ, Ravizza T, Perego C and De Simoni MG (2002). "Functional role of inflammatory cytokines and antiinflammatory molecules in seizures and epileptogenesis." *Epilepsia* 43 Suppl 5: 30-5.

Vukicevic V, Jauch A, Dinger TC, Gebauer L, Hornich V, Bornstein SR, Ehrhart-Bornstein M and Muller AM (2010). "Genetic instability and diminished differentiation capacity in long-term cultured mouse neurosphere cells." *Mech Ageing Dev* 131(2): 124-32.

Walton NM, Sutter BM, Laywell ED, Levkoff LH, Kearns SM, Marshall GP, 2nd, Scheffler B and Steindler DA (2006). "Microglia instruct subventricular zone neurogenesis." *Glia* 54(8): 815-25.

Wang Y, Sheen VL and Macklis JD (1998). "Cortical interneurons upregulate neurotrophins in vivo in response to targeted apoptotic degeneration of neighboring pyramidal neurons." *Exp Neurol* 154(2): 389-402.

Wegiel J, Wisniewski HM, Dziewiatkowski J, Tarnawski M, Kozielski R, Trenkner E and Wiktor-Jedrzejczak W (1998). "Reduced number and altered morphology of microglial cells in colony stimulating factor-1-deficient osteopetrotic op/op mice." *Brain Res* 804(1): 135-9.

Whitney NP, Eidem TM, Peng H, Huang Y and Zheng JC (2009). "Inflammation mediates varying effects in neurogenesis: relevance to the pathogenesis of brain injury and neurodegenerative disorders." *J Neurochem* 108(6): 1343-59.

Wollmer MA, Lucius R, Wilms H, Held-Feindt J, Sievers J and Mentlein R (2001). "ATP and adenosine induce ramification of microglia in vitro." *J Neuroimmunol* 115(1-2): 19-27.

Wong G, Goldshmit Y and Turnley AM (2004). "Interferon-gamma but not TNF alpha promotes neuronal differentiation and neurite outgrowth of murine adult neural stem cells." *Exp Neurol* 187(1): 171-7.

Wu J, Ohlsson M, Warner EA, Loo KK, Hoang TX, Voskuhl RR and Havton LA (2008). "Glial reactions and degeneration of myelinated processes in spinal cord gray matter in chronic experimental autoimmune encephalomyelitis." *Neuroscience* 156(3): 586-96.

Xu Y, Tamamaki N, Noda T, Kimura K, Itokazu Y, Matsumoto N, Dezawa M and Ide C (2005). "Neurogenesis in the ependymal layer of the adult rat 3rd ventricle." *Exp Neurol* 192(2): 251-64.

Yamamoto S, Yamamoto N, Kitamura T, Nakamura K and Nakafuku M (2001). "Proliferation of parenchymal neural progenitors in response to injury in the adult rat spinal cord." *Exp Neurol* 172(1): 115-27.

Yoshikawa K (2000). "Cell cycle regulators in neural stem cells and postmitotic neurons." *Neurosci Res* 37(1): 1-14.

Zhang J, Shi XQ, Echeverry S, Mogil JS, De Koninck Y and Rivest S (2007). "Expression of CCR2 in both resident and bone marrow-derived microglia plays a critical role in neuropathic pain." *J Neurosci* 27(45): 12396-406.

Zhang RL, Zhang L, Zhang ZG, Morris D, Jiang Q, Wang L, Zhang LJ and Chopp M (2003). "Migration and differentiation of adult rat subventricular zone progenitor cells transplanted into the adult rat striatum." *Neuroscience* 116(2): 373-82.

Zhao M, Momma S, Delfani K, Carlen M, Cassidy RM, Johansson CB, Brismar H, Shupliakov O, Frisen J and Janson AM (2003). "Evidence for neurogenesis in the adult mammalian substantia nigra." *Proc Natl Acad Sci U S A* 100(13): 7925-30.

Zigova T, Pencea V, Betarbet R, Wiegand SJ, Alexander C, Bakay RA and Luskin MB (1998). "Neuronal progenitor cells of the neonatal subventricular zone differentiate and disperse following transplantation into the adult rat striatum." *Cell Transplant* 7(2): 137-56.

19

Immune System Modulation of Germinal and Parenchymal Neural Progenitor Cells in Physiological and Pathological Conditions

Chiara Rolando, Enrica Boda and Annalisa Buffo
Department of Neuroscience
Neuroscience Institute Cavalieri Ottolenghi, NICO
University of Turin
Italy

1. Introduction

Historically, the Central Nervous System (CNS) was considered as an immune privileged site (Billingham and Boswell, 1953), being viewed as a territory physiologically out of the competence of immune cells. This notion has developed on initial studies showing that: (i) CNS unrelated antigens (i.e. foreign grafts, bacteria, viruses) evade an immune recognition when delivered to the brain parenchyma (Galea et al., 2007); (ii) no infiltrating immune cells nor antigen presenting cells (APCs, i.e. dendritic cells, DCs, see Table 1) can be detected in the CNS parenchyma in physiological conditions (Engelhardt and Ransohoff, 2005); (iii) CNS cells do not constitutively express major histocompatibility complex (MHC)I and MHCII molecules (Fabry et al., 1994); (iv) neural cells express apoptosis inductors for immune cells (Bechmann et al., 1999); (v) the CNS does not possess lymphatic vessels (Engelhardt and Ransohoff, 2005). The segregation between nervous and immune cells appeared tightly preserved by the anatomical separations offered by the Blood Brain Barrier (BBB) and the blood-cerebrospinal fluid barrier (Choi and Benveniste, 2004). Over time, on the basis of the association between immune inflammation and neurodegeneration, the concept of immune privilege further acquired the connotation of a defence mechanism against the detrimental effects of immune activation within the CNS.

However, during the last ten years evidence for an extensive and continuous bi-directional communication between the CNS and the immune system has accumulated, changing the traditional view of the CNS as an immune privileged site into an immune specialised site (Engelhardt and Ransohoff, 2005). Under physiological conditions, the CNS strictly controls circulating immune cell entry across its barriers by allowing a regulated exchange of factors between the nervous tissue and immune elements. Such exchange provides an incessant scavenging for self (host) and pathological antigens occurring in the CNS (immunosurveillance) and is at the basis of the newly recognised functions of immune cells in neural stem cell (NSC) activity, hippocampal neurogenesis, learning and stress-mediated responses (see below). In case of pathology, when the BBB is damaged or altered and immune attractive signals are released within the CNS, lymphocytes and macrophages penetrate into the CNS parenchyma. This invasion has now lost the former exclusively detrimental flavour in view of newly unveiled effects supportive for neuroprotection and reparative responses (Schwartz et al., 2009).

Studies on the interplay between the immune system and NSCs/progenitors in both health and disease have particularly contributed to this conceptual revolution in neuroimmunology. These studies are the actual focus of this chapter. We shall present them after having overviewed the main players and mechanisms involved in the CNS-immune system crosstalk.

1.1 Routes and modes of immunosurveillance in the healthy CNS

Main players in immunosurveillance are microglia cells residing within the CNS parenchyma, circulating monocytes and lymphocytes that mostly remain located at the outer anatomical borders of the CNS (Table 1, Schwartz and Shechter, 2010). These borders are defined by several structures: the outermost dural membrane, the arachnoid membrane and the innermost pial membrane. The subarachnoid space settles between the arachnoid and pial membrane and it is filled with the cerebrospinal fluid (CSF), which is continuously produced by the choroid plexus epithelium in the ventricular system. It circulates from the ventricle to the subarachnoid space and it is reabsorbed by the arachnoid villi that extend in the venous sinuses. The nervous artery supply follows the CNS surface in the subarachnoid space. As vessels enter the CNS parenchyma, they are surrounded by a perivascular space, the Virchow-Robin space, connected to the subarachnoid space. Moreover, the BBB separates the blood from the CNS parenchyma and is formed by highly specialized endothelial cells surrounded by basement membranes and astroglial end feet.

In the healthy CNS, immune reactivity is strictly controlled by limiting the presentation of neuroantigens outside the CNS and by tightly regulating the trafficking of immunocompetent cells. The BBB avoids leaking of neuroantigens into the systemic circulation, while within the nervous tissue microglia cells continuously survey the parenchyma with highly dynamic processes and protrusions that may clear accumulated metabolites and cell debris, thereby regulating microenvironmental homeostasis (Nimmerjahn et al., 2005). This microglial function is proposed to be directly influenced by T cell-derived soluble factors, at least at specific CNS sites (see section 2.1; Ziv et al., 2006a). T cells normally do not enter the healthy nervous tissue, but can reach the CSF together with monocytes (about 3000 leukocytes per ml can be found in the CSF of healthy individuals) from both vessels of the choroid plexi and post-capillary venules surrounded by the Virchow-Robin space. Soluble proteins and cells constitutively move from the CNS parenchyma into the CSF through the choroid plexi and ependymal cells and are transported up to peripheral lymph nodes, where they are presented to naïve CD4+ T lymphocytes (afferent arm of immunosurveillance) to achieve their first activation (see Table 1). In the efferent route of this loop, activated CD4+ T cells get to the CNS via the blood stream, and there, moving into the CSF, flow back to the systemic circulation (Ransohoff et al., 2003). The CSF is the site of secondary activation of CD4+ T cells that there encounter monocytes presenting neuroantigens and contribute to immunosurveillance without being encephalitogenic.

Very few if any leukocytes can directly access the healthy CNS parenchyma by crossing the BBB and the endothelial basal lamina. Of note, the few T cells entering the parenchyma have undergone the first activation with neuroantigens, while resting lymphocytes fail to penetrate even after stimulation with inflammatory cues (Ransohoff et al., 2003). The involvement of lymphocyte populations other than CD4+ T (CD8+ T, B cells and NKs) in CNS immunosurveillance has so far been considered poorly relevant, as they comprise neglectable fractions in the healthy CNS/CSF (Ransohoff et al., 2003). During pathology the scenario depicted so far can undergo dramatic changes that compromise the CNS specialised status. Dangerous antigens can be sensed locally within the CNS or directly drained to the periphery to stimulate further recruitment of immune cells. The accompanying massive release of cytokines and chemokines by immune and neural cells mediates the initiation of an immune reaction aimed at promoting CNS defence and restoring tissue homeostasis.

1.2 Innate and adaptive immune responses in the diseased CNS

To effectively exert its defensive functions, the immune system has developed two different reaction modes: a relatively fast and generic action against external agents (innate response), and a specific and targeted action that requires plasticity and memory (adaptive or acquired immunity). Innate and adaptive responses operate in strict collaboration and undergo distinct levels of activation depending on the type of pathology. The innate response is the first line of CNS defence, preceding and stimulating the adaptive reaction (Nguyen, 2002), and relies on microglia cells and on astrocytes. Both cell types constitutively express phagocytic and scavenger receptors (pattern-recognition receptors, PPRs) capable of distinguishing self (host) from non-self (i.e. pathogens, toxic agents and molecules released by damaged/dying cells). Peripheral macrophages/monocytes can also participate in this initial activation when lesions such as traumatic or vascular injury induce BBB breakdown and allow direct CNS parenchyma-blood interactions. Amongst PPRs are Toll-like receptors (TLRs, Table 1) that activate phagocytosis and, via the nuclear transcription factor NFkB pathway, promote the production of pro-inflammatory signals, including cytokines (Interleukin1β, IL1β; Tumor Necrosis Factor-α, TNFα; IL6) and chemokines (Becher et al., 2000; Farina et al., 2007), modulating the nervous tissue response to damage (Buffo et al., 2010) and triggering adaptive immunity (Becher et al., 2000). Notably, several CNS intrinsic mechanisms operate to avoid uncontrolled or hyperactive innate responses: (i) neurons, endothelial and ependymal cells express neuroimmune regulatory proteins (NIRegs) to protect CNS cells from the phagocytic activity of macrophages and microglia and attenuate inflammatory cytokine secretion by lymphocytes (Griffiths et al., 2007); (ii) gliotic astrocytes limit blood leukocyte infiltration (Voskuhl et al., 2009).

Initiation of an adaptive immune response requires time after the initial appearance of pathogenic signals, and implies the participation of numerous cell types and signalling molecules. Adaptive immunity can be either cell-mediated (major effectors are T cells), or humoral (with the involvement of B cells, see Table 1). Although humoral immune responses are most important for the organism's defence, their role in the regulation of neural stem cells and parenchymal progenitor activity is so far unknown. Thus, we shall leave them aside and focus on cell-mediated immune responses.

During CNS damage (e.g. traumatic and neurovascular injuries such as stroke) the integrity of the BBB is primarily disrupted, leading to increased and often deregulated communication between the CNS and the immune system, including the entry of immune cells. In other cases, however, an increased exchange of cellular elements and signals between the two systems occurs while the gross anatomy of the BBB is preserved, thereby implying more subtle functional alterations (Kaur and Ling, 2008). In some cases, such as the initial phase of Multiple Sclerosis (MS), deregulated autoimmune cellular elements find their way into the CNS where they trigger an acute inflammatory reaction to myelin components that can progress into a chronic phase of neurodegeneration. In general, after a CNS insult, resident microglia, astrocytes and DCs that can migrate from the perivascular space into the CNS parenchyma if the BBB is disrupted, present antigens to CD4+ and CD8+ T cells in association with co-stimulatory molecules. After priming with APCs, both T cell types become activated and proliferate: CD8+ T cells exert their cytotoxic activity inducing the apoptosis of antigen-bounded MHCI expressing cells, whereas CD4+ T cells produce pro- or anti-inflammatory cytokines, depending on their subtypes (Th1, Th2, see Table 1, Becher et al., 2000). Th1 cells release pro-inflammatory molecules that sustain and potentiate microglial activation in a feed-forward loop, and stimulate MHCII expression on astrocytes and endothelial cells (see Table 1). In turn, chemokines and cytokines from microglia and activated astrocytes such as interferon-γ (INFγ) and TNFα can both attract and activate immune cells (Carpentier et al., 2005). Conversely, Th2 cells exert anti-inflammatory effects through the production of IL4 and IL10, reducing macrophage and microglia activation (see Table 1). The balance between Th1/Th2 phenotypes is finely tuned by cytokines themselves (Goverman, 2009), and requires a tight regulation to avoid detrimental hyperinflammation: for instance, IL12 produced by activated microglia promotes Th1 type, while astrocytes are known to counteract this phenomenon (Becker et al., 2000).

A further modulatory mechanism of T cell activity involves a specific subclass of T cells, the CD4+ CD25+ Foxp3 T regulatory cells (Tregs, see Table1, Walsh and Kipnis, 2010). Tregs act by suppressing autoimmunity (T cells directed against self-antigens) and terminating immune responses. They exert their inhibitory action mostly through Transforming Growth Factor-β (TGFβ) signalling and IL10 production, which suppresses auto-reactive T cells (Vignali et al., 2008).

Further mechanisms participate in terminating immune reactions: (i) activated T cells themselves express receptors (i.e. CTLA-4) that reduce their proliferation and production of cytokines after interaction with microglia; (ii) T cells after cytokine exposure upregulate receptors that induce their apoptosis (i.e. CD95); (iii) IL2 potentiates CD95-mediated apoptosis. Thus, immune responses are self-limited and decline with time after antigenic stimulation, leaving functionally quiescent memory lymphocytes as indicators of previous antigen exposure (Parijis and Albas, 1998). Acute immune response and inflammation may therefore be soon resolved, and the damage circumscribed by astroglial scarring and microglia cells with poor replacement of lost cells and transected axons (Bush et al., 1999; Donnelly and Popovich, 2007). Yet, an involvement of abnormal autoimmunity or the persistence of pro-inflammatory stimuli can protract the inflammatory/immuno response into a chronic status and exacerbate the destructive effects of immune activation (McFarland and Martin, 2007).

Despite the described mechanisms of immune response have evolved primarily as a defence from infectious agents, they take place in all types of CNS injuries. In the following sections we will not deal with infectious diseases, but instead focus on traumatic, vascular, autoimmune damage and chronic neurodegeneration, where loss or malfunctioning of cellular elements is followed by activation and recruitment of NSCs and parenchymal progenitors, engaged for the most in replacing lost myelin rather then neurons, and in the production of scarring astrocytes (see below).

	Features	Functions	Molecular Signals	References
Microglia	• Invade the CNS parenchyma during late embryogenesis and perinatal stage • Myeloid cells • APC upon activation	• Immunosurveillance • Innate immunity • Cytokine secretion (IL1, IL6, IL12, TNFα) • Phagocytic and cytotoxic activity • T cell stimulation and apoptosis (Fas ligand mediated) • Neurogenesis control	• Upregulation MHCI and II and CD40, CD86 to activate T cells • TLR expression • Upregulation of complement receptor (CR1, CR3, CR4)	Yang et al., 2010 Aloisi et al., 2000
Astrocyte	• Most abundant glial cells in the CNS • Neuroectodermal origin	• Brain homeostasis • BBB formation • Scar formation • APC function induced by Th1 cytokines • Cytokine production (TNFα, IL6, IL12, IL1) • Polarization of T cell cytokine responses • B cell survival and differentiation • Microglia activation • Neurogenesis	• Upregulation MHCII, TLRs (TLRs1-6, TLR9), ICAM1 and VCAM1, chemokines (CCL2, CCL5) • Expression of BAFF (B cell activating factor) • Neurotrophic factor release (BDNF, NGF, IGF1, LIF)	Carpentier et al., 2005 Farina et al., 2007
CD4+ T	• Helper T cells (Th) • Antigen recognition bound to MHCII through T cell Receptor (TCR) • Th1 pro-inflammatory phenotype (INFγ, TNFα production) • Th2 anti-inflammatory phenotype (IL4 and IL10 production)	• Adaptive cell-mediated immune response • Autoimmunity • Activation of B cells	• BDNF production • IL2 production for T cell survival	Goverman, 2009 Dittel, 2008
CD8+ T	• Cytotoxic T cells • Antigen recognition bound to MHCI through TCR	• Cytolysis (perforin mediated) • CNS autoimmunity	• Cytokine production (INFγ and TNFα, IL10, IL17)	Goverman, 2009
Treg	• CD4 positive • Foxp3 positive (forkhead box P3) that controls their development and function and it is induced by TGFβ • CD25 expression (IL2R)	• Avoidance of autoimmune disease and tumoral autoimmunity (peripheral tolerance) • Suppression by cytokine inhibition, cytolysis, by metabolic disruption, by targeting DCs	• Release of TGFβ, IL10, IL35, Granzime B, Adenoside nucleoside, cAMP	Vignali et al., 2008; Walsh and Kipnis, 2010

	Features	Functions	Molecular Signals	References
		• Control CD8+ T cell invasion		
Th17	• Pro-inflammatory lineage • Autoimmune disease	• Cytokine production (TGFβ, IL6, IL23) • Suppression Treg differentiation (IL21 mediated)	• Activation of the STAT3 pathway • Activation of RORα/RORγ	Fabry et al., 2008
B cells	• Rarely detected in the healthy CSF • Four different developmental states (mature B cells, memory B cells, plasmablast, plasma cells)	• Humoral immune response • T cell activation • Ig production • APCs • Ig CNS autoantigen production that induce complement activation, MBP proteolysis, macrophages stimulation	• Upregulation of CXCR12/13 and CCR1,2,4 • Cytokine production (IL6, IL10, IL12, TGFβ) • Neurotrophic factor release (NGF and BDNF)	Meinl et al., 2006
DCs	• APCs • Lymphoid and myeloid origin • Localized in the meninges and chroid plexi in the healthy brain • During inflammation, autoimmune diseases and neurodegeneration they reach the CNS parenchyma	• Immunosurveillance • T cell stimulation, tolerance induction, T cell polarization, phagocytosis, cytokine secretion (IL1b, IL6, IL10, IL12, TNFα, INFγ)	• MHC class II, costimulatory proteins (CD40, CD80, CD86), chemokine receptor (CCR7)	McMahon et al., 2006
Toll like Receptors (TLR)	• Expressed on APCs including microglia and DCs • Surface recognition of PAMPs • Expression on resting and activated microglia and astrocytes (TLR3 on astrocytes, TLR2 and TLR4 on microglia) • Extracellular portion with multiple leucine-rich repeats	• Recognition of exogenous components of the bacterial membrane and flagella, bacterial DNA and viral dsRNA • Neurogenesis	• NFkB activation • Cytokine and chemokine production	Farina et al., 2007 Rolls et al., 2007

Table 1. Immunoplayers.

Main cellular types and molecular signals that regulate the interplay between the immune system and the CNS. This process involves the participation of CNS resident cells (microglia and astrocytes), immune system cells and numerous molecular signals.

2. Immune-based regulation of adult Neural Stem Cell activities and neurogenesis

In the adult brain NSCs displaying astrocytic features reside in two anatomically defined germinal niches, namely the subventricular zone (SVZ) of the lateral wall of the lateral ventricles and the subgranular zone of the hippocampus (SGZ). Adult NSCs of the SVZ (also termed Type-B cells) retain the capability to asymmetrically divide, giving rise to actively proliferating intermediate neural progenitors defined as transit amplifying cells or Type-C cells. These latter cells symmetrically divide to produce neuroblasts (Type-A cells) that migrate through the Rostral Migratory Stream (RMS) into the olfactory bulb, where they eventually differentiate in interneurons. *In vivo* type-B cells can also generate oligodendroglial cells of the corpus callosum and fimbria fornix, although to a lesser extent compared to neurons (Kriegstein and Alvarez-Buylla, 2009). NSCs residing within the hippocampal SGZ divide asymmetrically to give rise to neuroblasts that locally differentiate into mature granular neurons (Kriegstein and Alvarez-Buylla, 2009). Numerous studies indicate that continuous neurogenesis in the olfactory bulb and hippocampus is instrumental for memory acquisition, learning and mood regulation (Zhao et al., 2008), while both physiological (i.e. life experiences, such as learning, physical activity, environmental or olfactory enrichment, stress) and pathological (i.e. brain insults or pathologies, local or systemic inflammation) stimuli affect NSCs and their derivatives.

Surprisingly, a number of recent studies have provided evidence that local and systemic immune-mediated mechanisms, including both innate and adaptive factors, exert a key role in modulating neuro/oligodendrogenic events within the germinal niches in healthy and pathological conditions. Such immune-based regulation takes place at many levels, including (i) proliferation of NSCs and intermediate progenitors; (ii) neuronal vs. glial specification of NSC-derivatives; (iii) migratory ability of the new-born elements; and (iv) their survival, maturation and integration in the adult brain parenchyma. Since the identity of immune system players and the level of their recruitment/activation/production are dramatically different in distinct brain conditions (i.e. healthy vs. acutely injured vs. chronically diseased), immune modulation displays complex and context-dependent effects on the functioning and survival of NSCs and their derivatives (see below). For the sake of simplicity and brevity, in the following text we will refer to germinal functions as "neuro/oligodendrogenesis" or "neuro/oligodendrogenic activity", while the specific effects of immune elements on the diverse germinal components and activities will be dissected in Table 2 (see also Figure 1).

2.1 Immune regulation of adult germinal niche functioning under physiological conditions

Among all immune elements, microglial cells are reportedly crucial modulators of neurogenic niche activities in both the healthy and injured adult CNS. They populate both adult SVZ and SGZ, where they localize in close proximity to NSCs. Interestingly, germinal microglia displays phenotypes and behaviours (i.e. higher levels of activation, proliferation and phagocytic activity; Goings et al., 2006; Ziv et al., 2006a; Sierra et al., 2010) distinct from both their resting counterparts in the non-neurogenic CNS parenchyma, and fully activated and phagocytic microglia detected upon injury or in inflammatory conditions. *In vitro* experiments suggest that such basal germinal activated (BGA) phenotype is maintained through interactions with components of the niche environment, including matrix molecules (e.g. Tenascin-R; Liao et al., 2008), low levels of inflammatory molecules (e.g. IL4 and IFNγ), and elements of adaptive immunity (see below; Ziv et al., 2006a). Notably, various studies report a positive correlation between the persistence of the BGA microglial state and basal levels of neurogenesis and oligodendrogenesis both *in vivo* and *in vitro* (see also Table 2), with blockade of any microglial activation by minocycline resulting in decreased numbers of newborn neurons (Carpentier and Palmer, 2009 and references therein). *In vitro* studies suggest that, although microglial cells are abundantly distributed within the germinal niches in the adult CNS, direct contacts between microglia and NSCs/precursors/ neuroblasts may not be required for facilitating neurogenesis (Aarum et al., 2003; Walton et al., 2006). Notably, microglia cells residing within the germinal niches constitutively secrete a plethora of soluble mediators, including growth factors (e.g. Brain Derived Neurotrophic Factor BDNF and Insulin-like Growth Factor IGF1; Liao et al., 2008; Ziv et al., 2006a) and low levels of inflammatory molecules (e.g. TGFβ, TNFα, IL1β; Battista et al., 2006; Liao et al., 2008; Carpentier and Palmer, 2009; Yirmiya and Goshen, 2011). Both categories of mediators are importantly implicated in neurogenesis, as attested by the negative outcomes on constitutive NSC functioning of genetically- or pharmacologically-driven ablation of growth factor- or inflammatory cytokine-mediated signalling pathways (see also Table 2; Ziv et al., 2006a; Butovsky et al., 2006; Carpentier and Palmer, 2009 and references therein). Moreover, microglia-derived inflammatory molecules can directly influence basal NSC/progenitor functions, as these cells express a set of cytokine receptors, including those for IFNγ (Li et al., 2010), TNFα (Carpentier and Palmer, 2009), IL1β (Yirmiya and Goshen, 2011) and IL6 (Monje et al., 2003). Moreover, pioneer studies reported that adult NSCs basally display a set of features typical of immune cells, including the expression of MHC-related molecules, TLRs and complement receptors, whose activation may allow NSCs themselves to (i) act as antigen-presenting cells; (ii) directly sense alterations in tissue integrity and immune system activity upon injury; and (iii) plastically modulate their own neuro/oligodendrogenic activity in response to environmental alterations (see Table 2; Popa et al., 2011; Rolls et al., 2007; Moriyama et al., 2011; Rahpeymai et al., 2006).

Further novelty in the field has been provided by the unexpected discovery that, in addition to resident microglia and derived soluble factors, T cells contribute to maintain the neurogenic homeostasis in the adult CNS. By using nude or SCID (severe combined immune deficiency) mice, lacking respectively either only mature T cells or both T and B cell populations, Michal Schwartz and colleagues in 2006 demonstrated that T cell deficiency is correlated with impaired NSC/progenitor proliferation and neuronal differentiation of new born derivatives in both SGZ and SVZ, accompanied by a defective spatial learning ability (Ziv et al., 2006a). Such effect on SGZ progenitor cells is specifically exerted by T helper lymphocytes, as repopulation with CD4+, but not CD8+ or B cells, rescues defective neurogenesis (Wolf et al., 2009). Notably, antigenic specificity to CNS autoantigens and the consequent lymphocyte homing to the CNS appear required for the expression of these T cell supportive effects on neurogenesis. Consistently, transgenic (tg) mice in which the majority of the T cell population is directed to an irrelevant antigen (i.e. ovalbumin) show impaired hippocampal neurogenesis, while, conversely, tg mice in which the majority of the T cells is directed to a CNS-specific antigen, such as certain peptides of the myelin basic protein (MBP), display increased hippocampal neurogenesis (Ziv et al., 2006a). Interestingly, data collected so far allow to propose a model in which the homeostatic role of T cells on germinal niche functions includes both a direct action on NSCs and their derivatives via release of BDNF, and an indirect effect mediated by modulation of the microglial BGA state and increased BDNF production induced in surrounding neurons (Ziv et al., 2006a; Hohlfeld et al., 2006). In line with this scenario, minocycline treatment results in a reduced hippocampal neurogenesis even in tg mice where T cells are directed to MBP antigens, while BDNF levels are reduced in immune deficient mice and elevated in tg mice enriched with T cells directed to MBP antigens (Ziv et al., 2006a; Wolf et al., 2009).

The participation of immune cells in the regulation of hippocampal NSCs/progenitors is further supported by experiments showing that while wild type mice respond to enriched housing conditions (including social, sensory and motor stimulation) by increased neurogenesis, SCID animals do not show any change in NSC activity (Ziv et al., 2006a). Notably, in the same experimental condition it has been reported the appearance of T cells in the hippocampal hilus and an increased number of SGZ microglial cells upregulating MHCII molecules and IGF1 (Ziv et al., 2006a). Since this microglia phenotype is typically promoted by Th2-derived IL4 (Butovsky et al., 2005), it can be speculated that these changes are stimulated by defined T cell activities occurring as a consequence of the organism-environment interactions. In addition, based on the capability of microglia stimulated by either IL4 or low levels of INFγ (which is known to be produced by T cells) to promote neurogenesis and oligodendrogenesis in hippocampal progenitor cell/microglia co-cultures, (Butovsky et al., 2006), one may claim that the observed microglia changes contribute to the occurring increased neurogenesis. Stress and elevated levels of circulating glucocorticoid hormones can also affect the BGA microglial state (Song and Wang, 2011). Accordingly, the surgical removal of the adrenal gland and the consequent suppression of glucocorticoid production result in a moderately higher activation of microglial cells, whose density increase again correlates with a higher number of dividing cells and of newly generated neurons in the SGZ. Such microglial activation is accompanied by the upregulation of inflammatory cytokines, such as TGFβ (Battista et al., 2006). Taken together, these data indicate that T lymphocytes and microglia cells take part in the homeostatic regulation of adult neurogenesis, comprising the mediation of part of the pro-and anti-neurogenic effects of experience/emotional stimuli.

Another study confirmed the contribution of immune cells to the plastic regulation of adult neurogenesis. Wolf and colleagues in 2009 showed that when the effect of voluntary wheel running on neurogenesis is assessed in CD4 knock-out mice or in mice treated with anti-CD4 neutralizing antibodies, both wild type and CD4-depleted/deficient mice respond by increased hippocampal proliferation (although starting from different basal levels), while such expected effect is absent only in mice devoid of functional T, B and NK cells. This latter finding suggests that while CD4+ T cells are the major player in controlling constitutive neurogenesis, the entire pool of adaptive immune cells may contribute to induce a response to neurogenic stimuli. Only one study so far excluded a role for T cells and microglia activation in activity-induced increase of adult hippocampal neurogenesis (Olah et al., 2009). Whether differences in experimental paradigms applied (voluntary physical activity vs. enriched environment) or in animal models used (mice vs. rats) account for such different results remains to be assessed.

To sum up, data collected so far provide evidence that in non-pathological conditions both resident microglia and T cells have a prominent role in maintaining and plastically modulating the basal levels of neurogenesis within the adult CNS (see Figure 1). Microglia residing within the neurogenic niches displays a germinal-specific basally activated phenotype, characterized by the expression of defined pro-neurogenic inflammatory mediators (e.g. TGFβ, TNFα, IL1β) and growth factors (e.g. BDNF and IGF1). The maintenance of such BGA microglial phenotype depends on the interaction with local elements and with CNS-directed T cells, and positively correlates with the neurogenic activity of adult germinal niches.

2.2 Immune system regulation of germinal niche functioning after injury

The discovery of the retention of NSCs within the mature CNS has inspired two decades of intense investigations aimed at assessing whether such endogenous source of new neurons and oligodendrocytes could be exploited for the repopulation of lost neuronal populations and the restoration of damaged myelin sheaths. However, it is now well established that adult germinal niche activity has a very limited ability to mediate a long-lasting cell replacement and to support a complete repair of the injured CNS cytoarchitecture and functions. Many studies have shown that early after acute insults (e.g. stroke, trauma, single epileptic attacks, acute phase of relapsing-remitting experimental autoimmune encephalomyelitis EAE) endogenous neurogenesis and oligodendrogenesis are stimulated. However, in most cases reactive neurogenesis is eventually abortive, since the majority of the newly generated neurons are not recruited to the lesion site but remain within the germinal areas, nor are integrated into the parenchyma, and ultimately undergo cell death (Carpentier and Palmer, 2009). Moreover, in chronic pathologies (e.g. neurodegenerative diseases, recurrent seizures, chronic progressive EAE) and stress conditions (e.g. mouse isolation), neurogenesis and oligodendrogenesis appear impaired (Carpentier and Palmer, 2009; Pluchino et al., 2008; Rasmussen et al., 2011). Studies using ablation of single immune cell populations or molecular pathways have revealed that upon injury immune system elements exert a dual role, contributing to both the early neuro/oligodendrogenic reaction and the subsequent establishment of a milieu non-permissive for NSC activities. Major players in such regulation are again microglia and T cells (see below).

Early after acute damage, cell debris, nucleotides released from dying cells and reactive glial cells, and extracellular matrix protein fragments serve as ligands for the TLRs expressed by microglia residing within the germinal niches, and trigger its full activation, with subsequent release of high levels of pro-inflammatory cytokines (i.e. TNFα, INFγ, IL6 and IL1β) and growth factors (e.g. IGF1; Ekdahl, 2009 and references therein; Deierborg et al., 2010). Various studies reported that such early microglial activation is *per se* needed to induce the post-injury increase of neurogenesis. In fact, minocycline inhibition of activated microglial cells exposed to injury (i.e. stroke) abolishes the increase in NSC proliferation both *in vivo* and *in vitro* (Kim et al., 2010; Deierborg et al., 2010). Consistently, genetically- or pharmacologically-driven ablation of IL1β- or TNFα-mediated signalling pathways negatively affects neurogenesis after seizures and stroke (Spulber et al., 2008; Carpentier and Palmer, 2009 and references therein), indicating that, in addition to growth factors, these microglial-derived mediators contribute to post-injury reactive neurogenesis. Notably, similar to the healthy conditions, defined subpopulations of T cells contribute to modulate germinal niche functioning at early stages after injury. Removal of the whole CD4+ T cell population results in increased precursor/neuroblast generation few days after stroke, while depletion of the only Treg lymphocytes suppresses neurogenesis and reduces functional recovery (Saino et al., 2010). In contrast, promotion of Treg homing to the ischemic brain enhances NSC and neuroblast survival (Ishibashi et al., 2009). At variance with non-pathological conditions, post-injury T cell effects appear to modulate pro-inflammatory cytokine secretion by activated endothelial cells rather than being mediated by microglial functions (Saino et al., 2010; Ishibashi et al., 2009).

When persistent and uncontrolled microglial activation occurs, the same molecular players appear to switch their acute stimulating function into detrimental effect on germinal niche activity. For instance, when minocycline is administered in the chronic phase of CNS injuries or diseases, such treatment results in increased generation of neurons and oligodendrocytes (Carpentier and Palmer, 2009; Yirmiya and Goshen, 2011 and references therein; Rasmussen et al., 2011). Whether such phenomenon is due to a beneficial-to-detrimental switch in the microglial phenotype is still highly debated. It remains also unresolved whether changes in the intrinsic responsiveness to immuno/inflammatory mediators of NSCs and their derivatives may account for these harmful effects. Although data are not completely consistent (Ekdahl, 2009), transcriptional profiling of isolated SVZ microglia cells reveals that microglia exhibits disease phase-specific gene expression signatures (Starossom et al., 2011). Moreover, *in vitro* evidence suggests that while early-activated microglia displays pro-neurogenic features, it acquires a non-supportive phenotype at delayed time point after injury (Deierborg et al., 2010). Moreover, inflammatory cytokines can act through different receptors, thereby triggering distinct effects. This is the case of TNFα that can activate both the TNFα receptor 1 (TNFR1), mediating cytotoxic functions on NSCs and neuroblasts, and the TNFR2, activating pro-neurogenic pathways (Carpentier and Palmer, 2009). These data suggest that upon chronic damage both microglia and NSC can contribute to reduced neuro/oligodendrogenesis, by acquiring phenotypes non-supportive for germinal niche functioning and newborn cell survival.

In summary, early microglia activation is required to induce post-injury increase in NSC proliferation and neurogenesis. However, when microglial activation and inflammatory molecule secretion persist for long time, as in chronically injured CNS, adult neurogenesis and oligodendrogenesis are suppressed. At variance with what reported in physiological conditions, at early stages after injury CD4+ T cell activity negatively affects precursor/neuroblast generation through microglia-indipendent mechanisms. However, under the same conditions, the Treg subpopulation appear to exert beneficial effects on neurogenesis and functional recovery (see Figure 1).

Although still controversial, these data support the idea that anti-inflammatory treatments should be finely and temporarily calibrated in order to be beneficial for neuro/oligodendrogenesis and promote CNS regeneration following injury.

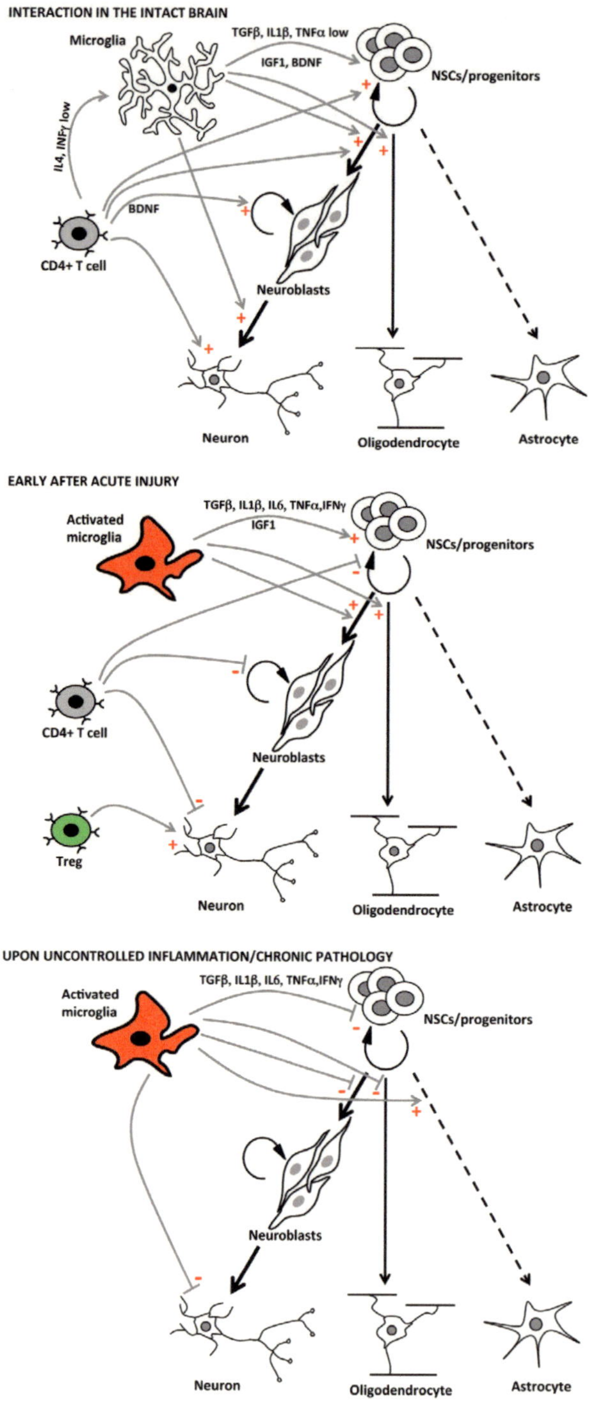

Fig. 1. Immune-based regulation of adult germinal niche activities in intact and pathological conditions. Curved arrows indicate proliferation; straight arrows indicate differentiation along cell lineages.

In the adult brain NSCs residing in the SVZ and SGZ asymmetrically divide and give rise to actively proliferating neuroblasts that eventually differentiate in mature neurons into the olfactory bulb or in the dentate gyrus. Within the adult germinal niches, oligodendrocytes and astrocytes are also produced, though to a much lesser extent. Depending on their state of activation, immune system elements can positively (+) or negatively (-) affect the generation of new neurons or glial cells within the mature CNS. Immune-mediated mechanisms include microglia- and T cell-derived soluble factors, and influence (i) NSC and neuroblast proliferation, (ii) neuronal vs. glial specification, (iii) survival and maturation of the newborn elements, thereby exerting an important role in modulating the germinal niche activities in both healthy and pathological conditions.

Immunoplayer	Proliferation	Specification	Survival	Neuroblast differentiation	Oligodendro genesis
Immune cells					
Microglia					
in physiological conditions	+ (NSCs/ precursors)	n.r.	n.r.	+	+
early after acute injury	+ (NSCs/ precursors)	+ neurogenesis + oligodendro-genesis	n.r.	n.r.	+
in chronic pathology/ uncontrolled immuno-activation	- (NSCs/ precursors)	- neurogenesis	- (neurobl.)	n.r.	-
CD4+ T cells					
in physiological conditions	+ (NSCs/ precursors/ neurobl.)	+ neurogenesis	+ (neurobl.)	+	n.r.
early after injury	- (NSCs/ precursors/ neurobl.)	n.r.	- (precursor/ neurobl.)	n.r.	n.r.
T regs	n.r.	n.r.	+ (neurobl.)	n.r.	n.r.
Inflammatory cytokines					
TGFβ	n.r.	+ neurogenesis	n.r.	+	
IL4	n.r.	n.r.	n.r.	+	+
IL1β	- (*in vitro*; NSCs/ precursors); + (*in vivo*; NSCs/ precursors)	n.r.	n.r.	n.r.	n.r.
IL6	- (NSCs/ precursors/ neurobl.)	+ astroglio-genesis - neurogenesis	-	n.r.	n.r.

Immunoplayer	Proliferation	Specification	Survival	Neuroblast differentiation	Oligodendro genesis
TNFα	- (>200ng/ml; precursors); + (1 ng/ml; NSCs/ precursors/ neurobl.)	n.r.	- (10-100 ng/ml; NSC/ precursors); + (1ng/ml; NSCs/ precursors/ neurobl.)	+ (1ng/ml); No effect (10ng/ml)	n.r.
IFNγ	- (NSCs/ precursors)	+ neurogenesis - oligodendro-genesis (*in vivo*); - astroglio-genesis (*in vitro*); - neurogenesis - oligodendro-genesis + astroglio-genesis (neurosphere assay)	- (neurobl.)	+ (20 ng/ml)	-
Receptors on NSCs and derivatives					
TLR2	n.r.	+ neurogenesis	n.r.	+	n.r.
TLR4	- (NSCs)	n.r.	+ (neurobl.)	+	n.r.
CR2 complement receptor	- (NSCs/ precursors/ neurobl.)	n.r.	n.r.	n.r.	n.r.
C3aR complement receptor	n.r.	n.r.	n.r.	+	n.r.
RAE-1 (MHCI – related)	+ (NSCs/ precursors)	n.r.	n.r.	n.r.	n.r.

Table 2. Major immune factors regulating the adult germinal niche activity.

Data were obtained from studies in which genetically- or pharmacologically-driven ablation of single cell populations or molecular pathways allows to unveil a causal relationship between the activity of a defined cell type or molecule and a specific effect on NSC or derivatives. References can be found in the text. Note that the effects of inflammatory cytokines are often context- or dose-dependent. Abbreviations: NSCs, Neural Stem Cells; neurobl., neuroblasts; +, increased; -, decreased; n.r., not reported.

3. Immune system regulation of parenchymal neural progenitors

Studies over the last decades have revealed that glia cells residing in the nervous parenchyma outside the neurogenic areas can display progenitor functions (Boda and Buffo, 2010), in addition to absolving supportive roles for neurons and contributing to information processing (Kettenmann and Verkhratsky, 2008; Bakiri et al., 2009). Typically, cells expressing the proteoglycan NG2 comprise the vast majority of cycling elements outside the germinal areas (Horner et al., 2000; Dawson et al., 2003) and respond to a variety of lesion conditions by an increased cytogenic activity and hypertrophy (Keirstead et al., 1998; Reynolds et al., 2002; Hampton et al., 2004). Conversely, mature parenchymal astrocytes remain quiescent in the healthy CNS, but can re-enter the cell cycle and assume features of progenitor cells upon injury (Buffo et al., 2008, 2010). Numerous approaches including proliferation studies, expression analysis, grafting experiments and Cre-lox based fate-mapping investigations (revised in Trotter et al., 2010; Richardson et al., 2011) have consolidated the view of NG2 positive cells as endogenous reservoir of mature and myelinating oligodendrocytes during development, adulthood and in most pathological conditions. Therefore, these cells are generally termed oligodendrocyte precursor cells (OPCs), despite the names 'polydendrocytes' or 'synanthiocytes' have been recently adopted in view of their morphology and contiguity to neurons. A controversial issue, dawned by seminal experiments showing that OPCs *in vitro* can revert to a stem cell-like state and differentiate along all the three neural lineages (Kondo and Raff, 2000), is whether *in vivo* these cells can undergo low levels of neurogenesis and generate glial cells other than oligodendrocytes at specific CNS sites or in specific conditions. Data on this issue are conflicting, although the prevailing view agrees that some astrogliogenesis (and generation of Schwann cells in the spinal cord) can occur in defined injury conditions and developmental ages (embryonic astrogliogenesis). Production of new neurons has also been reported, but remains to be further confirmed (see Boda and Buffo, 2010; Richardson et al., 2011; Fröhlich et al., 2011 for review). Recent studies on CNS lesions have also attributed precursor properties to reactive astrocytes and spinal cord ependymal cells. During anisomorphic gliosis, parenchymal astrocytes dedifferentiate and acquire progenitor features, which are not expressed *in vivo*, likely inhibited by a plethora of injury-evoked restrictive signals such as inflammatory molecules, but can be disclosed *ex vivo* (Buffo et al., 2008; 2010). Spinal cord ependymal cells appear instead able to undergo astrogliogenesis and oligodendrogenesis upon injury directly *in vivo* (Barnabè-Heider et al., 2010).

3.1 Protective and destructive effects of immune activation in the nervous tissue

As presented above (see also Table 2), poor survival of progenitor cells as well as restriction of their differentiation potentials to astrogliogenesis, blockade of maturational programs and induction of cell death have been long ascribed to immuno-mediated inflammatory signals released at sites of lesions. This purely negative view of immunity and inflammation has also extended to parenchymal progenitor functioning, based on the established detrimental inflammatory burden of immune (e.g. MS), traumatic, neurovascular and neurodegenerative (Alzheimer's Disease, Parkinson disease, Amyotrophic Lateral Sclerosis) pathologies. However, recent studies have highlighted a positive contribution of immunity to repair of neural damage. Thus, while nothing is known on whether and how both local microglia and peripheral immune cells physiologically modulate the proliferation and differentiation rates of OPCs and/or affect the progenitor potentials of other glial cells, it is increasingly clear that the concept of immune activation as purely harmful to CNS repair is too simplistic. Accordingly, well-defined features, levels and timing of immune activity appear to promote neuroprotection and post-injury plasticity in the forms of axon regrowth, replacement of degenerated cells and functional recovery.

This emerging view suggests that supportive functions for tissue repair and functional recovery can be exerted by defined populations or functional states of macrophages/microglia and T cells. For instance, infiltrating blood-derived macrophages have been shown to promote recovery at sub-acute stages in rodents with spinal cord injury (Rapalino et al., 1998; Shechter et al., 2009). This action appears related to the production of immunomodulatory (anti-inflammatory IL10) and neurotrophic molecules (BDNF), which is triggered by exposure to self-antigens or by the actions of T cells responding to neuroantigens ('protective autoimmunity', Schechter et al., 2009; see also Schwartz and Yoles, 2006; Schwartz and Shechter, 2010). A similar modulatory function would be exerted by T cells on local microglia that, upon proper stimulation, can become beneficial to the nervous tissue (Butovsky et al., 2005, 2006; Shaked et al., 2005). According to this view, autoimmune T cells sensibilised against CNS antigens (and in particular myelin components) (Moalem et al., 1999; Hauben et al., 2000; Kipnis et al., 2002; Fisher et al., 2001; Beers et al., 2008) have been proposed to play a crucial role in the recovery from acute CNS insults. These cells would enhance cellular and molecular mechanisms responsible of cleaning up the injured area and creating a milieu favourable to tissue remodelling and function restoration. Yet, in autoimmune neuropathologies such as EAE Tregs have been primarily implicated in neuroprotection and inflammation control (Liu et al., 2006; Huang et al., 2009; Reddy et al., 2004), where they would partly contribute to limiting the overactivation of cytotoxic autoimmune cells. A similar function for Tregs in non-autoimmune CNS damage has been confirmed by a further study on a stroke model (Liesz et al., 2009). The important novelty of these findings resides on the identification of physiological reparative mechanisms mediated by innate and adaptive immunity that, in the natural state may remain too weak or abortive to express their full neuroprotective and reparative potential, and could therefore be implemented for therapeutic purposes (Schwartz and Yoles, 2006; Walsh and Kipnis, 2011). In other words, with distinct timings and modalities, defined immune cell populations can be proposed as an endogenous therapeutic target to restrain or modulate self-checking mechanisms on the part of beneficial immunity activated spontaneously in response to CNS injury (Walsh and Kipnis, 2011). However, any immediate extension of this view to all types of CNS injury, including chronic neurodegenerative diseases, requires further confirmations and disclosure of the specific mechanisms of immune cell actions in distinct disease conditions (Walsh and Kipnis, 2011).

3.2 Immune-mediated control of parenchymal progenitor functioning

Whereas the above presented findings referred to nervous tissue protection and recovery from damage in general terms, in this section we will take a closer look on how and when innate and adaptive cells and inflammatory cues influence the activity and survival of OPCs and parenchymal progenitors. Oligodendrocyte produce myelin sheaths that allow fast conduction of electrical signals along axons. These cells undergo primary degeneration due to genetic causes (leukodistrophies) and are highly vulnerable to noxious signals produced during traumatic and ischemic events and inflammatory/autoimmune pathologies. In many instances their replacement with subsequent remyelination of temporary demyelinated axons occurs spontaneously. However, if oligodendrocyte death is particularly extended or in defined damage conditions such as traumatic compressive injuries, stroke and MS, this process remains incomplete or blocked. Neo-oligodendrogenesis and remyelination are not operated by spared mature oligodendrocytes but by OPCs. To attain remyelination, these progenitors have to become activated, undergo hyperthrophic changes, activate fast proliferation, migrate to the site of demyelination, start a complex differentiation process including the establishment of contacts with the denudated axons, expression of myelin genes, generation of the myelin membranes that wrap the axons and form the sheath. It is unquestionably true that OPCs respond to a variety of insults other than demyelination. Amongst these, compelling evidence supports a role for inflammatory/immune components in OPC proliferation, recruitment and differentiation.

In a mouse model of traumatic injury, Rhodes and colleagues have established that, amongst early factors capable to induce immediate reactivity in OPC in the form of NG2 upregulation, hypertrophy and increase in OPC cell number, are blood-derived macrophages in a defined activation state including the release of the inflammatory cytokines TNFα, IL1α, TGFβ, INFγ (Rhodes et al., 2006). Furthermore, specific macrophage/microglia activation phenotypes have been proposed to differentially affect OPC proliferation and regenerative capabilities through the selective activation of specific microglia/macrophage TLRs (Lehenardt et al., 2002; Glezer et al., 2006; Schonberg et al., 2007; Taylor et al., 2010). Despite data presented in distinct studies are not completely consistent (perhaps due to different experimental conditions), the consensus view is that defined microglia/macrophage activation states, correlated with specific pattern of cytokine production, act by either triggering or hampering OPC proliferation and differentiation. For instance, IL4-stimulated microglia has been shown to promote oligodendrogenesis from local progenitors in an autoimmune demyelination models, whereas INFγ-stimulated microglia had no or very limited effects (Butovsky et al., 2005). The role of innate immunity in OPC functioning in damage has further been substantiated by studies on non-immunity-mediated toxin-induced models of focal demyelination. In these models, genetic-based depletion or pharmacological inhibition of macrophages leads to an impairment of remyelination (Kotter et al., 2001, 2005), indicating a defective OPC response in an injury condition that normally leads to complete regeneration of myelinating oligodendrocytes by local reactive OPCs (Woodruff and Franklin, 1999). In the same experimental lesion, enhancing TLR4 mediated microglia activation by LPS infusion increases OPC reactivity, promotes a more efficient removal of myelin debris and triggers a faster appearance of remyelination markers (Glezer et al., 2006). It is clear that one key aspect of the innate immunity contribution to the full expression of the OPC regenerative potential is the removal of myelin debris. *In vitro* and *in vivo* data support the notion that myelin components dampen OPC differentiation (Miller, 1999). In line with these findings, anti-inflammatory drugs attenuating microglia/macrophage activity can affect OPC responses by delaying their differentiation in experimental demyelination (Li et al., 2005; Chari et al., 2005).

A similar role in the modulation of OPC reaction to demyelination has been attributed to T cells (both CD4+ and CD8+) indicating that also adaptive immunity is required for the correct OPC regenerative response (Bieber et al., 2003). Indeed, lack or depletion of either

CD4+ or CD8+ is associated with reduced remyelination in focal demyelination. Interestingly, the disease-delaying drug Glatiramer acetate (GA) adopted for therapy in MS, may promote remyelination by potentiating a specific T cell mediated effects. Indeed, it has been shown *in vitro* that GA increases the production of Th2 cells, IGF1, and that the conditioned medium from GA-reactive T cells promotes the formation of OPCs from embryonic brain-derived forebrain cell culture. These findings are confirmed *in vivo*, where GA increases the OPC number and the extent of remyelinaton in toxin-mediated focal demyelination (Skihar et al., 2009).

Moving from cells to molecular signals, a wide range of pro-inflammatory cytokines (e.g. IL1β and TNFα, along with lymphotoxin-β receptor and MHCII) have been implicated as mediators of remyelination in non-autoimmune remyelination, implying that they promote the reactivity and the reparative behaviour of OPCs (revised in Franklin and ffrench-Costant, 2008). Another cytokine, INFγ, has instead been shown to inhibit remyelination (Franklin and ffrench-Costant, 2008). In turn, upon INFγ stimulation glial precursors with features of OPCs have been shown to produce a variety of immunomodulators, trophic factors, microglia attractive factors, and activate the expression of specific TLRs (Cassiani-Ingoni et al., 2006), indicating that OPC participate in active and bidirectional interplay with immune cells. Finally, cytokines have also been proposed as capable to instruct alternative OPC fates: *in vitro* exposure to INFγ diverted glial progenitor from oligodendrogenesis to astrogliogenesis. Despite this finding is consistent with the capability of INFγ to block remyelination, astrogliogenesis from OPCs *in vivo* remains debated (see above).

Moving to injury models distinct from demyelination, spinal cord contusions offer an example of a traumatic injury where intense OPC proliferation is not accompanied by complete glial repopulation of the lesioned area. In this specific immuno-inflammatory condition, activated microglia/macrophages have been shown to secrete inhibitory factors (i.e. TNFα, and extracellular matrix modifiers) hampering survival and growth of OPC *ex vivo*, and impeding their migration into the lesioned demyelinated area (Wu et al., 2010). Opposite effects of activated microglia on tissue repair in different lesion models may indeed be explained by different timing of recruitment of T cells in this process ensuing distinct microglia activation states (following Schwartz and Yoles, 2006).

Immuno-inflammatory levels have also been suggested to affect the neurogenic potential of parenchymal precursors, independent on their identity. Low levels of inflammation or specific immuno/inflammatory states have been proposed to allow the disclosure of neurogenic potentialities. In the cerebral cortex, selective cortical neuron damage mediated by apoptotic events and very low levels of inflammatory/immune activation has been associated with the appearance of glial cells with radial progenitor traits and rare immature neurons, suggesting that injury-induced de-differentiatiation of resident astrocytes to a radial glia state may subserve local neurogenesis (Leavitt et al., 1999; Chen et al., 2004). Also mild ischemic damage has been reported to allow neurogenesis from parenchymal sources: viral-based tracing revealed that layer I cortical progenitors can give birth to a low number of GABAergic cortical interneurons (Ohira et al., 2010).

A further support to the contention that local immune response strongly influences the behaviour of local precursors was provided by the observation in a model of spinal cord lesion that combined modulation of T cell activation by myelin-derived peptide vaccination and transplantation of immunomodulatory adult NSCs correlated with the appearance of neurogenic attempts from local progenitors accompanied by modulation of parenchymal T cell response and microglia activation, and, increased BDNF and noggin expression (Ziv et al., 2006b). Pioneering studies have also started investigating the influence of T cells on astrocytes, showing that T-cell derived signals modify the astrocytic metabolic state *in vitro*. Namely, glutamate released by T cells promotes the acquisition of a neuroprotective phenotype and potentiates their capability to clear glutamate (Garg et al., 2008). Astroglial dysfunctions appear instead induced by LPS-activated microglia *in vivo*, resulting in defect of the BBB and subsequent myelin damage (Sharma et al., 2010). Astrocytes are obviously intensely involved in any kind of response to noxious stimuli, given their essential functions in the maintenance of tissue homeostatis, scavenging of toxic molecules, production of trophic support to neurons and oligodendrocytes, and cytogenic glial scarring to prevent the spreading of potential secondary damage to the healthy tissue (Buffo et al., 2010). The astrocytic reaction is directly or indirectly induced by various inflammatory cytokines and, in turn, reactive astrocytes produce pro-inflammatory molecules that modulate their own activation state and that of immune cells (Buffo et al., 2010; Kostianovsky et al., 2008). Whether and how inflammatory/immune factors specifically affect the progenitor potential of reactive astroglia is not known. What is well accepted is that extended damage is associated with high levels of inflammation and immune activation that are generally unfavourable to the disclosure of progenitor properties and regeneration (see also above). Accordingly, controlled microlesions to the CNS and associated low levels of inflammatory/immuno activation were reported to induce immature/progenitor phenotypes associated with rare neurogenic events as well as the establishment of a microenvironment more prone to support axon growth (Leavitt et al., 1999; Chen et al., 2004). It remains to be established whether specific components or modalities of innate/adaptive immune activation can boost such pro-reparative changes in resident astroglia in case of extended damage. On the whole, these data indicate that the expression of the reparative potentials of parenchymal progenitors can be supported by immune mechanisms directed at both removing debris and toxic molecules, and performing immunomodulation to avoid the overactivation of the immune response.

4. Concluding remarks

Recent discoveries have profoundly changed the perception of CNS–immune interactions. In particular, the novel roles of immune cells in the maintenance and plastic regulation of adult NSC functions have revealed an unexpected exchange of signals between the nervous and immune systems, opening the possibility that immune malfunction may have relevance in so far unsuspected CNS diseases. Furthermore, a decade of investigations has dissected components of the immune response to CNS injury that potentiate or dampen CNS reparative activities. While more research is needed to disclose the influence of immune factors on the properties of parenchymal sources of progenitor cells, on the whole immune cells can be proposed as an endogenous therapeutic target to modulate immune mechanisms on the part beneficial to foster CNS repair and function restoration.

5. Acknowledgments

Chiara Rolando' s fellowship and work in our laboratory are supported by Compagnia di San Paolo, GLIAREP project.

6. References

Aarum, J., Sandberg, K., Haeberlein, S.L.B., Persson, M.A.A. (2003). Migration and differentiation of neural precursor cells can be directed by microglia. *Proceedings of the National Academy of Sciences USA*, Vol.100, No.26 (Dec 2003), pp. 15983-15988, ISSN 1091-6490.

Aloisi, F., Ria, F., & Adorini, L. (2000). Regulation of T-cell responses by CNS antigen-presenting cells: different roles for microglia and astrocytes. *Immunology today*, Vol.21, No.3 (Mar 2000), pp. 141-147, ISSN 0167-5699.

Bakiri, Y., Attwell, D,. & Káradóttir, R. (2009). Electrical signalling properties of oligodendrocyte precursor cells. *Neuron Glia Bio*logy, Vol.5, No.1-2 (May 2009), pp. 3-11, ISSN 1740-925X.

Barnabé-Heider, F., Göritz, C., Sabelström, H., Takebayashi, H., Pfrieger, F.W., Meletis, K., & Frisén, J. (2010). Origin of new glial cells in intact and injured adult spinal cord. *Cell Stem Cell*. Vol.7, No.4 (Oct 2010), pp. 470-82, ISSN 19345909.

Battista, D., Ferrari, C.C., Gage, F.H., & Pitossi, F.J. (2006). Neurogenic niche modulation by activated microglia: transforming growth factor β increases neurogenesis in the adult dentate gyrus. *European Journal of Neuroscience,* Vol.23, No.1 (Jan 2006), pp. 83-93, ISSN 1460-9568.

Becher, B., Prat, A., & Antel, J.P. (2000). Brain-immune connection: immuno-regulatory properties of CNS-resident cells. *Glia*, Vol.29, No.4 (Feb 2000), pp. 293-304, ISSN 1098-1136.

Bechmann, I., Mor, G., Nilsen, J., Eliza, M., Nitsch, R., & Naftolin, F. (1999). FasL (CD95L, Apo1L) is expressed in the normal rat and human brain: evidence for the existence of an immunological brain barrier. *Glia*, Vol.27, No.1 (Jul 1999), pp.62-74, ISSN 1098-1136.

Beers, D.R., Henkel, J.S., Zhao, W., Wang, J., & Appel, S.H. (2008). CD4+ T cells support glial neuroprotection, slow disease progression, and modify glial morphology in an animal model of inherited ALS. *Proceedings of the National Academy of Sciences USA*, Vol.105, No.40 (Oct 2008), pp. 15558–15563, ISSN 0027-8424.

Bieber, A. J., Kerr, S. & Rodriguez, M. Efficient central nervous system remyelination requires T cells. (2003). *Annals of Neurology*, Vol.53, No.5 (May 2003), pp. 680–684, 531-8249.

Billingham, R.E., Boswell, T. (1953). Studies on the problem of corneal homografts. *Proceeding of the Royal Society Biological Science*, Vol.141, No.904, pp. 392-406, ISSN 1471-2954.

Boda, E., Buffo, A. (2010) Glial cells in non-germinal territories: insights into their stem/progenitor properties in the intact and injured nervous tissue. *Archives Italiennes de Biologie*, Vol. 148, No.2 (June 2010), pp 119-136, ISSN 0003-9829.

Buffo, A., Rite, I., Tripathi, P., Lepier, A., Colak, D., Horn, A.P., Mori, T., & Götz M. (2008). Origin and progeny of reactive gliosis: A source of multipotent cells in the injured brain. *Proceedings of the National Academy of Sciences U S A*. Vol.105, No.9 (Mar 2008), pp. 3581-3586, ISSN 0027-8424.

Buffo, A., Rolando, C., & Ceruti, S. (2010). Astrocytes in the damaged brain: molecular and cellular insights into their reactive response and healing potential. *Biochemical Pharmacology*, Vol.79, No.2 (Jan 2010), pp. 77-89, ISSN 0006-2952.

Bush, T.G., Puvanachandra, N., Horner, C.H., Polito, A., Ostenfeld, T., Svendsen, C.N., Mucke, L., Johnson, M.H., & Sofroniew, M.V. (1999). Leukocyte infiltration, neuronal degeneration, and neurite outgrowth after ablation of scar-forming, reactive astrocytes in adult transgenic mice. *Neuron*, Vol.23 No.2 (Jun 1999), pp. 297-308. ISSN 0896-6273.

Butovsky, O., Talpalar, A.E., Ben-Yaakov, K., & Schwartz, M. (2005). Activation of microglia by aggregated β-amyloid or lipopolysaccharide impairs MHC-II expression and renders them cytotoxic whereas IFN-γ and IL-4 render them protective. *Molecular and Cellular Neuroscience,* Vol.29, No.3 (Jul 2005), pp. 381-393, ISSN 1044-7431.

Butovsky, O., Landa, G., Kunis, G., Ziv, Y., Avidan, H., Greenberg, N., Schwartz, A., Smirnov, I., Pollack, A., Jung, S., & Schwartz, M. (2006). Induction and blockage of oligodendrogenesis by differently activated microglia in an animal model of multiple sclerosis. *Journal of Clinical Investigation,* Vol.116, No.4 (Mar 2006), pp. 905-915, ISSN 00219738.

Carpentier, P.A., Begolka, W.S., Olson, J.K., Elhofy, A., Karpus, W.J., & Miller, S.D. (2005). Differential activation of astrocytes by innate and adaptive immune stimuli. *Glia,* Vol.49, No.3 (Feb 2005), pp. 360-374, ISSN 1098-1136.

Carpentier, P.A., & Palmer, T.D. (2009). Immune Influence on Adult Neural Stem Cell Regulation and Function. *Neuron,* Vol.64, No.1 (Oct 2009), pp. 79-92, ISSN 0896-6273.

Cassiani-Ingoni, R., Coksaygan, T., Xue, H., Reichert-Scrivner, S.A., Wiendl, H., Rao, M.S., & Magnus, T. (2006). Cytoplasmic translocation of Olig2 in adult glial progenitors marks the generation of reactive astrocytes following autoimmune inflammation. *Experimental Neurology,* Vol.201, No.2 (Oct 2006), pp. 349-58, ISSN 0014-4886.

Chari, D. M., Zhao, C., Kotter, M. R., Blakemore, W. F. & Franklin, R. J. M. (2006). Corticosteroids delay remyelination of experimental demyelination in the rodent central nervous system. *Journal of Neuroscience Research,* Vol.83, No.4 (Mar 2006), pp. 594–605, ISSN 1097-4547.

Chen, J., Magavi, S.S., & Macklis, J.D. (2004). Neurogenesis of corticospinal motor neurons extending spinal projections in adult mice. *Proceedings of the National Academy of Sciences U S A*, Vol.101, No.46 (Nov 2004), pp. 16357-16362, ISSN 0027-8424.

Choi, C., Benveniste, E.N. (2004). Fas ligand/Fas system in the brain: regulator of immune and apoptotic responses. *Brain Research Review,* Vol.44, No.1 (Jan 2004), pp. 65-81, ISSN 0165-0173.

Dawson MR, Polito A, Levine JM, & Reynolds, R. (2003). NG2-expressing glial progenitor cells: an abundant and widespread population of cycling cells in the adult rat CNS. *Molecular Cellular Neuroscience,* Vol.24, No.2 (Oct 2003), pp.476-88, ISSN 1044-7431.

Deierborg, T., Roybon, L., Inacio, A.R., Pesic, J., & Brundin, P. (2010). Brain injury activates microglia that induce neural stem cell proliferation ex vivo and promote differentiation of neurosphere-derived cells into neurons and oligodendrocytes. *Neuroscience,* Vol.171, No.4 (Dec 2010), pp. 1386-1396, ISSN 03064522.

Dittel, B.N. (2008). CD4 T cells: Balancing the coming and going of autoimmune-mediated inflammation in the CNS. *Brain, Behavior, and Immunity,* Vol.22, pp. 421-430, ISSN 0889-1591.

Donnelly, D.J., Popovich, P.G. (2007). Inflammation and its role in neuroprotection, axonal regeneration and functional recovery after spinal cord injury. *Experimental Neurology,* Vol.209, No.2 (Feb 2008), pp. 378-388, ISSN 0014-4886.

Ekdahl, C.T., Kokaia, Z., & Lindvall, O. (2009). Brain inflammation and adult neurogenesis: The dual role of microglia. *Neuroscience,* Vol.158, No.3 (Feb 2009), pp. 1021-1029, ISSN 03064522.

Engelhardt, B., Ransohoff, R.M. (2005). The ins and outs of T-lymphocyte trafficking to the CNS: anatomical sites and molecular mechanisms. *Trends in Immunology,* Vol.26, No.9 (Sep 2005), pp. 485-495, ISSN 1471-4906.

Fabry Z, Raine CS, & MN, H. (1994). Nervous tissue as an immune compartment: the dialect of the immune response in the CNS. *Immunology Today,* Vol.15, No.5 (May 1994), pp. 218-224, ISSN 0167-5699.

Fabry, Z., Schreiber, H.A., Harris, M.G., & Sandor, M. (2008). Sensing the microenvironment of the central nervous system: immune cells in the central nervous system and their pharmacological manipulation. *Current Opinion in Pharmacology,* Vol.8, pp. 496-507, ISSN 1471-4892.

Farina, C., Aloisi, F., & Meinl, E. (2007). Astrocytes are active players in cerebral innate immunity. *Trends in Immunology,* Vol.28, No.3 (Feb 2007), pp. 138-145, ISSN 1471-4906.

Fisher, J., Levkovitch-Verbin, H., Schori, H., Yoles, E., Butovsky, O., Kaye, J.F., Ben-Nun, A., & Schwartz, M. (2001). Vaccination for neuroprotection in the mouseoptic nerve: implications for optic neuropathies. *Journal of Neuroscience,* Vol.21, No.1 (Jan 2001), pp. 136–142, ISSN 0270-6474.

Franklin, R.J., & Ffrench-Constant, C. (2008). Remyelination in the CNS: from biology to therapy. *Nature Reviews Neuroscience,* Vol.9, No.11 (Nov 2008), pp. 839-55, ISSN 1471-003X.

Fröhlich, N., Nagy, B., Hovhannisyan, A., & Kukley, M. (2011). Fate of neuron-glia synapses during proliferation and differentiation of NG2 cells. *Journal of Anatomy,* Vol.219, No.1 (Jul 2011), pp. 18-32, ISSN 1469-7580.

Galea, I., Bechmann, I., & Perry, V.H. (2007). What is immune privilege (not)? *Trends in Immunology,* Vol.28, No.1 (Jan 2007) pp. 12-18, ISSN 1471-4906.

Garg, S.K., Banerjee, R., & Kipnis, J. (2008). Neuroprotective immunity: T cell-derived glutamate endows astrocytes with a neuroprotective phenotype. *Journal of Immunology,* Vol.180, No.6 (Mar 2008), pp. 3866-3873, ISSN 0022-1767.

Glezer, I., Lapointe, A., & Rivest, S. (2006). Innate immunity triggers oligodendrocyte progenitor reactivity and confines damages to brain injuries. *FASEB Journal,* Vol.20, No.6 (Apr 2006), pp.750-2, ISSN 0892-6638.

Goings, G.E., Kozlowski, D.A., & Szele, F.G. (2006). Differential activation of microglia in neurogenic versus non-neurogenic regions of the forebrain. *Glia,* Vol.54, No.4 (Sept 2006), pp. 329-342, ISSN 1098-1136.

Goverman, J. (2009). Autoimmune T cell responses in the central nervous system. *Nature Review Immunology,* Vol.9, No.6 (Jun 2009), pp. 393-407, ISSN 1474-1733.

Griffiths, M., Neal, J.W., Gasque, P., Giacinto Bagetta, M.T.C., & Stuart, A.L. (2007). Innate Immunity and Protective Neuroinflammation: New Emphasis on the Role of Neuroimmune Regulatory Proteins. *International Review of Neurobiology*, Vol.82, pp. 29-55, ISSN 0074-7742.

Hampton, D.W., Rhodes, K.E, Zhao, C., Franklin, RJ, Fawcett, J.W. (2004) The responses of oligodendrocyte precursor cells, astrocytes and microglia to a cortical stab injury, in the brain. *Neuroscience*. Vol.127, No.4 (Jul 2004), pp. 813-20, ISSN 0306-4522.

Hauben, E., Butovsky, O., Nevo, U., Yoles, E., Moalem, G., Agranov, E., Mor, F., Leibowitz-Amit, R., Pevsner, E., Akselrod, S., Neeman, M., Cohen, I.R., &Schwartz, M. (2000). Passive or active immunization with myelin basic protein promotes recovery from spinal cord contusion. *Journal Neuroscience*, Vol.20, No.17 (Sep 2000), pp. 6421-30, ISSN 0270-6474.

Hohlfeld, R., Kerschensteiner, M., Stadelmann, C., Lassmann, H., & Wekerle, H. (2000). The neuroprotective effect of inflammation: implications for the therapy of multiple sclerosis. *Journal of Neuroimmunology*, Vol.107, No.2 (Jul 2000), pp.161-166, ISSN 0165-5728.

Horner ,P.J., Power, A.E., Kempermann. G., Kuhn, H.G., Palmer, T.D., Winkler, J., Thal, L.J., & Gage, FH. (2000). Proliferation and differentiation of progenitor cells throughout the intact adult rat spinal cord. *Journal of Neuroscience*, Vol.20, No.6 (Mar 2000), pp. 2218-2228, ISSN 0270-6474.

Huang, X.Y., Reynolds, A.D., Mosley, R.L., & Gendelman, H.E. (2009). CD 4+T cells in the pathobiology of neurodegenerative disorders. *Journal of Neuroimmunology*, Vol.211, No.1-2, pp. 3–15, ISSN 0165-5728.

Ishibashi, S., Maric, D., Mou, Y., Ohtani, R., Ruetzler, C., & Hallenbeck, J.M. (2008). Mucosal tolerance to E-selectin promotes the survival of newly generated neuroblasts via regulatory T-cell induction after stroke in spontaneously hypertensive rats. *Journal Cerebral Blood Flow Metabolism*, Vol.29, No.3 (Mar 2009), pp. 606-620, ISSN 0271-678X.

Kaur, C., Ling, E.A. (2008). Blood brain barrier in hypoxic-ischemic conditions. *Current Neurovascular Research*, Vol.5, No.1 (Feb 2008), pp. 71-81, ISSN 1567-2026.

Keirstead, H.S., Levine, J.M., & Blakemore, W.F. (1998). Response of the oligodendrocyte progenitor cell population (defined by NG2 labelling) to demyelination of the adult spinal cord. *Glia*, Vol.22, No.2 (Feb 1998), pp.161-170, ISSN 1098-1136.

Kettenmann H., & Verkhratsky, A. (2008). Neuroglia: the 150 years after. *Trends Neuroscience*, Vol.31, No.12 (Dec 2008), pp.653-659, ISSN 0166-2236.

Kim, D.H., Kim, J.M., Park, S.J., Lee, S., Yoon, B.H., & Ryu, J.H. (2010). Early-activated microglia play a role in transient forebrain ischemia-induced neural precursor proliferation in the dentate gyrus of mice. *Neuroscience Letters*, Vol.475, No.2 (May 2010), pp. 74-79, ISSN 0304-3940.

Kipnis, J., Mizrahi, T., Hauben, E., Shaked, I., Shevach, E., & Schwartz, M. (2002). Neuroprotective autoimmunity: naturally occurring CD4+CD25+ regulatory T cells suppress the ability to withstand injury to the central nervous system. *Proceedings of the National Academy of Sciences U S A*. Vol. 99, No. 24 (Nov 2002), pp. 15620-5, ISSN 0027-8424.

Kondo T., & Raff, M. (2000). Oligodendrocyte precursor cells reprogrammed to become multipotential CNS stem cells. *Science*, Vol.289, No.5485 (Sep 2000), pp. 1754-1757, ISSN 0036-8075.

Kostianovsky, A.M., Maier, L.M., Anderson, R.C., Bruce, J.N., & Anderson, D.E. (2008). Astrocytic regulation of human monocytic/microglial activation. *Journal of Immunology*, Vol.181, No.8 (Oct 2008), pp. 5425-5432, ISSN 0022-1767.

Kotter, M. R., Setzu, A., Sim, F. J., van Rooijen, N. & Franklin, R. J. M. (2001). Macrophage depletion impairs oligodendrocyte remyelination following lysolecithin induced demyelination. *Glia*, Vol.35, No.3 (Sep 2001), pp. 204–212, ISSN 1098-1136.

Kotter, M. R., Zhao, C., van Rooijen, N. & Franklin, R. J. M. (2005). Macrophage-depletion induced impairment of experimental CNS remyelination is associated with a reduced oligodendrocyte progenitor cell response and altered growth factor expression. *Neurobiology of Disease*, Vol.18, No.1 (Feb 2005), pp. 166–175, ISSN 0969-9961.

Kriegstein, A., & Alvarez-Buylla, A. (2009). The glial nature of embryonic and adult neural stem cells. *Annual Review of Neuroscience*, Vol.32, pp. 149-184, ISSN 1545-4126.

Leavitt, B.R., Hernit-Grant, C.S., & Macklis, J.D. (1999). Mature astrocytes transform into transitional radial glia within adult mouse neocortex that supports directed migration of transplanted immature neurons. *Experimenal Neurology*, Vol.157, No.1 (May 1999), pp. 43-57, ISSN 0022-3069.

Lehnardt, S., Lachance, C., Patrizi, S., Lefebvre, S., Follett, P.L., Jensen, F.E., Rosenberg, P.A., Volpe, J.J., & Vartanian, T. (2002). The toll-like receptor TLR4 is necessary for lipopolysaccharide-induced oligodendrocyte injury in the CNS. *Journal of Neuroscience*, Vol.22, No.7 (Apr 2002), pp. 2478-86, ISSN 0270-6474.

Li, L., Walker, T.L., Zhang, Y., Mackay, E.W., & Bartlett, P.F. (2010). Endogenous Interferon-γ Directly Regulates Neural Precursors in the Non-Inflammatory Brain. *The Journal of Neuroscience, Vol.30*, No.27 (Jul 2010) pp. 9038-9050, ISSN 0270-6474.

Li, W.W., Setzu, A., Zhao, C. & Franklin, R. J. M. (2005). Minocycline-mediated inhibition of microglia activation impairs oligodendrocyte progenitor cell responses and remyelination in a non-immune model of demyelination. *Journal of Neuroimmunology*, Vol.158, No.1-2 (Jan 2005), pp. 58–66, ISSN 0165-5728.

Liao, H., Huang, W., Niu, R., Sun, L., & Zhang, L. (2008). Cross-talk between the epidermal growth factor-like repeats/fibronectin 6–8 repeats domains of Tenascin-R and microglia modulates neural stem/progenitor cell proliferation and differentiation. *Journal of Neuroscience Research*, Vol.86, No.1 (Jan 2008), pp. 27-34, ISSN 0360-4012.

Liesz, A., Suri-Payer, E., Veltkamp, C., Doerr, H,. Sommer, C., Rivest, S., Giese, T., & Veltkamp, R. (2009). Regulatory T cells are key cerebroprotective immunomodulators in acute experimental stroke. *Nature Medicine*, Vol.15, No.2 (Feb 2009) pp. 192-199, ISSN 1078-8956.

Liu, Y.W., Teige, I., Birnir, B., & Issazadeh-Navika, S. (2006). Neuron-mediated generation of regulatory T cells from encephalitogenic T cells suppresses EAE. *Nature Medicine*, Vol.12, No.5 (May 2006), pp. 518–525, ISSN 1078-8956.

McFarland, H.F., Martin, R. (2007). Multiple sclerosis: a complicated picture of autoimmunity. *Nature Immunology*, Vol.8, No.9 (Sep 2007), pp. 913-919, ISSN 1529-2908.

McMahon, E.J., Bailey, S.L., & Miller, S.D. (2006). CNS dendritic cells: critical participants in CNS inflammation? *Neurochemistry International*, Vol.49, No.2, pp. 195-203, ISSN 0197-0186.

Meinl, E., Krumbholz, M., & Hohlfeld, R. (2006). B lineage cells in the inflammatory central nervous system environment: Migration, maintenance, local antibody production, and therapeutic modulation. *Annals of Neurology,* Vol.59, No.6 (Jun 2006), pp. 880-892, ISSN 0364-5134.

Miller, R.H. (1999). Contact with central nervous system myelin inhibits oligodendrocyte progenitor maturation. *Developmental Biology,* Vol.216, No.1 (Dec 1999), pp. 359–368, ISSN 0012-1606.

Moalem, G., Leibowitz-Amit, R., Yoles, E., Mor, F., Cohen, I.R., & Schwartz, M. (1999). Autoimmune T cells protect neurons from secondary degeneration after central nervous system axotomy. *Nature Medicine,* Vol.5, No.1 (Jan 1999), pp. 49-55, ISSN 1078-8956.

Monje, M.L., Toda, H., & Palmer, T.D. (2003). Inflammatory Blockade Restores Adult Hippocampal Neurogenesis. *Science,* Vol.302, No.5651 (Dec 2003), pp. 1760-1765, ISSN 1095-9203.

Moriyama, M., Fukuhara, T., Britschgi, M., He, Y., Narasimhan, R., Villeda, S., Molina, H., Huber, B.T., Holers, M., & Wyss-Coray, T. (2011). Complement receptor 2 is expressed in neural progenitor cells and regulates adult hippocampal neurogenesis. *The Journal of Neuroscience,* Vol.31, No.11 (Mar 2011), pp. 3981-3989, ISSN 0270-6474.

Nguyen, M.D., Julien, J.P., & Rivest, S. (2002). Innate immunity: the missing link in neuroprotection and neurodegeneration? *Nature Review Neuroscience,* Vol.3, No.3 (Mar 2002), pp. 216-227, ISSN 1471-0048.

Nimmerjahn, A., Kirchhoff, F., & Helmchen, F. (2005). Resting microglial cells are highly dynamic surveillants of brain parenchyma in vivo. *Science,* Vol.308, No.5726 (May 2005), pp. 271314-271318.

Ohira, K., Furuta, T., Hioki, H., Nakamura, K.C., Kuramoto, E., Tanaka, Y., Funatsu, N., Shimizu, K., Oishi, T., Hayashi, M., Miyakawa, T., Kaneko, T., & Nakamura, S. (2010) Ischemia-induced neurogenesis of neocortical layer 1 progenitor cells. *Nature Neuroscience,* Vol.13, No.2 (Feb 2010), pp. 173-179, ISSN 1097-6256.

Olah, M., Ping, G., De Haas, A.H., Brouwer, N., Meerlo, P., Van Der Zee, E.A., Biber, K., & Boddeke, H.W.G.M. (2009). Enhanced hippocampal neurogenesis in the absence of microglia T cell interaction and microglia activation in the murine running wheel model. *Glia,* Vol.57, No.10 (Aug 2010), pp. 1046-1061, ISSN 1098-1136.

Parijs, L.V., Abbas, A.K. (1998). Homeostasis and self-tolerance in the immune system: turning lymphocytes off. *Science,* Vol.280, No.5361 (Apr 1998), pp. 243-248, ISSN 1095-9203.

Pluchino, S., Muzio, L., Imitola, J., Deleidi, M., Alfaro-Cervello, C., Salani, G., Porcheri, C., Brambilla, E., Cavasinni, F., Bergamaschi, A., Garcia-Verdugo, J.M., Comi, G., Khoury, S.J., & Martino, G. (2008). Persistent inflammation alters the function of the endogenous brain stem cell compartment. *Brain,* Vol.131, No.10 (Oct 2010), pp. 2564-2578, ISSN 1460-2156.

Popa, N., Cedile, O., Pollet-Villard, X., Bagnis, C., Durbec, P., Boucraut, J.E. (2011). RAE-1 is expressed in the adult subventricular zone and controls cell proliferation of neurospheres. *Glia,* Vol.59, No.1 (Jan 2011), pp. 35-44, ISSN 1098-1136.

Rahpeymai, Y., Hietala, M.A., Wilhelmsson, U., Fotheringham, A., Davies, I., Nilsson, A.-K., Zwirner, J., Wetsel, R.A., Gerard, C., Pekny, M., & Pekna, M. (2006). Complement: a novel factor in basal and ischemia-induced neurogenesis. *EMBO Journal,* Vol.25, No.6 (Mar 2006), pp. 1364-1374, ISSN 1460-2075.

Ransohoff, R.M., Kivisakk, P., & Kidd, G. (2003). Three or more routes for leukocyte migration into the central nervous system. *Nature Review Immunology,* Vol.3, No.7 (Jul 2007), pp. 569-581, ISSN 1474-1733.

Rapalino, O., Lazarov-Spiegler, O., Agranov, E., Velan, G.J., Yoles, E., Fraidakis, M., Solomon, A., Gepstein, R., Katz, A., Belkin, M., Hadani, M., & Schwartz, M. (1998). Implantation of stimulated homologous macrophages results in partial recovery of paraplegic rats. *Nature Medicine,* Vol.4, No.7 (Jul 1998), pp. 814-21, ISSN 1078-8956.

Rasmussen, S., Imitola, J., Ayuso-Sacido, A., Wang, Y., Starossom, S.C., Kivisäkk, P., Zhu, B., Meyer, M., Bronson, R.T., Garcia-Verdugo, J.M., & Khoury, S.J. (2011). Reversible neural stem cell niche dysfunction in a model of multiple sclerosis. *Annals of Neurology,* Vol.69, No.5 (May 2011), pp. 878-891, ISSN 1531-8249.

Reddy, J., Illes, Z., Zhang, X., Encinas, J., Pyrdol, J., Nicholson, L., Sobel, R.A., Wucherpfennig, K.W., & Kuchroo, V.K. (2004). Myelin proteolipid protein-specific CD4+CD25+ regulatory cells mediate genetic resistance to experimental autoimmune encephalomyelitis. *Proceedings of the National Academy of Sciences U S A,* Vol.101, No.43 (Oct 2004), pp. 15434-15439, ISSN 0027-8424.

Reynolds, R., Dawson, M., Papadopoulos, D., Polito, A., Di Bello, I.C., Pham-Dinh, D., & Levine J. (2002) The response of NG2-expressing oligodendrocyte progenitors to demyelination in MOG-EAE and MS. *Journal Neurocytology,* Vol.31, No.6-7 (Jul-Aug 2002), pp.523-536, ISSN 0300-4864.

Rhodes, K.E., Raivich, G, & Fawcett, J.W. (2006). The injury response of oligodendrocyte precursor cells is induced by platelets, macrophages and inflammation-associated cytokines. *Neuroscience,* Vol.14, No.1 (Jun 2006), pp. 87-100, ISSN 03064522.

Richardson, WD, Young, K.M., Tripathi, R.B., & McKenzie, I. (2011). NG2-glia as multipotent neural stem cells: fact or fantasy? *Neuron.* Vol.70, No.4 (May 2011), pp. 661-673, ISSN 0896-6273.

Rolls, A., Shechter, R., London, A., Ziv, Y., Ronen, A., Levy, R., & Schwartz, M. (2007). Toll-like receptors modulate adult hippocampal neurogenesis. *Nature Cell Biology,* Vol.9, No.9 (Sep 2007), pp. 1081-1088, ISSN 1465-7392.

Saino, O., Taguchi, A., Nakagomi, T., Nakano-Doi, A., Kashiwamura, S.I., Doe, N., Nakagomi, N., Soma, T., Yoshikawa, H., Stern, D.M., Okamura, H., & Matsuyama, T. (2010). Immunodeficiency reduces neural stem/progenitor cell apoptosis and enhances neurogenesis in the cerebral cortex after stroke. *Journal of Neuroscience Research,* Vol.88, No.11 (Aug 2011), pp. 2385-2397, ISSN 0360-4012.

Schonberg, D.L., Popovich, P.G., & McTigue, D.M. (2007). Oligodendrocyte generation is differentially influenced by toll-like receptor (TLR) 2 and TLR4-mediated intraspinal macrophage activation. *Journal Neuropathology Experimental Neurology* Vol.66, No.12 (Dec 2007), pp. 1124-35, ISSN 0022-3069.

Schwartz, M., & Yoles, E. (2006). Immune-based therapy for spinal cord repair: autologous macrophages and beyond. *Journal Neurotrauma,* Vol.23, No.3-4 (Apr 2006), pp. 360-370, ISSN- 1557-9042.

Schwartz, M., A., & Shechter, R. (2009). Boosting T-cell immunity as a therapeutic approach for neurodegenerative conditions: the role of innate immunity. *Neuroscience,* Vol.158, No.3 (Febr 2009), pp. 1133-1142, ISSN 03064522.

Schwartz, M., & Shechter, R. (2010). Systemic inflammatory cells fight off neurodegenerative disease. *Nature Review in Neurology*, Vol.6, No.7 (Jul 2010) pp. 405-10, ISSN 1759-4758.

Shaked, I., Tchoresh, D., Gersner, R., Meiri, G., Mordechai, S., Xiao, X., Hart, R.P., & Schwartz, M. (2005). Protective autoimmunity: interferon-gamma enables microglia to remove glutamate without evoking inflammatory mediators. *Journal of Neurochemistry*, Vol.92, No.5 (Mar 2005), pp. 997-1009, ISSN 0022-3042.

Sharma, R., Fischer, M.T., Bauer, J., Felts, P.A., Smith, K.J., Misu, T, Fujihara, K., Bradl, M., & Lassmann, H. (2010). Inflammation induced by innate immunity in the central nervous system leads to primary astrocyte dysfunction followed by demyelination. *Acta Neuropathologica,* Vol.120, No.2 (Aug 2010), pp. 223-236, ISSN 0001-6322.

Shechter, R., London, A., Varol, C., Raposo, C., Cusimano, M., Yovel, G., Rolls, A., Mack, M., Pluchino, S., Martino, G., Jung, S., & Schwartz, M. (2009). Infiltrating blood-derived macrophages are vital cells playing an anti-inflammatory role in recovery from spinal cord injury in mice. *PLoS Medicine,* Vol.6, No.7 (Jul 2009), ISSN 1549-1676.

Sierra, A., Encinas, J.M., Deudero, J.J.P., Chancey, J.H., Enikolopov, G., Overstreet-Wadiche, L.S., Tsirka, S.E., & Maletic-Savatic, M. (2010). Microglia Shape Adult Hippocampal Neurogenesis through Apoptosis-Coupled Phagocytosis. (2010). *Cell Stem Cell,* Vol.7, No.4 (Oct 2010), pp. 483-495, ISSN 1934-5909.

Skihar, V., Silva, C., Chojnacki, A., Döring, A., Stallcup, W.B., Weiss, S., Yong, V.W. (2009). Promoting oligodendrogenesis and myelin repair using the multiple sclerosis medication glatiramer acetate. *Proceedings of the National Academy of Sciences U S A.* Vol.106, No.42 (Oct 2009), pp. 17992-17997, ISSN 0027-8424.

Song, C., & Wang, H. Cytokines mediated inflammation and decreased neurogenesis in animal models of depression. (2011). *Progress in Neuro-Psychopharmacology and Biological Psychiatry,* Vol.35, No.3 (Apr 2011) pp. 760-768, *ISSN* 0278-5846.

Spulber, S., Oprica, M., Bartfai, T., Winblad, B., & Schultzberg, M. (2008). Blunted neurogenesis and gliosis due to transgenic overexpression of human soluble IL-1ra in the mouse. *European Journal of Neuroscience,* Vol.27, No.3 (Feb 2008), pp. 549-558, ISSN 1460-9568.

Starossom, S.C., Imitola, J., Wang, Y., Cao, L., Khoury, S.J. (2011). Subventricular zone microglia transcriptional network. *Brain Behaviour Immunology,* Vol.25, No.5 (Nov 2010), pp. 991-999, ISSN 0889-1591.

Taylor, D.L., Pirianov, G., Holland, S., McGinnity, C.J., Norman, A.L., Reali, C., Diemel, L.T., Gveric, D., Yeung, D., & Mehmet, H. (2010). Attenuation of proliferation in oligodendrocyte precursor cells by activated microglia. *Journal of Neuroscience Research,* Vol.88, No.8 (Jun 2010), pp.1632-44, ISSN 1097-4547.

Trotter, J., Karram, K., & Nishiyama, A. NG2 cells: Properties, progeny and origin. (2003). *Brain Research Reviews, Vol.63,* No.1-2 (May 2003), pp. 72-82, *ISSN* 0165-0173.

Vignali, D.A.A., Collison, L.W., & Workman, C.J. (2008). How regulatory T cells work. *Nature Review Immunology,* Vol.8, No.7 (July 2008), pp. 523-532, ISSN 1474-1733.

Voskuhl, R.R., Peterson, R.S., Song, B., Ao, Y., Morales, L.B., Tiwari-Woodruff, S., & Sofroniew, M.V. (2009). Reactive astrocytes form scar-like perivascular barriers to leukocytes during adaptive immune inflammation of the CNS. *Journal of Neuroscience,* Vol.29, No.37 (Sep 2009), pp. 11511-11522, ISSN 0270-6474.

Walsh, J.T., Kipnis, J. (2010). Regulatory T cells in CNS injury: the simple, the complex and the confused. *Trends in Molecular Medicine, In Press, Corrected Proof,* ISSN 1471-4914.

Walton, N.M., Sutter, B.M., Laywell, E.D., Levkoff, L.H., Kearns, S.M., Marshall, G.P., Scheffler, B., & Steindler, D.A. (2006). Microglia instruct subventricular zone neurogenesis. *Glia,* Vol.54, No.8 (Dec 2006), pp. 815-825, ISSN 1098-1136.

Wolf, S.A., Steiner, B., Akpinarli, A., Kammertoens, T., Nassenstein, C., Braun, A., Blankenstein, T., & Kempermann, G. (2009). CD4-Positive T Lymphocytes Provide a Neuroimmunological Link in the Control of Adult Hippocampal Neurogenesis. *The Journal of Immunology,* Vol.182, No.7 (Apr 2009), pp. 3979-3984, ISSN 0165-5728.

Woodruff, R.H., & Franklin, R.J. (1999). Demyelination and remyelination of the caudal cerebellar peduncle of adult rats following stereotaxic injections of lysolecithin, ethidium bromide, and complement/anti-galactocerebroside: a comparative study. *Glia,* Vol.25, No.1 (Feb 1999), pp. 216-228, ISSN 1098-1136.

Wu, J., Yoo, S., Wilcock, D., Lytle, J.M., Leung, P.Y., Colton, C.A., & Wrathall, J.R. (2010). Interaction of NG2(+) glial progenitors and microglia/macrophages from the injured spinal cord. *Glia,* Vol.58, No.4 (Mar 2010), pp. 410-22, ISSN 1098-1136.

Yang, I., Han, S.J., Kaur, G., Crane, C., & Parsa, A.T. (2010). The role of microglia in central nervous system immunity and glioma immunology. *Journal of Clinical Neuroscience,* Vol.17, No.1, pp. 6-10, ISSN 0967-5868.

Yirmiya, R., & Goshen, I. Immune modulation of learning, memory, neural plasticity and neurogenesis. (2011). *Brain, Behavior, and Immunity,* Vol.25, No.2 (Feb 2011), pp. 181-213, ISSN 0889-1591.

Zhao, C., Deng, W., & Gage, F.H. (2008). Mechanisms and functional implications of adult neurogenesis. *Cell,* Vol.132, No.4 (Feb 2008), pp. 645-660.

Ziv, Y., Ron, N., Butovsky, O., Landa, G., Sudai, E., Greenberg, N., Cohen, H., Kipnis, J., & Schwartz, M. (2006a). Immune cells contribute to the maintenance of neurogenesis and spatial learning abilities in adulthood. *Nature Neuroscience,* Vol.9, No.2 (Feb 2006), pp. 268-275, ISSN 1097-6256.

Ziv, Y., Avidan, H., Pluchino, S., Martino, G., & Schwartz, M. (2006b). Synergy between immune cells and adult neural stem/progenitor cells promotes functional recovery from spinal cord injury. *Proceedings of the National Academy of Sciences U S A.* Vol.103, No.35 (Aug 2006), pp. 13174-13179, ISSN 0027-8424.